20th International Conference on Strangeness in Quark Matter (SQM 2022)

EPJ Web of Conferences Volume 276 (2023)

Busan, South Korea
13 – 18 June 2022

Editors:

Youngman Kim
Dong Ho Moon

ISBN: 978-1-7138-7050-0

Printed from e-media with permission by:

Curran Associates, Inc.
57 Morehouse Lane
Red Hook, NY 12571

Some format issues inherent in the e-media version may also appear in this print version.

For additional information, please contact EDP Sciences – Web of Conferences
at the address below.

EDP Sciences – Web of Conferences
17, Avenue du Hoggar
Parc d'Activité de Courtabœuf
BP 112
F-91944 Les Ulis Cedex A
France

Phone: +33 (0) 1 69 18 75 75
Fax: +33 (0) 1 69 28 84 91

contact-edps@webofconferences.org

Additional copies of this publication are available from:

Curran Associates, Inc.
57 Morehouse Lane
Red Hook, NY 12571 USA
Phone: 845-758-0400
Fax: 845-758-2633
Email: curran@proceedings.com
Web: www.proceedings.com

TABLE OF CONTENTS

edp sciences EPJ Web of Conferences

Statement of Peer review

In submitting conference proceedings to EPJ Web of Conferences, I certify to the Publisher that I adhere to the **Policy on Publishing Integrity** of the journal in order to safeguard good scientific practice in publishing.

1. All articles have been subjected to peer review administered by the proceedings editors.
2. Reviews have been conducted by expert referees, who have been requested to provide unbiased and constructive comments aimed, whenever possible, at improving the work.
3. Proceedings editors have taken all reasonable steps to ensure the quality of the materials they publish and their decision to accept or reject a paper for publication has been based only on the merits of the work and the relevance to the journal.

Title, date and place of the conference

Strangeness in Quark Matter 13-18 June 2022 Busan, Republic of Korea

Proceedings editor(s): Youngman Kim, Dong Ho Moon

Date and editor's signature

2023.02.03

Search for the Chiral Magnetic Wave in Pb–Pb collisions using the ALICE detector

Wenya Wu (for the ALICE Collaboration)[1,*]

[1]Key Laboratory of Nuclear Physics and Ion-beam Application (MOE), Institute of Modern Physics, Fudan University, Shanghai 200433, China

Abstract. The interplay of the chiral anomaly and the strong magnetic field ($\sim 10^{15}$ T) created in heavy-ion collisions could give rise to a collective excitation in the quark–gluon plasma called the Chiral Magnetic Wave (CMW). This effect can be experimentally sought by the charge asymmetry ($A_{\rm ch}$) dependence of elliptic flow v_2. However, non-CMW mechanisms such as local charge conservation (LCC) coupled with collective flow can also lead to a similar dependence of v_2 on $A_{\rm ch}$. The triangular flow (v_3) measurement serves as a reference as it is not expected to be affected by the CMW. The v_2 and v_3 of charged hadrons as a function of $A_{\rm ch}$ measured in Pb–Pb collisions at $\sqrt{s_{\rm NN}} = 2.76$ and 5.02 TeV are presented. In addition, the event-shape engineering (ESE) technique is adopted for the first time to quantitatively disentangle the CMW signal and the LCC background. The results indicate that the background effects dominate the CMW measurements.

1 Introduction

Vacuum topological fluctuation in the hot and dense matter created in heavy-ion collisions causes the chiral anomalies of quarks, which can be regarded as local parity violation in strong interactions. The interplay of the chiral anomaly and the strong external magnetic field is expected to produce various chiral anomalous phenomena [1]. One phenomenon, called the chiral magnetic wave (CMW), could lead to wave propagation of electric charge resulting in a charged quadrupole configuration in the dense matter [2].

The difference of the elliptic flow (v_2) between positively and negatively charged particles is expected to depend linearly on the charge asymmetry ($A_{\rm ch}$) for the measurement of the electric quadrupole moment from the CMW [2], i.e., $\Delta v_2 \equiv v_2^- - v_2^+ \approx rA_{\rm ch}$. Here the $A_{\rm ch} = (N^- - N^+)/(N^- + N^+)$ denotes the average net charge in an event. The v_2^+ (v_2^-) represents elliptic flow of positively (negatively) charged particles and the r is the extracted slope parameter. In general, slopes are estimated from Δv_2 normalized to v_2 (i.e. $r^{\rm Norm}$) to probe the intensity of CMW signals in different collision systems. In addition to the slope, there is another observable proposed in Ref. [3] for studying the CMW. It is defined as the integral covariance (Int. Cov.) between $v_2^\pm$ and $A_{\rm ch}$,

$$\text{Int. Cov.} \equiv \langle v_2^\pm A_{\rm ch}\rangle - \langle A_{\rm ch}\rangle \langle v_2^\pm\rangle \approx \mp r\left(\left\langle A_{\rm ch}^2\right\rangle - \langle A_{\rm ch}\rangle^2\right)/2 = \mp r\sigma_{A_{\rm ch}}^2/2. \quad (1)$$

A key advantage of Int. Cov. is that it does not require correction for detector acceptance or experimental tracking efficiency. The use of Int. Cov. also avoids the reduction of the data

*e-mail: wenya.wu@cern.ch

sample from dividing events into different bins of $A_{\rm ch}$ in the slope measurement. In the CMW measurement, the difference of Int. Cov. between positively and negatively charged particles (ΔInt. Cov.) is applied.

In previous analyses [4, 5], magnitudes of the slopes $r_{\Delta v_2}$ are consistent with theoretical predictions in semicentral collisions, which seems to confirm the CMW. On the other hand, theoretical calculations predict that the quadrupole moment is distorted by the third-order expansion of collision geometry, which would lead to a zero-magnitude slope of v_3 [2]. However, the similar values of $r_{\Delta v_2}$ and $r_{\Delta v_3}$ [6] imply that there are backgrounds contributing to the CMW measurement. In fact, local charge conservation (LCC) is widely regarded as the most important background [7]. This is further supported by the calculation from Ref. [8] which shows that $r_{\Delta v_2}$ stems from the intrinsic property of $A_{\rm ch}$ in the LCC scenario. The observed $A_{\rm ch}$ would automatically select particles from different pseudorapidity (η) and transverse momentum ($p_{\rm T}$) ranges leading to various kinds of dependence such as $A_{\rm ch} - \Delta v_2$, $A_{\rm ch} - \Delta p_{\rm T}$, and $A_{\rm ch} - \Delta|\eta|$ [8]. Note that the slopes between $A_{\rm ch}$ and v_2 cannot disentangle the CMW signal and the LCC background directly. Recently, an approach to quantitatively disentangle the CMW signal and the LCC background using the event-shape engineering (ESE) method [9] was introduced in Ref. [10]. It was shown that the ΔInt. Cov. is independent on v_2 in the pure CMW scenario, while in the background, it is proportional to v_2 for shape-selected events. In fact, ESE technique was successfully used in constraining the fraction of the Chiral Magnetic Effect signal [11].

2 Analysis details

The analysis is performed using Pb–Pb collisions at $\sqrt{s_{\rm NN}}$ = 5.02 TeV recorded in 2015 and 2018 by ALICE. Minimum bias trigger criteria, as well as central and semicentral trigger conditions, are required to define the data samples. Events with z−component of the reconstructed primary vertex (V_z) within ±10 cm are selected. The collision centrality is estimated from the amplitude of the signal measured by the V0 detector which consists of the V0A ($2.8 < \eta < 5.1$) and V0C ($-3.7 < \eta < -1.7$) sub-detectors. Inclusive charged particles are reconstructed using the Time Projection Chamber (TPC) and the Inner Tracking System (ITS) with kinematic ranges of $|\eta| < 0.8$ and $0.2 < p_{\rm T} < 2.0$ GeV/c. The results are corrected for the azimuthal acceptance of the TPC and for the $p_{\rm T}$ dependence of the tracking efficiency.

In this analysis, the two-particle Q-Cumulant method [12] is adopted to measure v_2 and v_3. This method expresses particle cumulants in the flow vectors (Q-vectors) with respect to the harmonic of order n, such as $Q_n = \sum_{i=0}^{M} e^{in\varphi_i}$, where M is the number of particles. In order to reduce the non-flow effect from short-range correlations such as resonance decay, jets, and quantum correlation, events are divided into two sub-events A ($-0.8 < \eta < -0.3$) and B ($0.3 < \eta < 0.8$).

The ESE technique is employed to disentangle LCC background contributions from the potential CMW signal as proposed in Ref. [10]. The event shape selection is performed based on the reduced flow vector given as $q_n = \frac{Q_n}{\sqrt{M}}$, where M is the multiplicity and Q_n is the magnitude of the flow vector. In this analysis, the Int. Cov. and v_2 are measured in ten classes of q_2 determined by the V0C detector. Note that the first and the last intervals of q_2 are discarded to remove impurities from neighboring classes.

3 Results

3.1 Slope parameters $r^{\rm Norm}_{\Delta v_2}$ and $r^{\rm Norm}_{\Delta v_3}$

The left panel of Fig. 1 shows the linear dependence of normalized Δv_2 on observed $A_{\rm ch}$ in the 40–50% centrality interval, while the right panel shows the extracted the $r^{\rm Norm}_{\Delta v_2}$ (red

https://doi.org/10.1051/epjconf/202327601001

circles) and $r^{\mathrm{Norm}}_{\Delta v_3}$ (blue crosses) after the efficiency correction. Clearly visible is a consistency between the second- and third-order results for most centrality intervals, confirming earlier measurements in Ref. [6]. The comparable magnitudes of $r^{\mathrm{Norm}}_{\Delta v_2}$ and $r^{\mathrm{Norm}}_{\Delta v_3}$ indicate that the background, which can be qualitatively explained by the LCC effect, dominates the measurements.

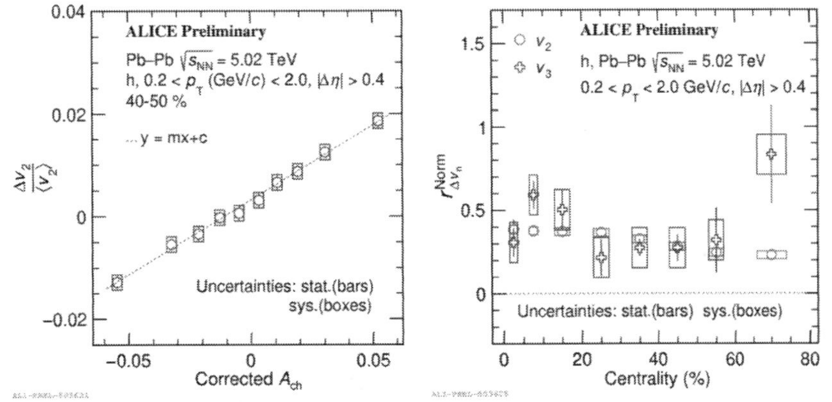

Figure 1. Left: normalized Δv_2 as a function of corrected A_{ch} for the 40–50% centrality interval, with the fit function represented by a dash line. Right: slope parameter $r^{\mathrm{Norm}}_{\Delta v_n}$ for v_2 (red circles) and v_3 (blue crosses), after efficiency correction, as a function of centrality.

3.2 Fraction of CMW with ESE

By implementing the ESE technique, one can estimate the fraction of the CMW signal [10]. The left panel of Fig. 2 presents the ΔInt. Cov. as a function of v_2 for shaped-selected events in the 30–40% centrality interval. The solid line represents a linear fit while the colored band indicates the confidence of the fit within 1σ (68%). The dependence of ΔInt. Cov. changing with shaped v_2 as well as the finite intercept indicate that the background from LCC plays an important role in the CMW measurement.

The right panel of Fig. 2 shows a moderate centrality dependence of the CMW fraction (f_{CMW}), which is defined as

$$f_{\mathrm{CMW}} = \frac{b}{a \times \bar{v_2} + b}, \tag{2}$$

where a and b are the slope and the intercept from the linear fit, respectively, illustrated in the left panel of Fig. 2. The total, centrality-independent systematical uncertainty is denoted by the shaded box on the edge of the line at zero. The fit over the 10–60% centrality interval yields a value of $f_{\mathrm{CMW}} \sim 0.338 \pm 0.084\,(\mathrm{stat.}) \pm 0.198\,(\mathrm{syst.})$. It can be seen that f_{CMW} is compatible with zero within uncertainties, implying that the CMW signal could be very small if it exits.

4 Summary

The $r^{\mathrm{Norm}}_{\Delta v_2}$ and $r^{\mathrm{Norm}}_{\Delta v_3}$ of charged hadrons have been measured in Pb–Pb collisions at $\sqrt{s_{\mathrm{NN}}}$ = 5.02 TeV. The v_3 results are comparable to those of v_2 which points to a large background contribution to the CMW. The ESE technique was applied to disentangle the CMW signal from the LCC background. The dependence of ΔInt. Cov. on v_2 also indicates that LCC dominates the CMW measurements. According to fit parameters, one can extract the f_{CMW} in different centrality classes. Averaging over the 10–60% centrality interval, the measured f_{CMW}

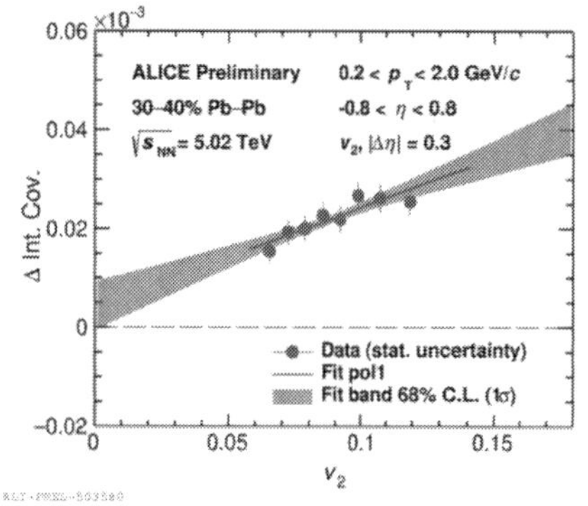

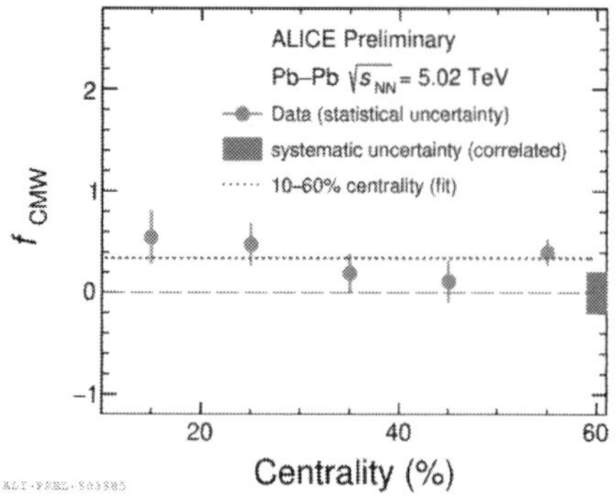

Figure 2. Left: Linear dependence of ΔInt. Cov. on v_2 of shape-selected events in 30–40% centrality class in Pb–Pb collisions at $\sqrt{s_{NN}}$ = 5.02 TeV. Right: Extracted CMW fraction as a function of centrality.

($f_{CMW} \sim 0.338 \pm 0.084$(stat.) $\pm$ 0.198(syst.)) is consistent with zero within experimental uncertainties in Pb–Pb collisions.

This work was supported in part by the National Key Research and Development Program of China (No. 2018YFE0104600), the National Natural Science Foundation of China (Nos. 11975078, 12061141008) and the Shanghai Rising-Star Program (20QA1401500).

References

[1] D. Kharzeev et al., Phys. Lett. B **633**, 260264 (2006)
[2] Y. Burnier et al., Phys. Rev. Lett. **107**, 052303 (2011)
[3] S. A. Voloshin et al., Nucl. Phys. A **931**, 992996 (2014)
[4] ALICE Collaboration, Phys. Rev. C **93**, 044903 (2016)
[5] STAR Collaboration, Phys. Rev. Lett. **114**, 252302 (2015)
[6] CMS Collaboration, Phys. Rev. C **100**, 064908 (2019)
[7] A. Bzdak et al., Phys. Lett. B **726** 239–243 (2013)
[8] W. Wu et al., Phys. Rev. C, **103**, 034906 (2021)
[9] J. Schukraft et al., Phys. Lett. B **719** 394–398 (2013)
[10] C. Wang et al., Phys. Lett. B **820**, 136580 (2021)
[11] ALICE Collaboration, Phys. Lett. B **777**, 151162 (2018)
[12] A. Bilandzic et al., Phys. Rev. C **83** 044913 (2011)

Collectivity in ultra-peripheral heavy-ion collisions

Chun Shen[1,2,*], *Wenbin* Zhao[1], and *Björn* Schenke[3]

[1]Department of Physics and Astronomy, Wayne State University, Detroit, Michigan 48201, USA
[2]RIKEN BNL Research Center, Brookhaven National Laboratory, Upton, NY 11973, USA
[3]Physics Department, Brookhaven National Laboratory, Upton, NY 11973, USA

Abstract. We present full (3+1)D dynamical simulations to study collective behavior in ultra-peripheral nucleus-nucleus collisions (UPC) at the Large Hadron Collider (LHC) with the 3DGlauber+MUSIC+UrQMD framework [1, 2]. By extrapolating from asymmetric p+Pb collisions, we simulate a quasi-real photon γ^* interacting with the Pb nucleus in an ultra-peripheral collision at the LHC, assuming strong final-state effects. We study the elliptic flow hierarchy between p+Pb and γ^*+Pb collisions, which is dominated by the difference in longitudinal flow decorrelations. Our theoretical framework provides a quantitative tool to study collectivity in small asymmetric collision systems at current and future collider experiments.

1 Introduction

Collective features of strongly-coupled systems have been observed in relativistic nuclear collisions with light and heavy nuclei, such as p+Au, d+Au, ^{3}He+Au at the Relativistic Heavy-Ion Collider (RHIC) [3, 4], and p+p and p+Pb collisions at the Large Hadron Collider (LHC) [5–7]. The theoretical interpretation of these flow-like signals has been a hot topic, driving our field to unravel how the collective behavior emerges depending on the collision system size [8, 9]. Recently, the ATLAS Collaboration measured the two-particle azimuthal correlations in ultra-peripheral Pb+Pb collisions (UPCs) at the LHC [10]. The high multiplicity UPC events created from the photo-nuclear interactions showed the persistence of collective phenomena with correlations comparable to those observed in p+p and p+Pb collisions at similar multiplicity [10].

Quantitative understanding of the many-body dynamics in these small collision systems requires the development and application of full (3+1)D simulations beyond Bjorken's boost-invariance paradigm in the high energy limit [11–18]. In photon-nucleus collisions, the quasi-real photon γ^*'s energy fluctuates event-by-event, and is much smaller than the energy of the incoming Pb nucleus. Such unbalanced and fluctuating kinematics leads to a highly asymmetric collision system, strongly violating the longitudinal boost invariance. In these asymmetric systems, the rapidity decorrelation of the collision geometry plays a crucial role when computing and measuring the anisotropic flow coefficients.

In this proceeding, we study the flow rapidity decorrelation in detail for γ^*+Pb and p+Pb collisions, providing complementary information to Ref. [1].

*e-mail: chunshen@wayne.edu

2 Fluctuations in collision kinematics of γ^*+A collisions in UPCs

The fast-moving Pb spectators in the UPC events generate strong fluxes of quasi-real photons. The emitted photons have the following energy spectrum [19, 20],

$$\frac{dN^\gamma}{dk_\gamma} = \frac{2Z^2\alpha}{\pi k_\gamma}\left[w_R^{AA} K_0(w_R^{AA}) K_1(w_R^{AA}) - \frac{(w_R^{AA})^2}{2}(K_1^2(w_R^{AA}) - K_0^2(w_R^{AA}))\right], \tag{1}$$

where $\alpha = 1/137$ and $w_R^{AA} = 2k_\gamma R_A/\gamma_L$, with the longitudinal Lorentz contraction factor $\gamma_L = \sqrt{s_{\rm NN}}/(2m_N)$. The functions $K_0(x)$ and $K_1(x)$ are the modified Bessel functions of the second kind. For the Pb nucleus, $R_A = 6.62$ fm and $Z = 82$. The kinematics for incoming photon projectile and nucleon target in the Pb nucleus are $P_\gamma^\mu \simeq (k_\gamma, 0, 0, k_\gamma)$ and $P_N^\mu \simeq (\sqrt{s_{\rm NN}}/2, 0, 0, -\sqrt{s_{\rm NN}}/2)$, where we neglect the photon's virtuality and nucleon's rest mass. The center of mass collision energy for the γ^*+A system is

$$\sqrt{s_{\gamma^* N}} = \sqrt{2k_\gamma \sqrt{s_{\rm NN}}}. \tag{2}$$

From Eqs. (1) and (2), we can compute the probability distribution for the center of mass energy in γ^*+A collisions,

$$P(\sqrt{s_{\gamma^* N}}) \propto \frac{\sqrt{s_{\gamma^* N}}}{k_\gamma}\left[w_R^{AA} K_0(w_R^{AA}) K_1(w_R^{AA}) - \frac{(w_R^{AA})^2}{2}(K_1^2(w_R^{AA}) - K_0^2(w_R^{AA}))\right], \tag{3}$$

with the photon momentum $k_\gamma = s_{\gamma^* N}/(2\sqrt{s_{\rm NN}})$. Because of the unequal incoming longitudinal momentum between the quasi-real photons and the target nucleon, the center of mass rapidity of the γ^*+Pb system differs from the lab frame rapidity by

$$\Delta y = y_{\rm beam}(\sqrt{s_{\gamma^* N}}) - y_{\rm beam}(\sqrt{s_{\rm NN}}), \tag{4}$$

where the beam rapidity for a given center-of-mass collision energy can be computed as $y_{\rm beam}(\sqrt{s}) = \text{arccosh}(\sqrt{s}/(2m_N))$.

The left panel of Figure 1 shows the probability distributions of the center-of-mass collision energies in Au+Au and Pb+Pb UPC events at RHIC and LHC. The center-of-mass

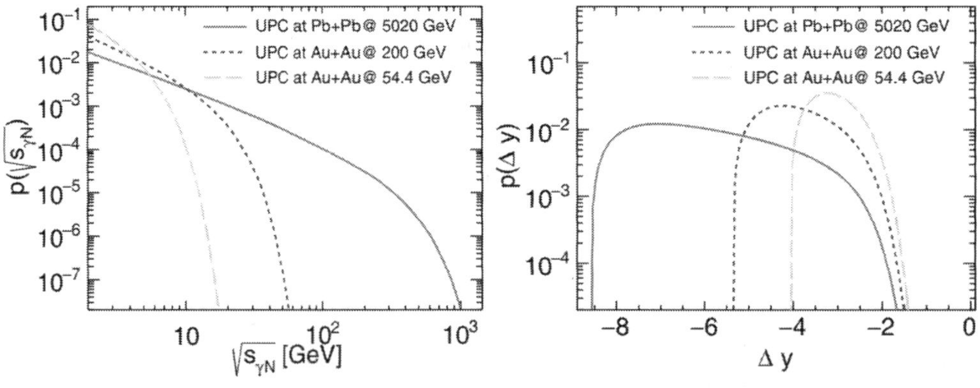

Figure 1. *Left Panel:* The probability distributions of the center-of-mass collision energies for photon-nucleus collisions in Au+Au and Pb+Pb UPC events at three collision energies. *Right Panel:* The probability distributions of the global rapidity shifts in γ^*+A collisions from the center-of-mass frame to the lab frame. Negative Δy represents the shift towards the nucleus-going direction.

collision energies in γ^*+A collisions are much smaller than in their corresponding heavy-ion collisions. The values of $\sqrt{s_{\gamma^* N}}$ fluctuate over wide ranges, which results in broad intervals for rapidity shifts between the center-of-mass frame and lab frame, as shown in the right panel of Figure 1. For UPC events in Pb+Pb collisions at 5020 GeV, the rapidity shifts fluctuate from −2 to −8.5. Therefore, it is important to include these kinematics fluctuations in γ^*+A collisions, which result in non-trivial effects in the rapidity direction. We note that small collision energy and large global rapidity shift result in little particle production at mid-rapidity in the lab frame. Therefore, triggering high multiplicity events at mid-rapidity effectively selects the γ^*+A collisions with large $\sqrt{s_{\gamma^* N}}$.

3 Flow rapidity decorrelation in γ^*+Pb and p+Pb collisions

In the work [1], we found the different amounts of longitudinal flow decorrelations in γ^*+Pb and p+Pb collisions led to the elliptic flow hierarchy observed by the ATLAS Collaboration [10]. The different flow rapidity decorrelations in γ^*+Pb and p+Pb collisions come from the difference in center-of-mass collision energy and the global rapidity shift in γ^*+Pb collisions [1].

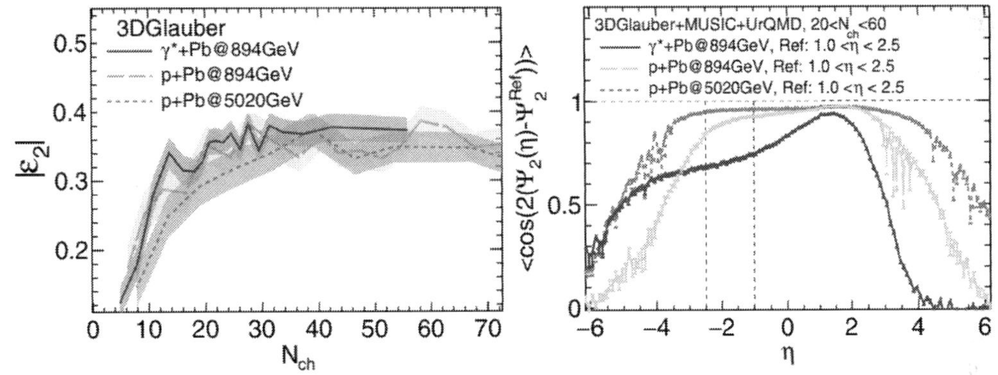

Figure 2. *Left Panel:* Initial-state ellipticities of the fireballs in $1 < |\eta_s| < 2.5$ at $\tau = 0.6$ fm/c for γ^*+Pb and p+Pb collisions at $\sqrt{s} = 894$ GeV and $\sqrt{s} = 5020$ GeV. *Right Panel:* Final-state elliptic flow event-plane correlation with respect to a reference flow angle $\psi_2^{\rm ref}$ at $1 < \eta < 2.5$ for γ^*+Pb and p+Pb collisions at two collision energies.

To investigate the flow rapidity correlations at a charged hadron multiplicity of 40 – 50, we focus on the following analysis with γ^*+Pb collisions at their highest energy, 894 GeV. The left panel of Figure 2 shows that the values of initial-state ellipticity are almost the same between γ^*+Pb and p+Pb collisions as a function of the particle multiplicity, which means that the shape fluctuations in the transverse plane are at the same level for the two collision systems. The right panel of Figure 2 shows the evolution of event-plane correlations in steps from p+Pb collisions at 5020 GeV to γ^*+Pb at 894 GeV collisions. The ATLAS Collaboration measured the two-particle correlation with a rapidity gap of $|\Delta\eta| > 2$. This analysis method computes the flow angular correlation between the two pseudorapidity intervals, namely $\eta_1 \in [-2.5, -1]$ and $\eta_2 \in [1, 2.5]$. The right panel of Figure 2 shows that the angular correlations of the elliptic flow vectors are strong in p+Pb collisions at 5020 GeV, very close to unity between these two η intervals. Reducing the collision energy to 894 GeV shortens the length of the produced strings in the rapidity space, weakening the event-plane correlation to ~ 0.9. The extra global rapidity shift in γ^*+Pb collisions further reduces the correlation strength to ~ 0.7.

4 Conclusion

We applied a newly developed (3+1)D dynamical framework to study the collectivity in highly asymmetric relativistic nuclear collisions, such as p+A collisions and γ^*+A in the ultra-peripheral A+A collisions at RHIC and LHC energies [1, 2]. In this proceeding, we present a detailed analysis of the flow rapidity decorrelation in γ^*+Pb and p+Pb collisions. We discuss how to include fluctuating collision energies in simulating the 3D dynamics of γ^*+A collisions in the UPC events. At LHC energies, the elliptic flow hierarchy between the γ^*+Pb and p+Pb collisions can be explained by the different amounts of flow rapidity decorrelations in these systems, demonstrating the necessity of full (3+1)D simulations for these asymmetric collision systems.

Acknowledgements

W.B.Z. is supported by the National Science Foundation (NSF) under grant numbers ACI-2004571 within the framework of the XSCAPE project of the JETSCAPE collaboration. B.P.S. and C.S. are supported by the U.S. Department of Energy, Office of Science, Office of Nuclear Physics, under DOE Contract No. DE-SC0012704, and Award No. DE-SC0021969, respectively. This research was done using resources provided by the Open Science Grid (OSG) [21, 22], which is supported by the National Science Foundation award #2030508.

References

[1] W. Zhao, C. Shen, B. Schenke (2022), 2203.06094
[2] C. Shen, B. Schenke, Phys. Rev. C **105**, 064905 (2022), 2203.04685
[3] C. Aidala et al. (PHENIX), Phys. Rev. Lett. **120**, 062302 (2018), 1707.06108
[4] C. Aidala et al. (PHENIX), Nature Phys. **15**, 214 (2019), 1805.02973
[5] W. Li, Mod. Phys. Lett. A **27**, 1230018 (2012), 1206.0148
[6] K. Dusling, W. Li, B. Schenke, Int. J. Mod. Phys. E **25**, 1630002 (2016), 1509.07939
[7] J.L. Nagle, W.A. Zajc, Ann. Rev. Nucl. Part. Sci. **68**, 211 (2018), 1801.03477
[8] C. Shen, L. Yan, Nucl. Sci. Tech. **31**, 122 (2020), 2010.12377
[9] B. Schenke, Rept. Prog. Phys. **84**, 082301 (2021), 2102.11189
[10] G. Aad et al. (ATLAS), Phys. Rev. C **104**, 014903 (2021), 2101.10771
[11] P. Bozek, A. Bzdak, G.L. Ma, Phys. Lett. B **748**, 301 (2015), 1503.03655
[12] W. Ke, J.S. Moreland, J.E. Bernhard, S.A. Bass, Phys. Rev. C **96**, 044912 (2017), 1610.08490
[13] B. Schenke, S. Schlichting, Phys. Rev. C **94**, 044907 (2016), 1605.07158
[14] C. Shen, B. Schenke, Phys. Rev. C **97**, 024907 (2018), 1710.00881
[15] C. Shen, S. Alzhrani, Phys. Rev. C **102**, 014909 (2020), 2003.05852
[16] X.Y. Wu, G.Y. Qin (2021), 2109.03512
[17] A. Schäfer, I. Karpenko, X.Y. Wu, J. Hammelmann, H. Elfner (2021), 2112.08724
[18] M.A. Lisa, J.a.G.P. Barbon, D.D. Chinellato, W.M. Serenone, C. Shen, J. Takahashi, G. Torrieri, Phys. Rev. C **104**, 011901 (2021), 2101.10872
[19] C.A. Bertulani, S.R. Klein, J. Nystrand, Ann. Rev. Nucl. Part. Sci. **55**, 271 (2005), nucl-ex/0502005
[20] A.J. Baltz, Phys. Rept. **458**, 1 (2008), 0706.3356
[21] R. Pordes et al., J. Phys. Conf. Ser. **78**, 012057 (2007)
[22] I. Sfiligoi, D.C. Bradley, B. Holzman, P. Mhashilkar, S. Padhi, F. Wurthwrin, WRI World Congress **2**, 428 (2009)

YITP-22-105, J-PARC-TH-0276

Anomalous enhancement of dilepton production due to soft modes in dense quark matter

Toru Nishimura[1,2,*], *Masakiyo* Kitazawa[2,3,1], and *Teiji* Kunihiro[2]

[1]Department of Physics, Osaka University, 560-0043, Toyonaka, Osaka, Japan
[2]Yukawa Institute for Theoretical Physics, Kyoto University, 606-8502, Kyoto, Japan
[3]J-PARC Branch, KEK Theory Center, Institute of Particle and Nuclear Studies, KEK, 203-1, 319-1106, Shirakata, Tokai, Ibaraki, Japan

Abstract. We explore how the dilepton production rate is modified near the critical temperature of color superconductivity and QCD critical point by the soft modes inherently associated with the phase transitions of second-order. It is shown that the soft modes affect the photon self-energy significantly through so called the Aslamasov-Larkin, Maki-Thompson and density of states terms, which are known responsible for the paraconductivity in the metalic superconductivity, and cause an anomalous enhancement of the production rate in the low energy/momemtum region.

1 Introduction

Rich physics is expected to exist in high baryon-density matter at finite temperature, and may be hopefully revealed by experimental programs in relativistic heavy-ion collisions (HIC) such as the beam energy scan program at RHIC, HADES and NA61/SHINE, as well as the future plans at FAIR, NICA and J-PARC-HI. In this report, we show a promising observability of the color superconducting (CSC) [1] phase transition and QCD critical point (CP) in these experiments through an anomalous enhancement of the dilepton production rate (DPR) caused by the soft modes inherently associated with these phase transitions of second-order; the CSC phase transion is assumed to be second-order [4]. We calculate the effects of the soft modes on the DPR by extending the theory of the paraconductivity in metals [5], and show that the soft modes lead to an anomalous enhancement of the DPR at low energy region near each phase transition.

2 Formalism

To explore the temperature and density region around the 2-flavor color superconductivity (2SC) and QCD CP, which is expected to be realized at relatively low densities, we employ the 2-flavor NJL model

$$\mathcal{L} = \bar{\psi} i(\gamma^\mu \partial_\mu - m)\psi + G_S[(\bar{\psi}\psi)^2 + (\bar{\psi} i\gamma_5 \vec{\tau}\psi)^2] + G_C(\bar{\psi} i\gamma_5 \tau_2 \lambda_A \psi^C)(\bar{\psi}^C i\gamma_5 \tau_2 \lambda_A \psi), \quad (1)$$

where τ_2 ($\gamma_{A=2,5,7}$) is the antisymmetric component of the Pauli (Gell-Mann) matrices for the flavor $SU(2)_f$ (color $SU(3)_c$), and $\psi^C(x) \equiv C\bar{\psi}^T(x)$ with $C = i\gamma_2\gamma_0$. We consider the chiral

*e-mail: nishimura@kern.phys.sci.osaka-u.ac.jp

limit for the analysis of the DPR near the critical temperature (T_c) of the 2SC, while we set to $m = 4$ MeV for investigating the QCD CP. The scalar coupling constant $G_S = 5.01$ MeV^{-2} and the three-momentum cutoff $\Lambda = 650$ MeV are used, which leads to the pion decay constant $f_\pi = 93$ MeV and the chiral condensate $\langle\bar{\psi}\psi\rangle = (-250\ \text{MeV})^3$, while the diquark coupling G_C is treated as a free parameter.

The DPR is related to the retarded photon self-energy $\Pi^{R\mu\nu}(k)$ as

$$\frac{d^4\Gamma}{d^4k} = -\frac{\alpha}{12\pi^4}\frac{1}{k^2}\frac{1}{e^{\omega/T}-1}g_{\mu\nu}\text{Im}\Pi^{R\mu\nu}(k), \tag{2}$$

where $k = (\mathbf{k}, \omega)$ is the four momentum of the photon and α is the fine structure constant. In this study, we calculate the modification of $\tilde{\Pi}^{\mu\nu}(k)$ due to the soft modes of the 2SC and QCD CP. In this section, we first calculate $\tilde{\Pi}^{\mu\nu}(k)$ including the diquark soft modes of 2SC [4] in Sec. 2.1, and then deal with the soft modes of the QCD CP that are the mesonic modes in Sec. 2.2.

2.1 Modification of $\Pi^{\mu\nu}$ by diquark soft modes

It is known that the diquark fluctuations form a collective mode with a significant strength above but near T_c of the 2SC [2, 3]. The collective mode associated with the second-order nature of phase transition is called the soft mode. The propagator of the diquark soft modes in the random-phase approximation (RPA) in the imaginary-time formalism is given by

$$\tilde{\Xi}_C(k) = \frac{1}{G_C{}^{-1} + Q_C(k)}, \quad Q_C(k) =$$  $$= -8\int_p \text{Tr}[\mathcal{G}_0(k-p)\mathcal{G}_0(p)], \tag{3}$$

where $Q_C(k) = Q_C(\mathbf{k}, i\nu_n)$ is the one-loop qq correlation function with $\mathcal{G}_0(p) = \mathcal{G}_0(\mathbf{p}, i\omega_m) = 1/[(i\omega_m+\mu)\gamma_0 - \mathbf{p}\cdot\gamma]$ the free quark propagator, ω_m (ν_n) the Matsubara frequency for fermions (bosons), and Tr the trace over the Dirac indices. The retarded propagator is obtained by the analytic continuation $\Xi_C^R(\mathbf{k}, \omega) = \tilde{\Xi}_C(\mathbf{k}, i\nu_n \to \omega + i\eta)$.

We remark that $\Xi_C^R(\mathbf{k}, \omega)$ satisfies the Thouless criterion, that is $[\Xi_C^R(\mathbf{0}, 0)]^{-1} = 0$ at $T = T_c$ for the second-order phase transition. The criterion shows that $\Xi_C^R(\mathbf{k}, \omega)$ has a pole at the origin at $T = T_c$, and hence the diquark fluctuations become soft near T_c [2]. This fact also allows us to adopt the time-dependent Ginzburg-Landau (TDGL) approximation

$$[\Xi_C^R(\mathbf{k}, \omega)]^{-1} = A(\mathbf{k}) + C(\mathbf{k})\omega, \tag{4}$$

near but above T_c, where $A(\mathbf{k}) = G_C^{-1} + Q_C^R(\mathbf{k}, 0)$ and $C(\mathbf{k}) = \partial Q_C^R(\mathbf{k}, \omega)/\partial\omega\,|_{\omega=0}$. It is known that Eq. (4) reproduces $\Xi_C^R(\mathbf{k}, \omega)$ over wide ranges of ω, $\mathbf{k}^2$ and $T (> T_c)$ [2].

To construct $\tilde{\Pi}^{\mu\nu}(k)$ that inolves the soft mode, we start from the one-loop diagram of $\tilde{\Xi}_C(\mathbf{k}, i\nu_n)$, which is the lowest contribution of the soft modes to the thermodynamic potential [4]. The photon self-energy is then constructed by attaching electromagnetic vertices at two points of quark lines in the thermodynamic potential. One then obtains four types of diagrams shown in Fig. 1. These diagrams are called the Aslamasov-Larkin (AL) (Fig. 1(a)), Maki-Thompson (MT) (Fig. 1(b)) and density of states (DOS) (Fig. 1(c, d)) [5] terms, respectively, in the theory of metallic superconductivity. Each contribution to the photon self-energy, $\tilde{\Pi}^{\mu\nu}_{\text{AL}}(k)$, $\tilde{\Pi}^{\mu\nu}_{\text{MT}}(k)$ and $\tilde{\Pi}^{\mu\nu}_{\text{DOS}}(k)$, respectively, is given by

$$\tilde{\Pi}^{\mu\nu}_{\text{AL}}(k) = 3\int_q \tilde{\Gamma}^{\mu}(q, q+k)\tilde{\Xi}_C(q+k)\tilde{\Gamma}^{\nu}(q+k, q)\tilde{\Xi}_C(q), \tag{5}$$

$$\tilde{\Pi}^{\mu\nu}_{\text{MT(DOS)}}(k) = 3\int_q \tilde{\Xi}_C(q)\mathcal{R}^{\mu\nu}_{\text{MT(DOS)}}(q, k), \tag{6}$$

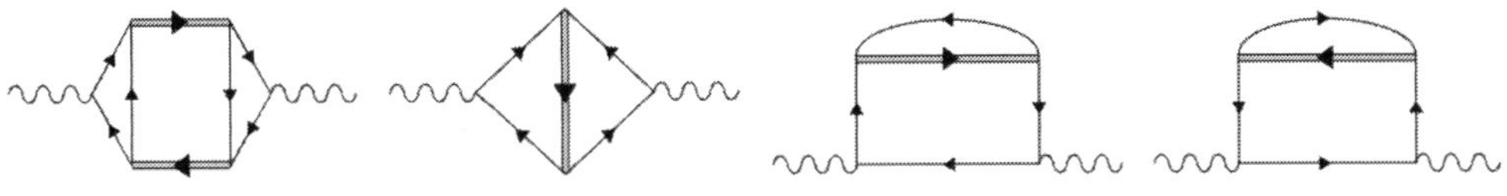

(a) (b) (c) (d)

Figure 1: Diagrammatic representations of the Aslamasov-Larkin (a), Maki-Thompson (b) and density of states (c,d) terms with the diquark soft mode. The wavy and double lines are photons and the diquark soft modes, respectively.

where $\tilde{\Gamma}^\mu(q,q+k)$ and $\mathcal{R}^{\mu\nu}(q,k) = \mathcal{R}^{\mu\nu}_{\rm MT}(q,k)+\mathcal{R}^{\mu\nu}_{\rm DOS}(q,k)$ satisfy the following Ward-Takahashi (WT) identities

$$k_\mu\tilde{\Gamma}^\mu(q,q+k) = (e_u+e_d)[Q_C(q+k)-Q_C(q)], \tag{7}$$

$$k_\mu\mathcal{R}^{\mu\nu}(q,k) = (e_u+e_d)[\tilde{\Gamma}^\nu(q-k,q)-\tilde{\Gamma}^\nu(q,q+k)], \tag{8}$$

where $e_u = 2|e|/3$ ($e_d = -|e|/3$) is the electric charge of up (down) quarks. The total photon self-energy is then given by $\tilde{\Pi}^{\mu\nu}(k) = \tilde{\Pi}^{\mu\nu}_{\rm free}(k)+\tilde{\Pi}^{\mu\nu}_{\rm AL}(k)+\tilde{\Pi}^{\mu\nu}_{\rm MT}(k)+\tilde{\Pi}^{\mu\nu}_{\rm DOS}(k)$. One can explicitly check that this photon self-energy satisfies the WT identity using Eqs. (7) and (8).

To obtain the DPR with Eq. (2), the calculation of the spatial components of $\Pi^{R\mu\nu}(k)$ is sufficient since $\Pi^{R00}(k)$ is expressed in terms of the longitudinal part from the WT identity as $\Pi^{R00}(k) = \mathbf{k}^2\Pi^{R11}(k)/k_0^2$ with $k = (k_0,|\mathbf{k}|,0,0)$. For the vertex functions (7) and (8), we approximate $\tilde{\Gamma}^\mu(q,q+k)$ and $\mathcal{R}^{\mu\nu}(q,k)$ using Eq. (4). For $\tilde{\Gamma}^i(q,q+k)$, we employ an ansatz

$$\tilde{\Gamma}^i(q,q+k) = -(e_u+e_d)\frac{Q_C(\mathbf{q}+\mathbf{k},0)-Q_C(\mathbf{q},0)}{|\mathbf{q}+\mathbf{k}|^2-|\mathbf{q}|^2}(2q+k)^i, \tag{9}$$

which is a real number and satisifes the WT identity (7) with Eq. (4). One can also obtain $\mathcal{R}^{ij}(q,k)$ in the same manner, and finds that it is a real number as well. Using this vertex and Eq. (4), the imaginary part of $\Pi^{Rij}_{\rm MT}(q)+\Pi^{Rij}_{\rm DOS}(q)$ vanishes. Therefore, the MT and DOS terms do not contribute to the calculation of the DPR. This result is in accordance with the case of the metallic superconductivity [5].

2.2 Modification of $\Pi^{\mu\nu}$ by mesonic soft modes near QCD CP

The photon self-energies including the mesonic soft modes of the QCD CP can also be calculated with a similar manner. The propagator of the mesonic mode is evaluated by the RPA as

$$\tilde{\Xi}_S(k) = \frac{1}{G_S{}^{-1}+Q_S(k)}, \quad Q_S(k) =$$  $$= 12\int_p {\rm Tr}[\mathcal{G}_0(p-k)\mathcal{G}_0(p)], \tag{10}$$

where $Q_S(k)$ is the one-loop $q\bar{q}$ correlation function. As described in Ref. [6], $\Xi^R_S(k)$ has a discontinuity at the light cone, and its strength is limited to the space-like region. This property is contrasted with the diquark soft mode (3), which does not have the discontinuity at the light cone, while the diquark soft mode also has a significant strength in the space-like region. The photon self-energy with the mesonic modes is given by

$$\tilde{\Pi}^{\mu\nu}_{\rm AL}(k) = \sum_f\int_q \tilde{\Gamma}^\mu_f(q,q+k)\tilde{\Xi}_S(q+k)\tilde{\Gamma}^\nu_f(q+k,q)\tilde{\Xi}_S(q), \tag{11}$$

$$\tilde{\Pi}^{\mu\nu}_{\rm MT(DOS)}(k) = \sum_f\int_q \tilde{\Xi}_S(q)\mathcal{R}^{\mu\nu}_{f\ \rm MT(DOS)}(q,k), \tag{12}$$

where $f = u,d$ are the indices of the flavors. $\tilde{\Gamma}^\mu_f(q,q+k)$ and $\mathcal{R}^{\mu\nu}_f(q,k) = \mathcal{R}^{\mu\nu}_{f\ \rm MT}(q,k) + \mathcal{R}^{\mu\nu}_{f\ \rm DOS}(q,k)$ satisfy the WT identity, similarly to Eqs. (7) and (8).

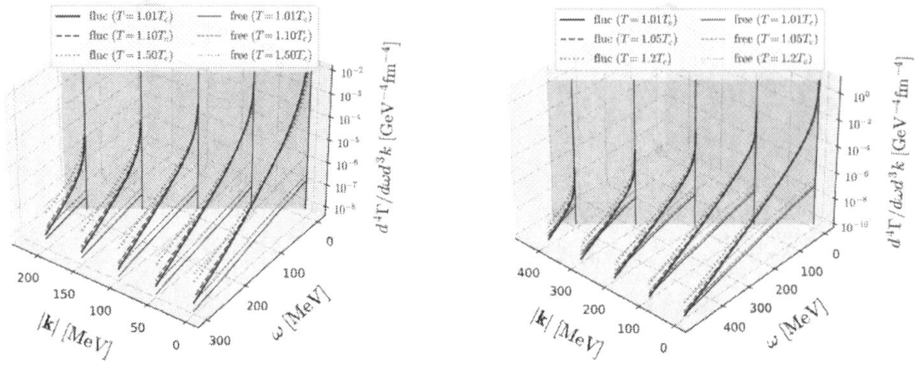

Figure 2: DPRs per unit energy ω and momentum $\mathbf{k}$ above T_c of the 2SC at $\mu = 350$ MeV (left) [4] and of the QCD CP at $\mu = \mu_{\rm CP}$ (right). The thick lines are the contribution of the soft modes, and the thin lines are the results for the free quark gas.

3 Numerical results and summary

In the left panel of Fig. 2, we show the DPR per unit energy and momentum $d^4\Gamma/d^4k$ near the 2SC phase transition for $T = 1.01T_c$, $1.1T_c$ and $1.5T_c$ at $\mu = 350$ MeV ($T_c = 42.94$ MeV) [4]. The thick lines are the contribution from the soft modes, while the thin lines are the ones of the free quark gas. One sees that the DPR from the soft modes is anomalously enhanced in the small ω and $\mathbf{k}$ region in comparison with the free quark gas for $T \lesssim 1.5T_c$, and this enhancement becomes more pronounced as T approaches T_c. This result is expected through the properties of the soft modes. Shown in the right panel of Fig. 2 is the DPR near the QCD CP for $T = 1.01T_{\rm CP}$, $1.05T_{\rm CP}$ and $1.2T_{\rm CP}$ at $\mu = \mu_{\rm CP}$, where $(T_{\rm CP}, \mu_{\rm CP}) = (35.07, 322.20)$ MeV is the location of the QCD CP. One finds that the anomalous enhancement of the DPR is also observed similarly to the case of the 2SC.

The low-energy part of the DPR is related to the electrical conductivity σ and relaxation time τ, which are obtained from differentiations of $\rho_L(\mathbf{0}, \omega) = \sum_i \mathrm{Im}\Pi^{R\ ii}_{AL}(\mathbf{0}, \omega)/\pi$ at $\omega = 0$. The analysis of σ and τ thus allows us to understand behavior of the DPR near T_c more quantitatively. We find that these quantities diverge at T_c with $\sigma \propto \epsilon^{-1/2}$ and $\tau \propto \epsilon^{-3/2}$ for the 2SC, while for the case of the QCD CP, $\sigma \propto \epsilon^{-2/3}$ and $\tau \propto \epsilon^{-1}$ with $\epsilon = |T - T_c|/T_c$, as will be discussed in the forthcoming publication. The origin of the difference of two cases is due to the T-dependence of $A(\mathbf{k})$ and the $\mathbf{k}$-dependence of $C(\mathbf{k})$ in Eq. (4).

In this report, we studied how the soft modes of the 2SC and QCD CP affect the DPR near but above T_c. The modification of the photon self-energy due to the soft modes are investigated through the analysis of the Aslamasov-Larkin, Maki-Thompson and density of states terms with the TDGL approximation for the soft mode propagator and vertices. The enhancement of the DPR found in this study would allow us to detect these signals of the 2SC and QCD CP in the future HIC experiments.

References

[1] M.G. Alford, A. Schmitt, K. Rajagopal and T. Schäfer, Rev. Mod. Phys. **80** (2008) 1455.
[2] M. Kitazawa, T. Koide, T. Kunihiro and Y. Nemoto, Phys. Rev. D **65** (2002) 091504; ibid., Phys. Rev. D **70** (2004) 056003; ibid., Prog. Theor. Phys. **114** (2005) 117.
[3] T. Kunihiro, M. Kitazawa and Y. Nemoto, PoS **CPOD07** (2007) 041.
[4] T. Nishimura, M. Kitazawa and T.Kunihiro, PTEP **2022** (2022) 093D02.
[5] A. I. Larkin and A. A. Varlamov, Springer Berlin Heidelberg (2008).
[6] H. Fujii and M. Ohtani, Phys. Rev. D **70** (2004) 014016.

Highlights from the NA61/SHINE

Szymon Pulawski[1,*] on behalf of NA61/SHINE

[1]Institute of Physics, University of Silesia, Katowice, Poland

Abstract. NA61/SHINE is a multipurpose fixed-target facility at the CERN Super Proton Synchrotron. The main goals of the NA61/SHINE strong-interactions program are to discover the critical point of strongly interacting matter as well as to determine the properties of produced particles relevant for the study of the onset of deconfinement - the transition between the state of hadronic matter and the quark-gluon plasma. An analysis of hadron production properties is performed in nucleus-nucleus, proton-proton, and proton-nucleus interactions as a function of collision energy and size of the colliding nuclei to achieve these goals.
The selected NA61/SHINE results from this program are presented. In particular, the latest results from different reactions p+p, Be+Be, Ar+Sc, and Pb+Pb on hadron spectra and fluctuations are discussed. The NA61/SHINE results are compared with worldwide experiments and predictions from various theoretical models, like EPOS, PHSD, UrQMD, and others.

1 The NA61/SHINE facility

The NA61/SHINE detector [1] is a large acceptance hadron spectrometer with excellent capabilities in charged particle momentum measurements and identification by a set of eight Time Projection Chambers as well as Time-of-Flight detectors. The high resolution forward calorimeter, the Projectile Spectator Detector (PSD), measures energy flow around the beam direction, which in nucleus-nucleus reactions is primarily a measure of the number of projectile spectator (non-interacted) nucleons and is thus related to the violence (centrality) of the collision. A set of beam detectors identifies beam particles and measures precisely their trajectories. The NA61/SHINE detector system was recently upgraded: readout rate was increased to 1kHz, new Vertex Detector and data acquisition and trigger systems were installed, extension of PSD was performed. The schematic view of upgraded system is shown in Fig. 1.

2 Study of the onset of deconfinement

NA61/SHINE performed a two-dimensional scan in collision energy (13*A*-150*A* GeV/c and system size (p+p, Be+Be, Ar+Sc, Xe+La, Pb+Pb) to study the phase diagram of strongly interacting matter. The main goals of NA61/SHINE are the search for the critical point and a study of the onset of deconfinemnet.

A plateau ("step") in the energy dependence of the inverse slope parameter T was observed by the NA49 experiment in Pb+Pb collisions for m_T spectra of $K^{\pm}$. It was expected

*e-mail: s.pulawski@cern.ch

https://doi.org/10.1051/epjconf/202327601004

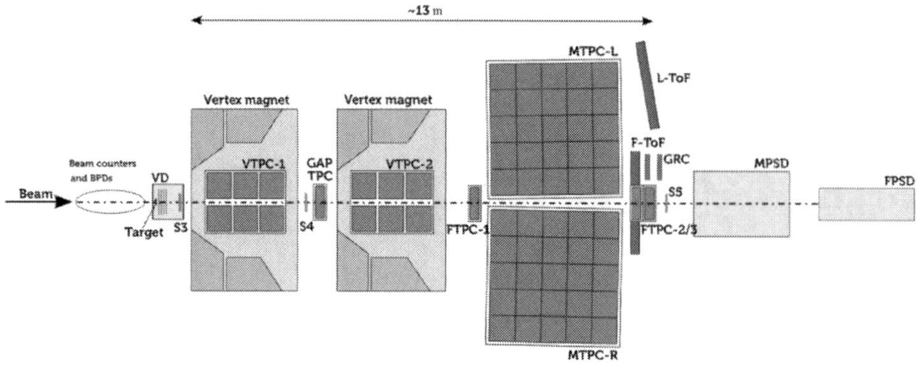

Figure 1. Schematic layout of NA61/SHINE detector system.

for the onset of deconfinement due to the presence of a mixed phase of hadron gas (HRG) and quark-gluon plasma (QGP) [2]. In p+p interactions at SPS energies the inverse slope parameter T of m_T spectra shows qualitatively similar enery dependence as in central Pb+Pb collisions ("step") and such a behaviour seems to emerge also in Be+Be reactions, as visible in Fig. 2. The values of the T parameter in Be+Be collisions are slightly above those in p+p interactions. The T parameter in Ar+Sc reactions is found between those in p+p/Be+Be and Pb+Pb collisions.

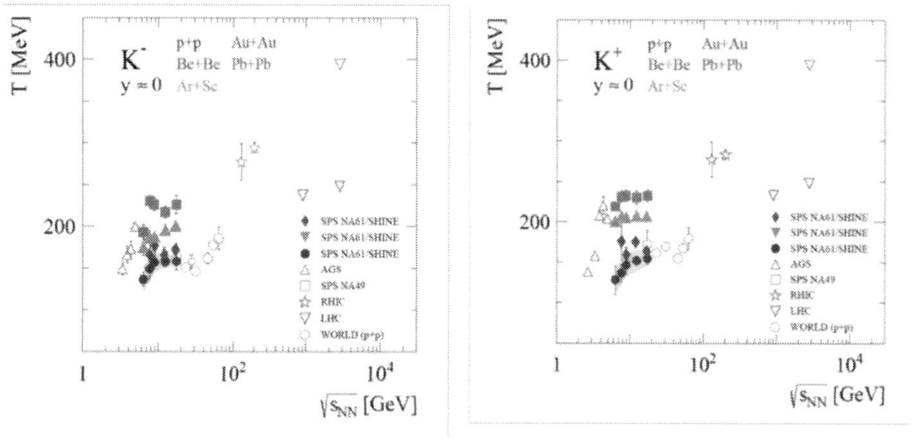

Figure 2. Inverse slope parameter T of m_T spectra of $K^{\pm}$ as function of collision energy. Most results are shown with statistical uncertainties only. For the p+p data the shaded band indicates systematic uncertainties.

Finally, rapid changes of the ratios K^+/π^+ at mid-rapidity and $\langle K^+ \rangle / \langle \pi^+ \rangle$ as function of collision energy ("horn") were observed in Pb+Pb collisions by the NA49 experiment. These were predicted by the SMES model [3] as a signature of the onset of deconfinement. These two ratios together with new NA61/SHINE results from Be+Be and Ar+Sc collisions are shown in Fig. 3. A plateau like structure is visible in p+p interactions. The ratio K^+/π^+ at mid-rapidity as well as the ratio of total yields from Be+Be collisions is close to the p+p measurements. For the five analysed energies of Ar+Sc collisions, the ratio K^+/π^+ at mid-rapidity and $\langle K^+ \rangle / \langle \pi^+ \rangle$ are higher than in p+p collisions but show a qualitatively similar energy dependence - no horn structure visible.

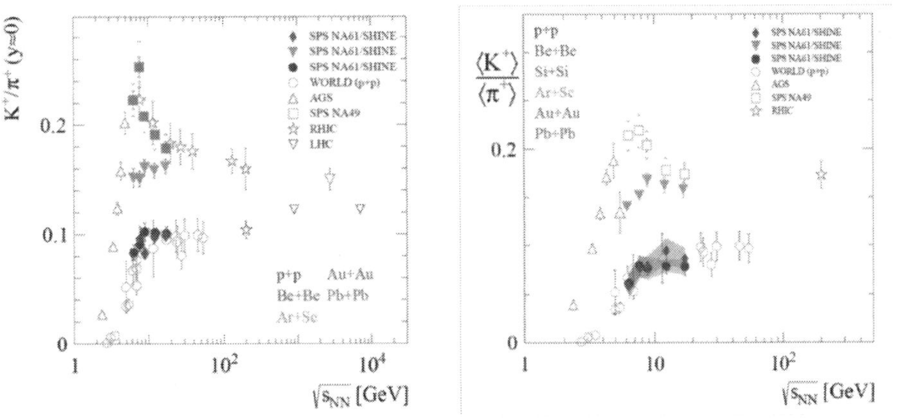

Figure 3. Yields ratio K^+/π^+ at mid-rapidity and total yields ratio $\langle K^+\rangle$ / $\langle\pi^+\rangle$ produced in p+p, Be+Be and Pb+Pb collisions as function of collision energy.

The Ξ mean multiplicities measured by NA61/SHINE in inelastic $p + p$ interactions are used to calculate the enhancement factors of Ξs observed in centrality selected Pb+Pb, in semi-central C+C, and in Si+Si collisions as measured by NA49 [4] at the CERN SPS. The results for mid-rapidity densities are shown in Fig. 4 (*left*) as a function of $\langle N_W\rangle$. The enhancement factor increases approximately linearly from 3.5 in C+C to 9 in central Pb+Pb collisions. With the advent of the NA61/SHINE results on Ξ production in $p + p$ interactions a new baseline reference becomes available and it is used here for the recalculation of the enhancement observed in the NA57 p+Be and A+A data. Figure 4 (*right*) shows the rapidity densities dn/dy of $\overline{\Xi}^+$ at mid-rapidity per mean number of wounded nucleons divided by the corresponding values for inelastic $p+p$ collisions as a function of $\langle N_W\rangle$. Apart from a slightly flatter rise the overall picture remains unchanged.

3 Search for critical point

The expected signal of a critical point (CP) is a non-monotonic dependence of various fluctuation or correlation measures in NA61/SHINE energy – system size scan. Fluctuations of conserved charges (electric, strangeness or baryon number) are of special interest [8–10]. To compare fluctuations in systems of different sizes, one should use quantities insensitive to system volume, i.e. intensive quantities. They are constructed by division of cumulants κ_i of the measured multiplicity distribution (up to fourth order), where i is the order of the cumulant. For second, third and fourth order cumulants intensive quantities are defined as: κ_2/κ_1, κ_3/κ_2 and κ_4/κ_2. Their reference values for no fluctuations are 0 and for independent particle production are 1. In case of net-charge, cumulant ratios are redefined to $\kappa_2/\left(\kappa_1\left[h^+\right]-\kappa_1\left[h^-\right]\right)$, $\kappa_3/\left(\kappa_1\left[h^+-h^-\right]\right)$ and $\kappa_4/\left(\kappa_2\left[h^+-h^-\right]\right)$ in order to keep the same references. Figure 5 shows the system size and energy dependence of second, third and fourth order cumulant ratio of net-electric charge in p+p as well as central Be+Be and Ar+Sc interactions. So far, there is no clear difference between systems for higher order moments. More detailed studies are needed.

Another possible tool for search of CP is a proton intermittency. In the proximity of CP a local power-law fluctuations of the baryon density should appear which can be searched for by studying second factorial moments with the cell size or, equivalently, with the number of cells in space of protons at mid-rapidity [11–13]. NA61/SHINE measures $F_2(M)$ using

 https://doi.org/10.1051/epjconf/202327601004

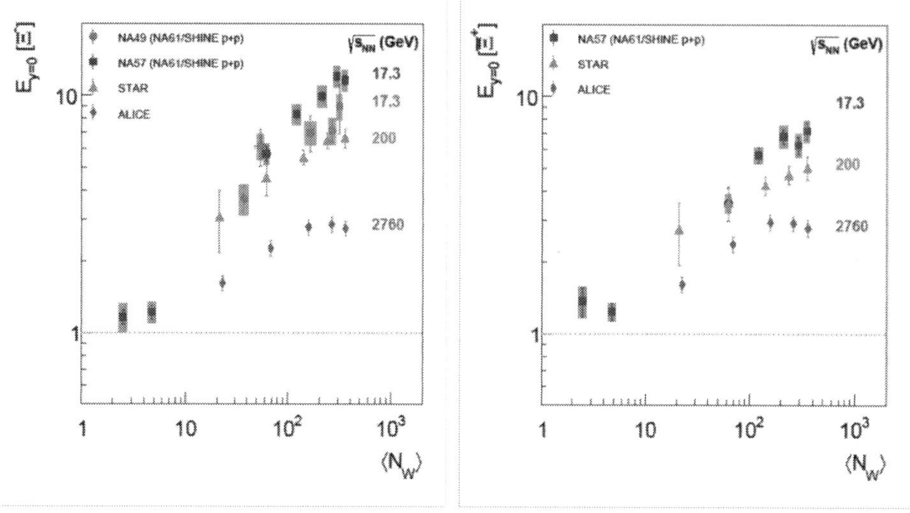

Figure 4. (Color online) The strangeness enhancement E at mid-rapidity as a function of average number of wounded nucleons $\langle N_W \rangle$ calculated as a ratio of rapidity density for Ξ^- production (*left*) and $\overline{\Xi}^+$ production (*right*) in nucleus-nucleus interactions per $\langle N_W \rangle$ divided by the corresponding value for $p+p$ interactions. The enhancement factor is defined as $E = \frac{2}{\langle N_W \rangle} \frac{dn/dy(A+A)}{dn/dy(p+p)}$. At SPS energies and above the number of wounded nucleons is close to or equal to the number of participating nucleons. Red circles – NA49 Pb+Pb at 158*A* GeV [4], blue squares - NA57 p+Be, p+Pb and Pb-Pb at the same center-of-mass energy $\sqrt{s_{NN}} = 17.3$ GeV [5], magenta triangles - STAR Au+Au at $\sqrt{s_{NN}} = 200$ GeV [6], gray diamonds - ALICE Pb+Pb at $\sqrt{s_{NN}} = 2.76$ TeV [7]. The systematic errors are represented by shaded boxes.

statistically independent points and cumulative variables. Preliminary results on $F_2(M)$ of mid-rapidity protons measured in 0-20% most central Ar+Sc collisions at 150*A* GeV/*c* and 0-10% most central Pb+Pb collisions at 30*A* and 13*A* GeV/*c* are presented in Fig. 6. The intermittency index ϕ_2 for a system freezing out at the QCD critical endpoint is expected to be $\phi_2 = 5/6$ assuming that the latter belongs to the 3-D Ising universality class [14]. Measured $F_2(M)$ of protons for Ar+Sc at 150*A* GeV/*c* and Pb+Pb at 30*A* and 13*A* GeV/*c* show no indication for power-law increase with a bin size which could indicate the presence of CP.

4 Strangeness production in p+p

Finally, NA61/SHINE provides new and unique results on strangeness particle production in p+p interactions. Figure 7 presents first rapidity distribution of $K^*(892)^0$ produced in inelastic p+p interactions at 40 and 80 GeV/*c* [15] and the only result on $\Xi(1530)^0$ and $\overline{\Xi}(1530)^0$ production in inelastic p+p interactions at 158 GeV/*c* [16].

The new measurements by NA61/SHINE of $\Xi(1530)^0$ and $\overline{\Xi}(1530)^0$ produced in inelastic $p+p$ interactions at 158 GeV/*c* as well as previously obtained results for π^+, π^-, K^+, K^-, p, $\overline{p}$, $K^*(892)^0$, Λ, $\phi(1020)$, Ξ^- and $\overline{\Xi}^+$ (see Refs. [17–23] were fitted by different variants of the Hadron Resonance Gas Model (HRG). The Canonical Ensemble with fixed $\gamma_s = 1$ and Canonical Ensemble with fitted strangeness saturation parameter γ_s configurations were used. Figure 8 compares the measured multiplicities of particles produced in inelastic $p+p$ interactions at 158 GeV/*c* with multiplicities of the same particles obtained from the HRG model in the CE formulation under two different model assumptions: $\gamma_s = 1$ and fitted γ_s. The

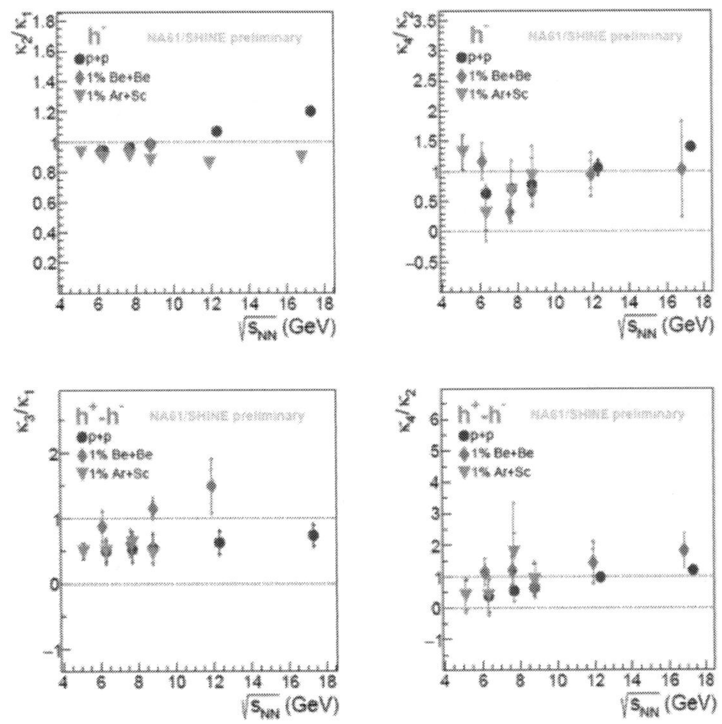

Figure 5. System size and energy dependence of κ_2/κ_1, κ_4/κ_2, $\kappa_3/(\kappa_1\,[h^+ - h^-])$ and $\kappa_4/(\kappa_2\,[h^+ - h^-])$. Statistical uncertainty was obtained with the bootstrap method and it is indicated as a dashed black bar. Systematic uncertainty/bias: p+p - corrected data with estimate on systematic uncertainty; Be+Be/Ar+Sc - uncorrected data with estimate of systematic bias. Systematic uncertainty/bias is indicated with arrows.

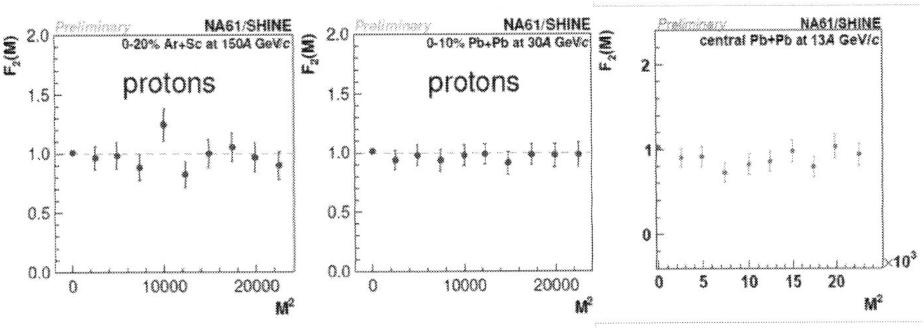

Figure 6. Preliminary results on $F_2(M)$ of mid-rapidity protons measured in 0-20% most central Ar+Sc collisions at 150*A* GeV/*c* and 0-10% most central Pb+Pb collisions at 30*A* and 13*A* GeV/*c*.

software package THERMAL-FIST 1.3 [24] was used for this purpose. For the small $p + p$ system, the appropriate statistical system is the Canonical Ensemble, which has as parameters the temperature T, the radius R of the system at chemical freezeout, and the strangeness suppression factor γ_s. The results plotted in Fig. 8 show significant discrepancies for the fitted parameters T and R between the fits with $\gamma_s = 1$ and free γ_s. Moreover, the fit with

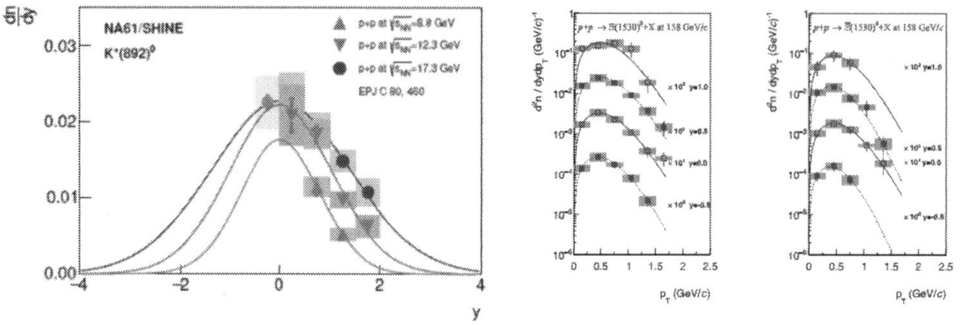

Figure 7. Rapidity distribution of $K^*(892)^0$ produced in inelastic p+p interactions at 40 and 80 GeV/c and transverse momentum spectra in rapidity intervals of $\Xi(1530)^0$ and $\overline{\Xi}(1530)^0$ production in inelastic p+p interactions at 158 GeV/c

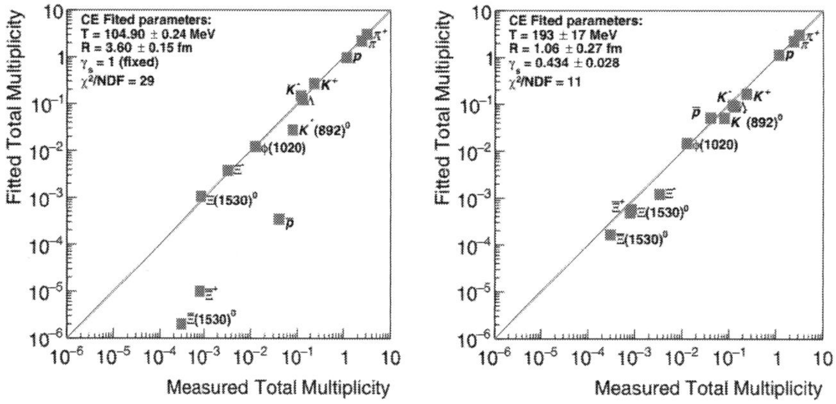

Figure 8. (Color online) Mean multiplicities of π^+, π^-, K^+, K^-, p, $\overline{p}$, $K^*(892)^0$, Λ, $\phi(1020)$, Ξ^-, $\overline{\Xi}^+$, $\Xi(1530)^0$ and $\overline{\Xi}(1530)^0$ produced in $p+p$ interactions at 158 GeV/c [17–23] measured by NA61/SHINE are compared with mean multiplicities obtained from the HRG model based on the Canonical Ensemble with fixed $\gamma_s = 1$ (*i*) and fitted γ_s (*ii*). Uncertainties of the measurement are smaller than the symbol size.

fixed γ_s shows unacceptably large χ^2/NDF = 29. This demonstrates that the statistical model fails when fixing γ_s to one. The fit with free γ_s finds $\gamma_s = 0.434 \pm 0.028$ and reproduces the measurements significantly better than with $\gamma_s = 1$. Thus the statistical model indicates a strong suppression of strange particle production in $p + p$ collisions at CERN SPS energies.

Acknowledgment

This work was supported by the Polish Minister of Education and Science (contract No. 2021/WK/10), the Norwegian Financial Mechanism 2014-2021 (grant 2019/34/H/ST2/00585), National Science Center, Poland Grant No. 2018/31/G/ST2/03910 - BEETHOVEN CLASSIC 3 and under the Research Excellence Initiative of the University of Silesia in Katowice.

References

[1] N. Abgrall et al. (NA61/SHINE), JINST **9**, P06005 (2014), 1401.4699
[2] R. Poberezhnyuk, M. Gazdzicki, M. Gorenstein (2015), 1502.05650
[3] M. Gazdzicki, M.I. Gorenstein, Acta Phys.Polon. **B30**, 2705 (1999), hep-ph/9803462
[4] T. Anticic et al. (NA49), Phys. Rev. C **80**, 034906 (2009), 0906.0469
[5] F. Antinori et al. (NA57), J. Phys. G **32**, 427 (2006), nucl-ex/0601021
[6] B. Abelev et al. (STAR), Phys. Rev. C **77**, 044908 (2008), 0705.2511
[7] B.B. Abelev et al. (ALICE), Phys. Lett. B **728**, 216 (2014), [Erratum: Phys.Lett.B 734, 409–410 (2014)], 1307.5543
[8] M.A. Stephanov, Phys. Rev. Lett. **102**, 032301 (2009), 0809.3450
[9] M. Asakawa, M. Kitazawa, Prog. Part. Nucl. Phys. **90**, 299 (2016), 1512.05038
[10] M.A. Stephanov, K. Rajagopal, E.V. Shuryak, Phys. Rev. Lett. **81**, 4816 (1998), hep-ph/9806219
[11] A. Bialas, R.B. Peschanski, Nucl. Phys. B **273**, 703 (1986)
[12] L. Turko, Phys. Lett. B **227**, 149 (1989)
[13] F.K. Diakonos, N.G. Antoniou, G. Mavromanolakis, PoS **CPOD2006**, 010 (2006)
[14] H. Satz, Nucl.Phys. **B326**, 613 (1989)
[15] A. Acharya et al. (NA61/SHINE), Eur. Phys. J. C **82**, 322 (2022), 2112.09506
[16] A. Acharya et al. (NA61/SHINE), Eur. Phys. J. C **81**, 911 (2021), 2105.09144
[17] N. Abgrall et al. (NA61/SHINE), Eur.Phys.J. **C74**, 2794 (2014), 1310.2417
[18] A. Aduszkiewicz et al. (NA61/SHINE), Eur. Phys. J. **C77**, 671 (2017), 1705.02467
[19] A. Aduszkiewicz et al. (NA61/SHINE), Phys. Rev. C **102**, 011901 (2020), 1912.10871
[20] A. Aduszkiewicz et al. (NA61/SHINE), Eur. Phys. J. C **80**, 460 (2020), 2001.05370
[21] A. Aduszkiewicz et al. (NA61/SHINE), Eur. Phys. J. C **80**, 199 (2020), 1908.04601
[22] A. Aduszkiewicz et al. (NA61/SHINE), Eur. Phys. J. **C76**, 198 (2016), 1510.03720
[23] A. Aduszkiewicz et al. (NA61/SHINE), Eur. Phys. J. C **80**, 833 (2020), 2006.02062
[24] V. Vovchenko, H. Stoecker, Comput. Phys. Commun. **244**, 295 (2019), 1901.05249

Fluctuations in heavy ion collisions and global conservation effects

Roman Poberezhnyuk[1,2,*], *Volodymyr* Vovchenko[3,4,2], *Oleh* Savchuk[1,5,2], *Volker* Koch[4], *Mark* Gorenstein[1,2], and *Horst* Stoecker[2,6,5]

[1]Bogolyubov Institute for Theoretical Physics, 03680 Kiev, Ukraine
[2]Frankfurt Institute for Advanced Studies, D-60438 Frankfurt am Main, Germany
[3]Institute for Nuclear Theory, University of Washington, Box 351550, Seattle, WA 98195, USA
[4]Lawrence Berkeley National Laboratory, 1 Cyclotron Road, Berkeley, CA 94720, USA
[5]GSI Helmholtzzentrum für Schwerionenforschung GmbH, Planckstr. 1, D-64291 Darmstadt, Germany
[6]Institut für Theoretische Physik, Goethe Universität Frankfurt, D-60438 Frankfurt am Main, Germany

Abstract. Subensemble is a type of statistical ensemble which is the generalization of grand canonical and canonical ensembles. The subensemble acceptance method (SAM) provides general formulas to correct the cumulants of distributions in heavy-ion collisions for the global conservation of all QCD charges. The method is applicable for an arbitrary equation of state and sufficiently large systems, such as those created in central collisions of heavy ions. The new fluctuation measures insensitive to global conservation effects are presented. The main results are illustrated in the hadron resonance gas and van der Waals fluid frameworks.

1 Introduction

Fluctuations are essential in studies of the quantum chromodynamics (QCD) phase diagram in first-principle lattice QCD simulations and heavy-ion collision (HIC) experiments. They are sensitive to fine details of interactions, for example to a phase structure including searched-for phase transition and a critical point (CP) of QCD matter [1–6]. Fluctuations are characterized by cumulants κ_n.

In QCD simulations and statistical models, the grand canonical ensemble (GCE) is mostly used. In GCE, cumulants of quantity B distribution can be calculated by taking the derivative of pressure p over the corresponding chemical potentials μ_B:

$$\kappa_n[B] = VT^3\chi_n^B, \quad \chi_n^B \equiv \partial^n(p/T^4)/\partial(\mu_B/T)^n. \tag{1}$$

GCE implies that the total system with volume V is separated in the coordinate space into the observed subsystem with volume V_1 and the rest of the system, which is called reservoir (or thermostat) with volume V_2, $V = V_1 + V_2$. The subsystem and thermostat can freely exchange particles (or charge B). In the total system, the charge is fixed, and fluctuations are absent. However, the charge fluctuates within the subsystem, and we measure fluctuations in this subsystem.

*e-mail: rpoberezhnyuk@bitp.kiev.ua, poberezhnyuk@fias.uni-frankfurt.de

For the observed subsystem to be in GCE, the following requirements must be satisfied: Firstly, V_1 must be much larger than the correlation length, $V_1 \gg \xi$, i.e., thermodynamic limit is realized. Secondly, the subsystem must be a small part of the total system, $V \gg V_1$. In this case, the effects of global charge conservation can be neglected.

The size of the created system in HIC is not large enough to satisfy these two requirements simultaneously. Thus, the results of experimental measurements of cumulants can't be compared directly with grand-canonical predictions [1] and the global charge conservation effects must be taken into account. [2] The developed subensemble acceptance method (AM) [8–10] allows to account for global conservation effects.

2 Subensemble Acceptance Method

Subensemble is a new type of statistical ensemble, a generalization of GCE. It corresponds to a finite subsystem in contact with a finite thermostat. We partition a thermal system with fixed conserved charge B and temperature T into two subsystems. [3] The Hamiltonian H can be written as $H = H_1 + H_2$. Then the canonical partition function of the total system can be written as

$$Z(T,V,B) = \sum_{B_1} Z(T,\alpha V,B_1)Z(T,\beta V,B-B_1), \qquad \alpha \equiv V_1/V, \qquad \beta \equiv 1-\alpha. \tag{2}$$

We express cumulants κ_n^B of fluctuations inside the subsystem through the corresponding grand canonical susceptibilities, $\chi_n^B \equiv \partial^n(p/T^4)/\partial(\mu_B/T)^n$. This can be done in the thermodynamic limit [8], yielding:

$$\kappa_1[B_1] = \alpha VT^3\chi_1^B, \quad \kappa_2[B_1] = \alpha VT^3\beta\chi_2^B, \quad \kappa_3[B_1] = \alpha VT^3\beta(1-2\alpha)\chi_3^B,$$
$$\kappa_4[B_1] = \alpha VT^3\beta\left[(1-3\alpha\beta)\chi_4^B - 3\,\alpha\,\beta\,\frac{(\chi_3^B)^2}{\chi_2^B}\right]. \tag{3}$$

The model-independent formulas (3) provide corrections for global charge conservation. In the case $\alpha \to 0$, the charge conservation effects are absent, and the cumulants κ_n reduce to their GCE limit (1). In the opposite case $\alpha \to 1$, cumulants approach the limit of the canonical ensemble where fluctuations are absent.

See Ref. [8] for results on cumulants up to 6th order, as well as on special cumulant ratios: scaled variance $\omega \equiv \kappa_2/\kappa_1$, skewness $S\sigma \equiv \kappa_3/\kappa_2$, and kurtosis $\kappa\sigma^2 \equiv \kappa_4/\kappa_2$.

3 Applications and Extensions

The subensemble acceptance analytic formulas have been checked [9] for a specific example of the van der Waals (vdW) mean-field model of interacting particles [11]. The model contains a first-order liquid-gas phase transition and a CP. Scaled variance and kurtosis of fluctuations inside the subsystem as functions of α are shown in Fig. (1).

[1] Other technical issues present in the experiment also hinder the direct comparison. Here we focus only on the effects of global conservation of QCD charges.

[2] Note that in HIC, event-by-event fluctuations are measured in momentum space rather than coordinate space. Therefore, for the method's applicability, the momentum-space correlations must be substantial due to collective flow. We claim that the strength of space-momentum correlations is sufficient, at least at high collision energies of the Large Hadron Collider, LHC. For the estimation of effects of thermal smearing and imperfect space-momentum correlations, see Ref. [7].

[3] Assuming the subsystem is large, we neglect surface effects.

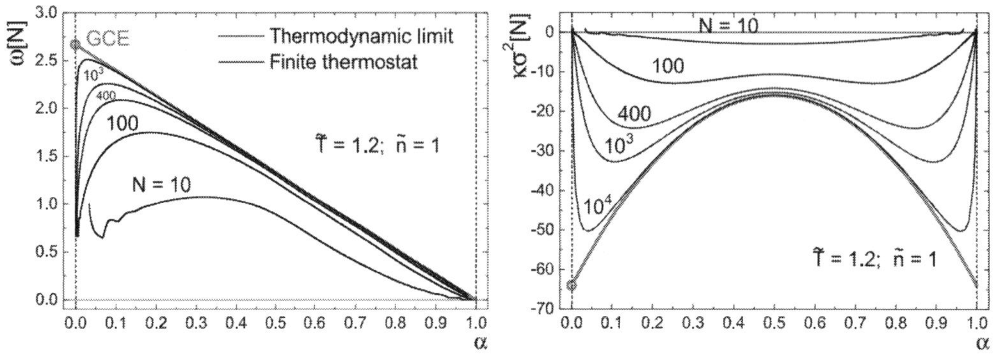

Figure 1. Scaled variance (left) and kurtosis (right) of particle number (conserved charge) fluctuations as functions of the parameter α. The red lines are obtained from the SAM formulas (3). The black lines represent the numeric results calculated explicitly from the vdW partition function [12] using Eq. (2) for different values of the total number of particles in the system N_0. The red dots at $\alpha = 0$ represent GCE values [13]. Taken from [9].

We see that observed fluctuations strongly depend on α, which displays the influence of global particle number conservation. One also sees that for a realistic number of baryon charge $B = 400$ encountered in HICS, the analytic formulas give a good approximation of the numeric results even for higher order cumulants, and even in the vicinity of the CP ($n = n_{crit}$, $T = 1.2\, T_{crit}$) where the correlation length is large. [4] [5]

The subensemble acceptance formulas were used to make lattice QCD-based predictions for baryon number fluctuations in HICs at LHC energies [8]. The rapidity acceptance window of unity, $\Delta Y_{acc} \approx 1$, was found to be a sweet spot where at the same time, the measurements of fluctuations are sensitive to the equation of state of QCD matter, and finite size effects are small.

The SAM formulas have been generalized to the presence of an arbitrary number of independent conserved charges [10]. As an example, here the 4th order cumulant of baryon number fluctuations inside the subsystem is presented in the case when there are two conserved charges (baryon number B and electric charge Q):

$$\kappa_4[B_1] = \alpha V T^3 \beta \left[(1 - 3\alpha\beta)\chi_4^B - 3\,\alpha\beta\, \frac{(\chi_3^B)^2\chi_2^Q - 2\chi_{21}^{BQ}\chi_{11}^{BQ}\chi_3^B + (\chi_{21}^{BQ})^2\chi_2^B}{\chi_2^B\chi_2^Q - (\chi_{11}^{BQ})^2} \right]. \quad (4)$$

Formulas are modified compared to the case of a single conserved charge (3), namely, the mixed (off-diagonal) grand-canonical susceptibilities χ^{BQ} appear. For the SAM formulas in case of arbitrary order of cumulant and number of conserved charges as well as for non-conserved quantities such as, for example, net proton number, see Ref. [10].

We show [10] that α-dependence cancels out in *(i)* any ratio of second order cumulants, *(ii)* any ratio of third order cumulants *(iii)* ratios of mixed second order cumulants involving non-conserved quantities and conserved charge, *(iv)* ratio Σ/Δ involving so-called strongly

[4] Also, SAM formulas were recently tested in a similar setting for molecular dynamics simulations of the classical Lennard-Jones fluid, where similar results were obtained [14].

[5] The present study does not consider fluctuations in the region below spinodals of the first-order phase transition. In Ref. [15], fluctuations in GCE below spinodals were addressed with a method similar to SAM. The subensemble results for the mixed phase region are in preparation [16].

intensive fluctuation measures [17] Σ and Δ. For example,

$$\frac{\kappa_2^Q}{\kappa_2^B} = \frac{\chi_2^Q}{\chi_2^B}, \quad \frac{\kappa_3^Q}{\kappa_3^B} = \frac{\chi_3^Q}{\chi_3^B}, \quad \frac{\kappa^{BQ}}{\kappa_2^S} = \frac{\chi^{BQ}}{\chi_2^S}, \quad \frac{\kappa_{p\hat{Q}_i}}{\kappa_{\hat{Q}_i\hat{Q}_j}} = \frac{\chi_{p\hat{Q}_j}}{\chi_{\hat{Q}_i\hat{Q}_j}}, \quad \frac{\kappa_{p\hat{Q}_j}}{\kappa_{p\hat{Q}_i}} = \frac{\chi_{p\hat{Q}_j}}{\chi_{p\hat{Q}_i}}, \quad \frac{\kappa_{p\hat{Q}_j}}{\kappa_{k\hat{Q}_i}} = \frac{\chi_{p\hat{Q}_j}}{\chi_{k\hat{Q}_i}}. \quad (5)$$

Here S is strangeness, p and k denote respectively net proton and net kaon numbers, and $\hat{Q}_i$, $\hat{Q}_j$ are arbitrary conserved charges. These ensemble-independent fluctuation measures are not sensitive to global conservation and are thus particularly convenient for experimental measurements.

4 Summary

The subensemble acceptance method provides general formulas to correct cumulants of distributions in heavy-ion collisions for the global conservation of all QCD charges. Explicit expressions for all diagonal and off-diagonal cumulants measured in the subsystem of the thermal system that relate them to the grand canonical susceptibilities are obtained for an arbitrary equation of state with an arbitrary number of different conserved charges. Presented fluctuation measures are insensitive to global conservation and, thus, can be particularly convenient for experimental measurement. The method can be extended for application at intermediate collision energies and supplemented with formulas for higher-order fluctuations of non-conserved quantities.

References

[1] M. A. Stephanov, K. Rajagopal and E. V. Shuryak, Phys. Rev. Lett. **81**, 4816-4819 (1998), hep-ph/9806219.

[2] M. A. Stephanov, K. Rajagopal and E. V. Shuryak, Phys. Rev. D **60**, 114028 (1999), hep-ph/9903292.

[3] C. Athanasiou, K. Rajagopal and M. Stephanov, Phys. Rev. D **82**, 074008 (2010) 1006.4636.

[4] M. A. Stephanov, Phys. Rev. Lett. **102**, 032301 (2009), 0809.3450.

[5] V. Vovchenko, D. V. Anchishkin, M. I. Gorenstein and R. V. Poberezhnyuk, Phys. Rev. C **92** (2015) no.5, 054901, 1506.05763.

[6] R. Poberezhnyuk, V. Vovchenko, A. Motornenko, M. I. Gorenstein and H. Stoecker, Phys. Rev. C **100**, no.5, 054904 (2019), 1906.01954.

[7] V. Vovchenko and V. Koch, Phys. Rev. C **103**, no.4, 044903 (2021), 2012.09954.

[8] V. Vovchenko, O. Savchuk, R. V. Poberezhnyuk, M. I. Gorenstein and V. Koch, Phys. Lett. B **811**, 135868 (2020), 2003.13905.

[9] R. V. Poberezhnyuk, O. Savchuk, M. I. Gorenstein, V. Vovchenko, K. Taradiy, V. V. Begun, L. Satarov, J. Steinheimer and H. Stoecker, Phys. Rev. C **102**, no.2, 024908 (2020), 2004.14358.

[10] V. Vovchenko, R. V. Poberezhnyuk and V. Koch, JHEP **10**, 089 (2020), 2007.03850.

[11] V. Vovchenko, D. V. Anchishkin and M. I. Gorenstein, Phys. Rev. C **91**, no.6, 064314 (2015), 1504.01363.

[12] W. Greiner, L. Neise, and H. Stoecker, Thermodynamics and statistical mechanics, Springer Science Business Media (2012).

[13] V. Vovchenko, R. V. Poberezhnyuk, D. V. Anchishkin and M. I. Gorenstein, J. Phys. A **49**, no.1, 015003 (2016), 1507.06537.

[14] V. A. Kuznietsov, O. Savchuk, M. I. Gorenstein, V. Koch and V. Vovchenko, Phys. Rev. C **105**, no.4, 044903 (2022), 2201.08486.

[15] R. V. Poberezhnyuk, O. Savchuk, M. I. Gorenstein, V. Vovchenko and H. Stoecker, Phys. Rev. C **103**, no.2, 024912 (2021), 2011.06420.

[16] V. A. Kuznietsov, O. Savchuk, R. V. Poberezhnyuk, V. Vovchenko, M. I. Gorenstein, in progress.

[17] M. I. Gorenstein and M. Gazdzicki, Phys. Rev. C **84**, 014904 (2011), 1101.4865.

Seventh and eighth-order cumulants of net-proton multiplicity distributions in heavy-ion collisions at RHIC-STAR

Ashish Pandav (for the STAR Collaboration)[1,2,*]

[1]School of Physical Sciences, National Institute of Science Education and Research, Jatni 752050, India
[2]Homi Bhabha National Institute, Training School Complex, Anushaktinagar, Mumbai 400094, India

Abstract. We report the first measurements of seventh and eighth-order cumulants of net-proton distributions in Au+Au collisions at $\sqrt{s_{NN}}$ = 27, 54.4, and 200 GeV. The measurements are performed at mid-rapidity $|y| < 0.5$ within $0.4 < p_T < 2.0$ GeV/c using the Time Projection Chamber and Time-of-Flight detector. Motivation for the measurements comes from lattice-QCD and QCD based model calculations that predict their negative signs for a crossover quark-hadron transition. While 0-40% centrality measurements at $\sqrt{s_{NN}}$ = 54.4 and 200 GeV are consistent with zero within large uncertainties, at $\sqrt{s_{NN}}$ = 27 GeV, they are negative with $\leq 1.4\sigma$ significance. The peripheral 70-80% measurements are either positive or consistent with zero for the three energies.

1 Introduction

The phase diagram of strongly interacting matter is called as the Quantum-Chromo-Dynamics (QCD) phase diagram [1]. It has at least two distinct phases: hadronic phase where quarks and gluons are confined within hadrons and the quark-gluon plasma (QGP) phase where they are deconfined. The transition between the two phases at vanishing baryonic chemical potential (μ_B) is shown to be a smooth crossover by first-principle lattice QCD calculations [2]. At large μ_B, several QCD-based model calculations indicate a first-order phase transition terminating at a QCD critical point [3, 4]. Experimental exploration of the QCD phase diagram forms one of the primary goals of the heavy-ion collision experiments. Higher-order cumulants of event-by-event net-particle distributions have been suggested as sensitive observables in this regard [5, 6]. Ratio of the measured cumulants are constructed to eliminate the system volume dependence and facilitate a direct comparison with ratio of susceptibilities (χ_n) calculated from lattice QCD, QCD-based models, and thermal models [7, 8].

The STAR experiment at RHIC has measured cumulants (C_n) of net-proton (as a proxy for net-baryon) distributions up to sixth-order in Au+Au collisions from $\sqrt{s_{NN}}$ = 3 – 200 GeV [9–13]. A non-monotonic collision energy dependence of net-proton cumulant ratio C_4/C_2 was observed, which is consistent with a model calculation that includes a critical point [9]. Furthermore, the net-proton C_6/C_2 shows an increasingly negative sign with decreasing collision energy in the range of $\sqrt{s_{NN}}$ = 7.7 – 200 GeV [13]. The negative values, albeit with large uncertainties, and the energy dependence are consistent with lattice QCD calculation (μ_B < 110 MeV) that includes a crossover quark-hadron transition [6].

*e-mail: ashish.pandav@niser.ac.in

The study presented in these proceedings extends the net-proton cumulant measurements to even higher orders, i.e. C_7/C_1 and C_8/C_2. The lattice-QCD and various QCD-based models predict negative seventh and eighth-order net-baryon susceptibilities for a crossover quark-hadron transition at vanishing baryonic chemical potential near the transition temperature $T = T_{pc}$ with larger magnitude compared to those at fifth- and sixth-orders [6, 14, 15]. Calculations from a model, called the Polyakov-loop extended quark-meson (PQM) model [15], presented in Fig 1 shows that χ_8^B is an order more negative than χ_6^B at $T = T_{pc}$ for vanishing chemical potential ($\mu_q \sim 0$). In addition to search for crossover in QCD phase diagram, net-proton cumulant ratios C_7/C_1 and C_8/C_2 can also be used to study the validity of hadron resonance gas (HRG) models: Canonical Ensemble (CE) vs. the Grand Canonical Ensemble (GCE) frameworks. In the CE approach [16], the cumulant ratios show a departure from unity and a strong collision energy dependence as compared to GCE [6], where they are unity across all collision energies.

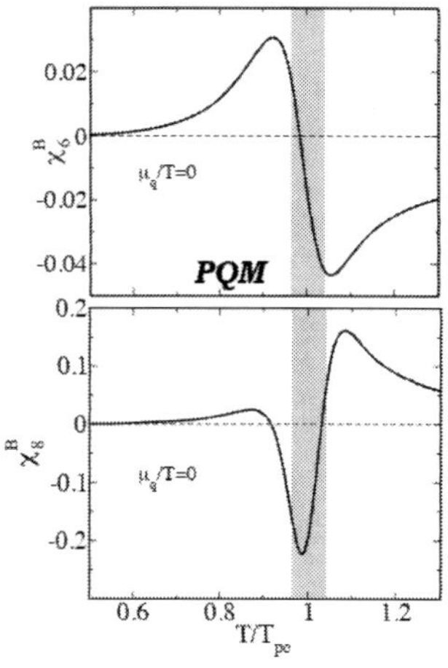

Figure 1. χ_6^B and χ_8^B from the PQM model as a function of T/T_{pc} [15].

2 Analysis Details

Data from Au+Au collisions at $\sqrt{s_{NN}} = 27$ (year 2018), 54.4, and 200 GeV recorded by the STAR detector were analysed. The number of minimum bias events for $\sqrt{s_{NN}} = 27$, 54.4, and 200 GeV are around 300, 550, and 900 millions, respectively. The detectors used for (anti-)proton identification are the Time-Projection-Chamber (TPC) and Time-of-Flight (TOF). The charged particle multiplicity in the psuedo-rapidity (η) range $|\eta| < 1$ excluding protons and anti-protons were used to define centrality to avoid self-correlation effects. The (anti-)protons at mid-rapidity ($|y| < 0.5$) within the transverse momentum (p_T) coverage of $0.4 < p_T < 2.0$ GeV/c were used for measurements. In the p_T range of $0.4 < p_T < 0.8$ GeV/c, only the TPC was used to select (anti-)protons whereas both TPC and TOF were required for (anti-)proton identification in the higher momentum region. To suppress the effect of initial system volume fluctuations on cumulants, Centrality-Bin-Width-Correction (CBWC) was applied [17]. To correct the cumulants for finite detection efficiency, an analytical correction was performed where the detector response was assumed to follow binomial distribution [18]. For estimating statistical uncertainties, bootstrap method was used [19, 20]. Systematic uncertainties on the measurements were estimated varying tracking efficiency, track selection, and particle identification criteria.

3 Results

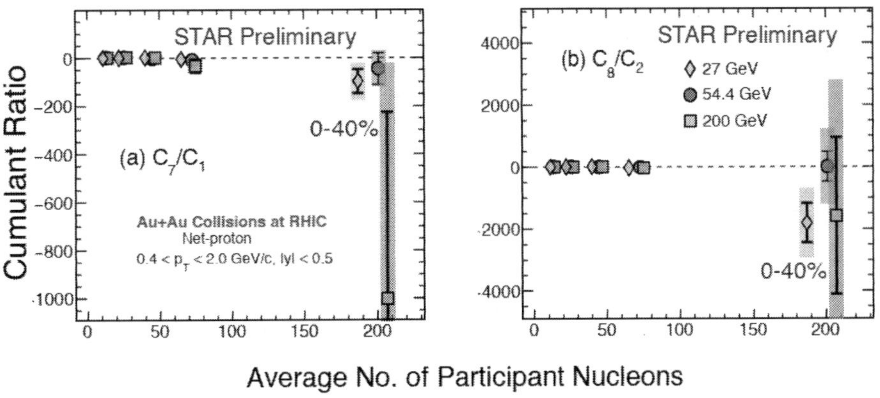

Figure 2. C_7/C_1 (a) and C_8/C_2 (b) of net-proton distributions in Au+Au collisions at $\sqrt{s_{NN}} = 27$ (diamonds), 54.4 (circles), and 200 GeV (squares) as a function of average number of participant nucleons. The bars and shaded bands on the data points represent the statistical and systematic uncertainties, respectively. Results from 70-80%, 60-70%, 50-60%, 40-50%, and 0-40% centrality bins are presented (from left to right).

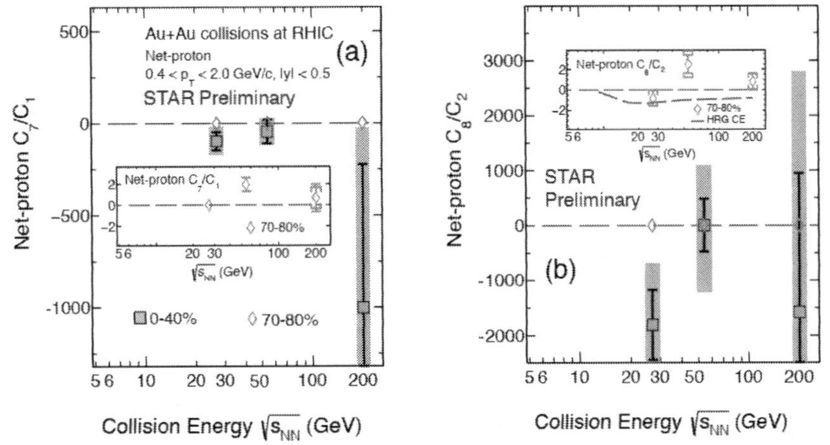

Figure 3. C_7/C_1 (a) and C_8/C_2 (b) of net-proton distributions from Au+Au collisions as a function of collision energy. The results for two centrality bins, 0-40% (filled squares) and 70-80% (open diamonds), are presented. The bars represent the statistical uncertainties. The shaded bands on 0-40% data points and caps on the 70-80% data represent systematic uncertainties. Insets are added in both panel containing peripheral 70-80% data for better visibility. The HRG CE calculations for C_8/C_2 [16] are also shown.

The collison centrality dependence of net-proton C_7/C_1 and C_8/C_2 for $\sqrt{s_{NN}} = 27$, 54.4, and 200 GeV is presented in Fig. 2. A weak dependence on collision centrality is observed within large uncertainties. Larger width of net-proton distributions at $\sqrt{s_{NN}} = 200$ GeV contributes to larger statistical uncertainties at 200 GeV as compared to other two energies. The cumulant ratios for 0-40% centrality at $\sqrt{s_{NN}} = 54.4$ and 200 GeV are consistent with zero within uncertainties. At $\sqrt{s_{NN}} = 27$ GeV, the results for 0-40% centrality are negative with a significance of $\leq 1.4\sigma$ (where σ is the quadrature sum of statistical and systematic uncertainties). The peripheral data, as seen more clearly in Fig. 3, are close to zero.

The energy dependence of net-proton C_7/C_1 and C_8/C_2 for 0-40% and 70-80% centralities are presented in Fig. 3. Within large uncertainties, no clear energy dependence could be observed for the 0-40% measurements. The peripheral data are found to be either positive or consistent with zero within uncertainties.

4 Summary

In summary, we presented the first measurements on seventh and eighth-order net-proton cumulant ratios: C_7/C_1 and C_8/C_2 in Au+Au collisions at $\sqrt{s_{NN}}$ = 27, 54.4, and 200 GeV. While lattice QCD and QCD-based model predicted negative sign of these observables for a smooth crossover, the experimental data at $\sqrt{s_{NN}}$ = 54.4 and 200 GeV are consistent with zero for central 0-40% collisions within large uncertainties. Both the cumulant ratios (for 0-40%) at 27 GeV deviate from unity and show negative values at a significance of $\leq 1.4\sigma$. The peripheral 70-80% data are either positive or consistent with zero for all the energies presented. Measurements in the lower collision energies from the upcoming high statistic BES-II data [21] will be interesting. In addition, $\sim$ 20 billion Au+Au minimum bias collision events at $\sqrt{s_{NN}}$ = 200 GeV which are scheduled to be collected at STAR in the year 2023+2025 [22], will greatly enhance the statistical precision of the measurements.

References

[1] A. Pandav, D. Mallick and B. Mohanty, Prog. Part. Nucl. Phys. **125** (2022), 103960
[2] Y. Aoki, G. Endrodi, Z. Fodor, S. D. Katz and K. K. Szabo, Nature **443**, 675-678 (2006)
[3] S. Ejiri, Phys. Rev. D **78**, 074507 (2008)
[4] E. S. Bowman and J. I. Kapusta, Phys. Rev. C **79**, 015202 (2009)
[5] M. A. Stephanov, Phys. Rev. Lett. 107, 052301 (2011).
[6] A. Bazavov, *et al.* Phys. Rev. D **101**, no.7, 074502 (2020)
[7] S. Gupta, X. Luo, B. Mohanty, H. G. Ritter and N. Xu, Science **332**, 1525 (2011)
[8] F. Karsch and K. Redlich, Phys. Lett. B **695**, 136-142 (2011)
[9] J. Adam et al. *et al.* [STAR], Phys. Rev. Lett. **126**, 092301 (2021)
[10] M. Abdallah *et al.* [STAR], Phys. Rev. C **104**, no.2, 024902 (2021)
[11] M. Abdallah *et al.* [STAR], Phys. Rev. Lett. **127**, no.26, 262301 (2021)
[12] M. S. Abdallah *et al.* [STAR], Phys. Rev. Lett. **128**, no.20, 202303 (2022)
[13] [STAR], [arXiv:2207.09837 [nucl-ex]].
[14] W. j. Fu *et al.* , Phys. Rev. D **104**, 094047 (2021)
[15] B. Friman *et al.*, Eur. Phys. J. C **71** (2011), 1694
[16] P. Braun-Munzinger *et al.*, Nucl. Phys. A **1008**, 122141 (2021)
[17] X. Luo, J. Xu, B. Mohanty and N. Xu, J. Phys. G **40**, 105104 (2013)
[18] T. Nonaka, M. Kitazawa and S. Esumi, Phys. Rev. C **95**, no. 6, 064912 (2017)
[19] X. Luo, J. Phys. G **39**, 025008 (2012)
[20] A. Pandav, D. Mallick and B. Mohanty, Nucl. Phys. A **991**, 121608 (2019)
[21] STAR BES-II White paper [STAR], https://drupal.star.bnl.gov/STAR/starnotes/public/sn0598
[22] STAR Beam Use Request [STAR], https://drupal.star.bnl.gov/STAR/starnotes/public/sn0773

Deciphering the phases of QCD matter with fluctuations and correlations of conserved charges

Anar Rustamov[1,2,*]

[1]GSI Helmholtzzentrum für Schwerionenforschung, Darmstadt, Germany
[2]National Nuclear Research Center, Baku, Azerbaijan

Abstract. A review is given on recent experimental and theoretical/phenomenological developments regarding the phase structure of the strongly interacting matter. Specifically, evolution with collision energy of net-proton number fluctuations as measured by several experiments are presented and their implications for the QCD phase diagram are outlined. In addition, theoretical calculations on correlations between conserved charges are presented and prospects for their experimental explorations are addressed.

1 Introduction

The quest for the phase structure of strongly interacting matter, hereinafter referred to as Quantum Chromodynamics (QCD) matter, remains at the focus of contemporary theoretical and experimental investigations. By know it is well established that, in nuclear collisions at cm energies from several TeV down to about 12 GeV per nucleon pair the matter created in head-on collisions of heavy ions freezes out close to the chiral phase transition line, taking place at a pseudo-critical temperature values of about 156 MeV [1, 2]. This opens unique experimental opportunities to probe the strongly interacting matter close to the phase boundary, such as experimental verification of its Equation of State (EoS). In this energy range the lattice QCD (lQCD) calculations predict crossover chiral transition [2], which still awaits experimental confirmations. Contrary to the high energy region, first principle calculations at low energies (high net-baryon densities) are still not available, however a number of effective models predict, that at reasonably large net-baryon densities the QCD matter undergoes a first order chiral phase transition [3–7]. This lends support to the existence of the chiral Critical End Point (CEP) at which the anticipated first order phase transition line terminates and smooth crossover transition sets in. Hence at the CEP the system would undergo a second order phase transition. One of the main goals of experiments at lower energies is to locate the position of the chiral CEP in the QCD phase diagram.

Fluctuations of conserved charges, such as baryon number, electric charge, strangeness and charm, as well as correlations between them are identified as promising tools to probe the chiral criticality. The objective is to characterise the response of the system to external perturbations, such as infinitesimal changes in the chemical potentials. In theoretical calculations, within the Grand Canonical Ensemble (GCE) of statistical mechanics, for a system of fixed volume V and temperature T this task is performed by evaluating the nth order susceptibilities

*e-mail: a.rustamov@gsi.de

χ_q^n, i.e., derivatives of the thermodynamic pressure P with respect to the chemical potentials μ_q responsible for the conservation of the corresponding charge q on average, evaluated at vanishing values of chemical potentials

$$\hat{\chi}_n^q = \left.\frac{\partial^n \hat{P}}{\partial \hat{\mu}_q^n}\right|_{\vec{\mu}=0} = \frac{\kappa_n(N_q)}{VT^3}, \tag{1}$$

where $\hat{\chi}_n^q$, $\hat{P} = P(T, V, \vec{\mu})/T^4$ and $\hat{\mu}_q = \mu_q/T$ denote reduced susceptibility, reduced pressure and reduced chemical potential, respectively. In experiments the cumulants are directly measured in a given detector acceptance either by first reconstructing the multiplicity distributions of the corresponding net-charge number (N_q) or by using probabilistic approaches based on the detector response functions for particle identification (cf. [8, 9] and references therein). The so obtained experimental cumulants $\kappa_n(N_q)$ can be directly compared to those computed via derivatives of thermodynamic pressure (cf. Eq. 1), provided that the system under consideration is in thermal equilibrium and all necessary instrumental effects and non critical signals, such as those stemming from detection efficiencies, participant fluctuations, suppression of fluctuations due to e-by-e conservation laws and resonance decays are properly accounted for [10–13].

2 Experimental Results

In this section I briefly discuss experimental measurements related to crossover and critical point studies, provide their comparison with the model calculations and implications for the QCD phase diagram.

2.1 Search for a critical point

Fig. 1 shows the evolution with collision energy of the fourth to seconder order cumulants of net-proton distributions[1] as measured by the STAR [14, 15] (black star symbols) and HADES [16] (magenta box) collaborations. The data points are contrasted with calculations within two different ensembles of statistical mechanics [13, 17, 18]: (i) Grand Canonical Ensemble (GCE) with the ideal gas EoS in the Boltzmann limit, represented with the dashed line along unity; (ii) Canonical Ensemble (CE), implemented in the full phase space, also using the ideal gas EoS. In the following these are referred to as the GCE or Hadron Resonance Gas (HRG) baseline and the CE baseline, respectively. Such treatment of cumulants within CE ensures that in each event the baryon number (number of baryons minus number of antibaryons) is exactly conserved in the full phase space, inducing strict correlations between baryon and anti-baryon numbers. As a consequence, since there are essentially no antibaryons at low energies, the baryon number does not fluctuates in the full phase space, leading to a binomial distributions in the finite acceptance at low energies (for more details see the Appendix C of Ref. [13])[2]. This, however, apples if wounded nucleons [19] do not fluctuate from event-to-event, while in practice they do [11, 20]. Thus, non-critical contributions to the finally measured (net-)proton cumulants, stemming from fluctuations of wounded nucleons, need to be corrected for. This is in particular essential at low energies because, in addition to the above mentioned arguments regarding binomial multiplicity distributions, most of these contributions scale with the net-proton number. For example, at LHC energies, where net-proton number at mid-rapidity is practically zero, contributions from wounded nucleon fluctuations

[1] At low collisions energies these are essentially cumulants of proton number.

[2] At high energies baryons and anti-baryons fluctuate in the full phase space, only net-baryon number does not.

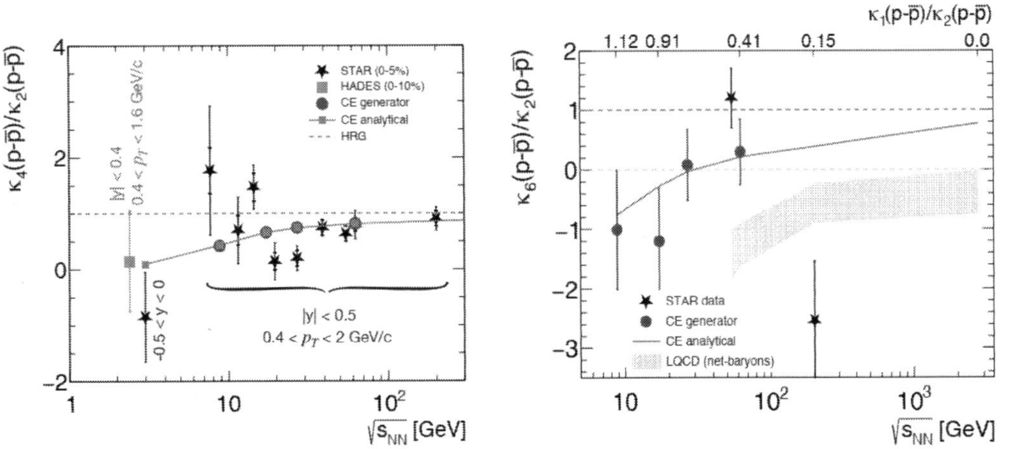

Figure 1. Left panel: evolution with the collision energy of κ_4/κ_2 measured for net-proton distributions. The STAR data, represented with black star symbols, are measured for |y|<0.5 (-0.5 <y < 0 for the lowest energy) and 0.4 < p_T[GeV/c] < 2. The HADES measurement (magenta square symbol) is for |y| < 0.4 and 0.4 < p_T[GeV/c] < 1.6. Also presented are two non-critical baselines; the HRG baseline: indicated with the dashed line at unity, the CE baseline: shown by solid the red line/boxes (analytical calculations) and blue symbols (generated values). Right panel: Similar to the left panel for κ_6/κ_2. The corresponding lQCD results for net-baryons are presented with the hashed cyan region.

are immaterial for the measured second order cumulants of net-protons, while at lower energies these contributions are substantial [11, 15, 16]. Coming back to Fig. 1, both analytic (red line/boxes) and generated (blue symbols) results of the CE baseline are presented there. It is worthwhile to mention that, ingredients of the CE calculations, such as single particle partition functions are estimated using the experimentally measured mean multiplicities of protons, which are "naturally" treated as CE multiplicities. Also, it is important to note that the analytic calculations and generated values fully incorporate experimental conditions, such as acceptances both in rapidity and transverse momentum coverage. Fig 1 shows that correlations induced by the baryon number conservation in full phase space survive in the finite experimental acceptance thereby suppressing, with respect to the GCE baseline, the net-proton cumulants. Clearly this kind of modifications of cumulants do not originate from searched for critical phenomena induced by phase transitions or existence of a critical point. The CE results represent more realistic ("proper") baseline. The critical behaviour due the presence of the critical point is predicted based on universality arguments and suggest non-monotonic behaviour of the order parameter as a function of collision energy, which should led to non-monotonic, with respect to a "proper" baseline, energy dependence of fourth order cumulants for proton or net-proton number distributions [21]. Such a non-monotonic behaviour with the significance of 3.1σ is reported by the STAR collaboration [14], albeit with respect to the GCE baseline. However, this conclusion changes if one considers the "proper" baseline. Indeed, within the uncertainties, the experimentally measured energy dependence of the κ_4/κ_2 ratio, presented in the left panel of Fig. 1, are consistent with the CE, i.e., noncritical baseline, as reported in Ref. [13]. The presented experimental measurements look promising, however the precision of the data does not allow to draw firm conclusions regarding the existence of the first order phase transition or related CEP. This situation will improve significantly with the available high statistics data from the STAR experiment, the ongoing analysis at NA61/SHINE and future measurements at the CBM experiment [22].

2.2 Search for a crossover transition

The search for the crossover transition has a clear advantage because of its closeness to the genuine second order phase transition belonging to the $O(4)$ universality class in three dimensions, realised in the limit of massless u and d quarks [23]. Indeed, lQCD calculations indicate that critical signals of the chiral limit survive also after accounting for physical light quark masses [24]. On the other hand, the cumulants of the $O(4)$ dynamics significantly differ from those calculated within GCE with the ideal gas EoS [25]. For example, the sixth (also eighth) order cumulants of net-baryons are negative, while in GCE with the ideal gas EoS they remain positive such that $\kappa_6/\kappa_2 = \kappa_8/\kappa_2 = 1$. Here one again needs to be careful, because the CE baseline for κ_6/κ_2 may also become negative starting from some collision energy [13, 26]. Hence deviations of these ratios from the CE baseline towards negative values would be a strong signal for the crossover transition [27–30], still to be confirmed in experiments. In the right panel of Fig. 1 the ratios of the six to second order cumulants of net-protons are presented, as measured by the STAR experiment. Qualitatively the 200 GeV STAR data are indeed consistent with the anticipated negative sign. However, the data point at 54.4 GeV of STAR is at odds with the lQCD calculations [30]. Another interesting measurements would be to test the hierarchy of cumulant rations as predicted by the lQCD calculations for the ratios of baryon number susceptibilities. Such measurements were recently reported by the STAR collaboration for net-protons [31], yet with sizable measurement uncertainties. Future as well as already available data from the STAR and ALICE collaborations will allow for systematic exploration of the crossover region of the QCD phase diagram with significantly enlarged statistics.

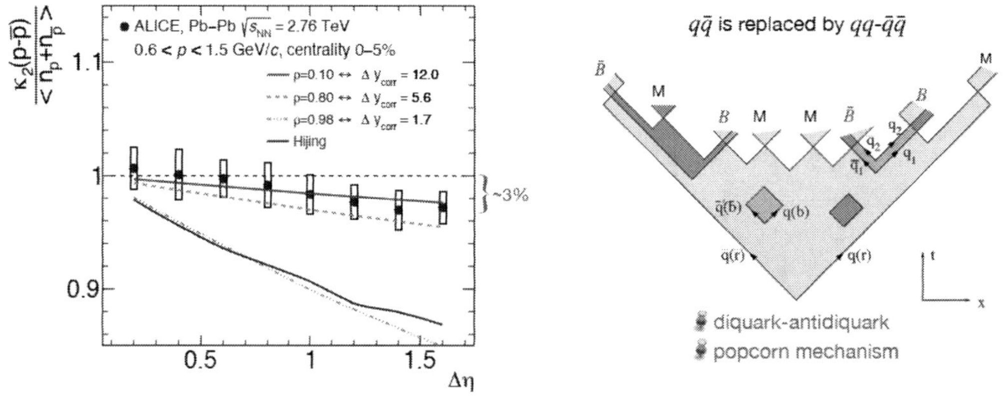

Figure 2. Left panel: scaled second order cumulants of net-proton distributions as a function of the acceptance coverage around mid-rapidity, measured by the ALICE experiment (black symbols) [32, 33]. The colored lines correspond to calculations accounting for baryon number conservation for different values of the correlation coefficient in the rapidity space [18, 34]. The solid black line represents the results from the HIJING event generator. Right panel: a schematic illustration of baryon production mechanisms via the Lund string model.

The acceptance dependence of the normalized second order cumulants of net-proton distributions as measured by the ALICE experiment are presented in the left panel of Fig. 2 [32, 33]. The ALICE data are compared to both GCE and CE baselines [13, 17]. Furthermore, the CE baseline is performed for different correlations between baryons and anti-baryons quantified with the Pearson correlation coefficient ρ and/or the correlation length

Δy_{corr} [18, 34]. In doing so either Cholesky factorisation [35] or the Metropolis algorithm [36] is used. The $\rho = 0$ case corresponds to the global baryon number conservation or long range correlations. As seen from Fig. 2 the experimental measurements are best described with $\rho = 0.1$, i.e., the data suggest long range correlations between baryons and antibaryons in the rapidity space. Thus, after accounting for baryon number conservation the ALICE data on second order cumulants of net-protons is consistent with the corresponding cumulants of the Skellam distribution. This is in line with the lQCD predictions of a Skellam behaviour for the second order cumulants of net-baryon distributions at a pseudo-critical temperature of about 156 MeV. The results from the HIJING event generator, shown with the black solid line in Fig. 2, grossly underestimate experimental measurements. The large discrepancy between the ALICE data and HIJING simulations calls into question the baryon production mechanism used in the Lund string model, namely by producing them via diquark-antidiquark pairs [37], schematically illustrated in the right panel of Fig. 2. The latter leads to a small correlation lengths in the rapidity space, which in turn, as presented in Fig. 2, suppresses fluctuations of net-baryons; but much stronger compared to experimental observations. Indeed, to catch the trend of the HIJING results one needs to introduce rather strong correlation coefficient of about ρ=0.9 (cf. Fig. 2). Such short-range rapidity correlations from models can be softened by exploiting popcorn mechanism for baryon production, in which additional meson (even several mesons) is produced in between the diquark-antidiquark pairs [38]. The presented ALICE measurements on net-proton fluctuations as a function of the rapidity coverage demonstrate that baryon production at high energies is not fully understood and are useful to discriminate between several baryon production mechanisms available in the Lund string model.

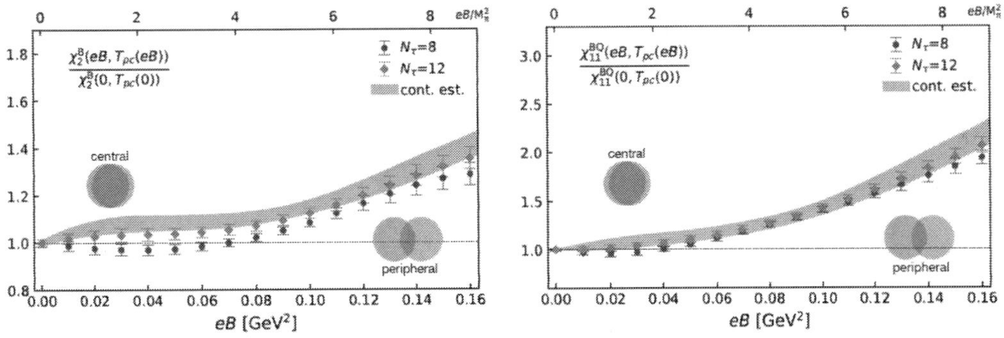

Figure 3. Left panel: normalised second order fluctuations of baryon number along the transition line as a function of the magnetic field strength. Right panel: similar to the left panel for the normalized second order correlations between baryon number and electric charge.

2.3 Fluctuations in the presence of a magnetic field

Fluctuations of conserved charges can serve as ideal probes of a magnetic field in nuclear collisions [39, 40]. First (2+1)-flavor lQCD calculations, with physical quark masses, on the second order susceptibilities for baryon number and correlations between baryon number and electric charge as a function of a magnetic field strength eB are presented in Fig. 3, where fluctuations without magnetic field are used as a reference [41, 42]. As seen from Fig. 3, from central towards peripheral collisions the ratio of susceptibilities progressively deviate from unity, induced by the presence of a magnetic field. The correlations between different conserved charges (right panel of Fig. 3) exhibit even more pronounced sensitivity to the presence of the magnetic field. In experiments the signal can be looked at by measuring net-baryon cumulants or mixed cumulants between different conserved charges as a function of centrality normalized to the results obtained for most central collisions.

References

[1] A. Andronic, P. Braun-Munzinger, K. Redlich, J. Stachel, Nature **561**, 321 (2018), 1710.09425
[2] A. Bazavov et al. (HotQCD), Phys. Lett. B **795**, 15 (2019), 1812.08235
[3] M. Asakawa, K. Yazaki, Nucl. Phys. A **504**, 668 (1989)
[4] K. Rajagopal, F. Wilczek, Nucl. Phys. B **399**, 395 (1993), hep-ph/9210253
[5] M.A. Stephanov, K. Rajagopal, E.V. Shuryak, Phys. Rev. Lett. **81**, 4816 (1998), hep-ph/9806219
[6] M.A. Stephanov, Phys. Rev. D **73**, 094508 (2006), hep-lat/0603014
[7] C. Sasaki, B. Friman, K. Redlich, Phys. Rev. Lett. **99**, 232301 (2007), hep-ph/0702254
[8] A. Rustamov, M.I. Gorenstein, Phys. Rev. C **86**, 044906 (2012), 1204.6632
[9] M. Arslandok, A. Rustamov, Nucl. Instrum. Meth. A **946**, 162622 (2019), 1807.06370
[10] M. Nahrgang, T. Schuster, M. Mitrovski, R. Stock, M. Bleicher, Eur. Phys. J. C **72**, 2143 (2012), 0903.2911
[11] P. Braun-Munzinger, A. Rustamov, J. Stachel, Nucl. Phys. A **960**, 114 (2017), 1612.00702
[12] A. Rustamov, Nucl. Phys. A **1005**, 121858 (2021), 2005.13398
[13] P. Braun-Munzinger, B. Friman, K. Redlich, A. Rustamov, J. Stachel, Nucl. Phys. A **1008**, 122141 (2021), 2007.02463
[14] J. Adam et al. (STAR), Phys. Rev. Lett. **126**, 092301 (2021), 2001.02852
[15] M.S. Abdallah et al. (STAR), Phys. Rev. Lett. **128**, 202303 (2022), 2112.00240
[16] J. Adamczewski-Musch et al. (HADES), Phys. Rev. C **102**, 024914 (2020), 2002.08701
[17] P. Braun-Munzinger, A. Rustamov, J. Stachel, Nucl. Phys. A **982**, 307 (2019), 1807.08927
[18] P. Braun-Munzinger, A. Rustamov, J. Stachel (2019), 1907.03032
[19] A. Bialas, M. Bleszynski, W. Czyz, Nucl. Phys. B **111**, 461 (1976)
[20] V. Skokov, B. Friman, K. Redlich, Phys. Rev. C **88**, 034911 (2013), 1205.4756
[21] M.A. Stephanov, Phys. Rev. Lett. **107**, 052301 (2011), 1104.1627
[22] B. Friman, C. Hohne, J. Knoll, S. Leupold, J. Randrup, R. Rapp, P. Senger, eds., *The CBM physics book: Compressed baryonic matter in laboratory experiments*, Vol. 814 (2011)
[23] R.D. Pisarski, F. Wilczek, Phys. Rev. D **29**, 338 (1984)
[24] S. Ejiri, F. Karsch, K. Redlich, Phys. Lett. B **633**, 275 (2006), hep-ph/0509051
[25] B. Friman, F. Karsch, K. Redlich, V. Skokov, Eur. Phys. J. C **71**, 1694 (2011), 1103.3511
[26] V. Vovchenko, V. Koch, C. Shen, Phys. Rev. C **105**, 014904 (2022), 2107.00163
[27] G.A. Almasi, B. Friman, K. Redlich, Phys. Rev. D **96**, 014027 (2017), 1703.05947
[28] W.j. Fu, X. Luo, J.M. Pawlowski, F. Rennecke, R. Wen, S. Yin, Phys. Rev. D **104**, 094047 (2021), 2101.06035

[29] S. Borsanyi, Z. Fodor, J.N. Guenther, S.K. Katz, K.K. Szabo, A. Pasztor, I. Portillo, C. Ratti, JHEP **10**, 205 (2018), 1805.04445
[30] A. Bazavov et al., Phys. Rev. D **101**, 074502 (2020), 2001.08530
[31] (2022), 2207.09837
[32] A. Rustamov (ALICE), Nucl. Phys. A **967**, 453 (2017), 1704.05329
[33] S. Acharya et al. (ALICE), Phys. Lett. B **807**, 135564 (2020), 1910.14396
[34] A. Rustamov, P. Braun-Munzinger, J. Stachel, *The imprint of conservation laws on correlated particle production*, in *29th International Conference on Ultra-relativistic Nucleus-Nucleus Collisions* (2022)
[35] A.L. Cholesky, Bulletin Géodésique **2**, 66–67 (1924)
[36] N. Metropolis, A.W. Rosenbluth, M.N. Rosenbluth, A.H. Teller, E. Teller, J. Chem. Phys. **21**, 1087 (1953)
[37] B. Andersson, G. Gustafson, G. Ingelman, T. Sjostrand, Phys. Rept. **97**, 31 (1983)
[38] P. Eden, G. Gustafson, Z. Phys. C **75**, 41 (1997), hep-ph/9606454
[39] D.E. Kharzeev, L.D. McLerran, H.J. Warringa, Nucl. Phys. A **803**, 227 (2008), 0711.0950
[40] V. Skokov, A.Y. Illarionov, V. Toneev, Int. J. Mod. Phys. A **24**, 5925 (2009), 0907.1396
[41] H.T. Ding, S.T. Li, Q. Shi, X.D. Wang, Eur. Phys. J. A **57**, 202 (2021), 2104.06843
[42] H.T. Ding, S.T. Li, J.H. Liu, X.D. Wang, *Fluctuations of conserved charges in strong magnetic fields from lattice QCD*, in *29th International Conference on Ultra-relativistic Nucleus-Nucleus Collisions* (2022), 2208.07285

SQM2022: Theory Overview in Heavy-Ion Physics

Joseph Kapusta[1,*]

[1]School of Physics and Astronomy, University of Minnesota, Minneapolis, Minnesota 55455 USA

Abstract. An overview is presented of the many open, interesting questions regarding the behavior of matter at large temperatures and chemical potentials.

1 Introduction

The main goal of colliding large nuclei at high energies is to measure the properties of matter under the conditions of extreme temperature and density. The first heavy ion collision experiments began approximately fifty years ago. Much has been learned, and much more will be learned, in the coming years. In a fictional world, one could produce hot dense matter in the accelerator and place it under a microscope for study. In the real world, theory and experiment must go hand-in-hand to interpet the data and extract physics from it.

It is not feasible to provide a complete review of the full theoretical effort here. Rather, the purpose of this overview is to alert the audience to many open physics issues and how they are addressed by the speakers in this conference.

2 Equation of State

The equation of state of QCD as a function of temperature and density is a fascinating subject with many unanswered questions.

• Is there a critical point in the phase diagram? Many models suggest so. Where is the critical curve?

• If a critical point exists it would be in the same universality class as the 3D Ising model and liquid-gas transitions. How would it be related to a chiral symmetry restoring phase transition, which is in a different universality class, if the up and down quarks were massless?

• To describe the matter created at RHIC beam energies and below, knowledge is required of the equation of state as a function of T, μ_B, μ_Q, and μ_S to conserve energy, baryon number, electric charge, and strangeness. This is nontrivial when there is critical behavior in the phase diagram.

• Such an equation of state is also needed for modeling neutron star mergers and closely related to the cold dense matter comprising neutron stars.

• Will hard working lattice practioners ever be able to provide the answers on account of the infamous sign problem in the partition function?

*e-mail: kapusta@umn.edu

3 Simulations are the Key

Analytic models were very useful at the beginning of the field. Numerical simulations are now essential to make contact between theory and experiment.

• Fluctuations are typically computed in coordinate space but measured experimentally in momentum space. Dynamical simulations are necessary to compare them.

• Photon and dilepton emission are computed theoretically as functions of the thermodynamic variables. They must be folded with the space-time evolution of the matter.

• Jet and charm evolution are computed theoretically as functions of the thermodynamic variables. They must be folded with the space-time evolution of the matter. Feedback on the matter should be taken into account.

• A sampling of publicly available codes includes
Initial state: Glauber, LEXUS, IP-Glasma, Trento
Relativistic hydrodynamics: MUSIC, iEBE-MUSIC, iEBE-VISHNU, SONIC & superSONIC, SPheRIO, CLVISC, vHLLE, ECHO-QGP
Transport theory: UrQMD, ZPC, AMPT, SMASH
Particlization of hydrodynamics: iSS, Microcanonical Cooper-Frye

4 Hydrodynamics and Transport Theory

The "standard model" of relativistic heavy ion collisions includes hydrodynamics, which has proven to be invaluable. However

• Perfect fluid dynamics is not realistic. The system is out of equilibrium. How to define flow velocity? Should it be via energy flow, baryon flow, electric charge flow, or something else? This is not an issue for transport models.

• Is the standard model of viscous hydrodynamics coupled with an hadronic afterburner the correct one for lower beam energies?

• What is the initial condition as it depends on beam energy?

• Transport theory with hadrons and mean fields is undoubtedly a better description at lower beam energies. It appears to be better able to handle longer nuclear transit times.

• How and when to transition from transport theory to hydrodynamics as a function of beam energy, impact parameter, or even at a fixed impact parameter? When does the assumption of (approximate) local equilibrium become reasonable?

• Can a theory of spin-hydrodynamics be realized in numerical simulations?

5 Strangeness

Strange quark and anti-quark production is the reason for the origin and existence of this conference series!

• How does strangeness first appear? As strange quarks and anti-quarks or as strange mesons and baryons? How does it depend on beam energy or impact parameter or even as a function of transverse position in a given collision?

• The differences between Λ and $\bar{\Lambda}$ and between K^+ and K^- are very small at the LHC but become increasingly large as the energy decreases. What interesting physics can we infer?

• The connection between polarized Λ and $\bar{\Lambda}$ baryons, aligned K^{*0} mesons, and vorticity of the matter is extremely interesting. Current models assume their spins are equilibrized but is that correct?

• How does strangeness evolve? Is the strange current directly proportional to the baryon current? If the quarks are polarized, how do they pass that on to the strange mesons and baryons?

6 Charm

Production of charmed mesons and baryons at RHIC and especially at the LHC is an exciting sub-field with many aspects to it. These include, but are not limited to the following:

• Charm production in p+p and p+Pb at the LHC is not well reproduced by fragmentation in PYTHIA, which is based on e^+e^- collisions. Some physics is missing.

• The charm diffusion coefficient is calculated as a function of T on the lattice and detailed modeling can connect it to experimental data. What about the chemical potential dependence at lower beam energies?

• Many new exotic charmed baryons and mesons have been detected at the LHC. Are they tetraquark and pentaquark states or are they molecular states of two hadrons?

• Do charmed nuclei exist?

• K^{*0} vector mesons in Pb+Pb collisions at the LHC are aligned but J/ψ mesons are not. What dynamics is involved?

7 Fluctuations

Relativistic heavy ion collisions can produce thousands, if not tens of thousands, of particles. Even so, fluctuations are important and can provide valuable information on the dynamics of the system and on the equation of state. Sources of fluctuations include:

• Initial state fluctuations from randomly chosen locations of nucleons.

• Initial state fluctuations from randomly chosen locations of quarks within nucleons.

• Initial state fluctuations of color gluon field at very high energy.

• Hydrodynamic fluctuations are intimately connected to transport coefficients due to the fluctuation-dissipation theorem. Inclusion of viscosity and heat conductivity requires them for consistency.

• Hydrodynamic fluctuations in the evolving matter is challenging to implement numerically in 3D.

8 Correlations

Particles can be correlated due to quantum mechanics, conservation laws, and dynamics. Some intriquing issues are:

• Hanbury-Brown and Twiss intensity interferometry provides information on the space-time evolution of the matter. Correlations are readily inferred from experiment, but a rigorous derivation for heavy ion collisions is subtle and lacking.

• Charge balance functions can provide information on the charge diffusivity, but extraction from data requires detailed knowledge and modeling of the space-time evolution of the matter.

• How can we understand theoretically the transition in dynamics from small to medium to large collision systems as defined by size of the beam and target or by multiplicity?

• Can Disoriented Chiral Condensates (DCC) be revived at the LHC?

9 Critical Point and First Order Phase Transition Dynamics

Whether or not QCD has a critical point and associated line of first order phase transition in the temperature - chemical potential plane is a major unanswered question. There are many issues that complicate the matter, including these:

• One may need to be within 10^{-2} or 10^{-3} of T_c or even less to see the true critical exponents. The distance is not universal. Beyond that one typically measures mean field exponents. This is quite a challenge for heavy ion collisions.

• Crossing a line of first order phase transition in a heavy ion collision ought to be more dramatic than passing near a critical point.

• It is uncertain how hydrodynamic models can deal with a first order phase transition. Nucleation is the statistical formation of a critical size bubble or droplet of the other phase in the metastable region, but the critical size may be too large for relevance in a heavy ion collision or may not happen fast enough.

• It is uncertain how hydrodynamic models can deal with a first order phase transition. Spinodal decomposition occurs when the system crosses into the unstable region where $c_T^2 < 0$. Subsequent evolution depends on pre-existing fluctuations and inhomogeneities.

• Transport theories can use mean fields to deal with a two phase system dynamically. But, fundamentally, does one use hadron or quark and gluon degrees of freedom? No matter whether a smooth crossover or a critical point, the issue remains.

10 Neutron Stars and their Mergers

Supernovae may reach temperatures of 50 MeV. Neutron stars may reach energy densities greater than 1 GeV/fm^3 with correspondingly high baryon densities. Mergers of neutron stars, crucially studied with gravitational waves, may equal or exceed both. This is a great synergy among fields of physics and astrophysics.

• The introduction of new degrees of freedom such as strangeness, in the form of quasi-free quarks or hyperons, softens the equation of state by lowering the pressure at the same energy density. This makes it challenging to produce neutron stars with two more solar masses.

• A common way to avoid this challenge is to introduce terms in the effective hadronic Lagrangian which makes it energetically unfavorable for strange degrees of freedom to appear. Can heavy-ion collisions provide quantitative information on these otherwise difficult to constrain terms?

• Interactions between or among nucleons and hyperons are important to constrain the effective Lagrangians. These interactions can be informed by correlations and HBT of those baryon species.

11 Conclusion

Asymptotic freedom was discovered almost fifty years ago. Within two years it was realized that matter should best be described in terms of quark and gluon degrees of freedom at high energy densities. Since then much analytical and numerical work has been done to elucidate the properties of matter at extreme energy densities. Theory and experiment have been driving each other to a state of the field that was almost unthinkable in the 1980's. Keep all this in mind during the conference. What will the theory summary talk [1] have to say?

I thank Sangyong Jeon, Che-Ming Ko, Volker Koch, Berndt Müller, Björn Schenke and Ramona Vogt for advice on preparing this talk. This work was supported by the U.S. DOE Grant No. DE-FG02-87ER40328.

References

[1] B. Müller, theory summary in this conference.

Particle production as a function of underlying-event activity and very forward energy with ALICE

Feng Fan * (for the ALICE Collaboration)

Central China Normal University, Wuhan, Hubei, 430079, China

Abstract. Measurements of charged-particle production in pp and p–Pb collisions at $\sqrt{s_{\rm NN}}$ = 5.02 TeV in the toward, away and transverse regions are discussed. These three regions are defined event by event relative to the track with the largest transverse momentum ($p_{\rm T}^{\rm trig}$). The transverse region is sensitive to the underlying event (UE), but it also includes contributions from initial- and final-state radiation (ISR-FSR). Therefore, it is further subdivided in two regions, defined according to their relative multiplicities: trans-max (the sub-transverse region with the larger multiplicity) and trans-min (the sub-transverse region with the smaller multiplicity) regions which have an enhanced sensitivity to ISR-FSR and UE, respectively. KNO-like scaling properties are explored in the three defined regions. Finally, the relationship between $p_{\rm T}^{\rm trig}$ and the energy detected in a region close to beam rapidity (very forward energy) is reported.

1 Introduction

In models incorporating multi-parton interactions (MPI), particles produced in the hard scattering (jet) are accompanied by particles from additional parton-parton interactions, as well as from the proton break-up [1]. This component of the collision makes up the underlying event (UE). The traditional UE analysis focuses on the study of particles in three topological regions depending on their azimuthal angle relative to the leading particle ($|\Delta\varphi| = |\varphi - \varphi^{\rm trig}|$), which is the one with the highest transverse momentum in the event ($p_{\rm T}^{\rm trig}$ or $p_{\rm T}^{\rm leading}$) at mid-pseudorapidity ($|\eta| < 0.8$). The toward region ($|\Delta\varphi| < \pi/3$ rad) contains the primary jet and UE, while the away region ($|\Delta\varphi| > 2\pi/3$ rad) contains the fragments of the recoil jet and UE. In contrast, the transverse region ($\pi/3$ rad $< |\Delta\varphi| < 2\pi/3$ rad) is dominated by the UE dynamics, but it also includes contributions from initial- and final-state radiation (ISR-FSR) [2, 3]. A study based on Monte Carlo (MC) simulations shows that at the LHC energies the multiplicity distributions in the transverse region obey a Koba-Nielsen-Olesen (KNO) scaling [4]. In this contribution, the transverse region is further subdivided into trans-max and trans-min, defined according to their relative multiplicities, in order to increase the sensitivity to ISR-FSR and UE effects, respectively. Measurements of multiplicity distributions in the trans-min region at different centre-of-mass energy allows for the study of the role of MPI to produce such scaling properties.

Another way to characterise the event is to look into a region in the very forward pseudorapidity ($|\eta| > 8.8$) using the ALICE zero degree calorimeters (ZDC), which measure the beam remnants. Moreover, the relationship between the very forward energy and $p_{\rm T}^{\rm trig}$ is

*e-mail: feng.fan@cern.ch

complementary to the UE measurements, and provides direct insights into the initial stages and the subsequent evolution of the collision. In this contribution, a set of new measurements aimed at understanding the role of MPI in hadronic interactions at the LHC is presented [5, 6].

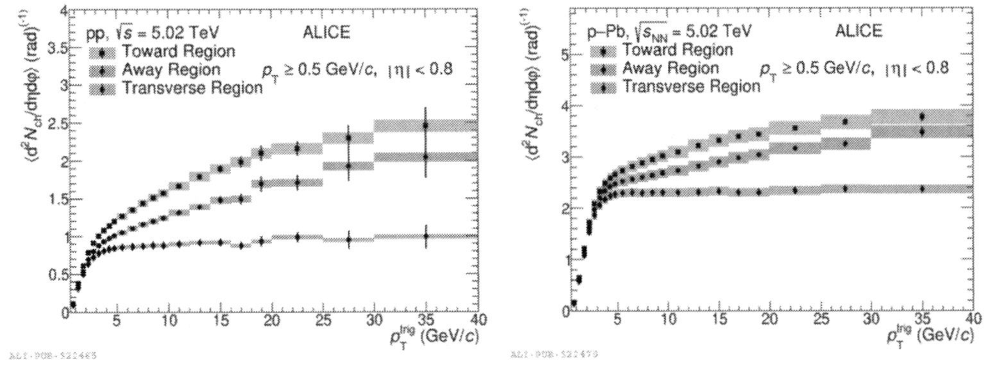

Figure 1: Multiplicity density as a function of $p_{\mathrm{T}}^{\mathrm{trig}}$ measured in pp (left) and p–Pb (right) collisions at $\sqrt{s_{\mathrm{NN}}} = 5.02\,\mathrm{TeV}$. Results for the toward, transverse, and away regions are displayed. The shaded boxes and the error bars represent the systematic and statistical uncertainties, respectively.

2 Results and discussion

Figure 1 shows the charged-hadron multiplicity density as a function of $p_{\mathrm{T}}^{\mathrm{trig}}$ measured in pp and p–Pb collisions at $\sqrt{s_{\mathrm{NN}}} = 5.02\,\mathrm{TeV}$. Results are presented for the toward, away and transverse regions. For both systems, the distributions behave in the same way. At low $p_{\mathrm{T}}^{\mathrm{trig}}$ ($< 5\,\mathrm{GeV}/c$), the multiplicity densities rapidly increase with $p_{\mathrm{T}}^{\mathrm{trig}}$ for those three topological regions, while at higher $p_{\mathrm{T}}^{\mathrm{trig}}$ ($> 5\,\mathrm{GeV}/c$), they still increase but less steeply in the toward and away regions, whereas in the transverse region they tend to saturate. This saturation is expected in models including an impact parameter dependence such that the requirement of the presence of a high-p_{T} particle in a pp collision biases the selection of collisions towards those with a smaller impact parameter [7]. Moreover, the increase observed in the toward and away regions at high p_{T} is due to the contributions from jets as well as ISR and FSR. The UE contributions to the toward and away regions are larger in p–Pb than in pp collisions.

Figure 2 shows the comparison between the multiplicity density in the transverse region [8] and the self-normalised neutral ZDC energy (ZN) as a function of $p_{\mathrm{T}}^{\mathrm{leading}}$ (at mid-pseudorapidity) in pp collisions at $\sqrt{s} = 13\,\mathrm{TeV}$. The very forward energy reaches its minimum value at $p_{\mathrm{T}}^{\mathrm{leading}} \approx 5\,\mathrm{GeV}/c$ and then remains constant for increasing $p_{\mathrm{T}}^{\mathrm{leading}}$. The saturation occurs at the same p_{T} value observed for UE quantities. The result corroborates the interpretation in terms of a bias towards collisions with small impact parameter, because the correlation between central and forward pseudorapidity can only be attributed to the initial stage of the collisions.

In a MC study of UE [4], it was shown that the multiplicity distributions in the transverse region ($|\eta| < 2.5, p_{\mathrm{T}} > 0\,\mathrm{GeV}/c$) at the plateau obey a KNO scaling at the LHC energies. The KNO scaling was expected in a model which assumes that a single pp collision results from the superposition of a given number of elementary partonic collisions emitting particles independently [9]. MPI could produce such an effect. However, the transverse region not

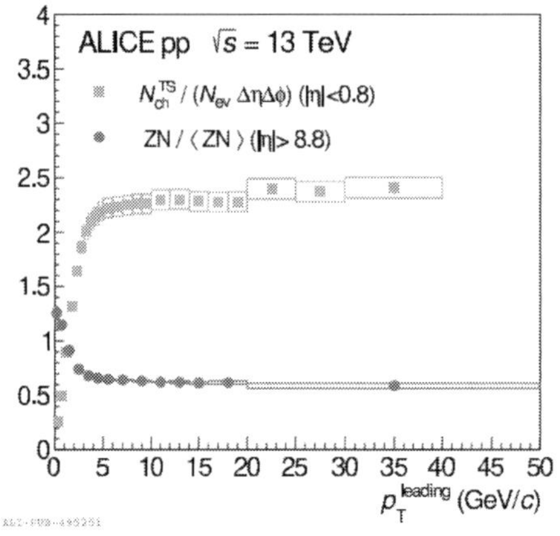

Figure 2: Multiplicity density (azure squares) in the transverse region [8] and self-normalised ZN signal (red circles) as a function of $p_{\mathrm{T}}^{\mathrm{leading}}$ (at mid-pseudorapidity) in pp collisions at $\sqrt{s} = 13\,\mathrm{TeV}$.

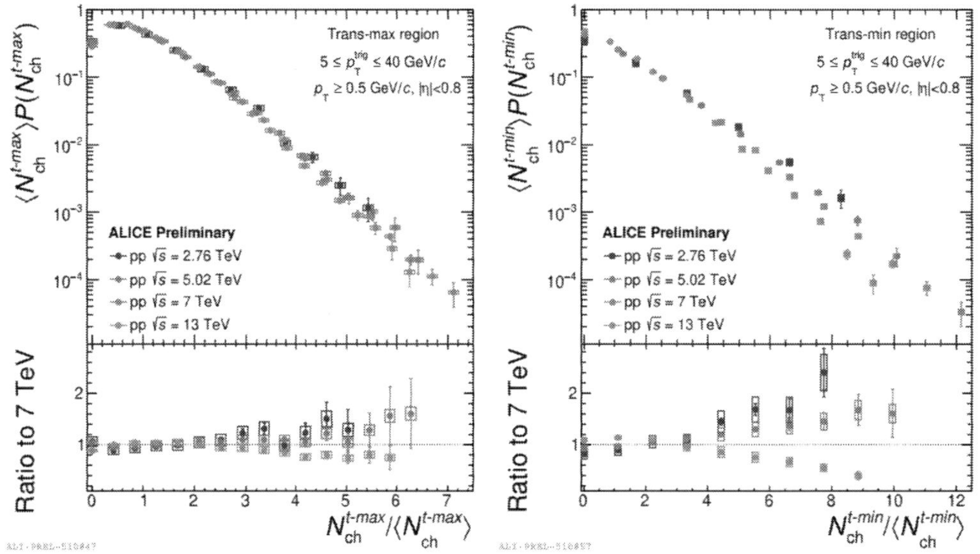

Figure 3: Top: multiplicity distributions in KNO variables in the trans-max (left) and trans-min (right) regions for pp collisions at $\sqrt{s} = 2.76$, 5.02, 7 and 13 TeV. Bottom: the KNO multiplicity distributions are normalized to that for pp collisions at $\sqrt{s} = 7\,\mathrm{TeV}$. The boxes and the error bars represent the systematic and statistical uncertainties, respectively.

only includes contributions from UE but also from ISR-FSR. In order to investigate the KNO-like scaling properties, a further treatment of the transverse side is implemented, where the transverse region is further divided in two regions:

- transverse-I: $\pi/3 < \Delta\varphi < 2\pi/3$
- transverse-II: $-\pi/2 < \Delta\varphi < -\pi/3$, $4\pi/3 < \Delta\varphi < 3\pi/2$

These two distinct regions are characterized in terms of their relative charged-particle multiplicities. Trans-max (trans-min) region refers to the sub-transverse region (I or II) with the larger (smaller) multiplicity which have an enhanced sensitivity to ISR-FSR (UE) [2, 10].

Figure 3 shows the charged-particle multiplicity distributions in KNO variables for pp collisions at $\sqrt{s}$ = 2.76, 5.02, 7 and 13 TeV. Results are presented for the trans-max and trans-min regions. In the trans-max region, within 20%, the KNO-like scaling is observed in a wider range of multiplicity ($0 < z =< 4$, where $z = N_{\rm ch}^{X}/\langle N_{\rm ch}^{X}\rangle$, X ='*t-max*' or '*t-min*' which corresponds to the 'trans-max' or 'trans-min' region) relative to the results reported in Ref. [3], while for higher z values ($z > 4$) the scaling is broken. It is worth noticing that for trans-max both contributions are considered: UE and ISR-FSR. If the effect of ISR-FSR is suppressed, i.e., exploiting the features of trans-min region, the KNO-like scaling also holds for $0 < z < 4$, and then for $z > 4$ the KNO-like scaling is still broken but a higher z reach is achieved, especially for $z > 6$, a lager violation is observed. Events with high-multiplicity jets can contribute to the large violation of the scaling properties. It was observed that for $z > 3$, the number of uncorrelated seeds (or MPI) deviate from the linear trend suggesting the presence of high-multiplicity jets [11, 12].

3 Conclusions

The multiplicity density as a function of $p_{\rm T}^{\rm trig}$ for both pp and p–Pb collisions at $\sqrt{s_{\rm NN}}$ = 5.02 TeV shows a remarkably similar behavior, such as in the transverse region the distributions increase with $p_{\rm T}^{\rm trig}$ until about 5 GeV/c where it saturates. On the other hand, the very forward energy as a function of $p_{\rm T}^{\rm leading}$ in pp collisions at $\sqrt{s}$ = 13 TeV decreases with increasing $p_{\rm T}^{\rm leading}$ and then saturates at the same $p_{\rm T}^{\rm leading}$ ($\approx$ 5 GeV/c). These two saturation effects are commonly interpreted as a bias towards collisions with small impact parameter. In addition, in the trans-max and trans-min regions the charged-particle multiplicity distributions exhibit a KNO-like scaling for $0 < z < 4$, which is expected for a single pp collision involving a large number of MPI emitting independently. And for $z > 4$ the scaling is broken. For the MPI-sensitive region, a higher z reach is achieved, especially for $z > 6$, a larger violation is observed, which may be attributed to high-multiplicity mini jets.

4 Acknowledgement

This work has been supported by CONACyT under the Grants CB No. A1-S-22917 and CF No. 2042.

References

[1] T. Sjöstrand, M. van Zijl, Phys. Rev. D **36**, 2019 (1987)
[2] G. Bencedi, A. Ortiz, A. Paz, Phys. Rev. D **104**, 016017 (2021), [arXiv]2105.04838
[3] A. Ortiz, Phys. Rev. D **104**, 076019 (2021), [arXiv]2108.08360
[4] A. Ortiz, L. Valencia Palomo, Phys. Rev. D **96**, 114019 (2017), [arXiv]1710.04741
[5] S. Acharya et al. (ALICE) (2022), [arXiv]2204.10389
[6] S. Acharya et al. (ALICE) (2022), [arXiv]2204.10157
[7] M. Strikman, Acta Phys. Polon. B **42**, 2607 (2011), [arXiv]1112.3834
[8] S. Acharya et al. (ALICE), JHEP **04**, 192 (2020), [arXiv]1910.14400
[9] J. Dias de Deus, C. Pajares, C.A. Salgado, Phys. Lett. B **408**, 417 (1997), hep-ph/9705425
[10] T.A. Aaltonen et al. (CDF), Phys. Rev. D **92**, 092009 (2015), [arXiv]1508.05340
[11] B. Abelev et al. (ALICE), JHEP **09**, 049 (2013), [arXiv]1307.1249
[12] A. Ortiz, E.A. Zepeda, J. Phys. G **48**, 085014 (2021), [arXiv]2101.10274

Conservation Laws and Hydrodynamics

Sangyong Jeon[1,*]

[1]Department of Physics, McGill University, 3600 rue University, Montreal, QC, Canada

Abstract. Recent years have seen much development in analyzing the structure of relativistic hydrodynamics. In this proceeding, some of the developments are highlighted including issues related to pseudo-gauge transformations and spin hydrodynamics.

1 Introduction

Hydrodynamics has a long history in describing heavy ion collisions starting with Landau's seminal paper [1] and Bjorken's [2]. In the era of RHIC and the LHC, hydrodynamics has become an indispensable tool to describe the bulk dynamics of relativistic heavy ion collisions.

In this 2022 Strangeness in Quark Matter conference, many of the recent developments in hydrodynamics were reported. On spin hydrodynamics and Λ polarization, Palermo [3], Liao [4] and Buzzegoli [5] presented their work. On the subject of nonlinear causality conditions, Plumberg [6] presented his and collaborators' work. On the development of hydro initial states, Shen [7] presented recent developments for ultra-peripheral collisions and Kanakubo [8] presented her recent work on the core-corona model. On applying hydrodynamics to systems near the QCD critical point, Pradeep [9], Wu [10], and Sogabe [11] presented their work in this conference.

As this is a brief introduction to the recent developments, inevitably only a limited number of topics can be discussed. Below, I will mainly focus on the pseudo-gauge transformation and its relevance in developing spin hydrodynamics. I should emphasize that this is a personal choice dictated by my familiarity with the subjects [12].

2 Is any current density observable?

Hydrodynamics starts with conservation laws

$$\partial_\mu J^{\mu,q} = 0 \tag{1}$$

where q is the index for different conserved quantities such as energy, momentum, electric charge and baryon number. Mathematically, a conserved current density is *not* uniquely defined. One can always add a "pseudo-gauge" term

$$J'^{\mu,q} = J^{\mu,q} + \partial_\lambda B^{\lambda\mu,q} \tag{2}$$

*e-mail: jeon@physics.mcgil.ca

where $B^{\lambda\mu,q} = -B^{\mu\lambda,q}$ is the pseudo-gauge potential. Adding $\partial_\lambda B^{\lambda\mu,q}$ neither changes the conservation law $\partial_\mu J'^{\mu,q} = 0$, nor the total charge provided that $B^{i0,q}$ vanishes at the boundary of the volume. But it does change the local current density since $J^{\mu,q}(x) \neq J'^{\mu,q}(x)$.

Does this mean that a current density is not observable? That depends on whether there exist other constraints that may select a particular form to be physical. For instance, in the classical particle dynamics, a point particle carries well-defined energy, momentum and other conserved charges. Hence, the classical current density

$$J^{\mu,q}_{\rm cl}(x) = \sum_{s={\rm ptcles}} \frac{q_s}{p^0_s} p^\mu_s \delta^{(3)}(\mathbf{x} - \mathbf{x}_s(t)) \tag{3}$$

is physical and observable as long as a particle is observable. Adding an arbitrary $\partial_\lambda B^{\lambda\mu,q}$ to $J^{\mu,q}_{\rm cl}$ is still a possibility, but the fact that a point particle carries definite amounts of energy, momentum and other charges clearly selects Eq.(3) to be the physical current density.

In reality, due to the uncertainty principle, particles cannot be localized as in Eq.(3). Particle positions are inevitably smeared so that

$$J^{\mu,q}_C(x) = \sum_{s={\rm ptcles}} \frac{q_s}{p^0_s} p^\mu_s g(\mathbf{x} - \mathbf{x}_s(t)) \tag{4}$$

with the normalization condition $\int d^3x\, g(\mathbf{x} - \mathbf{x}_s(t)) = 1$ that preserves the total charge. This expression still satisfies $\partial_\mu J^{\mu,q}_C = 0$ as long as $d\mathbf{x}_s/dt = \mathbf{p}_s/p^0_s$, but since there can be many choices for the smearing function, this does not define a unique value of the current density at x. For any change $\delta J^{\mu,q}_C$ introduced by a change of the smearing function, one can always find $B^{\lambda\mu,q}$ that satisfies

$$\delta J^{\mu,q}_C = \partial_\lambda B^{\lambda\mu,q} \tag{5}$$

since it has the same form as the Maxwell equation. Hence, without any further constraint on the form of the smearing function, the continuous current density $J^{\mu,q}_C$ can be ambiguous.

I have been so far careful to always refer to $J^{\mu,q}$ as a current density. This is to distinguish $J^{\mu,q}$ from the current. An electric current, for instance, is defined as

$$I_S = \int_S d\mathbf{S} \cdot \mathbf{J}^e \tag{6}$$

where $\mathbf{S}$ is the surface through which the electric current is measured and $\mathbf{J}^e$ is the spatial part of the electric current density. The current itself is a scalar quantity, not a vector quantity. One can easily show that as long as $\mathbf{B}$ vanishes at the boundary of the surface, I_S remains unchanged.

Does this mean that no current density is actually observable due to the pseudo-gauge freedom? In order to answer this question, consider the electric current density J^μ_e. As discussed above, the conservation law is not affected when a pseudo-gauge potential with the property $B^{\nu\mu}_e = -B^{\mu\nu}_e$ is added to J^μ_e. Using the new current density, The Maxwell equation becomes

$$\partial_\nu F'^{\nu\mu} = J'^\mu_e = J^\mu_e + \partial_\nu B^{\nu\mu}_e \tag{7}$$

The electric and magnetic field generated by J'^μ_e is then

$$F'^{\nu\mu} = F^{\nu\mu} + B^{\nu\mu}_e \tag{8}$$

where $F^{\nu\mu}$ is generated by J^μ_e. But this doesn't make much sense. The electric field and the magnetic field are measurable observables. They shouldn't be ambiguous. Hence, there must be a *physical* electric current density that generates the *measured* electric and magnetic fields.

For the energy-momentum tensor, a similar situation arises with the Einstein's equation where a symmetric Belinfante energy-momentum tensor (see below) sources the gravitational field. Since gravitational field is observable and measurable, the Belinfante tensor must be the physical energy-momentum tensor.

There are two lessons I would take from the discussions above. One is that as long as a current density generates an observable and measurable field (such as the electric and magnetic fields), one should be able to identify the unique physical and observable current density. The second lesson is that even though unambiguously identifying *classical* current density using the above arguement may be possible, identifying quantum current density could be still ambiguous (c.f. Eq.(5)). However, this ambiguity should represent only the $O(\hbar)$ correction to the classical current density.

3 Stress-energy tensors with spin

Hydrodynamics primarily deals with the energy-momentum tensor. Consider a field-theory Lagrangian $L(\varphi^A, \partial_\mu\varphi^A)$ where A is the field index. Using Noether's theorem, one can show that the conserved current densities for the translational symmetry in space-time is given by

$$\Theta^{\mu\nu} = \Pi^\mu_A \partial^\nu \varphi^A - g^{\mu\nu} L(\varphi_A, \partial_\mu\varphi^A) \tag{9}$$

where $\Pi^\mu_A = \partial L/\partial(\partial_\mu\varphi_A)$. This tensor satisfies

$$\partial_\mu\Theta^{\mu\nu} = \partial^\nu\varphi_A\left(\partial_\mu\Pi^\mu_A - \frac{\partial L}{\partial\varphi_A}\right) \tag{10}$$

Hence if φ_A satisfies the Euler-Lagrange equation $\partial_\mu\Pi^\mu_A = \partial L/\partial\varphi_A$, then $\Theta^{\mu\nu}$ is a conserved energy-momentum current density. The first index μ is the space-time index for the current density components and the second index ν indicates what is being conserved. Namely, $\nu = 0$ is the label for the energy current and $\nu = 1, 2, 3$ are for the momentum currents playing the role of the index q in the previous section. This canonical energy-momentum tensor is not symmetric unless it only contains scalar (spin $S = 0$) particles, and it is not gauge invariant when applied to gauge theories. As such, it cannot yet be directly interpreted as the measurable energy-momentum current density.

To see the consequences of having a non-symmetric $\Theta^{\mu\nu}$, define the orbital angular momentum current density

$$\mathcal{L}_O^{\lambda\mu\nu} = x^\mu\Theta^{\lambda\nu} - x^\nu\Theta^{\lambda\mu} \tag{11}$$

This tensor is in general not conserved $\partial_\lambda\mathcal{L}_O^{\lambda\mu\nu} = \Theta^{\mu\nu} - \Theta^{\nu\mu}$ unless all the fields have $S = 0$. If fields have non-zero spins, then the total angular momentum current density must include the spin angular momentum part as well as the orbital angular momentum part

$$\mathcal{L}_{\rm tot}^{\lambda\mu\nu} = x^\mu\Theta^{\lambda\nu} - x^\nu\Theta^{\lambda\mu} + \mathcal{S}^{\lambda\mu\nu} \tag{12}$$

where $\mathcal{S}^{\lambda\mu\nu}$ is the spin current density. The statement of the total angular momentum conservation becomes

$$\partial_\lambda\mathcal{S}^{\lambda\mu\nu} + 2\Theta^{\mu\nu}_{(a)} = 0 \tag{13}$$

In other words, the anti-symmetric part of the energy-momentum tensor, $\Theta^{\mu\nu}_{(a)} = (1/2)(\Theta^{\mu\nu} - \Theta^{\nu\mu})$, encodes the spin angular momentum part of the energy and momentum. That part is then necessarily of $O(\hbar)$ since the spin current density is $O(\hbar)$.

To find the pseudo-gauge transformation that results in the physical energy-momentum tensor, let

$$T^{\mu\nu} = \Theta^{\mu\nu} + \partial_\lambda B^{\lambda\mu,\nu} \tag{14}$$

where $T^{\mu\nu}$ is the symmetric tensor that we are after. Using $T^{\mu\nu} - T^{\nu\mu} = 0$, we get

$$0 = \Theta^{\mu\nu} - \Theta^{\nu\mu} + \partial_\lambda(B^{\lambda\mu,\nu} - B^{\lambda\nu,\mu}) \tag{15}$$

Using the conservation of total angular momentum, Eq.(13), one can identify

$$\mathcal{S}^{\lambda\mu\nu} = B^{\lambda\mu,\nu} - B^{\lambda\nu,\mu} \tag{16}$$

Equivalently,

$$B^{\lambda\mu,\nu} = \frac{1}{2}\left(\mathcal{S}^{\lambda\mu\nu} - \mathcal{S}^{\mu\lambda\nu} - \mathcal{S}^{\nu\lambda\mu}\right) \tag{17}$$

that yields

$$T^{\mu\nu} = \Theta^{\mu\nu}_{(s)} - \frac{1}{2}\partial_\lambda\left(\mathcal{S}^{\mu\lambda\nu} + \mathcal{S}^{\nu\lambda\mu}\right) \tag{18}$$

after explicit symmetrization. This form of the energy-momentum tensor is usually referred to as the Belinfante energy-momentum tensor and it can be shown that this is identical to the gravitational source term in the Einstein equation obtained by varying the space-time metric. As such, classically, this is the observable energy-momentum tensor. One should note that this form of energy-momentum tensor now incorporates the conservation of total angular momentum since $\partial_\mu T^{\mu\nu} = 0$ is satiafied only after using both the equations of motion and the total angular momentum conservation, Eq.(13). It also means that the energy density obtained from $T^{\mu\nu}$ contains the energy due to the spins [14].

The the total angular momentum current density is now just given by

$$\mathcal{L}^{\lambda\mu\nu}_{\text{tot}} = x^\mu T^{\lambda\nu} - x^\nu T^{\lambda\mu} \tag{19}$$

which could be further divided this into the orbital angular momentum term and the spin term although this division is not so clear-cut.[1] This also means that there is no separate spin current density which presents a bit of a problem when formulating spin hydrodynamics.

4 Spin hydrodynamics

STAR has measured Λ polarizations that showed a surprising azimuthal pattern [15] that may indicate a strong spin-orbit coupling between the polarization of a hadron and the underlying fluid motion. In order to figure out where such a strong correlation can originate from, it is clear that we need a formulation of hydrodynamics that includes the spin degrees of freedom. This is, however, not straightforward: If one insists that the energy-momentum tensor must be observable, then it must be the Belinfante tensor. But that means there is no separate spin current density. On the other hand, one can naturally get the spin current density via the anti-symmetric part of the canonical energy-momentum tensor. However, the canonical energy-momentum tensor is not physical.

There are currently two broad approaches to this problem. One is to approach the spin hydrodynamics from the semi-classical kinetic theory point of view [3, 5, 12, 16]. Another

[1]For instance, for a spin-1 gauge field, only the sum is gauge-invariant.

approach is to start with the general expression of the non-symmetric energy-momentum tensor and the spin current density tensor and develope a gradient expansion [14, 17]. One can then use the extended thermodynamic identities and the production of entropy to obtain the first order and the second order form of spin hydrodynamics. The relationships between various formulations of the spin hydrodynamics, however, is not yet fully settled and full spin hydrodynamics simulations are yet to be performed.

In all these approaches, a common task is to decompose the non-symmetric tensor into the physical symmetric part and the $O(\hbar)$ spin dependent part. This decomposition is necessary because quantities such as the energy density and the flow velocity must be defined to carry out any hydrodynamic calculations. In some sense, this largely deals with the pseudo-gauge problem. Namely, once we have identified the classical energy-momentum tensor (corresponding to the Belinfonte tensor), then only $O(\hbar)$ pseudo-gauge potentials are allowed in order not to spoil the classial part of the energy-momentum tensor.

5 Outlook

In this short review, I have tried to highlight some of the issues in formulating physical hydrodynamics that were raised in recent literatures. The two related topics I have chosen to highlight here, namely the issue of pseudo-gauge transformations and spin hydrodynamics, have been studied by several groups motivated by the Λ polarizations measured by the STAR collaboration at RHIC. Through such efforts, it has become clearer what role pseudo-gauge transformations can and cannot play in hydrodynamics in general and in spin hydrodynamics in particular.

Theoretically, the STAR Λ polarization measurement presents an interesting challenge. Although some resolutions of this puzzle[2] have been suggested [3], more studies, especially realistic spin hydrodynamics simulations, are needed to firmly settle it.

References

[1] L. D. Landau, Izv. Akad. Nauk SSSR **17**, 51 (1953).
[2] J. D. Bjorken, Phys. Rev. D **27**, 140-151 (1983)
[3] A. Palermo, F. Becattini, M. Buzzegoli, G. Inghirami and I. Karpenko, [arXiv:2208.09874 [nucl-th]].
[4] Y. Guo, J. Liao, E. Wang, H. Xing and H. Zhang, Phys. Rev. C **104**, no.4, L041902 (2021)
[5] M. Buzzegoli, [arXiv:2209.06879 [nucl-th]].
[6] C. Plumberg, D. Almaalol, T. Dore, J. Noronha and J. Noronha-Hostler, Phys. Rev. C **105**, L061901 (2021)
[7] W. Zhao, C. Shen and B. Schenke, [arXiv:2203.06094 [nucl-th]].
[8] Y. Kanakubo, Y. Tachibana and T. Hirano, [arXiv:2209.04110 [nucl-th]].
[9] M. Pradeep, K. Rajagopal, M. Stephanov and Y. Yin, Phys. Rev. D **106**, no.3, 036017 (2022)
[10] X. Y. Wu, G. Y. Qin, L. G. Pang and X. N. Wang, Phys. Rev. C **105**, no.3, 034909 (2022)
[11] N. Sogabe and Y. Yin, JHEP **03**, 124 (2022)
[12] S. Shi, C. Gale and S. Jeon, Phys. Rev. C **103**, no.4, 044906 (2021)
[13] . F. Becattini, [arXiv:2204.01144 [nucl-th]].
[14] K. Fukushima and S. Pu, Phys. Lett. B **817**, 136346 (2021)
[15] L. Adamczyk *et al.* [STAR], Nature **548**, 62-65 (2017)

[2]The measured azimuthal dependence of the Λ polarization has the opposite sign compared to the vorticity tensor.

[16] N. Weickgenannt, D. Wagner, E. Speranza and D. Rischke, [arXiv:2203.04766 [nucl-th]].
[17] A. Daher, A. Das, W. Florkowski and R. Ryblewski, [arXiv:2202.12609 [nucl-th]].
[18] C. Gale, S. Jeon, B. Schenke, P. Tribedy and R. Venugopalan, Phys. Rev. Lett. **110**, no.1, 012302 (2013)
[19] S. McDonald, C. Shen, F. Fillion-Gourdeau, S. Jeon and C. Gale, Phys. Rev. C **95**, no.6, 064913 (2017)

Polarization in heavy ion collisions: a theoretical review

Matteo Buzzegoli[1,*]

[1]Department of Physics and Astronomy, Iowa State University, Ames, Iowa 50011, USA

Abstract. In these proceedings I discuss the recent progress in the theory of spin polarization in relativistic fluids. To date, a number of studies have begun to examine the impact of the shear tensor on the local spin polarization and whether this contribution can restore agreement between the measurements and the predictions obtained from a polarization induced by the gradients of the plasma. I present the derivation of the spin polarization vector of a fermion at local thermal equilibrium and I discuss the role of pseudo-gauge transformations and of dissipative effects. I list what we can learn from the polarization measured at lower energies. Finally, I discuss possible applications of spin polarization measurements in relativistic heavy ion collisions.

1 Introduction

Particles in a rotating medium tends to align their spin in the direction of the total angular momentum due to the spin-rotation coupling. It was then expected that fermions produced in relativistic heavy-ion collisions will have a global spin polarization proportional to the vorticity of the fluid [1–3]. Based on this idea, one obtains that the spin polarization vector of a fermion is given by a Cooper-Frye like formula [4]:

$$S^\mu(p) = -\frac{1}{8m}\epsilon^{\mu\rho\sigma\tau} p_\tau \frac{\int_\Sigma \mathrm{d}\Sigma \cdot p\, n_F(1-n_F)\varpi_{\rho\sigma}}{\int_\Sigma \mathrm{d}\Sigma \cdot p\, n_F}, \tag{1}$$

where

$$\varpi_{\mu\nu} = -\frac{1}{2}\left(\partial_\mu\beta_\nu - \partial_\nu\beta_\mu\right) \tag{2}$$

is the thermal vorticity tensor and $n_F = \left(e^{\beta\cdot p-\mu/T}+1\right)^{-1}$ is the Fermi distribution function with $\beta^\mu = u^\mu/T$ the four-temperature vector and μ a chemical potential. The predictions for global polarization obtained from this formula were confirmed in 2017 with Λ hyperons produced in Au-Au collisions at $\sqrt{s_{NN}} = 200$ Gev [5, 6], see for instance [7] for a review.

However, models based on the vorticity induced spin polarization (1) are not able to reproduce later measurements of local Λ spin polarization, that is as a function of the particle momentum [8–10]. This is known as local polarization sign puzzle because the models predicted the opposite sign for the longitudinal (along the beam axis) component of the spin polarization vector. Once it was established that this discrepancy can not be explained with

*e-mail: mbuzz@iastate.edu
Presented at the 20th International Conference on Strangeness in Quark Matter (SQM 2022)

feed-down corrections [11, 12], the sign puzzle was a strong motivation for theoretical investigation as something relevant was clearly missing in the theoretical description. Further investigations, that I review in next section, indeed revealed that spin polarization can also be induced by the shear flow of the fluid and, for heavy-ion collisions, this turned out to be a relevant contribution.

2 Spin polarization at local thermal equilibrium

In this section I review the derivation of the spin polarization vector of a fermion in a fluid that reached the local thermal equilibrium. The spin polarization vector is obtained from the particle branch of the Wigner function using (see the review [13] for details)

$$S^\mu_\varpi(p) = \frac{1}{2}\frac{\int_{\Sigma_{FO}} \mathrm{d}\Sigma \cdot p \; \mathrm{tr}_4\left[\gamma^\mu\gamma^5 W_+(x,p)\right]}{\int_{\Sigma_{FO}} \mathrm{d}\Sigma \cdot p \; \mathrm{tr}_4\left[W_+(x,p)\right]}, \tag{3}$$

where, for heavy-ion collision applications, the hypersurface Σ_{FO} is the decoupling hypersurface, where the system can be described with the quasi-free hadronic effective fields. Hence, Λ particles being quasi-free, we can use the Wigner function of a free Dirac field:

$$W(x,k)_{AB} = \frac{1}{(2\pi)^4}\int \mathrm{d}^4y \, \mathrm{e}^{-\mathrm{i}k\cdot y}\langle : \bar{\Psi}_B(x+y/2)\Psi_A(x-y/2) : \rangle, \tag{4}$$

where the brackets denotes the thermal average with the statistical operator $\hat{\rho}$: $\langle \hat{X}\rangle = \mathrm{tr}\left(\hat{\rho}\hat{X}\right)$.

For a relativistic system whose underlying microscopic theory is a quantum field theory, the statistical operator describing the thermal states can be obtained using the Zubarev method [14]. By maximazing the entropy of the system at the moment of thermalization and imposing constrains on the momenta and energy densities, one obtains the covariant statistical operator at the time of decoupling

$$\widehat{\rho} = \frac{1}{Z}\exp\left[-\int_{\Sigma_{FO}} \mathrm{d}\Sigma_\mu \left(\widehat{T}^{\mu\nu}_B\beta_\nu - \zeta\widehat{j}^\mu\right) + \int_\Theta \mathrm{d}\Theta\left(\widehat{T}^{\mu\nu}_B\nabla_\mu\beta_\nu - \widehat{j}^\mu\nabla_\mu\zeta\right)\right], \tag{5}$$

where $\widehat{T}^{\mu\nu}_B$ is the Belinfante stress-energy tensor (SET) operator, $\widehat{j}$ is a conserved current, and $\zeta = \mu/T$. One can show that the second term in the exponent gives rise to dissipative effects, while the first term gives the non-dissipative local thermal equilibrium. I postpone the discussion of the dissipative part in a later section and here I focus on the local equilibrium statistical operator:

$$\widehat{\rho} \simeq \widehat{\rho}_{\mathrm{LE}} = \frac{1}{Z}\exp\left[-\int_{\Sigma_{FO}} \mathrm{d}\Sigma_\mu \left(\widehat{T}^{\mu\nu}_B\beta_\nu - \zeta\widehat{j}^\mu\right)\right]. \tag{6}$$

The derivation of the spin polarization at local thermal equilibrium from Eq. (3) reduces to the evaluation of the Wigner function at local equilibrium

$$W(x,k)_{LE} = \frac{1}{Z}\mathrm{tr}\left(\exp\left[-\int_{\Sigma_{FO}} \mathrm{d}\Sigma_\mu(y)\left(\widehat{T}^{\mu\nu}_B(y)\beta_\nu(y) - \zeta(y)\widehat{j}^\mu(y)\right)\right]\widehat{W}(x,k)\right). \tag{7}$$

For a fluid in the hydrodynamic regime, a good approximation of (7) is obtained expanding the thermodynamic fields as a series of gradients around the point x and using the linear response theory to obtain the Wigner function mean value. Keeping the first order in derivatives of the four-temperature and of the chemical potential

$$\beta_\nu(y) \simeq \beta_\nu(x) + \partial_\lambda\beta_\nu(x)(y-x)^\lambda, \quad \zeta(y) \simeq \zeta_\nu(x) + \partial_\lambda\zeta(x)(y-x)^\lambda, \tag{8}$$

one finds

$$\widehat{\rho}_{LE} \simeq \frac{1}{Z}\exp\left[-\beta_\nu(x)\widehat{P}^\nu + \frac{1}{2}\varpi_{\mu\nu}(x)\widehat{J}^{\mu\nu}_x - \frac{1}{2}\xi_{\mu\nu}(x)\widehat{Q}^{\mu\nu}_x + \partial_\lambda\zeta(x)\int \mathrm{d}\Sigma_\mu(y-x)^\lambda\widehat{j}^\mu(y) + \cdots\right], \quad (9)$$

where

$$\widehat{J}^{\mu\nu}_x = \int \mathrm{d}\Sigma_\lambda\left[(y-x)^\mu\widehat{T}^{\lambda\nu}_B(y) - (y-x)^\nu\widehat{T}^{\lambda\mu}_B(y)\right], \quad (10)$$

is the conserved angular momentum operator (the generator of boosts and rotations) and

$$\widehat{Q}^{\mu\nu}_x = \int \mathrm{d}\Sigma_\lambda\left[(y-x)^\mu\widehat{T}^{\lambda\nu}_B(y) + (y-x)^\nu\widehat{T}^{\lambda\mu}_B(y)\right], \quad (11)$$

is a non-conserved symmetric quadrupole like operator. Notice that $\widehat{J}^{\mu\nu}$ couples with the thermal vorticity ϖ, while $\widehat{Q}^{\mu\nu}$ couples with the thermal shear tensor defined as

$$\xi_{\mu\nu} = \frac{1}{2}\left(\partial_\mu\beta_\nu + \partial_\nu\beta_\mu\right). \quad (12)$$

The spin polarization vector obtained from (3) using the linear response theory on the operators appearing in (9) is

$$S^\mu(p) = -\epsilon^{\mu\rho\sigma\tau}p_\tau\frac{\int_\Sigma \mathrm{d}\Sigma\cdot p\, n_F(1-n_F)\left[\varpi_{\rho\sigma} + 2\,\hat{t}_\rho\frac{p^\lambda}{\varepsilon}\xi_{\lambda\sigma} - \frac{\hat{t}_\rho\partial_\sigma\zeta}{2\varepsilon}\right]}{8m\int_\Sigma \mathrm{d}\Sigma\cdot p\, n_F}, \quad (13)$$

where $\hat{t}$ is the time direction in the laboratory frame. The first term is the vorticity induced polarization (1), the second is the recently found shear induced polarization [15–19], and the last one is the contribution from the gradient of chemical potential obtained in [20, 21], sometimes called spin Hall effect term.

2.1 Thermal shear and solution of the sign puzzle

The effect of thermal shear in spin polarization was studied in ref. [16, 19, 22–27]. These analyses revealed the following facts. The thermal shear tensor at the freeze-out hypersurface can be larger than the thermal vorticity. Because of its momentum dependence (see Eq. (13)), the thermal shear does not contribute to global polarization but it was found to give a large contribution to local spin polarization. Predictions including the thermal shear are closer to the experimental data but they are still lacking a good quantitative agreement.

In order to reproduce the experimental data, no new effects are needed. It is sufficient to improve the approximation used to derive the spin polarization spectrum (13). Since the available experimental data for local spin polarization was taken at 200 GeV and at 5 TeV, at such high energies it is expected that the hadronization happens at the same temperature. In that case the statistical operator is given by the Isothermal Local Equilibrium operator:

$$\widehat{\rho}_{\rm ILE} = \frac{1}{Z}\exp\left[-\int_{\Sigma_{FO}} \mathrm{d}\Sigma_\mu\widehat{T}^{\mu\nu}_B\left(\frac{u_\nu}{T_{\rm FO}}\right)\right] = \frac{1}{Z}\exp\left[-\frac{1}{T_{\rm FO}}\int_{\Sigma_{FO}} \mathrm{d}\Sigma_\mu\widehat{T}^{\mu\nu}_B u_\nu\right], \quad (14)$$

because the temperature $T_{\rm FO}$ is constant along the decoupling hypersurface. In this setting, expanding the temperature in a Taylor series together with β as in Eq. (8) will only introduce unnecessary mistakes. The appropriate spin polarization spectrum at high energies involves only the fluid velocity gradients [16]:

$$S^\mu_{\rm ILE}(p) = -\epsilon^{\mu\rho\sigma\tau}p_\tau\frac{\int_\Sigma \mathrm{d}\Sigma\cdot p\, n_F(1-n_F)\left[\omega_{\rho\sigma} + 2\,\hat{t}_\rho\frac{p^\lambda}{\varepsilon}\Xi_{\lambda\sigma}\right]}{8mT_{\rm FO}\int_\Sigma \mathrm{d}\Sigma\cdot p\, n_F}, \quad (15)$$

where

$$\omega_{\rho\sigma} = \frac{1}{2}\left(\partial_\sigma u_\rho - \partial_\rho u_\sigma\right), \quad \Xi_{\rho\sigma} = \frac{1}{2}\left(\partial_\sigma u_\rho + \partial_\rho u_\sigma\right) \tag{16}$$

denote the kinematic vorticity and shear tensors. The predictions of (15) are in quantitative agreement with the data taken at 200 GeV [16]. Preliminary results presented at this conference [28] showed an improvement of the agreement when including the feed-down effects and that the predictions of (15) are also in quantitative agreement with the local spin polarization measured by ALICE [10]. Future analyses of high energy spin polarization should use the isothermal settings.

What also emerged from these analyses is that the local spin polarization is sensitive to the properties of the plasma, which ultimately determines the gradients of the thermo-hydrodynamic fields in (13). This feature allows to use spin polarization as a probe of the quark gluon plasma properties.

2.2 Low energies

With new data of spin polarization at very low energy [29], we can further test the models and see whether they can reproduce the maximum of spin polarization [30–34] expected to occurs around $\sqrt{S_{NN}} = 7$ GeV. This energy range is also important for polarization in the FAIR and NICA facilities. At lower energies we also observe a larger deviation between the polarization of Λ and $\bar{\Lambda}$, which allows to probe the gradients of baryonic chemical potential [20, 21] and the effect of the magnetic field [35]. Physics of rotating matter at these energies was also discussed in this conference [36].

2.3 Spin tensor and pseudo-gauge dependence

For the first time, the importance of the spin degrees of freedom in the determination of the spin polarization vector, opens the possibility to have a direct experimental test of a local description of spin, i.e. a spin tensor. Indeed one can include a spin tensor in hydrodynamic equations [37–43] and let it evolve with the rest. In the same way that the temperature is a thermodynamic potential for energy, the thermodynamic quantity associated with spin is called spin potential: $\Omega_{\mu\nu}$, and it might be present in spin polarization.

The definition of spin tensor is not unique in special relativity but is given up to a transformation called pseudo-gauge transformation. For instance, from the Belinfante pseudo-gauge with SET $\widehat{T}_B^{\mu\nu}$ and vanishing spin tensor, a generic pseudo-gauge Φ is obtained with:

$$\begin{aligned}\widehat{T}_\Phi^{\mu\nu} =& \widehat{T}_B^{\mu\nu} + \frac{1}{2}\nabla_\lambda\left(\widehat{\Phi}^{\lambda,\mu\nu} - \widehat{\Phi}^{\mu,\lambda\nu} - \widehat{\Phi}^{\nu,\lambda\mu}\right),\\ \widehat{\mathcal{S}}_\Phi^{\lambda,\mu\nu} =& -\widehat{\Phi}^{\lambda,\mu\nu} + \nabla_\rho \widehat{Z}^{\mu\nu,\lambda\rho},\end{aligned} \tag{17}$$

where $\widehat{\Phi}^{\lambda,\mu\nu} = -\widehat{\Phi}^{\lambda,\nu\mu}$ and $\widehat{Z}^{\mu\nu,\lambda\rho} = -\widehat{Z}^{\nu\mu,\lambda\rho} = -\widehat{Z}^{\mu\nu,\rho\lambda}$. This raises the question whether a pseudo-gauge transformation affects the results for spin polarization. It was indeed realized that the local equilibrium form of statistical operator is different for different pseudo-gauges [44]:

$$\widehat{\rho}_{\mathrm{LTE}}^{\Phi} = \frac{1}{\mathcal{Z}}\exp\left\{-\int \mathrm{d}\Sigma_\mu\left[\widehat{T}_B^{\mu\nu}\beta_\nu - \frac{1}{2}\left(\varpi_{\lambda\nu} - \Omega_{\lambda\nu}\right)\widehat{\Phi}^{\mu,\lambda\nu} - \xi_{\lambda\nu}\widehat{\Phi}^{\lambda,\mu\nu} - \frac{1}{2}\Omega_{\lambda\nu}\nabla_\rho\widehat{Z}^{\lambda\nu,\mu\rho} - \widehat{j}^\mu\zeta\right]\right\}. \tag{18}$$

Starting from this, the different predictions of the spin polarization spectrum for several pseudo-gauge were obtained [45]. This difference is not only related to the contribution of the spin potential but also of the thermal shear. This feature is not unique of the statistical mechanics approach and the role of pseudo-gauge was studied in different approaches [46–50]. This is an unique opportunity to study the physical meaning of pseudo-gauge transformations and possibly to shed light on the role of spin potential in gravitational phenomena.

3 Dissipative effects and relativistic kinetic theory with spin

The current data and the inclusion of the thermal shear support the local equilibrium picture with a small contribution of dissipative effects. However, one should check how much the dissipative phenomena affect the spin polarization. These corrections in the spin polarization formula have been found for the massless [51] and massive [52] Dirac field, but so far there are no analyses on the impact of these effects on the spin polarization of Λ particles in heavy-ion collisions. If the dissipative effects will be found to be relevant, this would imply a separation of scales between the interaction time and the time in which the spin degrees of freedom equilibrates [39, 40, 53]. In this scenario the inclusion of a spin tensor in hydrodynamics mentioned above will be crucial.

An estimate of the magnitude of the dissipative effect might not be easy, as it was realized that, in addition to usual transport coefficients, in a hydrodynamic with spin additional transport coefficients related to spin emerge [39, 40, 52, 53]. The values of these new transport coefficients for the QGP can be fixed by fitting the data of spin polarization, but in this way one inevitably loose the predictive power of the current theory.

A powerful approach to study dissipative effects and the role of spin in hydrodynamics is the relativistic kinetic theory. The Wigner-function formalism in particular is convenient to preserve all the covariant properties of the system and to find results as an expansion in $\hbar$. Motivated by spin polarization, this approach recently experienced a rich and fast development [54–59].

In this regard, I would like to draw attention to an often overlook assumption needed to solve the Wigner equation in kinetic theory with collisions:

$$\left[\gamma\cdot\left(p+\mathrm{i}\frac{\hbar}{2}\partial\right)-m\right]W_{\alpha\beta}=\hbar\, C_{\alpha\beta}. \tag{19}$$

Wigner equation itself is not enough to determine all the properties of the system. For instance, changing $\widehat{\Phi}$ in (18) results in different Wigner functions all solving Eq. (19). Indeed the determination of the non equilibrium, dissipative Wigner function from (19) requires the input of the equilibrium Wigner function. For a system in absence of vorticity, the equilibrium form is well known. But the general form of global equilibrium Wigner function in the presence of vorticity was obtained only recently [60, 61]. Usually the ansatz for the equilibrium form used in kinetic theory agrees with the correct form only at first order in thermal vorticity. For the current applications is enough, but one have to keep it in mind when deriving higher order results.

4 Applications of spin polarization in heavy ion collisions

Spin is a new element in heavy-ion collisions and as a commonly accepted theoretical framework is being built, several ideas on how to use it as a diagnostic tool for the QGP have been proposed. What makes spin polarization a good probe to the QGP properties is its dependence on the gradients of the thermo-hydrodynamics fields (fluid velocity, temperature and baryonic chemical potential).

For instance, reproducing the vorticity structure of the plasma requires knowledge of the properties and evolution of the QGP. It was then proposed to use spin correlations of two Λ hyperons to probe the vorticity structure of the fluid [23, 62, 63]. In this way one acquires more constraints on the properties of the QGP.

Correlations of helicity, that is the projection of spin along the momentum of the particle, of two hyperons in the same event can also reveal the presence of local parity violation in hot QCD matter [64–66]. Local parity violations should create an imbalance of right-handed and left-handed quarks in the plasma, described with an axial chemical potential μ_A. Currently the chiral magnetic effect is used to reveal the presence of μ_A [67]. But the axial imbalance should also add a contribution to spin polarization (13) as [65]

$$S^\mu(p) = S^\mu_{\text{hydro}}(p) + \frac{g_h}{2}\frac{\int_\Sigma d\Sigma \cdot p\,(\mu_A/T) n_F(1-n_F)}{\int_\Sigma d\Sigma \cdot p\, n_F}\frac{\varepsilon p^\mu - m^2\hat{t}^\mu}{m\varepsilon}, \tag{20}$$

resulting into a peculiar (parity breaking) contribution to helicity correlations.

Spin polarization can be used to study anything that produce large enough gradients. This is the case of a jet crossing the plasma. The vorticity field generated by it contains information about the energy lost by the jet and can be studied with spin polarization measurements in reference to the jet axis [68].

The results derived so far for spin polarization do not include interactions. Vorticity induced polarization is a consequence of spin-rotation coupling $g_\Omega \vec{S}\cdot\vec{\Omega}$, and Einstein Equivalence Principle (EEP) prevents spin-rotation coupling to receive radiative corrections: $g_\Omega \equiv 1$ [69]. However, the breaking of Lorentz covariance at finite temperature also breaks the EEP and this allows interactions to affect the spin-rotation coupling [70]. For strong interactions the leading order correction is expected to be

$$g_\Omega = 1 - \frac{N_c^2-1}{2}\frac{1}{6}\frac{g^2T^2}{m^2}, \quad T \ll m \tag{21}$$

which, in a typical event, results in a reduction of about 40% in the spin polarization of the strange quark. Whether we can detected this consequence of the breaking of EEP in the energy dependence of Λ polarization is still under investigation.

5 Conclusions

Spin polarization is a new topic in heavy ion collisions that just started to show its potential. It can be used to answer many questions left unanswered in relativistic hydrodynamics and to probe the properties of hot QCD matter. When all the relevant gradients are included and more accurate approximations adopted, the description of the QGP as a fluid close to local equilibrium with small viscous corrections is receiving confirmation as a solid model in heavy ion collisions. Right now, theory and experiments in this field are rapidly expanding and spin will be used as a tool to investigate fundamental physics in the QGP and beyond.

Acknowledgments. M.B. is supported by the US Department of Energy under Grant No. DE-FG02-87ER40371.

References

[1] Z.T. Liang, X.N. Wang, Phys. Rev. Lett. **94**, 102301 (2005), [Erratum: Phys.Rev.Lett. 96, 039901 (2006)], `nucl-th/0410079`

[2] F. Becattini, F. Piccinini, Annals Phys. **323**, 2452 (2008), 0710.5694
[3] F. Becattini, F. Piccinini, J. Rizzo, Phys. Rev. C **77**, 024906 (2008), 0711.1253
[4] F. Becattini, V. Chandra, L. Del Zanna, E. Grossi, Annals Phys. **338**, 32 (2013), 1303.3431
[5] L. Adamczyk et al. (STAR), Nature **548**, 62 (2017), 1701.06657
[6] J. Adam et al. (STAR), Phys. Rev. C **98**, 014910 (2018), 1805.04400
[7] F. Becattini, M.A. Lisa, Ann. Rev. Nucl. Part. Sci. **70**, 395 (2020), 2003.03640
[8] J. Adam et al. (STAR), Phys. Rev. Lett. **123**, 132301 (2019), 1905.11917
[9] T. Niida (STAR), Nucl. Phys. A **982**, 511 (2019), 1808.10482
[10] S. Acharya et al. (ALICE), Phys. Rev. Lett. **128**, 172005 (2022), 2107.11183
[11] X.L. Xia, H. Li, X.G. Huang, H.Z. Huang, Phys. Rev. C **100**, 014913 (2019), 1905.03120
[12] F. Becattini, G. Cao, E. Speranza, Eur. Phys. J. C **79**, 741 (2019), 1905.03123
[13] F. Becattini, Lect. Notes Phys. **987**, 15 (2021), 2004.04050
[14] F. Becattini, M. Buzzegoli, E. Grossi, Particles **2**, 197 (2019), 1902.01089
[15] F. Becattini, M. Buzzegoli, A. Palermo, Phys. Lett. B **820**, 136519 (2021), 2103.10917
[16] F. Becattini, M. Buzzegoli, G. Inghirami, I. Karpenko, A. Palermo, Phys. Rev. Lett. **127**, 272302 (2021), 2103.14621
[17] S.Y.F. Liu, Y. Yin, JHEP **07**, 188 (2021), 2103.09200
[18] Y.C. Liu, X.G. Huang, Sci. China Phys. Mech. Astron. **65**, 272011 (2022), 2109.15301
[19] C. Yi, S. Pu, D.L. Yang, Phys. Rev. C **104**, 064901 (2021), 2106.00238
[20] S.Y.F. Liu, Y. Yin, Phys. Rev. D **104**, 054043 (2021), 2006.12421
[21] B. Fu, L. Pang, H. Song, Y. Yin (2022), 2201.12970
[22] B. Fu, S.Y.F. Liu, L. Pang, H. Song, Y. Yin, Phys. Rev. Lett. **127**, 142301 (2021), 2103.10403
[23] S. Ryu, V. Jupic, C. Shen, Phys. Rev. C **104**, 054908 (2021), 2106.08125
[24] W. Florkowski, A. Kumar, A. Mazeliauskas, R. Ryblewski, Phys. Rev. C **105**, 064901 (2022), 2112.02799
[25] Y. Sun, Z. Zhang, C.M. Ko, W. Zhao, Phys. Rev. C **105**, 034911 (2022), 2112.14410
[26] S. Alzhrani, S. Ryu, C. Shen, arXiv:2203.15718 (2022), 2203.15718
[27] X.Y. Wu, C. Yi, G.Y. Qin, S. Pu, Phys. Rev. C **105**, 064909 (2022), 2204.02218
[28] A. Palermo, F. Becattini, M. Buzzegoli, G. Inghirami, I. Karpenko, *Local equilibrium and Lambda polarization in high energy heavy ion collisions*, in *These proceedings* (2022), 2208.09874
[29] M.S. Abdallah et al. (STAR), Phys. Rev. C **104**, L061901 (2021), 2108.00044
[30] Y.B. Ivanov, V.D. Toneev, A.A. Soldatov, Phys. Rev. C **100**, 014908 (2019), 1903.05455
[31] Y.B. Ivanov, A.A. Soldatov, Phys. Rev. C **105**, 034915 (2022), 2201.04527
[32] X.G. Deng, X.G. Huang, Y.G. Ma, S. Zhang, Phys. Rev. C **101**, 064908 (2020), 2001.01371
[33] X.G. Deng, X.G. Huang, Y.G. Ma (2021), 2109.09956
[34] Y. Guo, J. Liao, E. Wang, H. Xing, H. Zhang, Phys. Rev. C **104**, L041902 (2021), 2105.13481
[35] Y. Guo, S. Shi, S. Feng, J. Liao, Phys. Lett. B **798**, 134929 (2019), 1905.12613
[36] J. Liao, *Novel Effects of Rotational Polarization in Relativistic Nuclear Collisions*, in *These proceedings* (2022)

[37] W. Florkowski, A. Kumar, R. Ryblewski, Prog. Part. Nucl. Phys. **108**, 103709 (2019), 1811.04409
[38] K. Hattori, M. Hongo, X.G. Huang, M. Matsuo, H. Taya, Phys. Lett. B **795**, 100 (2019), 1901.06615
[39] M. Hongo, X.G. Huang, M. Kaminski, M. Stephanov, H.U. Yee, JHEP **11**, 150 (2021), 2107.14231
[40] M. Hongo, X.G. Huang, M. Kaminski, M. Stephanov, H.U. Yee (2022), 2201.12390
[41] S. Bhadury, W. Florkowski, A. Jaiswal, A. Kumar, R. Ryblewski, Phys. Rev. D **103**, 014030 (2021), 2008.10976
[42] S. Bhadury, J. Bhatt, A. Jaiswal, A. Kumar, Eur. Phys. J. ST **230**, 655 (2021), 2101.11964
[43] D. She, A. Huang, D. Hou, J. Liao (2021), 2105.04060
[44] F. Becattini, W. Florkowski, E. Speranza, Phys. Lett. B **789**, 419 (2019), 1807.10994
[45] M. Buzzegoli, Phys. Rev. C **105**, 044907 (2022), 2109.12084
[46] E. Speranza, N. Weickgenannt, Eur. Phys. J. A **57**, 155 (2021), 2007.00138
[47] K. Fukushima, S. Pu, Phys. Lett. B **817**, 136346 (2021), 2010.01608
[48] A. Das, W. Florkowski, R. Ryblewski, R. Singh, Phys. Rev. D **103**, L091502 (2021), 2103.01013
[49] S. Li, M.A. Stephanov, H.U. Yee, Phys. Rev. Lett. **127**, 082302 (2021), 2011.12318
[50] N. Weickgenannt, D. Wagner, E. Speranza, Phys. Rev. D **105**, 116026 (2022), 2204.01797
[51] S. Shi, C. Gale, S. Jeon, Phys. Rev. C **103**, 044906 (2021), 2008.08618
[52] N. Weickgenannt, D. Wagner, E. Speranza, D. Rischke (2022), 2203.04766
[53] Y. Hidaka, S. Pu, D.L. Yang, Phys. Rev. D **97**, 016004 (2018), 1710.00278
[54] N. Weickgenannt, X.L. Sheng, E. Speranza, Q. Wang, D.H. Rischke, Phys. Rev. D **100**, 056018 (2019), 1902.06513
[55] J.H. Gao, Z.T. Liang, Phys. Rev. D **100**, 056021 (2019), 1902.06510
[56] K. Hattori, Y. Hidaka, D.L. Yang, Phys. Rev. D **100**, 096011 (2019), 1903.01653
[57] Y.C. Liu, K. Mameda, X.G. Huang, Chin. Phys. C **44**, 094101 (2020), [Erratum: Chin.Phys.C 45, 089001 (2021)], 2002.03753
[58] N. Weickgenannt, E. Speranza, X.l. Sheng, Q. Wang, D.H. Rischke, Phys. Rev. Lett. **127**, 052301 (2021), 2005.01506
[59] N. Weickgenannt, E. Speranza, X.l. Sheng, Q. Wang, D.H. Rischke, Phys. Rev. D **104**, 016022 (2021), 2103.04896
[60] F. Becattini, M. Buzzegoli, A. Palermo, JHEP **02**, 101 (2021), 2007.08249
[61] A. Palermo, M. Buzzegoli, F. Becattini, JHEP **10**, 077 (2021), 2106.08340
[62] L.G. Pang, H. Petersen, Q. Wang, X.N. Wang, Phys. Rev. Lett. **117**, 192301 (2016), 1605.04024
[63] X.L. Xia, H. Li, Z.B. Tang, Q. Wang, Phys. Rev. C **98**, 024905 (2018), 1803.00867
[64] F. Du, L.E. Finch, J. Sandweiss, Phys. Rev. C **78**, 044908 (2008)
[65] F. Becattini, M. Buzzegoli, A. Palermo, G. Prokhorov, Phys. Lett. B **822**, 136706 (2021), 2009.13449
[66] J.H. Gao, Phys. Rev. D **104**, 076016 (2021), 2105.08293
[67] D. Kharzeev, Phys. Lett. B **633**, 260 (2006), hep-ph/0406125
[68] W.M. Serenone, J.a.G.P. Barbon, D.D. Chinellato, M.A. Lisa, C. Shen, J. Takahashi, G. Torrieri, Phys. Lett. B **820**, 136500 (2021), 2102.11919
[69] I.Y. Kobzarev, L. Okun, Zh. Eksperim. i Teor. Fiz. **43** (1962)

[70] M. Buzzegoli, D.E. Kharzeev, Phys. Rev. D **103**, 116005 (2021), 2102.01676
[71] S.K. Singh, J.e. Alam (2021), 2110.15604

Calculating QCD Phase Diagram Trajectories of Nuclear Collisions using a Semi-analytical Model

Todd Mendenhall[1] and *Zi-Wei* Lin[1]

[1]Department of Physics, East Carolina University, Greenville, NC 27858

Abstract. At low to moderate collision energies where the parton formation time $\tau_{\rm F}$ is not small compared to the nuclear crossing time, the finite nuclear thickness significantly affects the energy density $\epsilon(t)$ and net conserved-charge densities such as the net-baryon density $n_{\rm B}(t)$ produced in heavy ion collisions. As a result, at low to moderate energies the trajectory in the QCD phase diagram is also affected by the finite nuclear thickness. Here, we first discuss our semi-analytical model and its results on $\epsilon(t)$, $n_{\rm B}(t)$, $n_{\rm Q}(t)$, and $n_{\rm S}(t)$ in central Au+Au collisions. We then compare the $T(t)$, $\mu_{\rm B}(t)$, $\mu_{\rm Q}(t)$, and $\mu_{\rm S}(t)$ extracted with the ideal gas equation of state (EoS) with quantum statistics to those extracted with a lattice QCD-based EoS. We also compare the $T-\mu_{\rm B}$ trajectories with the RHIC chemical freezeout data. Finally, we discuss the effect of transverse flow on the trajectories.

1 Introduction

The Bjorken energy density formula [1] predicts the energy density in the central spacetime rapidity region produced in the initial state of heavy ion collisions assuming that partons originate at $(z_0, t_0) = (0, 0)$. After averaging over the transverse overlap area $A_{\rm T}$ (where we take the radius $R_A = 1.12A^{1/3}$ fm from the hard-sphere model), one needs to take a finite initial time because the Bjorken energy density formula diverges as $t \to 0$:

$$\epsilon^{Bj}(t) = \frac{1}{t A_{\rm T}} \frac{dE_{\rm T}}{dy}. \tag{1}$$

Here, $dE_{\rm T}/dy$ is the transverse energy rapidity density at mid-rapidity. A similar formula can be used to calculate the net-baryon density as a function of time [2]:

$$n_{\rm B}^{Bj}(t) = \frac{1}{t A_{\rm T}} \frac{dN_{\rm netB}}{dy}, \tag{2}$$

which depends on the net-baryon rapidity density at mid-rapidity $dN_{\rm netB}/dy$. In Eqs. (1)-(2), the peak density occurs at the earliest time, which we take as the parton proper formation time $\tau_{\rm F}$.

In our semi-analytical model [3], we neglect secondary parton interactions and consider that produced partons are free-streaming, like the Bjorken energy density formula of Eq. (1). However, we include the finite nuclear thickness by considering the finite time x and longitudinal width z_0 of the primary NN collisions, and obtain for the energy density:

$$\epsilon(t) = \frac{1}{A_{\rm T}} \iint_S \frac{dx\, dz_0}{t - x} \frac{d^3 m_{\rm T}}{dx\, dz_0\, dy} \cosh^3 y. \tag{3}$$

We then simplify the above integral by assuming that $d^3 m_{\rm T}/(dx\, dz_0\, dy) \propto dm_{\rm T}/dy$, i.e., the initial transverse mass rapidity density is uniformly distributed over the initial production area S in the $x - z_0$ plane [3]. Note that $dm_{\rm T}/dy = dE_{\rm T}/dy + m_{\rm N} dN_{\rm netB}/dy$ where $m_{\rm N}$ is the nucleon mass. Recently, we have further extended our semi-analytical model [2] to calculate the net conserved-charge densities including the net-baryon density $n_{\rm B}(t)$ as

$$n_{\rm B}(t) = \frac{1}{A_{\rm T}} \iint_S \frac{dx\, dz_0}{t - x} \frac{d^3 N_{\rm netB}}{dx\, dz_0\, dy} \cosh^2 y. \tag{4}$$

Since the initial net-charge comes from incoming protons and there is no net-strangeness in the incoming nuclei, the net-electric charge and net-strangeness densities in our semi-analytical model are respectively given by

$$n_{\rm Q}(t) = n_{\rm B}(t)\frac{Z}{A}, \text{ and } n_{\rm S}(t) = 0. \tag{5}$$

Using the densities from our semi-analytical model, the temperature $T(t)$ and chemical potentials $\mu(t)$ can then be extracted for the ideal gas EoS with quantum statistics with the following relations [2]:

$$\begin{aligned} \epsilon &= \frac{19\pi^2}{12}T^4 + 3\frac{(\mu_{\rm B} - 2\mu_{\rm S})^2 + \mu_{\rm S}^2}{2}T^2 + 3\frac{(\mu_{\rm B} - 2\mu_{\rm S})^4 + \mu_{\rm S}^4}{4\pi^2}, \\ n_{\rm B} &= \frac{\mu_{\rm B} - \mu_{\rm S}}{3}T^2 + \frac{(\mu_{\rm B} - 2\mu_{\rm S})^3 + \mu_{\rm S}^3}{3\pi^2}, \\ n_{\rm Q} &= \frac{2\mu_{\rm B} - 5\mu_{\rm S}}{3}T^2 + \frac{2(\mu_{\rm B} - 2\mu_{\rm S})^3 - \mu_{\rm S}^3}{3\pi^2}. \end{aligned} \tag{6}$$

In the above, we assume that the quark-gluon plasma (QGP) consists of massless gluons and quarks, and we have used the fact that $n_{\rm S}(t) = 0$ for the ideal gas EoS leads to $\mu_{\rm Q} = \mu_{\rm B} - 3\mu_{\rm S}$. Therefore, the problem of extracting a $T - \mu_{\rm B}$ trajectory in the QCD phase diagram is reduced from solving a system of four equations with four unknowns to solving the above system of three equations with three unknowns.

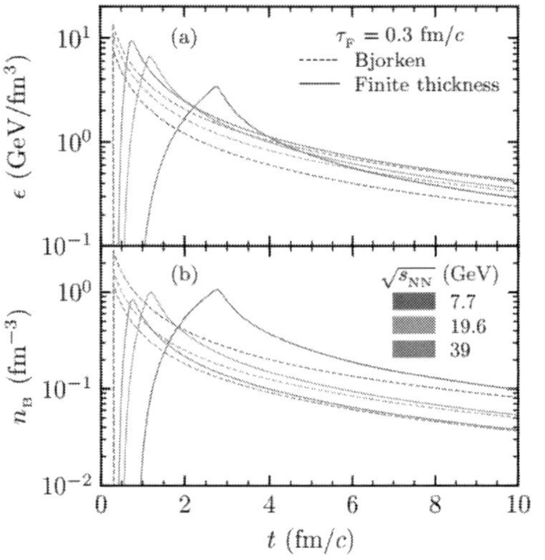

Figure 1. (a) $\epsilon(t)$ and (b) $n_B(t)$ at mid-spacetime-rapidity from the Bjorken formula (dashed) and our formula (solid) for central Au+Au collisions at $\sqrt{s_{NN}}$ = 7.7, 19.6, and 39 GeV with τ_F = 0.3 fm/c.

One can also use a lattice QCD-based EoS to extract the T, μ_B, μ_Q, μ_S from ϵ, n_B, n_Q, n_S, where these quantities are related with the standard thermodynamic relations [4]. Each of the conserved-charge densities n and the entropy density s are given by a derivative of the pressure p:

$$\frac{\epsilon}{T^4} = \frac{s}{T^3} - \frac{p}{T^4} + \frac{\mu_B}{T}\frac{n_B}{T^3} + \frac{\mu_Q}{T}\frac{n_Q}{T^3} + \frac{\mu_S}{T}\frac{n_S}{T^3},$$
$$\frac{n_B}{T^3} = \frac{1}{T^3}\frac{\partial p}{\partial \mu_B}\bigg|_{T,\mu_Q,\mu_S}, \quad \frac{n_Q}{T^3} = \frac{1}{T^3}\frac{\partial p}{\partial \mu_Q}\bigg|_{T,\mu_B,\mu_S},$$
$$\frac{n_S}{T^3} = \frac{1}{T^3}\frac{\partial p}{\partial \mu_S}\bigg|_{T,\mu_B,\mu_Q}, \quad \frac{s}{T^3} = \frac{1}{T^3}\frac{\partial p}{\partial T}\bigg|_{\mu_B,\mu_Q,\mu_S}. \quad (7)$$

In the above, the pressure p is defined by a Taylor series in powers of μ/T up to total power $i + j + k \leq 4$:

$$\frac{p(T, \mu_B, \mu_Q, \mu_S)}{T^4} = \sum_{i,j,k} \frac{1}{i!\,j!\,k!}\chi_{ijk}^{BQS}\left(\frac{\mu_B}{T}\right)^i\left(\frac{\mu_Q}{T}\right)^j\left(\frac{\mu_S}{T}\right)^k, \quad (8)$$

where the coefficients χ_{ijk}^{BQS} are parameterized as functions of T [4] based on lattice QCD results.

2 Results

In Fig. 1, we show the time evolution of the energy density $\epsilon(t)$ and net-baryon density $n_B(t)$ calculated with Eqs. (3)-(4) from our semi-analytical model compared to those calculated with the Bjorken formulas of Eqs. (1)-(2). The results are for central Au+Au collisions at $\sqrt{s_{NN}}$ = 7.7, 19.6, and 39 GeV for a parton formation time of τ_F = 0.3 fm/c. As $\sqrt{s_{NN}}$ increases, the maximum energy density ϵ^{max} in Fig. 1 increases while the maximum net-baryon

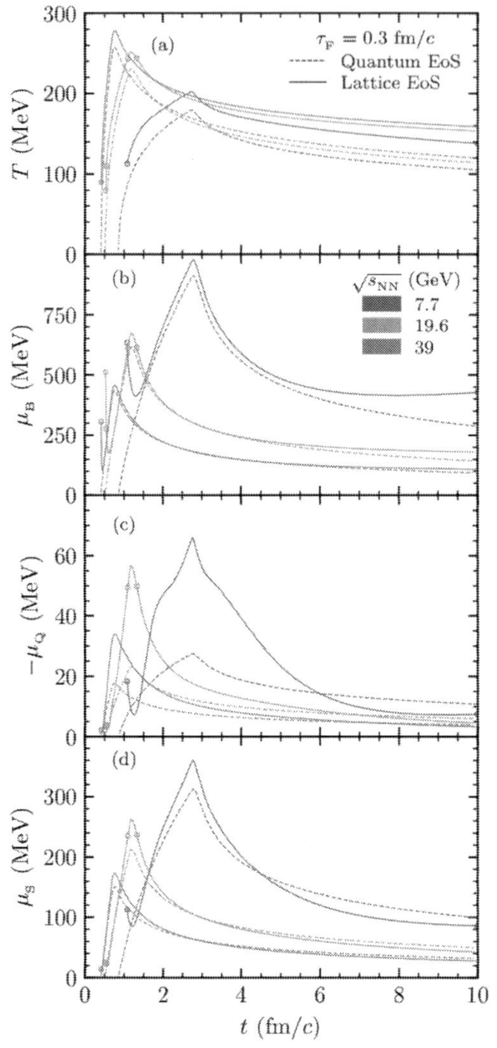

Figure 2. (a) $T(t)$, (b) $\mu_B(t)$, (c) $-\mu_Q(t)$, and (d) $\mu_S(t)$ extracted with the quantum EoS (dashed) and the lattice EoS (solid) for central Au+Au collisions at $\sqrt{s_{NN}}$ = 7.7, 19.6, and 39 GeV with τ_F = 0.3 fm/c. Open circles on the lattice EoS curves represent times when $\mu_B/T > 2.5$.

density n_B^{max} decreases according to our semi-analytical model. While ϵ^{max} also increases with $\sqrt{s_{NN}}$ according to the Bjorken formula, it occurs at $t = \tau_F$ whereas ϵ^{max} from our semi-analytical model occurs later, at time $t \in [t_a, t_2 + \tau_F]$ [2]. Note that t_1 and t_2 represent the starting and ending time of the nuclear overlap, respectively; the nuclear crossing time is $d_t = 2R_A/(\beta\gamma)$, and we choose $t_1 = d_t/6$ and $t_2 = 5d_t/6$ [2].

We also see in Fig. 1 that for both the Bjorken formula and our semi-analytical model, n_B^{max} decreases with $\sqrt{s_{NN}}$ and it is reached at the same time as ϵ^{max}. One major difference between the Bjorken formulas and our semi-analytical model is that our densities start at zero (at

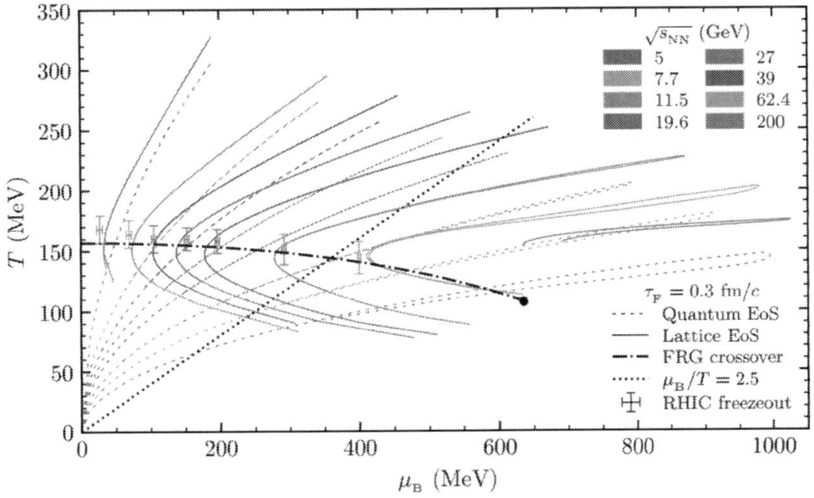

Figure 3. QCD phase diagram trajectories extracted with the quantum EoS (dashed) and the lattice EoS (solid) compared with the RHIC chemical freezeout data (symbol with error bars) for central Au+Au collisions at $\sqrt{s_{NN}}$ = 5.0, 7.7, 11.5, 19.6, 27, 39, 62.4, and 200 GeV with τ_F = 0.3 fm/c. The FRG crossover curve (dot-dashed) with the CEP and the μ_B/T = 2.5 line are also shown for reference.

$t = t_1 + \tau_F$), increase to their maximum values, then decrease thereafter. On the other hand, Bjorken densities start at their maximum values and decrease with time. We also find that the late time evolution of our densities approaches that of the Bjorken formula. This occurs because the formed partons in our model must have $y \sim 0$ in order to contribute to the densities in the mid-spacetime-rapidity region at late times, just like the Bjorken formula.

Figure 2 shows the time evolution of the temperature $T(t)$ and chemical potentials $\mu(t)$ for the quantum EoS and the lattice EoS extracted using our densities at $\sqrt{s_{NN}}$ = 7.7, 19.6, and 39 GeV with τ_F = 0.3 fm/c. For both equations of state, we extract the T, μ_B, μ_Q, and μ_S using the conditions in Eq.(5) from our semi-analytical model, which are relevant for heavy ion collisions and have also been used to constrain the lattice EoS [4]. As $\sqrt{s_{NN}}$ increases, the maximum temperature $T^{\max}$ increases, but the maximum baryon chemical potential $\mu_B^{\max}$ decreases in Fig. 2. The results for the lattice EoS show that μ_B first decreases before increasing with time, because the lattice EoS smoothly merges with the hadron resonance gas model at $T \lesssim 135$ MeV [4]. The open circles in Fig. 2 represent the times when the lattice trajectories are inside the region $\mu_B/T > 2.5$, where the lattice EoS is expected to break down [4]. We also observe in Fig. 2(c) that $\mu_Q^{\max}$ from the lattice EoS can be much larger than that from the quantum EoS (by a factor ~ 2), while in Fig.2(d) the $\mu_S^{\max}$ values extracted from the two EoS are reasonably close (within $\sim 20\%$ of each other). Note that a recent work [5] using the AMPT model, which includes secondary parton interactions, found similar results for the time dependences of T and μ as our results here.

In Fig. 3, we show the trajectories extracted from our densities using the quantum and lattice equations of state in comparison with the RHIC chemical freezeout data, which were obtained from grand canonical fits to the particle yields [6]. Trajectories for energies $\sqrt{s_{NN}}$ = 7.7, 11.5, 19.6, 27, 39, 62.4, and 200 GeV with τ_F = 0.3 fm/c cross the crossover curve and can thus be compared with the freezeout data, while the $\sqrt{s_{NN}}$ = 5.0 GeV lattice trajectory indicates a problem in finding the full $T-\mu_B$ solution. We can see the effect of using the more realistic lattice EoS on the extracted trajectories; e.g., the intersections with the crossover curve from the functional renormalization group (FRG) [7] shift to smaller μ_B and are closer to the RHIC chemical freezeout data.

The maximum temperature reached by the trajectories extracted with the lattice EoS are also larger than that extracted with the quantum EoS, which is also shown in Fig. 2. In addition, $\mu_B^{\max}$ extracted with the lattice EoS is significantly larger ($\sim 10\%$) at low to moderate collision energies than that with the quantum EoS. As $\sqrt{s_{NN}}$ increases, the difference in $\mu_B^{\max}$ between the two equations of state becomes smaller such that there is essentially no difference at $\sqrt{s_{NN}}$ = 200 GeV. Note that the lattice trajectories at late times below the FRG crossover curve do not approach the origin but instead go to a finite μ_B and low T in the QCD phase diagram. This behavior can also be seen in Fig. 2(b), where μ_B at late times can increase when using the lattice EoS but always decreases when using the quantum EoS.

In order to extract the lattice EoS trajectories in Fig. 3, we calculate the intersection points between the constant ϵ and n_B contours in the $T-\mu_B$ plane that correspond to the $\epsilon(t)$ and $n_B(t)$ values at a given time t from our model [2]. We find that the lattice EoS does not have $T-\mu_B$ solutions for low collision energies at very early or very late times; this usually happens when the trajectory is in the

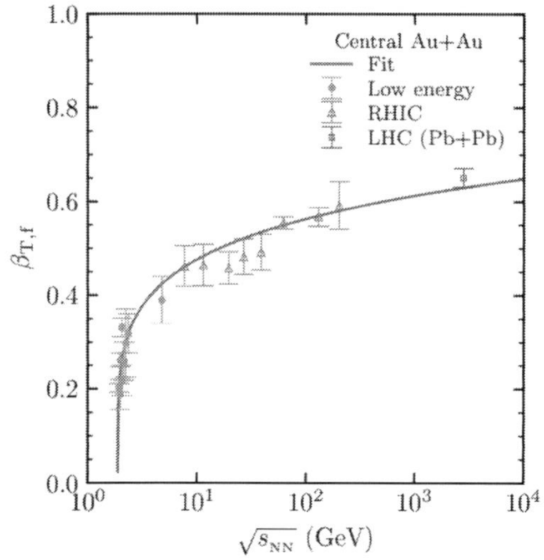

Figure 4. Our parameterization (solid) of the final transverse velocity $\beta_{T,f}$ in central Au+Au collisions as a function of collision energy compared to kinetic freezeout values from multiple low energy experiments (circles), the Beam Energy Scan program at RHIC (triangles), and the LHC (square).

large μ_B/T region where the lattice EoS is expected to be unreliable [4]. For example, the $\sqrt{s_{NN}}$ = 5.0 GeV lattice trajectory in Fig. 3 has no $T - \mu_B$ solution below the FRG crossover line for the densities from our semi-analytical model. Moreover, we find that no solution exists in the lattice EoS for our densities at any time in the evolution of central Au+Au collisions at $\sqrt{s_{NN}}$ = 2.0 GeV [2]. Therefore, the lattice EoS is an improvement over the ideal gas EoS at high collision energies where $\mu_B/T < 2.5$; however, it is expected to be unreliable at low energies.

We have also investigated the effect of transverse expansion by increasing the transverse overlap area A_T with time t in Eqs. (3)-(4). The transverse radius R_T of the overlap region increases according to a data-based parameterization of the transverse flow velocity $\beta_T(t)$. We assume that $\beta_T(t)$ starts at 0 and smoothly approaches a final value $\beta_{T,f}$. Using the kinetic freezeout data, which were obtained by fitting the transverse momentum spectra of central Au+Au collisions at various collision energies [6] to a blast-wave model, we parameterize $\beta_{T,f}$ as [2]:

$$\beta_{T,f} = \left[\frac{\ln\left(\sqrt{s_{NN}}/E_0\right)}{64.7 + \ln\left(\sqrt{s_{NN}}/E_0\right)} \right]^{0.202}, \quad (9)$$

where $E_0 = 2m_N$ is the threshold energy. Figure 4 shows the parameterization in comparison with the kinetic freezeout data [6]. The data at low energies (orange circles) and RHIC energies (green triangles) are for Au+Au collisions, while the data at the LHC energy (red square) is for Pb+Pb collisions at 2.76 TeV (where we have neglected the difference between Pb and Au for the kinetic freezeout data).

Further details regarding the implementation and effects of transverse expansion in our semi-analytical model can be found in the full study [2]. Overall, we find that including the transverse flow essentially does not change the path of the trajectory (at a given $\sqrt{s_{NN}}$ and τ_F), but it moves the trajectory endpoint, the $T - \mu_B$ point corresponding to ϵ^{max} and n_B^{max}, a bit closer to the origin. Importantly, the transverse expansion significantly decreases the time spent in the parton phase (i.e., the QGP lifetime) at all collision energies, and we also find that the QGP lifetime may have a local maximum below $\sqrt{s_{NN}} \sim 11.5$ GeV [2].

3 Summary and Outlook

In this proceeding, we have calculated the $T - \mu_B$ trajectories in the QCD phase diagram for central Au+Au collisions using our semi-analytical model, which includes the effect of the finite nuclear thickness. We have shown how the trajectories depend on the chosen equation of state and that the trajectories extracted with a lattice QCD-based EoS agree rather well with the chemical freezeout data from the RHIC Beam Energy Scan program. We also briefly discuss the implementation of transverse expansion and its effects on the trajectories. We have written a web interface [8], which currently calculates the densities and trajectories after the user specifies the colliding nuclei, $\sqrt{s_{NN}}$, τ_F, and the ideal gas EoS with quantum or Boltzmann statistics. We plan to further improve this web interface to include options to use the lattice EoS and/or consider transverse expansion. We hope that our semi-analytical model provides a useful tool for exploring the evolution of the dense matter in the QCD phase diagram.

This work has been supported by the National Science Foundation under Grant No. PHY-2012947.

References

[1] J.D. Bjorken, Phys. Rev. D **27**, 140 (1983)
[2] T. Mendenhall, Z.W. Lin (2021), `2111.13932`
[3] T. Mendenhall, Z.W. Lin, Phys. Rev. C **103**, 024907 (2021)
[4] J. Noronha-Hostler, P. Parotto, C. Ratti, J.M. Stafford, Phys. Rev. C **100**, 064910 (2019)
[5] H.S. Wang, G.L. Ma, Z.W. Lin, W.j. Fu, Phys. Rev. C **105**, 034912 (2022)
[6] L. Adamczyk et al. (STAR), Phys. Rev. C **96**, 044904 (2017)
[7] W.j. Fu, J.M. Pawlowski, F. Rennecke, Phys. Rev. D **101**, 054032 (2020)
[8] A web interface that performs our semi-analytical calculations is availabe at http://myweb.ecu.edu/linz/densities/ (2021)

Equation of State and Energy Loss of Hot and Dense Quark-Gluon matter from Holographic Black Holes

Joaquin Grefa[1,*], *Mauricio* Hippert[2], *Jorge* Noronha[2], *Jacquelyn* Noronha-Hostler[2], *Israel* Portillo[1], *Claudia* Ratti[1], and *Romulo* Rougemont[3]

[1]Physics Department, University of Houston, Houston, Texas 77204, USA
[2]Illinois Center for Advanced Studies of the Universe, Department of Physics, University of Illinois at Urbana-Champaign, Urbana, Illinois 61801, USA
[3]Instituto de Física, Universidade Federal de Goiás, Av. Esperança - Campus Samambaia, CEP 74690-900, Goiânia, Goiás, Brazil

Abstract. By using gravity/gauge correspondence, we construct a holographic model, constrained to mimic the lattice QCD equation of state at zero density, to investigate the temperature and baryon chemical potential dependence of the equation of state. We also obtained the energy loss of light and heavy partons within the hot and dense plasma represented by the heavy quark drag force, Langevin diffusion coefficients and jet quenching parameter at the critical point and across the first-order transition line predicted by the model.

1 Introduction

At vanishing density, strongly interacting matter undergoes a smooth but rapid crossover transition from hadrons at low temperature to a system of deconfined quarks and gluons at high temperature [1], a strongly interacting liquid called the quark-gluon plasma (QGP). This crossover is expected to evolve into a line of first order phase transition with a critical end point (CEP) at finite baryon chemical potential μ_B. Depending on the location of the CEP in the QCD phase diagram, its effects may be probed in relativist heavy ion collisions by looking at the fluctuations of the baryon charge as a function of the center of mass energy [2]. On the theory side, lattice simulations at finite baryon density are limited by the sign problem, and an effective field theory is needed to guide the experimental search for the QCD critical point.

Such effective approach to describe hot and dense quark-gluon matter must reproduce the lattice QCD equation of state (EoS) at vanishing chemical potential and exhibit nearly inviscid flow behavior, a feature of the QGP. In fact, Gauge/gravity correspondence [3–5] has successfully been employed to mimic the physics of the QGP around the crossover and allows calculations both in equilibrium as well as out of equilibrium. In this work, we summarize some of our results from Refs. [6, 7] where we considered a five-dimensional gravitational theory with a real scalar field ϕ (the dilaton field), and a potential of the scalar field $V(\phi)$, responsible for the dynamical breaking of conformal symmetry. Effects due to finite chemical potential are taken into account by adding a Maxwell field A_μ and a coupling function $f(\phi)$. This construction defines a holographic Einstein-Maxwell-dilaton (EMD) model.

*e-mail: vgrefaju@central.uh.edu

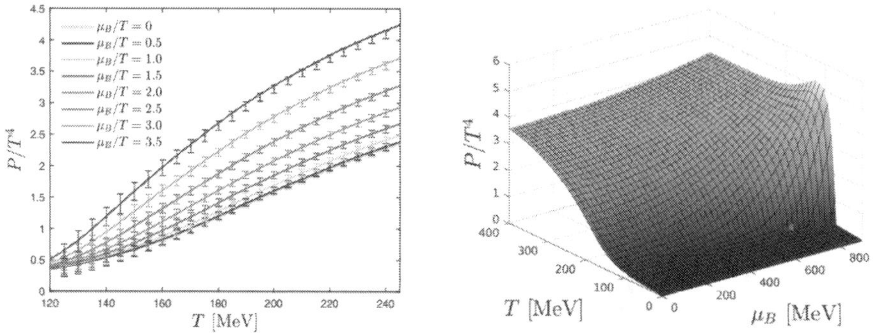

Figure 1. Left: normalized pressure as a function of the temperature for different values of μ_B/T and its comparison with the most recent lattice results. Right, the pressure in the phase diagram with the location of the CEP (red point).

2 The EMD holographic model and the EoS at finite density

The bulk EMD action is given by [8–10],

$$S = \frac{1}{2\kappa_5^2} \int_{\mathcal{M}_5} d^5x \sqrt{-g} \left[R - \frac{(\partial_\mu \phi)^2}{2} - V(\phi) - \frac{f(\phi) F_{\mu\nu}^2}{4} \right], \tag{1}$$

where κ_5^2 is the 5-dimensional gravitational constant, $g_{\mu\nu}$ is the metric tensor, and R is the Ricci scalar. We are interested here in charged isotropic and translationally invariant black hole backgrounds for the EMD fields. The two free functions in the holographic model, $V(\phi)$ and $f(\phi)$, can be dynamically fixed by matching the holographic EoS and the second order baryon susceptibility χ_2^B to the corresponding lattice QCD results with 2 + 1 flavours and physical quark masses at $\mu_B = 0$ from Refs. [11, 12].

The equations of motion obtained from the action (Eq. 1) are numerically solved given a pair of two initials conditions (ϕ_0, Φ_1) where ϕ_0 is the value of the dilaton field at the horizon, and Φ_1 is the electric field in the holographic direction evaluated at the horizon. Each black hole solution corresponds to a state in the QCD-like theory with a value for the entropy density (s), and baryon density (ρ_B) over the temperature and baryon chemical potential coordinates computed accordingly to the holographic dictionary [6, 10].

The comparison of our results for the holographic pressure with the most recent lattice QCD data up to the ratio of $\mu_B/T = 3.5$ [13] is shown in Fig. 1, where we also present this observable over a broad region in the phase diagram with the position of the CEP. The baryon susceptibilities, defined as $\chi_n(T, \mu_B) = \partial^n(P/T^4)/\partial(\mu_B/T)^n$, diverge in the vicinity of the critical point. In particular, the second order baryon susceptibility χ_2 develops a peak at larger chemical potential which evolves into a divergence at the CEP, located at $T^{CEP} = 89$ MeV and $\mu_B^{CEP} = 724$ MeV in this model.

3 Energy Loss

As a heavy quark moves through a hot and baryon dense medium, it loses energy and momentum through the drag force $F_{\text{drag}} = dp_x/dt$ which can be computed from the holographic trailing string approach [7]. The results for the heavy quark drag force for two different values of the quark velocity are shown in Fig. 2. One can infer that a very heavy quark (i.e. the bottom), which might not achieve a very high velocity within the plasma (the case for $v = 0.5$), is less sensitive to the in-medium effects in comparison with a less massive quark (i.e. the charm), which could attain higher velocities within the fluid (the case for $v = 0.99$). Additionally, one can observe that the magnitude of the energy loss associated with the heavy quark force increases by lowering the temperature or/and increasing μ_B.

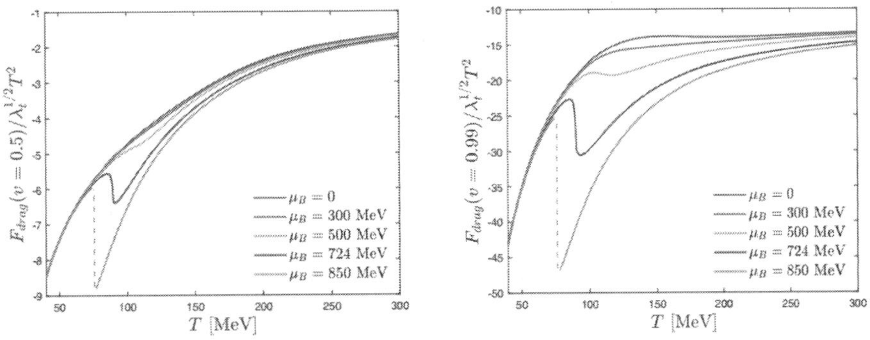

Figure 2. Heavy quark drag force at $v = 0.5$ (left) and $v = 0.99$ (right) as a function of the temperature for several values of μ_B

Langevin diffusion processes can be treated holographically. In particular, the Langevin diffusion coefficients, that describe the thermal fluctuations of a heavy quark trajectory under Brownian motion, can be computed from our EMD setup. The results for the parallel Langevin diffusion coefficient at $v = 0.99$ is shown in the left panel of Fig. 3. Analogously to the case of the heavy quark drag force, the Langevin diffusion coefficient is also enhanced with increasing baryon chemical potential.

Another transport coefficient that can be computed from the holographic EMD model is the jet quenching parameter ($\hat{q}$), defined as the rate for transverse momentum broadening. The jet quenching parameter characterizes the energy loss from collisional and radiative processes of high energy partons produced by the interaction with the hot and dense medium they travel through. The holographic results for the jet quenching parameter are shown in the right panel of Fig. 3. We observe that $\hat{q}/T^3$ displays a peak around the crossover that becomes sharper and increases in magnitude as the baryon chemical potential increases, which indicates more jet suppression and parton energy loss in the baryon rich regime.

In overall, the transport coefficients presented in this contribution exhibit a non trivial dependence with respect to the temperature and chemical potential, and they remain finite at the critical point as expected for holographic / large-N_c approaches (type-B dynamical univer-

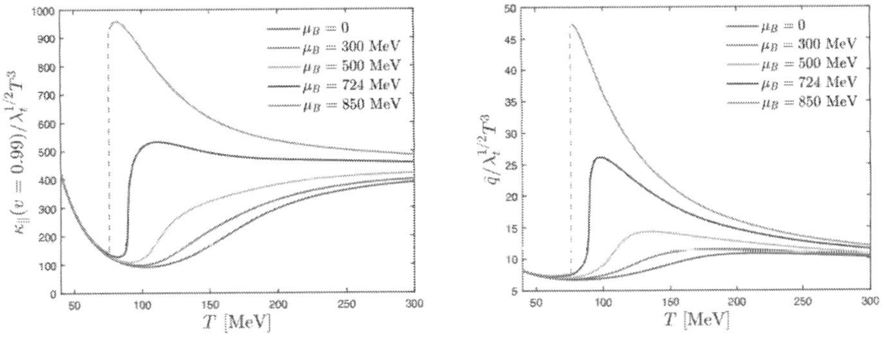

Figure 3. Parallel Langevin diffusion at coefficient $v = 0.99$ (left panel) and normalized jet quenching parameter $\hat{q}$ (right panel) as a function of the temperature for several values of μ_B.

sality class) [7]. While some possess a peak and others a local minimum, it is the inflection point, that gives rise to these extrema, that moves toward the CEP as μ_B is increased, acquiring an infite slope in the critical region. Since these transport coefficients also appear to be sensitive to the transition from the confined to the deconfined phase, its inflection points can be used to define pseudo-transition temperatures over the crossover region. Analogously to the equilibrium variables like the entropy density and baryon density, the transport coefficients display a discontinuity that corresponds to the line of first order phase transition.

4 Conclusions

By using the EMD model from Refs. [6, 10], we obtain the equilibrium and dynamical properties for a hot a dense QGP. Analogously to the EoS variables, the transport coefficients presented in this work display a temperature and baryon chemical potential dependence with different inflection points that can be used to describe the crossover transition region. They develop an infinite slope at the CEP that becomes a discontinuity gap at the first order transition line.

Acknowledgements:

This material is based upon work supported by the National Science Foundation under grants No. PHY-1654219, PHY-2208724 and PHY-2116686. This work was supported in part by the National Science Foundation (NSF) within the framework of the MUSES collaboration, under grant No. OAC-2103680, the US-DOE Nuclear Science Grant No. DE-SC0020633. J.N. is partially supported by the U.S. Department of Energy, Office of Science, Office for Nuclear Physics under Award No. DE-SC0021301.

References

[1] Y. Aoki, G. Endrodi, Z. Fodor, S.D. Katz, K.K. Szabo, Nature **443**, 675 (2006), hep-lat/0611014
[2] M.A. Stephanov, Phys. Rev. Lett. **107**, 052301 (2011)
[3] J.M. Maldacena, Adv. Theor. Math. Phys. **2**, 231 (1998), hep-th/9711200
[4] S.S. Gubser, I.R. Klebanov, A.M. Polyakov, Phys. Lett. B **428**, 105 (1998), hep-th/9802109
[5] E. Witten, Adv. Theor. Math. Phys. **2**, 253 (1998), hep-th/9802150
[6] J. Grefa, J. Noronha, J. Noronha-Hostler, I. Portillo, C. Ratti, R. Rougemont, Phys. Rev. D **104**, 034002 (2021), 2102.12042
[7] J. Grefa, M. Hippert, J. Noronha, J. Noronha-Hostler, I. Portillo, C. Ratti, R. Rougemont, Phys. Rev. D **106**, 034024 (2022), 2203.00139
[8] O. DeWolfe, S.S. Gubser, C. Rosen, Phys. Rev. D **83**, 086005 (2011), 1012.1864
[9] O. DeWolfe, S.S. Gubser, C. Rosen, Phys. Rev. D **84**, 126014 (2011), 1108.2029
[10] R. Critelli, J. Noronha, J. Noronha-Hostler, I. Portillo, C. Ratti, R. Rougemont, Phys. Rev. D **96**, 096026 (2017), 1706.00455
[11] S. Borsanyi, Z. Fodor, C. Hoelbling, S.D. Katz, S. Krieg, K.K. Szabo, Phys. Lett. B **730**, 99 (2014), 1309.5258
[12] R. Bellwied, S. Borsanyi, Z. Fodor, S.D. Katz, A. Pasztor, C. Ratti, K.K. Szabo, Phys. Rev. D **92**, 114505 (2015), 1507.04627
[13] S. Borsányi, Z. Fodor, J.N. Guenther, R. Kara, S.D. Katz, P. Parotto, A. Pásztor, C. Ratti, K.K. Szabó, Phys. Rev. Lett. **126**, 232001 (2021), 2102.06660

Lattice QCD equation of state at finite chemical potential from an alternative resummation: strangeness neutrality and beyond

Paolo Parotto[1,*], *Szabolcs* Borsányi[2], *Zoltan* Fodor[1,2,3,4,5], *Jana N.* Guenther[2], *Ruben* Kara[2], *Attila* Pásztor[3], *Claudia* Ratti[6], and *Kalman K.* Szabó[2,4]

[1]Pennsylvania State University, Department of Physics, University Park, PA 16802, USA
[2]Department of Physics, Wuppertal University, Gaussstr. 20, D-42119, Wuppertal, Germany
[3]Inst. for Theoretical Physics, ELTE Eötvös Loránd University, Pázmány P. sétány 1/A, H-1117 Budapest, Hungary
[4]Jülich Supercomputing Centre, Forschungszentrum Jülich, D-52425 Jülich, Germany
[5]Physics Department, UCSD, San Diego, CA 92093, USA
[6]Department of Physics, University of Houston, Houston, TX 77204, USA

Abstract. In this contribution we present a resummation of the Quantum Chromodynamics (QCD) equation of state from lattice simulations at imaginary chemical potentials. We generalize the scheme introduced in a previous work [1], to the case of non-zero strangeness chemical potential. We present continuum extrapolated results for thermodynamic observables in the temperature range 130 MeV $\leq T \leq$ 280 MeV, for chemical potentials up to $\mu_B/T = 3.5$, along the strangeness neutral line. Furthermore, we relax the constraint of strangeness neutrality, by extrapolating to small values of the strangeness-to-baryon-number ratio $R = n_S/n_B$.

1 Introduction

The knowledge of the QCD equation of state is of crucial importance, besides its intrinsic fundamental interest, for the modeling of heavy-ion collisions. At vanishing baryon density – where the transition is a crossover [2] – the equation of state has been known for about a decade from lattice QCD simulations [3–5]. On the other hand, results at finite baryon density are limited by the sign problem in lattice simulations. Though direct simulations have been performed recently with a novel reweighting approach [6, 7], on large lattices the use of indirect (extrapolation) techniques is necessary. These rely on Taylor expansion or analytical continuation from imaginary chemical potential [8–13].

Recently, different forms of reorganization of the Taylor series have been proposed [1, 14, 15], with the aim of improving its convergence. In particular, in [1] we introduced a scheme to extrapolate the equation of state of QCD based on the following ansatz:

$$F(T, \hat{\mu}_B) = F(T', 0) \; , \tag{1}$$

*e-mail: paolo.parotto@gmail.com

whereby the $\hat{\mu}_B$-dependence[1] of an observable $F(T, \hat{\mu}_B)$ is described by a rescaling of the temperature:

$$T' = T\left(1 + \kappa_2^F(T)\hat{\mu}_B^2 + \kappa_4^F(T)\hat{\mu}_B^4 + O(\hat{\mu}_B^6)\right) , \tag{2}$$

in turn described by the alternative coefficients $\kappa_n^F(T)$. In essence, we move to a framework in which we expand in the shift

$$\Delta T = T - T' = \left(\kappa_2^F(T)\hat{\mu}_B^2 + \kappa_4^F(T)\hat{\mu}_B^4 + O(\hat{\mu}_B^6)\right) . \tag{3}$$

In these proceedings we illustrate a generalization of this scheme, to the case of non-vanishing strangeness chemical potential, studying in particular the case of strangeness neutrality $n_S = 0$, which has phenomenological relevance for heavy-ion collisions. Moreover, we perform an initial extrapolation to non-zero strangeness density, through an expansion in the strangeness-to-baryon ratio $R = n_S/n_B$.

2 Observables

We consider three different observables:

$$c_1^B(\hat{\mu}_B, T), \qquad \mathrm{M}(\hat{\mu}_B, T) = \frac{\mu_S}{\mu_B}(\hat{\mu}_B, T), \qquad \chi_2^S(\hat{\mu}_B, T), \tag{4}$$

where

$$c_n^B = \frac{d^n}{d\hat{\mu}_B^n}\frac{p}{T^4} = \left(\frac{\partial}{\partial\hat{\mu}_B} + \frac{d\hat{\mu}_S}{d\hat{\mu}_B}\frac{\partial}{\partial\hat{\mu}_S}\right)^n \frac{p}{T^4} = \left(\frac{\partial}{\partial\hat{\mu}_B} - \frac{\chi_{11}^{BS}}{\chi_2^S}\frac{\partial}{\partial\hat{\mu}_S}\right)^n \frac{p}{T^4}$$

are the Taylor coefficients of the pressure along the strangeness neutral line, μ_S is such to realize strangeness neutrality, and $\chi_2^S(\hat{\mu}_B, T) = \partial^2(p/T^4)/\partial\hat{\mu}_S^2$. Note that, because of strangeness neutrality, $c_1^B = \chi_1^B = \partial(p/T^4)/\partial\hat{\mu}_B$. With respect to [1], we introduce an additional change in the case of c_1^B: for the observable F in Eq.(1), we consider the ratio $c_1^B(T, \hat{\mu}_B)/\bar{c}_1^B(\hat{\mu}_B)$, where $\bar{c}_1^B$ is the ($\hat{\mu}_B$-dependent) Stefan-Boltzmann limit of c_1^B. This is done in order to avoid the increase of κ_2 at large temperatures. With this modification, the expansion parameters are effectively changed, and we thus refer to the corrected ones by λ_n^F.

We show in the left panel of Fig.1 results for the ratio $c_1^B(T, \hat{\mu}_B)/\bar{c}_1^B(\hat{\mu}_B)$ at different values of the imaginary baryon chemical potential, in the case of strangeness neutrality (left). In the right panel, we show the same curves, with the x-axis replaced by $T \to T(1 + \lambda\hat{\mu}_B^2)$, with $\lambda = 0.0165$. We see in this case, that the difference between the different curves is minimal around the transition temperature, as well as in the high temperature limit. We also note that this value is different from the value $\kappa = 0.0205$ used in [1]. This is because the coefficients λ_n are defined after dividing by the Stefan-Boltzmann limit, but more importantly because the curvature in the case of strangeness neutrality seems to be smaller. A similar scenario appears in the curvature of the QCD transition line, where the numerical value in the case when $\mu_Q = \mu_S = 0$ is slightly larger [16].

The procedure to determine the coefficients λ_n^F makes use of results at both zero and non-zero chemical potential. The temperature T' for which Eq. (1) is satisfied is determined for different values of T, $\hat{\mu}_B$ and on lattices with different numbers of timeslices N_τ. A global fit for each temperature T is then performed, and continuum extrapolated results for the λ_n^F are obtained. For more details on the procedure, we refer the reader to [1, 15]. Once the coefficients are determined, the use of Eq. (1) for the observables in Eq. (4) allows us to determine the latter at finite (real) values of the chemical potential.

[1]We use the following notation for the dimensionless chemical potentials: $\hat{\mu}_i = \mu_i/T$.

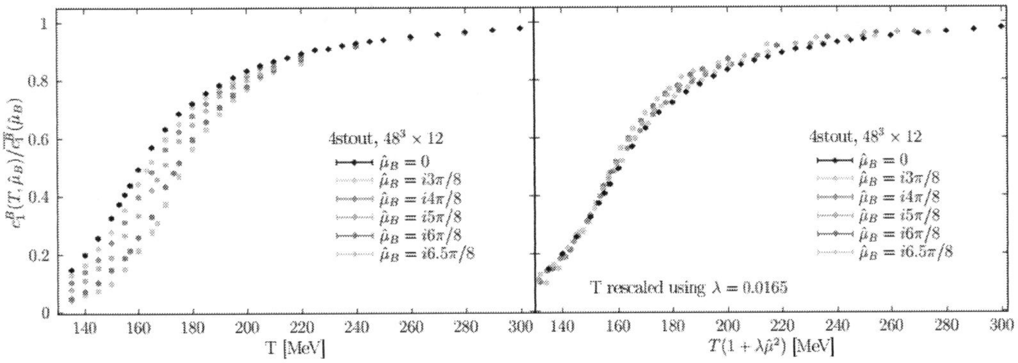

Figure 1. Left panel: results for the ratio $c_1^B(T,\hat{\mu}_B)/\overline{c}_1^B(\hat{\mu}_B)$ at different values of the imaginary baryon chemical potential, in the case of strangeness neutrality, on a $48^3 \times 12$ lattice. Right panel: same as on the left panel, with the x-axis rescaled according to $T \to T(1+\lambda\hat{\mu}_B^2)$, with $\lambda = 0.0165$.

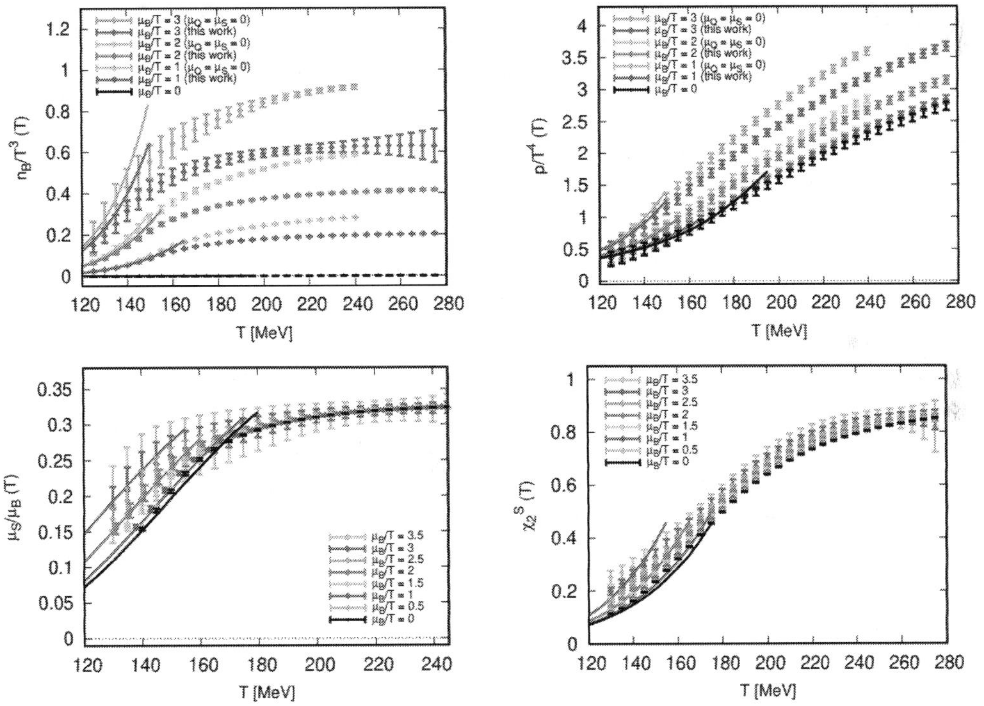

Figure 2. From top to bottom, left to right: baryon density, pressure, strangeness-to-baryon chemical potential ratio, and strangeness second susceptibility. The results are shown at increasing real values of $\hat{\mu}_B$. With solid lines we show the results from the hadron resonance gas (HRG) model. For baryon density and pressure, we also show (in lighter shades), the results for the non-strangeness neutral case.

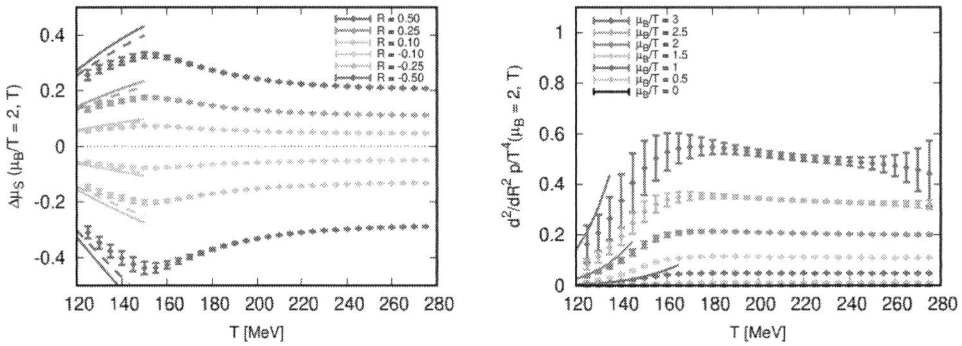

Figure 3. Left panel: shift in strangeness chemical potential for different values of the ratio R, at $\hat{\mu}_B = 2$. The lines indicate results from the HRG model obtained from Eq. (6) (dashed), or by solving $R = n_s/n_B$ for $\hat{\mu}_S$ (solid). Right panel: first non-vanishing coefficient for the expansion of the pressure in R around $n_S = 0$, for increasing values of $\hat{\mu}_B$. Solid lines show results from the HRG model.

We show in Fig. 2 the baryon density (top left), strangeness-to-baryon chemical potentials ratio (bottom left) and second strangeness susceptibility (bottom right) at increasing real values of $\hat{\mu}_B$. We also show (top right) the pressure, obtained from the baryon density through simple integration. For the baryon density and pressure, we also show the results in the case without strangeness neutrality. We see that all results, up to $\hat{\mu}_B = 3.5$ show very reasonable uncertainties, and do not hint at unphysical, non- monotonic behavior typical of traditional Taylor expansions.

3 Beyond strangeness neutrality

We wish to expand beyond strangeness neutrality in terms of the ratio $R = n_S/n_B$. Defining $\hat{\mu}_S^\star$ as the strangeness chemical potential realizing strangeness neutrality, we can write:

$$n_S \equiv \chi_1^S(\hat{\mu}_S) \approx \chi_2^S(\hat{\mu}_S^\star)\Delta\hat{\mu}_S \ , \qquad n_B \equiv \chi_1^B(\hat{\mu}_S) \approx \chi_1^B(\hat{\mu}_S^\star) + \chi_{11}^{BS}(\hat{\mu}_S^\star)\Delta\hat{\mu}_S \ , \tag{5}$$

where $\Delta\hat{\mu}_S \equiv \hat{\mu}_S - \hat{\mu}_S^\star$. Hence, we obtain:

$$R = \frac{\chi_1^S}{\chi_1^B} = \frac{\chi_2^S(\hat{\mu}_S^\star)\Delta\hat{\mu}_S}{\chi_1^B(\hat{\mu}_S^\star)\Delta\hat{\mu}_S + \chi_{11}^{BS}(\hat{\mu}_S^\star)} \ , \qquad \Delta\hat{\mu}_S = \frac{R\hat{\chi}_1^B(\hat{\mu}_S^\star)}{\chi_2^S(\hat{\mu}_S^\star) - R\chi_{11}^{BS}(\hat{\mu}_S^\star)} \ . \tag{6}$$

In the left panel of Fig. 3 we show the displacement $\Delta\hat{\mu}_S$ that realizes $n_S = Rn_B$, for $\hat{\mu}_B = 2$ and different values of R.

The correction to the pressure for small values of R vanishes at first order, because $n_S = 0$ by construction. The first non-zero contribution comes at the second order in R:

$$\frac{\mathrm{d}^2\hat{p}}{\mathrm{d}R^2}(T,\hat{\mu}_B) = \frac{\left(\chi_1^B(T,\hat{\mu}_B)\right)^2}{\chi_2^S(T,\hat{\mu}_B)} \ . \tag{7}$$

We show this correction coefficient for different values of $\hat{\mu}_B$ in the right panel of Fig. 3.

In these proceeding we have shown our continuum extrapolated results for thermodynamic quantities at finite real chemical potential, up to $\hat{\mu}_B = 3.5$. We obtained these with a generalization of the novel resummation scheme introduced in [1], applied to the case of strangeness neutrality. We showed that our results, for all chemical potentials, show no hints of pathological behavior, and display uncertainties well under control. Finally, we calculated the first terms for an expansion *around* the strangeness neutral line, in terms of the ratio $R = n_S/n_B$. These results can be improved systematically with larger statistics, and thus in the future provide guidance for the knowledge of the QCD equation of state for different chemical compositions of the QCD medium.

Acknowledgments

The project was supported by the BMBF Grant No. 05P18PXFCA and 05P21PXFCA. This work was also supported by the Hungarian National Research, Development and Innovation Office, NKFIH grant KKP126769. This material is based upon work supported by the National Science Foundation under grants no. PHY-1654219, PHY-2116686 and OAC-2103680. A.P. is supported by the J. Bolyai Research Scholarship of the Hungarian Academy of Sciences and by the ÚNKP-21-5 New National Excellence Program of the Ministry for Innovation and Technology. The authors gratefully acknowledge the Gauss Centre for Supercomputing e.V. (www.gauss-centre.eu) for funding this project by providing computing time on the GCS Supercomputers HAWK at HLRS, Stuttgart as well as the JUWELS/Booster and JURECA/Booster at FZ-Juelich. Part of the computation was performed on the QPACE3 funded by the DFG and hosted by JSC.

References

[1] S. Borsányi, Z. Fodor, J. N. Guenther, R. Kara, S. D. Katz, P. Parotto, A. Pásztor, C. Ratti and K. K. Szabó, Phys. Rev. Lett. **126**, no.23, 232001 (2021) doi:10.1103/PhysRevLett.126.232001 [arXiv:2102.06660 [hep-lat]].

[2] Y. Aoki, G. Endrodi, Z. Fodor, S. D. Katz and K. K. Szabo, Nature **443**, 675-678 (2006) doi:10.1038/nature05120 [arXiv:hep-lat/0611014 [hep-lat]].

[3] S. Borsanyi, G. Endrodi, Z. Fodor, A. Jakovac, S. D. Katz, S. Krieg, C. Ratti and K. K. Szabo, JHEP **11**, 077 (2010) doi:10.1007/JHEP11(2010)077 [arXiv:1007.2580 [hep-lat]].

[4] S. Borsanyi, Z. Fodor, C. Hoelbling, S. D. Katz, S. Krieg and K. K. Szabo, Phys. Lett. B **730**, 99-104 (2014) doi:10.1016/j.physletb.2014.01.007 [arXiv:1309.5258 [hep-lat]].

[5] A. Bazavov *et al.* [HotQCD], Phys. Rev. D **90**, 094503 (2014) doi:10.1103/PhysRevD.90.094503 [arXiv:1407.6387 [hep-lat]].

[6] S. Borsanyi, Z. Fodor, M. Giordano, S. D. Katz, D. Nogradi, A. Pasztor and C. H. Wong, Phys. Rev. D **105**, no.5, L051506 (2022) doi:10.1103/PhysRevD.105.L051506 [arXiv:2108.09213 [hep-lat]].

[7] S. Borsanyi, Z. Fodor, M. Giordano, J. N. Guenther, S. D. Katz, A. Pasztor and C. H. Wong, [arXiv:2208.05398 [hep-lat]].

[8] C. R. Allton, S. Ejiri, S. J. Hands, O. Kaczmarek, F. Karsch, E. Laermann, C. Schmidt and L. Scorzato, Phys. Rev. D **66**, 074507 (2002) doi:10.1103/PhysRevD.66.074507 [arXiv:hep-lat/0204010 [hep-lat]].

[9] P. de Forcrand and O. Philipsen, Nucl. Phys. B **642**, 290-306 (2002) doi:10.1016/S0550-3213(02)00626-0 [arXiv:hep-lat/0205016 [hep-lat]].

[10] M. D'Elia and M. P. Lombardo, Phys. Rev. D **67**, 014505 (2003) doi:10.1103/PhysRevD.67.014505 [arXiv:hep-lat/0209146 [hep-lat]].

[11] S. Borsanyi, Z. Fodor, J. N. Guenther, S. K. Katz, K. K. Szabo, A. Pasztor, I. Portillo and C. Ratti, JHEP **10**, 205 (2018) doi:10.1007/JHEP10(2018)205 [arXiv:1805.04445 [hep-lat]].

[12] A. Bazavov, D. Bollweg, H. T. Ding, P. Enns, J. Goswami, P. Hegde, O. Kaczmarek, F. Karsch, R. Larsen and S. Mukherjee, *et al.* Phys. Rev. D **101**, no.7, 074502 (2020) doi:10.1103/PhysRevD.101.074502 [arXiv:2001.08530 [hep-lat]].

[13] D. Bollweg *et al.* [HotQCD], Phys. Rev. D **105**, no.7, 074511 (2022) doi:10.1103/PhysRevD.105.074511 [arXiv:2202.09184 [hep-lat]].

[14] S. Mondal, S. Mukherjee and P. Hegde, Phys. Rev. Lett. **128**, no.2, 022001 (2022) doi:10.1103/PhysRevLett.128.022001 [arXiv:2106.03165 [hep-lat]].

[15] S. Borsanyi, Z. Fodor, J. N. Guenther, R. Kara, P. Parotto, A. Pasztor, C. Ratti and K. K. Szabo, Phys. Rev. D **105** (2022) no.11, 114504 doi:10.1103/PhysRevD.105.114504 [arXiv:2202.05574 [hep-lat]].

[16] P. Cea, L. Cosmai and A. Papa, Phys. Rev. D **93**, no.1, 014507 (2016) doi:10.1103/PhysRevD.93.014507 [arXiv:1508.07599 [hep-lat]].

Phase diagram of the QCD Kondo effect and inactivation of the magnetic catalysis

Koichi Hattori[1,2], *Daiki* Suenaga[2,3], *Kei* Suzuki[4], and *Shigehiro* Yasui[5]

[1]Zhejiang Institute of Modern Physics, Department of Physics, Zhejiang University, Hangzhou, 310027, China

[2]Research Center for Nuclear Physics (RCNP), Osaka University, Osaka 567-0047, Japan

[3]Strangeness Nuclear Physics Laboratory, RIKEN Nishina Center, Wako 351-0198, Japan

[4]Advanced Science Research Center, Japan Atomic Energy Agency (JAEA), Tokai 319-1195, Japan

[5]Research and Education Center for Natural Sciences, Keio University, Hiyoshi 4-1-1, Yokohama, Kanagawa 223-8521, Japan

Abstract. We investigate the QCD phase diagram in strong magnetic fields with heavy-quark impurities and determine the ground state within the mean-field analysis. The ground state is characterized by magnitudes of the pairing not only between the light quark and antiquark, i.e., chiral condensate, but also between the light quark and heavy-quark impurity, dubbed the Kondo condensate. We propose signatures of the interplay and/or competition between those two pairing phenomena reflected in the magnitude of the chiral condensate that is saturated with respect to the magnetic-field strength and anomalously increases with increasing temperature.

1 Introduction

Magnetic fields restrict motion of charged particles within the cyclotron orbits, and reduce the dimension of the phase space occupied by the low-energy particles. Pairing among the particles is enhanced in the dimensionally reduced systems. Two of such effects are the quark-antiquark pairing, i.e., chiral condensate, and the pairing between a light quark and heavy-quark impurity dubbed the Kondo condensate. It is important to note that the pairing occurs only for the dimensional reason no matter how small the interaction strength is: The occurrence of the pairing is inevitable when the magnetic-field strength is larger than the temperature scale. The former pairing has been intensively studied under the name of the magnetic catalysis [1] (see, e.g., Ref. [2] for a review), while the latter is proposed as the novel Kondo effect induced by background magnetic fields [3] particularly in QCD (which is analogous to that in finite-density QCD [4, 5]). We investigate interplay and competition between these two pairing phenomena that enrich the phase diagram in strong magnetic fields.

2 Formalism

We discuss the essence of the chiral symmetry breaking and the Kondo effect in magnetic fields with a four-Fermi interaction model [6] (see also Ref. [7])

$$\begin{aligned}\mathcal{L} &= \bar{\psi}(i\partial\!\!\!/_{\parallel} - m_l)\psi + \frac{G_{ll}}{2N}\left[(\bar{\psi}\psi)^2 + (\bar{\psi}i\gamma_5\psi)^2\right] \\ &+ \sum_{c=\pm}\left[c\bar{\Psi}_v^c i\partial_0 \Psi_v^c + \frac{G_{hl}}{N}\left\{(\bar{\psi}\Psi_v^c)(\bar{\Psi}_v^c\psi) + (\bar{\psi}i\gamma_5\Psi_v^c)(\bar{\Psi}_v^c i\gamma_5\psi)\right\}\right], \end{aligned} \tag{1}$$

where we have the number of colors $N = 3$, light-quark mass m_l, and the coupling strengths G_{hl} and G_{ll} for the interactions between the heavy and light quarks and between the light quarks, respectively. The light quark ψ in the lowest Landau level (LLL) has the dimensionally reduced kinetic term with $\partial\!\!\!/_{\parallel} = \gamma^0\partial_0 + \gamma^3\partial_3$ when the magnetic field is applied in the spatial third direction. Ψ_v^+ and Ψ_v^- are the heavy-fermion and its antifermion fields introduced as impurities. We took the leading-order terms in the heavy-quark expansion with respect to the inverse mass.

We introduce the chiral condensate $\langle\bar{\psi}\psi\rangle_{\rm LLL} \equiv -\frac{N}{G_{ll}}M$, and the Kondo condensate $\langle\bar{\psi}\Psi_v^{\pm}\rangle_{\rm LLL} \equiv \frac{N}{G_{hl}}\Delta$. In this work, we assume only the color-singlet condensates which are the dominant ones among all the color channels in the large N_c limit (see Ref. [8]). We then determine the magnitudes of those condensates self-consistently from the global minimum of the effective potential. At the linear order in the fluctuations around the mean fields, one can diagonalize the Lagrangian (1) to find four eigenvalues [6]

$$E_{\pm}(p_z) \equiv \pm\frac{1}{2}\left(\sqrt{E_{p_z}^2 + |2\Delta|^2} \pm E_{p_z}\right), \quad \tilde{E}_{\pm}(p_z) \equiv \pm\frac{1}{2}\left(\sqrt{E_{p_z}^2 + |2\Delta|^2} \mp E_{p_z}\right), \tag{2}$$

where $E_{p_z} \equiv \sqrt{p_z^2 + (m_l + M)^2}$. The thermodynamic potential is then written down as [6]

$$\tilde{\Omega}(M,\Delta) = \frac{1}{2\tilde{G}_{ll}}M^2 + \frac{1}{2\tilde{G}_{hl}}|2\Delta|^2 - \sum_i \int_{-\Lambda}^{\Lambda}\frac{dp_z}{2\pi}\frac{1}{2}|\mathcal{E}_i| - \sum_i \int_{-\infty}^{\infty}\frac{dp_z}{2\pi}\frac{1}{\beta}\ln\left(1 + e^{-\beta|\mathcal{E}_i|}\right), \tag{3}$$

where the last term is for the finite-temperature contribution with the inverse temperature $\beta = 1/T$. The sum runs over the four eigenmodes $\mathcal{E}_i = \{E_+, E_-, \tilde{E}_+, \tilde{E}_-\}$. Here, we have introduced normalized quantities $\tilde{\Omega} = \Omega/(N\rho_B)$, $\tilde{G}_{hl} = \rho_B G_{hl}$, and $\tilde{G}_{ll} = \rho_B G_{ll}$ with the Landau degeneracy factor $\rho_B = |q_l B|/(2\pi)$ that stems from the momentum integral over the degenerate two-dimensional phase space. As a result, we are left with the thermodynamic potential with the one-dimensional momentum integral. Λ is the momentum cutoff that may be determined by fitting physical quantities with the lattice QCD data in magnetic fields.

3 Zero-temperature results

By minimizing the thermodynamic potential, we determine the ground state. The numerical results at zero temperature are shown in the left panel of Fig. 1. The Kondo condensate is shown with the blue lines. The dotted and dashed lines show the Kondo condensate at zero and nonzero light quark mass m_l, respectively, when the chiral condensate is absent or formally $G_{ll} = 0$. While there is no phase transition at $m_l = 0$, a finite m_l changes it to a phase transition at a critical magnetic-field strength B_{cK}. Thus, the light-quark mass plays an important role for the Kondo effect.

The solid blue line shows the full result in the presence of competition with the chiral condensate at $G_{ll}\Lambda^2 = 1.0$ and $G_{hl}\Lambda^2 = 3.0$. We find a up-shift of the critical magnetic-field

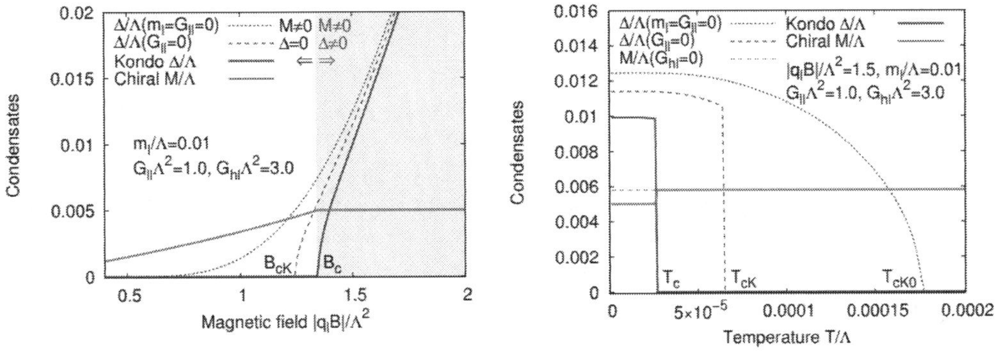

Figure 1. Magnitudes of the chiral and Kondo condensates at zero temperature (left) and at finite temperature (right) [6]. The solid lines show the full results with nonzero G_{hl} and G_{ll}, while the dotted and dashed lines show the results with the competition turned off with either $G_{ll} = 0$ or $G_{ll} = 0$.

strength to B_c, indicating that the Kondo effect is prohibited by the existence of the chiral condensate in the regime between B_{cK} and B_c. However, once the Kondo condensate appears at B_c, the chiral condensate, shown with the red line, saturates at a constant value. Such a *saturation of the chiral condensate* serves as a clear signal of the emergence of the Kondo condensate since it has been known that the chiral condensate monotonically increases by the magnetic catalysis [1] if the heavy-quark impurity is absent or formally $G_{hl} = 0$. Due to this saturation, the Kondo condensate eventually overwhelms the chiral condensate.

One can confirm those findings with analytic solutions to the gap equations [6]

$$\frac{M}{\tilde{G}_{ll}} + \frac{M + m_l}{2\pi} \ln \Xi = 0, \qquad \frac{\Delta}{\tilde{G}_{hl}} + \frac{\Delta}{2\pi} \ln \Xi = 0, \tag{4}$$

with $\Xi = \frac{(M+m_l)^2+4|\Delta|^2}{\left(\Lambda+\sqrt{(M+m_l)^2+4|\Delta|^2+\Lambda^2}\right)^2}$. They are derived from the stationary conditions $\partial\tilde{\Omega}/\partial M = 0 = \partial\tilde{\Omega}/\partial\Delta$. We immediately notice that the gap equations (4) have a trivial solution $\Delta = 0$ and a nontrivial solution M, indicating the prohibition of the Kondo condensate. This is the set of solutions below B_c which we call $(M_1, \Delta_1 = 0)$. By eliminating $\ln \Xi$ in Eq. (4), one can find another set of solutions (M_2, Δ_2) as [6]

$$M_2 = \frac{m_l G_{ll}}{G_{hl} - G_{ll}}, \qquad \Delta_2 = \frac{1}{2}\sqrt{\frac{\Lambda^2}{\sinh^2(2\pi^2/\tilde{G}_{hl})} - \frac{m_l^2 G_{hl}^2}{(G_{hl} - G_{ll})^2}}\,. \tag{5}$$

By solving $\Delta_2 = 0$, we obtain the critical magnetic-field strength [6]

$$q_l B_c = \frac{4\pi^3}{G_{hl}\,\mathrm{acrsinh}\left(\Lambda\frac{|G_{hl}-G_{ll}|}{m_l G_{hl}}\right)} \sim \frac{4\pi^3 m_l}{\Lambda|G_{hl} - G_{ll}|}, \tag{6}$$

where the last expression holds for a small value of $|G_{hl} - G_{ll}|$. The true vacuum switches over from $(M_1, \Delta_1 = 0)$ to (M_2, Δ_2) at B_c. The second set of solutions indicates the coexistence phase of both the condensates observed above B_c in Fig. 1, where the chiral condensate is saturated with the constant value M_2. Taking the cutoff scale Λ of the order of several hundred MeV, the above critical magnetic field is estimated to be sevelral GeV^2. This strength may be larger than those realized in LHC and RHIC, but are comparable to those that have been realized in recent lattice QCD simulations [9].

4 Finite-temperature results

Finally, we investigate the phase diagram at finite temperature and in strong magnetic fields. In the right panel of Fig. 1, we show the numerical results of temperature dependence of the condensates with the magnetic-field strength and the heavy-light coupling fixed at $|q_l B|/\Lambda^2 = 1.5$ and $G_{hl}\Lambda^2 = 3.0$. The Kondo condensate at $G_{ll} = 0$ and in the chiral limit ($m_l = 0$) is shown with the blue dotted line. We find a phase transition at the critical temperature T_{cK0}. When we switch on a light-quark mass $m_l/\Lambda = 0.01$, as shown by the blue dashed line, the magnitude of condensate is slightly suppressed, and the critical temperature decreases to T_{cK}. This m_l dependence is consistent with our result at zero temperature.

We now consider the chiral condensate with nonzero G_{ll}. First, without the competition at $G_{hl} = 0$, the chiral condensate M, shown with the red dashed line, is almost flat within the plot range and decreases at higher temperature. This is the result with the magnetic catalysis. The chiral condensate at nonzero m_l shows a crossover transition, and its pseudocritical temperature is located even in the higher-temperature region.

The full numerical results including both nonzero G_{ll} and G_{hl} are shown by the solid lines. We find that the chiral and Kondo condensates coexist at low temperature below T_c. Here, the critical temperature T_c for the Kondo condensate decreases from T_{cK} due to the competition effect. It is remarkable that the chiral condensate *increases abruptly* just above T_c to merge to the value in the absence of the competition. This anomalous increase with increasing temperature signals the end of competition and serves as an indirect evidence of the Kondo condensate below T_c even if measurement of the Kondo condensate itself is tough. There is no competition above T_c, and the chiral condensate never increases as we further increase temperature.

5 Conclusion and outlook

We reported the novel phase diagram emerging from the competition between the chiral and Kondo condensates in strong magnetic fields. At low temperature, the competition causes a *saturation behavior* of the chiral condensate above the critical magnetic-field strength. At finite temperature, the competition leads to a *steep increase* of the chiral condensate just above the critical temperature for the Kondo condensate. This anomalous increasing behavior signals the end of the competition. Our findings shed light on the intertwined dynamics with light and heavy quarks and may stimulate further developments of the nonperturbative many-body phenomena and of the lattice simulations in strong magnetic fields.

In fact, lattice QCD simulations in strong magnetic fields elucidated the magnetic catalysis in QCD with high precision, and they can be also applied to the counterpart in solid-state physics. The magnetically induced Kondo effect will be also confirmed by direct measurements of the Kondo condensate constructed from the correlation function of the heavy and light fermions or by indirect ones via the characteristic behaviors of the chiral condensate.

Acknowledgement.—This work was supported by Japan Society for the Promotion of Science (JSPS) KAKENHI under grant Nos. JP17K14277, JP20K14476, 20K03948, and 22H02316.

References

[1] V. P. Gusynin, V. A. Miransky, I.A. Shovkovy, Phys. Lett. B **349**, 477 (1995), hep-ph/9412257.

[2] I. A. Shovkovy, Lect. Notes Phys. **871**, 13 (2013), 1207.5081.

[3] S. Ozaki, K. Itakura, Y. Kuramoto, Phys. Rev. D **94**, 074013 (2016), 1509.06966.

[4] S. Yasui, K. Sudoh, Phys. Rev. C **88**, 015201 (2013), 1301.6830.

[5] K. Hattori, K. Itakura, S. Ozaki, S. Yasui, Phys. Rev. D **92**, 065003 (2015), 1504.07619.

[6] K. Hattori, D. Suenaga, K. Suzuki, S. Yasui, *in preparation.*

[7] Y. Araki, D. Suenaga, K. Suzuki, S. Yasui, Phys. Rev. Research **3**, 013233 (2021), 2008.08434.

[8] S. Yasui, K. Suzuki, and K. Itakura, Nucl. Phys. A **983**, 90–102 (2019), 1604.07208; S. Yasui, Phys. Lett. B **773**, 428–434 (2017), 1608.06450; S. Yasui, K. Suzuki, and K. Itakura, Phys. Rev. D **96**, 014016 (2017), 1703.04124.

[9] G. Endrődi, JHEP **07**, 173 (2015), 1504.08280; M. D'Elia, et al., Phys. Rev. D **105**, 034511 (2022), 2111.11237.

Femtoscopy measurements of two-kaon combinations in Au+Au collisions at the STAR experiment

Diana Pawłowska[1,*] for the STAR Collaboration

[1]Faculty of Physics, Warsaw University of Technology, Koszykowa 75, 00-662 Warsaw, Poland

Abstract. Relativistic heavy-ion collisions can study the properties of nuclear matter in high-energy experiments like the STAR experiment. One of the methods to learn about the properties of the bulk matter is the femtoscopy technique, which relies on information carried by the particles produced during the collisions. The emission source parameters, like space-time characteristics, can be extracted using the technique. High statistics data from RHIC make it possible to study the correlations between strange particles, like charged and neutral kaons. The pair-wise interactions between the identical kaons that form the basis for femtoscopy are quantum statistics and the Coulomb interaction for $K^{\pm}K^{\pm}$, and quantum statistics and the final-state interaction (FSI) through the $f_0(980)/a_0(980)$ threshold resonances for $K^0_S K^0_S$. The interactions between non-identical kaons pairs of $K^0_S K^{\pm}$ are essential, as the strong FSI is described only by the $a_0(980)$ resonance, which could be a four-quark state. These proceedings will present the femtoscopic measurements of strange particles with charged and neutral kaons correlations in Au+Au collisions at the RHIC energy.

1 Introduction

The main motivation to carry out the kaon femtoscopy study is that kaons are less affected by the feed-down from resonance decays. Kaons have a smaller cross-section on reaction with the hadronic matter and, what is also very important, they contain strange quarks. As it is known, one of the signatures of the Quark-Gluon Plasma (QGP) is the enhancement of the production of strange particles [1].
Comparison of femtoscopic results for all possible kaon combinations $K^{\pm}K^{\pm}$, $K^0_S K^0_S$, and $K^0_S K^{\pm}$ allows us to examine the a_0 resonance, which is suspected to be a four-quark state, e.g. a tetraquark [2]. All the systems should give the same emission source parameters of the homogeneity region and only final state interactions would cause the difference observed in correlations. Agreement between all kaon systems will shed light on the preceding assumption about the a_0 resonance.

2 Analysis methods

Experimentally, it is impossible to measure properties of the source directly with a size of about 1 fm and a lifetime of about 10^{-23} s. The relative momentum of particles produced during the heavy-ion collisions is used to learn about the source's characteristics. The equation (1) describes the correlation function:

*e-mail: diana.pawlowska.dokt@pw.edu.pl

$$CF(\vec{q}) = \int d^3r\, S(\vec{q},\vec{r})\, |\Psi(\vec{q},\vec{r})|^2, \quad (1)$$

where $S(\vec{q},\vec{r})$ is the emission function describing space-time properties of the source, $\Psi(\vec{q},\vec{r})$ is the pair wave function which contains information about all effects and interactions, and $\vec{q} = |\vec{p_1} - \vec{p_2}|$ is a relative momentum and $\vec{r} = |\vec{r_1} - \vec{r_2}|$ is relative distance between two particles. By measuring of pair relative momentum, space-time characteristics of the source like its size can be studied.
Theoretically, the $K^0_S K^0_S$ correlation function can be described analytically with a model derived by Lednicky-Lyuboshitz [3], which contains strong final state interaction (FSI):

$$CF(q_{inv}) = 1 + \lambda\left(e^{[-R^2_{inv}q^2_{inv}]} + \frac{1}{2}\left[\left|\frac{f(k^*)}{R_{inv}}\right|^2 + \frac{4\Re f(k^*)}{\sqrt{\pi}R_{inv}}F_1(q_{inv}R_{inv}) - \frac{2\Im f(k^*)}{R_{inv}}F_2(q_{inv}R_{inv})\right]\right), \quad (2)$$

where $F_1(z) = \int_0^z dx e^{x^2-z^2}/z$ and $F_2(z) = (1 - e^{-z^2})/z$ are analytic functions, $f(k^*)$ is s-wave $K^0_S K^0_S$ scattering amplitude, which depends on masses and decay couplings of near-threshold resonances $f_0(980)$ and $a_0(980)$ (Table 1), $q_{inv} = \sqrt{(\vec{p_1} - \vec{p_2})^2 - (E_1 - E_2)^2}$ is pair relative momentum (k^* is a first particle momentum in pair rest frame), R_{inv} is the size of the particle-emitting source and λ is the correlation strength.

Table 1. The most common parameters describing $f_0(980)$ and $a_0(980)$ resonances used in the Lednicky-Lyuboshitz model.

	m_{f_0}[GeV/c^2]	$\gamma_{f_0K\bar{K}}$	$\gamma_{f_0\pi\pi}$	m_{a_0}[GeV/c^2]	$\gamma_{a_0K\bar{K}}$	$\gamma_{a_0\pi\pi}$
Antonelli [4]	0.973	2.763	0.5283	0.985	0.4038	0.3711
Achasov2001 [5]	0.996	1.305	0.2684	0.992	0.5555	0.4401
Achasov2003 [6]	0.996	1.305	0.2684	1.003	0.8365	0.4580
Martin [7]	0.978	0.792	0.1990	0.974	0.3330	0.2220

To show the importance of considering the strong FSI in the parameterization, the experimental data was fitted by a correlation function that only considers the effects of quantum statistics (QS). This fit assumed a spherical source size described by the Gaussian density distribution.
In the case of $K^0_S K^{\pm}$ analysis, the parametrization was also performed using the Lednicky-Lyuboshitz model, where the strong FSI is due to one resonance, $a_0(980)$, and all quantities have the same description as in the formula for $K^0_S K^0_S$:

$$CF(k^*) = 1 + \frac{\lambda}{4}\left[\left|\frac{f(k^*)}{R}\right|^2 + \frac{4\Re f(k^*)}{\sqrt{\pi}R}F_1(2k^*R) - \frac{2\Im f(k^*)}{R}F_2(2k^*R)\right]. \quad (3)$$

3 Description of experiment and data selection

The femtoscopy method is successfully used in the case of the STAR experiment. Particle identification (PID) of charged particles is based on information from two detectors: the Time Projection Chamber (TPC) and Time of Flight (TOF) detector in the pseudorapidity range $|\eta| < 0.9$. TPC was used to measure the ionization energy loss of charged particles with a specific mass and momentum. For PID with TOF, the estimated time of flight was used to calculate the particle's mass and described as a function of the track length and momentum.
On the other hand, neutral particles like K^0_S were reconstructed using their decay products. Neutral kaons decay via the weak interaction into positive and negative daughters (π^+ and π^-). After reconstructing the secondary particles, neutral particles are identified based on the invariant mass applying cuts on their decay topology.

4 Results

4.1 Identical kaon femtoscopy

Figure 1 presents correlation functions for $K_S^0 K_S^0$ combination for 0-10% most central Au+Au collisions at $\sqrt{s_{NN}}$ = 39 GeV and 200 GeV. These correlations are dominated by the QS effects and strong FSI which gives a characteristic dip structure around q_{inv} = 0.1 GeV/c. The parameterization, including only QS effects (Gaussian), does not describe this shape well, so it is crucial to take into account the strong interaction in the fitting procedure as well.

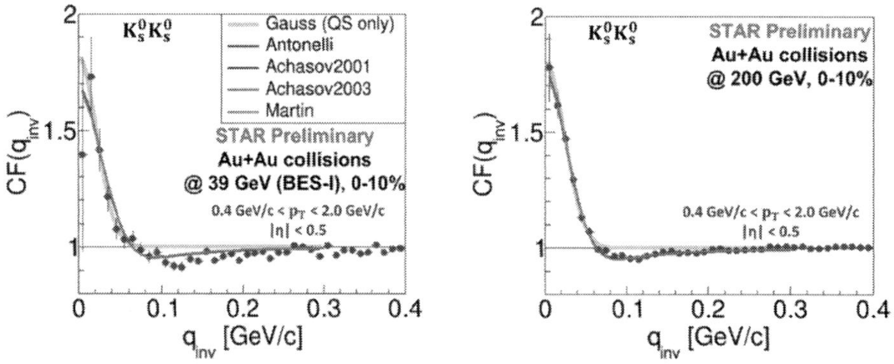

(a) (b)

Figure 1. $K_S^0 K_S^0$ correlation functions for $\sqrt{s_{NN}}$ = 39 GeV (a) and 200 GeV (b) with Gaussian and Lednicky-Lyuboshitz parameterizations. Only statistical uncertainties are shown.

4.2 Non-identical kaon femtoscopy

Correlation functions for $K_S^0 K^+$ system for central (0-10%) and noncentral (10-70%) Au+Au collisions at $\sqrt{s_{NN}}$ = 200 GeV are presented in the Figure 2. The correlation functions are

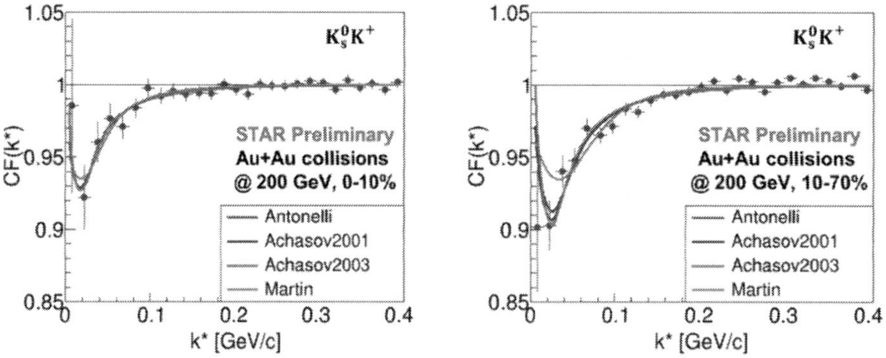

(a) (b)

Figure 2. $K_S^0 K^+$ correlation functions for $\sqrt{s_{NN}}$ = 200 GeV for central (a) and noncentral (b) collisions with Lednicky-Lyuboshitz parametrizations. Only statistical uncertainties.

characterized by dip structure due to near-threshold resonance $a_0(980)$, and the depth of this dip changes with the collision centrality. Three parametrizations Antonelli, Achasov2001, and Achasov2003 reflect this shape very well. Martin parametrization describes data worse for noncentral events.

4.3 Comparison of all kaon combinations

The comparison of the results from $K^\pm K^\pm$, $K_S^0 K_S^0$, and $K_S^0 K^\pm$ analyses for the top RHIC energy as a function of centrality is shown in Figure 3. The Antonelli parametrization gives the best agreement in the size of the particle emitting source among all combinations. Achasov's parametrizations lead to much larger source sizes for $K_S^0 K^{\pm}$ system, while the results with Martin are smaller for $K_S^0 K^{\pm}$. Antonelli parametrization favors the assumption that a_0 resonance could be the tetraquark.

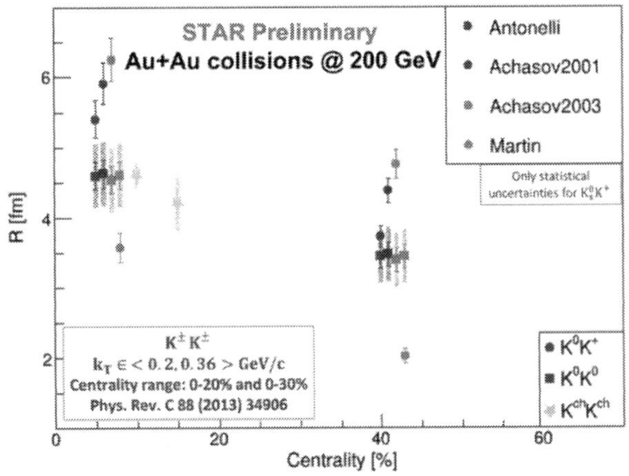

Figure 3. Centrality dependence of sizes of the $K^{\pm}K^{\pm}$ [8], $K_S^0 K_S^0$, and $K_S^0 K^{\pm}$ source at $\sqrt{s_{NN}} = 200$ GeV. For $K_S^0 K^{\pm}$ only statistical uncertainties are shown.

5 Summary

In summary, femtoscopic correlations with $K^{\pm}K^{\pm}$, $K_S^0 K_S^0$, and $K_S^0 K^{\pm}$ in Au+Au collisions at $\sqrt{s_{NN}} = 39$ and 200 GeV were presented. The extracted radii are compared among the three combinations for the first time. From the $K_S^0 K_S^0$ analysis, one can conclude that final state SI significantly affects these correlations due to two near-threshold resonances $a_0(980)$ and $f_0(980)$, and it is necessary to include this interaction in parametrization. Two-particle correlation in the $K_S^0 K^{\pm}$ system is reproduced by FSI due to the $a_0(980)$ resonance. Parametrization with $a_0(980)$ perfectly represents the signal region in the correlation function. The obtained source sizes are the biggest in the case of parametrization with the larger resonance mass (Achasov). For top RHIC energy, obtained source sizes for Antonelli parametrization are consistent among all the kaon combinations, which favors the assumption that $a_0(980)$ resonance could be a tetraquark.

Acknowledgement This work was supported by the Grant of National Science Centre, Poland, No: 2021/41/B/ST2/02409. Studies were funded by IDUB-POB-FWEiTE-1, IDUB-POB-FWEiTE-2 and IDUB-POB-FWEiTE-3 projects granted by Warsaw University of Technology under the program Excellence Initiative: Research University (ID-UB).

References

[1] M.I. Gorenstein, Heavy Ion Physics **14**, 141-148 (2001)
[2] E. Santopinto and G. Galatà, Physical Review C **75**, 045206 (2007)
[3] R. Lednicky and V.L.Lyuboshitz, Soviet Journal of Nuclear Physics **35**, 770 (1982)
[4] A. Antonelli, eConf **C020620**, THAT06 (2002)
[5] N.N. Achasov, V.V. Gubin, Physical Review D **63**, 094007 (2001)
[6] N.N. Achasov, A.V. Kiselev, Physical Review D **68**, 014006 (2003)
[7] A.D. Martin, E.N. Ozmutlu, E.J. Squires, Nuclear Physics B **121**, 514-530 (1977)
[8] STAR Collaboration, Physical Review C **88**, 34906 (2013)

Interplay between core and corona from small to large systems

Yuuka Kanakubo[1,*], *Yasuki* Tachibana[2,**], and *Tetsufumi* Hirano[1,***]

[1]Department of Physics, Sophia University, Tokyo 102-8554, Japan
[2]Akita International University, Yuwa, Akita-city 010-1292, Japan

Abstract. We present new results in p+p and Pb+Pb collisions at the LHC energies from the updated dynamical core–corona initialization framework (DCCI2). The fractions of final hadron yields originating from equilibrated and non-equilibrated components are extracted as functions of multiplicity. We find that the contributions from non-equilibrated components are non-negligible even in Pb+Pb collisions and affect p_T-integrated multi-particle correlations. These suggest the importance of non-equilibrated components for the sophisticated extraction of properties of the quark gluon plasma from comparisons between dynamical frameworks and experimental data.

1 Introduction

The properties of the quark-gluon plasma (QGP) have been investigated through dynamical frameworks based on relativistic hydrodynamics. Recently, several attempts from dynamical frameworks are reported [1–10] to quantitatively extract the QGP transport coefficients from the data-to-model comparison. This indicates that sophisticated description with dynamical frameworks is needed for a further investigation of transport coefficients. On the other hand, the possibility of the QGP formation in small colliding systems has been discussed following the report of strangeness enhancement in p+p collisions by the ALICE Collaboration [11]. Thus, the extension of the applicability of those dynamical frameworks to small colliding systems is indispensable.

Motivated by these backgrounds, we build the dynamical core–corona initialization as a state-of-the-art dynamical framework that is capable of describing from small to large colliding systems [12–14]. Under the core–corona picture [15], the system generated in high-energy nuclear collisions is described with two components: equilibrated matter (core) and non-equilibrated matter (corona). In our framework, we realize the dynamical description of the separation of the system into the core and corona by incorporating the picture with the novel dynamical initialization approach [16].

2 Dynamical Core–Corona Initialization framework

Under the dynamical initialization approach, initial conditions of QGP fluids are generated dynamically via source terms of relativistic hydrodynamic equations. We assume that the sys-

*e-mail: y-kanakubo-75t@eagle.sophia.ac.jp
**e-mail: ytachibana@aiu.ac.jp
***e-mail: hirano@sophia.ac.jp

tem is described with the two components, QGP fluids as core and non-equilibrated partons as corona, respectively, and consider the energy and momentum conservation for the entire system generated in a collision. Thus, the depositions of energy and momentum from non-equilibrated partons become sources of QGP fluids, and which makes it possible to model the generation of initial conditions of QGP fluids with keeping the energy and momentum conservation of the entire system. The dynamical realization of the core–corona picture is attained by modeling the energy and momentum deposition rates of non-equilibrated partons. Based on the core–corona picture, non-equilibrated partons with low transverse momentum, p_T, and in high-density regions are likely to deposit their energy and momentum while those with high p_T and in low-density regions are not.

A brief summary of the latest version of our framework [14, 17] is as follows: Initial conditions are phase-space distributions of partons after the final-state radiation obtained from Pythia8 [18] and Pythia8 Angantyr [19, 20] for p+p and A+A collisions, respectively. The dynamical initialization is performed with those initially produced partons and the system is dynamically separated into the core and corona. The space-time evolution of the core, QGP fluids, is described with (3+1)-D ideal hydrodynamics [21] by incorporating the equation of state $s95p$-v.1.1 [22]. Once temperatures of fluids become lower than a switching temperature, fluids are particlized with iS3D [23], a Monte-Carlo sampler of hadrons from hydrodynamic fields. The non-equilibrium partons are hadronized via the Lund string fragmentation with Pythia8. Finally, those directly emitted hadrons are handed to Jam [24], a hadronic transport model that performs hadronic rescatterings and resonance decays.

3 Results

We simulate p+p and Pb+Pb collisions at the LHC energies with our framework. Some major parameters control the fraction of generated core and corona components. We determine them so that the full simulation results can reasonably describe the multiplicity dependence of the yield ratios of omega baryons to charged pions in p+p and Pb+Pb collisions [14] [1] reported by the ALICE Collaboration [11]. By fixing those parameters, we extract the fractions of hadron yields originating from core and corona as functions of multiplicity near mid-pseudo-rapidity ($|\eta| < 0.5$) in p+p collisions at $\sqrt{s}$ = 7 and 13 TeV and Pb+Pb collisions at $\sqrt{s_{\mathrm{NN}}}$ = 2.76 TeV as shown in Fig. 1. One sees that the fractions of core and corona components show clear multiplicity scaling regardless of the system size or collision energies. Thus, within our framework, final hadron yields can be one of the reasonable estimators to evaluate the fraction of equilibrated and non-equilibrated components. We also find that the fraction of core components becomes dominant when multiplicity exceeds $\langle dN_{\mathrm{ch}}/d\eta\rangle \approx 20$. It should be also noted that the fraction of core (corona) components does not reach $\approx 100\%$ ($\approx 0\%$) even at the highest multiplicity bin in Pb+Pb collisions.

For a further investigation, we show p_T spectra of charged pions obtained in 20-40% centrality Pb+Pb collisions at $\sqrt{s_{\mathrm{NN}}}$ = 2.76 TeV in Fig. 2 (left). As an overall tendency, the core (corona) component dominates low (high) p_T regions, which is exactly the consequence of implementing the core–corona picture. However, against the naive expectation of the above tendency, there is a slight enhancement of the corona components at the very low p_T regions (p_T < 1 GeV). This can be intuitively understood just as a result of the interplay between the two different shapes of p_T spectra: exponential for the core components and power law for the corona components. The hadronic productions at such low p_T regions in the p_T spectra of corona components originate from the feed-down from string fragmentation of non-

[1]Note that some parameters are minor updated from Ref. [14]. For details, see Ref. [17].

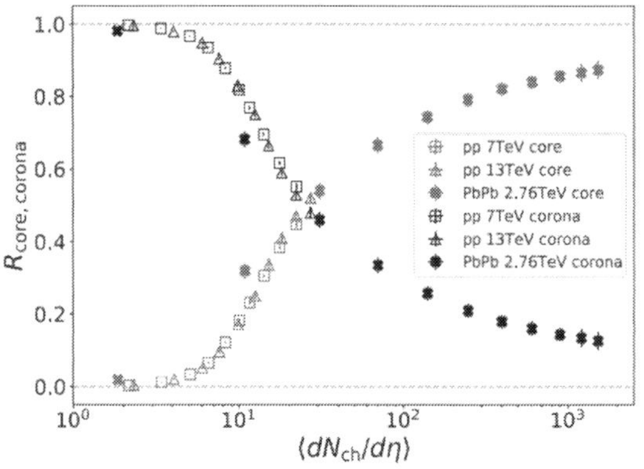

Figure 1. Fractions of hadron yields originating from the core and corona as functions of the number of charged particles produced at midrapidity $|\eta| < 0.5$. Results in p+p collisions at $\sqrt{s} = 7$ and 13 TeV (open squares and triangles, respectively) and Pb+Pb collisions at $\sqrt{s_{\text{NN}}} = 2.76$ TeV (closed circles) are shown.

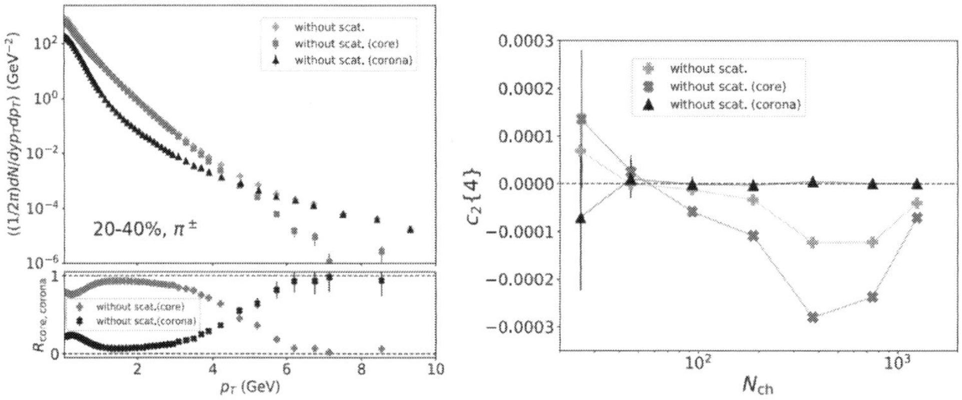

Figure 2. (Left) Charged pion p_T spectra obtained from 20-40% centrality class of Pb+Pb collisions at $\sqrt{s_{\text{NN}}} = 2.76$ TeV are shown in the upper panel. Fractions of core and corona contributions in each p_T bin are shown in the lower panel. (Right) Four-particle cumulant of charged particles as a function of the number of charged particles produced at midrapidity obtained in Pb+Pb collisions at $\sqrt{s_{\text{NN}}} = 2.76$ TeV. The solid lines are guides to the eyes. In both left and right figures, results with switching off hadronic rescatterings are shown in orange crosses and their breakdowns into the core and corona contributions are shown in red diagonal crosses and blue triangles, respectively.

equilibrated partons that lose their initial energy and momentum in dynamical initialization [14].

To see the influence of the existence of the corona components in the very low p_T on observables, four-particle cumulant of charged particles, $c_2\{4\}$, as a function of the number of charged particles, N_{ch}, produced near midrapidity obtained from Pb+Pb collisions at $\sqrt{s_{\text{NN}}} =$ 2.76 TeV is shown in Fig. 2 (right). For the kinematic ranges, $c_2\{4\}$ and N_{ch} are obtained from charged particles with $|\eta| < 0.8$ and $0.2 < p_T < 3.0$ GeV which is the same ranges used in Ref. [25]. The results from simulations switching off hadronic rescatterings and from the core components show the negative values of $c_2\{4\}$, while those from the corona components are zero-consistent in the entire range of N_{ch}. Notably, from the comparisons among those three results, one sees that the absolute values of $c_2\{4\}$ solely from the core components are diluted due to the existence of the corona contributions.

4 Summary

We reported new results from the updated version of the dynamical core–corona initialization framework. We found that the fractions of the core and corona components show clear scaling with multiplicity regardless of the system size or collision energies and that the core components become dominant above $\langle dN_{\mathrm{ch}}/d\eta \rangle \approx 20$. We also showed that there are non-negligible contributions from corona components, which significantly affect $c_2\{4\}$ as a function of N_{ch}. Our results suggest the importance of the non-equilibrium components even in heavy-ion collisions to extract the QGP properties quantitatively.

References

[1] J. Novak, K. Novak, S. Pratt, J. Vredevoogd, C. Coleman-Smith, R. Wolpert, Phys. Rev. C **89**, 034917 (2014), 1303.5769
[2] E. Sangaline, S. Pratt, Phys. Rev. C **93**, 024908 (2016), 1508.07017
[3] J.E. Bernhard, P.W. Marcy, C.E. Coleman-Smith, S. Huzurbazar, R.L. Wolpert, S.A. Bass, Phys. Rev. C **91**, 054910 (2015), 1502.00339
[4] J.E. Bernhard, J.S. Moreland, S.A. Bass, J. Liu, U. Heinz, Phys. Rev. C **94**, 024907 (2016), 1605.03954
[5] J. Auvinen, J.E. Bernhard, S.A. Bass, I. Karpenko, Phys. Rev. C **97**, 044905 (2018), 1706.03666
[6] J.E. Bernhard, J.S. Moreland, S.A. Bass, Nature Phys. **15**, 1113 (2019)
[7] D. Everett et al. (JETSCAPE), Phys. Rev. Lett. **126**, 242301 (2021), 2010.03928
[8] G. Nijs, W. van der Schee, U. Gürsoy, R. Snellings, Phys. Rev. Lett. **126**, 202301 (2021), 2010.15130
[9] J. Auvinen, K.J. Eskola, P. Huovinen, H. Niemi, R. Paatelainen, P. Petreczky, Phys. Rev. C **102**, 044911 (2020), 2006.12499
[10] J.E. Parkkila, A. Onnerstad, D.J. Kim, Phys. Rev. C **104**, 054904 (2021), 2106.05019
[11] J. Adam et al. (ALICE), Nature Phys. **13**, 535 (2017), 1606.07424
[12] Y. Kanakubo, M. Okai, Y. Tachibana, T. Hirano, PTEP **2018**, 121D01 (2018), 1806.10329
[13] Y. Kanakubo, Y. Tachibana, T. Hirano, Phys. Rev. C **101**, 024912 (2020), 1910.10556
[14] Y. Kanakubo, Y. Tachibana, T. Hirano, Phys. Rev. C **105**, 024905 (2022), 2108.07943
[15] K. Werner, Phys. Rev. Lett. **98**, 152301 (2007), 0704.1270
[16] M. Okai, K. Kawaguchi, Y. Tachibana, T. Hirano, Phys. Rev. **C95**, 054914 (2017), 1702.07541
[17] Y. Kanakubo, Y. Tachibana, T. Hirano (2022), 2207.13966
[18] T. Sjostrand, S. Mrenna, P.Z. Skands, Comput. Phys. Commun. **178**, 852 (2008), 0710.3820
[19] C. Bierlich, G. Gustafson, L. Lönnblad, JHEP **10**, 139 (2016), 1607.04434
[20] C. Bierlich, G. Gustafson, L. Lönnblad, H. Shah, JHEP **10**, 134 (2018), 1806.10820
[21] Y. Tachibana, T. Hirano, Phys. Rev. **C90**, 021902 (2014), 1402.6469
[22] P. Huovinen, P. Petreczky, Nucl. Phys. A **837**, 26 (2010), 0912.2541
[23] M. McNelis, D. Everett, U. Heinz, Comput. Phys. Commun. **258**, 107604 (2021), 1912.08271
[24] Y. Nara, N. Otuka, A. Ohnishi, K. Niita, S. Chiba, Phys. Rev. C **61**, 024901 (2000), nucl-th/9904059
[25] S. Acharya et al. (ALICE), Phys. Rev. Lett. **123**, 142301 (2019), 1903.01790

Elliptic and triangular flow of (multi-)strange hadrons and ϕ mesons in BES-II energies at STAR

Li-Ke Liu[1,*] (for the STAR Collaboration)

[1]Key Laboratory of Quark & Lepton Physics (MOE) and Institute of Particle Physics, Central China Normal University, Wuhan 430079, China

Abstract. In these proceedings, we present the measurements of elliptic (v_2) and triangular (v_3) flow of (multi-)strange hadrons and ϕ mesons in 19.6 and 14.6 GeV Au+Au collisions from the STAR. The number of constituent quark (NCQ) scaling of v_2 and v_3 holds well at $\sqrt{s_{\rm NN}}$ = 19.6 GeV, which indicates the collective flow is built up in the partonic stage. At these energies, the anti-particles show better NCQ scaling than the particles for both v_2 and v_3, which may be caused by the different contributions from the produced and transported quarks.

1 Introduction

At sufficiently high temperature and/or high density, quantum chromodynamics (QCD) predicts a transition from hadronic matter to deconfined partonic matter [1, 2]. Results from top-RHIC and LHC energies indicate a new form of matter with small viscosity and high temperature, created in high-energy heavy ion collisions [3, 4]. Lattice QCD calculations predict that, the phase transition from hadronic matter to the QGP phase is a smooth crossover at vanishing baryon chemical potential (μ_B) region. A first-order phase transition is expected at a finite baryon chemical potential region. Locating the first-order phase boundary with a critical point is essential in establishing the QCD phase diagram. It motivates the Beam Energy Scan (BES) program at RHIC, which covers energy $\sqrt{s_{\rm NN}}$ = 3.0 - 62.4 GeV corresponding baryon chemical potential 750 - 73 MeV. It can help us explore the QCD phase structure in the high baryon density region.

Anisotropies in particle momentum distributions relative to the reaction plane are referred to as anisotropic collective flow. Elliptic flow coefficient, v_2, is sensitive to the dynamics at the early stages of the system evolution in heavy-ion collisions and equation of state of the medium [5]. Triangular flow v_3 is particularly sensitive to the initial geometry fluctuations [6]. The hadronic interaction cross sections of multi-strange hadrons and ϕ mesons are expected to be small [7–9]. Hence, the anisotropic flow of these hadrons provides information on the early stages of the high-energy collisions. A hint of smaller v_2 for ϕ mesons compared to charged particles is observed for BES-I Au+Au collisions at $\sqrt{s_{\rm NN}}$ = 7.7 and 11.5 GeV [10].

In these proceedings, with the enhanced statistics datasets, we report the measurements of v_2 and v_3 of (multi-)strange hadrons ($K^{\pm}$, K^0_S, Λ, $\bar{\Lambda}$, Ξ, $\bar{\Xi}^{+}$, Ω, $\bar{\Omega}^{+}$) and ϕ mesons from BES-II Au+Au collisions at $\sqrt{s_{\rm NN}}$ = 14.6 and 19.6 GeV.

*e-mail: likeliu@mails.ccnu.edu.cn

2 Data sets and Analysis strategy

Data samples for Au+Au collisions at $\sqrt{s_{NN}}$ = 14.6 and 19.6 GeV taken in 2019 are used in the analysis. The number of good events analyzed are 270×10^6 and 440×10^6 for 14.6 and 19.6 GeV, respectively. The primary vertex position of each event along the beam direction, V_z, is selected to be within ± 70 cm from the center of the Time Projection Chamber (TPC). To eliminate possible beam interactions with the vacuum pipe, the vertex along the radial direction, V_r, is selected to be smaller than 2 cm. To select good quality tracks, p_T > 0.2 GeV/c, and a distance of closest approach (DCA) from vertex, DCA ≤ 3 cm, and at least 15 space points in the TPC acceptance are required. The particle identification at low transverse momentum is performed via their specific energy loss measured by the TPC. At intermediate and high transverse momenta, the particle identification is performed using the Time of Flight (TOF) detector. The strange hadrons such as K^0_S, Λ, $\bar{\Lambda}$, Ξ^-, Ξ^+, Ω^-, and Ω^+ are reconstructed using KF-Particle package [11]. The ϕ mesons are reconstructed via K^+K^- channel. The systematic uncertainties on the measurements are obtained by varying the above analysis cuts. The dominant sources of systematic uncertainties are the particle identification cuts and quality track selection cuts.

3 Results and Discussions

Figure 1 shows the p_T differential elliptic flow of the identified particles for mid-central (10-40%) Au+Au collisions at $\sqrt{s_{NN}}$ = 19.6 GeV. The black markers show baryon v_2 results, and the red markers show meson v_2 results. There is a clear mass ordering when p_T < 1.5 GeV/c. It could be due to the interplay of radial expansion and anisotropic flow, which is consistent with hydrodynamics predictions [12]. When p_T > 1.5 GeV/c, the v_2 of particles is grouped according to the hadron type, called baryon-meson splitting, which indicates that the coalescence could be dominant process of hadronization in this p_T-region.

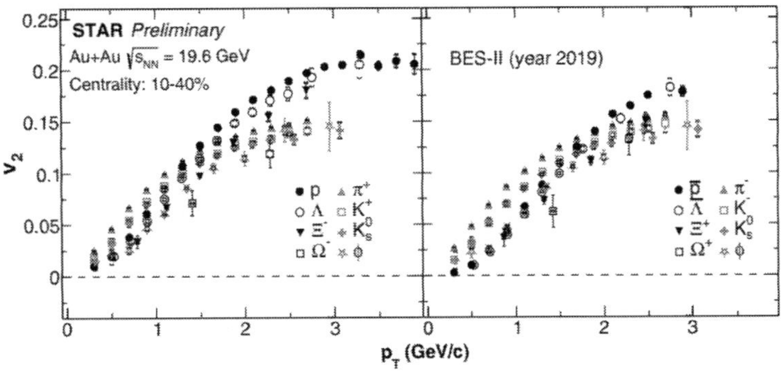

Figure 1. p_T differential elliptic flow of the identified particles in 10-40% Au+Au collisions at $\sqrt{s_{NN}}$ = 19.6 GeV.

The p_T differential triangular flow of the identified particles in 0-30% and 30-70% Au+Au collisions at $\sqrt{s_{NN}}$ = 19.6 GeV are shown in Fig. 2. The triangular anisotropy (ϵ_3) in the initial collision geometry originates from event-by-event fluctuations. The data of v_3 show a weak centrality dependence, same as ϵ_3 [6, 13]. It indicates the event-by-event fluctuations are the dominant source for v_3.

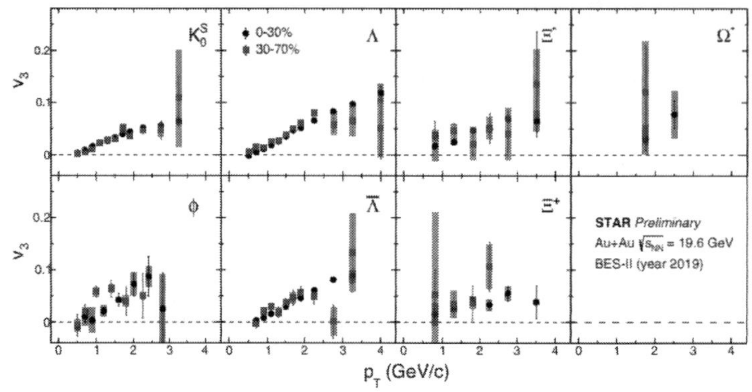

Figure 2. p_T differential triangular flow of the identified particles in 0-30% and 30-70% Au+Au collisions at $\sqrt{s_{\rm NN}}$ = 19.6 GeV. The vertical bars and shaded bands are the statistical and systematic uncertainties, respectively.

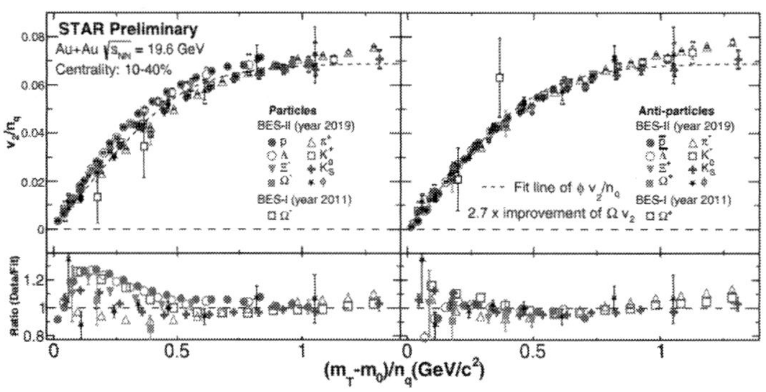

Figure 3. The number-of-constituent quark (NCQ) scaled elliptic flow, v_2/n_q versus $(m_T - m_0)/n_q$, for 10-40% central Au+Au collisions for identified particles (left pad) and corresponding anti-particles (right pad) at $\sqrt{s_{\rm NN}}$ = 19.6 GeV where dash lines show the fit of ϕ mesons v_2.

Figure 3 shows the test of NCQ scaling of v_2 in centrality 10-40% at $\sqrt{s_{\rm NN}}$ = 19.6 GeV. The NCQ scaling of v_2 holds within 10% for anti-particles, and 20% for particles. A similar scaling behavior is observed for v_3 and holds within 15% for anti-particles, and 30% for particles.

The NCQ scaling reflects that the quarks are the most effective degrees of freedom in determining hadron flow at intermediate p_T, which means that the collective flow has been built up in the partonic stage at this collision energy. Meanwhile, the NCQ scaling of anti-particles works better than that of particles both for v_2 and v_3, which may be caused by the different contributions from the produced and transported quarks. The strange hardons and ϕ meson v_2 are measured in Au+Au collisions at BES-II 14.6 GeV, the NCQ scaling of v_2 holds at 20% level shown in Fig. 4 and measurements in finer centrality bins are underway.

4 Summary

In summary, we report the elliptic and triangular flow of strange hadrons $K^{\pm}$, K^0_S, Λ, $\bar{\Lambda}$, Ξ^-, $\bar{\Xi}^+$, Ω^-, $\bar{\Omega}^+$, and ϕ mesons at 14.6 and 19.6 GeV from BES-II. At $\sqrt{s_{\rm NN}}$ = 19.6 GeV, the

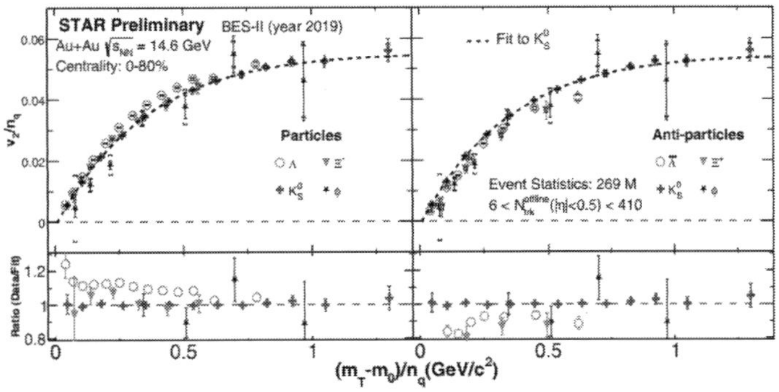

Figure 4. The number-of-constituent quark scaled elliptic flow, v_2/n_q versus $(m_T - m_0)/n_q$, in 0-80% Au+Au collisions for strange particles at $\sqrt{s_{NN}} = 14.6$ GeV.

NCQ scaling of v_2 (v_3) holds within 10 (15)% for anti-particles, and within 20 (30)% for particles. It indicates that the partonic collectivity could build up at this energy. Meanwhile, the NCQ scaling of anti-particles holds better than particles, which indicates the contribution of transported quarks in particles. At $\sqrt{s_{NN}} = 14.6$ GeV, the NCQ scaling of v_2 holds at 20% level in 0-80% central Au+Au collisions. The data taking for BES-II program (3-19.6 GeV) has completed during the year 2019 to 2021 with high statistics and detector upgrades. Such datasets will help scan the QCD phase diagram over a wide range of baryon chemical potential. It is expected to provide more precise differential measurements of v_2 and v_3 especially for less abundant particles: multi-strange hadrons and ϕ mesons. It will offer additional information to constrain the equation of state (EoS) and phase boundary of the produced QCD matter in the high baryon density region.

Acknowledgement This work was supported by the National Natural Science Foundation of China [Nos. 12175084, 11890710 (11890711)], the National Key Research and Development Program of China (No. 2020YFE0202002) and the Fundamental Research Funds for the Central Universities (No. CCNU220N003).

References

[1] D.J. Gross, R.D. Pisarski, L.G. Yaffe, Rev. Mod. Phys. **53**, 43 (1981).
[2] X. Luo, S. Shi, N. Xu, Y. Zhang, Particles **3**, 278 (2020).
[3] A. Bazavov et al., Phys. Rev. D **85**, 054503 (2012).
[4] K. Fukushima, C. Sasaki, Prog. Part. Nucl. Phys. **72**, 99 (2013).
[5] S.A. Voloshin, A.M. Poskanzer, R. Snellings, Landolt-Bornstein **23**, 293 (2010).
[6] B. Alver, G. Roland, Phys. Rev. C **81**, 054905 (2010).
[7] B. Mohanty, N. Xu, J. Phys. G **36**, 064022 (2009).
[8] M. Nasim, B. Mohanty, N. Xu, Phys. Rev. C **87**, 014903 (2013).
[9] S. Shi, Adv. High Energy Phys. **2016**, 1987432 (2016).
[10] L. Adamczyk et al. (STAR Collaboration), Phys. Rev. C **88**, 014902 (2013).
[11] A. Banerjee, I. Kisel, M. Zyzak, Int. J. Mod. Phys. A **35**, 2043003 (2020).
[12] N. Borghini, J.Y. Ollitrault, Phys. Lett. B **642**, 227 (2006).
[13] K. Xiao, N. Li, S. Shi, F. Liu, J. Phys. G **39**, 025011 (2012).

EPJ Web of Conferences **276**, 01019 (2023) https://doi.org/10.1051/epjconf/202327601019
SQM 2022

Higher-order moments of net-proton and the QCD phase structure

Ho San Ko[1,*]

[1] 1 Cyclotron Road Mailstop 70R0319 Berkeley, CA, 94720, U.S.A.

Abstract. Higher-order cumulants of the conserved quantities and their ratios are powerful tools to study the properties of QGP and explore the QCD phase structure, such as critical point and/or the first-order phase transition boundary. In this proceedings, we present the recent results of the higher-order cumulants and ratios at different collision energies. We compare the results with hadronic transport model, Hadron Resonance Gas model, and QCD thermodynamic calculations and discuss the physics implications. Additionally, we mention the near future cumulant analyses and an initial volume fluctuation issue.

1 Introduction

It has been reported from Lattice QCD calculations that it is an analytic crossover between the QGP and the hadron gas states at vanishing baryonic chemical potential (μ_B) [1] and at the temperature of 156.5 ± 1.5 MeV [2]. A critical point followed by a first-order phase transition at high μ_B is predicted from QCD-based model calculations. For example, see Ref. [3]. We search for the possible signature of the critical point and the first-order phase transition in the QCD phase diagram by scanning the temperature and μ_B. This is done by varying the collision energy [4, 5, 6].

One of the methods for the critical point search is measuring a fluctuation of conserved quantities: net-baryon number, net-charge, and net-strangeness. In this proceedings, we will discuss the net-proton analysis results. The net-proton number is experimentally used as a proxy to the net-baryon number [7, 8]. The moments of the distribution of this quantity are analysed by studying their cumulants.

1.1 Experimental observables

Cumulants from the first to the sixth-order can be written as:

$$\begin{aligned}
C_1 &= \langle N \rangle, \\
C_2 &= \langle (\delta N)^2 \rangle, \\
C_3 &= \langle (\delta N)^3 \rangle, \\
C_4 &= \langle (\delta N)^4 \rangle - 3\langle (\delta N)^2 \rangle^2, \\
C_5 &= \langle (\delta N)^5 \rangle - 10\langle (\delta N)^2 \rangle \langle (\delta N)^3 \rangle, \\
C_6 &= \langle (\delta N)^6 \rangle + 30\langle (\delta N)^2 \rangle^3 - 15\langle (\delta N)^2 \rangle \langle (\delta N)^4 \rangle - 10\langle (\delta N)^3 \rangle^2,
\end{aligned} \tag{1}$$

*e-mail: h-s.ko@lbl.gov

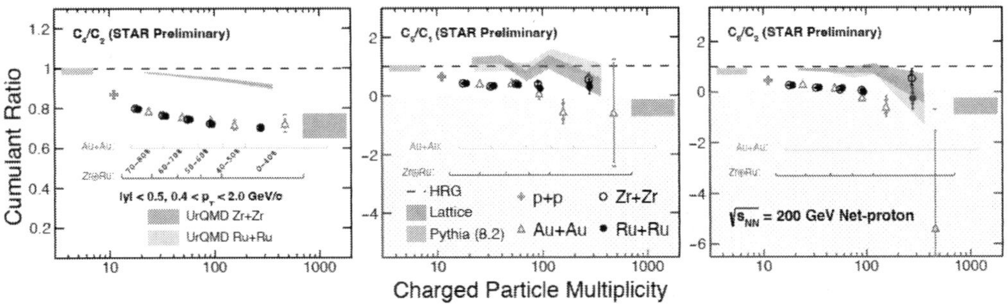

Figure 1: Cumulant ratios C_4/C_2, C_5/C_1, and C_6/C_2 of different collision systems: p+p, $^{96}_{40}$Zr+$^{96}_{40}$Zr, $^{96}_{44}$Ru+$^{96}_{44}$Ru, and Au+Au, as a function of charged particle multiplicity. Cumulant ratios for p+p are presented only in averaged charged particle multiplicity. The detector efficiencies for the charged particle multiplicity are not corrected but corrected for the cumulant ratios. The bars and brackets for each marker represent the statistical and systematic uncertainties, respectively. UrQMD calculations are shown in bands. HRG calculations are shown in dashed lines. Magenta bands represent Lattice QCD prediction for the formation of thermalized QCD matter. Pythia calculations shown in gold bands are for average charged particle multiplicity in p+p collisions.

where N represents the event-by-event conserved quantity distribution and $\delta N = N - \langle N \rangle$. The symbol $\langle N \rangle$ represents the average value of N of the events. The n-th order cumulants are related to the n-th order susceptibility:

$$\chi^{(n)} = \left(\frac{\partial^n p}{\partial \mu_B^n} \right)_T = \frac{1}{VT^3} \times C_n, \tag{2}$$

where p, V, and T represent the pressure, volume, and temperature of the system, respectively. Taking the ratio of the cumulants cancels out the volume dependence and the ratios can be directly compared to theoretical calculations. The higher the cumulant order, the more the cumulant is sensitive to the correlation length [9].

2 Experimental status

From the first principle Lattice QCD calculation, the C_5/C_1 and C_6/C_2 of net-baryon number are expected to be negative [10] where it is a smooth crossover from QGP to hadronic matter. STAR at RHIC measured C_6/C_2 of net-proton number of Au+Au collisions at three different collision energies: 27, 54.4, and 200 GeV [11]. While the results are mostly consistent being zero at other collision energies, the C_6/C_2 progressively become negative as the multiplicity increases and consistent with the Lattice QCD calculation at the most central collision at $\sqrt{s_{\rm NN}}$ = 200 GeV.

The same decreasing trend in multiplicity is observed, and presented in figure 1, in the higher-order cumulant ratios C_4/C_2, C_5/C_1, and C_6/C_2 regardless of the collision systems at $\sqrt{s_{\rm NN}}$ =200 GeV: p+p, $^{96}_{40}$Zr+$^{96}_{40}$Zr, $^{96}_{44}$Ru+$^{96}_{44}$Ru, and Au+Au [12, 11]. For better statistics, the collision centralities from 0% to 40% of $^{96}_{40}$Zr+$^{96}_{40}$Zr, $^{96}_{44}$Ru+$^{96}_{44}$Ru and Au+Au collisions are merged into one central collision bin. For p+p collisions, only the averaged values in multiplicity are shown. The decreasing trend in multiplicity makes the results deviate further from the Hadron Resonance Gas (HRG) model calculations in the Grand Canonical Ensemble picture [13, 14]. Due to the effects of decays, the p+p collisions results show below the Grand Canonical Ensemble prediction. Although the hadronic transport model (UrQMD)

[15] describes the overall multiplicity-dependent trend, they overpredict the results. At the most central Au+Au collision results, the results become consistent with the Lattice QCD prediction for the formation of the thermalized QCD matter and smooth crossover transition. PYTHIA 8.2 (Pythia) calculations in figure 1 represent the cumulant ratios averaged over charged particle multiplicity of the p+p collisions. All the higher-order cumulant ratios from Pythia are consistently positive which is inconsistent with the Lattice QCD results in the case of the fifth and the sixth-order.

For collision energy dependence, higher-order cumulant ratios of net-proton in Au+Au collisions from $\sqrt{s_{NN}}$ = 3 to 200 GeV for 0-40% and 50-60% centralities are presented in figure 2 [16]. For C_4/C_2, 0-5% centralities are shown instead of 0-40% [17]. For the 3 GeV results, a fixed Au target was used instead of colliding the two Au nuclei. Due to the experimental setup, the rapidity range is from -0.5 to 0 for the 3 GeV while it is ±0.5 for the others. Various theoretical calculations are described in the figure caption. The analysis results suggest that the cumulant ratios follow the ordering of $C_4/C_2 > C_5/C_1 > C_6/C_2$ in the range of collision energy of $\sqrt{s_{NN}}$ = 7.7 - 200 GeV. At 3 GeV however, the cumulant ratios show a reversed ordering. For 0-40% collision centrality, while C_5/C_1 does not show a systematic dependence on the collision energy, we observe a progressively negative sign with decreasing energy in the case of C_6/C_2 in the range of $\sqrt{s_{NN}}$ = 7.7 - 200 GeV. At 3 GeV, both cumulant ratios become positive. In the case of peripheral collision, 50-60% centrality, all three ratios are non-negative.

The hierarchy is expected from QCD thermodynamic calculations: Lattice QCD [10] and the QCD-based functional renormalization group (FRG) model [18]. They also show increasingly negative C_5/C_1 and C_6/C_2 as the collision energy decreases over a wide range of $\sqrt{s_{NN}}$ = 7.7 - 200 GeV (μ_B = 420 -20 MeV). One must, however, keep in mind the difference between net-baryon from the theoretical calculations and net-proton from the data with experimental acceptance. In contrast, the HRG canonical ensemble (HRG CE) model predictions [14] do not follow the ordering and remain positive up to the sixth-order. The UrQMD expectations are qualitatively consistent with the 50-60% centrality results. Additionally, the reversed ordering of the cumulant ratios is reproduced by the UrQMD calculations at 3 GeV. This suggests that at $\sqrt{s_{NN}}$ = 3 GeV, it is hadronic interaction dominant [16].

In the vicinity of a critical point, the fourth-order cumulant is expected to have a non-monotonic dependence on the collision energy [19, 20, 21, 22]. At STAR, this behavior for net-proton is observed (see the left plot of figure 2) in the collision energy of $\sqrt{s_{NN}}$ = 7.7 - 62.4 GeV with a significance 3.1σ for the 0-5% central Au+Au collisions [6, 12]. Recent analysis of Au+Au collisions at $\sqrt{s_{NN}}$ = 2.4 and 3 GeV at HADES [23] and STAR [17] respectively, show a suppression of net-proton C_4/C_2. The UrQMD reproduces the data at $\sqrt{s_{NN}}$ = 3 GeV indicating that it is hadronic interaction dominant at this energy.

We briefly mention here, two other measurements at $\sqrt{s_{NN}}$ = 3 GeV: anisotropic flow v_2 scaled by its number of constituent quarks [24] and ϕ-meson ratio over kaon [25, 26]. At higher collision energy, the scaled v_2 is positive. The positive v_2 along with the observation of its number-of-constituent-quarks (NCQ) scaling is a strong evidence of the formation of a hydrodynamically expanding QGP phase with partonic degrees of freedom. However at 3 GeV, it becomes negative and the scaling disappears. The 3 GeV result shows that it needs a description from a new equation of state, likely dominated by baryonic interactions. Also, strangeness production in different collision energy shows that it deviates strongly from Grand Canonical Ensemble at 3 GeV and favors Canonical Ensemble model. Also, a transport model describes the datapoint. This additionally indicates a change of medium. These measurements along with the fourth-order cumulants support that at 3 GeV and below, hadronic interaction is dominant.

 https://doi.org/10.1051/epjconf/202327601019

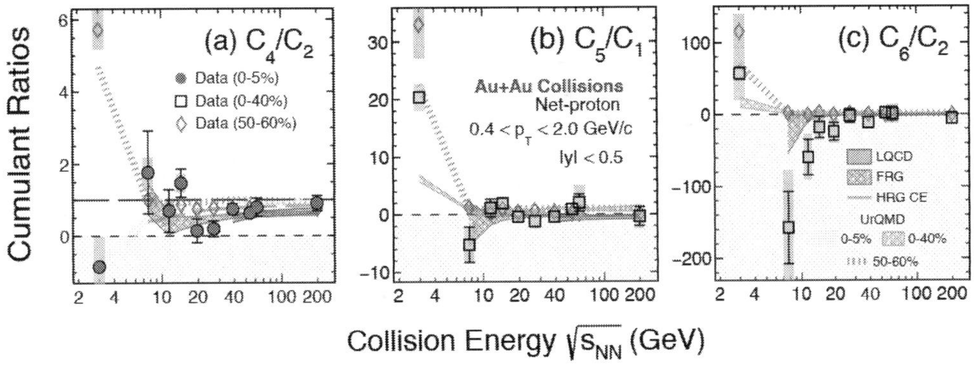

Figure 2: Energy dependence of the net-proton higher-order cumulant ratios of the Au+Au collisions. For C_4/C_2, two collision centralities, 0-5% and 50-60%, are shown while 0-40% and 50-60% are shown for the other cumulant ratios. The rapidity range for $\sqrt{s_{\rm NN}}$ = 3 GeV is from -0.5 to 0 while |y| < 0.5 is used for all other energies. The bars and bands represent the statistical and systematic uncertainties of the data points, respectively. Lattice QCD (39 - 200 GeV), FRG (7.7 - 200 GeV), UrQMD, and HRG CE (7.7 - 200 GeV) are shown as red, grey, brown and blue-dashed bands, and green dashed lines, respectively.

The higher-order cumulant analysis at $\sqrt{s_{\rm NN}}$ = 200 GeV strongly suggests that the thermalized QGP matter is formed and it is a smooth crossover between the QGP and hadron gas states. As the collision energy decreases, the fourth-order cumulant (C_4/C_2) exhibits non-monotonic energy dependence in the collision energy range of $\sqrt{s_{\rm NN}}$ = 7.7 - 62.4 GeV. At 3 GeV and below, it hints a hadronic interaction dominance. Therefore, if the QCD critical point exists, it is likely that it exists between 3 and 200 GeV.

3 Upcoming analyses

As described in equation 2, the n-th order cumulant is the n-th derivative of chemical potential. Therefore, the higher the order, the more cumulant contains the structure of the chemical potential and the more fluctuates to the variation. Cumulants of order higher than the sixth, up to the eighth, are currently under study. However, it needs attentiveness since the higher the order, the more statistic it needs and the more vulnerable to the background.

Analysis of cumulant ratios of mixed quantum numbers may enable us to measure the magnetic field created in heavy-ion collisions [27]. For baryon-strangeness correlation studies, strange baryon measurements become crucial [28]. Therefore, the detection efficiency and purity of the detected particles need to be carefully studied. See Ref. [29] as an example of the purity correction study.

The Phase II of the STAR beam energy scan collected higher statistics, compared to the first phase, on collision energies of $\sqrt{s_{\rm NN}}$ = 7.7 - 27 GeV. The focus of this energy range is to provide the most precise data to map the QCD phase diagram and pin down the critical point. Not only the statistics have been improved, but also the detector upgrades help reduce the systematic uncertainty of the measurement. The addition of the inner TPC (iTPC) to the TPC helps enlarge the acceptance. More rigorous rapidity scan can be performed.

4 Initial volume fluctuation at high baryon density

There are future experimental facilities that are focusing on the high baryon density region, with high statistics: NICA, HADES, HIAF, FAIR, and JPARC-HI. They will map out the

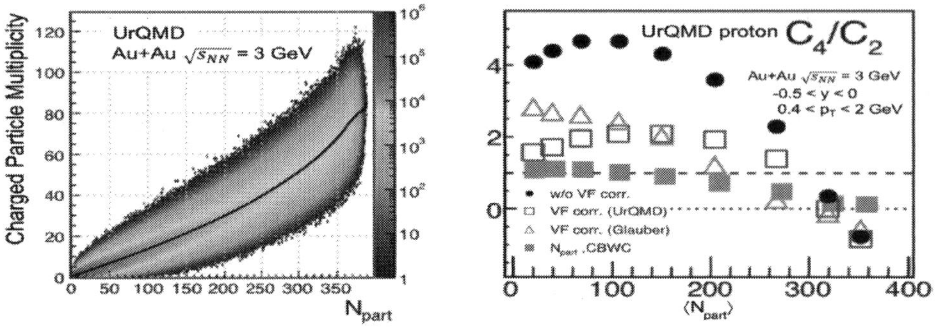

Figure 3: UrQMD model calculations. Left: charged particle multiplicity to N_{part} of Au+Au collisions at $\sqrt{s_{NN}} = 3$ GeV. Right: C_4/C_2 of the proton to $\langle N_{part} \rangle$ of Au+Au collisions at $\sqrt{s_{NN}} = 3$ GeV. Black dots represent the cumulant ratios calculated using the charged particle multiplicity. Red squares show when using the N_{part} for the calculation. The blue squares and the red triangles indicate the results using the charged particle multiplicity with initial volume fluctuation (IVF) corrections applied. Different model-based IVF corrections are applied for the blue squares and the red triangles results.

QCD phase diagram above the high baryon density region covered by RHIC. One of the challenges these facilities need to face is an initial volume fluctuation (IVF) issue.

We convert the charged particle multiplicity to the collision centrality or the number of participating nucleons (N_{part}) via a Glauber model coupled with a two-component particle production model. This method works well at the LHC energy and high RHIC energy, i.e. low baryon density region. However, as the collision energy decreases, the number of produced particles reduces. The plots in figure 3 are from the UrQMD model study. The left plot shows the charged particle multiplicity to N_{part} of Au+Au collisions at $\sqrt{s_{NN}} = 3$ GeV. Comparing the case of the charged particle multiplicity at 40 and to 100, for example, one can see that the corresponding range of N_{part} drastically increases. The non-one-to-one correspondence between the two is the IVF and it is more significant in the mid-central collisions due to no constraints by the minimum $N_{part} = 0$ nor the maximum $N_{part} = 394$. We find that IVF is significant for $\sqrt{s_{NN}} < 7.7$ GeV and further studies are ongoing.

The right plot of figure 3 shows the C_4/C_2 of proton of Au+Au collisions at $\sqrt{s_{NN}} = 3$ GeV. Using the traditionally obtained charged particle multiplicity gives different results from the ideal case of using the N_{part}. The difference is the largest for the mid-central collisions. This is because for the mid-central collisions, the IVF is the most enhanced as described above. In contrasts, the IVF is less significant for the most-central collisions due to the constraints of the maximum $N_{part} = 394$. The IVF corrections show model dependence. If one could measure the N_{part} independently, it would provide an additional constraints suppressing the IVF effect on the measured cumulants.

5 Summary

The cumulant analysis of conserved quantities probes the QCD phase structure. We presented in this proceedings the higher-order net-proton cumulant analysis results for different systems and energy range. At $\sqrt{s_{NN}} = 200$ GeV, we showed that the thermalized QGP is formed at the most central Au+Au collisions and it is an analytic crossover to the hadronic matter. In $\sqrt{s_{NN}} = 7.7$ - 200 GeV, the higher-order cumulant ratios of the central collisions are mostly consistent with the QCD thermodynamic calculations. We also observed the non-monotonic energy dependence of the fourth-order in the range of $\sqrt{s_{NN}} = 7.7$ - 62.4 GeV hinting that we are at the vicinity of the critical point. Additionally, we presented that at $\sqrt{s_{\mathrm{NN}}} = 3$ GeV, it is hadronic interaction dominant, implying that if the QCD critical point is created in heavy-ion collision, it likely exists at above 3 GeV. The upcoming experiments will provide better controlled systematic uncertainties and higher statistics to finely map out the QCD diagram to pin down the critical point. The cumulant analysis of mixed quantum numbers may allow us to measure the magnetic field created in heavy-ion collisions. In the high baryon density region, the initial volume fluctuation issue remains and needs to be dealt with.

References

[1] Y. Aoki et al. Nature 443 (2006), pp. 675–678.

[2] A. Bazavov et al., HotQCD. Phys. Lett. B 795 (2019), pp. 15–21.

[3] M. A. Halasz et al. Phys. Rev. D 58 (1998), p. 096007.

[4] Chi Yang. Nucl. Phys. A 967 (2017), pp. 800–803.

[5] L. Adamczyk et al., STAR. Phys. Rev. C 96.4 (2017), p. 044904.

[6] J. Adam et al., STAR. Phys. Rev. Lett. 126.9 (2021), p. 092301.

[7] Y. Hatta and M. A. Stephanov. Phys. Rev. Lett. 91 (2003). [Erratum: Phys.Rev.Lett. 91, 129901 (2003)], p. 102003.

[8] M. Kitazawa and M. Asakawa. Phys. Rev. C 86 (2012). [Erratum: Phys.Rev.C 86, 069902 (2012)], p. 024904.

[9] M. A. Stephanov. Phys. Rev. Lett. 102 (2009), p. 032301.

[10] A. Bazavov et al. Phys. Rev. D 101.7 (2020), p. 074502.

[11] M. S. Abdallah et al., STAR. Phys. Rev. Lett. 127.26 (2021), p. 262301.

[12] M. S. Abdallah et al., STAR. Phys. Rev. C 104.2 (2021), p. 024902.

[13] P. Garg et al. Phys. Lett. B 726 (2013), pp. 691–696.

[14] P. Braun-Munzinger et al. Nucl. Phys. A 1008 (2021), p. 122141.

[15] M. Bleicher et al. J. Phys. G 25 (1999), pp. 1859–1896.

[16] STAR. arXiv: 2207.09837v1 [nucl-ex].

[17] M. S. Abdallah et al., STAR. Phys. Rev. Lett. 128.20 (2022), p. 202303.

[18] Wei-jie Fu et al. Phys. Rev. D 104.9 (2021), p. 094047.

[19] M. A. Stephanov. Phys. Rev. Lett. 107 (2011), p. 052301.

[20] B. J. Schaefer and M. Wagner. Phys. Rev. D 85 (2012), p. 034027.

[21] J.-W. Chen et al. Phys. Rev. D 93.3 (2016), p. 034037.

[22] J.-W. Chen et al. Phys. Rev. D 95.1 (2017), p. 014038.

[23] J. Adamczewski-Musch et al., HADES. Phys. Rev. C 102.2 (2020), p. 024914.

[24] M. S. Abdallah et al. Phys. Lett. B 827 (2022), p. 137003.

[25] J. Adamczewski-Musch et al. Phys. Lett. B 778 (2018), pp. 403–407.

[26] M. S. Abdallah et al. Phys. Lett. B 831 (2022), p. 137152.

[27] H. -T. Ding et al. Eur. Phys. J. A 57.6 (2021), p. 202.

[28] Zhenzhen Yang, Xiaofeng Luo, and Bedangadas Mohanty. Phys. Rev. C 95.1 (2017), p. 014914.

[29] Toshihiro Nonaka. Nucl. Instrum. Meth. A 1039 (2022), p. 167171.

Light Nuclei and Hyper Nuclei Collectivity Measurements at High Baryon Density Region

Xionghong He[1,*]

[1]Institute of Modern Physics, Chinese Academy of Sciences, China

Abstract. High energy heavy-ion collisions produce large amounts of light nuclei and hype nuclei, especially at high baryon density around collision energy of several GeV. These light nuclei and hyper nuclei carry the information of nucleon-nucleon and hyperon-nucleon interactions and affect the chemical composition and properties of the collision system. This proceeding is a brief review for the recent measurements on light nuclei collective flow, including directed flow v_1, elliptic flow v_2, and high order flow coefficients, at finite and high baryon densities from different experiments. The light nuclei production mechanism is discussed based on comparisons of the measurements and model calculations. The first measurement for hyper nuclei v_1 at 3 GeV by STAR is also reported, which may imply the hyper nuclei are formed via the coalescence of nucleons and hyperon.

1 Introduction

The properties of strongly interacting matter under high temperatures and high densities are important research focuses of high energy heavy-ion collision experiments all over the world. It is still an open question whether the phase transition from hadronic matter to the quark-gluon plasma (QGP) is a first-order [1] and a critical point exists on the QCD phase diagram at finite baryon density [2]. Even if the QGP is not formed it is important to know the evolution of nuclear matter, which the thermodynamical properties are described by an equation-of-state (EoS). The researches on the EoS can also shed new light on the nuclear matter state existing in the dense stellar objects given their high baryon density, such as neutron star [3]. The azimuthal anisotropy in particle momentum distributions relative to the reaction plane, referred as collective flow, are thought be sensitive probes to the collision system properties at very stage of its evolution and to the EoS [4]. The collision energy dependence of the first- and second-order flow anisotropy parameters, v_1 and v_2, for different particle species provide valuable information on evolution of the nuclear matter. The high order flow parameters, $v_n (n > 2)$ are usually originated from the initial state fluctuations which are not correlated to the reaction plane at very high collision energies [5].

The light nuclei flow has been measured by different experiments from low to high energies [6–13]. These measurements suggest that compared to protons which is the lighted nuclei, the light nuclei (deuteron, triton, ^{3}He, and ^{4}He) flow have more pronounced dependencies on collision energies, and thus may be more sensitive to the collective motion and EoS of nuclear matter. Moreover, the light nuclei flow is expected to reveal their production

*e-mail: hexh@impcas.ac.cn

mechanisms in high energy heavy-ion collisions, which is still a question under debate. The statistical thermal model and nucleon coalescence model are two very popular but very different theoretical models for light nuclei production. The thermal model describes light nuclei production via the nucleon-nucleon or parton-parton interactions before the chemical freeze-out of the fireball [14, 15]. It is however difficult to explain how light nuclei can survive in the hot temperature given their small binding energy. The coalescence model describe the light nuclei is formed near kinetic freeze-out, when the system temperature is much lower, via combination of nucleons if these nucleons are near each other in coordinate and momentum space [16, 17]. One general feature of the nucleon coalescence model is that light nuclei flow is expected to follow an approximate atomic-mass-number (A) scaling under the assumption of small v_n

$$v_n^A(p_T, y)/A \approx v_n^p(p_T/A, y). \quad (1)$$

If one of the nucleon in light nuclei is replaced by a hyperon, a hyper nuclei is formed. In heavy-ion collisions, the production of hyper nuclei allows study of hyperon-nucleon (Y−N) interaction, which remains unclear. At high baryon densities, the EoS is expected to be changed by the appearance of hyperons and hyper nuclei [17], which is critical for understanding the inner structure of compact stellar objects [18]. The thermal model calculations have predicted that the $^3_\Lambda$H and $^4_\Lambda$H production get maximum at high baryon density around several GeV [19]. The measurement of hyper nuclei flow provides valuable information for understanding the Y−N interaction as well as their production mechanism in heavy-ion collisions.

2 Light Nuclei Flow

2.1 Light Nuclei v_1

The energy dependence of v_1 for proton and net-proton at $\sqrt{s_{NN}}$ = 7.7 to 62.4 GeV have been measured by STAR experiment [20]. The value of v_1 slopes at midrapidity, $dv_1/dy|_{y=0}$, change signs from positive to negative at $\sqrt{s_{NN}} \sim 10$ GeV with the increased energy and show a minimum at $\sqrt{s_{NN}} \sim 15$ GeV as shown in Fig. 1 [7], which may be a signature of first-order phase transition according to a hydrodynamical model [21]. According to the

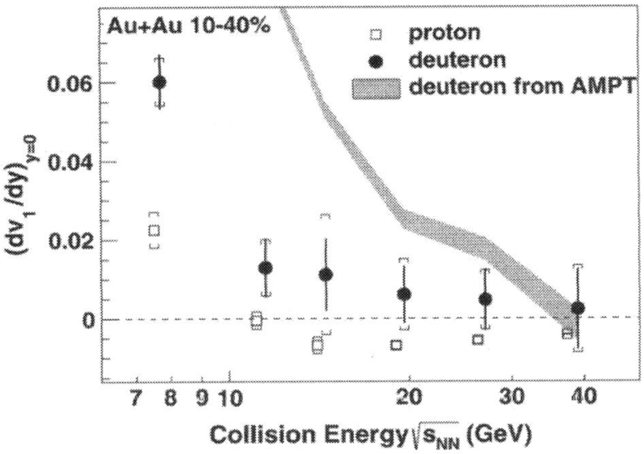

Figure 1. The v_1 slopes at the midrapidity as a function of collision energy in 10-40% Au+Au collisions from STAR experiment (figure from Ref. [7]). The open circles and solid circles represent the results for deuterons and protons, respectively.

picture of nucleon coalescence, the light nuclei v_1 should have a similar but more pronounced energy dependence behavior in the same energy region.

The deuteron v_1 has been measured by the STAR experiment and the extracted $dv_1/dy|_{y=0}$ as a function of collision energy is shown in Fig. 1 [7]. At $\sqrt{s_{NN}} = 7.7$ GeV, the value of $dv_1/dy|_{y=0}$ for deuteron is higher than that for proton and follows the A scaling considering the statistical and systematic uncertainties. Above 7.7 GeV, there is a hint that the deuteron $dv_1/dy|_{y=0}$ have a positive sign with large uncertainties, which is opposite to the corresponding proton $dv_1/dy|_{y=0}$. The p_T dependence of deuteron v_1 show an enhancement towards very low p_T at 7.7 GeV [7]. This measurement may suggest a more complicated mechanism than the simple coalescence picture for light nuclei production, while a stronger conclusion needs more event statistics which can be achieved with the second phase of STAR beam energy scan program.

Moving to lower energies, the light nuclei v_1 slopes reach maximum and have a weak energy dependence from $\sqrt{s_{NN}} = 2$ to 3 GeV [7–9], as shown in Fig. 2. There is a clear mass ordering and the value of $dv_1/dy|_{y=0}$ follow the A scaling approximately. Above 3 GeV, the v_1 slopes decrease with the increasing of collision energies. While the $dv_1/dy|_{y=0}$ has a steep drop with the decreasing energy below 2 GeV [9]. The p_T dependencies of light nuclei v_1 also show the A scaling at midrapidity [8]. The scaling, however, worsens for higher p_T and away from midrapidity; one of possible reasons is the increasing contamination of target/beam rapidity fragments. The proton v_1 at 3 GeV can be well reproduced by the baryon mean-field configuration of JAM transport model [8], while the cascade of the model underestimates the proton v_1. A nucleon coalescence using the proton and neutron phase space distributions from this JAM mean-field calculations can qualitatively describes the v_1 for all measured light nuclei species, which may suggest that the baryonic interactions dominate the evolution and light nuclei are likely formed via the coalescence of nucleons at energy of several GeV.

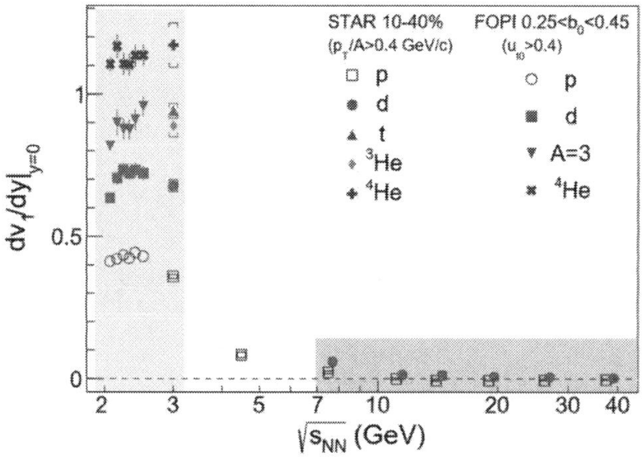

Figure 2. The v_1 slopes at the midrapidity as a function of collision energy in 10-40% Au+Au collisions from FOPI and STAR experiments [7–9]. The open markers represent proton and solid circles represent light nuclei. The results from STAR and FOPI are represented with different marker styles as they use different centrality definitions and transverse momentum cuts.

2.2 Light Nuclei v_2

The light nuclei v_2 as a function of p_T has been measured by the STAR experiment in $\sqrt{s_{NN}} = 7.7$ to 200 GeV Au+Au collisions [6]. The v_2 value shows a monotonic rise with

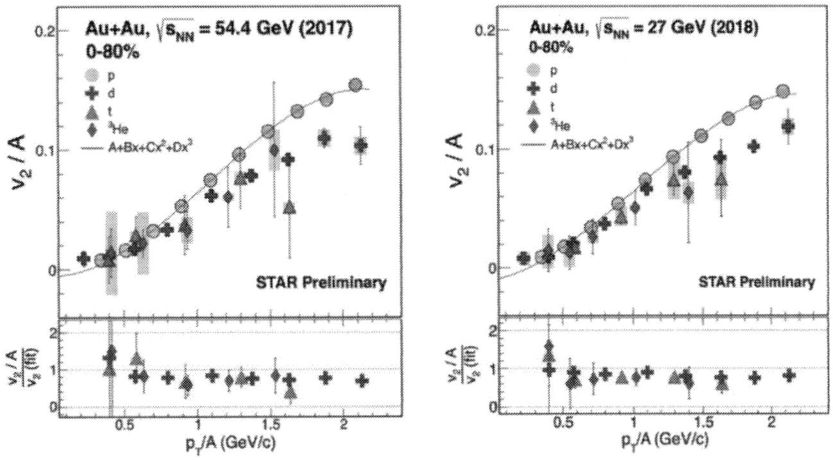

Figure 3. The v_2/A as a function of p_T/A in $\sqrt{s_{NN}}$ =27 GeV and 54.4 GeV Au+Au collisions from STAR [22]. The bottom panels show the ratio of light nuclei v_2 to the fit of proton v_2.

increasing p_T and a mass ordering with increasing mass number. The difference of v_2 between deuteron and anti-deuteron is qualitatively follow the difference between proton and anti-proton. For all measured collision energies, the A number scaling holds generally for all (anti-)light nuclei at low p_T (p_T/A < 1.5 GeV/c), while the scaling is broken, especially for low energies, towards high p_T. The results have been updated with more statistic events by STAR recently [22], as shown in Fig. 3. Compared to v_2 of protons, the v_2/A have a overall 20-30% deviations from A scaling for deuteron, triton, and ^{3}He at both $\sqrt{s_{NN}}$ = 27 GeV and $\sqrt{s_{NN}}$ = 54.4 GeV. The facts that both the deuteron v_1 and light nuclei v_2 do not have A scaling imply that light nuclei may not formed only via a simple coalescence of nucleons at the energy region $\sqrt{s_{NN}}$ = 7.7 – 200 GeV.

With decreasing of collision energies, the v_2 value at midrapidity for protons as well as light nuclei become negative at energy close to 3 GeV, as in Fig. 4. This negative v_2 may

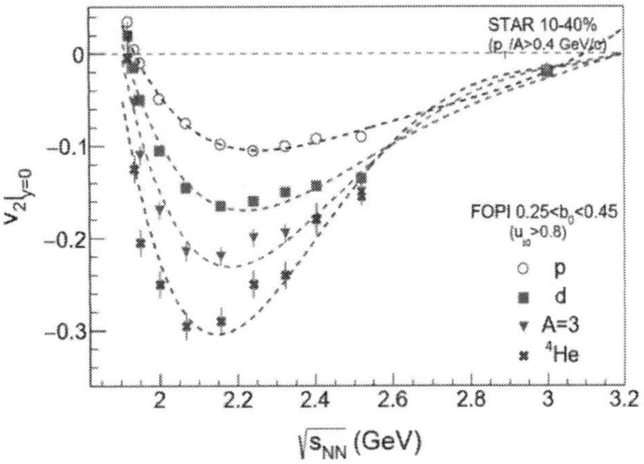

Figure 4. The v_2 in the midrapidity as a function of collision energy. The data points are extracted from STAR and FOPI measurements [8, 9]. The dashed lines represent fits with polynomial functions to guide eyes.

be caused by the squeeze-out and/or shadowing of the spectators due to their passage time becoming comparable with the expansion time of the fireball at $\sqrt{s_{NN}}$ = 2 to 3 GeV [9]. The light nuclei v_2 at the mid-rapidity have a clear mass hierarchy that increases with increasing collision energy from 1.9 GeV to 2.2 GeV and then decreases from 2.2 GeV to 3 GeV. The location of minimum value of v_2 varies with the nuclei mass. The mass hierarchy disappears at 3 GeV for all light nuclei. Similar to their v_1 slope behavior as a function of energy, the heavier nuclei have a stronger energy dependence which may suggest more sensitive to the EoS [9]. Although the light nuclei v_2 violate A scaling at 3 GeV, the JAM plus coalescence calculations can qualitatively describe the rapidity dependence [8]. The reason of the A scaling broken may partly be that the Eq. 1 is only valid under the assumption of small v_1 [8]. At 3 GeV, the v_1 is however not negligibly small as discussed in Section 2.1.

2.3 Light nuclei $v_n(n > 3)$

HADES experiment measure the light nuclei flow from v_1 to v_6 using the first-order event plane at $\sqrt{s_{NN}}$ = 2.4 GeV Au+Au collisions. It is expected that high order flow coefficients ($n > 3$) relative to the first-order event plane are zero if they originate from an initial state fluctuation. However, all the $v_3 - v_6$ are not zero but have strong dependencies on particle p_T and rapidity for all the measured light nuclei species. No mass hierarchy is observed for $v_3 - v_6$ in the midrapidity. Non-zero high order flow may imply these high order flows have more complicated causes at high baryon density and could be more sensitive to the EoS than v_1 and v_2 according to transport model calculations [23].

3 Hyper nuclei v_1

STAR experiment observed v_1 of ${}^3_\Lambda$H and ${}^4_\Lambda$H for the first time in $\sqrt{s_{NN}}$ = 3 GeV Au+Au collisions [24]. The extracted $dv_1/dy|_{y=0}$ at the midrapidity are similar to the corresponding mass light nuclei within the statistical uncertainties. This measurement suggest that the two hyper nuclei may be formed via the coalescence of nucleons and Λ hyperon.

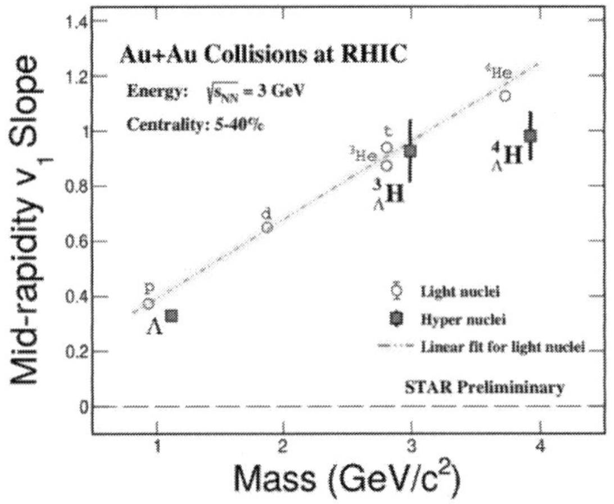

Figure 5. The v_1 slope at the midrapidity as a function of baryon mass number in 5-40% Au+Au collisions at $\sqrt{s_{NN}}$ = 3 GeV [24]. The solid markers represent the Λ baryon, ${}^3_\Lambda$H, and ${}^4_\Lambda$H. The open circles are the results for corresponding light nuclei. The dot-dashed line is a linear fit to the light nuclei v_1 slopes.

4 Summary

In summary, the light nuclei and hyper nuclei flow has been measured at finite and high baryon densities by several experiments. For collision energy above 7.7 GeV, there is a hint of opposite sign for deuteron v_1 slopes with the corresponding proton; the light nuclei v_2 show tensions with the expectations of coalescence model. The light nuclei v_1 slopes saturate and their v_2 at mid-rapidity are negative within 2–3 GeV. The data at 3 GeV can be qualitatively described by the coalescence model. Non-zero $v_3 - v_6$ have been observed at 2.4 GeV which may be more sensitive to the EoS than v_1 and v_2 at high baryon densities. The first flow measurement for hyper nuclei suggests they may be formed via the coalescence of nucleons and hyperon in heavy-ion collisions.

References

[1] A. Bzdak, S. Esumi, v. Koch, J, Liao, M. Stepanov, and N. Xu, Phys. Rept. **853**, 1 (2020).
[2] M. Nahrgang, Nucl. Phys. A **956**, 83 (2016).
[3] E. R. Most, *et al.*, arXiv:2201.13150 (2022).
[4] P. Danielewicz, R Lacey, and W. G. Lynch, Science **298**, 1592 (2002).
[5] J. Adam, *et al.*, ALICE Collaboration, JHEP **1609**, 164 (2016).
[6] L. Adamczyk, *et al.*, STAR Collaboration, Phys. Rev. C **94**, 034908 (2016).
[7] J. Adam, *et al.*, STAR Collaboration, Phys. Rev. C **102**, 044906 (2020).
[8] M. S. Abdallah, *et al.*, STAR Collaboration, Phys. Lett. B **827**, 136941 (2022).
[9] W. Reisdorf, *et al.*, FOPI Collaboration, Nucl. Phys. A **876**, 1 (2012).
[10] J. Adamczewski-Musch, *et al.*, HADES Collaboration, Phys. Rev. Lett. **125**, 262301 (2020).
[11] S. Wang, *et al.*, EOS Collaboration, Phys. Rev. Lett. **74**, 2646 (1995); M.D. Partlan, *et al.*, EOS Collaboration, Phys. Rev. Lett. **75**, 2100 (1995).
[12] J. Barrette, *et al.*, E877 Collaboration, Phys. Rev. C **59**, 884 (1999)
[13] S. Acharya, *et al.*, ALICE Collaboration, Eur. Phys. J. C **77**, 658 (2017); S. Acharya, *et al.*, ALICE Collaboration, Phys. Rev. C **102**, 055203 (2020).
[14] A. Z. Mekjian, Phys. Rev. C **17**, 1051 (1978).
[15] P. Braun-Munzinger and J. Stachel, J. Phys. G **21**, L17 (1995).
[16] H. Sato and K. Yazaki, Phys. Lett. B **98**, 153 (1981).
[17] J. Steinheimer, K. Gudima, A. Botvina, I. Mishustin, M. Bleicher, and H. Stöcker, Phys. Lett. B **714**, 85 (2012).
[18] D. Gerstung, N. Kaiser, and W. Weise, Eur. Phys. J. A **56**, 175 (2020).
[19] A. Andronic, P. Braun-Munzinger, J. Stachel, and H. Stöcker, Phys. Lett. B **697**, 203 (2011).
[20] L. Adamczyk, *et al.*, STAR Collaboration, Phys. Rev. Lett. **112**, 162301 (2014); L. Adamczyk, *et al.*, STAR Collaboration, Phys. Rev. Lett. **120**, 062301 (2018).
[21] H. Stöcker, Nucl. Phys. A **750**, 121 (2005).
[22] STAR Collaboration, talk in the Quark Matter 2022 conference (2022).
[23] P. Hillmann, J. Steinheimer, and M. Bleicher, arXiv:1802.01951 (2018).
[24] STAR Collaboration, talk in the 19th Strangeness in Quark Matter (2021).

Proton and Λ flow and the equation of state at high density

Jan Steinheimer[1,*], *Manjunath* Omana Kuttan[1,2,3], *Anton* Motornenko[1], *Agnieszka* Sorensen[4], *Yasushi* Nara[5], *Volker* Koch[6], *Marcus* Bleicher[2,7,8], and *Horst* Stoecker[1,2,7]

[1]Frankfurt Institute for Advanced Studies, Ruth-Moufang-Str. 1, D-60438 Frankfurt am Main, Germany
[2]Institut für Theoretische Physik, Goethe Universität Frankfurt, Max-von-Laue-Str. 1, D-60438 Frankfurt am Main, Germany
[3]Xidian-FIAS international Joint Research Center, Giersch Science Center, D-60438 Frankfurt am Main, Germany
[4]Institute for Nuclear Theory, University of Washington, Box 351550, Seattle, Washington 98195, USA
[5]Akita International University, Yuwa, Akita-city 010-1292, Japan
[6]Lawrence Berkeley National Laboratory, 1 Cyclotron Road, Berkeley, California 94720, USA
[7]GSI Helmholtzzentrum für Schwerionenforschung GmbH, Planckstr. 1, D-64291 Darmstadt, Germany
[8]Helmholtz Research Academy Hesse for FAIR (HFHF), GSI Helmholtzzentrum für Schwerionenforschung GmbH, Campus Frankfurt, Max-von-Laue-Str. 12, 60438 Frankfurt am Main, Germany

Abstract. Results on proton and Λ flow, calculated with the UrQMD model that incorporates different realistic density dependent equations of state, are presented. It is shown that the proton and hyperon flow shows sensitivity to the equation of state and especially to the appearance of a phase transition at densities below $4n_0$. Even though qualitatively hyperons and protons exhibit the same beam energy dependence of the flow, the quantitative results are different. In this context it is suggested that the hyperon measurements can be used to study the density dependence of the hyperon interaction in high density QCD matter.

1 Introduction

The properties of high density strongly interacting matter can be studied in large collider experiments where heavy nuclei are smashed at relativistic energies to create the densest and hottest matter observed in the universe. At the same time, new astrophysical observatories allow us to study the collisions of the densest macroscopic objects, neutron stars, as binary neutron star mergers (BNSM) which create matter that shows remarkable similarities with the matter created in heavy ion collisions [1–3]. In terrestrial experiments at the RHIC as well as the planned facilities at GSI/FAIR and HIAF, systems created in microscopic collisions of heavy nuclei can now be studied with unprecedented accuracy and statistics. One of the main open challenges in the physics of the strong interaction, Quantum Chromo Dynamics (QCD), is the possibility of a phase transition in matter at densities above twice nuclear saturation density. For systems with vanishing net baryon number and high temperatures, first principle lattice QCD calculations have converged to a picture of a smooth crossover from a hadronic system to a dense state of matter where chiral symmetry is restored [4, 5].

*e-mail: steinheimer@fias.uni-frankfurt.de

Since these simulations at finite baryon density suffer from the fermionic sign problem, no reliable predictions are yet possible for the range of moderate temperature up to $T < 150$ MeV and densities above twice saturation density. In this region of the phase diagram one has to rely on comparing model simulations with a wealth of experimental data which is and will become available. Many observables have been proposed in the past to be sensitive probes for a possible phase transition. To get a consistent picture of the high density QCD equation of state (EoS), using the now available wealth of data, requires a model which is able to incorporate any possible high density EoS and allow us to compute the various observables. In this talk we presented such an approach where it is now possible to incorporate any density dependent EoS in the UrQMD transport model. First results of proton and hyperon flow and its dependence on the EoS and specifically a phase transition were presented. It was found that a phase transition at densities below four times nuclear saturation density is not consistent with available flow data.

2 CMF in UrQMD

To extend the current version of UrQMD [6, 7] and allow the implementation of any density dependent EoS (also with a phase transition) a simple method is used. Besides the cascade part, including elastic and inelastic scattering of many hadronic species, resonance excitations and decays as well as string formation and fragmentation, the UrQMD model includes a quantum molecular dynamics (QMD) part to describe the real part of the nucleon interactions [8]. In the QMD part, the change in the momenta of the baryons are due to a density-dependent potential and are calculated using non-relativistic Hamilton equations of motion:

$$\dot{\mathbf{p}}_i = -\frac{\partial \mathbf{H}}{\partial \mathbf{r}_i} = -\frac{\partial \mathbf{V}}{\partial \mathbf{r}_i} = -\left(\frac{\partial V_i}{\partial n_i} \cdot \frac{\partial n_i}{\partial \mathbf{r}_i}\right) - \left(\sum_{j \neq i} \frac{\partial V_j}{\partial n_j} \cdot \frac{\partial n_j}{\partial \mathbf{r}_i}\right), \tag{1}$$

where $n_{\{i,j\}} \equiv n_B(\boldsymbol{r}_{\{i,j\}})$ is the local interaction density of baryon i or j. Thus, V_i corresponds to the average potential energy of a baryon at position $\boldsymbol{r}_i$, and the local interaction density n_B at position $\boldsymbol{r}_k$ is calculated by assuming that each particle can be treated as a Gaussian wave packet [6, 8].

To implement any realistic EoS in the QMD part of the UrQMD model, we need to calculate the density dependence of the average field energy per baryon $V(n_B)$ within each model, which then can be used in the QMD equations of motion given by Eq. (1). In particular, $V(n_B)$ and its derivative need to be provided in order to numerically calculate changes in momentum at a given time-step. In the following we will use the chiral mean field model (CMF) developed in Frankfurt [9] to provide this density dependent potential. The CMF is based on a chiral mean field Lagrangian in which nucleons and their parity partners as well as quarks interact with the scalar and vector fields. In this approach properties of nuclear matter as well as deconfinement aspects and lattice QCD constraints can be incorporated simultaneously [10]. In the CMF model, the nucleon interaction is described relativistically via scalar and vector mean fields which are not present in UrQMD. Fortunately, the effective field energy per baryon E_{field}/A can be used, i.e., the relevant quantity which enters the equations of motion is then defined as

$$V_{\text{CMF}} = E_{\text{field}}/A \equiv E_{\text{CMF}}/A - E_{\text{FFG}}/A\,, \tag{2}$$

where E_{CMF}/A is the total energy per baryon at $T = 0$ from the CMF model and E_{FFG}/A is the energy per baryon in a free non-interacting Fermi gas.

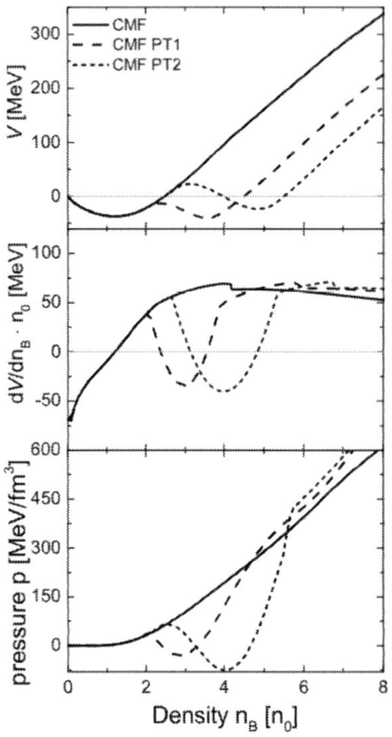

Figure 1. Upper panel: Potential energy per baryon as function of the net baryon density for the three QMD equations of state used in this study. PT1 and PT2 both contain a phase transition visible by the second minimum in the potential energy. Middle panel: Derivative of the potential energy with respect to the net baryon density. Lower panel: Pressure as function of baryon density, corresponding to the three equations of state used.

The resulting average field energy per baryon as a function of the baryon density, from the CMF model, is shown in the upper panel of Fig.1. Here, the original CMF model which incorporates only a crossover is compared with two extensions where an additional first order phase transition has been incorporated in the EoS, referred to as PT1 and PT2 (see [11] for a different method of expanding the potential that can incorporate a phase transition). The middle panel shows the derivative of the potential energy with respect to the baryon density as used in eq. 1 and the lower panel shows the thermodynamic pressure calculated with the given potential. The pressure represents the equation of state and the two phase transitions PT1 and PT2 can be clearly seen as dips in the pressure leading to unstable phases defined by the regions where the derivative of the pressure w.r.t. the density becomes negative. A detailed description of the implementation and details of the UrQMD+CMF model and how a phase transition can be included can be found in [12, 13].

To determine how well the UrQMD model with the CMF potential can describe the evolution of a hot fireball with given equation of state, we first compare the time dependence of bulk properties like the density and temperature of central collisions of Au nuclei at various beam energies. The UrQMD simulations with the standard CMF potential are contrasted with fully consistent relativistic fluid dynamic simulations that include also the CMF equation of state. The density can simply be calculated in the central volume of the collisions by averaging over many UrQMD events while the fluid dynamic simulation was initialized with a smooth initial condition of two cold and Lorenz contracted counterstreaming nuclei.

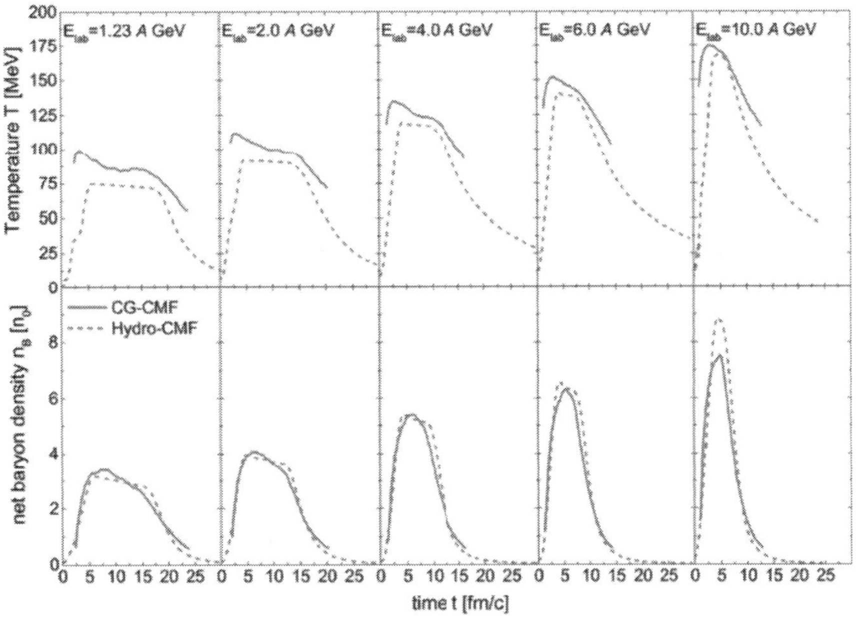

Figure 2. Upper panels: Time evolution of the temperature in a central volume for Au-Au collisions various beam energies calculated in a fully relativistic 3+1D 1-fluid dynamical simulation (dashed lines) and the UrQMD-CMF simulations (solid lines). Lower panels: Time evolution of the central net baryon density in the same systems. All simulations use the CMF-EoS as input [13].

The temperature can also be extracted directly from the CMF EoS included in the fluid simulation. In the UrQMD-CMF simulation, first the averaged energy and momentum density have to be calculated and can then be related to the corresponding temperature in the CMF EoS. The details on the procedure are described in [13]. Figure 2 shows the time evolution for the temperature (upper panels) and net baryon density (lower panels) for different kinetic beam energies in the laboratory frame. The UrQMD simulations (solid lines) are compared to the full fluid dynamic simulations (dashed lines). One can clearly see that the highest compression, which is strongly affected by the EoS used, is almost identical for the two different dynamical descriptions. Also the Temperature is similar. However, the non-equilbrium UrQMD approach leads to slightly larger temperatures. This result shows that indeed the UrQMD-CMF implementation leads to bulk matter compressions expected for this equation of state from a fully consistent fluid model. It is therefore expected that also other quantities can be consistently described by this non-equilibrium approach which has the advantage that no ad-hoc procedures for the fluidization and particlization process are required. In the following we will make use of this and calculate different flow coefficients using the CMF EoS with and without a phase transition and compare them to available data.

Figure 3 presents (from top to bottom) the excitation functions of the (mid-rapidity) mean transverse kinetic energy, the elliptic flow and the slope of the directed flow for protons as well as Λ's, simulated with the UrQMD model with 3 different equations of state. For all beam energies, we considered only Au-Au collisions at different centralities indicated in the figures. The solid lines correspond to the results obtained with the standard CMF-EoS and the dashed lines represent the two different phase transitions. All observables show a clear dependence on the equation of state. The mean transverse kinetic energy increases

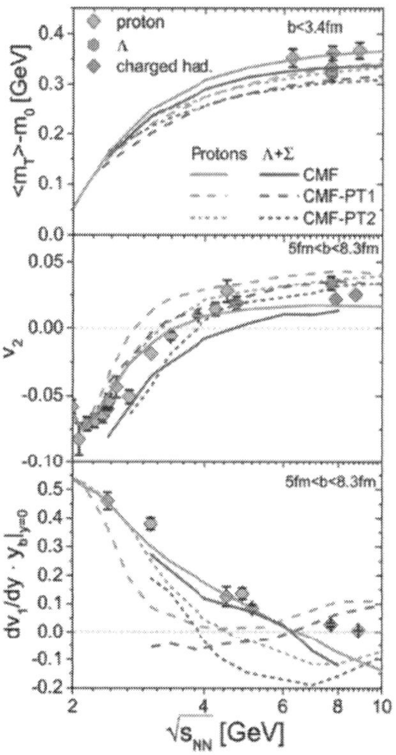

Figure 3. Upper panel: Mean transverse mass of protons (orange) and Λ's (green) in central Au-Au collisions at various beam energies compared to proton and Λ data [14–16]. Middle panel: Elliptic flow v_2 of protons (orange) and Λ's (green) in mid-central Au-Au collisions at various beam energies compared to proton data [17–23]. Lower panel: Slope of the directed flow v_1 of protons (orange) and Λ's (green) in mid-central Au-Au collisions at various beam energies compared to proton data [21–28].

for a stiffer EoS, the elliptic flow (calculated with respect to the reaction plane) shows an opposite behavior and increases for a softer EoS. The slope of the directed flow v_1 even shows a clear minimum for the scenarios with a phase transition as predicted from fluid dynamic simulations [29]. Regarding the hyperon flow we also observe the same effects as for the protons qualitatively, however, all hyperon flow components appear to be reduced compared to the protons. Compared to the available data on proton flow, the standard CMF-EoS provides the best description of the data.

3 Discussion

Results on the sensitivity of proton and hyperon flow on the equation of state in the UrQMD approach were presented. It was shown that the EoS and especially the presence of a phase transition at densities below $4n_0$ has significant impact on the observed flow of both protons and hyperons. Similar results have been obtained recently in [30]. In addition it was shown that, even though the QMD-potential used for protons and hyperons is identical, the flow is not. These findings are in line with a first study [31] where it was argued that Λ flow can be a sensitive probe for the hyperon interaction in dense nuclear matter. Our study, using a realistic density dependent potential consistent with astrophysical observations, confirms this statement and can be seen as a first step into the study of hyperon interactions with UrQMD-CMF. In the future more detailed studies on hyperon flow may be useful tools to understand the hyperon interaction in dense matter found also in compact stars and their mergers.

A.M. acknowledges the Stern–Gerlach Postdoctoral fellowship of the Stiftung Polytechnische Gesellschaft. M.O.K. thanks GSI for funding. A.S. acknowledges support by the U.S. Department of Energy, Office of Science, Office of Nuclear Physics, under Grant No. DE-FG02-00ER41132. Y.N. acknowledges support by JSPS KAKENHI Grant Number JP21K03577. V.K. acknowledges support by the U.S. Department of Energy, Office of Science, Office of Nuclear Physics, under contract number DE-AC02-05CH11231. M.B. acknowledges support by the EU–STRONG 2020 network. The computational resources for this project were provided by the Center for Scientific Computing of the GU Frankfurt and the Goethe–HLR.

References

[1] A. Bauswein, S. Goriely and H. T. Janka, Astrophys. J. **773**, 78 (2013).

[2] M. Hanauske, J. Steinheimer, L. Bovard, A. Mukherjee, S. Schramm, K. Takami, J. Papenfort, N. Wechselberger, L. Rezzolla and H. Stöcker, J. Phys. Conf. Ser. **878**, no.1, 012031 (2017).

[3] E. R. Most, A. Motornenko, J. Steinheimer, V. Dexheimer, M. Hanauske, L. Rezzolla and H. Stoecker, [arXiv:2201.13150 [nucl-th]].

[4] S. Borsanyi, Z. Fodor, C. Hoelbling, S. D. Katz, S. Krieg and K. K. Szabo, Phys. Lett. B **730**, 99-104 (2014).

[5] A. Bazavov *et al.* [HotQCD], Phys. Rev. D **90**, 094503 (2014).

[6] S. A. Bass, M. Belkacem, M. Bleicher, M. Brandstetter, L. Bravina, C. Ernst, L. Gerland, M. Hofmann, S. Hofmann and J. Konopka, *et al.* Prog. Part. Nucl. Phys. **41**, 255-369 (1998).

[7] M. Bleicher, E. Zabrodin, C. Spieles, S. A. Bass, C. Ernst, S. Soff, L. Bravina, M. Belkacem, H. Weber and H. Stoecker, *et al.* J. Phys. G **25**, 1859-1896 (1999).

[8] J. Aichelin, Phys. Rept. **202**, 233-360 (1991).

[9] A. Motornenko, S. Pal, A. Bhattacharyya, J. Steinheimer and H. Stoecker, Phys. Rev. C **103**, 054908 (2021).

[10] J. Steinheimer, S. Schramm and H. Stöcker, Phys. Rev. C **84**, 045208 (2011).

[11] A. Sorensen and V. Koch, Phys. Rev. C **104**, no.3, 034904 (2021).

[12] J. Steinheimer, A. Motornenko, A. Sorensen, Y. Nara, V. Koch and M. Bleicher, Eur. Phys. J. C **82** (2022) no.10, 911.

[13] M. Omana Kuttan, A. Motornenko, J. Steinheimer, H. Stoecker, Y. Nara and M. Bleicher, Eur. Phys. J. C **82**, no.5, 427 (2022).

[14] C. Alt *et al.* [NA49], Phys. Rev. C **73**, 044910 (2006).

[15] L. Adamczyk *et al.* [STAR], Phys. Rev. C **96**, no.4, 044904 (2017).

[16] J. Adam *et al.* [STAR], Phys. Rev. C **102**, no.3, 034909 (2020).

[17] C. Pinkenburg *et al.* [E895], Phys. Rev. Lett. **83**, 1295-1298 (1999).

[18] D. Adamova *et al.* [CERES], Nucl. Phys. A **698**, 253-260 (2002).

[19] A. Andronic *et al.* [FOPI], Phys. Lett. B **612**, 173-180 (2005).

[20] L. Adamczyk *et al.* [STAR], Phys. Rev. C **86**, 054908 (2012).

[21] J. Adam *et al.* [STAR], Phys. Rev. C **103**, no.3, 034908 (2021).

[22] J. Adamczewski-Musch *et al.* [HADES], Phys. Rev. Lett. **125**, 262301 (2020).

[23] M. S. Abdallah *et al.* [STAR], Phys. Lett. B **827**, 137003 (2022).

[24] M. S. Abdallah *et al.* [STAR], Phys. Lett. B **827**, 136941 (2022).

[25] J. Adamczewski-Musch *et al.* [HADES], [arXiv:2208.02740 [nucl-ex]].
[26] E. Kashirin *et al.* [NA61/Shine], J. Phys. Conf. Ser. **1690**, no.1, 012127 (2020).
[27] J. Barrette *et al.* [E877], Phys. Rev. C **55**, 1420-1430 (1997).
[28] C. Alt *et al.* [NA49], Phys. Rev. C **68**, 034903 (2003).
[29] H. Stoecker, Nucl. Phys. A **750**, 121-147 (2005).
[30] D. Oliinychenko, A. Sorensen, V. Koch and L. McLerran, [arXiv:2208.11996 [nucl-th]].
[31] Y. Nara, A. Jinno, K. Murase and A. Ohnishi, Phys. Rev. C **106** (2022) no.4, 4.

Measurements of collectivity in the forward region at LHCb

Imanol Corredoira[1,*]

[1]Instituto Galego de Física de Altas Enerxías (IGFAE, USC)

Abstract.
Due to its unique pseudorapidity coverage ($2 < \eta < 5$) and excellent performance at low p_T, the LHCb detector provides measurements of two-particle correlation in a complementary region to other LHC experiments. Ongoing studies in flow at different collision systems and energies, pPb collisions at $\sqrt{s_{NN}} = 8$ TeV, PbPb collisions at $\sqrt{s_{NN}} = 5$ TeV, and Bose-Einstein correlations of identical pions for pPb collisions at $\sqrt{s_{NN}} = 5$ TeV, are summarised.

1 Introduction

Particle correlations are a fundamental observable in heavy-ion studies. Since the discovery of the long-range near side ridge in pp collisions [1], the characterization of this effect in different collision systems and pseudorapidity ranges has become crucial. Due to the different acceptance of the LHCb experiment [2], it is possible to measure this observable in an unexplored region where the partons involved have a Bjorken-x value in the range $10^{-6} - 10^{-1}$. Previously at LHCb, two-particle correlations of charged hadrons have been measured in pPb collisions at $\sqrt{s_{NN}} = 5$ TeV providing the correlation function and the 1D yield for different activity classes [3]. This analysis will serve as a baseline to introduce ongoing analyses where the Fourier coefficients are extracted from the yield in order to measure directed (v_1), elliptic (v_2) and triangular (v_3) flow. Moreover, for PbPb collisions, the results will be displayed in different centrality classes. Finally, following a previous Bose-Einstein correlation measurement in pp collisions [4], an upcoming analysis on the same observable in pPb collisions will be detailed.

2 Near side long-range corrrelations

2.1 Two-particle angular correlations in pPb collisions at $\sqrt{s_{NN}} = 5$ TeV

Two-particle angular correlations of prompt charged particles have been measured in LHCb for both pPb and Pbp collisions [3]. This observable is widely motivated by the discovery of long-range near-side correlations ("ridge") in pp collisions by the CMS experiment [1]. The correlations are measured for events with one primary vertex and minimum bias trigger. The results have been displayed as a function of p_T and event-activity. The event activity is defined as percentiles of the number of VELO hits distribution, where VELO is the LHCb vertex detector. Five event-activity classes, from low (50-100%) to very high (0-3%) activity have been defined.

*e-mail: imanolcorredoira.fernandez@usc.es

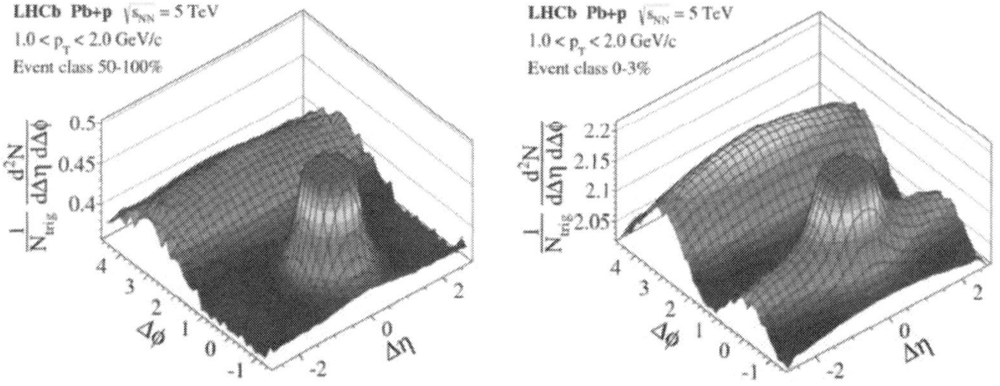

Figure 1. Correlation functions for Pbp collisions at low (left) and high (right) activity.

The correlation function is constructed as a ratio of signal over background,

$$\frac{1}{N_{trig}} \frac{d^2N_{pair}}{d\Delta\eta d\Delta\phi} = B(0, 0)\frac{S(\Delta\eta, \Delta\phi)}{B(\Delta\eta, \Delta\phi)}. \tag{1}$$

Where the signal, $S(\Delta\eta, \Delta\phi)$, is the density of particle pairs within the same event and the background, $B(\Delta\eta, \Delta\phi)$, the density of pairs from mixed events. Finally, this ratio is normalized by the value of the background contribution at the origin.

To quantify the correlation function, and study its evolution, a 1D-yield, $Y(\Delta\phi)$, is computed by integrating over a $\Delta\eta$ range,

$$Y(\Delta\phi) = \frac{1}{N_{trig}} \frac{dN_{pair}}{d\Delta\phi} = B(0, 0)\frac{S(\Delta\phi)}{B(\Delta\phi)}. \tag{2}$$

In order to exclude short-range correlations the integration is limited to $2.0 < |\Delta\eta| < 2.8$. The correlation functions for Pbp collisions can be seen in Fig. 1 for low and high activity events. The appearance of a ridge is visible in the high activity regime. This effect is consistent with the one observed by other experiments.

Furthermore, the underlying events have been subtracted by the zero yields at minimum (ZYAM) method. It is observed that the near side ridge yield increases when increasing event activity, as shown in Fig.2. Moreover, yields at pPb and Pbp collisions are compatible when comparing events with the same activity in the forward region. This is the first observation of near-side long-range correlation in the forward region.

2.2 Charged hadrons v_n in PbPb collisions at $\sqrt{s_{NN}} = 5$ TeV

The LHCb collaboration is going a step forward in measuring flow coefficients for charged hadrons in PbPb collisions. In this work, v_n coefficients will be extracted by fitting the 1D yield with a Fourier series. In that way, the directed, elliptic and triangular flow will be measured in the forward region.

In this study, a centrality classification is made according to Ref.[5] using calorimeter information and fitting to the Glauber model.

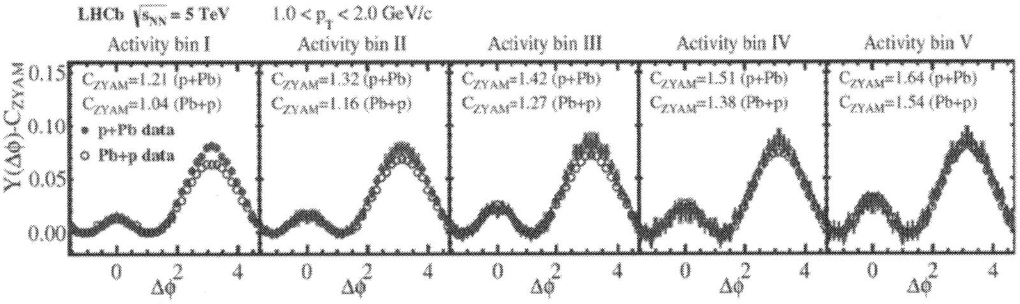

Figure 2. 1D yield from correlation function integration. The ZYAM normalization method is applied. An emergence of long-range near-side correlations can be observed in high activity events.

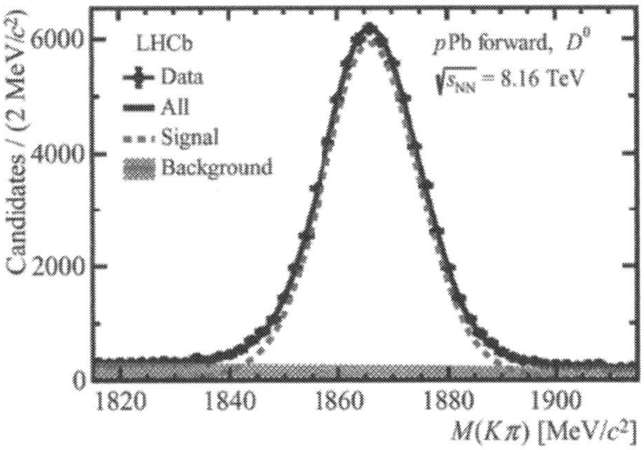

Figure 3. Fit to the invariant mass distribution of D^0 candidates from pPb collisions at $\sqrt{s_{NN}}$ = 8 TeV, [6].

2.3 Charged hadrons and charm meson v_n in pPb collisions at $\sqrt{s_{NN}}$ = 8 TeV

Using the same method as explained in Sec.2.2 to extract Fourier coefficients, a new analysis is being performed with pPb data at $\sqrt{s_{NN}}$ = 8 TeV. Additionally, the flow coefficients of charm mesons will be also measured. Thanks to LHCb's excellent hadron reconstruction capabilities, outstanding precision could be expected from this work. Fig.3 shows the excellent D^0 signals seen in the pPb dataset [6].

3 Bose-Einstein correlations of identical pions in pPb collisions

Following a previous LHCb work using pp collisions [4], a Bose-Einstein correlation analysis is ongoing at the LHCb collaboration using pPb data at $\sqrt{s_{NN}}$ = 5 TeV.

The correlation function, C_2, is computed by dividing the signal over the background. The signal corresponds to the pair density of the same sign pions in the same event while the background is constructed using pions from different events. An event-activity classification is also made. Moreover, the correlation function can be parameterized as a function of two parameters, R and Q,

$$C_2(Q) = 1 + e^{-|RQ|}. \quad (3)$$

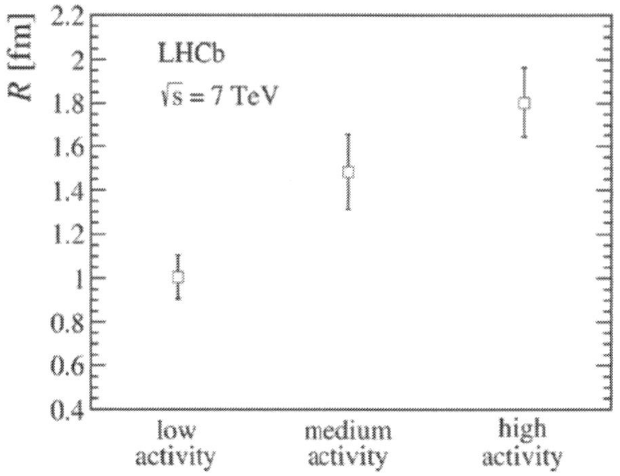

Figure 4. Correlation radius as a function of event-activity from *pp* data [4].

Where R is the correlation radius and $Q \equiv \sqrt{-q^2} = \sqrt{-(k_1 - k_2)^2}$, a Lorentz invariant quantity that gives a measure of the phase-space separation of the two-particle system. Fitting the correlation function with Eq.3 the correlation radius, R, can be extracted. In Fig.4 an example from the published measurement [4] can be seen.

4 Summary

Ongoing analyses in two-particle correlations at LHCb have been summarised. Thanks to its unique pseudorapidity coverage, LHCb detector can provide measurements in a complementary coverage to other LHC experiments. Three measurements have been detailed in this work. First, the already published long-range correlations functions of charged hadrons in *p*Pb/Pb*p* collisions, in which LHCb has confirmed the "ridge" effect emergence in high activity events. Secondly, two ongoing measurements, one with PbPb data at $\sqrt{s_{NN}} = 5$ TeV, and the other with *p*Pb data at $\sqrt{s_{NN}} = 8$ TeV, are extending the analysis method to the computation of the Fourier coefficients. Additionally, flow coefficients of charm mesons will also be measured. Finally, a measurement of the Bose-Einstein correlation of identical pions in *p*Pb collisions is being performed. In this work, the activity dependency of the correlation radius will be studied in *p*Pb collisions.

References

[1] V. Khachatryan et al. (CMS), Phys. Rev. Lett. **116**, 172302 (2016), 1510.03068
[2] J. Alves, A.Augusto et al. (LHCb), JINST **3**, S08005 (2008)
[3] R. Aaij, C. Abellán Beteta, B. Adeva, M. Adinolfi, A. Affolder, Z. Ajaltouni, S. Akar, J. Albrecht, F. Alessio, M. Alexander et al., Physics Letters B **762**, 473 (2016)
[4] R. Aaij et al. (LHCb), JHEP **12**, 025 (2017), 1709.01769
[5] R. Aaij, C.A. Beteta, T. Ackernley, B. Adeva, M. Adinolfi, H. Afsharnia, C. Aidala, S. Aiola, Z. Ajaltouni, S. Akar et al., Journal of Instrumentation **17**, P05009 (2022)
[6] Tech. rep., CERN, Geneva (2022), all figures and tables, along with any supplementary material and additional information, are available at https://cern.ch/lhcbproject/Publications/p/LHCb-PAPER-2022-007.html (LHCb public pages), 2205.03936, https://cds.cern.ch/record/2809400

EPJ Web of Conferences **276**, 01023 (2023) https://doi.org/10.1051/epjconf/202327601023
SQM 2022

Assessing critical point signatures through proton intermittency in NA61/SHINE

Nikolaos Davis[1,*,**]

[1]Institute of Nuclear Physics PAN, ul. Radzikowskiego 152, 31-342, Kraków, Poland

Abstract. The search for experimental signatures of the critical point (CP) of strongly interacting matter is one of the main objectives of the NA61/SHINE experiment at CERN SPS. One such candidate observable is local fluctuations of the proton density in transverse space, constituting an order parameter of the chiral phase transition, and expected to scale according to a universal power-law in the vicinity of the CP. Its scaling can be probed through an intermittency analysis of the proton second scaled factorial moments (SSFMs) in transverse momentum space. The first such analysis [1] revealed power-law behavior in NA49 Si+Si collisions at 158A GeV/c, with a fitted exponent value consistent with the critical prediction, within errors.
In the present work, we discuss known challenges posed by standard intermittency analysis in the face of low to moderate event statistics, and propose a novel technique that solves them. We perform a scan of Monte Carlo models simulating alternative critical and background scenarios, and weigh them against experimental results; we explore the method's efficacy on simulated data. Once adapted to the available NA61/SHINE systems, the method will allow us to obtain reliable confidence intervals for the intermittency index (power-law exponent) ϕ_2 compatible with the experimental data.

1 Introduction

NA61/SHINE is a fixed target particle and high-energy nuclear physics experiment at CERN SPS [2], colliding a variety of beams on hydrogen and nuclear targets. One of the physics goals of its strong interactions programme is to search for experimental signatures of the critical point (CP) of strongly interacting matter. To this purpose, a scan is performed of the phase diagram of strongly interacting matter, varying system size (p+p, Be+Be, Ar+Sc, Xe+La, Pb+Pb) and collision energy (13A – 150A GeV/c), thus probing different freeze-out conditions in temperature T and baryochemical potential μ_B (Fig. 1).

A characteristic feature of a second order phase transition (expected to occur at the CP) is the divergence of the correlation length, leading to a scale-invariant system effectively described by a universality class. Of particular interest are local fluctuations of the order parameter of the QCD chiral phase transition, the chiral condensate $\sigma(\mathbf{x}) = \langle \bar{q}(\mathbf{x}) q(\mathbf{x}) \rangle$. At finite baryochemical potential, the critical fluctuations of the chiral condensate are transferred to the net-baryon density [3]. For a critical system, we expect proton density fluctuations to

*e-mail: nikolaos.davis@ifj.edu.pl
**for the NA61/SHINE Collaboration

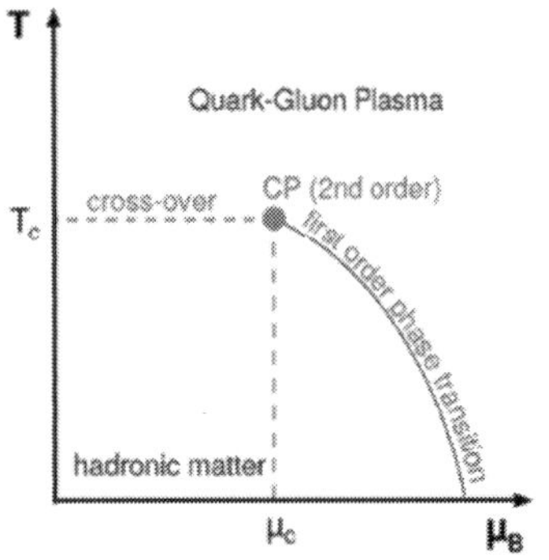

Figure 1. Hypothetical sketch of the phase diagram of strongly interacting matter with critical point, drawn as a function of baryochemical potential μ_B and temperature T.

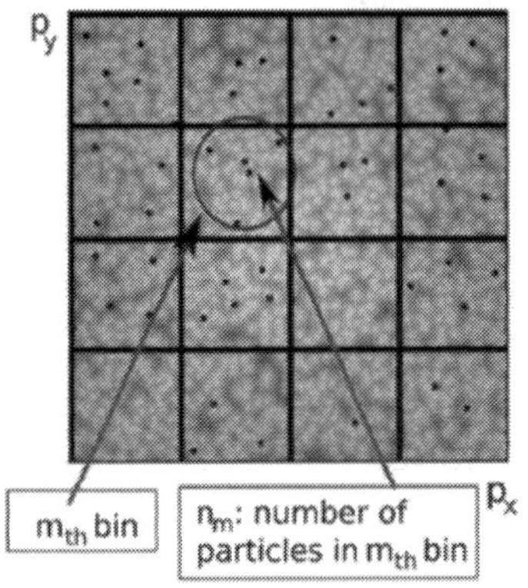

Figure 2. Counting proton pairs in a transverse momentum space partitioning of $M \times M$ equal size bins. *[Image by I. Sputowska]*

be self-similar [4], obeying power laws with critical exponents determined by the 3D Ising universality class [5–7]. Such fluctuations correspond to a power-law scaling of the proton density-density correlation function, which can be detected in transverse momentum space within the framework of an intermittency analysis [7, 8] of proton scaled factorial moments (SFMs). A detailed analysis can be found in Ref. [1], where we study various heavy nuclei collision datasets recorded in the NA49 experiment at the maximum energy (158A GeV/c, $\sqrt{s_{NN}} \approx 17$ GeV) of the SPS (CERN).

2 Methodology

Intermittency analysis examines how the Second Scaled Factorial Moments (SSFM) $F_2(M)$ of proton transverse momenta scale with the number of 2D bins M^2 at mid-rapidity (Fig. 2):

$$F_2(M) \equiv \left\langle \frac{1}{M^2} \sum_{i=1}^{M^2} n_i(n_i - 1) \right\rangle \Big/ \left\langle \frac{1}{M^2} \sum_{i=1}^{M^2} n_i \right\rangle^2 , \quad (1)$$

where n_i is the number of protons in the i-th bin, and $\langle \ldots \rangle$ denotes average over events.

For a pure critical system, $F_2(M)$ is predicted to follow a power-law [7]:

$$F_2(M) \sim M^{2 \cdot \phi_{2,cr}} \quad , \quad \phi_{2,cr}^{(p)} = 5/6. \quad (2)$$

For a noisy system, mixed event moments must be subtracted from the data moments in order to recover the critical component [1]. Thus, we define the correlator $\Delta F_2(M)$:

$$\Delta F_2(M) = F_2^{(d)}(M) - F_2^{(m)}(M). \quad (3)$$

SSFMs statistical errors are estimated by the bootstrap method [9], whereby the original set of events is resampled with replacement.

3 Results

Intermittency analysis of peripheral NA61/SHINE Ar+Sc collisions at 150A GeV/c (Fig. 3) reveals a non-trivial scaling effect; however, large uncertainties in $F_2(M)$ due to low multiplicity and event statistics, as well as M-bin error correlations [9–13] bias power-law fits of $F_2(M)$ and prevent an unbiased estimation of ϕ_2 confidence intervals.

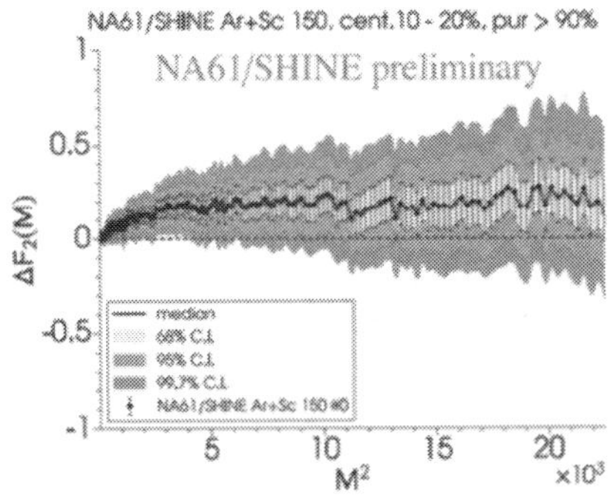

Figure 3. Correlated bin analysis results for $\Delta F_2(M)$ of NA61/SHINE Ar+Sc 10-20% most central collisions @ 150A GeV/c [14].

A better alternative to fitting for a power-law is to model the critical system, along with its background, for a variety of different power-law exponents and critical proton percentages; then, $F_2(M)$ values can be computed and compared to the corresponding experimental moments via a goodness-of-fit function (typically, a χ^2 metric). A scan performed for a large number of different models will hopefully reveal those compatible with the experimental results, thus allowing us to determine ϕ_2 confidence intervals.

We use a version of the Critical Monte Carlo (CMC) algorithm [7] adapted to proton cluster production, in order to simulate pure critical proton events. Subsequently, we "dilute" correlations by randomly replacing critical with uncorrelated protons at a fixed (adjustable) percentage. Fig. 4(a) shows $F_2(M)$ for 1000 iterations of the same critical + background model; Fig. 4(b) shows the correlation matrix between M-bins. Strong correlations between neighboring bins make the $F_2(M)$ distribution hard to understand and handle. We can untangle M-bins using the statistical technique of Principal Component Analysis (PCA), which determines linear combinations of the original M-bins such that their variations are maximal and statistically independent. Just the first few components give a fairly good reconstruction of the empirical $F_2(M)$ distribution.

Fig. 4(c) shows F_2 transformed by PCA, where the first 10 principal components have been kept. By construction, the new bins are uncorrelated , Fig. 4(d). We can now apply our goodness-of-fit function to the F_2 of the model vs the data, after we correspondingly transform the data to PCA-space. Fig. 5 shows the results of repeating this process on a grid of different models, each model compared to the one corresponding to Fig. 4. In this "exclusion plot", we see a narrow band of likely models (including our plug-in), surrounded by an "excluded" region. Event statistics (adjusted to that of Ref. [14]) is unfortunately insufficient to uniquely determine the optimal parameter set.

4 Conclusions

Proton intermittency analysis is a promising tool for detecting the critical point of strongly interacting matter. However, large uncertainties and bin correlations cannot be handled by the conventional analysis method. New techniques of scanning and evaluating Monte Carlo models against the data can handle statistical and systematic uncertainties without sacrificing event statistics.

A full model scan against NA61/SHINE experimental data, through a carefully calibrated Monte Carlo, is in progress; once completed, it will allow the estimation of ϕ_2 and critical component confidence intervals, for the Ar+Sc system as well as other NA61/SHINE available systems.

Acknowledgments: This work was supported by the National Science Centre, Poland (grant no. 2014/14/E/ST2/00018), and the Polish Minister of Education and Science (contract No. 2021/WK/10).

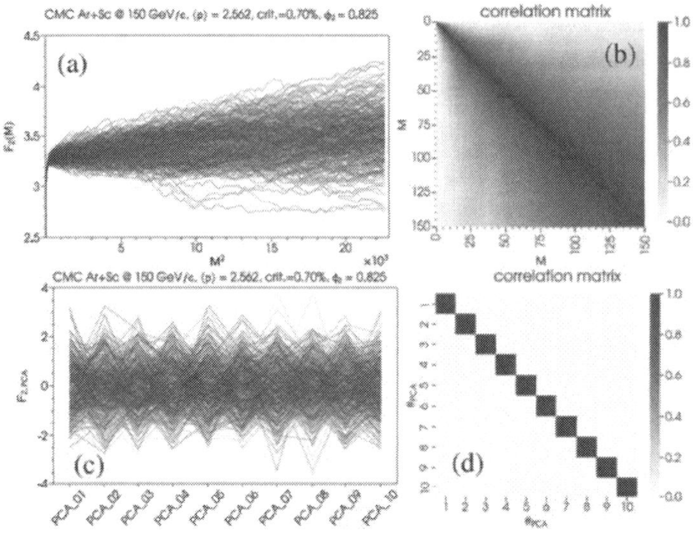

Figure 4. *(a)* $F_2(M)$ for 1000 iterations of CMC+background, for 0.70% critical protons and $\phi_2 = 0.825$, along with *(b)* the corresponding correlation matrix. *(c)* Principal Component Analysis (PCA) transformation of CMC $F_2(M)$; transformed correlation matrix *(d)* is diagonal by construction.

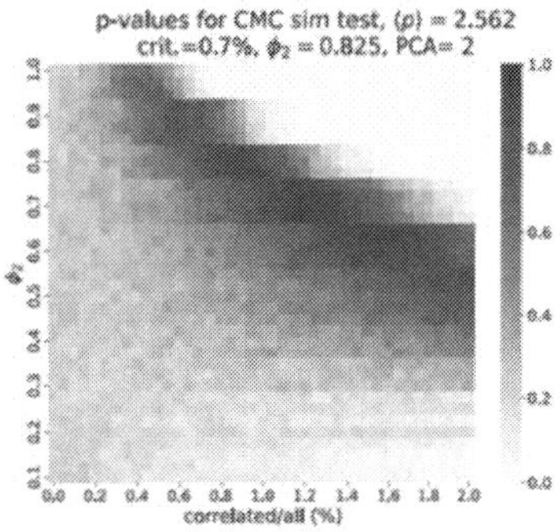

Figure 5. Exclusion plot comparing various models of ϕ_2 strength and critical proton percentage against the CMC model shown in Fig. 4. Color map corresponds to the p-value of each model vs the data, with high p-values corresponding to better fits. A narrow band of "best fit" models runs diagonally across the plot.

References

[1] T. Anticic et al. (NA49), Eur. Phys. J. C **75**, 587 (2015), 1208.5292
[2] N. Abgrall et al., Journal of Instrumentation **9**, P06005 (2014)
[3] Y. Hatta, M.A. Stephanov, Phys. Rev. Lett. **91**, 102003 (2003), hep-ph/0302002
[4] T. Vicsek, *Fractal Growth Phenomena* (World Scientific, 1989), ISBN 9789971508302, https://books.google.com/books?id=N8XSq7rTzTkC
[5] N.G. Antoniou et al., Nucl. Phys. A **693**, 799 (2001), hep-ph/0012164
[6] N.G. Antoniou et al., Nucl. Phys. A **761**, 149 (2005), hep-ph/0505185
[7] N.G. Antoniou et al., Phys. Rev. Lett. **97**, 032002 (2006), hep-ph/0602051
[8] A. Bialas, R.C. Hwa, Phys. Lett. B **253**, 436 (1991)
[9] W.J. Metzger (2004), HEN-455 (unpublished), https://repository.ubn.ru.nl/handle/2066/60397
[10] B. Efron, The Annals of Statistics **7**, 1 (1979)
[11] B. Wosiek, Acta Phys. Polon. B **21**, 1021 (1990)
[12] C. Michael, Phys. Rev. D **49**, 2616 (1994), hep-lat/9310026
[13] N.G. Antoniou et al., Nucl. Phys. A **1003**, 122018 (2020), 2004.12133
[14] N. Davis (NA61/SHINE), Acta Phys. Polon. Supp. **13**, 637 (2020), 2002.06636

YITP-22-107, J-PARC-TH-0277

From lattice to observables: Real and virtual experiments for exploring hot and dense QCD

Masakiyo Kitazawa[1,2,3,*]

[1]Yukawa Institute for Theoretical Physics, Kyoto University, 606-8502, Kyoto, Japan
[2]J-PARC Branch, KEK Theory Center, Institute of Particle and Nuclear Studies, KEK, 203-1, 319-1106, Shirakata, Tokai, Ibaraki, Japan
[3]Department of Physics, Osaka University, 560-0043, Toyonaka, Osaka, Japan

Abstract. Relativistic heavy-ion collisions and the lattice QCD Monte-Carlo simulations are important "experimental" tools for investigating the properties of the medium described by QCD under extreme conditions. After briefly examining their characteristics, I pick up the study of the critical points in QCD as an example of the research subjects that these two tools play complementary roles and discuss recent topics that have been achieved with the use of each tool.

1 Introduction

Exploring strongly-interacting medium under extreme conditions is one of the main topics of Quantum Chromodynamics (QCD) and hadron physics. The hadronic medium undergoes a phase transition to a new form of matter called the quark-gluon plasma (QGP) or the quark matter in which the quarks and gluons are liberated from the confinement into hadrons and act as basic degrees of freedom of the system. This phase transition is expected to occur when the temperature T or the baryon-number density ρ exceeds $T \simeq 150$ MeV $\simeq 1.5 \times 10^{12}$ K or $\rho \simeq 10^{15}$ g/cm^3 [1]. Understanding their properties is indispensable for various purposes, such as investigations of the form of the matter in the early Universe and inside neutron stars.

Although these conditions are far from the environments around us, we have experimental tools to investigate these extreme conditions. Roughly speaking, they are classified into two types. One is the relativistic heavy-ion collisions (HIC) that collide two heavy nuclei accelerated up to almost the speed of light. These experiments have been performed and are ongoing all over the world at the Large Hadron Collider (LHC), Relativistic Heavy Ion Collider (RHIC), and etc. Other experiments are the first-principle numerical simulations based on the lattice QCD, which are virtual experiments on supercomputers. Nowadays, numerical simulations with physical quark masses and their extrapolations to the continuum limit are possible for various observables. They thus are regarded as the "experimental" results. In the following, we denote experiments in this category as the LAT.

For exploring the hot and dense medium, the complementary use of these two experiments is highly effective, because the HIC and LAT have different characteristics, advantages and

*e-mail: kitazawa@yukawa.kyoto-u.ac.jp

Table 1. Short summary of the charcteristics of the HIC and LAT.

HIC	LAT
real	virtual
dynamical	thermal
$\mu_B \neq 0$	
possible	difficult

disadvantages. Now, let us discuss their properties and pros/cons in some more detail using Table 1 that shows a short summary.

First, needless to say, the HIC are real experiments, which is a clear advantage compared with the LAT. On the other hand, the fact that the LAT are simulations can become their unique advantage because the LAT can deal with experiments of the unphysical world, such as QCD with unphysical quark masses; such a study sometimes gives us novel insights into our world.

Second, the HIC produce highly-dynamical systems that expand rapidly after the creation. To extract physical information on the hot medium that exists only in the early stage of this expansion, therefore, a description of the whole dynamics of the space-time evolution is inevitable. In contrast, the LAT perform simulations of completely equilibrium systems. This allows the LAT to investigate thermal properties of the medium, such as the equation of states [2], in an ideal environment. These results are used as useful inputs for the dynamical modeling of the HIC. However, this fact can also be a disadvantage of the LAT, because it is difficult for the LAT to deal with non-equilibrium systems. For example, the quantitative analysis of the transport coefficients is still a challenging subject in the LAT, although they are important observables in describing the dynamics of the HIC.

Finally, the HIC are capable of investigating various T and ρ, since these parameters can be varied by changing the collision energy; for $\sqrt{s_{NN}} \gtrsim 3$ GeV, the baryon-number density of the medium produced by the collisions monotonically increases by reducing $\sqrt{s_{NN}}$. Series of the HIC with various collision energies is called the beam-energy scan, and active experimental activities are ongoing for exploring various regions on the T–ρ plane. For example, these experiments are ongoing at RHIC, NA61/SHINE and HADES, and they will be carried over to future experiments at FAIR and J-PARC-HI [3]. On the contrary, the conventional algorithm of the LAT is applicable only for the system at zero baryon chemical potential μ_B because of the so-called sign (complex-phase) problem. Therefore, investigation of the system with large ρ is difficult in the LAT.

2 QCD Critical points

In the following part of this manuscript, as an example of the research subjects where the HIC and LAT play useful roles in a complementary way, I would like to report on recent progress in the analysis of the critical points in QCD. Since the same mathematical concept for the probability distribution is used in these studies, their simultaneous explanation will also be useful for the mutual understanding.

An end-point of the first-order transition line to crossover is called the critical point (CP). The CP is a second-order transition point and singularities of various observables manifest themselves there. In the QCD phase diagram on the T–μ_B plane, the existence of the first-order transition line is expected at the large μ_B region as shown in Fig. 1. If this is the case, the CP does exist and it is called the QCD CP. The experimental search for the CP is one of the central subjects of the beam-energy scan in the HIC [5, 6].

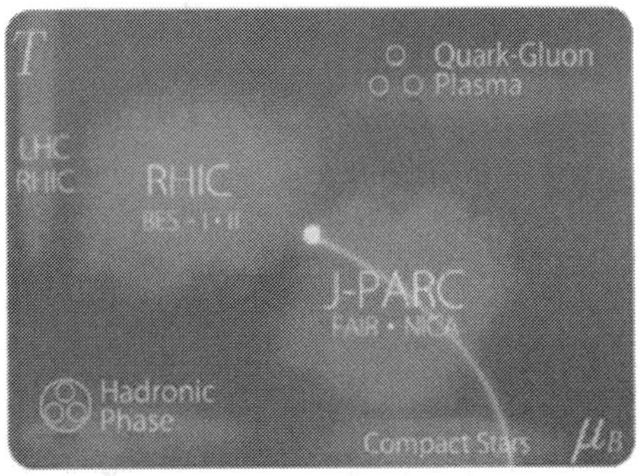

Figure 1. Conceptual phase diagram of QCD on the T–μ_B plane [4].

The appearance of CPs in QCD is not limited to the one shown in Fig. 1. Besides the possible existence of other CP on the T–μ_B plane [7], CPs can appear with variations of the quark masses and other external parameters. For example, whereas the phase transition of QCD at $\mu_B = 0$ is crossover at the physical quark masses, the phase transition becomes first order in the massless three-flavor QCD and the heavy-mass limit. This means that CPs do exist in the phase diagram of quark masses [8]. Their attractive feature is that they exist at $\mu_B = 0$, and hence their investigation in the LAT is feasible with the conventional method. For this reason, these critical points have been actively investigated by the LAT over the past several decades [8–11].

3 Cumulants

Since a CP is a second-order transition point, fluctuations of observables having a coupling with the order parameter show a divergence. Fluctuations are described by the probability distribution function, and the cumulants are useful quantities to denote its property [5]. In this section, we give a brief review of the cumulants as an introduction to the later sections.

Let us consider a real variable x and its probability distribution function $P(x)$. A set of quantities to characterize $P(x)$ is the moments

$$\langle x^n \rangle = \int dx x^n P(x) = \partial_\theta^n \langle e^{x\theta} \rangle|_{\theta=0}. \tag{1}$$

Cumulants are another useful set of quantities to describe $P(x)$ that are defined by

$$\langle x^n \rangle_c = \partial_\theta^n \ln\langle e^{x\theta} \rangle|_{\theta=0}. \tag{2}$$

For lower orders, they are related to each other as

$$\langle x \rangle_c = \langle x \rangle, \quad \langle x^2 \rangle_c = \langle (\delta x)^2 \rangle, \quad \langle x^3 \rangle_c = \langle (\delta x)^3 \rangle, \quad \langle x^4 \rangle_c = \langle (\delta x)^4 \rangle - 3\langle (\delta x)^2 \rangle^2, \tag{3}$$

Compared with the moments, the cumulants have various useful properties [5]. For example, cumulants in thermal systems are extensive variables, whereas the moments are not. For the Gauss distribution, the cumulants at the third and higher orders vanish; therefore, nonzero cumulants higher than the second order represent non-Gaussianity of $P(x)$. The third- and fourth-order cumulants are related to skewness S and kurtosis κ as

$$S = \frac{\langle x^3 \rangle_c}{\langle x^2 \rangle_c^{3/2}}, \qquad \kappa = \frac{\langle x^4 \rangle_c}{\langle x^2 \rangle_c^2}. \tag{4}$$

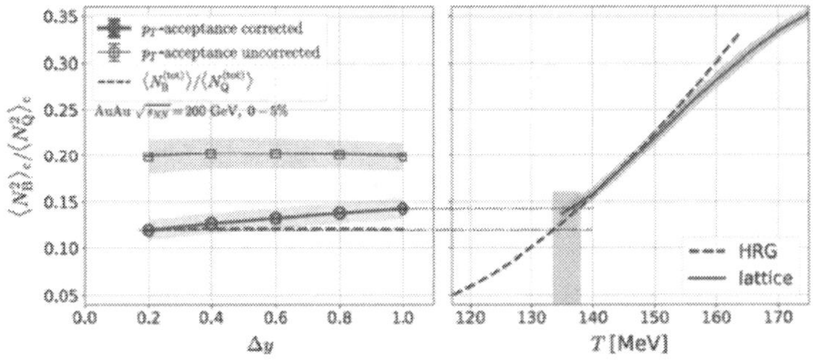

Figure 2. Ratio of the second-order cumulants of the baryon number and electric charge $\langle N_B^2\rangle_c/\langle N_Q^2\rangle_c$ constructed from the experimental data by the STAR Collaboration at $\sqrt{s_{NN}} = 200$ GeV and its comparison with the lattice results [21].

Also, for the Poisson distribution, all the cumulants are equivalent; $\langle x^n\rangle_c = \langle x\rangle_c$.

In Sec. 5, we use the quantity called the (fourth-order) Binder cumulant B_4, which is defined by

$$B_4 = \kappa + 3. \tag{5}$$

For the Gauss distribution, one has $B_4 = 3$.

4 Search for the QCD CP in the HIC

One of the most important subjects of the beam-energy scan in the HIC is the search for the QCD CP on the T–μ_B plane. Fluctuations of particle numbers are observable by the event-by-event analysis in the HIC, and it is believed that an enhancement of the fluctuations associated with the QCD CP is observable in these experiments. Besides the enhancement of the second-order cumulant, it is pointed out that yet higher order ones have more characteristic behavior around the CP, and thus are clearer signals of the CP [12, 13]. Active experimental analyses on the cumulants have been performed recently [14, 15].

Among the fluctuation observables, those of conserved charges have particularly useful properties. First, the cumulants of conserved charges, such as the baryon number N_B and electric charge N_Q, are well-defined and calculable unambiguously in QCD, whereas those of non-conserved quantities are not [5]. This also means that the conserved-charge cumulants are observable on the LAT [16, 17]. This allows one to make a direct comparison between the LAT and HIC in terms of the conserved-charge cumulants. Another advantage of the conserved-charge fluctuations is that their time evolution is typically slow, and hence the fluctuations developed near the CP can survive until the final state [5, 18].

In the HIC, however, a direct measurement of the conserved-charge cumulants is difficult. In particular, the measurement of the baryon number N_B is problematic since typical detectors in the HIC cannot observe neutrons that are charge neutral. For this reason, the proton number cumulants are measured and compared with theoretical results on the baryon number cumulants. As pointed out in Ref. [19], however, they are different quantities that should not be compared directly. Similarly, imperfect efficiencies and acceptance of experimental detectors affect the measurement. Although the effects of the former are usually corrected [20], the latter's effect has not been considered seriously in the literature.

To make a solid comparison between the LAT and HIC, in Ref. [21] we constructed the ratio of the second-order cumulants of the baryon number and electric charge $\langle N_B^2\rangle_c/\langle N_Q^2\rangle_c$

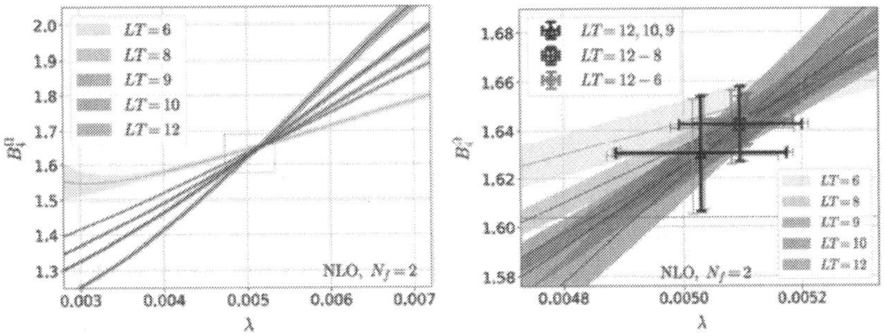

Figure 3. Finite-size scaling analysis of the Binder cumulant B_4 [11].

from the experimental data of STAR Collaboration for $\sqrt{s_{NN}} = 200$ GeV [14] by correcting the effect of the acceptance cut on the transverse momentum. The corrected result for $\langle N_{\rm B}^2\rangle_c/\langle N_{\rm Q}^2\rangle_c$ is shown in the left panel of Fig. 2 by the red circles [21]. By assuming that the cumulant ratio reflects a single temperature, the result can be compared with the result obtained on the LAT [17]. This comparison is made in Fig. 2, and it shows that the value of T extracted from the comparison is $T \simeq 134 - 138$ MeV. Although it is sometimes assumed that the fluctuations observed in the HIC are generated at the chemical freezeout temperature $T_{\rm chem} \simeq 155$ MeV, Fig. 2 shows that this assumption is not justified.

5 Study of QCD critical point in LAT

The direct analysis of the CP on the T–μ_B plane in the LAT is difficult due to the sign problem. On the other hand, the CPs that appear with the variations of the quark masses at $\mu_B = 0$ can be investigated in detail with the conventional algorithms. For this reason, these CPs and the phase structure on the quark-mass plane have been investigated in the LAT over the past decades [8–11].

One of the conventional methods to study the critical point in numerical simulations is the Binder cumulant analysis. In this method, one focuses on the fourth-order Binder cumulant B_4 defined in Eq. (5). On the transition line, this quantity behaves as

$$B_4 = 1 \ \text{(first order side)}, \qquad B_4 = 3 \ \text{(crossover side)}, \tag{6}$$

in the thermodynamic limit. Therefore, the discontinuous change of the value of B_4 can be used for locating the critical point. However, on the finite volume the relation (6) is violated and the discontinuity is smoothen. Even in this case, from the finite-size scaling (FSS) argument it is known that B_4 of the magnetic observable behaves as

$$B_4(t) = b_4 + c(t - t_c)L^{1/\nu} + O((t - t_c)^2), \tag{7}$$

where t is a parameter along the critical line and t_c represents the location of the CP, L is the spatial extent of the system, and b_4 and ν are quantities having common values on a universality class. In the Z_2 universality class to which the QCD CPs belong, their values are

$$b_4 \simeq 1.604, \qquad \nu \simeq 0.63. \tag{8}$$

In Ref. [11], the behavior of the Binder cumulant around the CP in the heavy-quark region has been investigated in the LAT with large spatial volumes up to the aspect ratio $LT = 12$.

To realize the simulations with these LT, the lattice spacing is fixed to a relatively coarse value $a = 1/(4T)$ and the quark determinant is evaluated by the hopping parameter expansion up to the next-to-leading order.

In Fig. 3, we show the result of the Binder cumulant of the Polyakov loop B_4^{Ω} for various values of LT as a function of $\lambda \sim \kappa^4$ with κ the hopping parameter [11]. These numerical results are then fitted by an ansatz inspired by Eq. (7). When the results with large spatial volume $LT \geq 9$ are used for the fit, it gives $b_4 = 1.630(24)$ and $\nu = 0.614(48)$, which are almost consistent with Eq. (8) within 1σ. The scaling of the Z_2 universality class thus is confirmed with high precision. We also found that the values of b_4 and ν become inconsistent with Eq. (8) when the results for smaller spatial volumes are included. This result shows that the large spatial volumes are needed to apply the FSS analysis of the Binder cumulant.

References

[1] K. Yagi, T. Hatsuda and Y. Miake, Camb. Monogr. Part. Phys. Nucl. Phys. Cosmol. **23**, 1 (2005).

[2] S. Borsanyi, *et al.*, Phys. Lett. B **730**, 99 (2014); A. Bazavov, *et al.* [HotQCD], Phys. Rev. D **90**, 094503 (2014).

[3] T. Galatyuk, Nucl. Phys. A **982**, 163 (2019).

[4] H. Sako, *et al.* [J-PARC-HI], https://asrc.jaea.go.jp/soshiki/gr/hadron/jparc-hi/documents.html (2016).

[5] M. Asakawa and M. Kitazawa, Prog. Part. Nucl. Phys. **90**, 299 (2016).

[6] M. Bluhm, *et al.*, Nucl. Phys. A **1003**, 122016 (2020).

[7] M. Kitazawa, T. Koide, T. Kunihiro and Y. Nemoto, Prog. Theor. Phys. **108**, 929 (2002).

[8] S. Gavin, A. Gocksch and R. D. Pisarski, Phys. Rev. D **49**, R3079 (1994).

[9] Y. Kuramashi, Y. Nakamura, H. Ohno and S. Takeda, Phys. Rev. D **101**, 054509 (2020); F. Cuteri, O. Philipsen, A. Schön and A. Sciarra, Phys. Rev. D **103**, 014513 (2021).

[10] S. Ejiri, *et al.* [WHOT-QCD], Phys. Rev. D **101**, 054505 (2020).

[11] A. Kiyohara, M. Kitazawa, S. Ejiri and K. Kanaya, Phys. Rev. D **104**, 114509 (2021).

[12] M. A. Stephanov, Phys. Rev. Lett. **102**, 032301 (2009).

[13] M. Asakawa, S. Ejiri and M. Kitazawa, Phys. Rev. Lett. **103**, 262301 (2009).

[14] J. Adam, *et al.* [STAR], Phys. Rev. C **100**, 014902 (2019); M. Abdallah, *et al.* [STAR], Phys. Rev. C **104**, 024902 (2021).

[15] M. Arslandok [ALICE], Nucl. Phys. A **1005**, 121979 (2021); M. Maćkowiak-Pawłowska [NA61/SHINE], Nucl. Phys. A **1005**, 121753 (2021); J. Adamczewski-Musch, *et al.* [HADES], Phys. Rev. C **102**, no.2, 024914 (2020).

[16] R. Bellwied, *et al.*, Phys. Rev. D **101**, 034506 (2020); A. Bazavov, *et al.* Phys. Rev. D **101**, 074502 (2020).

[17] D. Bollweg, *et al.* [HotQCD], Phys. Rev. D **104**, 074512 (2021).

[18] M. Kitazawa, M. Asakawa and H. Ono, Phys. Lett. B **728**, 386 (2014); M. Kitazawa, Nucl. Phys. A **942**, 65 (2015); M. Sakaida, M. Asakawa, H. Fujii and M. Kitazawa, Phys. Rev. C **95**, 064905 (2017);

[19] M. Kitazawa and M. Asakawa, Phys. Rev. C **85**, 021901 (2012); Phys. Rev. C **86**, 024904 (2012).

[20] T. Nonaka, M. Kitazawa and S. Esumi, Phys. Rev. C **95**, 064912 (2017); M. Kitazawa and X. Luo, Phys. Rev. C **96**, 024910 (2017); T. Nonaka, M. Kitazawa and S. Esumi, Nucl. Instrum. Meth. A **906**, 10 (2018).

[21] M. Kitazawa, S. Esumi and T. Nonaka, Nucl. Phys. A **1030**, 122591 (2023).

EPJ Web of Conferences **276**, 01025 (2023) https://doi.org/10.1051/epjconf/202327601025
SQM 2022

Exploration of the phase diagram within a transport approach

Olga Soloveva[1,2,*], *Pierre* Moreau[3], *Lucia* Oliva[2,4,5], *Taesoo* Song[9], *Ilia* Grishmanovskii[2], *Vadym* Voronuyk[1,8], *Viktar* Kireyeu[1,8], *Jörg* Aichelin[6,7], and *Elena* Bratkovskaya[9,1,2]

[1]Helmholtz Research Academy Hesse for FAIR (HFHF), GSI Helmholtz Center for Heavy Ion Physics, Campus Frankfurt, 60438 Frankfurt, Germany

[2]Institut für Theoretische Physik, Johann Wolfgang Goethe-Universität, Max-von-Laue-Str. 1, 60438 Frankfurt am Main, Germany

[3]Department of Physics, Duke University, Durham, NC 27708, USA

[4]Department of Physics and Astronomy "Ettore Majorana ", University of Catania, Via S. Sofia 64, I-95123 Catania, Italy

[5]INFN Sezione di Catania, Via S. Sofia 64, I-95123 Catania, Italy

[6]SUBATECH, University of Nantes, IMT Atlantique,IN2P3/CNRS 4 rue Alfred Kastler, 44307 Nantes cedex 3, France

[7]Frankfurt Institute for Advanced Studies, Ruth-Moufang-Strasse 1, 60438 Frankfurt am Main, Germany

[8]Joint Institute for Nuclear Research, Joliot-Curie 6, 141980 Dubna, Moscow region, Russia

[9]GSI Helmholtzzentrum für Schwerionenforschung GmbH, Planckstrasse 1, D-64291 Darmstadt, Germany

Abstract. We study equilibrium as well as out-of-equilibrium properties of the strongly interacting QGP medium under extreme conditions of high temperature T and high baryon densities or baryon chemical potentials μ_B within a kinetic approach. We present the thermodynamic and transport properties of the QGP close to equilibrium in the framework of effective models with N_f=3 active quark flavours such as the Polyakov extended Nambu-Jona Lasinio (PNJL) and dynamical quasiparticle model with the CEP (DQPM-CP). Considering the transport coefficients and the EoS of the QGP phase, we compare our results with various results from the literature. Furthermore, out-of equilibrium properties of the QGP medium and in particular, the effect of a μ_B- dependence of thermodynamic and transport properties of the QGP are studied within the Parton-Hadron-String-Dynamics (PHSD) transport approach, which covers the full evolution of the system during HICs. We find that bulk observables and flow coefficients for strange hadrons as well as for antiprotons are more sensitive to the properties of the QGP, in particular to the μ_B - dependence of the QGP interactions.

1 Introduction

It is known that the evolution of the deconfined QCD phase in ultra-relativistic heavy-ion collisions has been successfully described within hydrodynamic simulations and hybrid methods [1–4]. However, only a microscopic treatment can provide a proper non-equilibrium

*e-mail: soloveva@itp.uni-frankfurt.de

description of the entire dynamics through possibly different phases up to the final asymptotic hadronic states. Here we report on a recent progress made within the PHSD transport approach [5–9], which is an off-shell transport approach based on the Kadanoff-Baym equations in first-order gradient expansion. This approach sequentially describes the full evolution of relativistic heavy-ion collisions from the initial hard collisions and formation of strings, the deconfinement with a dynamic phase transition to a strongly interacting QGP, to hadronization and subsequent interactions in the expanding hadronic phase.
While the hadronic part is essentially equivalent to the conventional HSD approach [10], the microscopic properties of the QGP phase are described by the DQPM, which is based on the lQCD data and allows to interpret the equations of state (EoS) in terms of dynamical degrees of freedom and furthermore allows to evaluate the cross sections of the corresponding in/elastic reactions.The PHSD transport approach well describes observables from p+A and A+A collisions from SPS to LHC energies including electromagnetic probes [9].
In order to tackle the new challenge – i.e. the evolution of the partonic systems at finite μ_B – the PHSD approach has been extended to incorporate partonic quasiparticles and their differential cross sections that depend not only on T as in the previous PHSD studies, but also on μ_B explicitly [11]. Within this extended approach, the 'bulk' observables in HICs for different energies – from AGS to RHIC – for symmetric and asymmetric Au+Au/Pb+Pb collisions have been studied. Only a small influence of the μ_B modification of the parton properties (masses and widths) and their interaction cross sections has been found in bulk observables.

Furthermore, in Ref. [12] we extended our study to more sensitive observables, such as collective flow coefficients and a manifestation of the μ_B dependencies of partonic cross sections in the flow coefficients. In addition, we explore the relations between the in- and out-of equilibrium QGP by means of transport coefficients and collective flows.

2 Transport properties of the QGP at finite μ_B

We present transport coefficients of the QGP medium at finite μ_B where the phase transition is possibly changing from a crossover to a 1st order one. Due to the notoriuos difficulty for the estimation of transport coefficients at finite μ_B in lattice QCD it is necessary to resort to effective models which describe the chiral phase transition. It is important to note that while most of the models have a similar EoS, which agrees well with available lattice data, predictions for transport coefficients of the QGP can vary significantly already at $\mu_B = 0$ [11, 13–16].

In Refs. [15, 16] we have evaluated transport coefficients of the QGP medium for a wide range of μ_B for two models with a similar phase structure: the extended $N_f = 3$ PNJL model and the dynamical quasiparticle model with a hypothetical CEP (DQPM-CP) located at $\mu_B = 0.96$ GeV.
The shear and bulk viscosities for quasiparticles with medium-dependent masses $m_i(T, \mu_q)$ can be derived using the Boltzmann equation in the relaxation-time approximation (RTA) through the relaxation time:

$$\eta(T,\mu_q) = \frac{1}{15T} \sum_{i=q,\bar{q},g} \int \frac{d^3p}{(2\pi)^3} \frac{\mathbf{p}^4}{E_i^2} \tau_i(\mathbf{p}, T, \mu_q) \times d_i(1 \pm f_i) f_i, \tag{1}$$

$$\zeta(T,\mu_q) = \frac{1}{9T} \sum_{i=q,\bar{q},g} \int \frac{d^3p}{(2\pi)^3} \tau_i(\mathbf{p}, T, \mu_q) \frac{d_i(1 \pm f_i) f_i}{E_i^2} \left[\mathbf{p}^2 - 3c_s^2 (E_i^2 - T^2 \frac{dm_i^2}{dT^2}) \right]^2, \tag{2}$$

where $q = (u, d, s)$, $d_q = 6$ and $d_g = 16$ are the spin and color degeneracy factors for quarks and gluons, respectively, whereas τ_i are the corresponding relaxation times. c_s is the speed of sound for fixed μ_B. The relaxation times are evaluated from the interaction rates by calculating the partonic differential cross-sections as a function of T and μ_B for the leading tree-level diagrams [11]:

$$1/\tau(p_i, T, \mu_q) = \Gamma_i^{on}(p_i, T, \mu_q) = \sum_{j=q,\bar{q},g} \int \frac{d^3 p_j}{(2\pi)^3} d_j\, f_j\, v_{rel} \int d\sigma_{ij\to cd}^{on}\, (1 \pm f_c)(1 \pm f_d), \quad (3)$$

where $v_{rel} = \frac{\sqrt{(p_i \cdot p_j)^2 - m_i^2 m_j^2}}{E_i E_j}$ denotes the relative velocity in the c.m. frame, d_j is the degeneracy factor for spin and color. The specific shear viscosity of the QGP matter is shown in Fig.

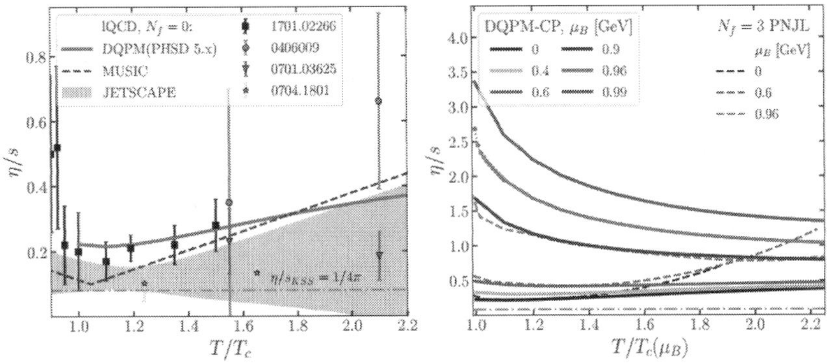

Figure 1. (Left) η/s as a function of T/T_c at $\mu_B = 0$. The symbols corresponds to the lQCD results for pure SU(3) gauge theory (black squares) [17], (green triangles and magenta circles) [18], (cyan stars) [19]. The red line corresponds to the DQPM results [16], while the dashed blue line displays the η/s parametrisation used in hydrodynamic simulations within MUSIC in [22]. The dash-dotted gray line demonstrates the Kovtun–Son–Starinets bound $(\eta/s)_{KSS} = 1/(4\pi)$ [20]. The grey area represents the model-averaged results from a Bayesian analysis of experimental heavy-ion data [21]. (Right) η/s as a function of $T/T_c(\mu_B)$ at finite μ_B: DQPM-CP results [16] (solid lines) are compared to the estimates from the $N_f = 3$ PNJL model (dashed lines) [15].

1 as a function of scaled temperature T/T_c at $\mu_B = 0$ (left) and at finite μ_B (right). At $\mu_B = 0$ we show results from the DQPM [16] (solid red line), in comparison with the lQCD results for pure SU(3) gauge theory [17–19], model-averaged results from a Bayesian analysis of the experimental heavy-ion data [21] (grey area) and η/s as employed in hydrodynamic simulations in [22] (dashed blue line). For finite $\mu_B \geq 0$ we compare the results from the PNJL and DQPM-CP models obtained by the RTA approach with the interaction rate. The estimates from both models show an increase of the specific shear viscosities η/s and electric conductivities σ_{QQ}/T with μ_B. While results for η/s are in agreement for moderate μ_B in the vicinity of the phase transition, there is a clear difference in σ_{QQ}/T essentially due to the different description of the partonic degrees of freedom [16].

3 Evolution of the QGP in the PHSD transport approach

To investigate the sensitivity of 'bulk' observables as well as flow coefficients of different hadrons produced in HICs to the modification of the partonic interactions and their transport properties at non-zero baryon density we have considered the following two settings for the transport simulations:

- **PHSD5.0 -** $\mu_B = 0$ (dashed blue lines): the pole masses and widths of quarks and gluons depend only on T; however, the differential and total partonic cross sections are obtained by calculations of the leading order Feynman diagrams employing the effective propagators and couplings $g^2(T/T_c)$ from the DQPM at $\mu_B = 0$ [11]. Thus, the cross sections depend explicitly on the invariant energy of the colliding partons $\sqrt{s}$ and on T. This is realized in the PHSD5.x by keeping $\mu_B = 0$.
- **PHSD5.0 -** μ_B (solid red lines): the pole masses and widths of quarks and gluons depend on T and μ_B explicitly; the differential and total partonic cross sections are obtained by calculations of the leading order Feynman diagrams from the DQPM and explicitly depend on $\sqrt{s}$, T and μ_B. This is realized in the full version of PHSD5.x [11].

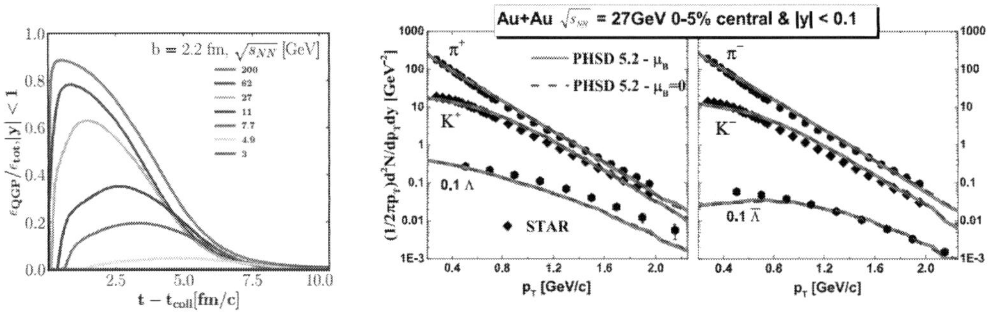

Figure 2. (Left) The QGP energy fraction from the PHSD as a function of time in central (impact parameter $b = 2.2$ fm) Au+Au collisions for different $\sqrt{s_{NN}} = 200 - 3$ GeV at the midrapidity region $|y| < 1$. (Right) The transverse momentum distributions for 0-5% central Au+Au collisions at $\sqrt{s_{NN}} =$ 27 GeV and midrapidity ($|y| < 1$) for PHSD5.2 with partonic cross sections and parton masses calculated for $\mu_B = 0$ (dashed blue lines) and with cross sections and parton masses evaluated at the actual μ_B in each individual space-time cell (solid red lines) in comparison to the experimental data from the STAR collaboration [23].

Fig. 2 (Left) displays the actual results for transverse momentum distributions for 0-5% central Au+Au collisions at $\sqrt{s_{NN}} = 27$ GeV and midrapidity ($|y| < 0.1$) for PHSD5.1 with partonic cross sections and parton masses calculated for $\mu_B = 0$ (dashed blue lines) and with cross sections and parton masses evaluated at the actual chemical potential μ_B in each individual space-time cell (solid red lines) in comparison to the experimental data from the STAR collaboration [23].

4 Summary

We find that HIC results from the extended PHSD transport approach, where in the QGP phase we found that transport coefficients have noticeable T and μ_B dependence, have been in agreement with the BES STAR data in case of bulk observables [11] and elliptic flow of charged particles [12], and reasonably agrees with the results from the hybrid approach [22]. It is important to note that, η/s used for hydrodynamic evolutions is close to the DQPM estimates as shown in Fig. 1 (left). However, results from the PHSD transport approach have shown a rather small influence of the μ_B-dependence of the QGP interactions on the elliptic flow than hybrid simulations. This small sensitivity of final observables to the influence of baryon density on the QGP dynamics can be explained by the fact that at high energies, where the matter is dominated by the QGP phase, one probes the QGP at a very small μ_B, whereas at lower energies, where μ_B becomes larger, the fraction of the QGP drops rapidly (see Fig. 2 (left)). Therefore, the final observables for lower energies of order of $1-10$ GeV are in total dominated by the hadrons which participated in hadronic rescattering and thus the information about their QGP origin is washed out or lost.

The authors acknowledge inspiring discussions with W. Cassing, J. Torres-Rincon, C. Ratti. Furthermore, we acknowledge support by the Deutsche Forschungsgemeinschaft (DFG, German Research Foundation) through the grant CRC-TR 211 "Strong-interaction matter under extreme conditions" - project number 315477589 - TRR 211. I.G. also acknowledges support from the "Helmholtz Graduate School for Heavy Ion research". This work is supported by the European Union's Horizon 2020 research and innovation program under grant agreement No 824093 (STRONG-2020) and by the COST Action THOR, CA15213. The computational resources have been provided by the Goethe-HLR Center for Scientific Computing.

References

[1] H. Song and U. W. Heinz, Phys. Rev. C **78** (2008), 024902

[2] C. Shen, Z. Qiu, H. Song, J. Bernhard, S. Bass and U. Heinz, Comput. Phys. Commun. **199** (2016), 61-85

[3] I. A. Karpenko, P. Huovinen, H. Petersen and M. Bleicher, Phys. Rev. C **91** (2015) no.6, 064901

[4] G. S. Denicol, C. Gale, S. Jeon, A. Monnai, B. Schenke and C. Shen, Phys. Rev. C **98** (2018) no.3, 034916

[5] W. Cassing and E. L. Bratkovskaya, Phys. Rev. C **78** (2008), 034919

[6] W. Cassing, Eur. Phys. J. ST **168** (2009), 3-87

[7] W. Cassing and E. L. Bratkovskaya, Nucl. Phys. A **831** (2009), 215-242

[8] E. L. Bratkovskaya, W. Cassing, V. P. Konchakovski and O. Linnyk, Nucl. Phys. A **856** (2011), 162-182

[9] O. Linnyk, E. L. Bratkovskaya and W. Cassing, Prog. Part. Nucl. Phys. **87** (2016), 50-115

[10] W. Cassing and E. L. Bratkovskaya, Phys. Rept. **308** (1999), 65-233

[11] P. Moreau, O. Soloveva, L. Oliva, T. Song, W. Cassing and E. Bratkovskaya, Phys. Rev. C **100** (2019) no.1, 014911

[12] O. Soloveva, P. Moreau, L. Oliva, V. Voronyuk, V. Kireyeu, T. Song and E. Bratkovskaya, Particles **3** (2020) no.1, 178-192

[13] R. Marty, E. Bratkovskaya, W. Cassing, J. Aichelin and H. Berrehrah, Phys. Rev. C **88** (2013), 045204

[14] J. Grefa, M. Hippert, J. Noronha, J. Noronha-Hostler, I. Portillo, C. Ratti and R. Rougemont, Phys. Rev. D **106** (2022) no.3, 034024

[15] O. Soloveva, D. Fuseau, J. Aichelin and E. Bratkovskaya, Phys. Rev. C **103** (2021) no.5, 054901

[16] O. Soloveva, J. Aichelin and E. Bratkovskaya, Phys. Rev. D **105** (2022) no.5, 054011

[17] N. Astrakhantsev, V. Braguta and A. Kotov, JHEP **04** (2017), 101

[18] A. Nakamura and S. Sakai, Phys. Rev. Lett. **94** (2005), 072305

[19] H. B. Meyer, Phys. Rev. D **76** (2007), 101701

[20] P. Kovtun, D. T. Son and A. O. Starinets, Phys. Rev. Lett. **94** (2005), 111601

[21] D. Everett *et al.* [JETSCAPE], Phys. Rev. Lett. **126** (2021) no.24, 242301
[22] C. Shen and S. Alzhrani, Phys. Rev. C **102** (2020) no.1, 014909
[23] L. Adamczyk *et al.* [STAR], Phys. Rev. C **96** (2017) no.4, 044904

Local equilibrium and Lambda polarization in high energy heavy ion collisions

Andrea Palermo[1,2,*], *Francesco* Becattini[1], *Matteo* Buzzegoli[3], *Gabriele* Inghirami[4], and *Iurii* Karpenko[5]

[1]Università di Firenze and INFN Sezione di Firenze, Via G. Sansone 1, I-50019 Sesto Fiorentino (Florence), Italy
[2]Institut für Theoretische Physik, Johann Wolfgang Goethe-Universität, Max-von-Laue-Straße 1, D-60438 Frankfurt am Main, Germany
[3]Department of Physics and Astronomy, Iowa State University, Ames, Iowa 50011, USA
[4]GSI Helmholtzzentrum für Schwerionenforschung GmbH, Planckstr. 1, 64291 Darmstadt , Germany
[5]Faculty of Nuclear Sciences and Physical Engineering, Czech Technical University in Prague, Břehová 7, 11519 Prague 1, Czech Republic

Abstract. The polarization of the Λ hyperon has become an important probe of the Quark-Gluon Plasma produced in relativistic heavy-ion collisions. Recently, it has been found that polarization receives a substantial contribution from a local equilibrium term proportional to the symmetric derivative of the four-temperature vector, the thermal shear tensor. We show that, at very high energies, this term can restore the agreement between the experimental measurements and the predictions of the hydrodynamic model, provided that the hadronization hypersurface is isothermal. We review the theoretical derivation of this new term, discuss numerical computations at RHIC and LHC energies, and compare them with the experimental data. We also present the effect of feed-down corrections.

1 Introduction

In non-central heavy ion collisions, the Quark-Gluon Plasma (QGP) is produced with a tremendous orbital angular momentum, which in turn polarizes particles produced in the hadronization stage. The first quantitative relation for the expectation value of the spin-vector of Dirac fermions was found to be [1]:

$$S^\mu_\varpi(p) = -\frac{1}{8m}\epsilon^{\mu\nu\sigma\tau}p_\tau\frac{\int_\Sigma \mathrm{d}\Sigma\cdot p\, n_F(1-n_F)\varpi_{\nu\sigma}}{\int_\Sigma \mathrm{d}\Sigma\cdot p\, n_F}, \tag{1}$$

where m is the mass of the particle, n_F is the Fermi-Dirac distribution function and the integration hypersurface Σ is taken to be the decoupling (or freeze-out) hypersurface. Given the four-temperature vector $\beta^\mu = u^\mu/T$ where u^μ is the four-velocity and T the proper temperature, the antisymmetric thermal vorticity tensor $\varpi_{\mu\nu}$ reads:

$$\varpi_{\mu\nu} = -\frac{1}{2}(\partial_\mu\beta_\nu - \partial_\nu\beta_\mu). \tag{2}$$

*e-mail: andrea.palermo@unifi.it
Presented at the 20th International Conference on Strangeness in Quark Matter (SQM 2022).

The formula (1) was successful in explaining the data concerning the global polarization, meaning the polarization integrated over all momenta [2, 3] but, unexpectedly, it could not explain the data of polarization as a function of momentum, the so-called local polarization [4, 5]. This inconsistency between global and local polarization has been known in the literature as the polarization sign puzzle.

Recently, a new contribution coming from the symmetric derivative of the four-temperature was found [6–9]. Adopting the results of [6], the additional term reads:

$$S^\mu_\xi(p) = -\frac{1}{4m}\epsilon^{\mu\nu\sigma\tau}\frac{p_\tau p^\rho}{\varepsilon}\frac{\int_\Sigma \mathrm{d}\Sigma\cdot p\, n_F(1-n_F)\hat{t}_\nu\xi_{\sigma\rho}}{\int_\Sigma \mathrm{d}\Sigma\cdot p\, n_F}. \tag{3}$$

In the above formula, $\varepsilon = \sqrt{m^2+\boldsymbol{p}^2}$ is the particle's energy, the vector $\hat{t}$ is given by $\hat{t}^\mu = (1,0,0,0)$ in the laboratory frame, and its origin will be briefly commented in the next section, and the symmetric tensor $\xi_{\mu\nu}$ is the so-called thermal shear:

$$\xi_{\mu\nu} = \frac{1}{2}(\partial_\mu\beta_\nu + \partial_\nu\beta_\mu). \tag{4}$$

Equation (3) represents a non-dissipative contribution to spin polarization because it originates from the local equilibrium statistical operator (see section 2). Nonetheless, it is a non-equilibrium effect, as in global equilibrium the thermal shear is necessarily vanishing due to the Killing condition.

To account for both effects, the total spin vector is the sum of equation (1) and (3), that is $S^\mu(p) = S^\mu_\varpi(p) + S^\mu_\xi(p)$.

2 Polarization in high energy nuclear collisions: isothermal decoupling

We comment on some noteworthy steps of the theoretical derivation, having in mind applications to heavy ion collisions at very high energy, e.g $\sqrt{s_{NN}} \geq 200$ GeV. More details about the derivation can be found in [6, 7].

The polarization vector is computed using the formula [10]:

$$S^\mu(k) = \frac{1}{2}\frac{\int_\Sigma \mathrm{d}\Sigma\cdot k\,\mathrm{tr}\left[\gamma^\mu\gamma^5 W^+(x,k)\right]}{\int_\Sigma \mathrm{d}\Sigma\cdot k\,\mathrm{tr}\left[W^+(x,k)\right]}, \tag{5}$$

where the integration is performed on the decoupling hypersurface. The function $W^+(x,k)$ is the particle component of the so-called Wigner function, to which we can associate a Wigner operator $\widehat{W}$ in such a way that the Wigner function is its expectation value, $W(x,k) = \mathrm{Tr}\left(\widehat{\rho}\,\widehat{W}(x,k)\right)$. The particle component of the Wigner operator is:

$$\widehat{W}^+_{ab}(x,k) = \theta(k^0)\theta(k^2)\frac{1}{(2\pi)^4}\int \mathrm{d}^4s\, e^{-ik\cdot s} : \bar{\Psi}_b(x+s/2)\Psi_a(x-s/2) :, \tag{6}$$

For what concerns the density operator $\widehat{\rho}$, it is assumed to be the so-called local equilibrium density operator:

$$\widehat{\rho}_{LE} = \frac{1}{Z}e^{-\int \mathrm{d}\Sigma_\mu \widehat{T}^{\mu\nu}\beta_\nu}, \tag{7}$$

where $\widehat{T}^{\mu\nu}$ is the stress-energy tensor operator, the integration hypersurface is the decoupling hypersurface and we have neglected the presence of conserved charges. It is known that the density operator (7) describes an ideal fluid out of equilibrium, hence the results obtained from (7) are non-dissipative [11, 12].

In the hydrodynamic regime, when dealing with expectation values of local operators such as the Wigner operator $\widehat{W}(x,k)$, a Taylor expansion of the four-temperature about the point x is possible, and the density operator is approximated as:

$$\widehat{\rho}_{LE} = \frac{1}{Z_{LE}} \exp\left[-\beta(x)\cdot\widehat{P} + \frac{1}{2}\varpi_{\nu\lambda}(x)\widehat{J}^{\nu\lambda}_x - \frac{1}{2}\xi_{\nu\lambda}\widehat{Q}^{\nu\lambda}_x\right]. \tag{8}$$

The thermal vorticity couples with the angular momentum operator $\widehat{J}^{\lambda\nu}$, whereas the termal shear couples with the operator $\widehat{Q}^{\lambda\nu}$:

$$\begin{aligned}\widehat{J}^{\lambda\nu}_x &= \int_\Sigma \mathrm{d}\Sigma_\mu \left[(y-x)^\lambda \widehat{T}^{\mu\nu}(y) - (y-x)^\nu \widehat{T}^{\mu\lambda}(y)\right],\\ \widehat{Q}^{\lambda\nu}_x &= \int_\Sigma \mathrm{d}\Sigma_\mu \left[(y-x)^\lambda \widehat{T}^{\mu\nu}(y) + (y-x)^\nu \widehat{T}^{\mu\lambda}(y)\right] = \int_\Sigma \mathrm{d}\Sigma_\mu\, \widehat{Q}^{\mu,\lambda\nu}.\end{aligned} \tag{9}$$

The remarkable difference between $\widehat{J}^{\lambda\nu}_x$ and $\widehat{Q}^{\lambda\nu}_x$ is that the latter depends on the integration hypersurface because $\partial_\mu \widehat{Q}^{\mu,\lambda\nu} \neq 0$. That's why the vector $\hat{t}$ appears in (3): $\hat{t}$ is to be interpreted as the average normal vector to the decoupling hypersurface.

If we use linear response theory up to first order in gradients with the operator (8), we obtain the results (1) and (3). For very high energy collisions, however, we can do better. In fact, if the collision happens at very high energy the chemical potential is negligible, and the only intensive thermodynamic parameter is the temperature T. It follows that the only possible parametrization of the freeze-out hypersurface is $T = T_{\mathrm{dec}}$ for some constant decoupling temperature T_{dec}. Therefore, since $\beta^\mu = u^\mu/T$, (7) reduces to the isothermal local equilibrium density operator:

$$\widehat{\rho}_{ILE} = \frac{1}{Z}\exp\left[-\int_{T=T_{\mathrm{dec}}} \mathrm{d}\Sigma_\mu \widehat{T}^{\mu\nu}\beta_\nu\right] \simeq \frac{1}{Z}\exp\left[-\frac{1}{T_{\mathrm{dec}}}\int_{T=T_{\mathrm{dec}}} \mathrm{d}\Sigma_\mu \widehat{T}^{\mu\nu}u_\nu\right]. \tag{10}$$

Although $\widehat{\rho}_{LE}$ and $\widehat{\rho}_{ILE}$ are completely equivalent in an isothermal decoupling scenario, linear response theory leads to different results depending on which operator we use. On the one hand equation (8) is expanded in terms of gradients of the four-temperature vector β^μ. On the other, for the operator (10), the expansion in gradients would involve only the gradients of u^μ and not of the temperature T. If we could compute polarization exactly, it would be irrelevant to use (7) or (10) as the density operator for an isothermal decoupling hypersurface, but having to resort to linear response theory it is paramount to choose the best density operator before computing the gradient expansion, and in the case of high energy nuclear collisions, the best choice is (10).

Using the isothermal local equilibrium density operator the mean spin vector reads:

$$S^\mu_{ILE}(k) = -\frac{1}{8m}\epsilon^{\mu\nu\sigma\tau}k_\tau \frac{\int_\Sigma \mathrm{d}\Sigma\cdot k\, n_F(1-n_F)\left(\omega_{\nu\sigma} + 2\hat{t}_\nu \Xi_{\sigma\rho}\frac{p^\rho}{\varepsilon}\right)}{T_{\mathrm{dec}}\int_\Sigma \mathrm{d}\Sigma\cdot k\, n_F}, \tag{11}$$

where we have defined the kinematic vorticity $\omega_{\mu\nu}$ and the kinematic shear $\Xi_{\mu\nu}$:

$$\omega_{\mu\nu} = -\frac{1}{2}(\partial_\mu u_\nu - \partial_\nu u_\mu), \qquad \Xi_{\mu\nu} = \frac{1}{2}(\partial_\mu u_\nu + \partial_\nu u_\mu). \tag{12}$$

In [7] it was found that for a decoupling temperature $T_{\mathrm{dec}} \sim 155$ MeV the predictions of (11) are in very good agreement with the data, thus solving the polarization sign puzzle. Such a result confirms that, for high-energy applications, the use of (10) and (11) is more appropriate.

3 Feed-down corrections

Hitherto, only the contribution of primary Λs has been considered, whereas feed-down may have some significance. We report for the first time the effect of feed-down on polarization including the thermal shear term at $\sqrt{s_{NN}} = 200$ GeV and $\sqrt{s_{NN}} = 5020$ GeV, corresponding to RHIC and ALICE top energies. The full detail of the simulation will appear in a forthcoming publication [13]. There have been already precious studies on the feed-down correction to polarization, and we refer the reader to them for a detailed theoretical analysis [14, 15].

Considering only the main channels of production of Λ hyperons, namely the primary Λs, and those produced in the processes $\Sigma^* \to \Lambda + \pi$ and $\Sigma_0 \to \Lambda + \gamma$, total spin vector of the Λ particle in its rest frame is given by:

$$S_{\Lambda,tot}(p) = \frac{n^{(FO)} S^{(FO)}(p) + n^{(\Sigma^*)} S^{(\Sigma^*)}(p) + n^{(\Sigma_0)} S^{(\Sigma_0)}(p)}{n^{(FO)} + n^{(\Sigma^*)} + n^{(\Sigma_0)}}. \tag{13}$$

where $S^{(X)}$ are the spin vector computed in each production channel, where the primaries have been denoted by (FO), and the other channels have the label of the mother particle. The vector $S^{(FO)}$ corresponds to the back-boost of (11) to the rest frame of the Λ, and $S^{(\Sigma^*)}$ and $S^{(\Sigma_0)}$ are calculated using the formula (31) of ref. [15].

In equation (13), $n^{(X)}$ represent the fraction of the total Λs produced in the channel X. An estimation has been given in [15], namely $n_\Lambda^{(FO)} = 0.243$, $n_\Lambda^{(\Sigma^*)} = 0.359$ and $n_\Lambda^{(\Sigma_0)} = 0.165$. These fractions are used for both 200 and 5020 GeV calculations.

The hydrodynamic stage of the evolution of the QGP is simulated using the 3+1 D hydrodynamical code vHLLE [16]. The initial state has been generated with averaged entropy density profile from the Monte Carlo Glauber model, generated by GLISSANDO v.2.702 code [17].

In figures 1 and 2, we show the dependence of $\langle P_z \sin 2\phi \rangle$ as a function of p_T and P_z (z being the beam axis) as a function of ϕ, for $\sqrt{s_{NN}} = 200$ GeV and $\sqrt{s_{NN}} = 5020$ GeV respectively, using the isothermal decoupling result for the spin vector (11). We remind the reader that the polarization vector for fermions is defined as $P^\mu = 2S^\mu$ where S^μ is the spin vector.

A good agreement with the data is found for a decoupling temperature of $T_{\rm dec} = 160$ MeV. For 200 GeV calculations, the simulation refers to $20 - 60\%$ centrality, and we used $\eta/s = 0.08$ and $\zeta/s = 0$ where η and ζ are the shear and bulk viscosities and s is the entropy density. The simulation at 5020 GeV, on the other hand, refers to $30 - 50\%$ centrality and $\eta/s = 0.12$. Interestingly, the experimental data cannot be reproduced if we keep $\zeta/s = 0$, so that a non-vanishing bulk viscosity is used. In fig. 2, the results are reported using ζ/s as in the parametrization III of ref. [18].

4 Conclusion and outlook

We have studied the polarization of Λ hyperons at local equilibrium, discussing both the contributions coming from thermal vorticity and from thermal shear. We have argued that the isothermal local equilibrium density operator should be used when dealing with high-energy nuclear collisions and we have used it to calculate polarization including the feed-down correction. The feed-down reduces the signal of about 20% compared to the case of primary Λs and a good fit with the data is obtained for a decoupling temperature of 160 MeV. We have found that at 5020 GeV a non-vanishing bulk viscosity is needed to explain the experimental data. The origin of this effect and the importance of bulk viscosity for spin physics represent an interesting avenue for future research.

 https://doi.org/10.1051/epjconf/202327601026

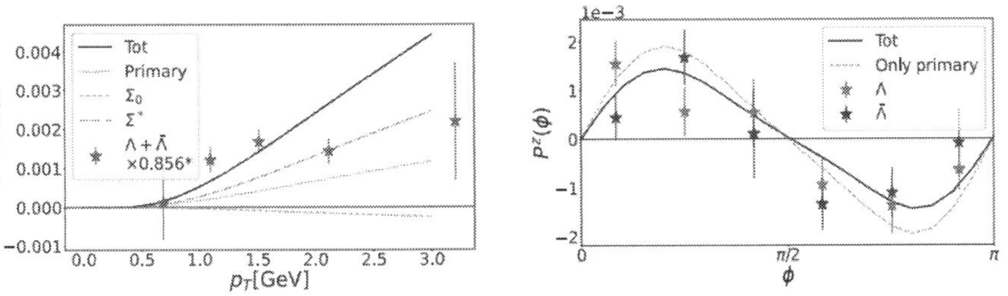

Figure 1. Calculations of the polarization vector at 200 GeV. Left panel: second Fourier harmonic of P_z. The contribution of each channel is plotted after multiplication for the respective weight $n^{(x)}$. Right panel: azimuthal dependence of the feed-down corrected P_z compared to the primary-only channel. The data points are taken from [4], and we used $\langle \cos^2 \theta_p^* \rangle = 1/3$ and $\alpha_H = 0.732$.

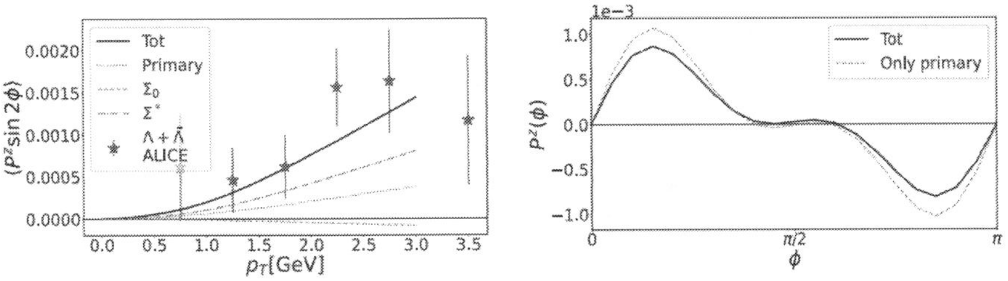

Figure 2. Calculations of the polarization vector at 5020 GeV. Left panel: second Fourier harmonic of P_z, similar as figure 1. The data points are taken from [19]. Right panel: comparison between P_z before and after the inclusion of feed down, much like fig. 1. In this simulation, a non-vanishing bulk viscosity has been used.

References

[1] F. Becattini, V. Chandra, L. Del Zanna and E. Grossi, "Relativistic distribution function for particles with spin at local thermodynamical equilibrium", Annals Phys., **338**, pp 32-49 (2013)

[2] L. Adamczyk et al., "Global Λ hyperon polarization in nuclear collisions: ev- idence for the most vortical fluid", Nature, **548**, pp. 62–65, 2017.

[3] J. Adam et al., "Global polarization of Λ hyperons in Au+Au collisions at $\sqrt{s_{NN}} = 200$ GeV", Phys. Rev. C, **98**, p. 014910, 2018.

[4] J. Adam et al., "Polarization of $\Lambda(\bar{\Lambda})$ hyperons along the beam direction in Au+Au collisions at $\sqrt{s_{NN}} = 200$ GeV", Phys. Rev. Lett., **123**, no. 13, p. 132301, 2019.

[5] T. Niida, "Global and local polarization of Λ hyperons in Au+Au collisions at 200 GeV from STAR", Nucl. Phys. A, **982**, pp. 511–514, 2019.

[6] F. Becattini, M. Buzzegoli, and A. Palermo, "Spin-thermal shear coupling in a relativistic fluid", Phys. Lett. B, **820**, p. 136519, 2021.

[7] F. Becattini, M. Buzzegoli, G. Inghirami, I. Karpenko, and A. Palermo, "Lo- cal Polarization and Isothermal Local Equilibrium in Relativistic Heavy Ion Collisions", Phys. Rev. Lett., **127**, no. 27, p. 272302, 2021.

[8] S. Y. F. Liu and Y. Yin, "Spin polarization induced by the hydrodynamic gradients", JHEP, **07**, p. 188, 2021.

[9] B. Fu, S. Y. F. Liu, L. Pang, H. Song, and Y. Yin, "Shear-Induced Spin Polarization in Heavy-Ion Collisions", Phys. Rev. Lett., vol. **127**, no. 14, p. 142301, 2021.
[10] F. Becattini, "Polarization in relativistic fluids: a quantum field theoretical derivation", Lect. Notes Phys., **987**, pp. 15–52, 2021.
[11] F. Becattini, M. Buzzegoli, and E. Grossi, "Reworking the Zubarev's approach to non-equilibrium quantum statistical mechanics", Particles, **2**, no. 2, pp. 197–207, 2019.
[12] C. van Weert, "Maximum entropy principle and relativistic hydrodynamics", Annals of Physics, **140**, no. 1, pp. 133–162, 1982
[13] A. Palermo, I. Karpenko, F. Becattini, in preparation.
[14] X.-L. Xia, H. Li, X.-G. Huang, and H. Z. Huang, "Feed-down effect on Λ spin polarization", Phys. Rev. C **100**, 014913 (2019)
[15] F. Becattini, G. Cao, and E. Speranza,"Polarization transfer in hyperon decays and its effect in relativistic nuclear collisions", Eur. Phys. J. C 79, **741**, p. 741 (2019)
[16] I. Karpenko, P. Huovinen, and M. Bleicher, "A 3+1 dimensional viscous hydrodynamic code for relativistic heavy ion collisions", Comput. Phys. Commun. **185**, 3016 (2014)
[17] M. Rybczynski, G. Stefanek, W. Broniowski, and P. Bozek, Comput. Phys. Commun. **185**, 1759 (2014)
[18] B. Schenke, C. Shen and P. Tribedy, "Running the gamut of high energy nuclear collisions", Phys.Rev.C **102**, 044905 (2020)
[19] Shreyasi Acharya *et al.*, "Polarization of Λ and $\bar{\Lambda}$ Hyperons along the Beam Direction in Pb-Pb Collisions at $\sqrt{s_{NN}}$ = 5.02 TeV", Phys.Rev.Lett. **128**, 172005 (2022)

Fluctuations of conserved charges in hydrodynamics and molecular dynamics

Volodymyr Vovchenko[1,2,3,*]

[1]Institute for Nuclear Theory, University of Washington, Seattle, WA 98195, USA
[2]Frankfurt Institute for Advanced Studies, D-60438 Frankfurt am Main, Germany
[3]Nuclear Science Division, Lawrence Berkeley National Laboratory, Berkeley, CA 94720, USA

Abstract. We present an overview of recent theoretical results on fluctuations of conserved charges in heavy-ion collisions obtained in relativistic hydrodynamics and molecular dynamics frameworks. In particular, we discuss the constraints on the location of the QCD critical point based on comparisons of experimental data on proton number cumulants with precision calculations of non-critical contributions. We also cover recent developments on critical fluctuations in molecular dynamics simulations.

1 Introduction

Fluctuations of conserved charges are sensitive probes of QCD matter. In thermodynamic equilibrium within the grand-canonical ensemble, the cumulants of a conserved charge distribution are linked to the susceptibilities, i.e., to the chemical potential derivatives of the partition function,

$$\kappa_n \propto \frac{\partial^n (\ln Z^{\rm gce})}{\partial \mu^n}. \qquad (1)$$

Therefore, fluctuations probe the finer details of the equation of state. In particular, the conserved baryon number plays the role of the order parameter for the hypothetical first-order QCD phase transition at finite baryon densities, and its fluctuations exhibit singular behavior near the critical point (CP) of the transition [1, 2]. Critical opalescence is a classical example of such a phenomenon when a usually transparent substance with respect to laser irradiation becomes opaque due to large density fluctuations of macroscopic scales near the CP. The location (and even existence) of the QCD CP and the associated phase structure of QCD matter are one of the most important open questions tackled by relativistic heavy-ion collisions at various energies [3].

In contrast to classical fluids, it is impossible to trap a droplet of hot and dense QCD fluid to try to observe critical opalescence. The blob of QCD matter created in heavy-ion collisions quickly expands and hadronises, producing at most a few thousands of (anti)baryons and typically even less. On the other hand, the relative "smallness" of the number of particles created in heavy-ion collisions makes it possible to track each particle in each event (modulo detector efficiency and acceptance limitations), and compute the event-by-event distribution and the associated fluctuation measures directly. In this regard, the cumulants of the proton

*e-mail: vlvovch@uw.edu

number – a proxy for the baryon number in the experiment – are sensitive probes of critical behavior [4], in particular the high-order non-Gaussian cumulants [2, 5]. The cumulants of proton number are also used to extract other information, and recent works have studied, for instance, the speed of sound [6] or freeze-out temperature [7].

Measurements of proton number fluctuations are being performed by several experiments, including ALICE [8], STAR [9–11], NA61/SHINE [12], and HADES [13]. Much attention is devoted to measuring the net-proton kurtosis $\kappa\sigma^2$ at RHIC beam energy scan, which indicated a possible non-monotonic collision energy dependence of this quantity [9]. Such a feature has earlier been predicted as a potential signal of the QCD critical point [14]. It should be noted, however, that experimental uncertainties are significant, and some of the observed features could be attributed to baryon conservation as opposed to critical point [15]. More robust conclusions will become possible once the improved data on $\kappa\sigma^2$ from BES-II becomes available. In the meantime, however, one could explore what can be learned from the more precise available measurements of the second and third-order proton cumulants. These analyses, though, require quantitative comparisons between theory and experiment, which for event-by-event fluctuations involve many caveats (see e.g. [16] for an overview). Addressing these caveats requires extensive dynamical modeling of heavy-ion collisions.

Several strategies are being considered to tackle the search for the QCD critical point in heavy-ion collisions within dynamical models:

- *Precision calculation of non-critical proton number fluctuations and its deviations from experimental data.*

  This approach, recently developed in Ref. [17], incorporates essential non-critical contributions to proton number fluctuations (conservation laws, hadronic interactions, momentum cuts) on top of the hydrodynamical background. Recent results are summarized in Sec. 2.

- *Molecular dynamics with a critical point.*

  Molecular dynamics is a microscopic approach that can address many of the caveats associated with fluctuations. Insights on critical fluctuations using classical molecular dynamics have recently been explored in Ref. [18] and covered in Sec. 3.

- *Hydrodynamics with critical fluctuations.*

  Ultimately, critical fluctuations should be incorporated into the hydrodynamic framework for heavy-ion collisions [19] as well the particlization procedure [20], allowing one to make testable predictions based on the location of the CP. Such a framework has been under development, for instance, within the BEST Topical Collaboration [21].

2 Hydrodynamics based analysis of (net-)proton fluctuations

One way to search for critical behavior in proton number cumulants is to study deviations of experimental data from baseline predictions that do not incorporate critical fluctuations. The simplest baseline corresponds to uncorrelated proton production, which would yield proton number cumulants consistent with Poisson statistics. However, additional non-critical mechanisms such as baryon number conservation and the repulsive core in baryon-baryon interaction break this assumption and make the non-critical baseline considerably more involved. Relativistic hydrodynamics is considered to be the standard model of heavy-ion collisions and provides a realistic background for calculating the aforementioned non-critical contributions to proton number cumulants. More specifically, the effects of baryon conservation and repulsion are implemented at the Cooper-Frye particlization stage, which is performed either analytically [17, 22] or through Monte Carlo sampling [23]. Baryon repulsion is modeled by utilizing the excluded volume model, with baryon excluded volume parameter $b = 1$ fm^3 as follows from fits to lattice QCD susceptibilities [24].

LHC. Net-proton number cumulants have been measured up to third order by the ALICE Collaboration at the LHC [8, 25]. Although fluctuations at the LHC are not expected to be sensitive to the possible presence of the QCD critical point in the baryon-rich regime, the available measurements of a normalized net-proton number variance $\kappa_2[p-\bar{p}]/\langle p+\bar{p}\rangle$ appear to establish the relevance of non-critical effects, in particular, that of baryon number conservation [8, 26]. Recently, the authors of Ref. [27] pointed out that precision measurements of $\kappa_2[p-\bar{p}]/\langle p+\bar{p}\rangle$ at the LHC are also sensitive both to the range of baryon conservation as well as to baryon annihilation in the hadronic phase. Namely, a surplus of annihilation over regeneration leads to an increase of $\kappa_2[p-\bar{p}]/\langle p+\bar{p}\rangle$ while reducing the range of correlations associated with baryon conservation brings this quantity down. The data can be described equally well by employing either (1) global baryon conservation and no baryon annihilation or (2) local baryon conservation across $\Delta y_{\rm cor} \sim 3$ units of rapidity and baryon annihilation without regeneration modeled by UrQMD afterburner. The two effects can be constrained experimentally by a combined measurement of $\kappa_2[p-\bar{p}]/\langle p+\bar{p}\rangle$ and $\kappa_2[p+\bar{p}]/\langle p+\bar{p}\rangle$, while additional constrains can come from fluctuation measurements involving light nuclei [28].

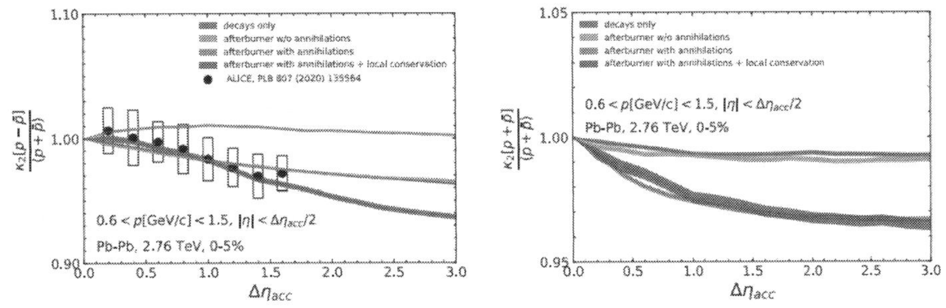

Figure 1. Dependence of proton cumulant ratios $\kappa_2[p-\bar{p}]/\langle p+\bar{p}\rangle$ (left) and $\kappa_2[p+\bar{p}]/\langle p+\bar{p}\rangle$ (right) on the pseudorapidity acceptance in 2.76 TeV Pb-Pb collisions calculated in different scenarios regarding the treatment of baryon annihilation and the range of baryon conservation. Taken from [27].

RHIC-BES. Proton number cumulants at RHIC-BES energies ($\sqrt{s_{\rm NN}} = 7.7 - 200$ GeV) have been measured by the STAR Collaboration [9, 10]. The non-critical contributions were calculated in Ref. [17] based on hydrodynamic simulations of 0-5% central Au-Au collisions within the MUSIC code [29], where both the ordinary as well as factorial cumulants [30] of (net-)proton number distribution were analyzed. It was shown that multi-proton ($n > 3$) correlations are small in the non-critical scenario. Thus, the behavior of all the cumulants is driven by two-proton correlations, which are found to be negative. This behavior contrasts critical fluctuations, which would generate sizeable multi-particle correlations [31].

The experimental data of the STAR Collaboration is quantitatively consistent with simultaneous effects of baryon conservation and repulsion at $\sqrt{s_{\rm NN}} \gtrsim 20$ GeV (Fig. 2), i.e. with non-critical physics. At lower collision energies, however, the data indicate an excess proton correlation over the non-critical baseline. This excess becomes even more prominent when new data from STAR fixed target programme at $\sqrt{s_{\rm NN}} = 3$ GeV [11] and HADES experiment at GSI at $\sqrt{s_{\rm NN}} = 2.4$ GeV [13] is considered.

Additional non-critical contributions such as volume fluctuations and electric charge conservation can influence proton number cumulants. In particular, adding volume fluctuations via an additional parameter can improve the data description at lower collision energies, but

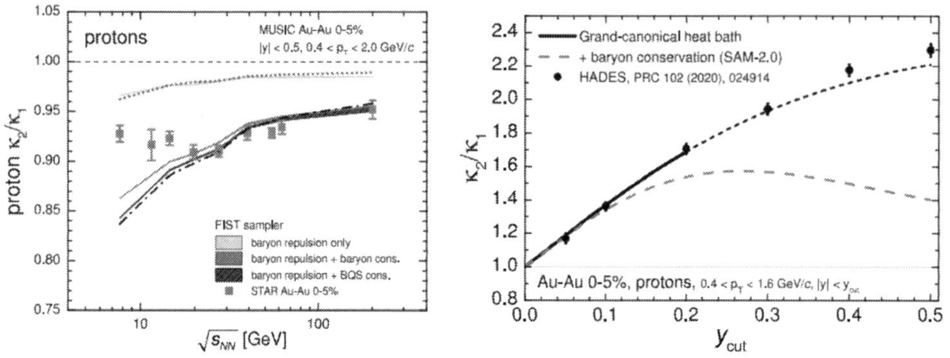

Figure 2. Scaled variance κ_2/κ_1 of the proton number distribution in central Au-Au collisions as a function of (left panel) collision energy and (right panel) acceptance in rapidity at $\sqrt{s_{\rm NN}} = 2.4$ GeV, incorporating various non-critical contributions. Adapted from [23, 32].

it will spoil the agreement at higher collision energies. The effect of exact conservation of electric charge was explored in Ref. [23] and did not yield any improvement in the description of the data at $\sqrt{s_{\rm NN}} \lesssim 20$ GeV.

Lower energies. One can conclude that it is challenging to understand the data at $\sqrt{s_{\rm NN}} \lesssim$ 20 GeV in terms of non-critical physics. In particular, the HADES data point shows a large proton scaled variance of $\kappa_2/\kappa_1 > 2$, warranting a closer look at these data. Proton number cumulants in Au-Au collisions at $\sqrt{s_{\rm NN}} = 2.4$ GeV have recently been analysed in [32] in the framework of Siemens-Rasmussen-like fireball model, with freeze-out parameters based on Refs. [33, 34]. In contrast to the non-critical baseline calculation, this analysis requires no assumptions on baryon number susceptibilities characterizing the emitting source. Instead, one extracts their values from the data by fitting the first four proton cumulants at different rapidity acceptance cuts. In order to minimize the effects of baryon conservation, the cuts were considered up to $y_{\rm cut} \leq 0.2$.

The experimental data of HADES Collaboration is described by assuming a thermal emission of nucleons from a grand-canonical heat bath (Fig. 2), provided that the corresponding baryon number susceptibilities of QCD matter are highly non-Gaussian and exhibit the following hierarchy: $\chi_4^B \gg -\chi_3^B \gg \chi_2^B \gg \chi_1^B$. This kind of hierarchy of conserved charge susceptibilities can appear in the vicinity of a critical point [2, 35]. Therefore, naively, this observation could point to a presence of the QCD critical point close to the HADES chemical freeze-out at $T \sim 70$ MeV and $\mu_B \sim 850-900$ MeV.

However, the behavior of proton cumulants at $y_{\rm cut} > 0.2$ is challenging to describe even qualitatively when one incorporates the effect of exact baryon conservation into the calculations (dashed red line). The above statement applies for κ_2 shown in Fig. 2 as well as for κ_3 and κ_4 not shown here, indicating that more theoretical and experimental effort is required to reach a firm conclusion. We emphasize that the challenge in understanding the results in the context of baryon conservation persists already at the second-order, κ_2/κ_1, and should be resolved at this level before turning to third- and fourth-order cumulants.

3 Critical point particle number fluctuations from molecular dynamics

Molecular dynamics (MD) is a microscopic approach to studying (non-)equilibrium evolution of dynamical systems and provides an alternative approach to model heavy-ion collisions.

Most MD codes such as UrQMD or SMASH are purely hadronic, given that it is challenging to implement both hadron and quark degrees of freedom. Nevertheless, such an approach is better suited to describe non-equilibrium evolution and is arguably preferable for intermediate collision energies. MD simulations already provide useful insights into various non-critical contributions to particle number fluctuations [11, 36, 37].

Critical fluctuations can also be studied in MD simulations, as long as one uses an appropriate interaction potential that leads to the presence of a first-order phase transition and a CP. One famous and well-studied example of such a potential is the Lennard-Jones (LJ) potential, which describes fluids exhibiting a liquid-gas type transition. The LJ fluid corresponds to a system of non-relativistic particles with attractive and repulsive interactions. This system is quite different from QCD matter, as QCD contains different degrees of freedom – hadron and partons – on the different sides of the QCD transition. Nevertheless, due to the universality of critical behavior, simulations of the LJ fluid can provide helpful insight into the microscopic development of critical fluctuations.

Recent work [18] used MD simulations of the LJ fluid to study the behavior of particle number fluctuations near and away from the CP through a numerical solution of Newton's equations of motion (classical N-body problem). Simulations were performed in a box with periodic boundary conditions and within the microcanonical ensemble. Since the total particle number N is conserved, fluctuations were studied in subsystems by performing cuts either in longitudinal coordinate z or longitudinal velocity v_z.

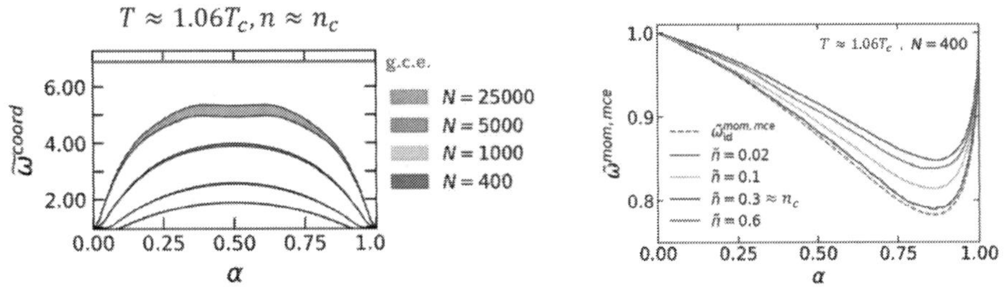

Figure 3. Dependence of the scaled variance of particle number corrected for global particle number conservation on the subvolume fraction α in coordinate (left panel) and momentum (right panel) space calculated using molecular dynamics simulations of the Lennard-Jones fluid near the critical point. Adapted from [18].

The left panel of Fig. 3 depicts the results for the scaled variance $\tilde{\omega} = \omega/(1-\alpha)$ of particle number fluctuations corrected for total particle number conservation by $1 - \alpha$ factor [38], as function of acceptance fraction α along the z-coordinate. The calculations were performed near the critical point ($T \simeq 1.06T_c$, $n \simeq n_c$), and for different values of the total particle number N. The results are consistent with an approach toward the grand-canonical scaled variance of $\omega^{\rm gce} \sim 7$ as the particle number (system size) is increased. These observations indicate that the presence of the CP manifests itself in large particle number fluctuations and finite system-size effects, as expected.

However, when one studies the fluctuations in the momentum subspace instead, via a cut $|v_z| < v_z^{\rm cut}$ in longitudinal velocity (right panel of Fig. 3), the effect of critical fluctuations is completely washed out. This observation reflects that coordinates and momenta in uniform LJ fluid are uncorrelated. Thus, the significant correlations in coordinate space due to CP do not translate into the momentum space. The situation may be different in expanding systems encountered in heavy-ion collisions, where collective velocities generate correlations between coordinates and momenta of emitted particles and may preserve CP signals even in momentum space which is accessible experimentally. Extensions of MD simulations with critical fluctuations to expanding systems will be the subject of future works.

4 Summary

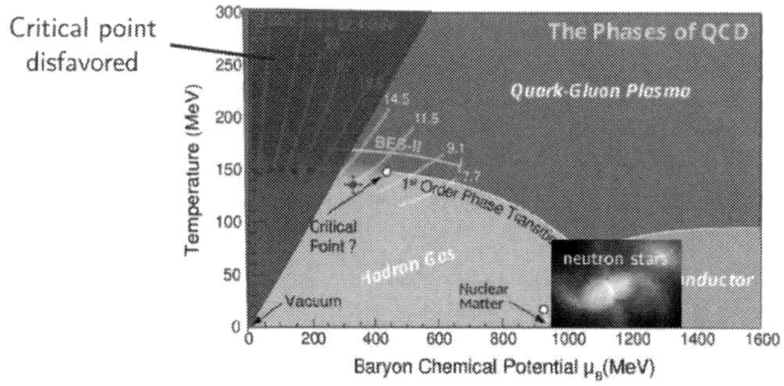

Figure 4. The figure shows the disfavored region for the QCP CP location on the QCD phase diagram based on the analysis of proton number cumulants in heavy-ion collisions. Modified from [3].

The analysis of (net-)proton number cumulants at different collision energies indicates that experimental data are consistent with non-critical physics such as baryon number conservation and short-range repulsion at $\sqrt{s_{\rm NN}} \gtrsim 20$ GeV. In contrast, the data at lower energies indicate significant excess proton correlations over various non-critical baselines, which require further analysis. These observations thus disfavor the existence of QCD CP at small baryon densities, $\mu_B/T < 2-3$, consistent with observations from lattice QCD [39–41]. They also indicate that the CP, if it exists, is located in the baryon-rich matter probed by heavy-ion collisions at intermediate energies $\sqrt{s_{\rm NN}} \lesssim 20$ GeV, see Fig. 4 for the summary of the available CP constraints. Future efforts in the search for the CP at intermediate collision energies will require improved modeling of CP effects in baryon-rich matter, and new developments in molecular dynamics simulations of critical fluctuations will play a valuable role there.

Acknowledgments

This work was supported by the U.S. Department of Energy, Office of Science, Office of Nuclear Physics, under contract number DE-FG02-00ER41132.

References

[1] M. A. Stephanov, Prog. Theor. Phys. Suppl. **153**, 139-156 (2004).
[2] M. A. Stephanov, Phys. Rev. Lett. **102,** 032301 (2009).
[3] A. Bzdak, S. Esumi, V. Koch, J. Liao, M. Stephanov, N. Xu, Phys. Rept. **853**, 1-87 (2020).
[4] Y. Hatta and M. A. Stephanov, Phys. Rev. Lett. **91**, 102003 (2003).
[5] M. A. Stephanov, Phys. Rev. Lett. **107,** 052301 (2011).
[6] A. Sorensen, D. Oliinychenko, V. Koch, L. McLerran, Phys. Rev. Lett. **127**, 042303 (2021).
[7] M. Kitazawa, S. Esumi and T. Nonaka, [arXiv:2205.10030 [hep-ph]].

[8] S. Acharya *et al.* [ALICE], Phys. Lett. B **807**, 135564 (2020).
[9] J. Adam *et al.* [STAR], Phys. Rev. Lett. **126**, 092301 (2021).
[10] M. Abdallah *et al.* [STAR], Phys. Rev. C **104**, 024902 (2021).
[11] M. S. Abdallah *et al.* [STAR], Phys. Rev. Lett. **128**, 202303 (2022).
[12] M. Gazdzicki [NA61/SHINE], PoS **CPOD2017**, 012 (2018) [arXiv:1801.00178 [nucl-ex]].
[13] J. Adamczewski-Musch *et al.* [HADES], Phys. Rev. C **102**, 024914 (2020).
[14] M. A. Stephanov, K. Rajagopal and E. V. Shuryak, Phys. Rev. D **60**, 114028 (1999).
[15] P. Braun-Munzinger, B. Friman, K. Redlich, A. Rustamov and J. Stachel, Nucl. Phys. A **1008**, 122141 (2021).
[16] V. Vovchenko, PoS **CPOD2021**, 013 (2022) [arXiv:2110.02446 [nucl-th]].
[17] V. Vovchenko, V. Koch and C. Shen, Phys. Rev. C **105**, 014904 (2022).
[18] V. A. Kuznietsov, O. Savchuk, M. I. Gorenstein, V. Koch and V. Vovchenko, Phys. Rev. C **105**, 044903 (2022).
[19] M. Stephanov and Y. Yin, Phys. Rev. D **98**, 036006 (2018).
[20] M. Pradeep, K. Rajagopal, M. Stephanov and Y. Yin, Phys. Rev. D **106**, 036017 (2022).
[21] X. An, *et al.*, Nucl. Phys. A **1017**, 122343 (2022).
[22] V. Vovchenko, Phys. Rev. C **105**, 014903 (2022).
[23] V. Vovchenko, [arXiv:2208.13693 [hep-ph]].
[24] J. M. Karthein, V. Koch, C. Ratti and V. Vovchenko, Phys. Rev. D **104**, 094009 (2021).
[25] ALICE Collaboration, [arXiv:2206.03343 [nucl-ex]].
[26] V. Vovchenko and V. Koch, Phys. Rev. C **103**, 044903 (2021).
[27] O. Savchuk, V. Vovchenko, V. Koch, J. Steinheimer and H. Stoecker, Phys. Lett. B **827**, 136983 (2022).
[28] ALICE Collaboration, [arXiv:2204.10166 [nucl-ex]].
[29] C. Shen and S. Alzhrani, Phys. Rev. C **102**, 014909 (2020).
[30] A. Bzdak, V. Koch and N. Strodthoff, Phys. Rev. C **95**, 054906 (2017).
[31] B. Ling and M. A. Stephanov, Phys. Rev. C **93**, 034915 (2016).
[32] V. Vovchenko and V. Koch, Phys. Lett. B **833**, 137368 (2022).
[33] S. Harabasz, W. Florkowski, T. Galatyuk, M. Gumberidze, R. Ryblewski, P. Salabura and J. Stroth, Phys. Rev. C **102**, 054903 (2020).
[34] A. Motornenko, J. Steinheimer, V. Vovchenko, R. Stock and H. Stoecker, Phys. Lett. B **822**, 136703 (2021).
[35] V. Vovchenko, D. V. Anchishkin, M. I. Gorenstein and R. V. Poberezhnyuk, Phys. Rev. C **92**, 054901 (2015).
[36] M. Nahrgang, T. Schuster, M. Mitrovski, R. Stock and M. Bleicher, Eur. Phys. J. C **72**, 2143 (2012).
[37] J. Hammelmann and H. Elfner, [arXiv:2202.11417 [nucl-th]].
[38] V. Vovchenko, O. Savchuk, R. V. Poberezhnyuk, M. I. Gorenstein and V. Koch, Phys. Lett. B **811**, 135868 (2020).
[39] A. Bazavov, *et al.*, Phys. Rev. D **95**, 054504 (2017).
[40] V. Vovchenko, J. Steinheimer, O. Philipsen, H. Stoecker, Phys. Rev. D **97**, 114030 (2018).
[41] S. Borsányi, Z. Fodor, J. N. Guenther, R. Kara, S. D. Katz, P. Parotto, A. Pásztor, C. Ratti and K. K. Szabó, Phys. Rev. Lett. **126**, 232001 (2021).

Insights into the initial conditions and evolution of hadronic collisions with flow observables.

Vytautas Vislavicius[1,*]

[1]Lund University, Sweden

Abstract. Studies of anisotropic flow in heavy-ion collisions are one of the critical elements in understanding the production and evolution of the Quark Gluon Plasma. These proceedings review the most recent results on anisotropic flow measurements at the LHC energies. In particular, recent measurements of p_{T}-differential $v_2\{4\}$ in Pb–Pb collisions for a plethora of light-flavour particles are discussed and compared to models with and without partonic coalescence. The measurements of $v_2\{2\}$ and $v_2\{4\}$ allow for the calculation of relative flow fluctuations, which for the first time are also reported for (multi)strange baryons. The study is further extended to higher moments of the underlying charged-particle v_2 PDF, where a transition from non-Gaussian to Gaussian fluctuations is observed. Finally, the correlation between v_2 and mean transverse momentum $[p_{\mathrm{T}}]$ is studied in terms of the Pearson Correlation Coefficient. This opens new venues to study nuclear deformation in ultra-central heavy-ion collisions and could provide a tool to identify processes in the pre-QGP phase.

1 Introduction

Studies of anisotropies in azimuthal distributions of particles produced in heavy-ion collisions have significantly contributed to the understanding of the hot and dense quark-gluon plasma (QGP). In particular, the anisotropies of energy density in the initial state give rise to large pressure gradients, which drive the expansion of the QGP. On a macroscopic level, the evolution of the QGP is described by relativistic hydrodynamics, typically studied in terms of anisotropic flow coefficients v_n, which measure the momentum anisotropy of final state particles. Experimentally the hydrodynamical prescription has been observed to be dominant at low transverse momenta ($p_{\mathrm{T}} \lesssim 3\,\mathrm{GeV}/c$).

One of the outcomes of the hydrodynamical evolution of the system is the (inverse) mass-ordering of flow coefficients v_n at low p_{T}. This has been experimentally observed as, for example, $v_2^{\pi} \gg v_2^{p}$ [1] and is due to heavier particles receiving a larger boost from the flowing medium. On the other hand, at large momenta ($p_{\mathrm{T}} \gtrsim 10\,\mathrm{GeV}/c$) where no mass-ordering has been observed [2] the dominant source for non-zero v_n is jet quenching. The transition from hydrodynamics and jet-dominated v_n occurs in the intermediate p_{T} region, $3 \lesssim p_{\mathrm{T}}/(\mathrm{GeV}/c) \lesssim 10$, where other processes such as partonic coalescence might contribute.

A common way to measure anisotropic flow coefficients in hadronic collisions is using Q-cumulants calculated from multi-particle correlations [3]. The advantage of this method is that different orders of correlations exhibit different sensitivity to the moments of underlying

*e-mail: vytautas.vislavicius@cern.ch

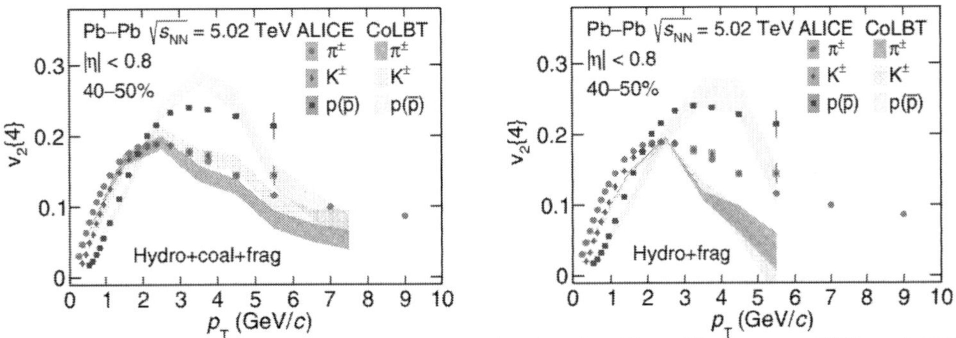

Figure 1: $v_2\{4\}$ of π, K, and p in 40-50% centrality Pb–Pb collisions at $\sqrt{s_{NN}} = 5.02$ TeV. Shaded bands show predictions from CoLBT model [8] with (left) and without (right) coalescence. Figure from [7].

v_n probability density function (PDF). As a result, from the measurement of v_n with 2- and 4-particle correlations ($v_n\{2\}$ and $v_n\{4\}$, respectively) one can calculate relative flow fluctuations as prescribed in [3]. On the other hand, measurements of particle correlations also receive contributions from non-flow effects such as jet fragmentation, jet quenching, and decays of resonance hadrons. To suppress the short-range correlations, one usually introduces a minimum separation in pseudorapidity, $|\Delta\eta|$, between the correlated particles. Alternatively, measurements of higher-order cumulants have also been shown to be less sensitive to non-flow effects [4].

2 Anisotropic flow in Pb–Pb collisions at $\sqrt{s_{NN}} = 5.02$ TeV

Recently, the ALICE Collaboration [5, 6] has reported on p_T-differential measurements of v_2 for identified particles, shown in Fig. 1 [7]. This is the first measurement of $v_2\{4\}$ that includes (multi) strange hyperons, although, in these proceedings, only π, K, and p are considered for brevity. At lower transverse momenta, $p_T \lesssim 2\,\text{GeV}/c$ $v_2\{4\}$ exhibits an inverse mass ordering, $v_2^{\pi}\{4\} > v_2^{K}\{4\} > v_2^{p}\{4\}$. The same holds for hyperons as reported in [7]. This is expected from hydrodynamics, where heavier particles receive a larger boost in p_T from the common velocity field. Around $p_T \approx 2\,\text{GeV}/c$, we see a characteristic crossing, above which $v_2\{4\}$ is grouped between baryons and mesons. The grouping is understood in terms of coalescence, where a combination of three partons results in a larger momentum of a hadron than a combination of two partons.

The measured $v_2\{4\}$ for different particle species is also compared to predictions of CoLBT model [8]. In particular, two cases are considered: one where the evolution of QGP follows hydrodynamics and fragmentation, and the other where quarks are allowed to coalesce before hadronization. At low p_T we see that hydrodynamics alone underestimates the $v_2\{4\}$ for all particle species. The description of the data is significantly better once coalescence is considered. Note that even with coalescence, the model underestimates the $v_2\{4\}$ of π; this could be due to a large number of π coming from the decays of short-lived heavy resonances, where a fraction of flow builds up before the decay.

The crossing between baryons and mesons at $p_T \approx 2\,\text{GeV}/c$ is present in both models. While it comes out naturally from coalescence, in the hydro+frag model, this crossing is facilitated by a mass-dependent p_T value where fragmentation starts to dominate over hydrodynamics. It therefore suggests that the observed baryon/meson crossing is not unique to quark coalescence. Nevertheless, at larger transverse momenta ($p_T \gtrsim 3\,\text{GeV}/c$) we find that fragmentation alone predicts a sharp decrease in $v_2\{4\}$, a trend which is not observed in the data.

On the other hand, the model including quark coalescence provides a significantly better description, although the absolute values for π and p are still underestimated. Finally, it is also worth noting that the CoLBT model does not consider jet quenching, which could further modify $v_2\{4\}$ at larger p_{T} values.

2.1 Flow fluctuations

Measurements of $v_2\{2\}$ and $v_2\{4\}$ allow for the estimation of relative flow fluctuations $F(v_2) = \sigma(v_2)/\langle v_2\rangle$. For the first time these have been reported for identified particles in [7] and are shown for several centrality classes in Fig. 2. No apparent species dependence is observed in central collisions, where $F(v_2)$ shows little evolution with p_{T}. In more peripheral collisions starting from 30-40% centrality, a species dependence emerges and, in particular, $F(v_2)$ of baryons and mesons show a non-monotonic evolution with minima at different p_{T}. The baryon/meson grouping seen in v_2 is also present in $F(v_2)$, but occurs at lower transverse momenta. The emerging species and p_{T}-dependence of $F(v_2)$ suggest that $\sigma(v_2)$ is also sensitive to final state interactions.

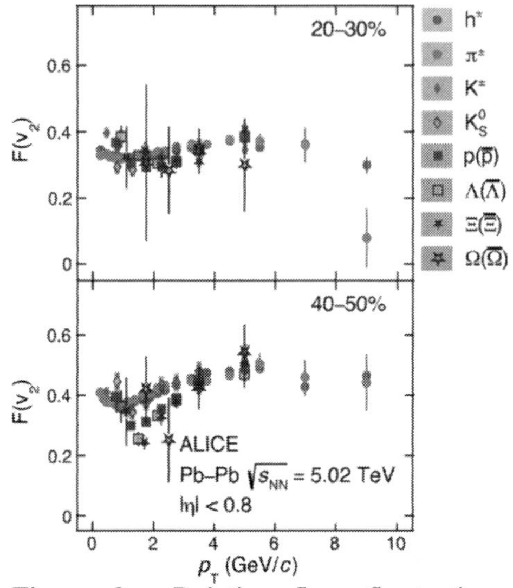

Figure 2: Relative flow fluctuations of identified hadrons in two centrality classes in Pb–Pb collisions at $\sqrt{s_{\mathrm{NN}}} = 5.02\,\mathrm{TeV}$. Figure from [7].

Measuring the underlying PDF of v_2 allows us to probe the initial geometry of the collision. The higher moments of the v_2 PDF can be calculated from the measurements of v_2 using higher orders of particle correlations as prescribed in [10]. The CMS Collaboration has reported on p_{T}-integrated skewness (γ_1) and kurtosis (γ_2) of charged hadrons as a function of centrality in Pb–Pb collisions at $\sqrt{s_{\mathrm{NN}}} = 5.02\,\mathrm{TeV}$ [9]. In addition, the contributions coming from the higher moments have been removed from the reported γ_1 and γ_2, see Fig. 3. This contribution appears to be larger for γ_1 and generally results in steeper slopes of γ_n with centrality.

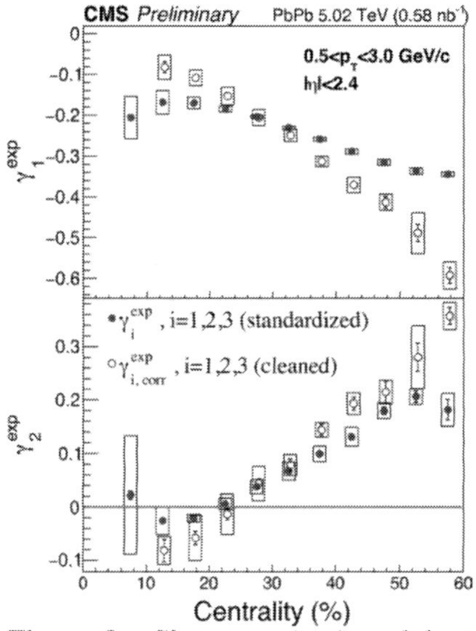

Figure 3: Skewness (top) and kurtosis (bottom) of the underlying v_2 PDF in Pb–Pb collisions at $\sqrt{s_{\mathrm{NN}}} = 5.02\,\mathrm{TeV}$. Figure from [9].

The skewness of the v_2 PDF is negative in the whole centrality range and reduces in more peripheral collisions. The negative values of γ_1 originate from a larger tail of the v_2 distribution at lower values [11], and the data suggests that this effect is enhanced with increasing centrality. The kurtosis γ_2 is small up to 20% centrality and rises as collisions get more peripheral. Overall, the non-zero γ_1 and γ_2 suggest non-Gaussian fluctuations of v_2.

To first order, v_2 is a linear response to the eccentricity ϵ_2 of the initial state. In practice, the hydrodynamical expansion of the QGP could also modify the underlying PDF. This is studied by the ALICE Collaboration, where γ_1 and γ_2 are measured as a function of p_T, see Fig. 4. Interestingly, the negative (positive) values of γ_1 (γ_2) are only seen at $p_T \lesssim$ 3 GeVc. The observed p_T-dependence suggests that the v_2 PDF is modified in a non-trivial way and thus the exact linear response $v_n = \kappa\epsilon_n$ breaks down. At higher transverse momenta, γ_1 and γ_2 are consistent with zero and thus v_2 fluctuations become Gaussian. Finally, it is worth noting that the observed p_T-dependence of skewness and kurtosis is observed in the same region where $F(v_2)$ exhibits baryon/meson grouping, and it would therefore be interesting to measure these moments for identified particles.

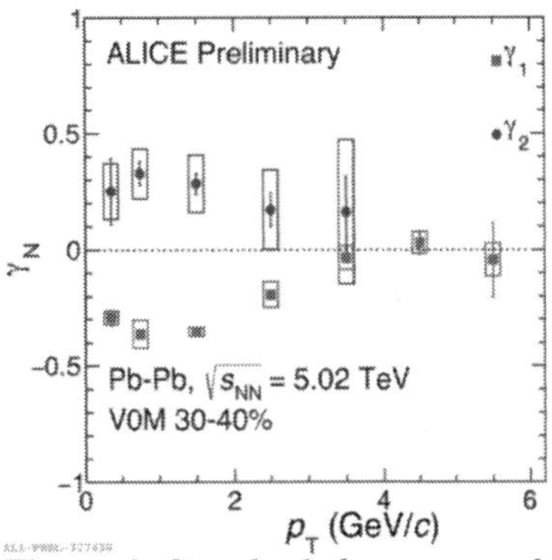

Figure 4: Standard skewness and kurtosis as a function of p_T in Pb–Pb collisions at $\sqrt{s_{NN}}$ = 5.02 TeV and 30-40% centrality.

3 Correlations between v_2 and $[p_T]$

Recently, the correlation between the mean transverse momentum $[p_T]$ and v_2 gained popularity as a tool to study the initial state of heavy-ion collisions. It is studied in terms of the Pearson Correlation Coefficient, $\rho_2 \equiv \rho\left(v_2^2, [p_T]\right)$, where $[p_T]$ is calculated on an event-by-event basis and in a limited kinematic region. The $[p_T]$ in the hydrodynamic regime is driven by the radial flow and thus is expected to increase as collisions become more central. At the same time, the elliptic component driving v_2 becomes less pronounced and in the most central collisions originates predominantly from local density fluctuations and the deformation of projectile nuclei. This can be used to study the nuclear deformation of heavy ions, as discussed in [12].

The measured ρ_2 in Pb–Pb and Xe–Xe collisions at $\sqrt{s_{NN}}$ = 5.02 and 5.44 TeV, respectively, is reported by ALICE in [13] and shown in Fig. 5 together with comparisons to different models. The sharp drop in ρ_2 in the most central collisions is smeared out by choice of a wider centrality class, 0-5%, but the decreasing trend at low centralities is seen nevertheless. In semi-central collisions (10-60%) the correlation shows little evolution with centrality and is higher in Pb–Pb collisions compared to Xe–Xe. Comparing the ρ_2 in Pb–Pb to different models, we find that in the 0-15% centrality range the data is described by all the models considered. However, Trajectum [14], Jetscape [15], and v-USPhydro [16] predict a decreasing trend beyond 15% centrality, which clearly is not in agreement with the Pb–Pb data. On the other hand, the IP-Glasma [17] initial conditions, where gluon distributions are estimated from Color Glass Condensate (CGC) calculations, coupled to hydrodynamics code MUSIC [18, 19] and UrQMD [20, 21] provides a qualitatively correct description of the measurement in Pb–Pb collisions. A similar conclusion can be drawn for measurements in Xe–Xe, but here the Xe nucleus in the IP-Glasma prescription is modelled as a spherical (β_2 = 0) or as a deformed (β_2 = 0.16) nucleus. The deformation parameter β_2 = 0.16 for the Xe nucleus has been previously predicted in [22] and tested in [23] using measurements of anisotropic flow. While β_2 does not have any significant effect on ρ_2 in semi-central collisions, in the 0-5% centrality class where deformation effects are most pronounced, the IP-Glasma model with β_2 = 0.16 is preferred.

On the other hand, ρ_2 can also be used to probe the fundamental processes leading to the formation of the QGP. From purely geometrical considerations, with decreasing multiplicity, the transverse size of the initial state reduces while the eccentricity grows. This drives ρ_2 towards lower values, ultimately changing the sign of the correlation. As proposed in [24], at sufficiently low multiplicity ρ_2 could probe initial momentum correlations. These can originate from novel processes leading to the pre-QGP phase, such as the Color Glass Condensate. In particular, for a fixed multiplicity, events with larger $[p_T]$ typically have smaller transverse sizes. In the CGC picture, this reduces the number of color domains, which in turn enhances the initial momentum anisotropy [25] and results in positive values of ρ_2. The initial momentum correlations are weak and could only dominate over geometry at very low multiplicities. This would result in ρ_2 being positive at the lowest multiplicity, decreasing steeply as collisions become more central (competition between initial momentum- and geometry-dominated ρ_2), then changing the slope from negative to positive and following geometry-only driven ρ_2.

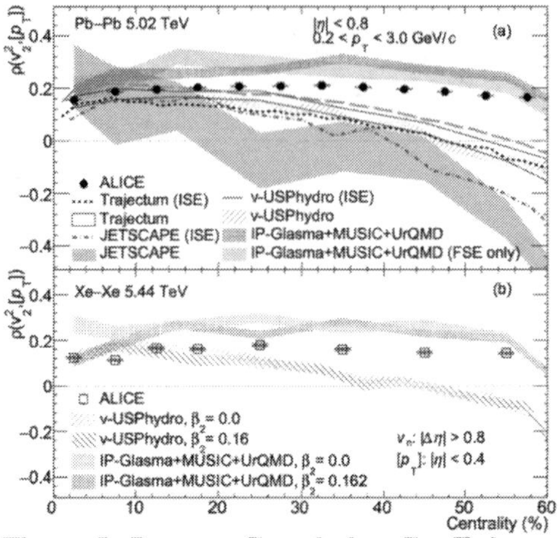

Figure 5: Pearson Correlation Coefficient of v_2 and $[p_T]$ as a function of centrality in Pb–Pb (top) and Xe–Xe collisions at $\sqrt{s_{NN}}$ = 5.02 and 5.44 TeV, respectively. Figure from [13].

By now, multiple collaborations have reported on measurements of ρ_2. The measurement by ALICE is shown in Fig. 6, where both Pb–Pb and pp collisions at $\sqrt{s_{NN}}$ = 5.02 and 13 TeV, respectively, are compared to different MC models. The measured ρ_2 is positive in the whole multiplicity region considered and is very similar in Pb–Pb and pp collisions. The change of slope in Pb–Pb collisions is observed at significantly larger multiplicity, N_{ch} ($|\eta| < 0.8$) $\approx$ 100, as compared to the CGC-inspired model where the slope changes at around N_{ch} ($|\eta| < 0.8$) $\approx$ 20. Moreover, IP-Glasma + MUSIC + UrQMD predicts a rapid change of ρ_2 at low multiplicity while the evolution seen in data is less pronounced. The similarities between pp and Pb–Pb data at comparable N_{ch} suggest that the mechanism giving rise to $[p_T]$-v_2 correlations in the two systems might be the same. It is also interesting to see that AMPT predicts the decreasing trend of ρ_2 at N_{ch} ($|\eta| < 0.8$) $\lesssim$ 130, suggesting that the change of slope is not unique to CGC initial conditions. In both data and the models considered, the rise of ρ_2 at the lowest multiplicities could result from non-flow, but this requires further studies.

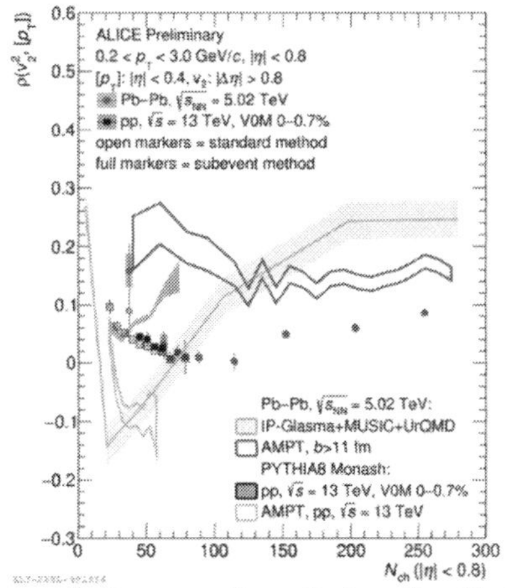

Figure 6: Pearson Correlation Coefficient of v_2 and $[p_T]$ as a function of $N_{ch}(|\eta| < 1.6)$ in Pb–Pb and high multiplicity pp collisions at $\sqrt{s_{NN}}$ = 5.02 and 13 TeV, respectively, in comparison to multiple model predictions.

Finally, the ATLAS Collaboration reported on sensitivity to the kinematic region where ρ_2 is measured[1] [26]. As shown in Fig. 7, lower p_T results in systematically smaller values of ρ_2, which is most pronounced in semi-central collisions. Reducing the η acceptance from 5 to 2 units of pseudorapidity modifies ρ_2 in a non-trivial way. Most notably, the change from

[1]Note that the ATLAS measurements are reported as a function of centrality and are performed in a different kinematic range as compared to ALICE, and so direct comparison of the results is currently not possible.

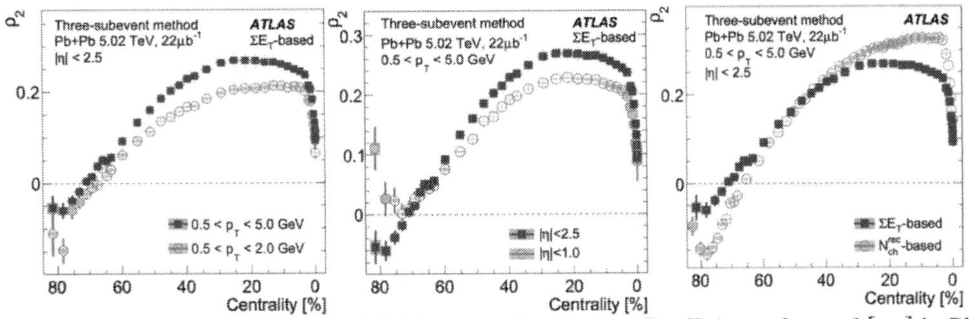

Figure 7: Centrality dependence of the Pearson Correlation Coefficient of v_2 and $[p_{\mathrm{T}}]$ in Pb–Pb collisions at $\sqrt{s_{\mathrm{NN}}} = 5.02\,\mathrm{TeV}$, measured in different p_{T} regions (left), different η regions (middle), and using different centrality estimators (right). Figure from [26].

decreasing to increasing trend is observed at lower centrality values for a smaller acceptance. This is possibly due to larger non-flow contamination due to a smaller $|\Delta\eta|$ separation. At the same time, it is not obvious whether initial momentum correlations can survive longitudinal decorrelation, and thus a large $|\Delta\eta|$ gap could also destroy them. Finally, the ρ_2 measured using two different centrality estimators by ATLAS are also compared in Fig. 7, where the ΣE_{T}-based estimator classifies events into centrality classes based on the total energy deposited into forward calorimeters and the N_{ch}-based classification depends on the total number of charged particles reconstructed at midrapidity. It is found that the N_{ch}-based estimator results in a steeper slope of ρ_2 and a more rapid change of behaviour within 75%-80% centrality. This is likely due to the finite resolution of the event activity, where midrapidity-based estimators are more sensitive to event-by-event fluctuations. Similar observations were previously reported in a model-inspired study [27], albeit at significantly lower center-of-mass energies and collision systems. These findings indicate that a more thorough model study is required to understand whether the observed signal in the data is unique to new phenomena or is an artefact of more basic kinematic considerations.

4 Summary

Measurements of the anisotropic flow at the LHC have significantly expanded our understanding of the production and evolution of the QGP. First measurements of $v_2\{4\}$ for a plethora of identified particles show a mass ordering at low p_{T} and baryon/meson grouping at higher momenta. The trends observed in the data are qualitatively described by hydrodynamic models that consider partonic coalescence. The baryon/meson grouping is also observed in the relative flow fluctuations, but it occurs at significantly lower p_{T}, suggesting it is of a different origin. The higher moments of the inclusive particle v_2 distribution indicate non-Gaussian fluctuations in the same p_{T} region where baryon/meson grouping is observed for $F(v_2)$, and they become Gaussian at $p_{\mathrm{T}} \gtrsim 3\,\mathrm{GeV}/c$. This calls for future measurements of γ_1 and γ_2 for identified particles. Finally, measurements of $v_2 - [p_{\mathrm{T}}]$ correlations open new venues to study nuclear deformation and processes leading to the formation of the QGP. Measurements of ρ_2 in central Pb–Pb and Xe–Xe collisions suggest the deformation parameter of Xe nucleus $\beta_2 = 0.16$, which different methods have previously confirmed. The evolution of ρ_2 with centrality shows trends that are qualitatively better described by models that utilize IP-Glasma in the initial state. In ultraperipheral heavy-ion collisions, indications of a ρ_2 slope change are observed. However, this effect cannot be uniquely attributed to initial momentum correlations, and a better understanding of biases originating from the kinematic selections is required.

References

[1] B.B. Abelev et al. (ALICE), JHEP **06**, 190 (2015), 1405.4632
[2] B. Abelev et al. (ALICE), Phys. Lett. B **719**, 18 (2013), 1205.5761
[3] A. Bilandzic, C.H. Christensen, K. Gulbrandsen, A. Hansen, Y. Zhou, Phys. Rev. C **89**, 064904 (2014), 1312.3572
[4] S. Acharya et al. (ALICE), JHEP **07**, 103 (2018), 1804.02944
[5] K. Aamodt et al. (ALICE), JINST **3**, S08002 (2008)
[6] B.B. Abelev et al. (ALICE), Int. J. Mod. Phys. A **29**, 1430044 (2014), 1402.4476
[7] S. Acharya et al. (ALICE) (2022), 2206.04587
[8] W. Chen, S. Cao, T. Luo, L.G. Pang, X.N. Wang, Phys. Lett. B **810**, 135783 (2020), 2005.09678
[9] Tech. rep., CERN, Geneva (2022), https://cds.cern.ch/record/2806158
[10] R.S. Bhalerao, G. Giacalone, J.Y. Ollitrault, Phys. Rev. C **99**, 014907 (2019), 1811.00837
[11] G. Giacalone, L. Yan, J. Noronha-Hostler, J.Y. Ollitrault, Phys. Rev. C **95**, 014913 (2017), 1608.01823
[12] G. Giacalone, Phys. Rev. C **102**, 024901 (2020), 2004.14463
[13] S. Acharya et al. (ALICE), Phys. Lett. B **834**, 137393 (2022), 2111.06106
[14] G. Nijs, W. van der Schee, U. Gürsoy, R. Snellings, Phys. Rev. Lett. **126**, 202301 (2021), 2010.15130
[15] D. Everett et al. (JETSCAPE), Phys. Rev. Lett. **126**, 242301 (2021), 2010.03928
[16] J. Noronha-Hostler, J. Noronha, M. Gyulassy, Phys. Rev. C **93**, 024909 (2016), 1508.02455
[17] B. Schenke, P. Tribedy, R. Venugopalan, Phys. Rev. C **86**, 034908 (2012), 1206.6805
[18] B. Schenke, S. Jeon, C. Gale, Phys. Rev. C **82**, 014903 (2010), 1004.1408
[19] J.F. Paquet, C. Shen, G.S. Denicol, M. Luzum, B. Schenke, S. Jeon, C. Gale, Phys. Rev. C **93**, 044906 (2016), 1509.06738
[20] S.A. Bass et al., Prog. Part. Nucl. Phys. **41**, 255 (1998), nucl-th/9803035
[21] M. Bleicher et al., J. Phys. G **25**, 1859 (1999), hep-ph/9909407
[22] P. Möller, A.J. Sierk, T. Ichikawa, H. Sagawa, Atom. Data Nucl. Data Tabl. **109-110**, 1 (2016), 1508.06294
[23] S. Acharya et al. (ALICE), Phys. Lett. B **784**, 82 (2018), 1805.01832
[24] G. Giacalone, B. Schenke, C. Shen, Phys. Rev. Lett. **125**, 192301 (2020), 2006.15721
[25] T. Lappi, B. Schenke, S. Schlichting, R. Venugopalan, JHEP **01**, 061 (2016), 1509.03499
[26] G. Aad et al. (ATLAS) (2022), 2205.00039
[27] J. Jia, S. Huang, C. Zhang, Phys. Rev. C **105**, 014906 (2022), 2105.05713

Global spin alignment of ϕ and K^{*0} vector mesons in Au+Au collisions from RHIC BES-II program

Gavin Wilks[1,*], for the STAR Collaboration

[1]The University of Illinois at Chicago, 1200 W Harrison St, Chicago, IL 60607

Abstract. Global spin alignment is a preferential alignment of a particle's spin along the orbital angular momentum produced in heavy-ion collisions. The global spin alignment of vector mesons ($J^P = 1^-$) ϕ and K^{*0} may be sensitive to the vorticity in the medium and hadronization mechanisms. The second phase of RHIC Beam Energy Scan (BES-II) program provides higher statistics data sets for Au+Au collisions at $\sqrt{s_{NN}}$ = 7.7-19.6 GeV. From this data, we can make high precision measurements of ϕ and K^{*0} global spin alignment, allowing for more differential studies not possible with the BES-I data. We can also compare global spin alignment between ϕ and K^{*0}, where the lifetime of ϕ is roughly ten times larger than that of K^{*0} and the latter is more sensitive to hadronic re-scattering. We report high precision measurements for the global spin alignment of ϕ at $\sqrt{s_{NN}}$ = 19.6 GeV from BES-II.

1 Introduction

In non-central heavy-ion collisions, a large orbital angular momentum (OAM) is generated. This OAM may preferentially align a particle's spin projection on the spin quantization axis through spin-orbit couplings, a phenomenon known as global spin alignment [1].

Preliminary results from the first phase of the RHIC Beam Energy Scan program (BES-I) showed a positive global spin alignment ($\rho_{00} > 1/3$) for ϕ-mesons and zero spin alignment ($\rho_{00} \sim 1/3$) for K^{*0} mesons over the $\sqrt{s_{NN}}$ of 11.5 to 200 GeV for Au+Au collisions [2]. The $\sqrt{s_{NN}}$ dependence of the ϕ-meson ρ_{00} can be accommodated by a theoretical model invoking a ϕ-meson mean field, which is supported by the existence of this field [2,3,4]. In these proceedings, we present the ϕ-meson global spin alignment from the BES-II Au+Au 19.6 GeV data set, which contains 478 million events for $\sqrt{s_{NN}}$ = 19.6 GeV compared to 19 million in BES-I.

2 Analysis Details

As discussed, the global spin alignment of vector mesons is a measurement of the 00th component of the spin density matrix called ρ_{00}. This quantity can be extracted from a fit to the vector meson yield as a function of the daughter's angular distribution,

$$\frac{dN}{d(\cos\theta^*)} \propto (1-\rho_{00}) + (3\rho_{00}-1)\cos^2\theta^*, \qquad (1)$$

*e-mail: gwilks3@uic.edu

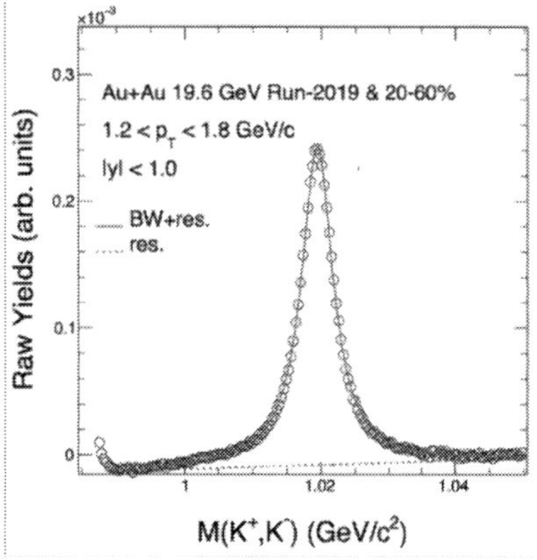

Figure 1. Invariant mass of K^+K^- pairs integrated over $\cos\theta^*$ for 20-60% central Au+Au collisions at $\sqrt{s_{\rm NN}} = 19.6$ GeV. The solid line represents the combined fit of a Breit-Wigner and first order polynomial. The dashed line is the first order polynomial, which is subtracted to extract the final yields.

where N is the number of ϕ-mesons and θ^* is the angle of the daughter K^+ with respect to the polarization axis in the parent's rest frame [5]. The polarization axis is taken to be the orbital angular momentum direction, which is estimated by the normal of the second order event plane. The determination of the second order event plane in this analysis is achieved through the use of the Time Projection Chamber at the STAR experiment. To avoid self correlation and short range non-flow contributions, we use the sub-event plane method for second order event plane reconstruction [6]. Two sub-event windows were defined η_{east} $(-1.0 < \eta < -0.05)$ and η_{west} $(0.05 < \eta < 1.0)$ with an η-gap of 0.1. The second order sub-event plane resolution achieved for mid-central collisions (20-60% centrality) was roughly 0.4. The acceptance inefficiency of the detectors in STAR leads to a non-uniform event plane distribution and we must therefore apply a recentering correction to the flow vector run-by-run for each centrality and vertex z position range [7]. After the recentering correction a shift correction is applied on an event-by-event basis to further correct for this non-uniformity [8].

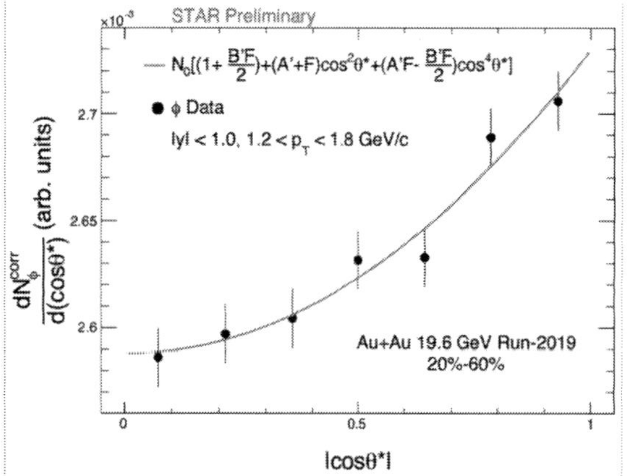

Figure 2. Efficiency corrected ϕ meson yields as a function of $\cos\theta^*$ for 20-60% central Au+Au collisions at $\sqrt{s_{\rm NN}} = 19.6$ GeV. The red line is the fit to the data used to extract ρ_{00}, given in detail by Eqs. (2-4). Only statistical uncertainties are shown.

In addition to the efficiency correction, we need to correct for the STAR detector's finite acceptance in $|\eta|$, which can artificially contribute to the ρ_{00} signal. Also, since we are unable to know the exact reaction plane, we must therefore correct for the effect of the event plane resolution. The final fit function for the $\cos\theta^*$ dependent yields is

$$\frac{dN^{corr}}{d(\cos\theta^*)} = N_0[(1 + \frac{B'F}{2}) + (A' + F)\cos^2\theta^* + (A'F - \frac{B'F}{2})\cos^4\theta^*], \quad (2)$$

where

$$A' = \frac{A(1+3R)}{4+A(1-R)}, B' = \frac{A(1-R)}{4+A(1-R)}, \quad (3)$$

and

$$A = \frac{3\rho_{00} - 1}{1 - \rho_{00}}. \quad (4)$$

In the above equations, N^{corr} is the efficiency corrected yield, N_0 is the normalization factor, R is the event plane resolution and F is the acceptance parameter found through STAR detector simulations. Figure 2 shows an example of ϕ-meson yield as a function of $\cos\theta^*$. Details of the acceptance and resolution corrections can be found in [9].

3 Results and Discussion

In Figure 3, we show the ϕ-meson ρ_{00} calculated for Au+Au collisions from BES-II at $\sqrt{s_{\mathrm{NN}}}$ = 19.6 GeV in four p_T bins at 20-60% centrality. The transverse momentum integrated global spin alignment for ϕ-meson at $\sqrt{s_{\mathrm{NN}}}$ = 19.6 GeV from BES-II mid-central collisions was measured to be ρ_{00} = 0.3622 ± 0.0026 (stat.) ± 0.0049 (sys.), which is larger than 1/3 with 5.2σ significance. Further theoretical progress is needed to understand this p_T dependence and is currently being investigated with a model invoking a ϕ-meson mean field [3,4,10].

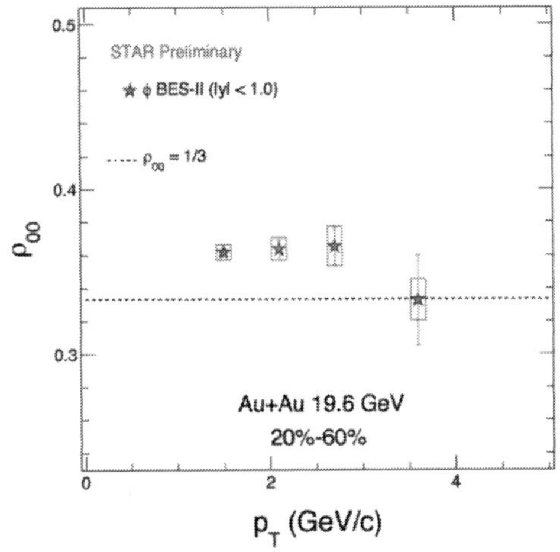

Figure 3. p_T dependent ρ_{00} for 20-60% centrality Au+Au collisions at $\sqrt{s_{\mathrm{NN}}}$ = 19.6 GeV. The p_T dependence is currently being investigated using a ϕ-meson mean field model [10]. The vertical lines are statistical uncertainties and boxes represent systematic uncertainties.

The analysis for ρ_{00} of K^{*0} is currently ongoing for the BES-II $\sqrt{s_{\mathrm{NN}}}$ = 19.6 GeV data set. In the BES-I results, K^{*0} was found to have no global spin alignment. This could be due to K^{*0} having a roughly ten times smaller lifetime than that of ϕ, resulting in different in medium interactions and a different late stage hadronic reaction [11]. Additionally, the fluctuating vector meson fields for the d and $\bar{s}$ quarks are expected to be weaker than s and $\bar{s}$, leading to a negligible contribution to ρ_{00} for K^{*0} [4].

4 Conclusion

We presented results for the ϕ-meson ρ_{00} from the BES-II 19.6 GeV Au+Au data set. The measured global spin alignment is significantly positive ($\rho_{00} > 1/3$). Our precision is greatly improved by the high-statistics BES-II data sets. This will allow us to perform differential studies at these energies that were previously unachievable with the BES-I data sets, such as centrality and rapidity dependent studies. These precise studies of vector meson ρ_{00} aim to further theoretical progress towards understanding of this quantity and how it can be used to probe other physical phenomena, such as spin polarization by a possible vector meson strong force field.

Acknowledgements

The author is supported in part by The United States Department of Energy.

References

[1] Z. T. Liang and X. N. Wang, Phys. Lett. B **629**, 20 (2005).
[2] M. S. Abdallah et al. (STAR Collaboration), arXiv:2204.02302 (2022).
[3] X. L. Sheng et al., Phys. Rev. D **101**, 096005 (2020).
[4] X. L. Sheng et al., Phys. Rev. D **102**, 056013 (2020).
[5] K. Schilling et al., Nucl. Phys. B **18**, **332** (1970).
[6] A. M. Poskanzer and S. A. Voloshin, Phys. Rev. C **58**, 1671 (1998).
[7] P. Danielewicz et al., Phys. Rev. C **38**, 120 (1988).
[8] B. I. Abelev et al. (STAR Collaboration), Phys. Rev. C **81**, 044902 (2010).
[9] A. Tang et al., Phys. Rev. C **98**, 044907 (2018).
[10] X. L. Sheng et al., arXiv:2205.15689 (2022).
[11] I. Karpenko and F. Becattini, Eur. Phys. J. C **77**, 213 (2017).

Charmonium, B_c and X(3872) Transport at the LHC

Biaogang Wu[1,*], *Zhanduo* Tang[1,**], *Min* He[2,***], and *Ralf* Rapp[1,****]

[1]Cyclotron Institute and Department of Physics and Astronomy, Texas A&M University, College Station, TX 77843-3366, USA

[2]Department of Applied Physics, Nanjing University of Science and Technology, Nanjing 210094, China

Abstract. We deploy a kinetic-rate equation to evaluate the transport of J/ψ, $\psi(2S)$, B_c and $X(3872)$ in ultrarelativistic heavy-ion collisions and compare their production yields to experimental data from the Large Hadron Collider. The rate equation has two main transport parameters: the equilibrium limit and reaction rate for each state. The temperature-dependent equilibrium limits include charm- and bottom-quark fugacities based on their initial production. The reaction rates for charmonia, bottomonia and B_c rely on charm- and bottom-quark masses and binding energies from a thermodynamic T-matrix approach. For the $X(3872)$ particle, internal structure information is encoded in reaction rates and initial conditions in the hadronic phase via two different scenarios: a loosely bound hadronic molecule vs. a compact tetraquark.

1 Introduction

The production of charm and bottom hadrons in ultrarelativistic heavy-ion collisions (URHICs) has been intensely studied for several decades to help unravel the structure of the strongly-interacting matter at high temperature [1]. Recent data on heavy quarkonia, *i.e.*, charmonia and bottomonia, as well as B_c and exotic particles, have further fueled these efforts [2–4]. In addition, the internal structure of the $X(3872)$ particle – a compact tetraquark with cq anti/-diquark components vs. a loosely bound hadronic molecule of D and D^* mesons – has been conjectured to affect its yields in URHICs [5]. Here, we report comprehensive results for the nuclear modification factor for various quarkonia using our rate-equation approach [6–8], as a function of both collision centrality and transverse momentum (p_T), and compare to pertinent data from the Large Hadron Collider (LHC).

2 Transport Approach in Thermal Medium

We focus on Pb-Pb collisions at 5 TeV, simulating the medium evolution via a cylindrically expanding fireball with a transverse flow of blastwave type [6–8]. In this background, we solve rate equations including dissociation and regeneration for each quarkonium state, Q,

$$\frac{dN_Q(\tau)}{d\tau} = -\Gamma(T(\tau))\left[N_Q(\tau) - N_Q^{\rm eq}(T(\tau))\right] . \quad (1)$$

*e-mail: bgwu@tamu.edu
**e-mail: zhanduotang@tamu.edu
***e-mail: minhephys@gmail.com
****e-mail: rapp@comp.tamu.edu

This equation is governed by two transport parameters. The first one is the equilibrium limit,

$$N_Q^{\rm eq}(T(\tau)) = V_{\rm FB} d_Q \gamma_{Q_1} \gamma_{Q_2} \int \frac{d^3k}{(2\pi)^3} \exp(-\sqrt{k^2 + m_Q^2}/T)\,, \quad (2)$$

with the fireball volume, $V_{\rm FB}$, and the quarkonium degeneracy, d_Q. The heavy-quark (HQ) fugacities, γ_{Q_i} (Q_i=b, c), are calculated assuming heavy-flavor (HF) conservation within the thermal statistical model,

$$N_{Q\bar{Q}} = \frac{1}{2}\gamma_Q n_{\rm op} V_{\rm FB} \frac{I_1(\gamma_Q n_{\rm op} V_{\rm FB})}{I_0(\gamma_Q n_{\rm op} V_{\rm FB})} + \gamma_Q^2 n_{\rm hid} V_{\rm FB}\,, \quad (3)$$

where $N_{Q\bar{Q}}$ is the total number of charm-anticharm or bottom-antibottom pairs in the fireball, as determined by hard production in primordial NN collisions at given collision centrality; $I_{0,1}$ are modified Bessel functions of zeroth and first order, and $n_{\rm op}$ and $n_{\rm hid}$ are the total densities of open and hidden HF particles, respectively (quarks in the QGP or HF hadrons in hadronic matter).

The inelastic reaction rates, Γ_Q, for charmonia, bottomonia and B_c are computed in quasi-free approximation [9], with HQ masses, m_Q, and binding energies taken from in-medium T-matrix calculations [10]. For the $X(3872)$, we employ a schematic parameterization [11],

$$\Gamma(T) = \Gamma_0 \left(\frac{T}{T_0}\right)^n\,, \quad (4)$$

akin to hadronic reaction rates for charmonia [8], focusing on n=3. Taking guidance from existing literature [12–14], Γ_0 at an initial temperature of T_0=180 MeV is set to $\Gamma_0^{\rm mol}$≃300-500 MeV and $\Gamma_0^{\rm tet}$≃50-80 MeV for the molecule and tetraquark, respectively, representing a hierarchy where the latter can survive in the QGP phase while the former cannot. This leads to a natural assumption for the initial conditions where the $X(3872)$ abundance is at its chemical-equilibrium value for the tetraquark and zero for the molecule scenario [11].

3 Time Evolution and Observables

Let us start with our calculations for charmonia using the most recent input charm/onium production cross sections from pp collisions and shadowing estimates, cf. Fig. 1. The p_T-dependent $R_{\rm AA}$ for both charmonia show the characteristic signature of a maximum at low

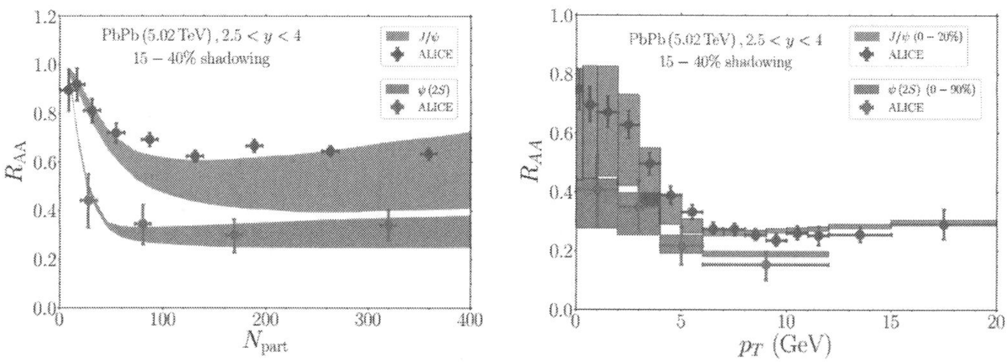

Figure 1. Centrality (left) and p_T (right) dependence for J/ψ (blue bands) and $\psi(2S)$ (red bands) production in forward-rapidity 5.02 TeV Pb-Pb collisions, compared to ALICE data [2, 15]. The bands indicate uncertainties due to the input $c\bar{c}$ cross section, $d\sigma_{c\bar{c}}^{pp}/dy$=0.72±0.7 mb, and 15-40% shadowing.

https://doi.org/10.1051/epjconf/202327602001

p_T. The centrality and the p_T dependence show fair agreement with the most recent ALICE data [2, 15, 16], in particular given that the $\psi(2S)$ results were theoretical predictions based on Ref. [8] (with only the input charm/onium cross sections from pp collisions updated).

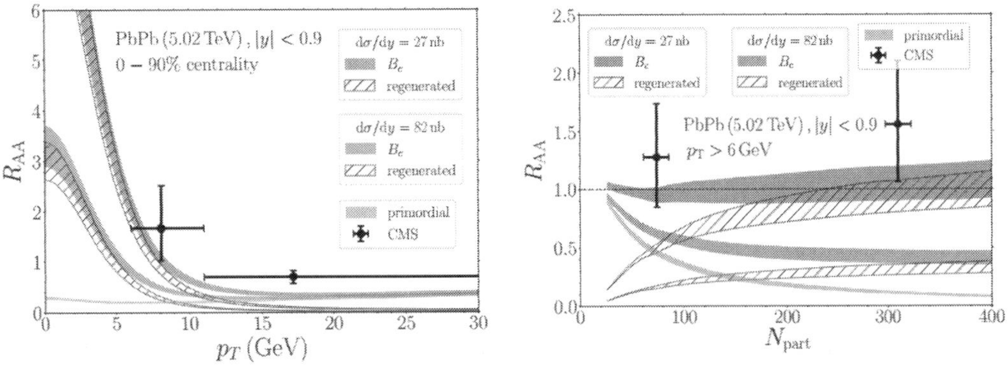

Figure 2. Transverse-momentum spectra (left) and centrality dependence (right, with p_T>6 GeV) of B_c production in Pb-Pb (5.02 TeV) collisions, compared to CMS data [4]. The red (blue) curves correspond to a pp cross section of dσ_{B_c}/dy=27 (82) nb; the primordial B_c component (orange curves) is independent of the cross section in pp collisions. The band widths reflect 10-30% nuclear shadowing.

Next, we turn to our calculations for B_c production. The input pp cross section (figuring in the denominator of the R_{AA}) is currently not well known; we have estimated it following Ref. [17] at dσ^{pp}/dy=27-82 nb, with p_T spectra extrapolated from 8 TeV forward-rapidity data [18]. We first compute the centrality dependence of the inclusive yields and then obtain the p_T spectra for the regeneration contribution from recombining b- and c-quark spectra from Langevin transport calculations [19], at average formation temperatures for $B_c(1S)$ and $B_c(1P)$ of T=220 MeV and T_c, respectively (the suppressed primordial spectra are obtained from Boltzmann simulations). The regenerated B_c dominate the spectra up to $p_T \simeq 2m_{B_c}$, cf. Fig. 2 left. The comparison to CMS data [4], especially for the centrality dependence with a $p_T > 6$ GeV cut (Fig. 2 right), shows better agreement for smaller pp cross sections.

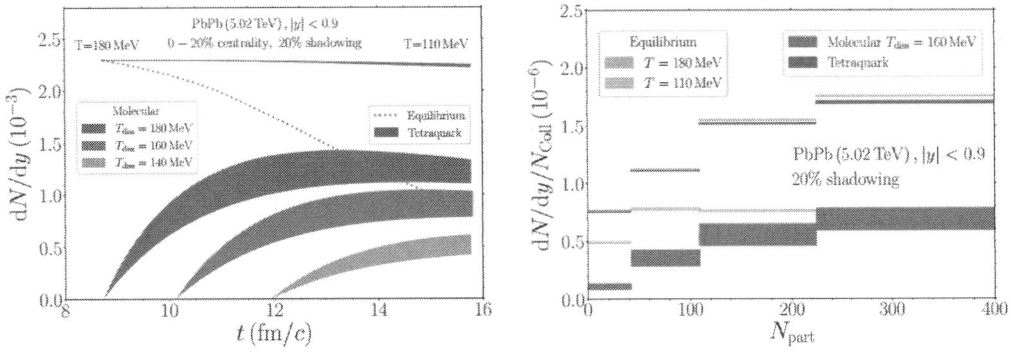

Figure 3. $X(3872)$ production in 0-20% Pb-Pb (5.02 TeV) collisions. Left: time evolution in hadronic matter for the molecule (lower 3 bands), tetraquark (red) and the equilibrium limit (green line). Right: centrality dependence for molecule (purple), tetraquark (red) and equilibrium limits at chemical (blue) and thermal freezeout (green); band widths reflect rates of $\Gamma_0^{\rm mol}$=300-500 MeV and $\Gamma_0^{\rm tet}$=50-80 MeV.

Finally we study $X(3872)$ production; the time evolution of its yield in the hadronic phase (cf. left panel of Fig. 3) shows little variation for the tetraquark scenario, remaining near its level at chemical freezeout. On the other hand, for the loosely bound hadronic molecule, assumed to come into existence only in the hadronic phase, appreciable regeneration occurs, quantitatively depending on its dissociation temperature. In any case, its final yield is significantly lower than for the tetraquark scenario. The calculated centrality dependencies (normalized to the number of binary nucleon-nucleon collisions), displayed in right panel of Fig. 3, show a rising trend for both scenarios – a clear-cut signature of regeneration, which, however, is a factor of 2-5 larger for the (early produced) tetraquark compared to the (later produced) molecule.

4 Conclusions

We have conducted transport calculations of J/ψ, $\psi(2S)$, B_c and $X(3872)$ production in URHICs within a well-tested kinetic rate-equation approach. Our predictions for the $\psi(2S)$ turn out to be in good agreement with recent ALICE data. For B_c production we find regeneration to dominate while pertinent $R_{\rm AA}$'s show a strong sensitivity to the (currently uncertain) input cross section from pp collisions. For the $X(3872)$, we encoded different structure scenarios in a scale hierarchy of the reaction rates, $\Gamma^{\rm mol} \gtrsim T_c \gtrsim \Gamma^{\rm tet}$. The resulting yields for the tetraquark are 2-5 times higher than for the molecule, which differs from most coalescence model predictions.

Acknowledgments. This work is supported by the U.S. NSF under grant nos. PHY-1913286 and PHY-2209335.

References

[1] M. He, H. van Hees, R. Rapp (2022), 2204.09299
[2] S. Acharya et al. (ALICE), JHEP **02**, 041 (2020)
[3] A.M. Sirunyan et al. (CMS), Phys. Rev. Lett. **128**, 032001 (2022)
[4] A. Tumasyan et al. (CMS), Phys. Rev. Lett. **128**, 252301 (2022)
[5] A. Esposito, A. Pilloni, A. Polosa, Phys. Rept. **668**, 1 (2017)
[6] L. Grandchamp, R. Rapp, G.E. Brown, Phys. Rev. Lett. **92**, 212301 (2004)
[7] X. Zhao, R. Rapp, Nucl. Phys. A **859**, 114 (2011)
[8] X. Du, R. Rapp, Nucl. Phys. A **943**, 147 (2015)
[9] L. Grandchamp, R. Rapp, Phys. Lett. B **523**, 60 (2001)
[10] S.Y.F. Liu, R. Rapp, Phys. Rev. C **97**, 034918 (2018)
[11] B. Wu, X. Du, M. Sibila, R. Rapp, Eur. Phys. J. A **57**, 122 (2021), [Erratum: Eur.Phys.J.A 57, 314 (2021)]
[12] M. Cleven, V.K. Magas, A. Ramos, Phys. Lett. B **799**, 135050 (2019)
[13] F. Brazzi, B. Grinstein, F. Piccinini, A.D. Polosa, C. Sabelli, Phys. Rev. D **84**, 014003 (2011)
[14] E.G. Ferreiro, J.P. Lansberg, JHEP **10**, 094 (2018), [Erratum: JHEP 03, 063 (2019)]
[15] S. Acharya et al. (ALICE) (2022), 2210.08893
[16] H. Hushnud, *Overview of quarkonium production with ALICE at the LHC* (2022), 2208.14757
[17] R. Aaij et al. (LHCb), Phys. Rev. D **100**, 112006 (2019)
[18] R. Aaij et al. (LHCb), Phys. Rev. Lett. **114**, 132001 (2015)
[19] M. He, R.J. Fries, R. Rapp, Nucl. Phys. A **910-911**, 409 (2013)

ψ(2S) production and nuclear modification factor in nucleus–nucleus collisions with ALICE

H. Hushnud[1,*] on behalf of the ALICE Collaboration

[1]Department of Physics, Aligarh Muslim University, Aligarh, India

Abstract. Charmonium production is sensitive to deconfinement in nucleus-nucleus collisions. The production via regeneration within the QGP or at the phase boundary has been identified as an important ingredient for the description of the centrality and transverse momentum dependence (p_{T}) of the J/ψ nuclear modification factor (R_{AA}) at the LHC. ψ(2S) production relative to J/ψ is one possible discriminator between the two different regeneration scenarios. At the LHC, there is so far no significant observation of the ψ(2S) in central nucleus-nucleus collisions at low-p_{T}, where regeneration is expected to play an important role. The combined Run 2 data set of ALICE allows one to extract a significant ψ(2S) signal in this kinematic region at forward rapidity, in the dimuon decay channel. In this contribution, we present for the first time results on the ψ(2S)-to-J/ψ single ratio and double ratio (to pp collisions) as well as the ψ(2S) nuclear modification factor, in Pb–Pb collisions at $\sqrt{s_{\mathrm{NN}}}$ = 5.02 TeV. The ψ(2S)-to-J/ψ double ratio and ψ(2S) R_{AA} are calculated using a new proton-proton reference with improved precision. Results are compared with model calculations.

1 Introduction

Charmonia, a bound state of $c\bar{c}$ pairs, are considered unique probes of the deconfined hot and dense medium made of free quarks and gluons, known as quark-gluon plasma (QGP), created in ultra-relativistic heavy-ion collisions [1]. In such medium, charmonium production yield is expected to be significantly suppressed with respect to the yield measured in proton-proton (pp) collisions at the same centre-of-mass energy, scaled by the number of binary nucleon-nucleon collisions, due to color screening of the $q\bar{q}$ potential [1] or dissociation [2]. The temperature required for dissociating a specific charmonium state depends on its binding energy, or equivalently on its radius. Hence, the strongly bound charmonium states, such as J/ψ, should melt at higher temperatures compared to more loosely bound states, namely ψ(2S) and χ_c. This is known as sequential dissociation. As a consequence, the in-medium dissociation probability of such states should provide an estimate of the medium temperature [3], assuming that the charmonium dissociation is the main mechanism at play. At the LHC energies, a large number of $c\bar{c}$ pairs is expected to be produced in central Pb–Pb collisions, leading to the possibility to form charmonia via recombination of c and $\bar{c}$ quarks, either in medium [4] or at the phase boundary [5, 6]. This new additional source of charmonium production is couterbalancing the suppression mechanism. The regeneration

*e-mail: hushnud.hushnud@cern.ch

mechanism has been identified as an important ingredient for the description of the observed centrality, rapidity (y) and p_{T} dependence of the J/ψ production in Pb–Pb collisions at the LHC [7, 8]. The measurement of the single (double) ratio between the ψ(2S) and J/ψ cross sections in Pb–Pb collisions (with respect to pp collisions), is predicted to be very sensitive to the details of the recombination mechanism. Experimentally, the single ratio is interesting as most of the systematic uncertainties cancel, with the remaining systematic uncertainties being only due to the signal extraction and some uncorrelated components related to the acceptance times efficency evaluation. On the theory side, this ratio is also weakly dependent on the total charm production cross section employed as inputs to the models.

2 Experimental set-up and data analysis

The ALICE collaboration has studied the ψ(2S) production in Pb–Pb collisions down to zero transverse momentum through its dimuon decay channel. The details of the ALICE detector are described in Ref. [9]. Muons coming from quarkonium decays are recontructed in the muon spectrometer, covering a pseudorapidity range -4 < η < -2.5. The two innermost layers of the inner tracking system (ITS), which consist of silicon pixel detectors, provide primary vertex reconstruction. The VZERO detectors, two scintillator arrays covering the pseudo-rapidity intervals 2.8 ≤ η ≤ 5.1 and -3.7 ≤ η ≤ -1.7, provide the minimum-bias trigger, the determination of the collision centrality and help to remove the beam-induced background. Two sets of zero degree calorimeters (ZDC) are used to suppress the background from electromagnetic processes in Pb–Pb collisions.

3 Results

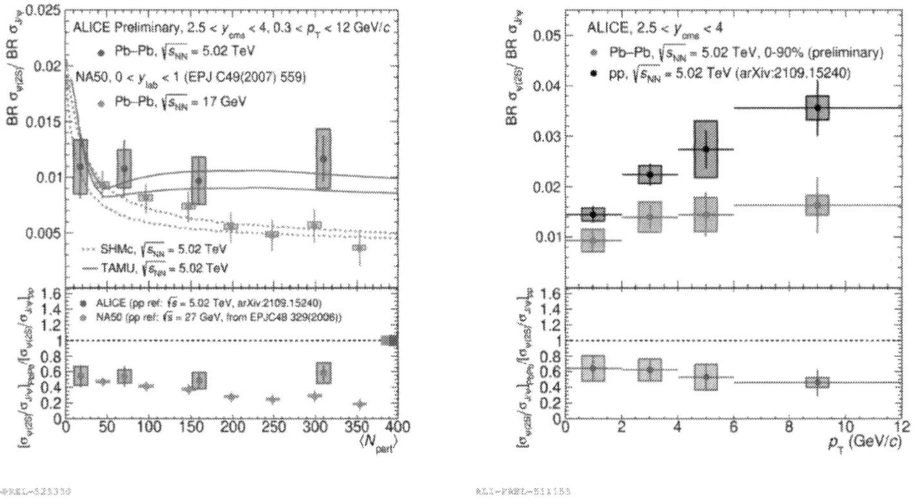

Figure 1. ψ(2S)-to-J/ψ cross section ratio measured by the ALICE collaboration in Pb–Pb collisions at $\sqrt{s_{\mathrm{NN}}}$ = 5.02 TeV as a function of the average number of participant nucleons ($\langle N_{\mathrm{part}}\rangle$) and p_{T}, in the left and right panel, respectively. In the left panel, NA50 measurements at SPS carried out at $\sqrt{s_{\mathrm{NN}}}$ = 17 GeV are also shown. The results, are compared with theoretical predictions from TAMU [10] and SHMc [11, 12]. Bottom panels show the ψ(2S)-to-J/ψ ratio normalized to the corresponding pp value (double ratio).

The left panel of Fig. 1 shows the ψ(2S)-to-J/ψ cross section ratio measured by the ALICE collaboration in Pb–Pb collisions at $\sqrt{s_{\rm NN}}$ = 5.02 TeV at forward rapidity as function of centrality (expressed in terms of average number of participant nucleons $\langle N_{\rm part}\rangle$). The bottom panel of Fig. 1 left shows the values of the ψ(2S)-to-J/ψ double ratio, indicating a suppression effect by 40% in Pb–Pb with respect to pp collisions. No significant centrality dependence is observed within uncertainties. The ALICE results in the left panel are also compared with NA50 ones in Pb–Pb collisions at $\sqrt{s_{\rm NN}}$ = 17 GeV in $0 < y_{Lab} < 1$ [13]. Both the ψ(2S)-to-J/ψ single and double cross section ratios measured by NA50 exhibit a stronger centrality dependence, reaching smaller values in central collisions. The ψ(2S)-to-J/ψ single ratio measured by ALICE is compared with theoretical calculations based on a transport approach (TAMU) [10] and with the Statistical Hadronization Model (SHMc) [11, 12]. The TAMU [10] model well reproduces the ψ(2S)-to-J/ψ cross section ratio as a function of centrality, while SHMc [11, 12] tends to underestimate the data in central Pb–Pb collisions.

In the right panel of Fig. 1, the ψ(2S)-to-J/ψ ratio as a function of $p_{\rm T}$ in Pb–Pb collisions is compared with the corresponding ratio in pp collisions. The ψ(2S)-to-J/ψ ratio in Pb–Pb collisions is systematically larger compared to the one measured in pp. The corresponding double ratio shown in the bottom panel, indicates a significant relative suppression in Pb–Pb with respect to pp, with no strong $p_{\rm T}$ dependence and reaching a value of ~ 0.5 at high $p_{\rm T}$.

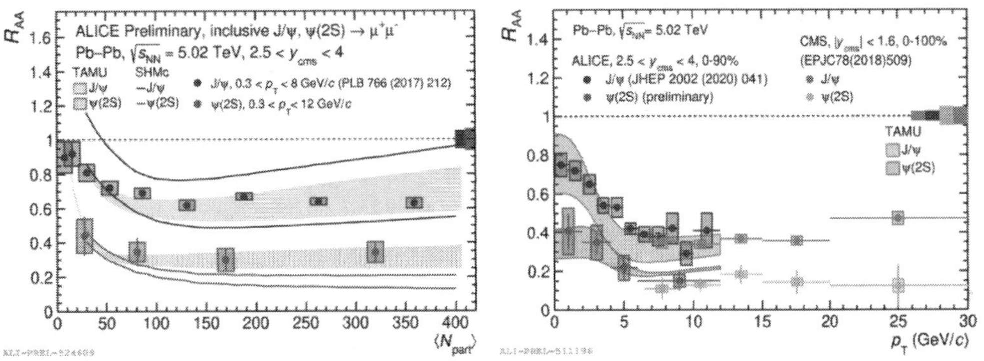

Figure 2. The $R_{\rm AA}$ of ψ(2S) and J/ψ as a function of the average number of participant nucleons ($\langle N_{\rm part}\rangle$) and $p_{\rm T}$, in the left and right panel, respectively. In the right panel, the ALICE data are compared with CMS results [14] for $|y| < 1.6$, $6.5 < p_{\rm T} < 30$ GeV/c and centrality 0 - 100 %. The results are also compared with theoretical predictions from TAMU [10] (left and right plots) and SHMc [11, 12] (left plot).

Figure 2 shows the nuclear modification factor $R_{\rm AA}$ of J/ψ and ψ(2S) measured by the ALICE collaboration as a function of $\langle N_{\rm part}\rangle$ (left panel) and $p_{\rm T}$ (right panel). The ψ(2S) $R_{\rm AA}$ shows significant suppression and no strong centrality dependence (assuming an almost constant value of about 0.4). It is significantly smaller compared to the $R_{\rm AA}$ of the J/ψ, both as a function of $p_{\rm T}$ and centrality. It also hints at less suppression at low-$p_{\rm T}$ with respect to higher $p_{\rm T}$, as also observed with more significance for the J/ψ. This could be a first indication for ψ(2S) production via recombination of $c\bar{c}$ pairs. The (TAMU) model calculation [10] reproduces both the centrality and $p_{\rm T}$ dependence of the $R_{\rm AA}$ for both charmonium states. On the other hand, the SHMc model [11, 12] reproduces the centrality dependence of the J/ψ $R_{\rm AA}$, while it overestimates the ψ(2S) suppression in central events.

The charmonium $R_{\rm AA}$ as a function of $p_{\rm T}$ is compared with CMS measurements [14] carried out for $|y| < 1.6$, $6.5 < p_{\rm T} < 30$ GeV/c and centrality 0 - 100 %. A strong suppression of the ψ(2S) persists up to 30 GeV/c, as shown by the CMS data which agree very well with those from ALICE in the common p_{T} range, in spite of the different rapidity coverages.

4 Summary

The first accurate measurement of the ψ(2S) production in central Pb–Pb collisions at $\sqrt{s_{\mathrm{NN}}}$ = 5.02 TeV and low-p_{T} has been reported by ALICE at forward rapidity. The ψ(2S)-to-J/ψ double ratio shows a relative suppression of ~40 %. No significant p_{T} or centrality dependence is observed within the uncertainties. The double ratio measurements from NA50 show a more pronounced centrality dependence compared to ALICE. The ψ(2S) R_{AA} hints at a decrease as a function of p_{T} similar to the J/ψ one and connected with charm quarks recombination processes. As a function of centrality, the value of the ψ(2S) R_{AA} is almost constant and reaches ~0.4. The ψ(2S) shows more suppression with respect to the J/ψ, both as a function of centrality and p_{T}. The transport model, which includes recombination of charm quarks through the QGP medium and is already able to describe the J/ψ data, shows a fair agreement with ψ(2S) data, especially for central events.

References

[1] T. Matsui *et al.*, Phys. Lett. **B178**, 416 (1986).
[2] A. Rothkopf *et al.*, Phys. Rev. Lett. **108**, 162001 (2012).
[3] S. Digal *et al.*, Phys. Rev. **D64**, 094015 (2001).
[4] R. L. Thews *et al.*, Phys. Rev. **C63**, 054905 (2001).
[5] P. Braun-Munzinger *et al.*, Phys. Lett. **B490**, 196 (2000).
[6] A. Andronic *et al.*, Jour. of Phys. **G38**, 124081 (2011).
[7] J. Adam *et al.* [ALICE Collaboration], Phys. Lett. **B 766**, 212 (2017).
[8] S. Acharya *et al.* [ALICE Collaboration], Phys. Lett. **B 805**, 135434 (2020).
[9] K. Aamodt *et al.* [ALICE Collaboration], JINST **3**, S08002 (2008).
[10] X. Du and R. Rapp, Nucl. Phys. **A943**, 147 (2015).
[11] A. Andronic *et al.*, Phys. Lett. **B797**, 134836 (2019).
[12] A. Andronic *et al.*, Nature **561** no. 7723, 321 (2018).
[13] B. Alessandro *et al.* [NA50 Collaboration], Eur. Phys. J.**C 49** 559 (2007).
[14] A. M. Sirunyan *et al.* [CMS Collaboration], Eur. Phys. J. **C78** no. 6, 509 (2018).

Measurement of quarkonium elliptic flow in pPb collisions at 8.16 TeV with the CMS detector

Kisoo Lee on behalf of the CMS Collaboration[1,*]

[1]Korea University

Abstract. The second-order Fourier coefficients (v_2) of Υ(1S) and J/ψ mesons in high-multiplicity pPb collisions are studied using data collected by the CMS experiment at a nucleon-nucleon center-of-mass energy of 8.16 TeV. The dimuons used to reconstruct the quarkonium states are correlated with charged hadrons using the long-range two-particle correlation technique. The measurement of the Υ(1S) v_2 is reported for the first time in small collision systems. The results are discussed in terms of collectivity and modification of heavy quarks.

1 Introduction

Strong azimuthal correlations in nucleus-nucleus collisions are observed at the BNL RHIC [1, 2] and the CERN LHC facilities [3, 4]. These correlations are understood to arise from the creation of a strongly interacting medium, called quark-gluon plasma (QGP), that exhibits nearly ideal hydrodynamic behavior [5]. Similarly, long-range correlations have been also observed in high particle multiplicity events in smaller collision systems, such as proton-lead (pPb) collisions [6, 7].

Heavy quarks (charm and bottom) are particularly useful to probe such QGP-like signatures as they are produced in the early stages of heavy ion collisions and experience the entire medium evolution [8]. In lead-lead (PbPb) collisions, significant elliptic flow (v_2) signals have been measured for J/ψ mesons [9] while the v_2 values are found to be consistent with zero for Υ(1S) and Υ(2S) mesons [10], indicating different collective behavior for charmonia and bottomonia.

This contribution reports the measurements of the J/ψ mesons v_2 based on long-range, two-particle correlations in high-multiplicity pPb collisions at a center-of-mass energy per nucleon pair of $\sqrt{s_{NN}}$ = 8.16 TeV. At the same $\sqrt{s_{NN}}$, the Υ(1S) meson v_2 is measured for the first time in small collision systems. The results provide unique information to the collective dynamics of heavy quarks in small collision systems.

2 Event Selection

This analysis uses pPb collision data collected with the CMS detector in 2016, with an integrated luminosity of 186 nb^{-1} [11]. The detailed description of the CMS detector can be found in Ref. [12]

*e-mail: kilee@cern.ch

The results are reported for high-multiplicity events $185 \leq N_{\mathrm{trk}}^{\mathrm{offline}} < 250$ ($70 \leq N_{\mathrm{trk}}^{\mathrm{offline}} < 300$) for J/ψ (Υ(1S)) mesons, where $N_{\mathrm{trk}}^{\mathrm{offline}}$ is the number of charged particle tracks with $|\eta| < 2.4$ and $\mathrm{p_T} > 0.4$ GeV/c. Low-multiplicity events $N_{\mathrm{trk}}^{\mathrm{offline}} < 35$ ($N_{\mathrm{trk}}^{\mathrm{offline}} < 50$) for J/ψ (Υ(1S)) are also used to estimate the nonflow contribution, e.g., residual back-to-back jet-like correlations.

In this analysis, muons are reconstructed by extrapolating tracks from the silicon tracker to match a hit on at least one segment of the muon detectors. The muons are selected to be in $|\eta^{\mu}| < 2.4$ and $p_T^{\mu} > 3.0$ (3.5) GeV/c for J/ψ (Υ(1S)), to ensure the high efficiency for the reconstructed dimuons.

3 Analysis

In this analysis, the long-range ($|\Delta\eta| > 1$) two-particle correlation technique [6, 13] is used. The dimuons are used as the "trigger" particles, and are matched with "associated" charged particles with $|\eta| < 2.4$ and $0.3 < p_T < 3$ GeV/c.

The per-trigger-particle associated yield distribution is defined as

$$\frac{1}{N_{\mathrm{trig}}}\frac{\mathrm{d}^2N^{\mathrm{pair}}}{\mathrm{d}\Delta\eta\mathrm{d}\Delta\phi\Delta m_{\mu+\mu-}\Delta\mathrm{p_T}} = B(0,0) \times \frac{S(\Delta\eta, \Delta\phi, \Delta m_{\mu+\mu-}, \Delta\mathrm{p_T})}{B(\Delta\eta, \Delta\phi, \Delta m_{\mu+\mu-}, \Delta\mathrm{p_T})}, \quad (1)$$

where $\Delta\eta$ and $\Delta\phi$ are differences in η and ϕ of the pair, N_{trig} and N^{pair} are the number of trigger particles and the trigger-associated pairs in the event, respectively. S($\Delta\eta$, $\Delta\phi$) and B($\Delta\eta$, $\Delta\phi$) represents the yield of same-event pairs and yield of mixed-event pair.

The two-dimensional (2-D) distributions are projected onto the $\Delta\phi$ axis and the azimuthal anisotropy harmonics are determined from a Fourier decomposition,

$$\frac{1}{N_{\mathrm{trig}}}\frac{\mathrm{d}^2N^{\mathrm{pair}}}{\mathrm{d}\Delta\eta\mathrm{d}\Delta\phi} = \frac{N_{\mathrm{assoc}}}{2\pi}\left\{1 + \sum_n 2V_{n\Delta}\cos(n\Delta\phi)\right\}, \quad (2)$$

where $V_{n\Delta}$ are the Fourier coefficients and N_{assoc} represents the total number of pairs per trigger particle for a given ($\mathrm{p_T}^{\mathrm{trig}}$, $\mathrm{p_T}^{\mathrm{assoc}}$) bin.

Signals and background components are used to fit the dimuon invariant mass ($m_{\mu^+\mu^-}$) distribution, and the fraction α of signal, as a function of $m_{\mu^+\mu^-}$, is defined as

$$\alpha(m_{\mu^+\mu^-}) = \mathrm{Sig}(m_{\mu^+\mu^-})/(\mathrm{Sig}(m_{\mu^+\mu^-}) + \mathrm{Bkg}(m_{\mu^+\mu^-})). \quad (3)$$

The V_2 of dimuon as a function of the invariant mass $V_2^{\mathrm{Sig+Bkg}}(m_{\mu^+\mu^-})$ is fitted using

$$V_2^{\mathrm{Sig+Bkg}}(m_{\mu^+\mu^-}) = \alpha(m_{\mu^+\mu^-})V_2^{\mathrm{Sig}} + (1 - \alpha(m_{\mu^+\mu^-}))V_2^{\mathrm{Bkg}}(m_{\mu^+\mu^-}). \quad (4)$$

Here, the signal (V_2^{Sig}) represents the V_2 of each dimuon resonance. The V_2 of background $V_2^{\mathrm{Bkg}}(m_{\mu^+\mu^-})$ is described by a polynomial function.

Contributions from jets are subtracted using the low-multiplicity subtraction method developed in Ref. [14] as

$$V_2^{\mathrm{sub}} = V_2^{\mathrm{Sig}}(\mathrm{high}) - V_2^{\mathrm{Sig}}(\mathrm{low}) \times \frac{N_{\mathrm{assoc}}(\mathrm{low})}{N_{\mathrm{assoc}}(\mathrm{high})} \times \frac{J_{\mathrm{jet}}(\mathrm{high})}{J_{\mathrm{jet}}(\mathrm{low})}, \quad (5)$$

where J_{jet} represents the near-side jet yield as the integral in the short-range region ($|\Delta\eta| < 1$). The ratio $J_{\mathrm{jet}}(\mathrm{high})/J_{\mathrm{jet}}(\mathrm{low})$, is introduced to account for the enhanced jet correlations resulting from the multiplicity selection. The associated ratio, $N_{\mathrm{assoc}}(\mathrm{low})/N_{\mathrm{assoc}}(\mathrm{high})$ accounts for the enhanced jet yield due to the difference of the associated track yield.

Finally, to determine the dimuon v_2 values, factorization is assumed where the V_2 value is taken as the product of single-particle v_2 value for the trigger particle and the associated charge hadrons, with

$$v_2^{sub}(p_T{}^{trig}) = \frac{V_2^{sub}(p_T{}^{trig}, p_T{}^{assoc})}{\sqrt{V_2^{sub}(p_T{}^{assoc}, p_T{}^{assoc})}}. \tag{6}$$

4 Results

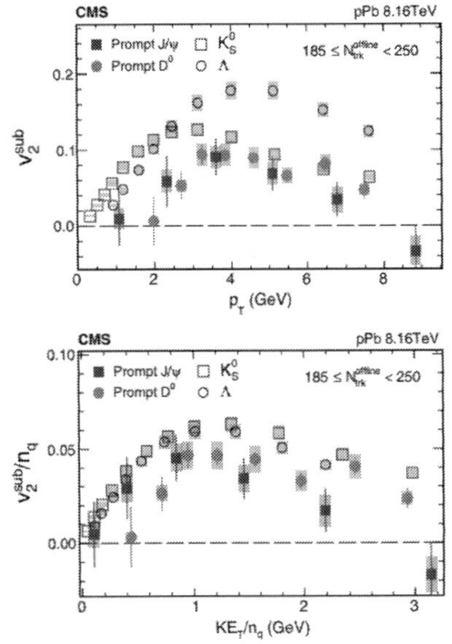

Figure 1. (Upper) The p_T-dependent v_2^{sub} values for prompt J/ψ mesons at forward rapidities ($1.4 < |y_{lab}| < 2.4$) are compared to K_S^0 and Λ hadrons and prompt D^0 mesons at midrapidity ($|y_{lab}| < 1.0$) [15] for pPb collisions at 8.16 TeV for $180 \leq N_{trk}^{offline} < 250$ where a low-multiplicity range with $N_{trk}^{offline} < 35$ to used estimate and correct for the dijet contribution [17]. (Lower) The n_q-normalize v_2^{sub} results. The vertical bars denote statistical uncertainties and the rectangular boxes systematic uncertainties.

Figure 1 shows the measured v_2^{sub} values in upper panel and nq-normalized v_2^{sub} in lower panel for prompt J/ψ meson where a clear positive v_2^{sub} is observed in $2 < p_T < 8$ GeV/c, comparing to D^0, K_S^0 and Λ particles. This results indicate weaker collective dynamics for charm quarks than for light quarks in small systems. On the other hand, Υ(1S) meson v_2^{sub} values are within one standard deviation of zero over the measured p_T range similar as the results in PbPb collisions shown in Fig. 2 (left). The consistency of v_2^{sub} values with zero over the entire p_T region suggests that the collectivity of bottom quarks does not strongly depends on the difference of the medium path length. In the comparison with J/ψ mesons, as shown in Fig. 2 (right), a discrepancy has been found in $3 < p_T < 6$ GeV/c. The different v_2^{sub} values for the two quarkonium states indicates different collective behavior of charm and bottom quarks in pPb collisions.

5 Summary

The elliptic flow (v_2^{sub}) of J/ψ and Υ(1S) mesons is measured as a function of transverse momentum (p_T) in high-multiplicity proton-lead (pPb) collision events at a center-of-mass

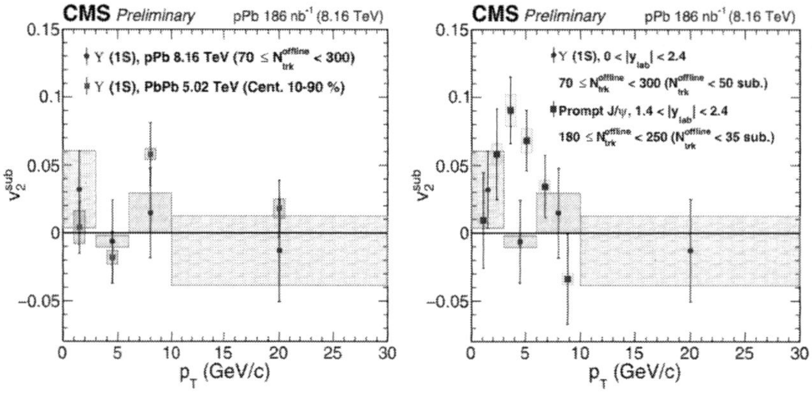

Figure 2. (Left) The p_T dependent v_2^{sub} values for $\Upsilon(1S)$ mesons [16] are compared to the corresponding results from PbPb collisions at 5.02 TeV, measured within the 10–90% centrality range [10]. (Right) The same distribution is also compared with the v_2^{sub} values for prompt J/ψ mesons within $1.4 < |y_{lab}| < 2.4$ in pPb collisions at 8.16 TeV for $180 \leq N_{trk}^{offline} < 250$, where a low-multiplicity range of $N_{trk}^{offline} < 35$ is used estimate and account for the dijet contribution [17]. The vertical bars denote statistical uncertainties and the rectangular boxes systematic uncertainties, while the widths of the boxes represent the p_T bin ranges.

energy per nucleon pair of $\sqrt{s_{NN}} = 8.16$ TeV. The v_2^{sub} values are reported for $0.2 < p_T < 10$ GeV$/c$ and $0 < p_T < 30$ GeV$/c$ for J/ψ and $\Upsilon(1S)$ mesons, respectively. The $\Upsilon(1S)$ meson v_2^{sub} is reported for the first time in pPb collisions. A sizable v_2^{sub} has been found for the J/ψ meson in mid-p_T region, while the $\Upsilon(1S)$ meson v_2^{sub} is consistent with zero in the overall studied p_T region. This finding hints to different in-medium effects for charm and bottom quarks in pPb collisions which provides important constraints to the study of heavy flavor dynamics in small collision systems.

References

[1] STAR Collaboration, Phys. Rev. Lett. **95**, 152301 (2005)
[2] PHOBOS Collaboration, Phys. Rev. Lett. **104**, 062301 (2010)
[3] ALICE Collaboration, Phys. Rev. Lett. **105**, 252302 (2010)
[4] CMS Collaboration, Phys. Rev. C **87**, 014902 (2013)
[5] C. Gale, S. Jeon, B. Schenke, Int. J. Mod. Phys. A **28**, 1340011 (2013)
[6] CMS Collaboration, Phys. Lett. B. **718**, 795 (2013)
[7] ALICE Collaboration, Phys. Lett. B **719**, 29 (2013)
[8] P. Braun-Munzinger, J. Phys. G **34**, S471 (2007)
[9] CMS Collaboration, Eur. Phys. J. C **77**, 252 (2017)
[10] CMS Collaboration, Phys. Lett. B **819**, 136385 (2021)
[11] CMS Collaboration, CMS-PAS-LUM-17-002, (2018)
[12] CMS Collaboration, JINST **3**, S08004 (2008)
[13] CMS Collaboration, JHEP **07**, 076 (2011)
[14] CMS Collaboration, Phys. Lett. B **724**, 213 (2013)
[15] CMS Collaboration, Phys. Rev. Lett. **121**, 082301 (2018)
[16] CMS Collaboration, CMS-PAS-HIN-21-001, (2022)
[17] CMS Collaboration, Phys. Lett. B **790**, 172 (2019)

Charm production: constraint to transport models and charm diffusion coefficient with ALICE

Martin Völkl[1,*] for the ALICE collaboration

[1]Physikalisches Institut der Universität Heidelberg

Abstract. New measurements of D mesons with the ALICE detector, interpreted with theoretical models, can give additional information about the spatial diffusion coefficient, a characteristic of the QGP medium. The relevant ingredients to the models can be assessed by comparing the predictions with diverse measurements of charmed hadrons. This contribution discusses the comparison of measurements with a large set of predictions and the role of the different ingredients implemented in the theoretical models, as well as the conclusions about the spatial diffusion coefficient.

1 Introduction

Heavy quarks are particularly useful probes for the properties of the quark-gluon plasma (QGP). Produced dominantly in the initial hard scatterings of a heavy-ion collision, they experience the evolution of the medium and can still be identified afterwards when measuring hadrons with heavy valence quarks. Due to the excellent primary vertex resolution and particle identification capabilities, ALICE is well suited to measure a range of different charmed hadrons. With the help of theoretical models, these measurements provide information about the interaction of heavy quarks with the QGP and thus its properties.

In most available models, the interaction of heavy quarks with the medium is treated as a series of scatterings of the heavy quarks. They can be considered as interactions with the quarks and gluons or with effective scattering centers. Here, a distinction can be made between elastic $2 \rightarrow 2$ scatterings and inelastic $2 \rightarrow 3$ scatterings producing an additional gluon. Elastic (or collisional) interactions happen independently, while the gluons from inelastic (or radiative) interaction might interfere. This can mean that the energy exchange is not necessarily proportional to the path length. For particles much faster than the surrounding medium, the radiative interaction is expected to be dominant.

The interaction via individual scatterings leads to a description of the propagation through the medium with the Boltzmann-, Fokker-Planck-, and Langevin transport equations in the different models. These make use of different parameter sets, which can however be related to each other. A useful parameter here is the spatial diffusion coefficient D_s, which is related to the mean squared displacement by $\langle \vec{r}^2 \rangle = (2d)D_s t$, where d is the number of spatial dimensions and t the time. This quantity at a reference temperature (e.g. 155 MeV) is a characteristic of the medium and can be constrained by the measurements of charmed hadrons.

*e-mail: martin.andreas.volkl@cern.ch

2 Measurements of charmed hadrons

The main observables are the nuclear modification factor R_{AA} and the flow coefficients v_n. However, these are not only dependent on the interaction of heavy quarks with the medium but can also be affected by other effects, such as modifications due to nuclear parton distribution functions (nPDFs), or the hadronization process. These effects need to be understood to disentangle them.

2.1 Nuclear Parton Distribution Functions

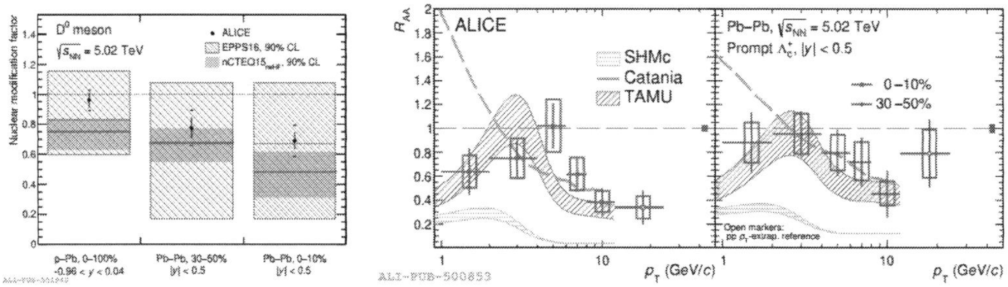

Figure 1. Left: p_T-integrated R_{AA} for D^0 mesons [1]. Right: R_{AA} of Λ_c compared with models [2].

In the most recent measurement of ALICE [1], D^0 mesons are measured at midrapidity in Pb–Pb collisions down to $p_T = 0$. This makes it possible to calculate the integrated yield without extrapolation uncertainties. The p_T-integrated nuclear modification factor is sensitive to differences in the production of charm quarks between pp and Pb–Pb collisions and thus to the nPDFs, in particular to the shadowing effects that reduce the charm production in Pb–Pb and in p–Pb with respect to the production in pp collisions. Figure 1 (left) shows that the measurements performed in Pb–Pb collisions indeed deviate from one and does so in agreement with calculations based on nPDFs. However, deviations from pure binary collision scaling can also appear due to the charm quarks ending up in different hadron types in Pb–Pb collisions compared to the pp reference.

2.2 Charm hadrochemistry

To test the contribution from different charmed hadron species, several measurements were performed. Figure 1 (right) shows the nuclear modification factor for Λ_c^+ baryons [2] compared with theoretical models. The TAMU calculations are in fair agreement with the measurement while the result for the Catania model somewhat overestimates the low-p_T values.

The measurement of D_s^+ mesons [3] leads to a similar result as shown in figure 2 (left) with many models describing the shape of the R_{AA} well. The peak of the R_{AA} seems to be shifted towards higher p_T compared to the non-strange D mesons and even further for the Λ_c^+. This would be consistent with the generic expectation that participation in the flow of the medium should push heavier particles towards higher transverse momenta.

This connection to the collective expansion of the medium can be seen in the measurements of the elliptic and triangular flow coefficients of D_s^+. The measurement of the elliptic flow is shown in figure 2 (right). The values for v_2 are similar to those of the non-strange D mesons. This provides important constraints about the interplay of the elliptic flow of

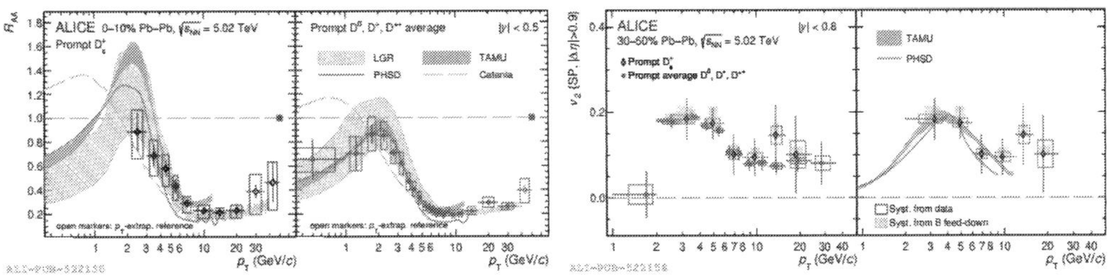

Figure 2. R_{AA} (left) and v_2 (right) of D_s^+ mesons compared with models [3].

the charm quarks themselves and the surrounding medium where they hadronize. The two models, which include coalescence effects describe the v_2 well.

A comparison of the results with the statistical hadronization model [4] shows that the yields are consistent with thermal production with the exception of Λ_c^+. This result could be explained by the existence of additional charmed baryon resonances decaying into Λ_c^+.

3 Discussion of model comparisons and the spatial diffusion coefficient

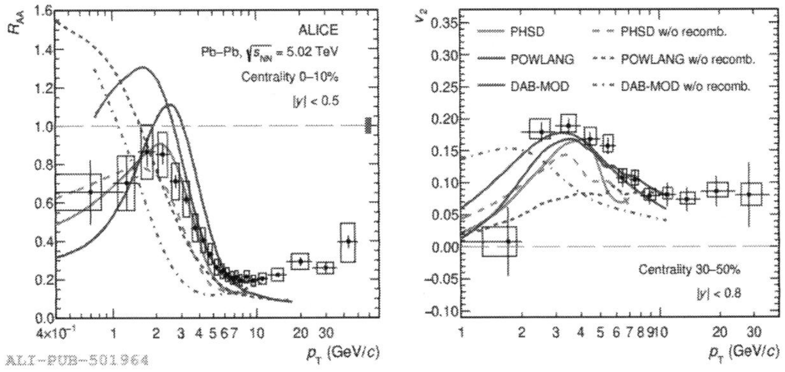

Figure 3. Comparison of R_{AA} and v_2 measurements with models with recombination processes turned on and off [1].

Further knowledge about the role of the different ingredients implemented in theoretical models is gained comparing the p_T-differential R_{AA} [1] and v_2 [5] of non-strange D mesons with models, removing key ingredients. As shown in figure 3, recombination effects are an important ingredient in the models to describe production and flow. The radiative contributions to the interaction are particularly important at higher p_T when attempting to describe a large p_T-range.

The path-length dependence of heavy-quark interactions with the medium was investigated by measuring leptons from heavy-flavor decays in Xe–Xe collisions [6]. Here, the path length scales differently with centrality and multiplicity compared with Pb–Pb collisions. Interestingly, the resulting R_{AA} takes on very similar values between Pb–Pb and Xe-Xe collisions at the same multiplicity. However, comparison of the measurement with a larger number of models is necessary for more thorough conclusions about the scaling with the path length.

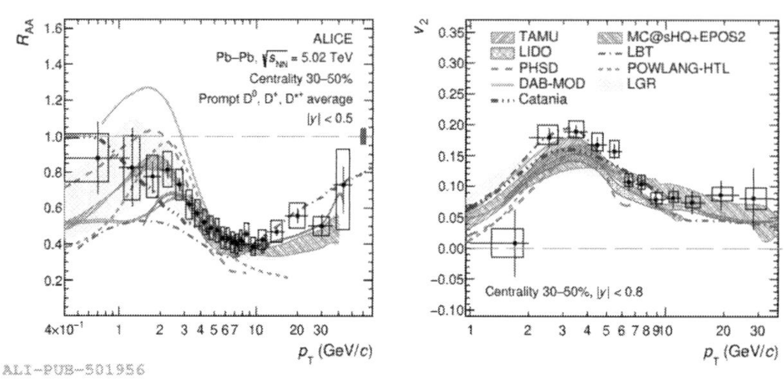

Figure 4. Comparison of R_{AA} and v_2 measurements in semiperipheral events with different models [1].

To understand the influence of the spatial diffusion coefficient, several models were compared to the R_{AA} and v_2 measurements in central and peripheral collisions as in figure 4. The comparison was done for $p_{\mathrm{T}} < 8$ GeV$/c$, where observables are more sensitive to the diffusion and hadronization processes of charm and in order to have a more direct comparison for models covering different p_{T} ranges. As the nuclear modification factor is in general influenced by additional ingredients to the models other than the spatial diffusion coefficient, models were qualified by a reasonable agreement with data, i.e. a reduced $\chi^2 < 5$ for the R_{AA} and $\chi^2 < 2$ for the v_2. Thereby, uncertainties both in the model and the measurements are considered.

The models, which provide a reasonable description of the data by this criterion are TAMU [7], MC@sHQ+EPOS2 [8], LIDO [9], LGR [10], and Catania [11]. The corresponding values for the spatial diffusion coefficient at T_c are $1.5 < 2\pi D_s T_c < 4.5$. These values correspond to relaxation time for charm quarks of $\tau_c \approx 3 - 8$ fm$/c$, i.e. during the lifetime of the QGP.

References

[1] S. Acharya et al. (ALICE), JHEP **01**, 174 (2022), 2110.09420
[2] S. Acharya et al. (ALICE) (2021), 2112.08156
[3] S. Acharya et al. (ALICE), Phys. Lett. B **827**, 136986 (2022), 2110.10006
[4] A. Andronic, P. Braun-Munzinger, M.K. Köhler, A. Mazeliauskas, K. Redlich, J. Stachel, V. Vislavicius, JHEP **07**, 035 (2021), 2104.12754
[5] S. Acharya et al. (ALICE), Phys. Lett. B **813**, 136054 (2021), 2005.11131
[6] S. Acharya et al. (ALICE), Phys. Lett. B **819**, 136437 (2021), 2011.06970
[7] M. He, R. Rapp, Phys. Rev. Lett. **124**, 042301 (2020), 1905.09216
[8] M. Nahrgang, J. Aichelin, P.B. Gossiaux, K. Werner, Phys. Rev. C **89**, 014905 (2014), 1305.6544
[9] W. Ke, Y. Xu, S.A. Bass, Phys. Rev. C **100**, 064911 (2019), 1810.08177
[10] S. Li, J. Liao, Eur. Phys. J. C **80**, 671 (2020), 1912.08965
[11] F. Scardina, S.K. Das, V. Minissale, S. Plumari, V. Greco, Phys. Rev. C **96**, 044905 (2017), 1707.05452

Constraining hadronization processes with charmed baryons in pp and p–Pb collisions with ALICE

Jinjoo Seo[1,*] for the ALICE collaboration

[1]Inha University, Incheon, 22212, South Korea

Abstract. The latest measurements of D^0, D^+ and $D_s{}^+$ mesons with the final measurements of Λ_c^+, $\Xi_c^{0,+}$, $\Sigma_c^{0,++}$, and the first measurement of Ω_c^0 baryons performed with the ALICE detector at midrapidity in pp collisions at $\sqrt{s}$ = 5.02 TeV and $\sqrt{s}$ = 13 TeV are presented. Furthermore, the recent Λ_c^+/D^0 yield ratio, measured down to $p_T = 0$ in p–Pb collisions are discussed. Finally, the first measurements of charm fragmentation fractions and charm production cross section at midrapidity per unit of rapidity are shown for both pp and p–Pb collisions using all measured charm ground state hadrons.

1 Introduction

The production cross sections of heavy flavor mesons are well described by theoretical models based on perturbative QCD calculations that assume fragmentation functions of charm to be universal among different collision systems. Recent measurements of charm-baryon production at midrapidity in pp and p–Pb collisions systems show a baryon-to-meson ratio significantly higher than that in e^+e^- collisions [1, 2], suggesting that the fragmentation of charm is not universal. Therefore, measurements of charm-baryon production are crucial to study the charm quark hadronization in a partonic-rich environment like the one produced in pp collisions at LHC energies. Thanks to the excellent particle identification capabilities, and optimal vertex resolution, ALICE is well suited to measure various charmed hadrons. These measurements provide information about the hadronization process of heavy quarks.

2 Charmed baryon measurements

2.1 Λ_c^+ and $\Sigma_c^{0,++}$measurements

ALICE measured the Λ_c^+/D^0 ratio in pp collisions at $\sqrt{s}$ = 5.02 TeV [3, 4] and 13 TeV [5] down to $p_T = 0$, showing that they are compatible within uncertainties. The Λ_c^+/D^0 ratio, shown in Fig. 1 (left), presents for $p_T < 10$ GeV/c an enhancement with respect to the values $\Lambda_c^+/D^0 \sim 0.11$ [1, 2] measured in e^+e^- and ep collisions, as well as strong p_T dependence not expected by models, which by assuming the fragmentation functions in pp collisions to be the same as those extracted from e^+e^- and ep collisions values. The p_T dependence of charmed baryon-to-meson ratio may suggest that the assumption of universality of charm fragmentation among collision systems is not valid.

*e-mail: jseo@cern.ch

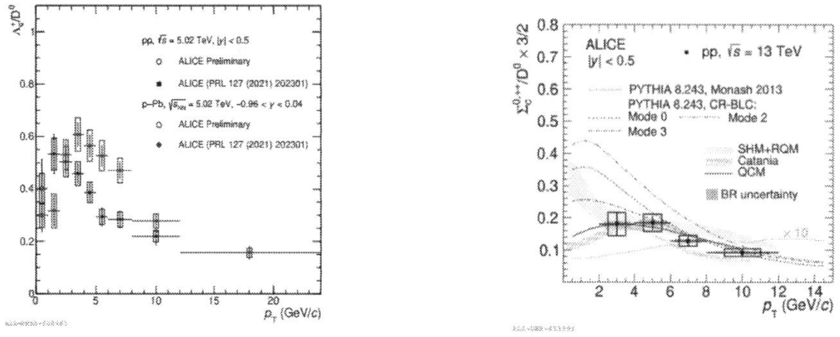

Figure 1. Left panel: Λ_c^+/D^0 as a function of p_T in pp and p–Pb collisions at $\sqrt{s} = 5.02\,\text{TeV}$. Right panel: $\Sigma_c^{0,++}/D^0$ ratio as a function of p_T at $\sqrt{s} = 13\,\text{TeV}$.

The production of Λ_c^+ in p–Pb collisions is also measured down to $p_T = 0$. The Λ_c^+/D^0 ratio measured in p–Pb collisions is higher than the one measured in pp collisions for $p_T > 3\,\text{GeV}/c$, and lower for $p_T < 2\,\text{GeV}/c$. This could be interpreted as a consequence of radial flow [6] or different hadronization processes between pp and p–Pb collisions [7]. The p_T integrated yields of Λ_c^+/D^0 ratio in both collision systems are compatible within uncertainties [4].

ALICE measured $\Sigma_c^{0,++}$to investigate further the charmed baryon production in pp collisions at $\sqrt{s} = 13\,\text{TeV}$. The ratio shown in Fig. 1 (right) presents an enhancement in the measured p_T region with respect to the e^+e^- predictions of models based on e^+e^- data and assuming hadronization to be unaltered in pp collisions. The PYTHIA8 Monash [8] uses the Lund string model and the fragmentation functions tuned to reproduce e^+e^- data. The PYTHIA8 Mode 0,2,3 tunes implement the new Color Reconnection topologies beyond leading color approximation (CR-BLC) [9], including junctions. The junctions (three-leg topology) increase the chance to connect partons from different multi-parton interactions, enhancing the formation of baryons. Therefore, the PYTHIA 8 with CR-BLC can describe the enhancement of Λ_c^+ and $\Sigma_c^{0,++}$. The Catania model [10] assumes that the hadronization occurs via vacuum fragmentation and coalescence of charm quarks with light quarks in a hot QCD matter in pp collisions. A blast wave parametrization is used for the light quark spectra, and the FONLL calculation [11, 12] is used for the heavy quark spectra. The Statistical Hadrobnization Model (SHM) [13], which assumes that the yields of the various hadron species are determined by thermal weights depending mainly on the system temperature and hadron masses, can reproduce the Λ_c^+/D^0 and $\Sigma_c^{0,++}/D^0$ ratios by assuming the existence of exciting charm baryon states which are not yet observed. All three CR modes and the SHM describe the observed p_T dependence, while the PYTHIA8 Monash tune significantly underestimates the measurement by a factor of 15-20 at low p_T and a factor of 5 at high p_T [5].

2.2 Ξ_c^0 and Ω_c^0 measurements

The charm-strange baryons, $\Xi_c^{0,+}$ and Ω_c^0, were measured by ALICE in pp collisions at $\sqrt{s} = 13\,\text{TeV}$. The $\Xi_c^{0,+}/D^0$ ratios [14] are shown in the left panel of Fig. 2, and the Ω_c^0/D^0 [15] ratio is shown in the right panel. The branching ratio (BR) of this Ω_c^0 decay is not yet measured, and a theoretical calculation [16] is used to scale the model predictions. Most models significantly underestimate the measurements, except for Catania model, which is close to the data in the measured p_T interval, both for the $\Xi_c^{0,+}/D^0$ and the Ω_c^0/D^0 ratios.

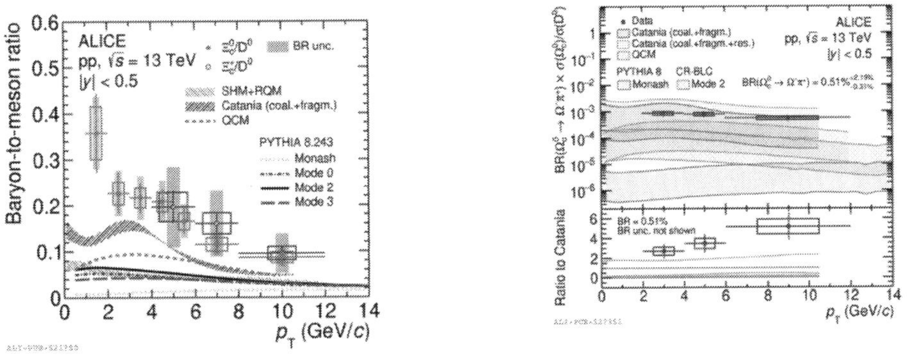

Figure 2. Left panel: $\Xi_c^{0,+}/D^0$ as a function of p_T in pp collisions at $\sqrt{s} = 13$ TeV. Right panel: BR × Ω_c^0/D^0 ratio as a function of p_T in pp collisions at $\sqrt{s} = 13$ TeV.

ALICE also measured $\Xi_c^{0,+}/\Sigma_c^{0,++}$ ratio yield. Most models underestimate the $\Xi_c^{0,+}/\Sigma_c^{0,++}$ ratio measurement, as observed for the charmed baryon-to-meson ratio measurements. However, the PYTHIA8 Monash describes the measurement of $\Xi_c^{0,+}/\Sigma_c^{0,++}$, and the value shows a similar enhancement with respect to e^+e^- results. A possible explanation for this behavior may be the similar masses of the light diquarks in the two baryons ($m(uu, ud, dd)_1 \approx m(ds)_0$), which implies a similar suppression with respect to Λ_c^+ yield due to the lower probability of forming the needed diquark-pair in a string breaking with respect to the lighter diquark ($m(ud)_0$) necessary to form a Λ_c^+ .

2.3 Charm fragmentation fractions and production cross section

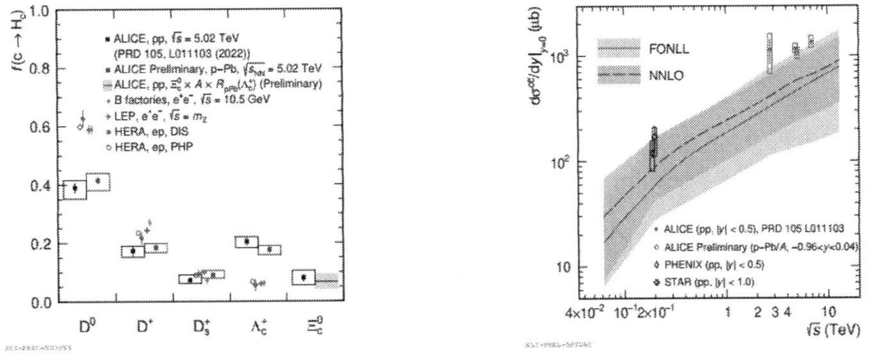

Figure 3. Left panel: Charm-quark fragmentation fractions into charm hadrons in pp and p–Pb collisions at $\sqrt{s} = 5.02$ TeV . Right panel: charm production cross section at midrapidity as a function of the collision energy.

The charm fragmentation fractions in pp [17] and p–Pb collisions at $\sqrt{s} = 5.02$ TeV are calculated summing the production cross section of all all single charm hadron ground states measured by ALICE. Thanks to measurements of D^0 [18, 19], D^+, and Λ_c^+ [3] performed down to $p_T = 0$ in pp and p–Pb collisions, and the measurement of D_s^+ and Ξ_c^0 down to very low p_T (extrapolated to $p_T = 0$ with theoretical predictions), ALICE was could determine the charm fragmentation fractions in pp and p–Pb collisions for the first time.

The Ω_c^0 was not measured in pp and p–Pb collisions at $\sqrt{s} = 5.02$ TeV . Its contribution is treated as a systematic uncertainty, taking into consideration the measurement performed in pp collisions at $\sqrt{s} = 13$ TeV. Charm fragmentation fractions in pp and p–Pb collisions are presented in the left panel of Fig. 3, showing significant baryon enhancement with respect to the fragmentation fractions measured in e^+e^- and ep collisions.

The total charm production cross section per unit of rapidity at midrapidity measured in pp [17] and p–Pb collisions at $\sqrt{s} = 5.02$ TeV are shown in Fig. 3 (right). They are compatible within uncertainties. Using the new charm fragmentation fractions [17], the charm production cross section in pp collisions at $\sqrt{s} = 2.76$ TeV [20] and 7 TeV [21] are about 40% higher than the previously published results. The updated charm production cross section lies on the upper edge of the pQCD calculation uncertainty bands [11].

3 Conclusions

ALICE measured the production cross section of all charm hadron ground states in pp collisions. The productions of Λ_c^+ was measured down to $p_T = 0$ in pp and p–Pb collisions. A large enhancement of all charm baryon-to-meson ratios is observed in pp collisions with respect to e^+e^- collisions. None of the models describes the enhancement for all charm-baryons. Reproducing the measurements of charm baryons with a strange quark is particularly challenging. The charm fragmentation fractions measured in pp and p–Pb collisions at $\sqrt{s} = 5.02$ TeV show that the assumption of the universality of charm fragmentation fractions across colliding systems is not valid already from e^+e^- to pp collisions.

References

[1] H. Albrecht et al. (ARGUS), Phys. Lett. B **207**, 109 (1988)
[2] P. Avery et al. (CLEO), Phys. Rev. D **43**, 3599 (1991)
[3] S. Acharya et al. (ALICE), Phys. Rev. C **104**, 054905 (2021), 2011.06079
[4] S. Acharya et al. (ALICE), Phys. Rev. Lett. **127**, 202301 (2021), 2011.06078
[5] S. Acharya et al. (ALICE), Phys. Rev. Lett. **128**, 012001 (2022), 2106.08278
[6] K. Werner, B. Guiot, I. Karpenko, T. Pierog, Phys. Rev. C **89**, 064903 (2014), 1312.1233
[7] J. Adolfsson et al., Eur. Phys. J. A **56**, 288 (2020), 2003.10997
[8] P. Skands, S. Carrazza, J. Rojo, Eur. Phys. J. C **74**, 3024 (2014), 1404.5630
[9] J.R. Christiansen, P.Z. Skands, JHEP **08**, 003 (2015), 1505.01681
[10] V. Minissale, S. Plumari, V. Greco, Phys. Lett. B **821**, 136622 (2021), 2012.12001
[11] M. Cacciari, M. Greco, P. Nason, JHEP **05**, 007 (1998), hep-ph/9803400
[12] M. Cacciari, S. Frixione, N. Houdeau, M.L. Mangano, P. Nason, G. Ridolfi, JHEP **10**, 137 (2012), 1205.6344
[13] M. He, R. Rapp, Phys. Lett. B **795**, 117 (2019), 1902.08889
[14] S. Acharya et al. (ALICE), Phys. Rev. Lett. **127**, 272001 (2021), 2105.05187
[15] S. Acharya et al. (ALICE), arXiv **10.48550** (2022), 2205.13993
[16] Y.K. Hsiao, L. Yang, C.C. Lih, S.Y. Tsai, Eur. Phys. J. C **80**, 1066 (2020), 2009.12752
[17] S. Acharya et al. (ALICE), Phys. Rev. D **105**, L011103 (2022), 2105.06335
[18] S. Acharya et al. (ALICE), JHEP **05**, 220 (2021), 2102.13601
[19] J. Adam et al. (ALICE), Phys. Rev. C **94**, 054908 (2016), 1605.07569
[20] B. Abelev et al. (ALICE), JHEP **07**, 191 (2012), 1205.4007
[21] S. Acharya et al. (ALICE), Eur. Phys. J. C **77**, 550 (2017), 1702.00766

Beauty production in heavy–ion collisions with ALICE at the LHC

Stefano Politanò[1,*]

[1]Politecnico and INFN Torino

Abstract. In this contribution, the measurement of the centrality dependence of the nuclear modification factor (R_{AA}) of non-prompt D^0 meson in Pb–Pb collisions at $\sqrt{s_{\mathrm{NN}}} = 5.02$ TeV is presented. This measurement provides important constraints to the in-medium mass dependence of energy loss and hadronization of the beauty quark. In addition, the first measurements of non-prompt D_s^+ meson production in central and semi-central Pb–Pb collisions at $\sqrt{s_{\mathrm{NN}}} = 5.02$ TeV are discussed. The non-prompt D_s^+ meson measurements provide further information on the production and hadronization of B_s^0 mesons. Finally, the first measurement of non-prompt D-meson elliptic flow in Pb–Pb collisions at $\sqrt{s_{\mathrm{NN}}} = 5.02$ TeV is shown. This measurement can help to investigate the degree of thermalization of beauty quarks in the hot and dense QCD medium.

1 Introduction

The formation of a colour-deconfined state, called quark-gluon plasma (QGP) in ultra-relativistic collisions of heavy ions is expected from quantum chromodynamics calculations on the lattice [1]. Heavy quarks (i.e. charm and beauty) are produced in hard-scattering processes occurring in the early stage of the collision. As the medium expands in presence of pressure gradients, the heavy quarks interact with its constituents via elastic and inelastic (gluon radiation) scatterings. The measurement of the nuclear modification factor (R_{AA}) of heavy-flavor hadrons allows us to study the mechanisms of energy loss, which provides information on the medium properties and expansion. For heavy-flavour hadrons, in absence of in-medium effects, the R_{AA} is expected to be equal to unity and be independent from transverse momentum (p_{T}). At low momenta, heavy-quark propagation through the medium can be described by transport models as a diffusion process [2]. Heavy-quarks undergo multiple low-energy-transfer interactions that favour the participation of heavy quarks in the collective expansion of the medium, which can alter their azimuthal distribution. This can be probed by measuring their elliptic flow (v_2). The study of the energy loss and the possible thermalization of heavy quarks in the medium provide constraints on the transport coefficient (D_s) of the QGP.

2 Beauty in-medium energy loss and thermalization

Prompt and non-prompt D^0-meson R_{AA} were measured by ALICE in central Pb–Pb collisions at $\sqrt{s_{\mathrm{NN}}} = 5.02$ TeV. Figure 1 shows their ratio as a function of p_{T} [3]. The non-prompt to prompt D^0 R_{AA} ratio is compared with predictions from various models, namely

*e-mail: stefano.politano@cern.ch

MC@sHQ+EPOS2 [4], LGR [5, 6], TAMU [7], and CUJET3.1 [8]. The predictions for $p_T < 5$ GeV/c are similar among the different models: they show a depletion close to unity in $2 < p_T < 3$ GeV/c and an increase at lower and higher p_T. The data show a flat trend as a function of p_T, about 3.9 σ above unity, for $p_T > 6$ GeV/c. Only the TAMU model predicts a decrease of the R_{AA} ratio in this p_T region. This different trend could derive from the absence in the TAMU model of radiative processes, which dominate energy loss in the high p_T region in the other models. A differential study of the LGR model predictions is shown in the bottom panel of Fig 1. In particular, the following configurations are implemented to estimate the effect of the different assumptions on the final prediction: the mass of the beauty is set to the one of the charm quark in the energy-loss (I) and in the hadronization (II) processes, the shadowing effect (III) and the hadronization via coalescence (IV) are excluded. In the low-p_T region, the absence of the hadronization via coalescence (IV) largely overestimates the data. The main difference observed in the high-p_T region is related to the change of the beauty mass in the energy-loss processes supporting the interpretation of the deviation of the measured ratio from unity as a result of the mass-dependence of quark in-medium energy-loss, which is expected to be smaller for more massive quarks.

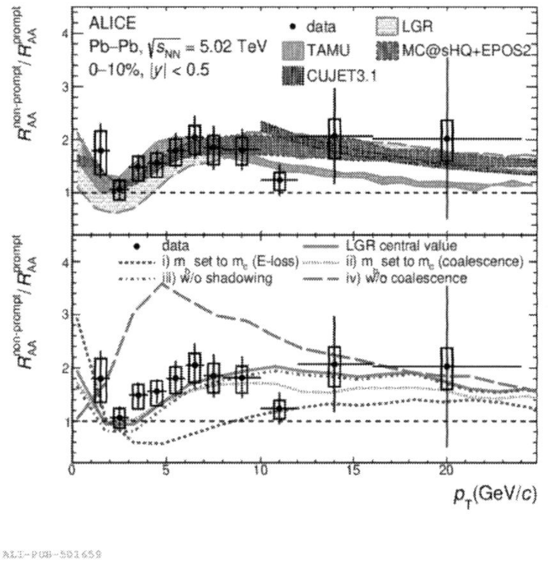

Figure 1. Non-prompt to prompt D^0-meson R_{AA} ratio [3] as a function of p_T in central Pb–Pb collisions at $\sqrt{s_{NN}}$ = 5.02 TeV, compared to model predictions [4–8] (top), and to different modifications of LGR calculations (bottom).

Measurements of strange D mesons provide additional information to investigate the in-medium energy loss and the degree of thermalization of beauty quarks. Figure 2 shows the comparison of the non-prompt to prompt D_s^+-meson R_{AA} ratio (left panel) and non-prompt D_s^+ to D^0-meson R_{AA} ratio (right panel) measured in central and semi-central Pb–Pb collisions at $\sqrt{s_{NN}}$ = 5.02 TeV. The non-prompt to prompt D_s^+ (non-prompt D_s^+ to D^0) R_{AA} ratio in central collisions shows a hint of enhancement of about 1.6 σ (1.7 σ) with respect to unity in the $p_T < 12$ GeV/c interval. This effect could be caused by a smaller energy-loss for beauty than charm quarks, similarly to the non-prompt-to-prompt D^0-meson R_{AA} ratio. The enhancement observed for non-prompt strange to non-strange D mesons R_{AA} ratio could be due to a substantial contribution of hadronization via recombination of the beauty quark with light quarks in the strangeness-rich QGP environment. The amount of strange quarks relative to that of light quarks is expected to increase from pp to Pb–Pb collisions. Thus, the production of strange to non-strange beauty mesons is expected to increase in Pb–Pb collisions in the low transverse momentum interval. Given the large uncertainties affecting both measurements, a firm conclusion on the size of this effect is not possible. The R_{AA} ratios are also compared with the predictions from the TAMU model. The TAMU model qualitatively describes the trend of the data indicating an enhancement of the R_{AA} ratios in the low-p_T region with respect to unity and a decrease at higher transverse momentum. However, it overestimates the measurements.

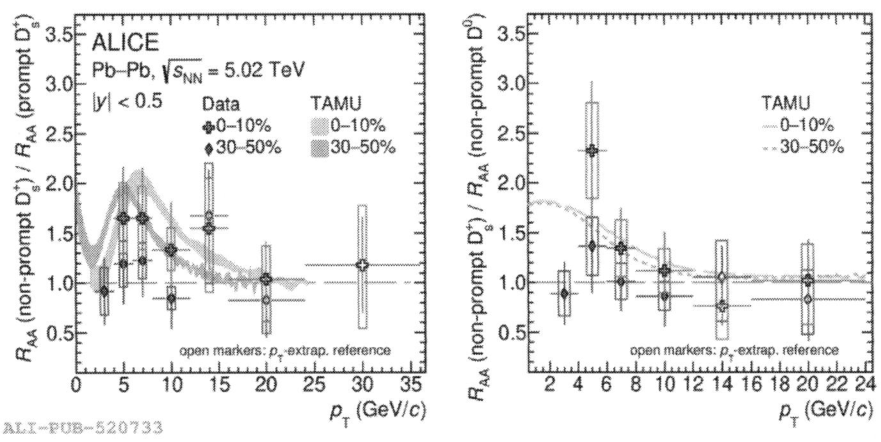

Figure 2. The R_{AA} of non-prompt D_s^+ meson divided by the one of prompt D_s^+ meson (left panel) and non-prompt D^0 meson (right panel) for central and semi-central Pb–Pb collisions at $\sqrt{s_{NN}}$ = 5.02 TeV [9]. The measurements are compared with TAMU model predictions [7].

The degree of thermalization of beauty quarks in the QGP can be further studied by measuring the elliptic flow of beauty hadrons. Figure 3 shows the v_2 of prompt and non-prompt D^0 mesons in semi-central Pb–Pb collisions at $\sqrt{s_{NN}}$ = 5.02 TeV. The non-prompt D^0 mesons show a positive v_2 with a 2.7 σ significance in the $2 < p_T < 12$ GeV/c region. A larger v_2 for prompt D^0 with respect to the one measured for non-prompt D^0 is measured with a significance of 3.2 σ in the low p_T interval. Since the beauty quark is more massive than the charm one, a larger relaxation time is expected. If the relaxation time of beauty quarks is comparable with or larger than the lifetime of the QGP, then the beauty quarks would be able to thermalize only partially.
Finally, the R_{AA} and v_2 of non-prompt D^0 mesons together can set constraints on heavy-quark D_s. Figure 4 shows the R_{AA} and v_2 of non-prompt D^0 compared to theoretical predictions based on Langevin calculations adopting two different values of D_s [10].

3 Conclusion

Heavy quarks are unique probes of the QGP that allow us to characterize its properties. ALICE results show that heavy quarks experience different in-medium energy loss and partially reach thermal equilibrium with the QGP medium. To be able to describe the data, models have to include both the elastic and the radiative energy loss processes. A significant dependence of energy loss on the mass of the quark is observed. The discrepancy observed between prompt and non-prompt D^0 meson v_2 suggests that beauty quarks are less coupled to the medium. More precise measurements, also of fully-reconstructed beauty-hadron decay, will be available in the next data-taking period at the LHC, helping us to further constrain the spatial diffusion coefficient of heavy quarks.

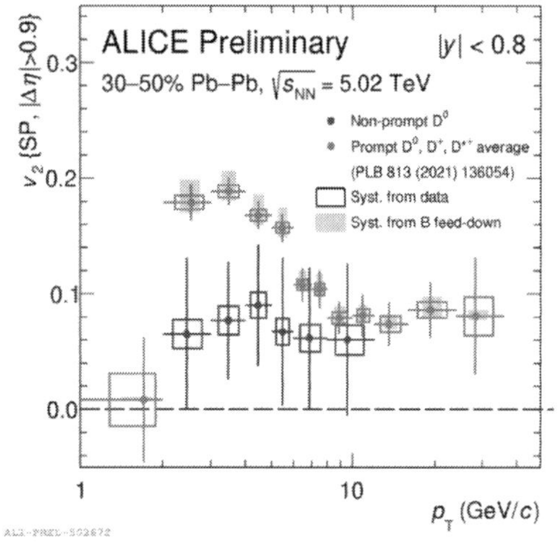

Figure 3. Prompt and non-prompt D^0 meson v_2 in semi-central (30–50%) Pb–Pb collisions at $\sqrt{s_{NN}}$ = 5.02 TeV.

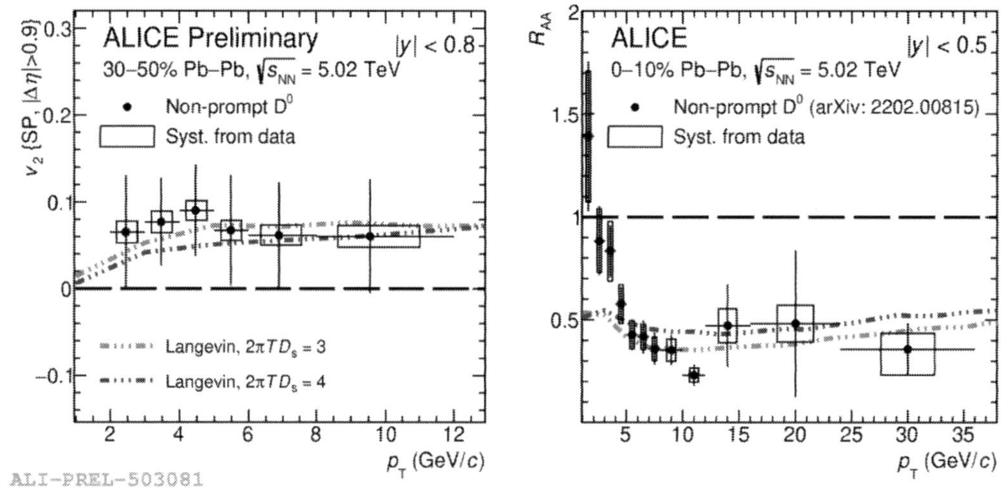

Figure 4. Non-prompt D^0 meson v_2 (left) and R_{AA} (right) in central and semi-central Pb–Pb collisions at $\sqrt{s_{NN}}$ = 5.02 TeV, respectively. The results are compared with theoretical predictions based on Langevin calculations that adopt different values for the diffusion coefficient [10].

References

[1] F. Karsch, J. Phys. Conf. Ser. 46 (2006) 122–131.
[2] M. He, H. van Hees, R. Rapp, arXiv.2204.09299.
[3] ALICE Collaboration, arXiv.2202.00815.
[4] M. Nahrgang, J. Aichelin, P. B. Gossiaux, K. Werner, PRC 89 (2014) 014905.
[5] S. Li and J. Liao, Eur. Phys. J. C 80 no. 7, (2020) 671.
[6] S. Li, W. Xiong, and R. Wan, Eur. Phys. J. C 80 no. 12, (2020) 1113.
[7] M. He, R. J. Fries, and R. Rapp, PLB 735 (2014) 445–450.
[8] S. Shi, J. Liao, and M. Gyulassy, Chin. Phys. C 43 no. 4, (2019) 044101.
[9] ALICE Collaboration, arXiv:2204.10386.
[10] S.Q. Li et al., EPJC 81 (2021) 11, 1035.

Heavy-flavour jet properties and correlations from small to large systems measured by ALICE

Antonio Carlos Oliveira da Silva[1]*, on behalf of the ALICE Collaboration

[1]University of Tennessee - Knoxville

Abstract. The early production of heavy-flavour partons makes them an excellent probes for investigating the evolution of QCD systems. Jets tagged by the presence of a heavy-flavour hadron give access to the kinematics of the heavy partons and allow for comparisons of their production, propagation and fragmentation across different systems. Whilst traversing the quark-gluon plasma (QGP), charm and bottom quarks lose energy through interactions with the medium, at a different rate relative to light quarks and gluons. To constrain the energy loss in the QGP, the nuclear modification factor of D^0-tagged jets is measured in the 0–10% most central Pb–Pb collisions at $\sqrt{s_{NN}} = 5.02$ GeV/c. The properties of charm fragmentation are also investigated in pp collisions at $\sqrt{s} = 5.02$ GeV/c through measurements of the production and the momentum fraction of the D^0 with respect to its jet.

1 Introduction

Heavy-flavour quarks are created in hard-scatterings in high-energy collisions. These heavy quarks can lose energy via gluon radiation forming a parton shower. The original heavy quark produced in this process finally hadronizes into a heavy hadron. ALICE is capable of fully reconstructing the heavy hadron and using it to tag jets that originated from the hard-scatterings and heavy quark parton shower.

In heavy-ion collisions, since heavy quarks are produced in hard scattering and, therefore, in the early stages of the collision, they will experience the whole evolution of the system. This makes heavy quarks ideal tools to study properties of the QGP in large systems.

2 Analysis methods and results

The D^0 meson is fully reconstructed through the hadronic decay channel $D^0 \rightarrow K^-\pi^+$ exploiting the particle identification provided by the ALICE detector[1]. The main strategy for D^0-jet reconstruction is the replacement of the D^0 decay daughters by the 4-momentum vector of the reconstructed D^0. After that, the D^0 and charged particles in the event are processed by the FastJet package for jet finding using the anti-k_T algorithm [2].

D^0-jets are fully corrected for reconstruction efficiency and detector effects. The contribution of non-prompt D^0, originating from the decay of hadrons containing bottom quarks, is subtracted using POWHEG [3] simulations.

*e-mail: antonio.silva@cern.ch

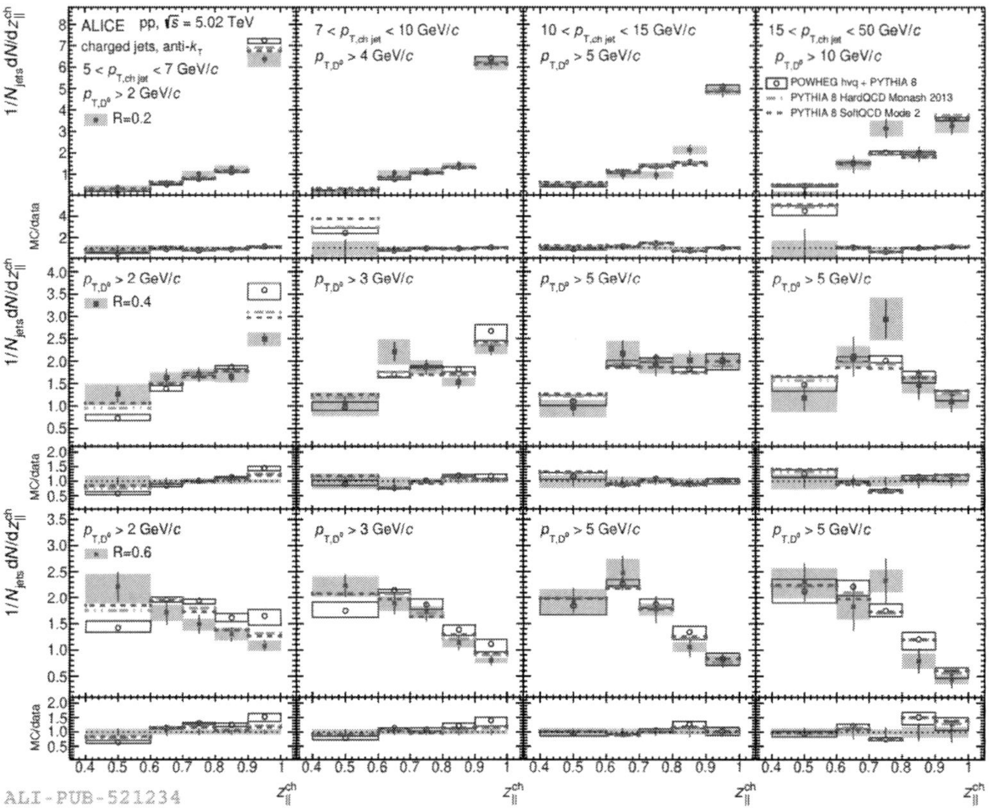

Figure 1. D^0-jet momentum fraction distribution measured in pp collisions at $\sqrt{s}$ = 5.02 TeV. Vertical lines and boxes are, respectively, statistical and systematic uncertainties [4]. Jet and D^0 p_T increase in panels from left to right and R increases in panels from top to bottom. Data are compared to predictions from models [3, 7–9].

The fraction of the D^0 momentum with respect to its jet is defined as $z_{||}^{\mathrm{ch,jet}} = \frac{\vec{p}_{\mathrm{D}} \cdot \vec{p}_{\mathrm{ch,jet}}}{||\vec{p}_{\mathrm{ch,jet}}||^2}$, where $\vec{p}_D$ and $\vec{p}_{ch,jet}$ are the momenta of the D^0 and the jet, respectively. Fig. 1 presents the D^0-jet momentum fraction in pp collisions at $\sqrt{s}$ = 5.02 TeV for jet resolution parameters R = 0.2, 0.4, and 0.6 in four jet p_T intervals.

The shape of the D^0-jet momentum fraction for R = 0.2 is peaked around $z_{||}^{\mathrm{ch,jet}} \approx 1$ for the first three jet p_T intervals and minimum D^0 p_T. This changes in the largest jet p_T interval ($15 < p_{\mathrm{T,ch\ jet}} < 50$ GeV/c), where the peak is not clearly defined. This is also a consequence of the dead-cone effect [5, 6], which suppresses gluon emission at small angles. When the heavy-quark energy becomes larger, the dead-cone angle becomes smaller and the effect also gets smaller. For larger jet resolution parameters, the distributions are not dominated by larger $z_{||}^{\mathrm{ch,jet}}$ because the jet recovers a large fraction of the gluon emissions. The measurements are compared with POWHEG-hvq + PYTHIA 8, POWHEG-hvq is a generator for heavy-flavour hadroproduction at next-to-leading order in QCD [7]; PYTHIA 8 HardQCD Monash 2013 [8]; and PYTHIA 8 SoftQCD Mode 2, which includes colour reconnection beyond the leading-colour approximation [9]. The bottom panels in Fig. 1 present the ratios of the Monte Carlo simulations to data and all models presented are in agreement with the measurements.

The charm interaction with the QGP was studied by measuring the p_T-differential yields and nuclear modification factor (R_{AA}) of D^0-jets in central (0–10%) Pb–Pb collisions at $\sqrt{s_{NN}} = 5.02$ TeV.

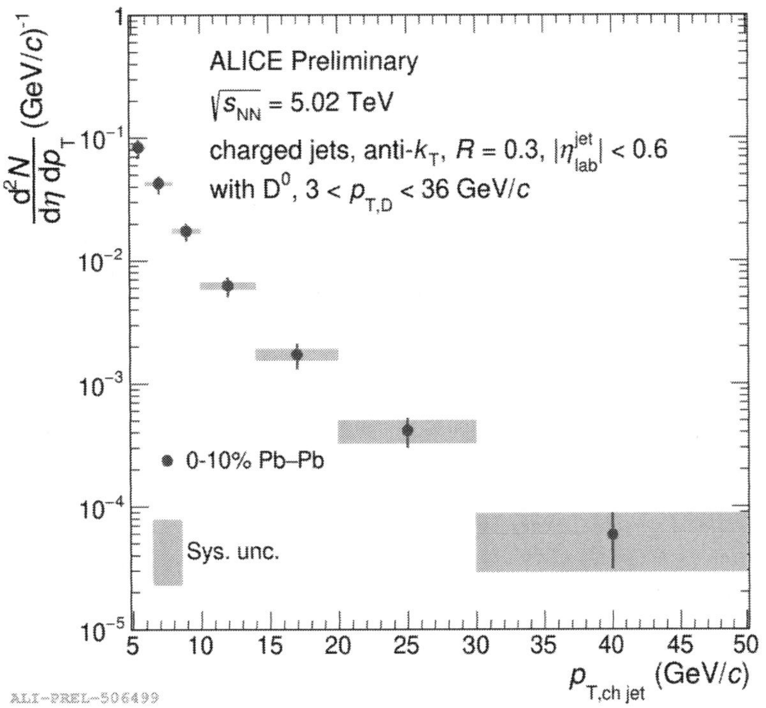

Figure 2. D^0-tagged jet p_T-differential yields in central (0–10%) Pb–Pb collisions at $\sqrt{s_{NN}} = 5.02$ TeV. The red vertical lines and gray boxes are, respectively, the statistical and systematic uncertainties.

The D^0-jet p_T-differential yields are presented in Fig. 2. The D^0 p_T interval inside the jets is $3 < p_{T,D} < 36$ GeV/c. The jet resolution parameter is $R = 0.3$ and all D^0-jets within absolute pseudorapidity $|\eta| < 0.6$ were accepted.

The R_{AA} is defined as the ratio of the p_T-differential yields of jets in Pb–Pb and pp collisions scaled by the number of binary collisions [10]. The D^0-jet R_{AA} is presented in Fig. 3 and varies from 0.3 at jet transverse momentum p_T between 5 to 8 GeV/c and grows as a function of the jet p_T. The D^0-jet R_{AA} is compared to that of inclusive charged-particle jets with R = 0.2, which is also compatible to the same measurement using R = 0.4. In the jet transverse momentum region where both measurements are available ($20 < p_{T,jet} < 50$ GeV/c), the inclusive jet R_{AA} is systematically lower than D^0-jet R_{AA}, although compatible within uncertainties in the interval $30 < p_{T,jet} < 50$ GeV/c.

The comparison between D^0-jet and inclusive charged-particle can be sensitive to different contributions of quarks and gluons energy loss mechanisms. Mass dependent effects, such as the dead-cone effect, can also play a role.

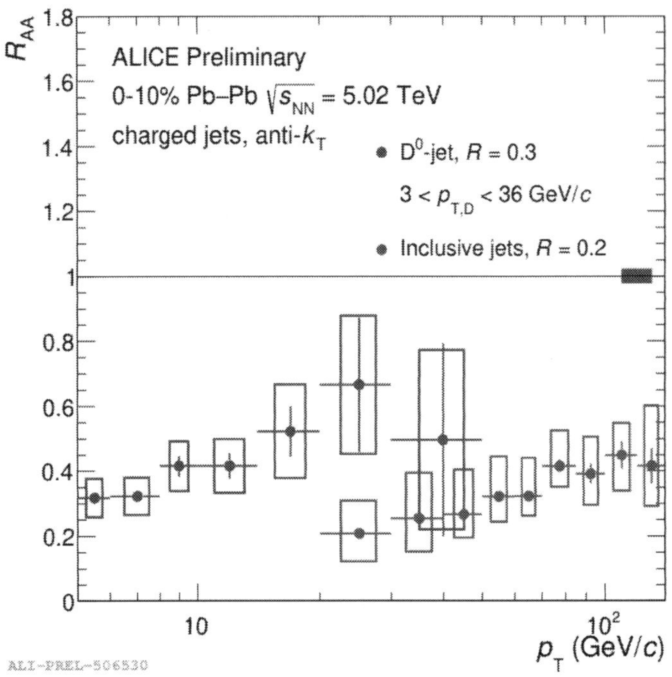

Figure 3. Nuclear modification factor of D^0-jets (red markers) and inclusive charged-particle jets (green markers) in central (0–10%) Pb–Pb collisions at $\sqrt{s_{NN}}$ = 5.02 TeV. Statistical and systematic uncertainties are, respectively, presented as vertical lines and boxes.

References

[1] The ALICE experiment at the CERN LHC, ALICE Collaboration, K. Aamodt et al., Journal of Instrumentation **3** 08 (2008)

[2] FastJet user manual, Cacciari, Matteo and Salam, Gavin P. and Soyez, Gregory, The European Physical Journal C **72**, 3 (2012)

[3] Matching NLO QCD computations with Parton Shower simulations: the POWHEG method, S. Frixione, P. Nason, and C. Oleari, JHEP **11**, 070 (2007)

[4] Measurement of the production of charm jets tagged with D^0 mesons in pp collisions at $\sqrt{s}$ = 5.02 and 13 TeV. ALICE collaboration, ALICE-PUBLIC-2022-017 (2022).

[5] On specific QCD properties of heavy quark fragmentation ('dead cone'). Y. L. Dokshitzer, V. A. Khoze, and S. I. Troian, J. Phys. **G17**, 1602–1604 (1991).

[6] Direct observation of the dead-cone effect in quantum chromodynamics. ALICE Collaboration. Nature **605**, 440–446 (2022)

[7] A Positive-weight next-to-leading-order Monte Carlo for heavy flavour hadroproduction, S. Frixione, P. Nason, and G. Ridolfi, JHEP **09**, 126 (2007)

[8] An introduction to PYTHIA 8.2, T. Sjöstrand, S. Ask, J. R. Christiansen, et al., Comput. Phys. Commun. **191**, 159–177 (2015)

[9] String Formation Beyond Leading Colour, J. R. Christiansen and P. Z. Skands, JHEP **08**, 003 (2015)

[10] Glauber modeling in high-energy nuclear collisions. M. Miller, K. Reygers, S. Sanders, and P. Steinberg. Annual Review of Nuclear and Particle Science **57** 205–243 (2007).

Investigation of in-medium effects of charmonia using azimuthal anisotropy and jet fragmentation function in PbPb collisions at $\sqrt{s_{\mathrm{NN}}} = 5.02$ TeV with the CMS experiment

Gyeonghwan Bak on behalf of the CMS Collaboration[1,*]

[1]Chonnam National University

Abstract. Quarkonia have been long proposed as the golden probes to study quark-gluon plasma (QGP) in heavy-ion collisions. In this presentation, we present the second- and third-order Fourier coefficients, v_2 and v_3, for prompt and nonprompt J/ψ mesons in PbPb collisions as functions transverse momentum (p_{T}) and PbPb collision centrality. Also, we report the first measurements of v_2 and v_3 for the prompt ψ(2S) mesons in PbPb collisions. The results provide v_2 and v_3 values over the wide studied kinematics regions of transverse momentum and collision centrality.

1 Introduction

Quarkonia (bound states of a heavy quark and its antiquark such as J/ψ, $\psi(2S)$, $\Upsilon(1S, 2S, 3S)$, etc.) are useful probes to study the properties of the quark-gluon plasma (QGP) in heavy ion collisions [1, 2]. Since heavy quarks are produced via hard parton scattering at early stages of the collisions, they experience the whole space-time evolution of the medium. Azimuthal correlation of these particles can be characterized by the Fourier coefficients (v_n) of the azimuthal distribution. In particular, the second (v_2) and the third (v_3) order components are sensitive to the initial collision geometry and event-by-event fluctuations, respectively. The azimuthal correlation of quarkonium states are useful in particular to study the quarkonium dynamics and its collectivity inside the QGP [3]. In addition, the path-length dependent suppression of quarkonium states could be captured by measurements of quarkonium v_2 values [4].

This conference proceeding reports the measurement of the v_2 and v_3 values for prompt and nonprompt (decayed from b-hadrons) J/ψ mesons, as well as for the prompt ψ(2S) mesons using the scalar product method [5]. The data set used in this analysis was collected with the CMS detector in PbPb collisions at $\sqrt{s_{\mathrm{NN}}} = 5.02$ TeV, corresponding to an integrated luminosity of 1.6 nb^{-1}.

2 Analysis procedure

The charmonia are identified using their dimuon decay channel. The separation of the prompt and the nonprompt charmonia relies on the displacement of a secondary $\mu^+\mu^-$ vertex relative to the primary collision vertex. The pseudo-proper decay length is defined as $\ell_{J/\psi} = L_{xyz}\, m_{J/\psi}\, c/|p_{\mu\mu}|$, where L_{xyz} is the distance between the primary and dimuon vertices,

*e-mail: gyeonghwan.bak@cern.ch

$m_{J/\psi}$ is the world average value of the J/ψ meson mass, and $p_{\mu\mu}$ is the dimuon momentum. To extract prompt and nonprompt J/ψ meson yields, the invariant mass spectrum of $\mu^+\mu^-$ pairs and their $\ell_{J/\psi}$ distribution are fitted using a two-dimensional (2D) extended unbinned maximum likelihood fit. For ψ(2S) mesons, the nonprompt component is reduced by placing tight constraints on the $\ell_{J/\psi}$ distributions. The detailed procedure of the fitting can be found in [6].

The v_n ($n = 2$ and 3) values of J/ψ and ψ(2S) candidates are determined using the scalar product method [5]. The event plane angle for the second- and third-order harmonics are defined using Q-vectors in the complex plane $Q_n = \sum_{k=1}^{M} \omega_k e^{in\phi_k}$, and are obtained using the tracker or hardron forward (HF) calorimeters. Here, k runs over the particles sampled by the subdetector; M is the multiplicity of particles in the tracker or the number of towers for HF; ϕ is the azimuthal angle of the particle or the tower; ω is the weighting factor given by the p_T of a particle for the tracker or the transverse energy deposited in an HF tower. In this analysis, three event plane Q-vectors are calculated using the tracker at the mid-pseudorapidity ($|\eta| < 0.75$) and the two HF calorimeters covering the forward ($3 < \eta < 5$, HF+) and backward ($-5 < \eta < -3$, HF−) regions. The v_n coefficients for J/ψ or ψ(2S) mesons are obtained as:

$$v_n\,\{\mathrm{SP}\} \equiv \frac{\left\langle Q_n^{J/\psi,\psi(2S)} Q_{nA}^{*} \right\rangle}{\sqrt{\frac{\langle Q_{nA} Q_{nB}^{*}\rangle \langle Q_{nA} Q_{nC}^{*}\rangle}{\langle Q_{nB} Q_{nC}^{*}\rangle}}}. \quad (1)$$

The subscripts A and B refer to either HF+ or HF−, depending on the rapidity of the J/ψ or ψ(2S) candidate. To avoid autocorrelations, the η gap between the J/ψ or ψ(2S) candidate and the detector used for event plane determination is required to be at least three units of rapidity [5, 7, 8]. For this purpose, HF+ is selected for A (B) when J/ψ and ψ(2S) candidates are produced at negative (positive) rapidity. The $\langle\rangle$ indicates the real component of the average Q-vector product of all the candidates in a given invariant mass bin. The subscript C denotes the event plane taken from the tracker. The denominator in Eq.(1) is the correction factor to remove finite resolution effect of the detectors.

To extract the prompt and nonprompt J/ψ meson v_n coefficients, the J/ψ candidates are sampled into fine v_n intervals. Then, a simultaneous fit of the invariant mass and the $\ell_{J/\psi}$ distributions, as described above, is performed in each interval.

A different method is used to extract the prompt ψ(2S) meson v_n values. The v_n values are extracted by a simultaneous fit on the invariant mass and v_n distribution of ψ(2S) mesons using a binned χ^2 fit method. The v_n distribution is fitted to the following formula:

$$v_n{}^{\mathrm{Sig+Bkg}}(m_{\mathrm{inv}}) = \alpha(m_{\mathrm{inv}}) v_n{}^{\psi(2S)} + [1 - \alpha(m_{\mathrm{inv}})] v_n{}^{\mathrm{Bkg}}(m_{\mathrm{inv}}), \quad (2)$$

3 Results

The measured v_2 and v_3 values for prompt and nonprompt J/ψ mesons in PbPb collisions are shown in Fig.1 as functions of p_T and $\langle N_{\mathrm{part}}\rangle$, which is the average number of participating nucleons [9]. Results for the prompt J/ψ mesons show a flat v_2 for $p_T > 9$ GeV/c, while no clear dependence on p_T is found for nonprompt J/ψ mesons. Previous studies from CMS found the J/ψ being produced in large surrounding jet-activity, which suggested the importance of incorporating jet quenching in describin J/ψ suppression in PbPb collisions at high-p_T [10]. The sizable v_2 of prompt J/ψ mesons at high-p_T in Fig.1 implies the large contribution from path length dependent jet energy loss into the final measured v_2. The v_2 values for nonprompt J/ψ mesons are found to be smaller than those for prompt J/ψ mesons in the studied kinematic region. This finding indiciates different medium effects for charm and bottom quarks induced by the interaction with the QGP.

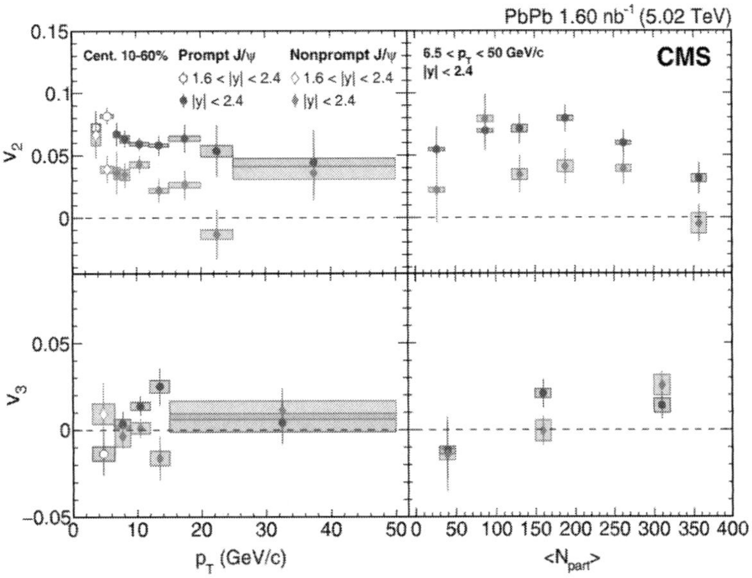

Figure 1. The v_2 (upper) and v_3 (lower) values as functions of p_T (left) and $\langle N_{part} \rangle$ (right) for prompt and nonprompt J/ψ mesons. The results for $3 < p_T < 6.5$ and $6.5 < p_T < 50$ GeV/c are studied in the rapidity range of $1.6 < |y| < 2.4$ and $|y| < 2.4$, respectively (left panel). The results for p_T bins for the right panel is 6.5–50 GeV/c and rapidity of $|y| < 2.4$. The vertical bars denote the statistical uncertainties, and the rectangular bands show the systematic uncertainties.

Figure 2 shows the first v_n measurements of the azimuthal anisotropy for prompt ψ(2S) mesons in heavy ion collisions together with the results for prompt J/ψ mesons. The v_2 values are found to be slightly larger for the prompt ψ(2S) than for the prompt J/ψ mesons, especially at higher p_T and in peripheral PbPb collisions, although the large uncertainties prevent from making a firm conclusion, with p-values of at best 10%. The v_3 values are found to be consistent with zero in the studied kinematic range.

4 Summary

The second-order (v_2) and third-order (v_3) Fourier coefficients of the azimuthal distributions for prompt and nonprompt J/ψ mesons and prompt ψ(2S) mesons are measured in PbPb collisions as functions of transverse momentum (p_T) and $\langle N_{part} \rangle$. The v_2 values for prompt and nonprompt J/ψ mesons both indicate a decreasing trend from mid-central towards central collision events. On the other hand, the v_2 is found to be flat for $p_T > 9$ GeV/c for prompt J/ψ mesons, while the dependence on p_T is found to be weak for nonprompt J/ψ mesons. The prompt J/ψ meson v_2 values are found to be larger than that of nonprompt J/ψ mesons throughout the studied kinematic region, suggesting different in-medium effects for charm and bottom quarks. The observation of sizable and flattened v_2 values at high p_T for prompt J/ψ mesons provides a hint of the contribution of jet quenching to prompt J/ψ meson suppression. The J/ψ v_3 values are reported for the first time separately for prompt and nonprompt components, which are found to be consistent with zero in the measured p_T and centrality intervals. The v_2 and v_3 values are also measured for prompt ψ(2S) mesons for the first time

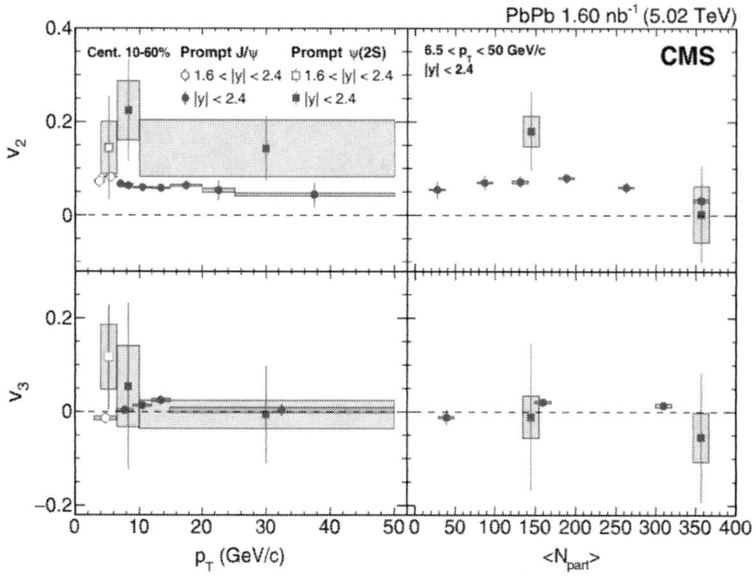

Figure 2. The v_2 (upper) and v_3 (lower) values as functions of p_T (left) and $\langle N_{part}\rangle$ (right) for prompt J/ψ (blue circles) and prompt ψ(2S) (green squares). The results for $p_T < 6.5$ and $p_T > 6.5$ GeV/c are diplayed in the rapidity range of $1.6 < |y| < 2.4$ and $|y| < 2.4$, respectively (left panel). The results for p_T bins for the right panel is 6.5–50 GeV/c and rapidity of $|y| < 2.4$. The vertical bars denote the statistical uncertainties, and the rectangular bands show the systematic uncertainties.

in heavy ion collisions. These J/ψ and ψ(2S) meson measurements provide new insights in the dynamics and in-medium effects of charmonia in heavy ion collisions.

References

[1] T. Matsui and H. Satz, Phys. Lett. B, **178**, 416 (1986)
[2] Digal, S. and Petreczky, P. and Satz, H., Phys. Rev. D **64**, 094015 (2001)
[3] A. Bzdak, B. Schenke, P. Tribedy, and R. Venugopalan, Phys. Rev. C **87** 064906 (2013)
[4] Yunpeng Liu and Nu Xu and Pengfei Zhuang, Nucl. Phys. A **834**, 317C (2010)
[5] CMS Collaboration, Phys. Lett. B **819**, 136385 (2021)
[6] CMS Collaboration, Phys. Lett. B **825**, 136842 (2022)
[7] CMS Collaboration, Phys. Rev. Lett. **120**, 202301 (2018)
[8] M. Luzum and J. Ollitrault, Phys. Rev. C **87**, 044907 (2013)
[9] Loizides, Constantin and Kamin, Jason and d'Enterria, David, Phys. Rev. C **97**, 054910 (2018)
[10] CMS Collaboration, Phys. Lett. B **825**, 136842 (2022)

Heavy-flavor meson and baryon production in high-energy nucleus-nucleus collisions

A. Beraudo[1,*], *A.* De Pace[1], *D.* Pablos Alfonso[1], *F.* Prino[1], *M.* Monteno[1], and *M.* Nardi[1]

[1]INFN - Sezione di Torino, via Pietro Giuria 1, I-10125 Torino

Abstract. We present a new model for the description of heavy-flavor hadronization in relativistic heavy-ion collisions. We explore its effect on the charmed hadron yields and kinematic distributions once the latter is applied at the end of transport calculations used to simulate the propagation of heavy quarks in the deconfined fireball produced in the collision. The model is based on the formation of color-singlet clusters through the recombination of charm quarks with light antiquarks or diquarks from the same fluid cell. This local mechanism of color neutralization leads to a strong space-momentum correlation, which provides a substantial enhancement of charmed baryon production and of the collective flow of all charmed hadrons.

1 Introduction

The diffusion of Brownian particles can be used to access microscopic properties of the medium in which their propagation takes place. Historically, the diffusion of small grains in water was used by Perrin to prove the discontinuous structure of matter and to provide a first estimate of the Avogadro number, finding $N_A \approx 5.5 - 7.2 \cdot 10^{23}$ [1]. More than one century later one of the goals of heavy-ion collisions is to exploit the *relativistic* brownian motion of heavy quarks to get an estimate of comparable accuracy of medium properties like the *momentum broadening coefficient* κ. An important difference in this case is that the nature of the Brownian particle changes during its propagation, since after a few fm/c it undergoes hadronization. This introduces a source of systematic uncertainty in the extraction of transport coefficients; on the other hand it can be considered an issue of interest in itself, in particular to study how hadronization changes in the presence of a medium acting as a color reservoir. This, in the case of charm quarks, will be the subject of our study.

2 The model

Any hadronization model must start grouping colored partons into color-singlet structures which will give rise to the final hadrons. In "elementary" hadronic collisions partons are taken from the hard process, from the shower stage, from the underlying event and from the beam remnants. In heavy-ion collisions recombining partons are taken from the hot medium produced in the collision and they are *close in space*: since the fireball undergoes a collective expansion, this last detail has deep phenomenological consequences.

*e-mail: beraudo@to.infn.it

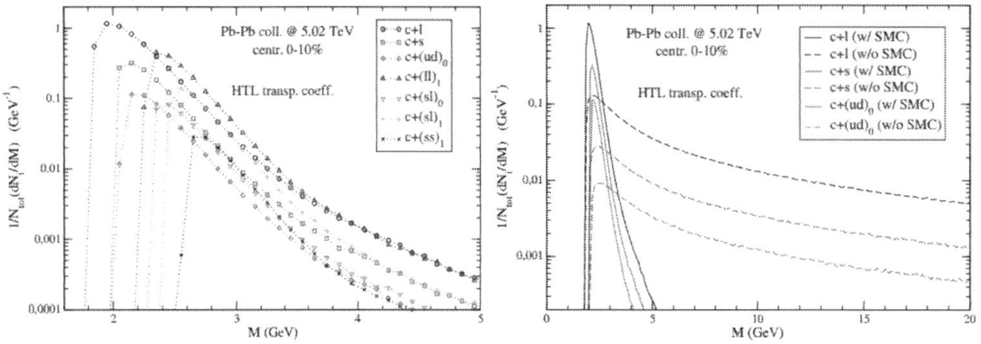

Figure 1. Left panel: the invariant-mass distribution of the $Q\overline{q}$ and $Q(qq)$ clusters at the hadronization hypersurface in the different isospin, strangeness and spin channels. Right panel: cluster invariant-mass distributions in case recombination occurs locally (continuous curves, default implementation), with strong correlation between the parton momenta and position, or non-locally (dashed curves), with no space-momentum correlation. All results refer to central Pb-Pb collisions at $\sqrt{s_{\rm NN}} = 5.02$ TeV, with charm quark propagation before hadronization described by HTL transport coefficients.

We now briefly describe our model for in-medium charm hadronization. For more details we refer the reader to our original publication [2]. Once a c quark, during its stochastic propagation through the fireball, reaches a fluid cell at $T_H = 155$ MeV it recombines with a light antiquark or diquark from the same fluid element. Both the species and the momentum of the medium particle are sampled assuming they obey a thermal distribution in the local rest frame (LRF) of the fluid. The thermal particle is then boosted to the laboratory frame and recombined with the charm quark, leading to the formation of the cluster C. Since the c quark and the thermal particle are taken from the same fluid cell typically their momenta are quite collinear (Space-Momentum Correlation) and this leads to the production of quite low invariant-mass clusters, as one can see from the left panel of Fig. (1). We also check that, suppressing SMC's by randomly redistributing the HQ's on the decoupling hypersurface (see right panel of Fig. 1), the production of heavier clusters is favored. As in Herwig [3], one treats differently light and heavy clusters. Clusters with $M_C < M_{\rm max} \approx 4$ GeV undergo an isotropic two-body decay in their LRF, giving rise to a charmed hadron and a second soft particle; the very rare heavier clusters are fragmented into multiple hadrons as Lund strings [4].

3 Results and discussion

Interfacing our new hadronization model to numerical simulations of heavy-quark transport in the deconfined fireball one can obtain a satisfactory description of important and partially unexpected experimental findings, in particular the strong enhancement of charmed-baryon production recently observed in heavy-ion [5] and also in proton-proton collisions [6], which cannot be explained by any hadronization model tuned to reproduce e^+e^- data. An example of our results is given in Fig. 2, where we plot as a function of p_T the D^+/D^0, D_s^+/D^0 and Λ_c^+/D^0 ratios in Pb-Pb collisions at $\sqrt{s_{\rm NN}} = 5.02$ TeV, comparing our predictions to recent ALICE data [5, 7, 8]. Notice that if hadronization occurred as in e^+e^- collisions one would get $\Lambda_c^+/D^0 \approx 0.1$. As one can see the strong enhancement of the Λ_c^+/D^0 ratio for intermediate values of p_T is correctly reproduced. Analogous results are obtained when our model is compared to STAR data for Au-Au collisions at $\sqrt{s_{\rm NN}} = 200$ GeV [9]. Integrating the momentum distributions one gets the *fragmentation fractions* into the different charmed hadrons, which

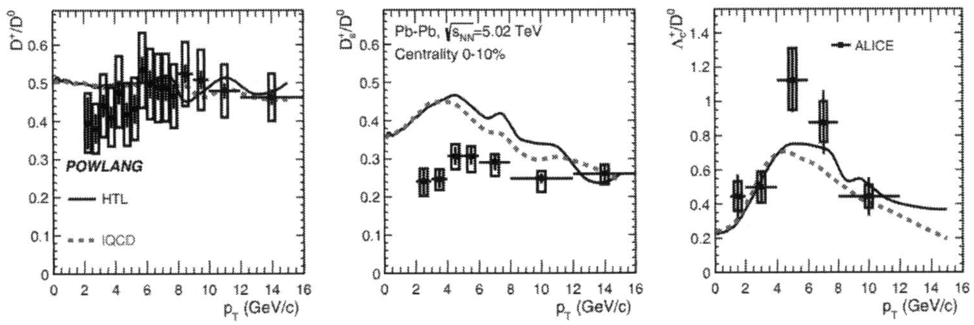

Figure 2. Predictions for the relative yields of charmed hadrons (relative to D^0 mesons) in central Pb-Pb collisions at $\sqrt{s_{\rm NN}} = 5.02$ TeV for different transport coefficients compared to recent ALICE data [5, 7, 8].

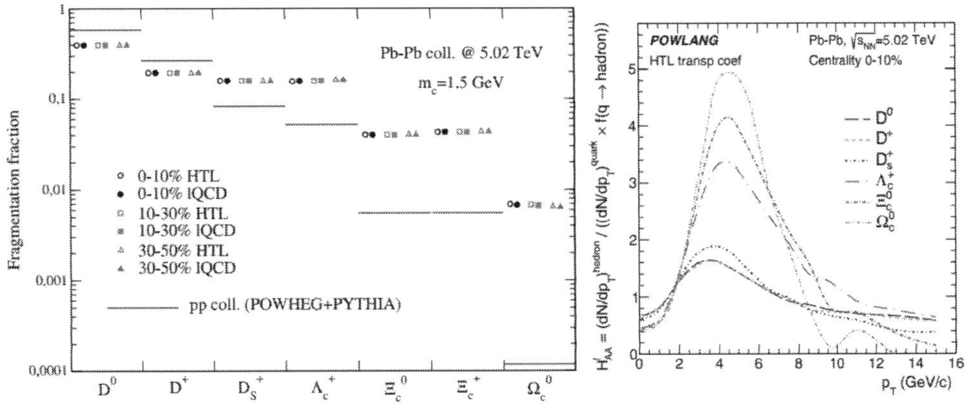

Figure 3. Left panel: charm fragmentation fractions in Pb-Pb collisions for different centrality classes and transport coefficients; also shown, in blue, the results obtained for proton-proton collisions at the same center-of-mass energy obtained with the POWHEG-BOX package [10]. Right panel: hadron-to-quark ratio of the normalized p_T-distributions for the different species of charmed-hadrons.

we display in the left panel of Fig. 3, comparing them to the predictions for proton-proton collisions provided by a pQCD event generator [10]. Our results in Pb-Pb collisions are pretty independent from the centrality class and from the heavy-quark transport coefficients. The most relevant prediction is the huge enhancement of the production of all charmed baryons, in agreement with recent experimental observations. In the right panel of Fig. 3 we show how hadronization, beside the yields, affects the kinematic distributions. To this purpose we plot the ratio of the p_T spectra of the different charmed hadrons (normalized to the corresponding fragmentation fractions) relative to the one of the parent quarks. As one can see, for all hadron species such a ratio displays a bump at intermediate values of p_T, the enhancement being particularly relevant for charmed baryons and increasing with the mass of the particle.

The above results can be interpreted as follows within our model. First of all the possibility for c quarks to recombine with a light diquark enhances the fraction of clusters carrying baryon number $B = 1$, which will decay into a charmed baryon. Secondly, SMC favors the recombination of c quarks with a thermal particle with a quite collinear momentum, parallel to the velocity of the fluid cell. Thus, the resulting clusters typically have a quite low invariant

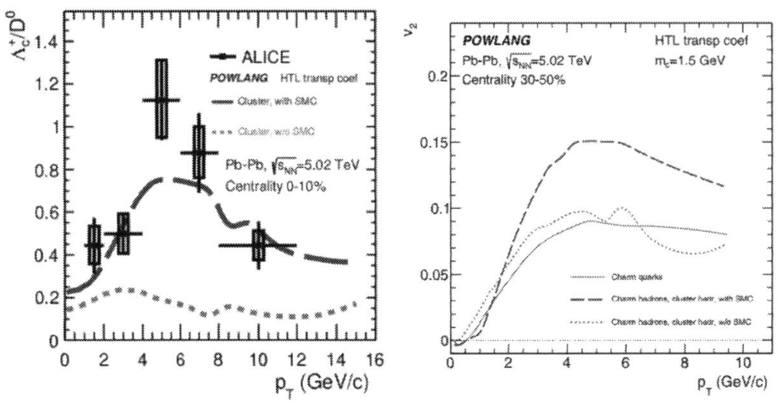

Figure 4. Charmed hadron ratios (left panel) and elliptic flow (right panel) with (dashed blue curves) and without (dotted brown curves) space-momentum correlation, the last case being obtained mixing momentum and position of heavy quarks from different events at hadronization.

mass and, furthermore, receive a boost due to the collective flow of the fireball. The effect is particularly important for the heaviest diquarks, due to the mass dependence of the radial flow. Furthermore, the light invariant mass of most clusters usually leaves room only for a two-body decay, in which the daughter charmed hadron carries a sizable momentum fraction of the ancestor. As a result, hadronization via local recombination is an efficient mechanism to transfer the collective flow of the fireball to the final charmed hadrons and this holds not only for their p_T-distributions, but also for their azimuthal anisotropies, as can be appreciated from the right panel of Fig. 4).

The importance of SMC's can be accessed by artificially breaking the connection between the position and the momentum of the hadronizing particles. This can be implemented by randomly redistributing the quarks undergoing recombination over the hadronization hypersurface. The effect on the final charmed hadrons is shown in Fig. 4: the Λ_c^+/D^0 ratio is no longer enhanced, but one finds a value around 0.1, as in e^+e^- collisions; furthermore the charmed hadron v_2 is lower, very similar to the one of the parent heavy quarks. Recombining partons are no longer collinear: this suppresses the collective flow of the clusters and increases their invariant mass, leading them to hadronize via string fragmentation, with no enhanced baryon production.

References

[1] J.B. Perrin, Nobel Lecture (1926)
[2] A. Beraudo, A. De Pace, M. Monteno, M. Nardi, F. Prino, Eur. Phys. J. C **82**, 607 (2022), 2202.08732
[3] B.R. Webber, Nucl. Phys. B **238**, 492 (1984)
[4] B. Andersson, G. Gustafson, G. Ingelman, T. Sjostrand, Phys. Rept. **97**, 31 (1983)
[5] S. Acharya et al. (ALICE) (2021), 2112.08156
[6] S. Acharya et al. (ALICE), Phys. Rev. Lett. **127**, 202301 (2021), 2011.06078
[7] S. Acharya et al. (ALICE) (2021), 2110.10006
[8] S. Acharya et al. (ALICE) (2021), 2110.09420
[9] J. Adam et al. (STAR), Phys. Rev. Lett. **124**, 172301 (2020), 1910.14628
[10] S. Alioli, P. Nason, C. Oleari, E. Re, JHEP **06**, 043 (2010), 1002.2581

Quarkonium production and elliptic flow in small systems measured with ALICE

Rita Sadek Finot[1,*], on behalf of the ALICE collaboration

[1]SUBATECH, IMT Atlantique, Nantes Université, CNRS-IN2P3, Nantes, France

Abstract. Quarkonium production in hadronic collisions provides a unique testing ground for understanding quantum chromodynamics (QCD) since it involves both perturbative and non-perturbative regimes of this theory. A variety of experimental observables, such as the production cross-section, polarization, particle correlations, serve as insights into the phenomenology of the quarkonium production and help constrain the theoretical models. Measurements of quarkonia as a function of multiplicity probe multiple parton interactions (MPI), i.e., several parton-parton interactions occurring in a single hadron-hadron collision. Finally, measurements of the azimuthal correlation structure of emitted particles in high multiplicity pp collisions can probe the existence of collective behaviour in small systems. The ALICE detector can reconstruct inclusive quarkonia over a broad kinematical range, spanning from mid-rapidity up to forward rapidity, and down to zero transverse momentum. In addition, at midrapidity, the non-prompt charmonium contribution can be separated from the prompt contribution. In these proceedings, we present new results on the inclusive, prompt and non-prompt J/ψ production cross sections measured by ALICE in pp collisions at different collision energies. The self-normalized ψ(2S)-to-J/ψ yield ratio measurement is investigated as a function of the charged particle multiplicity in pp collisions at $\sqrt{s}$ = 13 TeV. Finally, the first measurement of the J/ψ elliptic flow (v_2) in high multiplicity pp collisions at $\sqrt{s}$ = 13 TeV is presented. Results will be compared to theoretical models.

1 Introduction

The ALICE detector at the LHC is designed to study ultra-relativistic heavy-ion collisions (A–A). In order to understand the quarkonium production mechanisms in A–A, it is important to study its production in pp collisions. In particular, these measurements are considered a reference for nucleus-nucleus and proton-nucleus measurements, the former being used to probe the quark-gluon plasma formation and the latter to investigate cold nuclear matter effects. Results in pp collisions are also crucial for studying perturbative and non-perturbative quantum chromodynamics (QCD) aspects in vacuum [1]. Notably, pp collisions allow constraining theoretical models describing charmonium production, including the non-prompt contribution, originating from beauty hadron decays [1–3]. In addition, measurements of multiplicity-dependent quarkonium production shed light on multiple parton-parton interactions (MPI). MPI are proposed as a possible explanation for the observed collective behavior

*e-mail: rita.sadek@cern.ch

in high-multiplicity pp collisions [4, 5]. The ALICE detector can measure inclusive quarkonium down to zero p_T in two rapidity regions. At midrapidity ($|y| < 0.9$), quarkonium measurements are performed through their dielectron decay channel and separation of prompt and non-prompt contributions is achievable. At forward rapidity ($2.5 < y < 4$), inclusive quarkonium states are measured through their dimuon decay channel. The main detectors used at midrapidity for the measurements presented in these proceedings are the Inner Tracking System (ITS), providing tracking, primary and secondary vertexing determination. The ITS innermost layers compose the Silicon Pixel Detectors (SPD), used for charged particle multiplicity measurements. Finally, the Time Projection Chamber (TPC) provides tracking and particle identification. The measurements at forward rapidity are performed with the muon spectrometer, which provides reconstruction and identification of muon tracks. A description of the ALICE apparatus can be found in [6].

2 Results

The p_T-differential cross section measurements for the inclusive J/ψ are presented in Fig. 1 (top left panel) at forward rapidity and for various collision energies ranging from $\sqrt{s} = 5.02$ to 13 TeV [7]. An increase of the cross section is observed with increasing collision energy. Data are compared with calculations using NRQCD [8] to describe the prompt J/ψ production summed with FONLL [3] calculation for the non-prompt J/ψ contribution. The measured J/ψ cross section is in good agreement with the NRQCD + FONLL models within uncertainties, for all colliding energies. In addition, a hardening of the inclusive J/ψ p_T-differential spectra at $\sqrt{s} = 13$ TeV is noticed with respect to lower energies. This hardening can arise from the increase of the prompt J/ψ mean p_T with increasing energy, and from the increase of the non-prompt J/ψ contribution at high p_T (as predicted by FONLL). In Fig. 1 (bottom left panel), the cross sections ratios for different collision energies to the results at $\sqrt{s} = 13$ TeV are presented. In these ratios, partial cancellation of theoretical and experimental uncertainties occur, leading to stronger constraints on models when compared to data. An agreement within uncertainties between data and the models for the 8-to-13 TeV and 5.0-to-13 TeV ratios is observed. The 7-to-13 TeV ratio is slightly overestimated by NRQCD+FONLL model, the latter expecting values closer to the 8-to-13 TeV ratio.

The self-normalized ψ(2S)-to-J/ψ yield ratio at forward rapidity is presented as a function of the self-normalized charged particle multiplicity at midrapidity in Fig. 1 (right panel) [9]. The self-normalized ratio is compatible with unity, indicating that the multiplicity dependence of the charmonium production at forward rapidity is similar for the J/ψ and ψ(2S) states. The data is compared with PYTHIA8 [10], which considers MPI for heavy-flavour productions, with and without colour reconnections. A comparison is also made with the comovers model [11]. The latter considers a quarkonia dissociation effect, particularly affecting the less bound ψ(2S) excited state, due to its interactions with the final state comoving particles. The PYTHIA8 model describes data fairly considering the uncertainties. However, there is tension at low multiplicity, where the ratio is underestimated. As for comovers calculations, an agreement between data and the model can be observed within the large experimental uncertainties. The model suggests a decreasing ratio with the increasing multiplicity, which is not strongly supported by the current data.

The p_T-differential cross section for prompt and non-prompt J/ψ, measured at midrapidity in pp collisions at $\sqrt{s} = 13$ TeV are presented in Fig. 2 [12]. The prompt J/ψ p_T-differential cross section (Fig. 2 left) is compared with ICEM [2] and several NRQCD based models (see corresonding Refs. in [12]). Compatibility within uncertainties is noticed between data and models, except for the NRQCD Lipatov calculations [13], which slightly overestimate data at

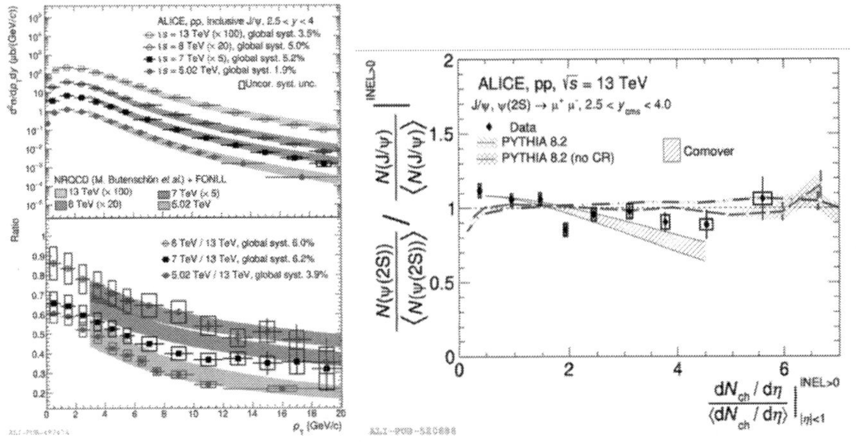

Figure 1. (left) The inclusive J/ψ p_T-differential cross section (top), and the ratio of the 5.02, 7, and 8 TeV cross sections to the 13 TeV one (bottom), at forward rapidity, in pp collisions [7]. (right) The self-normalised ψ(2S)-to-J/ψ yield ratio measured at forward rapidity as a function of the charged particle multiplicity measured at midrapidity, in pp collisions at $\sqrt{s}$ = 13 TeV [9].

low p_T. An agreement is observed between the non-prompt J/ψ p_T-differential cross section at $\sqrt{s}$ = 13 TeV and the FONLL model (Fig. 2 right).

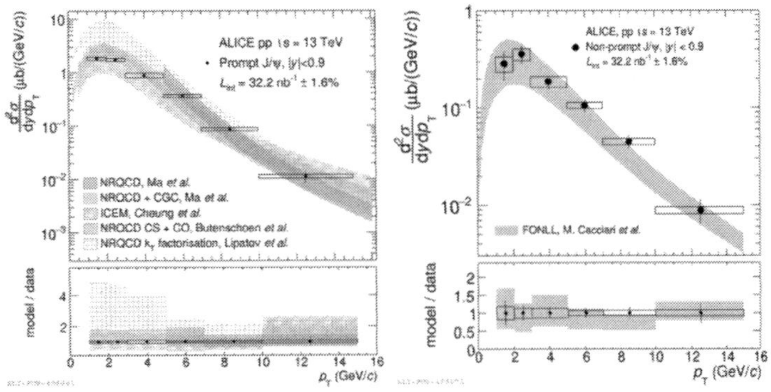

Figure 2. The prompt (left) and non-prompt (right) J/ψ cross section as a function of p_T, in pp collisions at $\sqrt{s}$ = 13 TeV and midrapidity [12].

Finally, to investigate eventual collective effects on quarkonium production in pp collisions, the J/ψ v_2 coefficient is extracted in high multiplicity pp collisions at $\sqrt{s}$ = 13 TeV, using the two-particle correlation technique. The resulting v_2 is presented in Fig. 3 down to p_T = 0 GeV/c, and is compared to similar results in p–Pb [14], and Pb–Pb [15] collisions at forward rapidity. In p–Pb and Pb–Pb, the inclusive J/ψ exhibits a positive v_2, which is interpreted as a sign of collectivity. As a first comparison, the following ordering of the J/ψ v_2 across systems is observed: $v_{2,J/\psi}^{pp} < v_{2,J/\psi}^{pPb} < v_{2,J/\psi}^{PbPb}$. No significant deviation from zero of the J/ψ v_2 is noted in pp collisions. The same observation can be made for the p_T-integrated J/ψ v_2, as its deviation from zero is within about 1 σ. Thus, there is no indication of collective behavior for the J/ψ within uncertainties, in pp collisions.

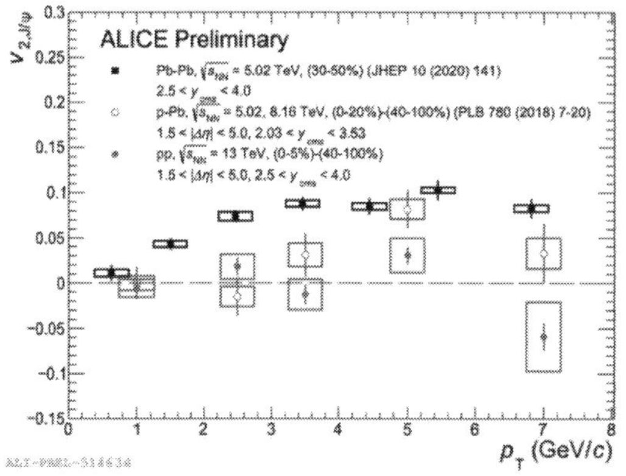

Figure 3. The inclusive J/ψ v_2 measured at forward rapidity in pp, p–Pb [14] and Pb–Pb [15] collisions.

3 Summary

The inclusive J/ψ p_T-differential cross section is presented in pp collisions at $\sqrt{s}$ = 5.02, 7, 8, and 13 TeV and at forward rapidity. NRQCD + FONLL describe well the data within uncertainties. Measurement of multiplicity-dependent self-normalized ψ(2S)-to-J/ψ yield ratio at $\sqrt{s}$ = 13 TeV is also shown. This is found to be compatible with unity within uncertainties, pointing that the multiplicity dependence of the charmonium production is similar for J/ψ and ψ(2S) states. Data is well described within uncertainties by PYTHIA8 and comovers, with some tensions observed notably at very low or high multiplicity, depending on the model. The p_T-differential cross section measurements at midrapidity are also presented separately for prompt and non-prompt J/ψ in pp collisions at $\sqrt{s}$ = 13 TeV. A good agreement within uncertainties is observed between NRQCD and prompt J/ψ data, as well as FONLL and non-prompt J/ψ data. Finally, the inclusive J/ψ elliptic flow at forward rapidity is presented in high multiplicity pp collisions. No significant collective behavior can be observed.

References

[1] Bodwin et al., Phys. Rev. D **51**, 1125–1171 (1995)
[2] Vincent Cheung and Ramona Vogt, Phys. Rev. D **98**, 11 (2018)
[3] Matteo Cacciari and Mario Greco and Paolo Nason, JHEP **1998**, 007–007 (1998)
[4] ALICE collaboration, Phys. Lett. B **719**, 29–41 (2013)
[5] CMS collaboration, JHEP **2010**, 9 (2010)
[6] ALICE collaboration, JINST **3**, S08002 (2008)
[7] ALICE Collaboration, arXiv:2109.15240
[8] Butenschoen, Mathias and Kniehl, Bernd A., arXiv:2207.09366
[9] ALICE Collaboration, arXiv:2204.10253
[10] T. Sjostrand et al., Comput. Phys. Commun. **191**, 159–177 (2015)
[11] E. G. Ferreiro and C. Pajares, Phys. Rev. C **86**, 034903 (2012)
[12] ALICE Collaboration, JHEP **03**, (2022) 190
[13] S. Baranov and A. Lipatov, Phys. Rev. D **100**, 11 (2019)
[14] ALICE Collaboration, JHEP **10**, 141 (2020)
[15] ALICE Collaboration, Phys. Lett. B **780**, 7–20 (2018)

EPJ Web of Conferences **276**, 02011 (2023) https://doi.org/10.1051/epjconf/202327602011
SQM 2022

Measurement of nonprompt and prompt D^0 azimuthal anisotropy in Pb-Pb collisions at $\sqrt{s_{\mathrm{NN}}} = 5.02$ TeV

Milan Stojanovic[1,*] *for the CMS Collaboration*

[1]Purdue University, USA

Abstract. Heavy quarks are primarily produced via initial hard scatterings, and thus carry information about the early stages of the Quark-Gluon Plasma (QGP). Measurements of the azimuthal anisotropy of the final-state heavy flavor hadrons provide information about the initial collision geometry, its fluctuation, and more importantly, the mass dependence of energy loss in QGP. Due to the larger bottom quark mass as compared to the charm quark mass, separate measurements of charm and bottom hadron azimuthal anisotropy can shed new light on understanding the dependence of the heavy quark and medium interaction. Because of the high branching ratio and large D^0 mass, measurements of D^0 meson coming from bottom hadron decay (nonprompt D^0) can cover a broad kinematic range and be a good proxy of the parent bottom hadrons results. In this talk, we report on the prompt D^0 and the first nonprompt D^0 measurements of the azimuthal anisotropy elliptic (v_2) and triangular (v_3) coefficients of nonprompt D^0 in PbPb collisions at $\sqrt{s_{\mathrm{NN}}} = 5.02$ TeV. The measurements are performed as functions of transverse momentum p_{T}, in three centrality classes, from central to midcentral collisions. Compared to the prompt D^0 results, the nonprompt D^0 v_2 flow coefficients are systematically lower but have a similar dependence on p_{T} and centrality. A non-zero v_3 coefficient of the nonprompt D^0 is observed. The obtained results are compared with theoretical predictions. The comparison could provide new constraints on the theoretical description of the interaction between heavy quarks and the medium.

1 Introduction

Azimuthal angle distribution of particles emitted in heavy ion collisions can be described via Fourier decomposition, where Fourier harmonics, v_n, represent the magnitude of the n-th order azimuthal anisotropy. Thermally produced particle anisotropy can be successfully described with relativistic hydrodynamics [1]. Unlike light quarks, heavy quarks are dominantly produced in the initial hard scattering [2], and their azimuthal anisotropy is expected to arise from the interaction with the medium. Indeed, significant positive charm (c) quark values of v_2 and v_3 have been measured and remarkable similarity is observed in transverse momentum (p_{T}) and centrality dependence compared to light hadrons [3]. The magnitude of flow coefficients, however, is smaller in the case of c quarks, particularly in the low- and intermediate-p_{T} range ($p_{\mathrm{T}} < 15$ GeV/c). The bottom (b) quark mass is even larger than the c quark mass, hence b hadron azimuthal anisotropy measurement is the next step in understanding the dependence of the heavy quark and medium interaction.

*e-mail: mstojano@purdue.edu

In this proceedings we report the first measurement of the second, v_2, and third, v_3, azimuthal anisotropy coefficients of D^0 mesons that come from b hadron decays, in lead-lead (PbPb) collisions. Results are shown as a function of D^0 meson transverse momentum, in three centrality intervals.

2 Dataset and analysis details

This analysis uses PbPb collision data at a center-of-mass energy per nucleon pair of $\sqrt{s_{NN}}$ = 5.02 TeV collected with the CMS detector [4] in 2018 with a total integrated luminosity of 0.58 pb^{-1}. Inclusive D^0 ($\overline{D}^0$) meson candidates are reconstructed via decay channel $D^0 \rightarrow \pi^+ + K^-$ ($\overline{D}^0 \rightarrow \pi^- + K^+$). No particle identification is applied and the D^0 meson candidates are formed by combining pairs of oppositely charged detected particles in the event where both possible particle assignments are considered for each pair. The background is partially suppressed (both combinatorial background and prompt D^0 component) by applying a boosted decision tree (BDT) algorithm from the TMVA package [5].

Inclusive D^0 meson yield, in each bin of scalar product (v_n^i), is extracted from the fit of the invariant mass spectrum [6]. Nonprompt fraction is further obtained by fitting the distributions of the distance of closest approach (DCA) between the D^0 meson momentum vector and the collision point with a linear combination of prompt and nonprompt D^0 DCA simulated distributions. Finally, the v_n coefficients are measured by taking the mean of the nonprompt D^0 mesons scalar product distribution in each bin of p_T and centrality.

3 Results

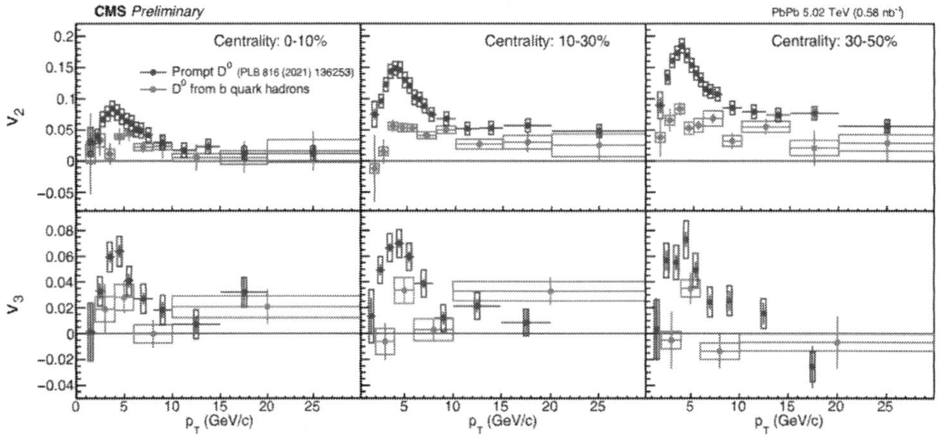

Figure 1. The elliptic, v_2 (top panel), and the triangular, v_3 (bottom panel), flow coefficients of nonprompt and prompt D^0 mesons as functions of p_T and in three bins of centrality. The bars and the boxes represent statistical and systematic uncertainties, respectively. The figure is taken from Ref. [6].

Figure 1 shows the results of v_2 and v_3 azimuthal anisotropy coefficients for nonprompt D^0 mesons, compared with previously published values for prompt D^0 from Ref. [7]. The results show non-zero values of the elliptic flow of b hadron daughters, although the flow magnitude is significantly smaller than for the prompt D^0 case. This is particularly evident in the low- and mid-p_T range (p_T < 15 GeV/c) and it is consistent with the mass ordering observed in pPb collisions [8], as well as mass ordering observed in leptonic channels in PbPb collisions [9, 10]. At larger D^0 transverse momentum, v_2 values from charm and bottom quarks start converging. Elliptic flow of nonprompt D^0 mesons exhibits very weak centrality and transverse momentum dependence, unlike flow of prompt D^0 mesons. The v_3 Fourier coefficients are strongly affected by statistical fluctuations so that neither the p_T, nor the centrality dependence can be determined. However, results indicate a non-zero value in the $4 < p_T < 6$ GeV/c range for all centralities. Similar as in the case of elliptic flow, the magnitude of v_3 is lower for nonprompt than for prompt D^0.

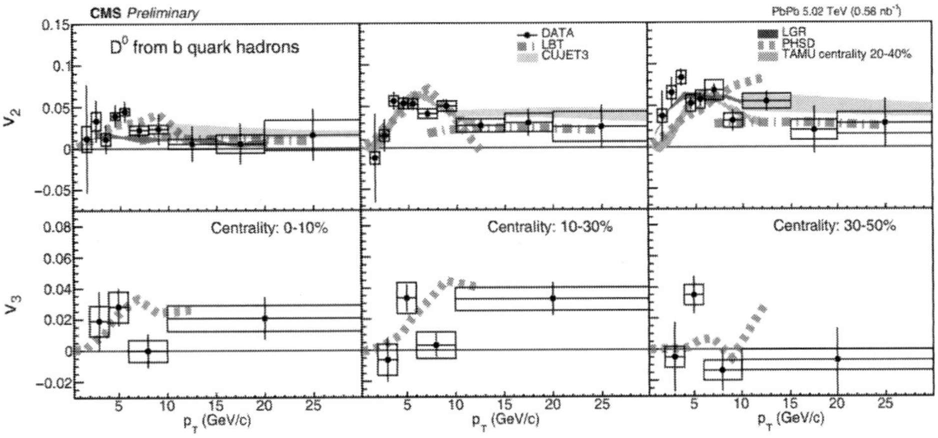

Figure 2. The elliptic, v_2 (top panel), and the triangular, v_3 (bottom panel), flow coefficients of nonprompt D^0 mesons as functions of transverse momentum and in three bins of centrality. The bars and the boxes represent statistical and systematic uncertainties, respectively. The colored bands show theoretical predictions [11–18]. The figure is taken from Ref. [6].

Figure 2 shows the measured nonprompt D^0 v_n coefficients compared with theoretical calculations that have different modeling of the b quark flow. All models show qualitative agreement with the measured p_T dependence of the elliptic flow, but there are quantitative discrepancies which could provide constraints on the description of bottom quark interactions with the medium. In the low- and mid- p_T regime, for the centrality range 30–50%, the measurements favor the LGR model [11, 12], while for the most central events (0–10%) the same model does not predict a peak structure seen in data. The PHSD model [13] provides a good description of the data, for more central collisions, 0–10% and 10–30%, but shows higher discrepancies in the 30–50% range. The TAMU model [14] provides predictions in the centrality range 20–40% only, so, direct comparison with data is not possible, but the model describe well the measurement performed in the centrality range 10–30%. At higher p_T, where anisotropy is driven by the path-length dependence of parton energy loss, the LBT [15, 16] model appears to be slightly closer to the data than CUJET3.0 [17, 18], which is systematically above the experimental results, however still within uncertainties. The PHSD is the only model that has predictions for v_3 coefficients. While uncertainties are rather large, both model and data reach similar maximal values in all centralities.

4 Summary

In summary, the second (v_2) and third (v_3) Fourier harmonics of D^0 mesons that come from b hadron decays (nonprompt D^0) are measured in PbPb collisions at $\sqrt{s_{NN}} = 5.02$ TeV. The

v_2 results exhibit a weak transverse momentum dependence and increase from central to midcentral collisions. The measured flow magnitudes are lower compared to the prompt D^0 mesons. The v_3 coefficient shows no centrality dependence, but an indication of non-zero triangular flow is found in the transverse momentum range $4 < p_T < 6$ GeV/c. The new measurements place additional constraints on theoretical understanding of heavy quark interactions with the medium.

References

[1] U. Heinz and R. Snellings, Ann. Rev. Nucl. Part. Sci **63** 123 (2013)
[2] F. Prino and R. Rapp, Journal of Physics G: Nuclear and Particle Physics **43** 093002 (2016)
[3] CMS Collaboration, Phys. Rev. Lett. **120** 202301 (2018)
[4] CMS Collaboration, JINST **3** S08004 (2008)
[5] H. Voss et al., XIth International Workshop on Advanced Computing and Analysis Techniques in Physics Research (ACAT) 40 (2007)
[6] CMS Collaboration, CMS Physics Analysis Summary, CMS-PAS-HIN-21-003 (2022), https://cds.cern.ch/record/2806157
[7] CMS Collaboration, Phys. Lett. B **816** 136253 (2021)
[8] CMS Collaboration, Phys. Lett. B **813** 136036 (2021)
[9] ATLAS Collaboration, Phys. Lett. B **807** 135595 (2020)
[10] ALICE Collaboration, Phys. Rev. Lett **126** 162001 (2021)
[11] S. Li, C. Wang, and J. Liao, Phys. Rev. C **99** 054909 (2019)
[12] S. Li and J. Liao, Eur. Phys. J. C **80** 671 (2020)
[13] T. Song et al., Phys. Rev. C **92** 014910 (2015)
[14] M. He, R. J. Fries, and R. Rapp, Phys. Lett. B **735** 445 (2014)
[15] S. Cao et al., Phys. Rev. C **94** 014909 (2016)
[16] W. Xing et al., Phys. Lett. B **805** 135424 (2020)
[17] S. Shi, J. Liao, and M. Gyulassy, Chin. Phys. C **42** 104104 (2018)
[18] S. Shi, M. Gyulassy, and J. Chin. Phys. C **43** 044101 (2019)

Quark susceptibilities, transport properties and heavy quark production in an extended Quasi-Particle Model with $N_f = 2 + 1 + 1$ flavors

Maria Lucia Sambataro[1,2,*], *Salvatore* Plumari[1,2], and *Vincenzo* Greco[1,2]

[1]Department of Physics and Astronomy 'E. Majorana', University of Catania, Via S. Sofia 64, 1-95123 Catania, Italy
[2]Laboratori Nazionali del Sud, INFN-LNS, Via S. Sofia 62, 1-95123 Catania, Italy

Abstract. The quark susceptibilities are a very useful tool to understand the nature of the degrees of freedom in the vicinity of the QCD phase transition while Heavy-Quarks (HQs) transport coefficients give us information on their thermalization time in the Quark-Gluon Plasma (QGP). Recently, new lattice results for the equation of state of QCD with 2 + 1 + 1 dynamical flavors have become available. Therefore, we extend our QPM approach for $N_f = 2 + 1$ to $N_f = 2 + 1 + 1$ where the charm quark is included. We also explore the extension of QPM approach to a more realistic model in which partonic propagators explicitly depend on the three-momentum with respect to the partonic matter at rest in order to match pQCD at high momenta following Dyson-Schwinger studies in the vacuum. In this context, we evaluate and correctly reproduce both EoS and quark susceptibilities which are understimated in the simple QPM approach. Therefore, we study the impact of the extended QPM approach on the spatial diffusion coefficient D_s of charm quark making a comparison with the results in the standard QPM approach.

1 Introduction

The description of heavy quarks diffusion in the QGP through the lower order perturbation theory for quantum chromodynamics has been extensively studied in the literature [1]. The use of pQCD was initially considered a reference point for the study of the heavy quarks dynamics since the coupling constant in the typical energy range of the formed system assumes intermediate values: perturbation analysis has allowed to describe QGP as a system of weakly interacting particles. The experimental results of the last decades have however shown that the QGP, produced in collisions between heavy ions at high energies, is a system characterized by a greater interaction than the one predicted by a perturbative treatment so that the created medium can therefore be described as a strongly many-body system influenced by non-perturbative effects even at very high temperatures ($\sim 3 - 4\ T_c$). In particular, lattice QCD calculations predict that the energy density of the system created deviates from the Stefan-Boltzmann limit of a perfect gas by about $15 - 20$ % resulting in a value of the interaction measures $< \Theta_{\mu\mu} >= \epsilon - 3P$ significantly different from zero [2]. In this scenario of course perturbative QCD treatment fails in reproducing lattice QCD (lQCD) expectations

*e-mail: sambataro@lns.infn.it

and the observables related to the dynamics of heavy quarks such as R_{AA} and v_2, are not well reproduced by the model in comparison with the experimental data. In particular, these calculations considerably underestimate the energy loss and the elliptic flow of heavy quarks and, therefore, in general, the strenght of the HQs interaction with the bulk partons especially in the low p_T region. A successful way to account the non-perturbative QCD effects at non-zero temperature is to encode the lQCD thermodynamical expectations within effective temperature dependent particle masses by prescription of a Quasi-Particle Model (QPM) [3] [4]. The origin of quark and gluon masses can be explained as the energy contribution coming out from the particle correlations within the strongly interacting QCD medium: if a large part of the interaction can be included into the effective masses, quasi-particles move freely or interact only weakly. We provide the expressions of these partonic masses according to their thermal pQCD evaluation:

$$m_g^2 = \frac{1}{6}g(T)^2\left[\left(N_c + \frac{1}{2}N_f\right)T^2 + \frac{N_c}{2\pi^2}\sum_q \mu_q^2\right],$$
$$m_{u,d}^2 = \frac{N_c^2-1}{8N_c}g(T)^2\left[T^2 + \frac{\mu_{u,d}^2}{\pi^2}\right] \tag{1}$$

where N_c is the number of colors and N_f is the number of flavors. The presence of temperature dependent masses give rises an extra term in the energy density of the system which does not have the ideal gas form. Therefore, in order to ensure the thermodynamic consistency, pressure and energy density contain additional medium temperature dependent contribution, which we will call $B(T)$: this bag contribution takes into account further non-perturbative effects and preserve the entropy expression of an ideal gas. In this approach, the strength coupling g is a general T-dependent function whose parametrized form is obtained by fitting the energy density ϵ of lQCD thermodynamics. For more details see Ref.[3].

2 The extended Quasi-Particle Model

Recently, new lattice results for the equation of state of QCD with $2+1+1$ dynamical flavors have become available [5]. Therefore, we extend our quasiparticle model(QPM) approach for $N_f = 2+1$ to $N_f = 2+1+1$ where the charm quark is included. We also explore the extension of QPM approach to a more realistic model (in the following QPM^*) in which partonic propagators explicitly depend on the three-momentum with respect to the partonic matter at rest in order to match pQCD at high momenta. In fact, the masses in the QPM increase with the temperature due to interactions with the medium and this is a good approximation if the particle has a low momentum but, if the momentum is large compared to the temperature, perturbation theory becomes accessible and the partons should behave like almost massless particles. This is not taken into account in the standard QPM approach. Following the Ref. [6], the essentially new elements in this extended approach with respect to the standard QPM are the momentum-dependent factor in the masses motivated by Dyson-Schwinger studies in the vacuum [7] which yields the limit of pQCD for $p \to \infty$.

The functional forms for the partons masses at finite temperature T, quark chemical potential μ_q, and three-momentum squared p^2 are assumed to be given by:

$$
\begin{aligned}
M_g(T, \mu_q, p) = & \left(\frac{3}{2}\right)\left(\frac{g^2(T^*/T_c)}{6}\left[(N_c + \tfrac{1}{2}N_f)T^2 + \frac{N_c}{2}\sum_q \frac{\mu_q^2}{\pi^2}\right]\left[\frac{1}{1+\Lambda_g(T_c/T^*)p^2}\right]\right)^{1/2} + m_{\chi g}, \\
M_{q,\bar{q}}(T, \mu_q, p) = & \left(\frac{N_c^2-1}{8N_c} g^2(T^*/T_c)\left[T^2 + \frac{\mu_q^2}{\pi^2}\right]\left[\frac{1}{1+\Lambda_q(T_c/T^*)p^2}\right]\right)^{1/2} + m_{\chi q} \qquad (2)
\end{aligned}
$$

where $T^{*2} = T^2 + \frac{\mu_q^2}{\pi^2}$ is the effective temperature used to extend the QPM to finite μ_q, $\Lambda_g(T_c/T^*) = 5(T_c/T^*)^2 GeV^{-2}$ and $\Lambda_q(T_c/T^*) = 12(T_c/T^*)^2 GeV^{-2}$. Here $m_{\chi g} \approx 0.5\ GeV$ is the gluon condensate and $m_{\chi q}$ is the light quark chiral mass ($m_{\chi q} = 0.003\ GeV$ for up and down quark and $m_{\chi q} = 0.06\ GeV$ for s quark. In Eq. 2, $m_{\chi g}$, ($m_{\chi q}$) gives the finite gluon (light quark) mass in the limit $p \to 0$ and $T = 0$ or for $p \to \infty$. In our QPM^*calculations, likely in standard QPM approach, we consider $\mu_q = 0$. In the following we discuss the main results on the impact of the extended model on the EoS and charm quark coefficients.

3 Results

In the this section we present our reproduction of EoS and quark susceptibility but also a discussion on the spatial diffusion coefficient D_s making a comparison between the standard QPM with 2 + 1 flavours, the standard QPM with 2 + 1 + 1 flavours including charm and the extended one with momentum-dependent propagators. In the left panel of Fig.1 we show the lattice data of Wuppertal-Budapest collaboration [5] for the pressure and trace anomaly as functions of the temperature, together with the quasi-particle model curves. Meanwhile, in the right panel the susceptibility $\chi_s = \frac{T}{V}\frac{\partial^2 lnZ}{\partial \mu_s^2}$ is shown as function of temperature T both for standard QPM and extended QPM^*.

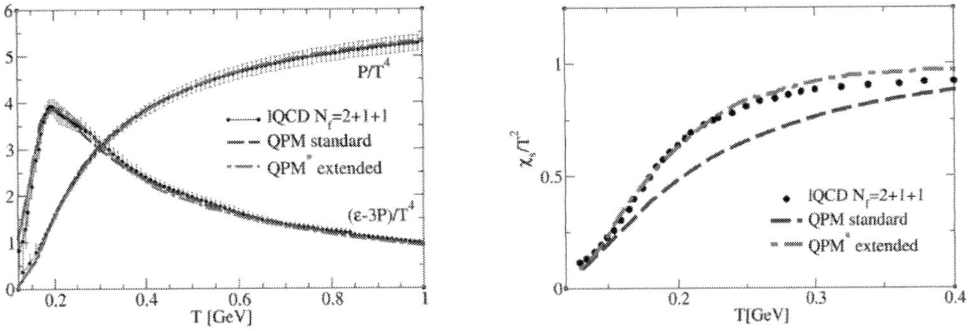

Figure 1. Pressure P/T^4 and interaction measure $(\epsilon - 3p)/T^4$ (left panel) and susceptibility $\chi_s = \frac{T}{V}\frac{\partial^2 lnZ}{\partial \mu_s^2}$ (right panel) for standard QPM and extended QPM^* with $N_f = 2 + 1 + 1$. The lattice data of the Wuppertal-Budapest collaboration are taken from Ref. [5].

Differently from the standard QPM approach, in which the quark susceptibility is underestimated [3], in the QPM^* approach in which the parton masses have a momentum dependence, we are able to reproduce correctly both the lQCD quark susceptibility and the EoS. We finally discuss the impact of the extension of our QPM approach on the spatial diffusion coefficient D_s of charm quark. In fig. 2 we show the D_s of charm quark for standard QPM and the extended one both for $N_f = 2 + 1$ and $N_f = 2 + 1 + 1$. The D_s for the extended QPM

 https://doi.org/10.1051/epjconf/202327602012

is modified when $T/T_c \leq 2$ due to the enhancement of the $g(T)$ at low temperatures which corresponds to a decrease of the D_s coefficient of about 35%. This means that QPM describes a strong non-perturbative behaviour near to T_c similar to the one achieved in strongly coupled theory as AdS/CFT and in the high T region the D_s reaches the natural pQCD limit quickly than the standard QPM.

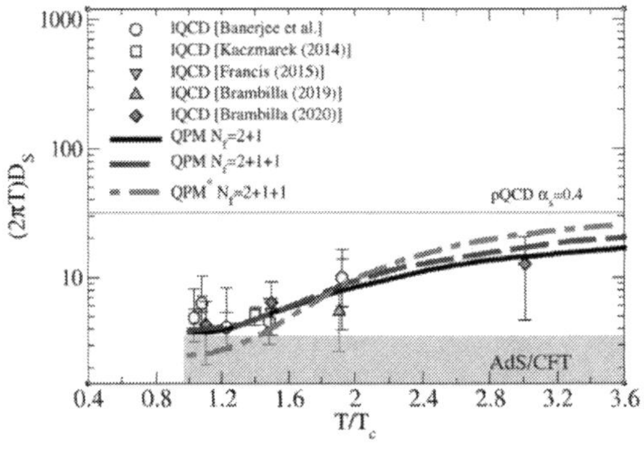

Figure 2. Spatial diffusion coefficient $D_s(T)$ of charm quark for standard QPM with $N_f = 2+1$ flavours (black solid line), standard QPM with $N_f = 2+1+1$ flavours (blue dashed line) and extended QPM^* with $N_f = 2+1+1$ (red dot-dashed line).

4 Conclusions

In this study we have upgraded our QPM approach with the recent lQCD calculations which include the charm quark. Therefore, we have also studied the extension of our QPM approach to a more realistic model in which partonic masses explicitly depend on the three-momentum of plasma particle in order to reach the pQCD limit in the high momentum regime. In this extended approach we reproduce correctly both the EoS and quark susceptibility which are understimated in the standard QPM approach. Since this extended model better describes the most recent lattice QCD data and in particular the susceptibilities, it will important to study if it leads to a better description of these dynamical observables lowering the global χ^2 coming from the comparison to the experimental data. An explicit treatment of the moment dependence of quasi-particle masses and coupling has been often postponed, but it significantly affects the p dependence of R_{AA}, v_2, v_3. Given the new upgrade of ALICE and CMS will allow to access low p_T obsevarbles with high precision, the corresponding development of the QPM approach employed for realistic simulation is a necessary step toward a more solid and precise determination of D_s.

References

[1] Roger Cutler and Dennis Sivers, Phys. Rev. D 17, 196 (1978).
[2] Claudia Ratti, Rept. Prog. Phys. 81, 084301 (2018).
[3] S. Plumari,W. M. Alberico, V. Greco and C. Ratti, Phys. Rev. D 84, 094004 (2011).
[4] M.L. Sambataro, S. Plumari and V. Greco, Eur.Phys.J.C 80 (2020) 12, 1140.
[5] Sz. Borsanyi et al., Nature 539 (2016) 7627, 69-71.
[6] Hamza Berrehrah, Wolfgang Cassing, Elena Bratkovskaya, and Thorsten Steinert, Phys. Rev. C 93, 044914 (2016).
[7] Christian S. Fischer, J. Phys. G 32, R253 (2006).

Coalescence plus fragmentation approach for the hadronization mechanism of heavy hadrons from AA to pp collisions

Vincenzo Minissale[1,*], *Salvatore* Plumari[1,2], and *Vincenzo* Greco[1,2]

[1]INFN - Laboratori Nazionali del Sud , via S.Sofia 62, 95123 Catania (Italy),
[2]Dipartimento di Fisica e Astronomia "Ettore Majorana" Università di Catania Via S. Sofia, 64 95123 Catania (Italy)

Abstract. One of the present challenge for the theoretical understanding of heavy-quark hadronization is represented by the description of the measurements of heavy hadron production in pp, pA and AA collisions from RHIC to top LHC energies. The Λ_c/D^0 ratio observed in AA collisions has a value of the order of the unity, and experimental measurements in pp collisions at both $\sqrt{s} = 5.02\,\text{TeV}$ and $\sqrt{s} = 13\,\text{TeV}$ have shown ratios for charm baryons Λ_c, Ξ_c^0 and Ω_c^0 respect to D^0 meson larger than that measured and expected in e^+e^-, ep collisions. We present an hadronization mechanism based on the coalescence and fragmentation processes, and the results obtained in AA collisions for D^0 and Λ_c related baryon to meson ratios at RHIC and LHC. We present moreover results obtained for the charmed hadron production in pp collisions at LHC energies assuming the formation of an hot QCD matter at finite temperature.We calculate the heavy baryon/meson ratio and the p_T spectra of charmed hadrons D^0, Λ_c^+ and the recently measured Ξ_c baryon, finding an enhancement in comparison with the ratio observed for e^+e^-, ep collisions; with this approach we also predict a significant production of Ω_c.

1 Introduction

A new state of matter composed of deconfined quark and gluons that strongly interact, called Quark-Gluon Plasma (QGP), has been studied, over the years, in the ultra-relativistic heavy ion collision at Large Hadron Collider (LHC) and at Relativistic Heavy-Ion Collider (RHIC). The bulk properties of this state of matter is characterized by the characteristics of light-quark and gluons. Instead, because of their large mass, the heavy quarks, i.e. charm and bottom, can be used as probes of the QGP properties Recently, the relation between the heavy-quark dynamics in the QGP and the bulk properties has been studied with a lot of theoretical efforts [1–10]. The experimental data have shown, for heavy ion collisions in the intermediate region of momenta, a $\Lambda_c/D^0 \sim 0.8 \div 1.5$ at RHIC and $\Lambda_c/D^0 \sim 0.2 \div 1.2$ at LHC [11–13], with an enhancement of the baryon/meson ratio in the heavy flavor sector similar to the one observed for light and strange hadrons, and larger than the one expected from the simple fragmentation in collision systems as e^+e^-, $e^\pm p$ [14]. We study the heavy hadrons production using for the hadronization process a coalescence plus fragmentation model. [15–17]. The

*e-mail: vincenzo.minissale@lns.infn.it

different heavy hadron species produced manifests into a baryon-to-meson enhancement for charmed hadrons. [18–21] Measurements of heavy baryon production in pp collisions, at LHC energies, have recently attracted interest [22]. It represents, at the moment, a challenge for the heavy-quark hadronization theoretical understanding. In fact, at the energy of $\sqrt{s} = 5TeV$ and $13TeV$, the Λ_c/D^0 ratio has been measured with a value of about 0.6 in the low p_T region that is larger than the one evaluated in [14]. Furthermore the measured production of other single-charmed baryons with content of strangeness, i.e. Ξ_c and Ω_c, and their ratios to D^0 and Λ_c show an unexpected behavior in pp collision [23].

2 Coalescence plus Fragmentation hadronization model

The hadron production that comes from coalescence, in the model that we use, is based on the Wigner formalism, and is evaluated with the following coalescence integral:

$$\frac{d^2N_H}{dP_T^2} = g_H \int \prod_{i=1}^{n} \frac{d^3p_i}{(2\pi)^3E_i} p_i \cdot d\sigma_i \; f_{q_i}(x_i, p_i) f_H(x_1...x_n, p_1...p_n)\, \delta^{(2)}\left(P_T - \sum_{i=1}^{n} p_{T,i}\right) \quad (1)$$

where $d\sigma_i$ denotes an element of a space-like hypersurface, g_H is the statistical factor to form a colorless hadron while f_{q_i} are the quark (anti-quark) phase-space distribution functions for i-th quark (anti-quark). $f_H(x_1...x_n, p_1...p_n)$ is the Wigner function and describes the distribution of quarks, in space and momentum, within an hadron. In the case of charm hadrons we use a Wigner function that is a Gaussian distribution in relative space and momentum, $f_H(x_1...x_{N_q}; p_1...p_{N_q}) = \prod_{i=1}^{N_q-1} A_W \exp(-\frac{x_{r_i}^2}{\sigma_{r_i}^2} - p_{r_i}^2\sigma_{r_i}^2)$ where x_{r1} and p_{r1} are the 4-vectors for the relative coordinates of the N_q quarks. The covariant width parameter σ_r can be related to the oscillator frequency ω by $\sigma = 1/\sqrt{\mu\omega}$ where $\mu = (m_1m_2)/(m_1 + m_2)$ is the reduced mass, and is related to the root mean square charge radius of the meson. For D^+ meson $\langle r^2\rangle_{ch} = 0.184\,fm^2$ corresponds to a $\sigma_p = \sigma_r^{-1} = 0.283\,\mathrm{GeV}$; for Λ_c^+ the mean square charge radius is $\langle r^2\rangle_{ch} = 0.15\,fm^2$ with the related widths $\sigma_{p_1} = \sigma_{r_1}^{-1} = 0.251\,\mathrm{GeV}$ and $\sigma_{p_2} = \sigma_{r_2}^{-1} = 0.424\,\mathrm{GeV}$. We assign a fragmentation probability as $P_{frag}(p_T) = 1 - P_{coal}(p_T)$, after the evaluation of the coalescence probability P_{coal} for each charm quark. Based on the constraint of confinement and with the physical motivation that a quark with zero momentum cannot fragment into an hadron at lower momentum; we determine the normalization constant A_W in order to have, in the limit $p \to 0$, that all the charm quarks hadronize via coalescence. Then we can evaluate the final hadron fragmentation momentum spectra, coming from the charm quarks; it is given by the convolution of the parton spectrum that do not undergo to coalescence with the fragmentation function $D_{had}(z, Q^2) \propto 1/\left[z\left[1 - \frac{1}{z} - \frac{\epsilon_c}{1-z}\right]^2\right]$. In this case we employ the Peterson fragmentation function [24], where ϵ_c is a parameter determined in order to well reproduce, with the fragmentation, the high momenta tails of the D and Λ_c production in $p + p$ collisions. [4, 21]. The relative fragmentation fractions of charm quarks into different hadron channels are used according to the fractions evaluated in [14]. In our model we use a fireball formed by a thermalized system of gluons and u, d, s quarks and anti-quarks at a temperature of $T_C = 165\,\mathrm{MeV}$, that form the bulk of particles. The fireball is considered at $\tau = 8.\,fm/c$ for LHC Pb+Pb collisions at $\sqrt{s_{NN}} = 5.02\,TeV$, and $\tau = 4.5\,fm/c$ for RHIC Au+Au collisions at $\sqrt{s_{NN}} = 200$ GeV. In $p + p$ collisions at LHC at $\sqrt{s_{NN}} = 5.02\,TeV$ we fix the parameter according to hydro-dynamical simulations [25] with $\tau = 2.5\,\mathrm{fm}/c$. The collective flow is considered assuming a radial flow profile as $\beta_T(r_T) = \beta_{max}\frac{r_T}{R}$, where R is the transverse radius of the fireball and β_{max} is the radial flow value on the external surface of the fireball. For partons at low transverse momentum, $p_T < 2\,\mathrm{GeV}$, hence we consider a thermal distribution, instead for $p_T > 2.5\,\mathrm{GeV}$, we consider the minijets that have undergone the jet

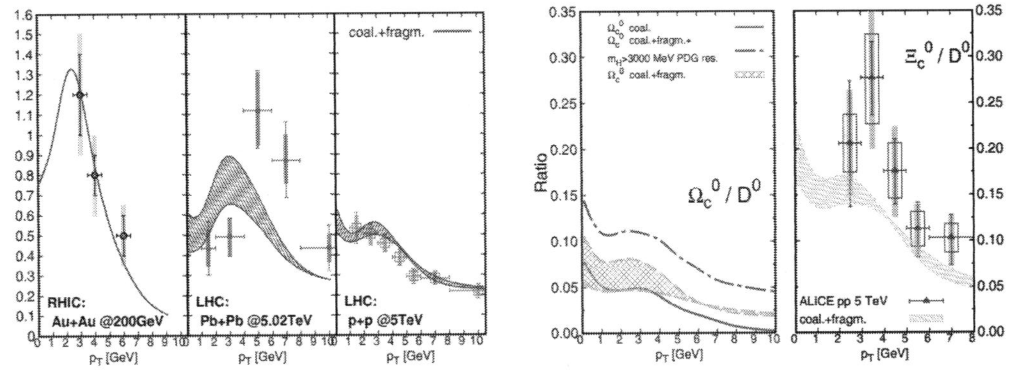

Figure 1: Λ_c^+ to D^0 ratio as a function of p_T and at mid-rapidity for (left panel) $Au + Au$ collisions at $\sqrt{s} = 200\,\mathrm{GeV}$ [11] for (middle panel) $Pb + Pb$ collisions at $\sqrt{s} = 5.02\,\mathrm{TeV}$ [13] and for (right panel) pp collisions at $\sqrt{s} = 5\,\mathrm{TeV}$ [22].
(left) Ratios at mid-rapidity for pp collisions at $\sqrt{s} = 5.02\,\mathrm{TeV}$. Ω_c/D^0 ratio considering Ω_c with: only coalescence (purple solid line), coalescence plus fragmentation (pink band), resonances (black dot-dashed lines). D^0 coming from coalescence plus fragmentation. (right) $\Xi_c^{0,+}/D^0$ with both coalescence and fragmentation, data from [23]

quenching mechanism. For heavy quarks, in heavy ion collisions, the distribution is obtained by solving the relativistic Boltzmann equation [4] giving a good description of R_{AA} and v_2 for D mesons. For $p + p$ collisions, the charm quark spectrum have been taken according to Fixed Order + Next-to-Leading Log (FONLL) distribution [26]

3 Results

Our model with coalescence plus fragmentation predicts a rise and fall of the baryon/meson ratio. The left and middle panel in Fig.1 show the comparison between RHIC and LHC for the Λ_c^+/D^0 ratio. The production ratio between coalescence and fragmentation is smaller at LHC than at RHIC. At LHC the larger contribution in particle production from fragmentation leads to a final ratio that is smaller than at RHIC [19]. Experimental data of the Λ_c^+ to D^0 ratio have been released, in the past years, for pp collisions at LHC [22, 29], and show an unexpected excess of production of Λ_c with respect to the simple fragmentation, with values of the ratio of ~ 0.6 in the region at low momenta.We have applied our model in the case of pp collisions at mid-rapidity at $\sqrt{s} = 5.02\,TeV$, assuming the formation of a QGP medium like the one simulated in hydrodynamics calculations [25] and we obtain a good description of the Λ_c/D^0 ratio, see Fig.1 (right panel), with the band determined including the uncertainties given by a 20% variation of the Wigner function widths. Our calculations describe the disappearance of the peak, and an enhancement that is significantly different from the ratio obtained with the only fragmentation [21]. In Fig. **??**, we show the Ξ_c and Ω_c to D^0 ratios in pp collisions at $\sqrt{s} = 5.02\,TeV$. In the left panel we show the Ω_c^0/D^0 ratio having considered, for the D^0 meson, the contribution that comes from both coalescence and fragmentation; while, for the Ω_c^0, we show three cases: only coalescence contribution (purple line), coalescence plus fragmentation (pink band) and including some resonance states in addition to the ground state (i.e. $\Omega_c(3000)^0$, $\Omega_c(3005)^0$, $\Omega_c(3065)^0$, $\Omega_c(3090)^0$, $\Omega_c(3120)^0$.) (black dash-dotted line). In the right panel, we show the Ξ_c/D^0 ratio (green band) in comparison with recent data from ALICE collaboration [23], where we have included the contribution from the main existing Ξ_c resonance states.

Acknowledgments

S.P. acknowledges the funding from UniCT under 'Linea di intervento 3' (HQsmall Grant).

References

[1] He M, Fries R J and Rapp R 2013 *Phys. Rev. Lett.* **110** 112301
[2] Uphoff J, Fochler O, Xu Z and Greiner C 2012 *Phys. Lett.* **B717** 430–435
[3] Cao S, Qin G Y and Bass S A 2015 *Phys. Rev.* **C92** 024907
[4] Scardina F, Das S K, Minissale V, Plumari S and Greco V 2017 *Phys. Rev.* **C96** 044905
[5] S. K. Das, M. Ruggieri, F. Scardina, S. Plumari and V. Greco, J. Phys. G **44** (2017) no.9, 095102
[6] Das S K, Plumari S, Chatterjee S, Alam J, Scardina F and Greco V 2017 *Phys. Lett.* **B768** 260–264
[7] Das S K, Scardina F, Plumari S and Greco V 2015 *Phys. Lett.* **B747** 260–264
[8] Das S K, Torres-Rincon J M, Tolos L, Minissale V, Scardina F and Greco V 2016 *Phys. Rev.* **D94** 114039
[9] S. Plumari, G. Coci, V. Minissale, S. K. Das, Y. Sun and V. Greco, Phys. Lett. B **805** (2020), 135460
[10] M. L. Sambataro, Y. Sun, V. Minissale, S. Plumari and V. Greco, [arXiv:2206.03160 [hep-ph]].
[11] J. Adam *et al.* [STAR], Phys. Rev. Lett. **124** (2020) no.17, 172301
[12] S. Acharya *et al.* [ALICE], Phys. Lett. B **793** (2019), 212-223
[13] S. Acharya *et al.* [ALICE], [arXiv:2112.08156 [nucl-ex]].
[14] Lisovyi M, Verbytskyi A and Zenaiev O 2016 *Eur. Phys. J.* **C76** 397
[15] Fries R J, Muller B, Nonaka C and Bass S A 2003 *Phys. Rev. Lett.* **90** 202303
[16] Greco V, Ko C and Levai P 2003 *Phys.Rev.* **C68** 034904
[17] Minissale V, Scardina F and Greco V 2015 *Phys. Rev.* **C92** 054904
[18] Oh Y, Ko C M, Lee S H and Yasui S 2009 *Phys. Rev.* **C79** 044905
[19] Plumari S, Minissale V, Das S K, Coci G and Greco V 2018 *Eur. Phys. J.* **C78** 348
[20] S. Cho, K. J. Sun, C. M. Ko, S. H. Lee and Y. Oh, Phys. Rev. C **101** (2020) no.2, 024909
[21] V. Minissale, S. Plumari and V. Greco, Phys. Lett. B **821** (2021), 136622
[22] S. Acharya *et al.* [ALICE], [arXiv:2011.06079 [nucl-ex]].
[23] S. Acharya *et al.* [ALICE], [arXiv:2105.05616 [nucl-ex]].
[24] Peterson C, Schlatter D, Schmitt I and Zerwas P M 1983 *Phys. Rev.* **D27** 105
[25] R. D. Weller and P. Romatschke, Phys. Lett. B **774** (2017), 351-356
[26] M. Cacciari, S. Frixione, N. Houdeau, M. L. Mangano, P. Nason and G. Ridolfi, JHEP **10** (2012), 137
[27] Adamczyk L *et al.* (STAR) 2014 *Phys. Rev. Lett.* **113** 142301 [Erratum: Phys. Rev. Lett.121,no.22,229901(2018)]
[28] Abelev B *et al.* (ALICE) 2012 *JHEP* **09** 112
[29] A. M. Sirunyan *et al.* [CMS], Phys. Lett. B **803** (2020), 135328

Heavy-flavor jets in heavy-ion collisions

Saehanseul Oh[1,2,*]

[1]Sejong University, Seoul, South Korea
[2]Lawrence Berkeley National Laboratory, Berkeley, CA, USA

Abstract. Heavy-flavor jets are a unique tool to study dynamical evolution of QCD systems via heavy-flavor particles, allowing one to access to the original parton kinematics and to separate out production and fragmentation effects. In these proceedings, various heavy-flavor jet results in *pp*, *p*A, and AA collisions are presented, extending the scope of heavy-flavor studies discussed at this conference series, the Strangeness in Quark Matter.

1 Introduction

The measurement of heavy-flavor production has long been considered as one of the ideal probes to investigate the properties of the Quark-Gluon Plasma created in relativistic heavy-ion collisions. The large mass of charm (c) and bottom (b) quarks requires its production to occur at the early stage of the collision. This leads heavy quarks to interact with the medium as they traverse it, and to be sensitive to the transport properties of the medium. Recently, there have been significant efforts to study heavy-flavor jets, which can provide further information with respect to heavy-flavor single-hadron measurements. Heavy-flavor jets have better access to the original parton kinematics, and they enable to separate out production and fragmentation effects.

Jets themselves have been widely used to understand the interaction of hard-scattered partons with the quark-gluon plasma [1]. The study of heavy-flavor jets can particularly shed light on the dependence of jet-medium interaction on the parton flavor, which largely attributes to mass dependent effects and color-charge effects [2–5]. Compared to the inclusive jet measurements, the main complication in heavy-flavor jet measurements arises from how to identify heavy-flavor jets. There are mainly two distinguished methods, i) reconstructing jets first with jet reconstruction algorithm, e.g. anti-k_T algorithm [6], then using the properties of jet constituents, such as impact parameter of constituents or secondary vertex reconstructed with constituents, for identification of the original parton flavor [7], and ii) reconstructing heavy-flavor single-hadron first, then running the jet finding algorithm including the heavy-flavor hadron as a constituent, and identifying the jet containing the heavy-flavor hadron [8]. Each method has pros and cons considering statistics, kinematic limit in detectors, and identification purity. Heavy-flavor jets identified with method i) are referred to as c-jet or b-jet, and those with method ii) are D^0-jet or J/ψ-jet depending on the heavy-flavor hadron used for the identification.

*e-mail: saehanseul.oh@sejong.ac.kr

In these proceedings, selected heavy-flavor jet results are discussed including results in small collision systems, such as proton-pronton (pp) and proton-ion (pA) collisions, as well as in heavy-ion (AA) collisions.

2 Heavy-flavor jets in pp and pA collisions

Attributing to its large mass compared to the non-perturbative QCD scale (Λ_{QCD}), heavy-quark production cross section has been calculated with the perturbative QCD framework [9–11]. In order to further test the validity of theoretical predictions and to provide constraints on perturbative QCD calculations, a variety of heavy-flavor jet cross-section measurements have been performed in pp collisions at several center of mass energies from different experiments at the LHC [8, 12–16]. Figure 1 shows cross-section ratios of D^0-jets to inclusive jets [15] (left), and of b-jets to inclusive jets [16] (middle) in $\sqrt{s}$ = 5.02 TeV pp collisions reported by ALICE and ATLAS Collaborations, respectively. Throughout the $p_{\mathrm{T,jet}}$ range, which is significantly different in two results, the ratios agree well with the PYTHIA 8 predictions, while the POWHEG + PYTHIA 8 calculations are observed to slightly disagree with the data at low $p_{\mathrm{T,jet}}$.

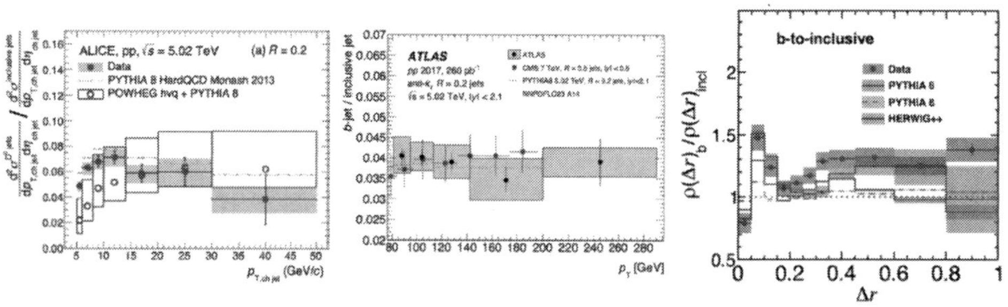

Figure 1. Cross-section ratios of D^0-jets to inclusive charged-particle jets [15] (left) and of b-jets to inclusive jets [16] (middle). b-to-inclusive jet shape ratio as a function of radial distance from the jet axis (Δr) with $p_{\mathrm{T,jet}} > 120$ Ge/c and $p_{\mathrm{T,trk}} > 1$ GeV/c is shown on the right panel [17]. All results are from pp collisions at $\sqrt{s}$ = 5.02 TeV.

In addition to the cross-section measurements, several other observables for heavy-flavor jets have been measured in pp collisions, not only providing constraints to theoretical calculations, but also serving as a baseline for the corresponding measurements in heavy-ion collisions. The right panel in Fig. 1 presents the b-to-inclusive jet shape ratio as a function of radial distance, Δr, with respect to the jet axix for jets with $p_{\mathrm{T,jet}} > 120$ Ge/c and $p_{\mathrm{T,trk}} > 1$ GeV/c [17]. The result indicates a shift of transverse momentum from small to large radial distance for b-jets compared to inclusive jets, and it gives a hint of the dead cone effect, which predicts the suppression of radiation from an emitter of mass m and energy E in the angular region, $\theta \leq m/E$ [18].

Furthermore, the dead cone effect in QCD has been directly observed using D^0-jets and experimental declustering techniques [19]. The observable of this measurement is defined by the ratio of the splitting angle (θ) distributions for D^0-jets to inclusive jets in bins of E_{Radiator},

$$R = \frac{1}{N^{D^0\mathrm{jet}}}\frac{\mathrm{d}n^{D^0\mathrm{jet}}}{\mathrm{d}\ln(1/\theta)} \Big/ \frac{1}{N^{\mathrm{inclusivejet}}}\frac{\mathrm{d}n^{\mathrm{inclusivejet}}}{\mathrm{d}\ln(1/\theta)}\Bigg|_{k_{\mathrm{T}},E_{\mathrm{Radiator}}},$$

where θ, k_{T}, and E_{Radiator} are the splitting angle between two prongs, relative transverse momentum of the splitting, and sum of the energy of two prongs at each declustering step, respectively. In Fig. 2, a significant suppression in the rate of small-angle splittings is observed in D^0-jets with respect to inclusive jets, which directly contrasts to the no dead-cone limit from theoretical calculations. The suppression magnitude increases with decreasing E_{Radiator}, agreeing with the expectation from the equation for dead-cone effect region, $\theta \leq m/E$.

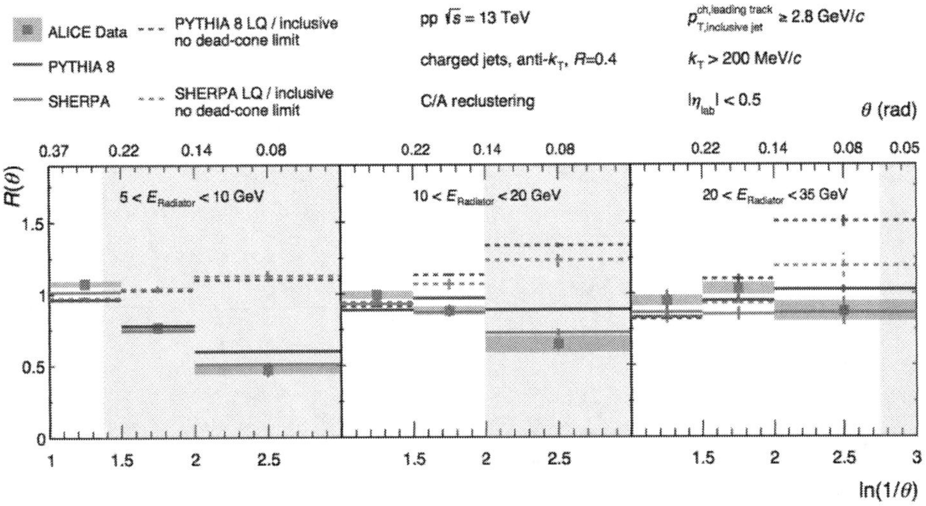

Figure 2. Ratios of the splitting-angle probability distributions for D^0 jets to inclusive jets in $\sqrt{s} = 13$ TeV pp collisions with $5 < E_{\text{Radiator}} < 10$ GeV (left), $10 < E_{\text{Radiator}} < 20$ GeV (middle), and $20 < E_{\text{Radiator}} < 35$ GeV (right) [19]. No dead-cone limit from event generators are shown with dashed lines.

Heavy-flavor jet cross section measurements have also been performed in pA collisions to investigate contributions of cold nuclear matter effects on jet production and jet quenching in heavy-ion collisions [21]. Figure 3 shows results of nuclear modification factor in pA collisions, $R_{p\text{A}}$, i.e. the ratio of cross section in pA collisions to that in pp collisions scaled by the ion mass number, with c-jets (left) and b-jets (middle and right) [14, 22, 23]. The $R_{p\text{A}}$ value with its uncertainties demonstrates no significant jet modification in p-Pb collisions at $\sqrt{s_{\text{NN}}} = 5.02$ TeV. This leads to the conclusion that cold nuclear matter effects do not play a significant role in jet energy modification in heavy-ion collisions in the corresponding collision energy.

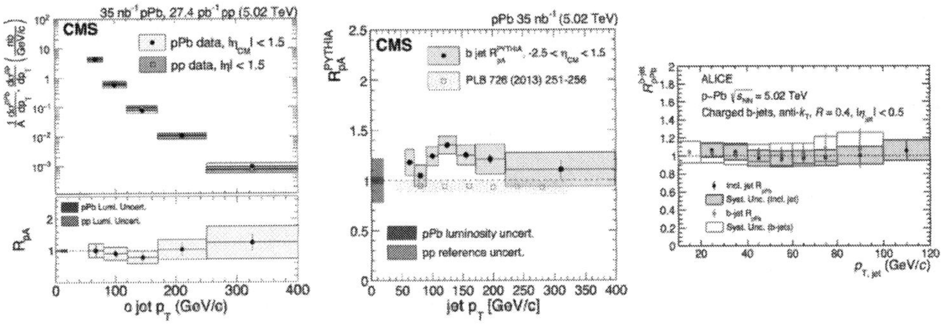

Figure 3. Cross-section and $R_{p\text{A}}$ with c-jets [23] (left) and $R_{p\text{A}}$ with b-jets (middle and right) [14, 22] in $\sqrt{s_{\text{NN}}} = 5.02$ TeV p-Pb collisions.

3 Heavy-flavor jets in AA collisions

With baseline results in pp and pA collisions, heavy-flavor jets are measured in heavy-ion collisions in order to study the parton flavor dependence of jet quenching. The nuclear modification factor, $R_{\rm AA}$, indicates the deviation of inclusive spectra in heavy-ion collisions with respect to those in pp collisions, and is defined as

$$R_{\rm AA} = \frac{\frac{1}{N_{\rm event}}(\mathrm{d}^2 N_{\rm AA})/(\mathrm{d}p_{\rm T}\,\mathrm{d}\eta)}{\langle T_{\rm AA}\rangle(\mathrm{d}^2\sigma_{pp})/(\mathrm{d}p_{\rm T}\,\mathrm{d}\eta)},$$

where $N_{\rm AA}$, $\sigma_{\rm pp}$, and $T_{\rm AA}$ corespond to the yield in AA collisions, corss section in pp collisions, and the nuclear overlap function, respectively. Figure 4 shows $R_{\rm AA}$ of b-jets [16]

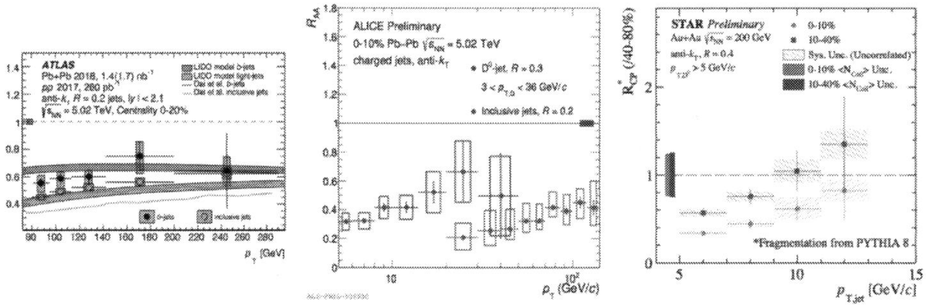

Figure 4. $R_{\rm AA}$ for b-jets and inclusive jets in 0–20% central collisions [16] (left), and for D^0-jets and inclusive jets in 0-10% central collisions [24] in $\sqrt{s_{\rm NN}}$ = 5.02 TeV Pb+Pb collisions at the LHC, and $R_{\rm CP}$ for D^0-jets in $\sqrt{s_{\rm NN}}$ = 200 GeV Au+Au collisions at RHIC (right) [25].

(left) and D^0-jets [24] (middle) in Pb+Pb collisions at $\sqrt{s_{\rm NN}}$ = 5.02 TeV, and $R_{\rm CP}$, nuclear modification factor with respect to peripheral heavy-ion collisions instead of pp collisions, in Au+Au collisions at $\sqrt{s_{\rm NN}}$ = 200 GeV [25]. Although only the most central collision results (0–20%) are presented at $\sqrt{s_{\rm NN}}$ = 5.02 TeV due to the page limit in these proceedings, the $R_{\rm AA}$ values of b-jets and inclusive jets show a gradual decrease in more central collisions, and $R_{\rm CP}$ results follow the similar trend. In central collisions, both b-jets and D^0-jets have larger $R_{\rm AA}$ values compared to those of inclusive jets. These results along with theoretical predictions provide a hint that the difference may attribute to larger gluon contribution in inclusive jet sample, as gluon jets are expected to lose more energy than quark jets. However, the current systematic uncertainties do not allow more definitive conclusion.

There have also been measurements regarding internal structure of heavy-flavor jets, such as jet shape, fragmentation function, and radial profile of heavy-flavor meson within jets [25–28]. Among such results, the radial distributions of D^0 meson within D^0-jets in pp and Pb+Pb collisions are shown in Fig. 5. These distributions are sensitive not only to the production mechanisms of the meson, but also to the energy loss and diffusion processes in the interaction between parton and hot QCD medium created in heavy-ion collisions. In Pb+Pb collisions, the D^0 meson distribution for low transverse momentum ($4 < p_{\rm T}^{D^0} < 20$ GeV/c) is enhanced at a larger distance with respect to the jet axis compared to the corresponding result in pp collisions, and this is an indication of charm quark diffusion in heavy-ion collisions. On the other hand, this modification is not observed at higher transverse momentum ($p_{\rm T}^{D^0} > 20$ GeV/c), and the overall results well agree with the CCNU energy loss model [29]. Such results as well as other heavy-flavor jet internal structure measurements provide constraints on physics mechanisms relevant to the heavy quark propagation inside the quark gluon plasma.

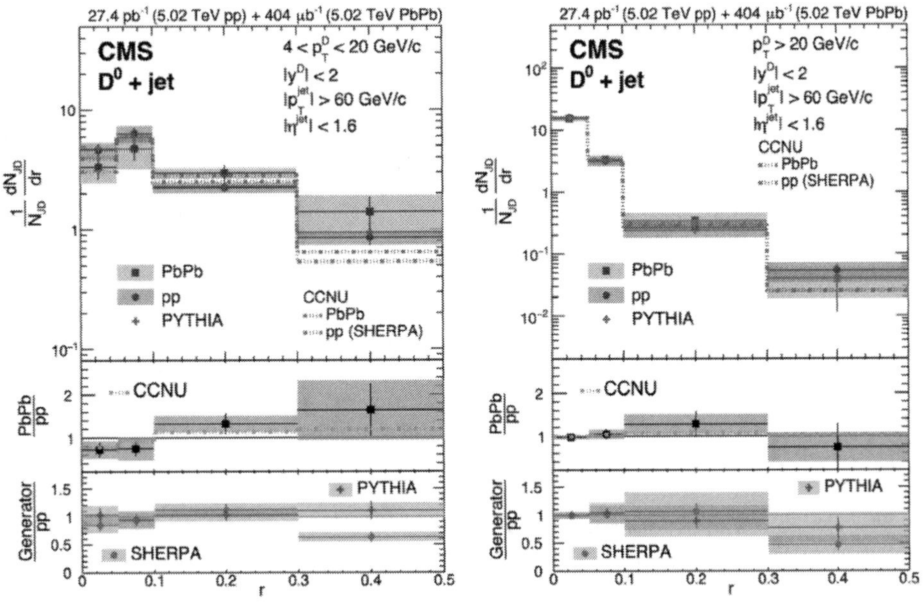

Figure 5. Radial distributions of D^0 mesons within D^0-jets in pp and Pb+Pb collisions at $\sqrt{s_S}$ = 5.02 TeV. r is defined by the radial distance between D^0 meson and the jet axis. D^0-jets have $p_{\mathrm{T,jet}}$ larger than 60 GeV/c, and D^0 mesons are selected with $4 < p_{\mathrm{T}}^{D^0} < 20$ GeV/c (left) and $p_{\mathrm{T}}^{D^0} > 20$ GeV/c [28].

Heavy-flavor jet measurements are one of the key points where the upcoming LHC Run 3 and 4, as well as sPHENIX and STAR runs in years of 2023-2025 will play a crucial role with significantly larger statistics and improved detector performances. Different kinematic ranges from different experiments can provide comprehensive picture of the flavor dependence of jet quenching, for example, verifying the statement that low-$p_{\mathrm{T,jet}}$ results are more sensitive to mass-dependent effects.

References

[1] G.-Y. Qin and X.-N. Wang, Int. J. Mod. Phys. E **24**, 1530014 (2015)
[2] H. van Hees, R. Rapp, Phys. Rev. C **71**, 034907 (2005)
[3] G. D. Moore, D. Teaney, Phys. Rev. C **71** 064904 (2005)
[4] S. Wang, W. Dai, B.-W. Zhang, E. Wang, Eur. Phys. J. C **79**, 789 (2019)
[5] Y. L. Dokshitzer, D. E. Kharzeev, Phys. Lett. B **519**, 199 (2001)
[6] M. Cacciari, G. P. Salam, and G. Soyez, JHEP **04**, 063 (2008)
[7] CMS Collaboration, JINST **8**, P04013 (2013)
[8] ALICE Collaboration, JHEP **1908** (2019) 133
[9] J.C. Collins, D.E. Soper, Annu. Rev. Nucl. Part. Sci. **37**, 383 (1987)
[10] S. Frixione, M. L. Mangano, P. Nason, G. Ridolfi, Adv. Ser. Direct. High Energy Phys. **15**, 609–706 (1998)
[11] M. Cacciari, M. Greco, P. Nason, JHEP **05**, 007 (1998)
[12] CMS Collaborartion, JHEP **04**, 084 (2012)
[13] ATLAS Collaboration, Eur. Phys. J. C **71**, 1846 (2011)
[14] ALICE Collaboration, JHEP **01**, 178 (2022)

[15] ALICE Collaboration, arXiv:2204.10167, submitted to JHEP (2022)
[16] ATLAS Collaboration, arXiv:2204.13530, submitted to Eur. Phys. J. C (2022)
[17] CMS Collaboration, JHEP **05**, 054 (2021)
[18] Y. L. Dokshitzer, V. A. Khoze, S. I. Troian, J. Phys. G **17**, 1602-1604 (1991)
[19] ALICE Collaboration, Nature **605**, 7910, 440–446 (2022)
[20] ALICE Collaboration, Phys. Lett. B **827**, 136984 (2022)
[21] C. A. Salgado et al., J. Phys. G **39**, 015010 (2012)
[22] CMS Collaborartion, Phys. Lett. B **754**, 59 (2016)
[23] CMS Collaborartion, Phys. Lett. B **772**, 306 (2017)
[24] M. Mazzilli (for the ALICE Collaboration), arXiv:2208.10908, Quark Matter 2022 proceedings (2022)
[25] D. Roy (for the STAR Collaboration), arXiv:2207.14434, Quark Matter 2022 proceedings (2022)
[26] X. Wang (for the CMS Collaboration), Quark Matter 2022 (2022)
[27] CMS Collaboration, Phys. Lett. B **825**, 136842 (2021)
[28] CMS Collaboration, Phys. Rev. Lett. **125**, 102001 (2020)
[29] S. Wang, W. Dai, B.-W. Zhang, E. Wang, Eur. Phys. J. C **79** 789 (2019)

Measurements of quarkonium production and polarization in Pb–Pb collisions with ALICE

Xiaozhi Bai on behalf of the ALICE Collaboration

State Key Laboratory of Particle Detection and Electronics, University of Science and Technology of China, Hefei 230026, China

Abstract. Quarkonia are excellent probes of deconfinement in heavy-ion collisions. For J/ψ, a bound state of $c\bar{c}$ quarks, the (re-)generation is found to be the dominant production mechanism at the LHC energies. Production measurements of non-prompt J/ψ, originating from beauty-hadron decays, allow one to access the interaction of beauty-quarks with the quark-gluon plasma (QGP). Polarization and spin alignment measurements can be used to investigate the characteristics of the formed medium. Moreover, it has been hypothesized that quarkonium states can be polarized by the strong magnetic field generated in the initial state of the collision and by the large angular momentum of the medium in non-central heavy-ion collisions.

In these proceedings, the measurements of the inclusive, prompt, and non-prompt J/ψ nuclear modification factor R_{AA} in Pb–Pb collisions at $\sqrt{s_{NN}} = 5.02$ TeV are shown. The measured non-prompt J/ψ fraction extends down to very low p_T with a significantly improved precision compared to previous publications. The results from the first publication on the J/ψ polarization with respect to the event-plane in Pb–Pb collisions at $\sqrt{s_{NN}} = 5.02$ TeV at forward rapidity are presented as well. The results are compared with available calculations.

1 Introduction

Heavy quarks (charm and beauty) are an excellent probe to study the quark-gluon plasma (QGP), the strongly interacting medium created in high-energy heavy-ion collisions. They are mainly produced via initial hard partonic scatterings and thus experience the entire QGP evolution. Quarkonia are bound states composed of a heavy quark and its corresponding anti-quark. In the QGP, the color screening of the surrounding medium may prevent the heavy quark and anti-quark from forming their bound states depending on the QGP temperature and binding energy of the state [1]. A new charmonium production mechanism, known as (re-)generation [2–4], is found to be significant at the LHC energies due to the increase of heavy quark densities in the QGP [5, 6]. Moreover, the measurement of non-prompt J/ψ production can provide insights into the beauty quark energy loss mechanisms in the QGP. The nuclear modification factor (R_{AA}) is an interesting observable to study the medium influence on particle production yields and it is defined as the ratio of the yield in heavy-ion collisions to that in pp collisions at the same center-of-mass energy, scaled by the number of binary nucleon-nucleon collisions. Finally, it has been argued that the strong electromagnetic field created at the very early stage of the heavy-ion collision [7], as well as the angular momentum of the medium in non-central collisions [8], may affect vector meson polarization.

2 Analysis and Results

In ALICE, inclusive J/ψ-mesons are reconstructed by employing the e^+e^- decay channel at midrapidity ($|y| < 0.9$) and the $\mu^+\mu^-$ decay channel at forward rapidity ($2.5 < y < 4$). The analysis techniques are discussed in detail in Refs. [5, 9]. The ALICE detector is described in Ref. [10].

The left panel of Fig. 1 shows the $p_{\rm T}$-integrated inclusive J/ψ $R_{\rm AA}$ measured at midrapidity as a function of the average number of participant nucleons ($\langle N_{\rm part} \rangle$) in Pb–Pb collisions at $\sqrt{s_{\rm NN}}$ = 5.02 TeV. In order to exclude J/ψ photoproduction processes [11], a selection of the dielectron with $p_{\rm T} > 0.15$ GeV/c is applied. The $R_{\rm AA}$ exhibits a sizeable increase in moving from semi-central to central collisions, described by models including (re-)generation. The right panel of Fig. 1 compares the $p_{\rm T}$-differential inclusive J/ψ $R_{\rm AA}$ in 0-10% and 0-20% centrality intervals at midrapidity and forward rapidity, respectively. A larger $R_{\rm AA}$ is observed at midrapidity and low $p_{\rm T}$ as a consequence of the higher $c\bar{c}$ density, which leads to larger J/ψ yields being produced via (re-)generation. The statistical hadronization model [12] describes data at low $p_{\rm T}$ and underestimates them at high $p_{\rm T}$. The transport model agrees with data in the whole measured $p_{\rm T}$ region [3].

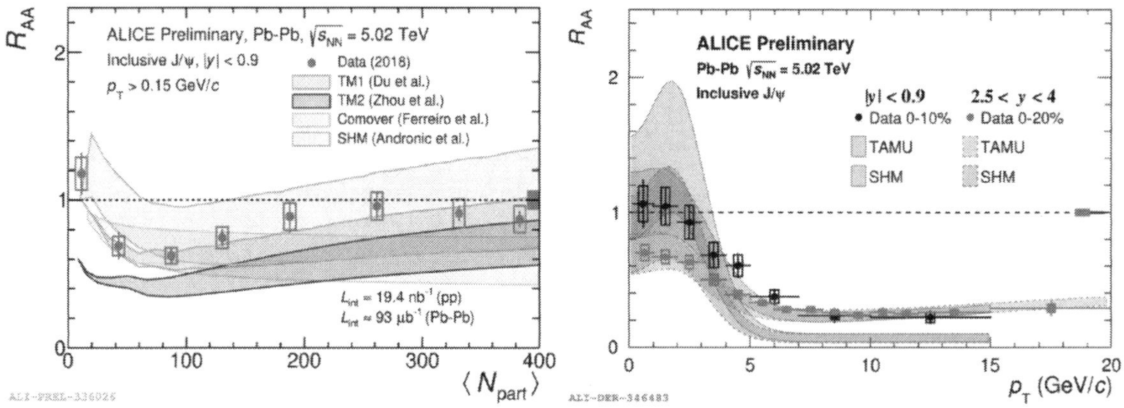

Figure 1. Left panel: Inclusive J/ψ $R_{\rm AA}$ at midrapidity, integrated over $p_{\rm T}$, as a function of $\langle N_{\rm part} \rangle$ in Pb–Pb collisions at $\sqrt{s_{\rm NN}}$ = 5.02 TeV. Right panel: $p_{\rm T}$ dependence of the inclusive J/ψ $R_{\rm AA}$ at midrapidity and forward rapidity in the 0–10% and 0–20% centrality intervals, respectively. The results are compared to model calculations from Refs. [3, 4, 12, 13].

The left panel of Fig. 2 shows the non-prompt J/ψ fraction as a function of $p_{\rm T}$, measured in Pb–Pb collisions at $\sqrt{s_{\rm NN}}$ = 5.02 TeV, in three different centrality classes. The results are compatible with CMS in the common $p_{\rm T}$ interval, which refers to the centrality class 0-100% [14]. The right panel of Fig. 2 shows the $p_{\rm T}$ differential non-prompt J/ψ $R_{\rm AA}$ in the 0-10% centrality class in Pb–Pb collisions at $\sqrt{s_{\rm NN}}$ = 5.02 TeV, compared to similar results from other experiments as well as with model calculations. The new ALICE data extends the non-prompt J/ψ $R_{\rm AA}$ measurement down to very low $p_{\rm T}$ and are consistent with CMS [14] and ATLAS [15] data in the overlapping $p_{\rm T}$ region. The non-prompt J/ψ $R_{\rm AA}$ is comparable to the non-prompt D^0 $R_{\rm AA}$ measurement performed by the ALICE collaboration [16] (small differences could arise from the decay kinematics). The suppression of the non-prompt J/ψ $R_{\rm AA}$ at high $p_{\rm T}$ can be described by model calculations from Refs. [17, 18]. Both calculations include beauty quark energy loss in the QGP via both radiative [19] and collisional [20] processes.

 https://doi.org/10.1051/epjconf/202327602015

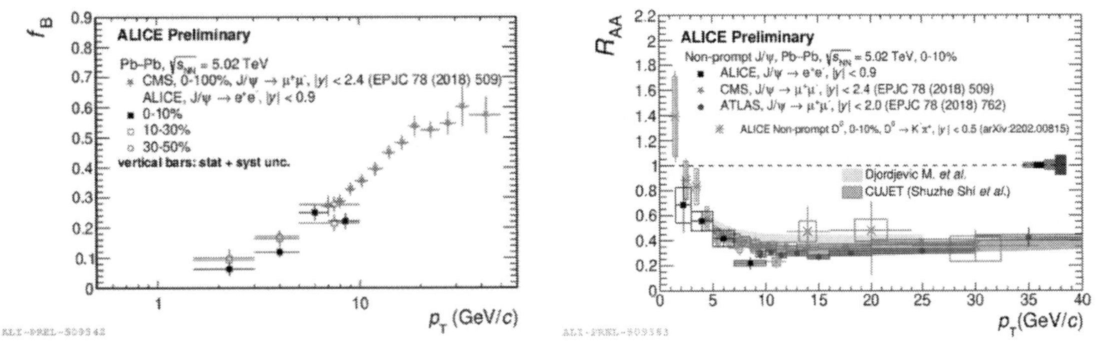

Figure 2. Left panel: p_T dependence of the non-prompt J/ψ fraction in Pb–Pb collisions at $\sqrt{s_{NN}}$ = 5.02 TeV and midrapidity. The results are compared to the corresponding CMS measurements in the centrality range of 0–100% [14]. Right panel: Non-prompt J/ψ R_{AA} as a function of p_T compared with similar measurements from CMS [14] and ATLAS [15]. Non-prompt D^0 R_{AA} measurements from ALICE [16] are shown as well. Results are compared with model calculations from Refs. [17, 18].

The polarization of the inclusive J/ψ production with respect to the event plane is measured for the first time in Pb–Pb collisions at $\sqrt{s_{NN}}$ = 5.02 TeV via the dimuon decay channel at forward rapidity [21]. The left panel of Fig. 3 shows the p_T integrated λ_θ parameter measured in the range $2 < p_T < 6$ GeV/c in different centrality classes. A significant non-zero polarization is found in central and semi-central collisions, in particular in the 40–60% centrality interval, where a 3.5σ effect is observed. The p_T dependence of λ_θ in central (0–20%) and semi-central (30–50%) collisions is shown on the right panel of Fig. 3. The maximum deviation from zero polarization is observed for $2 < p_T < 4$ GeV/c and in the 30–50% centrality range, where the significance of the effect reaches 3.9σ when considering the total uncertainties.

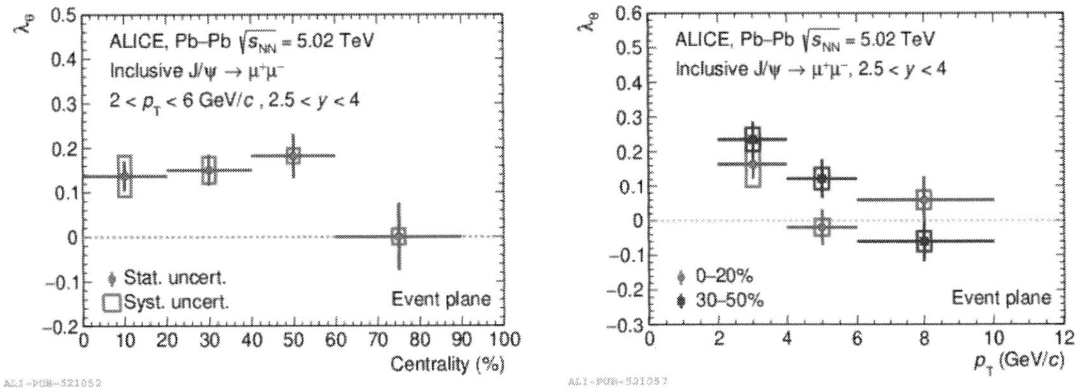

Figure 3. Centrality (left panel) and p_T dependence (right panel) of the λ_θ polarization parameter. The vertical bars and open boxes represent the statistical and systematic uncertainties, respectively [21].

3 Summary

In this contribution, recent measurements of the inclusive and non-prompt J/ψ production in Pb–Pb collisions at $\sqrt{s_{NN}} = 5.02$ TeV are reported. The R_{AA} of inclusive J/ψ shows an increasing trend at low p_T and towards more central collisions, which indicates a large contribution of J/ψ production from the (re-)generation process. Non-prompt J/ψ R_{AA} is measured down to $p_T = 1.5$ GeV/c in 0-10% central Pb–Pb collisions. The strong suppression of non-prompt J/ψ at high p_T is well reproduced by models implementing both collisional and radiative energy loss processes of beauty quarks inside the QGP. The first measurement of inclusive J/ψ polarization with respect to the event plane is shown as a function of centrality and p_T in Pb–Pb collisions at $\sqrt{s_{NN}} = 5.02$ TeV. A small but significant polarization effect, reaching 3.9σ for $2 < p_T < 4$ GeV/c in the 30-50% centrality interval, is presented.

4 Acknowledgement

The author is supported in part by the National Key R&D Program of China under Grant No. 2018YFE0104900, the National Natural Science Foundation of China under Grant No. 12105277

References

[1] T. Matsui, H. Satz, Phys. Lett. B **178**, 416 (1986)
[2] P. Braun-Munzinger, J. Stachel, Nature **448**, 302–309 (2007)
[3] X. Du, R. Rapp, Nucl. Phys A **943**, 147 (2015)
[4] K. Zhou, N. Xu, Z. Xu, P. Zhuang, Phys. Rev. C **89**, 054911 (2014)
[5] ALICE Collaboration, Phys. Lett. B **766**, 212 (2017)
[6] ALICE Collaboration, Phys. Rev. C **94**, 054908 (2016)
[7] W.T. Deng, X.G. Huang, Phys. Rev. C **85**, 044907 (2012)
[8] Z.T. Liang, X.N. Wang, Phys. Rev. Lett. **94**, 102301 (2005)
[9] ALICE Collaboration, Phys. Lett. B **805**, 135434 (2020)
[10] ALICE Collaboration, JINST **3**, S08002 (2008)
[11] ALICE Collaboration, arXiv: **2204.10684** (2022)
[12] A. Andronic et al., Phys. Lett. B **797**, 134836 (2019)
[13] E. Ferreiro, Phys. Lett. B **731**, 57 (2014)
[14] CMS Collaboration, Eur. Phys. J. C **78**, 1 (2018)
[15] ATLAS Collaboration, Eur. Phys. J. C **78**, 762 (2018)
[16] ALICE Collaboration, arXiv: **2202.00815** (2022)
[17] S. Shi, J. Liao, M. Gyulassy, Chin. Phys. C **43**, 044101 (2019)
[18] S. Stojku et al., Phys. Rev. C **105**, L021901 (2022)
[19] A. Adil, I. Vitev, Phys. Lett. B **649**, 139 (2007)
[20] N. Armesto, C.A. Salgado, U.A. Wiedemann, Phys. Rev. D **69**, 114003 (2004)
[21] ALICE Collaboration, arXiv: **2204.10171** (2022)

Heavy-flavor anisotropic flow at RHIC and LHC energies within a full transport approach

Lucia Oliva[1,2,*], *Maria Lucia* Sambataro[1,3], *Yifeng* Sun[4,1,3], *Vincenzo* Minissale[1,3], *Salvatore* Plumari[1,3], and *Vincenzo* Greco[1,3]

[1]Department of Physics and Astronomy "Ettore Majorana", University of Catania, Via S. Sofia 64, I-95123 Catania, Italy
[2]INFN Sezione di Catania, Via S. Sofia 64, I-95123 Catania, Italy
[3]Laboratori Nazionali del Sud, INFN-LNS, Via S. Sofia 62, I-95123 Catania, Italy
[4]School of Physics and Astronomy, Shanghai Key Laboratory for Particle Physics and Cosmology, and Key Laboratory for Particle Astrophysics and Cosmology (MOE), Shanghai Jiao Tong University, Shanghai 200240, China

Abstract. The propagation of heavy quarks (HQs) in the quark-gluon plasma (QGP) is described by means of a full Boltzmann transport approach. The non-perturbative dynamics and the interaction between HQs and the bulk is taken into account by means of a Quasi-Particle Model. Including the intense electromagnetic and vortical fields, we discuss their impact on the directed flow of neutral D mesons at RHIC energy, clarifying the role of this observable in giving information on QGP transport properties. We also show a novel study of the correlations between different flow coefficients of D mesons and soft hadrons at LHC energy within a coalescence plus fragmentation hadronization scheme and including event-by-event initial state fluctuations. This investigations can put further constraints on HQ transport coefficients towards a solid determination in comparison to the lattice QCD calculations.

1 Introduction

Heavy quarks (HQs), namely charm and bottom quarks, are excellent probes of the dynamical evolution of ultra-Relativistic Heavy Ion Collisions (uRHICs). Being produced in hard binary processes in the early off-equilibrium stage, their formation time is very small compared to that of light quarks. Due to the large masses, their thermalization time is comparable or larger than the lifetime of the Quark-Gluon Plasma (QGP). Hence, the HQs witness and keep a better memory of both initial stage and QGP evolution. Key observables of the HQ final states, such as D mesons, are the anisotropic flow coefficients v_n that characterize the azimuthal particle distribution in momentum space. The first coefficient, the directed flow v_1, is a collective sidewards deflection of particles on the event plane with respect to the beam axis; the v_1 of neutral D-meson has proven to give important evidences of the non-perturbative dynamics, the longitudinal bulk matter asymmetry and the electromagnetic (EM) fields [1–5]. Among the observables connected to higher-order flow coefficients, the Symmetric Cumulant (SC) [6] gives important evidences of the event-by-event fluctuations coming from the random nucleon positions as well as of the correlations between different-order flow anisotropies.

*e-mail: lucia.oliva@dfa.unict.it

2 Heavy quark transport approach

We describe the QGP and HQ evolution in relativistic heavy-ion collisions by solving the relativistic Boltzmann transport equations [7–11]:

$$\left[p_\mu \partial_x^\mu + q_j F_{\mu\nu}(x) p^\nu \partial_p^\mu\right] f_j(x,p) = C[f_j, f_k](x,p) \quad (1)$$

$$\left[p_\mu \partial_x^\mu + q_Q F_{\mu\nu}(x) p^\nu \partial_p^\mu\right] f_Q(x,p) = C[f_j, f_k, f_Q](x,p) \quad (2)$$

where $f_{j,k}(x,p)$ and $f_Q(x,p)$ are the phase-space one-body distribution function of, respectively, the light parton $j,k = g, q, \overline{q}$ (gluon, quark or antiquark) and the HQ $Q = c, \overline{c}$ (charm or anticharm); $F_{\mu\nu}$ is the electromagnetic strength tensor and $C = C_{22}$ is the relativistic collision integral that accounts for $2 \rightarrow 2$ scattering processes. In the collision integral of Eq. 1 the total cross section is determined in order to have a fixed value of the viscosity over entropy ratio, $\eta/s = 1/(4\pi)$, during the QGP evolution [9]. In Eqs. (2) we neglect collisions between HQs and describe the HQ interaction with the QGP medium by means of scattering matrices calculated at Leading-Order in pQCD but accounting for the non-perturbative effects by means of the Quasi-Particle Model (QPM) [10, 12] in which quarks and gluons are dressed with thermal masses that allow to reproduce lattice QCD thermodynamics [13]. When the fireball temperature goes below the quark-hadron transition temperature, $T_c = 155$ MeV, the charm quarks are hadronized to D-meson by means of a coalescence plus fragmentation model [14]. For the calculation of the directed flow we adopt a non boost-invariant initial condition for the bulk medium in the coordinate space able to describe the onset of vorticity in the fireball caused by the angular momentum of the colliding nuclear system. This is obtained by means of an asymmetric initial distribution of the QGP density in the $x - \eta_s$ plane that gives rise to a tilted fireball on the reaction plane [5]. For the calculations of the higher flow harmonics we assume longitudinal boost invariance and employ a Monte-Carlo Glauber model in order to take into account the initial event-by-event fluctuations. In momentum space the initial conditions for the bulk medium are given by a Boltzmann-Jüttner distribution function up to transverse momentum $p_T = 2$ GeV while at larger momenta we adopt mini-jet distributions as calculated by pQCD at NLO order in [15]. The charm quark and antiquark distributions are initialized in coordinate space in accordance with the number of binary nucleon-nucleon collisions. In momentum space we use charm quark production calculated at the Fixed Order + Next-to-Leading Log (FONLL) [16] [10]. The EM fields appearing in Eqs. (1) and (2) are computed with the method of Ref. [1, 17], where the QGP electrical conductivity σ_{el} is assumed constant in order to have analytic solutions of the Maxwell equations.

3 Results and discussion

The initial QGP density asymmetry is translated to the final azimuthal asymmetry of final hadrons measured by the charged particle v_1 shown in the left panel of Fig. 1. We show results for two values of the parameter η_m, which determines the degree of tilt of the fireball and, consequently, the directed flow magnitude. In the right panel of Fig. 1 we show our result for the v_1 of D mesons in comparison to the STAR data [19]. The values of the combined v_1 of D^0 and $\overline{D}^0$ mesons in the central rapidity range $|y| < 1$ are about 20-30 times larger than those of charged particle shown in left panel. This is a result of the shift in the transverse profiles of the tilted QGP fireball and the symmetrically-distributed HQ emission points from initial hard scatterings [2]. This shift leads to a pressure push of the bulk medium on the HQs, which is however effective only because the HQ interaction in QGP is largely non-perturbative, corresponding to a very small space diffusion coefficient $2\pi T D_s$ of charm quarks in satisfying agreement with the lattice QCD data; this generate an amplification of the D-meson v_1. While

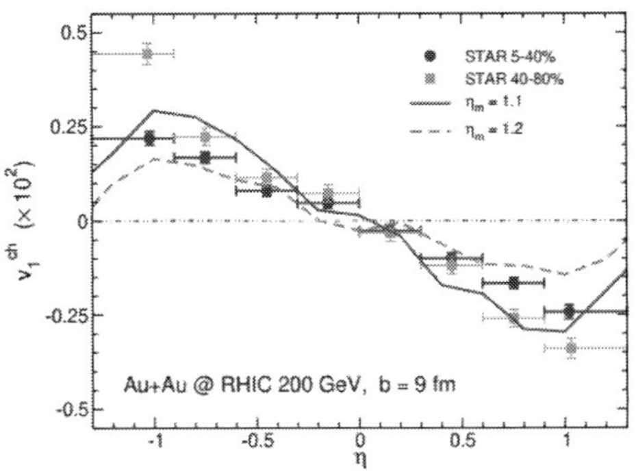

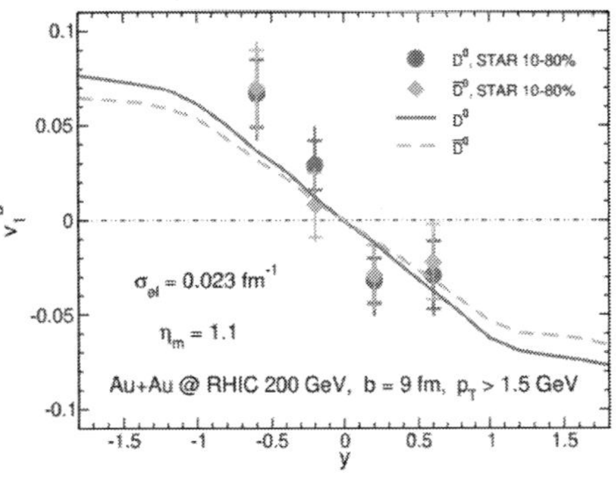

Figure 1. (Pseudo)rapidity dependence of the directed flow of charged particles (left) and D mesons (right) in Au+Au collisions at RHIC energy $\sqrt{s_{NN}} = 200$ GeV with impact parameter $b = 9$ fm. The experimental data are from the STAR Collaboration [18, 19]. Figures from Ref. [5].

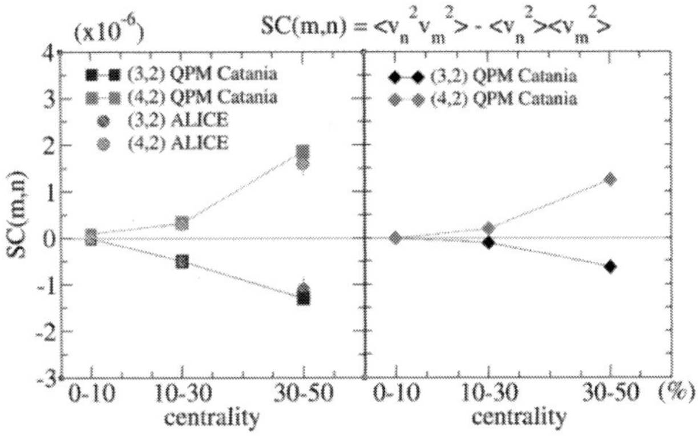

Figure 2. Centrality dependence of the Symmetric Cumulant Correlator for charged particles (left) and *D* mesons (right) with $|\eta| < 0.8$ and $0.5 < p_T < 5.0\,GeV$ in *PbPb* collisions at $\sqrt{s_{NN}} = 5.02$ TeV. The circles in the left panel are the ALICE experimental data [22]. Figure from Ref. [23].

the v_1 magnitude is associated with the HQ diffusion coefficient, the v_1 splitting is connected to the electric conductivity of the QGP medium. In simulations without the EM fields the two lines of the v_1 of D^0 and $\overline{D}^0$ lie one on top of each other; we see clearly from Fig. 1 (right) that a splitting of the two curves is generated when the EM fields are considered, in agreement with Refs. [1, 3]. In this calculation we assume that D^0 and $\overline{D}^0$ are produced only by fragmentation of c and $\overline{c}$ quarks respectively, thus the v_1 of neutral D mesons is driven by the v_1 of charm quarks and antiquarks. Our result indicates that the total effect coming from B_y and E_x which push positively and negatively charged particles in opposite directions is slightly dominated by the electric field, consistently also with studies on the light sector in large [17, 20] and small [21] systems at top RHIC energy.

The event-by-event transport approach allows to analyse higher-order v_n through the SC correlator that is defined as $SC(m,n) =< v_n^2 v_m^2 > - < v_n^2 >< v_m^2 >$, where the brackets indicate average over all events for a given centrality class. The SC correlator vanishes for fixed values of v_n and v_m over all events as well as when event-by-event fluctuations of v_n and v_m are uncorrelated. Therefore, a nonzero SC indicates that the event-by-event fluctuations of v_n and v_m are correlated. In Fig. 2 we show $SC(4,2)$ (red lines) and $SC(3,2)$ (black lines) for charged particles (left panel) and D mesons (right panel). The result for charged particles is in good agreement with the experimental data from the ALICE Collaboration [22]. The $SC(4,2)$ is positive, indicating that for a given centrality interval, events with larger v_2 have on average larger v_4. On the other hand, the $SC(3,2)$ is negative indicating an anti-correlation between v_2 and v_3 at fixed centrality. This confirm what was observed within the event-shape selected v_n–v_m correlations [23] and extend a previous study on the heavy-light flavor correlations of anisotropic flows [24]. The predictions for the SC correlators of D mesons indicate a weaker but similar behaviour with the centrality with respect to light hadrons. In both panels we observe that $SC(4,2)$ and $SC(3,2)$ decrease considerably in magnitude from peripheral to central events. This large decrease does not imply that the correlations between these harmonics becomes negligible for centrality class smaller then $(0-10)\%$, but occurs since $SC \sim v_n^2 v_m^2$ and for very central events the flow coefficients (especially v_2) become quite small.

References

[1] S.K. Das, S. Plumari, S. Chatterjee, J. Alam, F. Scardina, V. Greco, Phys. Lett. **B768**, 260 (2017)

[2] S. Chatterjee, P. Bożek, Phys. Rev. Lett. **120**, 192301 (2018)

[3] S. Chatterjee, P. Bozek, Phys. Lett. **B798**, 134955 (2019)

[4] L. Oliva (2020), 2007.00560

[5] L. Oliva, S. Plumari, V. Greco, JHEP **05**, 034 (2021)

[6] A. Bilandzic, C.H. Christensen, K. Gulbrandsen, A. Hansen, Y. Zhou, Phys. Rev. C **89**, 064904 (2014)

[7] S. Plumari, A. Puglisi, F. Scardina, V. Greco, Phys. Rev. C **86**, 054902 (2012)

[8] M. Ruggieri, F. Scardina, S. Plumari, V. Greco, Phys. Rev. C **89**, 054914 (2014)

[9] S. Plumari, G.L. Guardo, F. Scardina, V. Greco, Phys. Rev. C **92**, 054902 (2015)

[10] F. Scardina, S.K. Das, V. Minissale, S. Plumari, V. Greco, Phys. Rev. C **96**, 044905 (2017)

[11] S. Plumari, Eur. Phys. J. C **79**, 2 (2019)

[12] S. Plumari, W.M. Alberico, V. Greco, C. Ratti, Phys. Rev. **D84**, 094004 (2011)

[13] S. Borsanyi, G. Endrodi, Z. Fodor, A. Jakovac, S.D. Katz, S. Krieg, C. Ratti, K.K. Szabo, JHEP **11**, 077 (2010), 1007.2580

[14] S. Plumari, V. Minissale, S.K. Das, G. Coci, V. Greco, Eur. Phys. J. **C78**, 348 (2018)

[15] V. Greco, C.M. Ko, P. Levai, Phys. Rev. Lett. **90**, 202302 (2003)

[16] M. Cacciari, S. Frixione, N. Houdeau, M.L. Mangano, P. Nason, G. Ridolfi, JHEP **10**, 137 (2012)

[17] U. Gursoy, D. Kharzeev, K. Rajagopal, Phys. Rev. **C89**, 054905 (2014)

[18] B.I. Abelev et al. (STAR), Phys. Rev. Lett. **101**, 252301 (2008), 0807.1518

[19] J. Adam et al. (STAR), Phys. Rev. Lett. **123**, 162301 (2019), 1905.02052

[20] U. Gürsoy, D. Kharzeev, E. Marcus, K. Rajagopal, C. Shen, Phys. Rev. **C98**, 055201 (2018)

[21] L. Oliva, P. Moreau, V. Voronyuk, E. Bratkovskaya, Phys. Rev. **C101**, 014917 (2020)

[22] S. Acharya et al. (ALICE), Phys. Lett. B **818**, 136354 (2021)

[23] M.L. Sambataro, Y. Sun, V. Minissale, S. Plumari, V. Greco, Eur. Phys. J. C **82**, 833 (2022)

[24] S. Plumari, G. Coci, V. Minissale, S.K. Das, Y. Sun, V. Greco, Phys. Lett. B **805**, 135460 (2020)

Understanding mass hierarchy in different energy loss mechanisms through heavy flavor data

Bojana Ilic[1,*] and *Magdalena* Djordjevic[1]

[1]Institute of Physics Belgrade, University of Belgrade, Belgrade, Serbia

Abstract. The theoretical analysis of experimental observations, such as the mass hierarchy effect, often neglects some ingredients, which may have a significant impact. The forthcoming measurements at RHIC and LHC will generate heavy flavor data with unprecedented precision, providing an opportunity to utilize high-$p_\perp$ heavy flavor data to analyze the interaction mechanisms in QGP. To this end, we use our recently developed DREENA framework based on the dynamical energy loss formalism. We present: i) How to disentangle the signature of different interaction mechanisms (radiative and collisional energy losses) at the same dataset. ii) Novel observables susceptible to these different mechanisms to be tested by future high-precision measurements. iii) Analytical and numerical extraction of the mass hierarchy effect in energy losses through this observable.

1 Introduction

One of the fundamental properties of parton's energy loss is the flavor dependence, that is, the mass ordering. The experimental observation of suppression mass hierarchy [1] is attributed and analyzed within radiative models [2–5], while collisional interactions are neglected. On the other hand, at the intermediate-$p_\perp$ range ($p_\perp \lesssim 10$ GeV) the collisional energy loss for heavy flavor is comparable to, or even larger, than the radiative energy loss [6–10]. However, neither a direct relation between collisional energy loss and heavy quark mass is established, nor an observable which quantifies this effect. Additionally, the upcoming high-precision RHIC and LHC measurements present an opportunity to utilize high-$p_\perp$ heavy flavor data to study the interaction mechanisms in QGP. Note also that D-mesons suppression is practically indistinguishable [1] from $h^\pm$ R_{AA}, which we will exploit in the first part of this document.

Thus, first, we discuss how nonintuitive suppression patterns can be utilized to qualitatively disentangle collisional interaction from radiative energy loss mechanism on the same dataset. In the second part, we focus on the lower-$p_\perp$ range (below 50 GeV) and present: A pursuit for an observable, which could unravel collisional interactions from radiative energy loss. And finally, analytical and numerical extraction of the mass ordering effect in collisional energy loss (for the first time) through this observable, to be more rigorously tested by the upcoming high-luminosity measurements. Note that, in this document, only the main results are displayed, while for more details on the first and second parts, we refer the reader to [4] and [11], respectively.

*e-mail: bojanab@ipb.ac.rs

2 Methods

We employ our full-fledged DREENA-C framework [12], which is based on our state-of-the-art dynamical energy loss formalism [13] (for more details on the model and its reliability see [11]). It also assumes a medium modeled with a constant average temperature. We here choose DREENA-C instead of hydrodynamically-wise more sophisticated versions of DREENA-B or -A [14], to exclude complications originating from details of medium evolution and thus, to provide analytical tractability. This can be achieved without significant loss of accuracy, as we already showed [12, 14, 15] that energy loss-sensitive observable R_{AA} (considered here) is barely sensitive to the medium evolution model. Therefore, DREENA-C provides an optimal framework for these studies, through observable R_{AA}, as it assumes a state-of-the-art energy loss model.

3 Results

In the first part of this paper, we address experimentally observed non-intuitive suppression patterns for both: light or D probes as well as heavy B probes. For light probes (upper left plot of Fig. 1) we see that R_{AA} as a function of N_{part} curves become flatter with increasing $p_\perp$, and that the difference between the curves decreases with increasing $p_\perp$. We call this effect saturation in R_{AA} vs. N_{part} curves. From the lower left plot of Fig. 1, we see flattening of R_{AA} at very high $p_\perp$, to which we refer to as saturation in R_{AA} vs. $p_\perp$ dependence.

Experimental measurements show qualitatively different R_{AA} vs. N_{part} pattern for B probes, compared to the light ones in the upper right plot of Fig. 1. Namely, for two opposite momentum ranges (purple stars vs. blue triangles) the experimental data are practically overlapping. Moreover, R_{AA} vs. N_{part} is flatter across the entire $p_\perp$ range, all of this indicating the saturation of this observable, not only at high $p_\perp$. Measurements of R_{AA} vs. $p_\perp$ (in lower right panel of Fig. 1) display slower change with $p_\perp$ compared to light probes. All these experimental observations are in good agreement with our DREENA-C predictions. So, the question is: which energy loss mechanism is responsible for the observations from Fig. 1?

To qualitatively explain the observations for light or D probes, in Fig. 2 we provide predictions for total (i.e., radiative plus collisional), only collisional, and only radiative R_{AA} vs. $p_\perp$, for a family of curves corresponding to different centralities. The equidistant in $p_\perp$ arrows (at 10, 100, and 190 GeV) indicate the density changes of these curves. From the left plots in first and second rows of Fig. 2, we observe that the leftmost arrow (at lower $p_\perp$) spans a much larger total R_{AA} range compared to the remaining two arrows (at higher $p_\perp$) and which are of a similar span. This explains much steeper R_{AA} vs. N_{part} curves at lower $p_\perp$, and their saturation with increasing $p_\perp$. From the central and right plots in first and second rows of Fig. 2, we see that collisional contribution is only important at lower $p_\perp$, where it increases steeply, while radiative contribution is significant across the entire $p_\perp$ range, but increases slowly. Consequently, the interplay between collisional and radiative contributions is responsible for R_{AA} vs. $p_\perp$ pattern, where steep R_{AA} increase is due to collisional, while saturation is due to radiative contributions.

Regarding B, mostly uniform R_{AA} vs. $p_\perp$ curves' density across the entire $p_\perp$ range (left plot in third row of Fig. 2) results in R_{AA} vs. N_{part} curves' overlap regardless of $p_\perp$. Central and right plots in the third row of Fig. 2 indicate that at lower $p_\perp$ both collisional and radiative contributions are important but significantly smaller than for light probes. Consequently, the mass hierarchy in collisional and radiative energy losses are responsible for R_{AA} vs. $p_\perp$ shape at lower $p_\perp$, while the flat radiative contribution plays a decisive role at higher $p_\perp$.

In the second part of this paper, we dive deeper into energy loss mechanisms and focus on the $p_\perp \lesssim 50$ GeV region. First, by comparing DREENA-C patterns in energy losses of charm

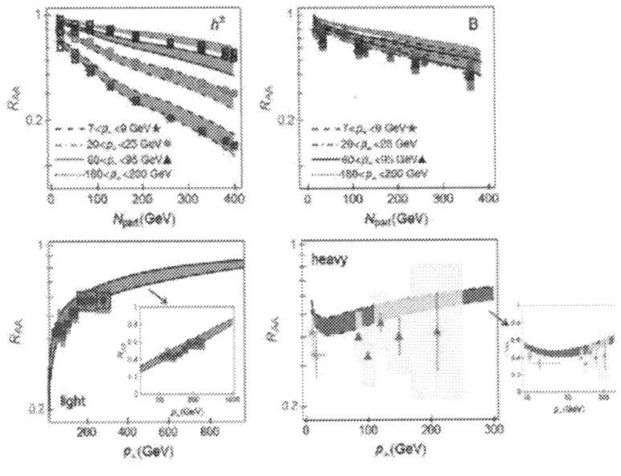

Figure 1. Comparison of suppression patterns data and predictions (gray bands) for light (left column) and B-probes (right column). For experimental data see [4]. Figure adapted from [4].

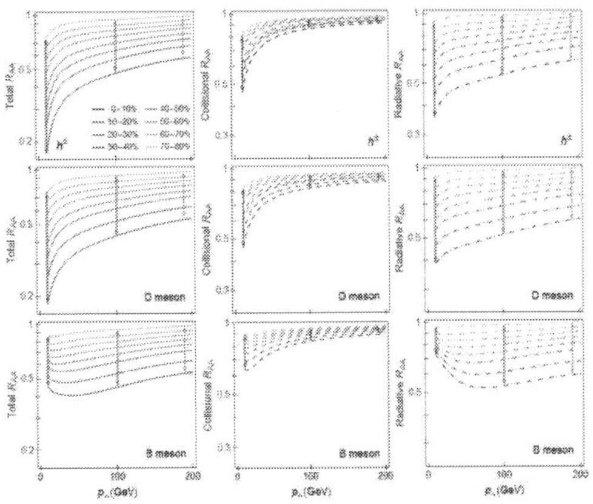

Figure 2. Explanation of suppression patterns for light (upper), D (middle) and B (lower row) probes. Left, central and right plots show, respectively, total, collisional and radiative R_{AA}. Figure adapted from [4].

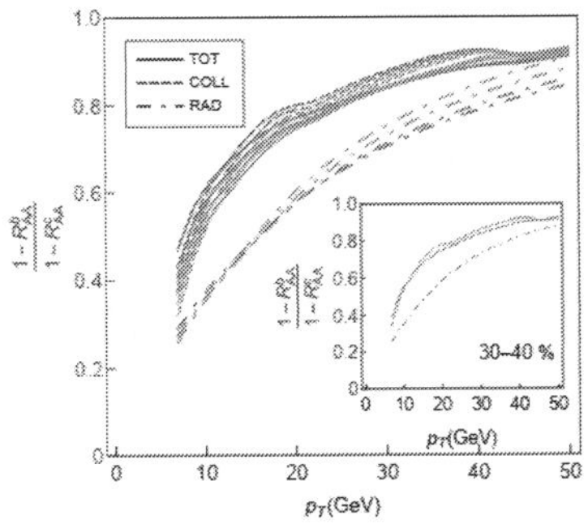

Figure 3. DREENA-C $(1 - R^b_{AA})/(1 - R^c_{AA})$ total (full), collisional (dashed) and radiative (dot-dashed curves) cases for centralities (as in Fig. 4 legend) are displayed. For clarity, $30 - 40\%$ centrality is singled out. Figure adapted from [11].

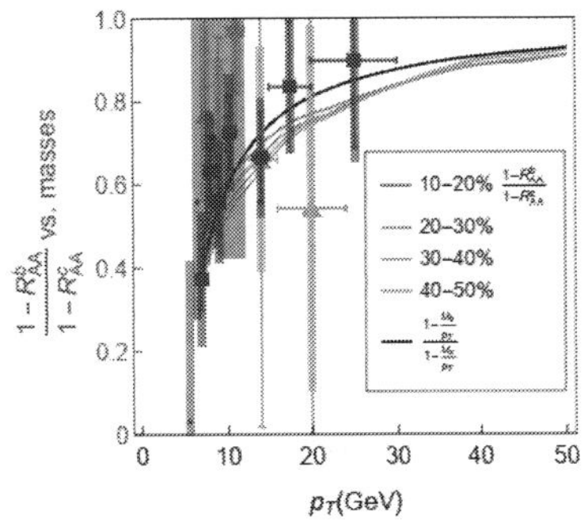

Figure 4. Numerical verification of $(1-R^b_{AA})/(1-R^c_{AA})$ as an adequate observable for mass extraction from collisional contribution, through comparison with experimental data (for more details see [11]. Figure adapted from [11].

and bottom (see Fig. 1 from [11]) we observe that we reproduced the dead-cone effect [16] in radiative energy loss, and clear mass ordering in $\Delta E_{coll}/E$ also. Compared to the $\Delta E_{rad}/E$, this effect in $\Delta E_{coll}/E$ is slightly less pronounced, but it is a significant observation. The question is, which observable could quantify this effect? To this end, we recall [11] that it should be a function of high-$p_\perp$ $1 - R_{AA}$, since it carries explicit information on the parton's energy loss [15, 17] while being practically insensitive to the details of medium evolution.

Guided by this idea, in Fig. 3 we compare $1-R_{AA}$ bottom to charm ratios (for different centralities), when only collisional, only radiative, and total interactions are considered. Surprisingly, we find that $(1-R^{b,tot}_{AA})/(1-R^{c,tot}_{AA})$ is essentially overlapping with $(1-R^{b,coll}_{AA})/(1-R^{c,coll}_{AA})$, which is true for all centralities. This is a nontrivial and significant insight. Consequently, Fig. 3 shows that collisional contribution is in the origin of $(1 - R^{b,tot}_{AA})/(1 - R^{c,tot}_{AA})$. Thus, in paper [11] we proposed $(1 - R^b_{AA})/(1 - R^c_{AA})$ as a new observable which may quantify the mass hierarchy in collisional energy loss. Moreover, after some algebraic manipulations (for more details see [11]), where collisional suppression definition [10, 18] as well as mass parameterization of heavy quarks initial distribution is applied [19], we infer that the $(1-R^{b}_{AA})/(1-R^{c}_{AA}) \simeq (1-\frac{M_b}{p_\perp})/(1-\frac{M_c}{p_\perp})$ reflects the mass hierarchy in collisional energy loss. This provides analytical support to $(1-R^{b}_{AA})/(1-R^{c}_{AA})$ as the appropriate new observable.

Furthermore, we get that $(1-R^{b}_{AA})/(1-R^{c}_{AA})$ is robust to the collision centrality (see Fig. 3), collision system, and collision energy (see Fig. 4). Therefore, we propose that the new observable should have general applicability to both the RHIC and the LHC experiments, independently of collision centrality, as long as QGP is formed.

4 Conclusions and outlook

Complex and significantly different suppression patterns for different flavors inspired us to distinguish between radiative and collisional contributions at the same dataset. Then, focused on the $p_\perp \lesssim 50$ GeV range, where both energy loss mechanisms are important for heavy probes, we proposed an observable, which could disentangle collisional from radiative energy loss. Also, by comparing with scarce heavy flavor data, we numerically verified the adequacy of the proposed observable for extracting mass hierarchy from collisional energy loss.

As an outlook, our analysis provides specific guidelines for future experiments. For instance, regarding non-intuitive suppression patterns, efforts should be concentrated on single-particle measurements to much higher $p_\perp$, while regarding the new observable, at the lower $p_\perp$ region, which is accessible at both RHIC and LHC. Note that current error bars are still large, so we expect that the upcoming high-luminosity measurements will provide a more rigorous test for this new observable. Finally, the suppression measurements of both b and c probes in the same centrality bins would be highly useful.

References

[1] A. M. Sirunyan *et al.* [CMS], Phys. Rev. Lett. **123**, no.2, 022001 (2019)
[2] N. Armesto, C. A. Salgado and U. A. Wiedemann, Phys. Rev. D **69**, 114003 (2004)
[3] M. Djordjevic and U. Heinz, Phys. Rev. C **77**, 024905 (2008)
[4] M. Djordjevic, Phys. Lett. B **763**, 439-444 (2016)
[5] B. W. Zhang, E. Wang and X. N. Wang, Phys. Rev. Lett. **93**, 072301 (2004)
[6] S. Wicks, W. Horowitz, M. Djordjevic and M. Gyulassy, Nucl. Phys. A **784**, 426 (2007)
[7] M. Djordjevic, Phys. Rev. C **74**, 064907 (2006)
[8] B. Blagojevic and M. Djordjevic, J. Phys. G **42**, no.7, 075105 (2015)
[9] M. H. Thoma, Phys. Lett. B **273**, 128-132 (1991)
[10] M. G. Mustafa, Phys. Rev. C **72**, 014905 (2005)
[11] Bojana Ilic and Magdalena Djordjevic, Phys. Rev. C **1**06, 014902 (2022)
[12] D. Zigic, I. Salom, J. Auvinen, M. Djordjevic and M. Djordjevic, J. Phys. G **46**, 085101 (2019)
[13] M. Djordjevic and M. Djordjevic, Phys. Lett. B **734**, 286-289 (2014)
[14] D. Zigic, I. Salom, J. Auvinen, M. Djordjevic and M. Djordjevic, Phys. Lett. B **791**, 236-241 (2019); [arXiv:2110.01544 [nucl-th]]
[15] M. Djordjevic, D. Zigic, M. Djordjevic and J. Auvinen, Phys. Rev. C **99**, 061902 (2019)
[16] Y. L. Dokshitzer and D. E. Kharzeev, Phys. Lett. B **519**, 199-206 (2001)
[17] S. Stojku, B. Ilic, M. Djordjevic and M. Djordjevic, Phys. Rev. C **103**, 024908 (2021)
[18] M. H. Thoma and M. Gyulassy, Nucl. Phys. B **351**, 491-506 (1991)
[19] G. D. Moore and D. Teaney, Phys. Rev. C **71**, 064904 (2005)

Medium-modifications of $g \to c\bar{c}$ splitting

Sohyun Park[1,*]

[1]Theoretical Physics Department, CERN, 1211 Geneva 23, Switzerland

Abstract. We study medium-modifications of the gluon splitting into a quark and anti-quark pair. Applying the Baier-Dokshitzer-Mueller-Peigné-Schiff and Zakharov (BDMPS-Z) formalism, we derive a path-integral formula for the in-medium $g \to q\bar{q}$ splitting function in the close-to-eikonal limit. Our analysis shows that there are two qualitatively different medium effects: transverse momentum broadening of $q\bar{q}$ pairs and enhanced production of such pairs. We note that both effects are numerically sizeable if the average momentum transfer from the medium to the parton is at the quark mass scale. In ultra-relativistic heavy-ion collisions, this condition is realized by charm quarks, therefore we focus our numerical analysis on the medium-modifications of $g \to c\bar{c}$ splitting.

1 Introduction

Heavy quarks are produced by high-momentum transfer processes in hadronic collisions. In particular for a $c\bar{c}$ pair to be produced, the invariant mass $Q^2 = (p_c + p_{\bar{c}})^\mu (p_c + p_{\bar{c}})_\mu$ should satisfy $4m_c^2 \leq Q^2 \leq \hat{s}$, where m_c is the charm quark mass and $\sqrt{\hat{s}}$ is the partonic center of mass energy. Since $m_c \gg \Lambda_{\rm QCD}$, this process is perturbatively calculable in QCD. Moreover, most $c\bar{c}$ pairs emerge back-to-back with $Q^2 \sim O(\hat{s})$ at short distances, where the Quark Gluon Plasma (QGP) has not formed yet. That is, the total charm cross-section is mainly determined by such high-Q^2 short-distance processes and therefore almost unmodified by the QCD medium.

In heavy-ion collisions, charm quarks produced at short distances enter the QGP, and the dominant in-medium process is gluon radiation off a charm, $c \to c\,g$ at long distances. The BDMPS-Z calculation for the $c \to c\,g$ splitting function indicates that charm quarks traversing the QGP lose energy due to medium-induced gluon radiation [1–6]. This effect has been observed as the suppression of high-p_T spectra of charmed hadrons. It should be noted however that this dominant process $c \to c\,g$ does not affect the total charm yield but only modifies the charm transverse momentum distribution.

The process we are focusing on is the medium-induced charm production with $Q^2 \ll \hat{s}$ at long distances via a gluon splitting into a $c\bar{c}$ pair inside the QGP. This is much rarer compared to the charm production at short distances and sub-dominant compared to the medium-induced gluon radiation, but it is a dominant process for the in-medium charm production. We also remark two features making this process useful. First, the splitting rate depends on the degree of momentum transfer from the medium, hence it provides a way to probe the QGP properties. Second, the $c\bar{c}$ pair emerges collinear to the gluon direction, in contrast to back-to-back emergence in vacuum, so they are likely to remain inside the jet. Hence, splittings are traceable with jets.

*e-mail: sohyun.park@cern.ch

2 Medium-modified $g \to c\bar{c}$ splitting function

In the collinear limit $Q^2 \ll \hat{s}$, partonic cross-sections factorize. For example, the cross-section for the collision of two gluons producing a $c\bar{c}$ pair plus something else, say X, can be factorized by the cross-section for producing a parent gluon and the probability that this gluon splits into a $c\bar{c}$ pair,

$$\hat{\sigma}^{gg\to c\bar{c}X} \xrightarrow{Q^2\ll\hat{s}} \hat{\sigma}^{gg\to gX}\,\frac{\alpha_s}{2\pi}\,\frac{1}{Q^2}\,P_{g\to c\bar{c}}(z)\,. \tag{1}$$

Here the probability $P_{g\to c\bar{c}}(z)$ is called the $g \to c\bar{c}$ splitting function. In the collinear limit, the virtuality is given by $Q^2 = \frac{m_c^2+\kappa^2}{z(1-z)}$, where $\kappa = \frac{1}{2}\left(\mathbf{k}_c - \mathbf{k}_{\bar{c}}\right)$ is the relative momentum of the $c\bar{c}$ pair transverse to the gluon direction, and z is the longitudinal momentum fraction carried by the charm. The in-vacuum $g \to c\bar{c}$ splitting function to leading order in α_s reads [7]

$$\left(\frac{1}{Q^2}\,P_{g\to c\bar{c}}\right)^{\text{vac}}(\kappa, z) = \frac{1}{2}\frac{1}{Q^4}\left[\left(m_c^2 + \kappa^2\right)\frac{z^2 + (1-z)^2}{z(1-z)} + 2m_c^2\right]. \tag{2}$$

The in-medium $g \to c\bar{c}$ splitting function in the close-to-eikonal limit with the multiple soft scattering approximation in the framework of the BDMPS-Z formalism reads [8]

$$\begin{aligned}\left(\frac{1}{Q^2}\,P_{g\to c\bar{c}}\right)^{\text{med}}(E_g, \kappa, z, \hat{q}, L) &= 2\,\Re\mathrm{e}\,\frac{1}{8\,E_g^2}\int_0^L dt\int_t^\infty d\bar{t}\,\exp\left[i\frac{m_c^2}{2E_g z(1-z)}(t-\bar{t})\right]\\ &\times\int d\mathbf{r}_{\text{out}}\exp\left[-\frac{1}{4}\int_{\bar{t}}^\infty d\xi\,\hat{q}(\xi,z)\,\mathbf{r}_{\text{out}}^2\right]\exp\left[-i\,\kappa\cdot\mathbf{r}_{\text{out}}\right]\\ &\times\left[\left(m_c^2 + \frac{\partial}{\partial\mathbf{r}_{\text{in}}}\cdot\frac{\partial}{\partial\mathbf{r}_{\text{out}}}\right)\frac{z^2+(1-z)^2}{z(1-z)} + 2m_c^2\right]\mathcal{K}\left[\mathbf{r}_{\text{in}} = 0, t; \mathbf{r}_{\text{out}}, \bar{t}\right],\end{aligned} \tag{3}$$

where E_g is the gluon energy and L is the longitudinal size of the medium, and

$$\mathcal{K}[\mathbf{x}, t; \mathbf{r}, \bar{t}] = \int_{\rho(t)}^{\rho(\bar{t})}\mathcal{D}\rho\,\exp\left[i\int_t^{\bar{t}} d\xi\left(\frac{E_g z(1-z)}{2}\,\dot{\rho}^2 - \frac{\hat{q}(\xi,z)\,\rho^2}{4\,i}\right)\right] \tag{4}$$

is the path-integral of a harmonic oscillator. This formula (3) describes the rate at which a gluon born at $t = 0$ splits into a $c\bar{c}$ pair at longitudinal positions t in amplitude and $\bar{t}$ in complex conjugate amplitude. The medium effect is encoded in the so-called quenching parameter $\hat{q}(\xi, z)$ entering the absorption factor $\exp[\cdots]$ in the second line and the path-integral $\mathcal{K}$ in the last line, which measures the squared momentum transferred per unit path-length from the medium to a parton.

Medium-induced shifts in transverse momentum arise in two occasions in the evolution, one from back-propagating the phase $\exp\left[-i\,\kappa\cdot\mathbf{r}_{\text{out}}\right]$ from $\bar{t}$ to ∞ in the second line, and the other from acting the differential operator $\frac{\partial}{\partial\mathbf{r}_{\text{in}}}\cdot\frac{\partial}{\partial\mathbf{r}_{\text{out}}}$ on the path-integral in the last line of (3). Evaluating (3), we find $P^{\text{med}}_{g\to c\bar{c}} \sim O\left(\frac{\langle\mathbf{q}^2\rangle_{\text{med}}}{Q^2}\right)$. We recall that the mass correction of $P^{\text{vac}}_{g\to c\bar{c}}$ is of order $O\left(\frac{m_c^2}{Q^2}\right)$. Therefore, medium-modifications become comparable to this mass correction at the scale

$$\langle\mathbf{q}^2\rangle_{\text{med}} = \int_{\tau_i}^{\tau_f} d\tau\hat{q}(\tau) \sim O\left(m_c^2\right). \tag{5}$$

From model extraction in central PbPb collisions at $\sqrt{s_{NN}} = 5.02\,\text{TeV}$, we obtain $4\,\text{GeV}^2 < \langle q^2\rangle_{\text{med}} < 8\,\text{GeV}^2$ [9]. That is, the condition (5) is realized for the charm mass of $m_c = 1.27\,\text{GeV}$. This is indeed why we focus on the $g \to c\bar{c}$ splitting although the formula (3) applies to gluons splitting into any quark-anti-quark pairs.

3 Numerical results

There are two signatures of medium-modification: transverse momentum broadening of $c\bar{c}$ pairs and enhanced production of such pairs. Left panel (from [8]) of Figure 1 plots numerical results of the medium-modification of (3) when the average squared momentum transferred from the medium is $\langle q^2 \rangle_{\rm med} = \hat{q}L = 4\,{\rm GeV}^2$. It shows the depletion of low κ^2 splittings due to the in-medium broadening and enhanced production rate at the scale $\kappa^2 \sim \hat{q}\,L$. Qualitatively these two phenomena can be understood as follows: A gluon originally having a virtuality lower than the mass threshold $Q^2 < 4m_c^2$ acquires transverse momentum as it traverses the medium, and hence the virtuality reaches or exceeds the threshold $Q^2 \geq 4m_c^2$ so that the gluon can produce a $c\bar{c}$ pair, which results in the enhanced production. The charm and anti-charm produced also get momentum transfers in the medium, which leads to the momentum broadening.

As an experimental strategy we propose to measure the ratio of the number of $D^0\bar{D}^0$-tagged gluon jets to the number of inclusive gluon jets as a function of jet p_T. Due to the enhanced production of $c\bar{c}$ pairs in the QGP, a larger fraction of $D^0\bar{D}^0$-tagged jets is expected in heavy-ion collisions than in pp collisions. Right panel (from [10]) of Figure 1 plots Monte Carlo simulation (using Pythia) results, which reveal $10\% - 40\%$ enhancement of $D^0\bar{D}^0$-tagged jets in heavy-ion collisions with respect to pp collisions [10]. See the small box in the right corner. One remark is that for a very high p_T correspondingly a very high gluon energy E_g, the enhancement is suppressed. This feature can be understood in terms of the formation time $\tau_{\rm form}$ of a $c\bar{c}$ pair. In the rest frame of the parent gluon, $\tau_{\rm form,\,rest} = \frac{1}{Q}$. In the lab frame, the gluon is boosted with respect to the medium by a Lorentz factor $\gamma = E_g/Q$, and hence dilated to $\tau_{\rm form,\,boosted} = \frac{E_g}{Q^2}$. If the gluon energy E_g is too high, the formation time gets too delayed such that the splitting occurs after the gluon leaves the medium and hence no momentum transfers from the medium, i.e., no medium-induced production of $c\bar{c}$ pairs.

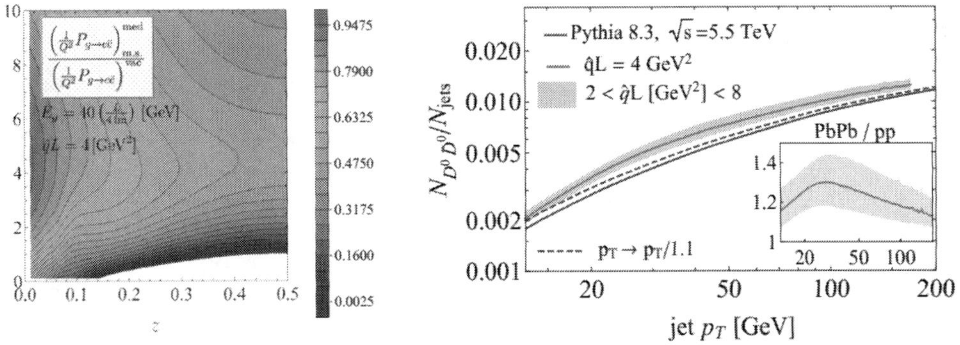

Figure 1. Left (from [8]): The ratio of the medium-modified splitting function to the vacuum one shows the characteristic momentum broadening and enhanced production of $c\bar{c}$ pairs. Right (from [10]): The fraction of jets carrying a $D^0\bar{D}^0$-tag in $\sqrt{s} = 5.5\,{\rm TeV}$ mid-central pp collisions (blue line), and in central PbPb collisions (red band). The blue dashed line is included to take into account jet energy-loss estimated as $\Delta p_T/p_T \approx 10\%$ of inclusive jets.

Another remark concerns other mechanisms, besides the medium-modified $g \to c\bar{c}$ splitting, that can affect the ratio $N_{D^0\bar{D}^0}/N_{\rm jets}$: First, jets lose energy to the QGP via the medium-induced gluon radiation $g \to gg$ and $q \to qg$, and as a result $p_T^{\rm jet}$ decreases. We estimate the size of this jet energy-loss by the average fractional energy-loss $\Delta p_T/p_T \approx 10\%$ of inclusive jets in central PbPb collisions [11] (blue dashed line in Figure 1 Right). Second, the medium-induced gluon radiation processes $g \to gg$ and $q \to qg$ change the distribution of gluons that can split into $c\bar{c}$ pairs and hence affect the rate of $c\bar{c}$ production inside jets. This effect certainly exists, however we have checked that it is numerically smaller and does not significantly affect our main conclusion of medium-enhanced $c\bar{c}$ production [10].

4 Conclusion

We have calculated the medium-modification of the QCD leading order gluon splitting function into a charm and anti-charm pair in the BDMPS-Z formalism. The result of $P^{\text{med}}_{g\to c\bar{c}}$ shows broadening of the relative transverse momentum of $c\bar{c}$ pairs and enhancement of $c\bar{c}$ production in the QGP, which are sizeable if the average momentum transfer from the medium is at the charm mass scale. As an experimental strategy for testing the enhanced $g \to c\bar{c}$ splittings, we have made a Monte Carlo study for the fraction of $D^0\bar{D}^0$-tagged jets over inclusive jets, which shows 10% – 40% enhancement for such measurements in heavy-ion collisions with respect to elementary collisions.

5 Acknowledgment

I thank Maximilian Attems, Jasmine Brewer, Gian Michele Innocenti, Aleksas Mazeliauskas, Wilke van der Schee and Urs Achim Wiedemann for collaboration and for critical readings of this manuscript. I also would like to thank Dániel Barta, Sangyong Jeon, Min-Jung Kweon, Saehanseul Oh and Jaebeom Park for useful questions and discussions during SQM 2022.

References

[1] R. Baier, Y. L. Dokshitzer, A. H. Mueller, S. Peigne and D. Schiff, Nucl. Phys. B **483**, 291-320 (1997) [arXiv:hep-ph/9607355 [hep-ph]].
[2] B. G. Zakharov, JETP Lett. **63**, 952-957 (1996) [arXiv:hep-ph/9607440 [hep-ph]].
[3] Y. L. Dokshitzer and D. E. Kharzeev, Phys. Lett. B **519**, 199-206 (2001) [arXiv:hep-ph/0106202 [hep-ph]].
[4] U. A. Wiedemann, Nucl. Phys. B **588**, 303-344 (2000) doi:10.1016/S0550-3213(00)00457-0 [arXiv:hep-ph/0005129 [hep-ph]].
[5] M. Gyulassy, P. Levai and I. Vitev, Nucl. Phys. B **594**, 371-419 (2001) doi:10.1016/S0550-3213(00)00652-0 [arXiv:nucl-th/0006010 [nucl-th]].
[6] X. N. Wang and X. f. Guo, Nucl. Phys. A **696**, 788-832 (2001) doi:10.1016/S0375-9474(01)01130-7 [arXiv:hep-ph/0102230 [hep-ph]].
[7] R. K. Ellis, W. J. Stirling and B. R. Webber, *QCD and collider physics* (Cambridge University Press, 1996) 1-435.
[8] M. Attems, J. Brewer, G. M. Innocenti, A. Mazeliauskas, S. Park, W. van der Schee and U. A. Wiedemann, [arXiv:2203.11241 [hep-ph]].
[9] A. Huss, A. Kurkela, A. Mazeliauskas, R. Paatelainen, W. van der Schee and U. A. Wiedemann, Phys. Rev. C **103**, no.5, 054903 (2021) [arXiv:2007.13758 [hep-ph]].
[10] M. Attems, J. Brewer, G. M. Innocenti, A. Mazeliauskas, S. Park, W. van der Schee and U. Wiedemann, [arXiv:2209.13600 [hep-ph]].
[11] S. Chatrchyan *et al.* [CMS], Phys. Rev. C **84**, 024906 (2011) [arXiv:1102.1957 [nucl-ex]].

Impact of the electromagnetic field on Z^0 leptonic invariant mass in ultrarelativistic heavy ion collisions

Yifeng Sun[1,2,*], *Vincenzo* Greco[2,3], *Xin-Nian* Wang[4], and *Salvatore* Plumari[2,3]

[1]School of Physics and Astronomy, Shanghai Key Laboratory for Particle Physics and Cosmology, and Key Laboratory for Particle Astrophysics and Cosmology (MOE), Shanghai Jiao Tong University, Shanghai 200240, China

[2]Department of Physics and Astronomy, University of Catania, Via S. Sofia 64, 1-95125 Catania, Italy

[3]Laboratori Nazionali del Sud, INFN-LNS, Via S. Sofia 62, I-95123 Catania, Italy

[4]Nuclear Science Division, Lawrence Berkeley National Laboratory, Berkeley, CA 94720, USA

Abstract. The extraordinary strong magnetic field generated in ultrarelativistic heavy ion collisions makes it possible to investigate many novel effects in the hot QCD matter, and a direct probe of it thus becomes an urgent task in the field. Here we report a new way to probe it, where the leptonic invariant mass distribution of Z^0 boson will be modified in the presence of the magnetic field due to the Lorentz force on the lepton pairs from the decay of Z^0 boson. It is found that the magnetic field can decrease the mean value of Z^0 leptonic invariant mass and increase its width, and both shifts approximately depend on the integral of the magnetic field over its time duration quadratically.

1 Introduction

The presence of a strong magnetic field in ultrarelativistic heavy ion collisions can induce many novel effects, such as the chiral magnetic effect (CME) which can probe the local parity (P) and charge conjugation parity (CP) violations of QCD [1, 2] and the magnetic field dependence of the chiral condensate [3]. However, there are quite a lot of disagreements in the calculation of the time evolution of the magnetic field due to the uncertainty of the electrical conductivity of QGP [4, 5] as well as the difficulty in numerically solving magnetohydrodynamics or transport approach coupled to the Maxwell equations. This inspired the studies to search for a direct probe of the strong electromagnetic (e.m.) fields, and it is proposed that the directed flow $v_1 = \langle p_x/p_T \rangle$ splitting between positively and negatively charged light hadrons [6, 7] as well as heavy meson pairs [8–13] can serve the purpose. Here we report a new and cleaner probe of the strong e.m. fields proposed in our recent manuscript [14] via the invariant mass distribution of Z^0 boson reconstructed from its decaying lepton pairs, since the momenta of these leptons should be modified by e.m. fields. It is significantly easier to measure the leptonic invariant mass distribution of Z^0 boson than the v_1 splitting, thus opening up a more accessible experimental probe.

*e-mail: sunyfphy@sjtu.edu.cn

2 Z^0 boson in ultrarelativistic heavy ion collisions and the transport of leptons from the decay of Z^0 boson

The initial conditions of the lepton pairs decaying from Z^0 boson in heavy ion collisions inherit from the distribution of Z^0 boson in the 4-D momentum and coordinate spaces. In momentum space, the transverse momenta $\mathbf{p}_T$ and rapidity y_z dependences of Z^0 boson are given by fitting the experimental measurements on Z^0 boson in pp collisions at the same collision energies, and the energy of Z^0 boson is given by $E_p = \sqrt{M^2 + \mathbf{p}^2}$. Here the mass distribution of Z^0 is given by a Breit-Wigner form:

$$\rho(M) = \frac{1}{\pi}\frac{\Gamma/2}{(M - M_0)^2 + \Gamma^2/4}, \tag{1}$$

with $M_0 = 91.1876$ GeV and $\Gamma = 2.4952$ GeV[15]. In coordinate space, the transverse coordinate distribution of Z^0 boson is given by the profile of the binary nucleon-nucleon collisions, while the longitudinal coordinate and time are given by $z = \tau_{Z^0} \sinh y_z$ and $t = \tau_{Z^0} \cosh y_z$ with $\tau_{Z^0} = \hbar/m_{Z^0} = 0.0022$ fm$/c$. This indicates that the space-time rapidity η_S of Z^0 boson is assumed to be equal to its rapidity. Finally one can obtain the spacetime coordinate of lepton pairs from their mother Z^0 that moves in a straight line, where the decay time follows a distribution $\rho(\Delta t) \propto e^{-\frac{\Gamma\Delta t}{\gamma_v}}$, with γ_v being the Lorentz contraction factor.

The lepton pairs not only interact with e.m. fields after their production, but also interact with the charged quarks due to the lepton-quark scattering in the quark-gluon plasma (QGP) phase, though the effect of the latter one may be small. To consider both effects, the Langevin equations are employed:

$$d\mathbf{x} = \frac{\mathbf{p}}{E}dt, \tag{2}$$

$$d\mathbf{p} = -\gamma\mathbf{p}dt + \boldsymbol{\xi}\sqrt{2D_p dt} + q(\mathbf{E} + \mathbf{v}\times\mathbf{B}), \tag{3}$$

where the momentum diffusion coefficient D_p is related to the drag coefficient γ, energy of leptons E, and the local temperature T by $D_p = \gamma ET$, and each component of $\boldsymbol{\xi}$ is a real number randomly sampled from a normal distribution with $\langle\xi_i\rangle = 0$ and $\langle\xi_i\xi_j\rangle = \delta_{ij}$. The value of D_p can be calculated by the lepton-quark scattering, see details in Ref. [14].

For the space and time dependence of e.m. fields, a general parametrization of them used in several studies is adopted in this study, knowing the large uncertainty in calculating e.m. fields:

$$eB_y(x, y, \tau) = -eB(\tau)\rho_B(x, y), \rho_B(x, y) = \exp[-\frac{x^2}{2\sigma_x^2} - \frac{y^2}{2\sigma_y^2}], eB(\tau) = eB_0/(1 + (\tau/\tau_B)^a), \tag{4}$$

where eB_0, σ_x and σ_y are usually given by the estimates of e.m. fields in the vacuum in AA collisions at $t = 0$ [16], and τ_B and a are the parameters that control the lifetime and the specific manner of decay of e.m. fields, respectively. The electric field eE_x is then determined by solving the Faraday's Law $\nabla \times \mathbf{E} = -\partial\mathbf{B}/\partial t$:

$$eE_x(t, x, y, \eta_S) = \rho_B(x, y)\int_0^{\eta_S} d\chi B'\left(\frac{t}{\cosh\chi}\right)\frac{t}{\cosh\chi}. \tag{5}$$

After the modification on the momenta by e.m. fields and lepton-quark scattering, the invariant mass of Z^0 boson is reconstructed from these lepton pairs.

The modification of Z^0 boson leptonic invariant mass is investigated primarily on systems of PbPb collisions at 5.02 TeV due to the large number of Z^0 production events. In this colliding system at centrality 20-30%, corresponding to impact parameter $b = 7.5$ fm, the

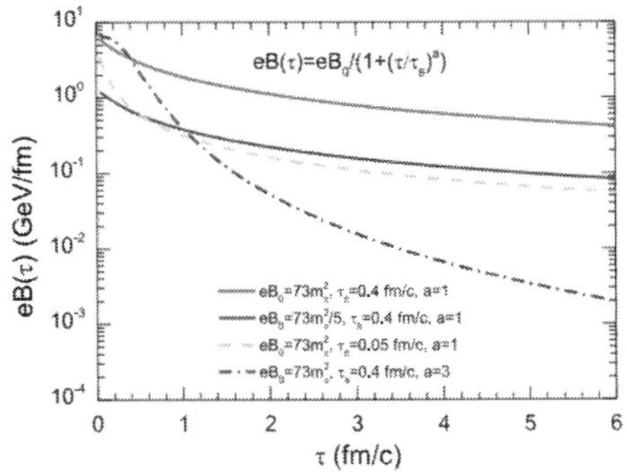

Figure 1. (Color online) The time dependence of $eB(\tau)$ with different parameter sets of eB_0, τ_B and a.

parameters in Eq. (4) are found to be $eB_0 = 73\,m_\pi^2$, $\sigma_x = 3$ fm and $\sigma_y = 4$ fm [11], where eB_0 is the magnitude of magnetic field estimated in the vacuum at $t = 0$. To have a general investigation eB_0 is varied by 5 times, τ_B is varied by 8 times and a is varied by 3 times in this study. In Fig.1 we show the time evolution of $eB(\tau)$ with four parameter sets of eB_0, τ_B and a,and it should be noted that the parameter set $eB_0 = 73\,m_\pi^2$, $\tau_B = 0.4$ fm$/c$ and $a = 1$ can reproduce the ALICE measurement on the v_1 splitting between $(D^0,\overline{D}^0)$.

3 Numerical results

To the effect of the lepton-quark scattering in the QGP, the modification is found to be extremely small. To quantitatively characterize the modification of Z^0 leptonic invariant mass, we define its mean value shift $\Delta\langle M\rangle$ and width shift $\Delta\sigma$ as

$$\Delta\langle M\rangle = \langle M_f\rangle - \langle M_i\rangle, \Delta\sigma = \sigma_f - \sigma_i = \sqrt{\frac{\sum(M_f - \langle M_f\rangle)^2}{N-1}} - \sqrt{\frac{\sum(M_i - \langle M_i\rangle)^2}{N-1}}. \quad (6)$$

It is found that $\Delta\langle M\rangle$ is about -1.9 MeV due to the drag of QGP and $\Delta\sigma$ is about 0.2 MeV, though both shifts are beyond the experimental determination. On the other hand, as shown by the red solid lines in the left, middle and right panels of Fig.2, the parameter set $eB_0 = 73\,m_\pi^2$, $\tau_B = 0.4$ fm$/c$ and $a = 1$ can modify largely Z^0 leptonic invariant mass compared to the black solid lines, where $\langle M\rangle$ decreases by 246 MeV and σ increases by 305 MeV. Moreover, if one decreases the accumulation of e.m. fields by decreasing eB_0 or τ_B or by increasing a, the modification becomes smaller as shown in Fig. 2.

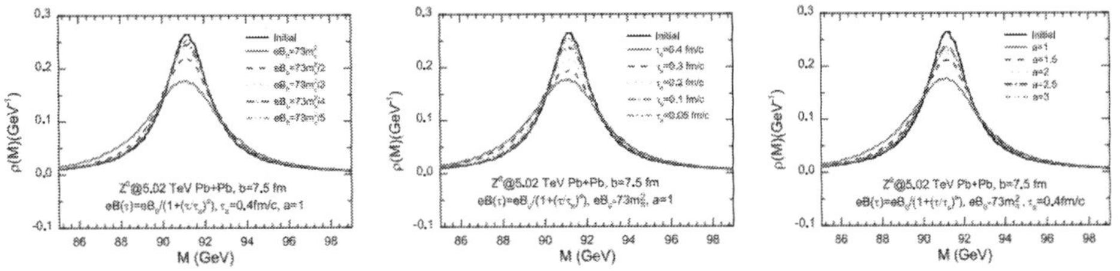

Figure 2. (Color online) The eB_0, τ_B and a dependences of Z^0 invariant mass distribution at midrapidity $|y_z| \leq 0.5$ reconstructed by lepton pairs after interacting with e.m. fields. Left: eB_0 varies from $73\,m_\pi^2/5$ to $73\,m_\pi^2$; Middle: τ_B varies from 0.05 fm$/c$ to 0.4 fm$/c$; Right: a varies from 1 to 3.

Since the momenta change of leptons is directly related to the accumulation of e.m. fields $\chi_B = \int_{\tau_0}^{\tau_1} d\tau eB(\tau)$, where τ_0 and τ_1 are the formation time of leptons and the time when they leave e.m. fields, we further check the relation between $\Delta\langle M\rangle$ ($\Delta\sigma$) and χ_B. It is found [14] that $\Delta\sigma$ depends on χ_B quadratically no matter one varies eB_0, τ_B or a. Since the invariant mass is symmetric with charge conjugation, it should depend on the square of χ_B at leading order if one varies the electric and magnetic fields by the same factor. $\Delta\langle M\rangle$ thus should depend on χ_B quadratically when one varies eB_0, which is what is found in Ref. [14]. On the other hand, though $\Delta\langle M\rangle$ does not depend on the χ_B quadratically when one varies τ_B or a, it still depends on the power of it. The power is found to be 2.12 or 2.33, which is still close to 2.

4 Conclusion

The Z^0 leptonic invariant mass in ultrarelativistic heavy ion collisions is investigated with Langevin equations at LHC energy. It is found that lepton-quark scattering modifies Z^0 leptonic invariant mass extremely small while e.m. fields can modify both its mean value and width by few hundred MeV, thus providing a cleaner probe of e.m. fields in ultrarelativistic heavy ion collisions. With the current statistics on the Z^0 boson in PbPb collisions, it is possible to experimentally detect this modification of a few hundred MeV in the invariant mass of Z^0 boson. It is also found that both its mean value and width depend approximately on the integral of the magnetic field over its time duration quadratically, which makes a measurement on the modification of Z^0 leptonic invariant mass a direct probe of the time accumulation of magnetic field.

References

[1] D. Kharzeev, R. D. Pisarski and M. H. G. Tytgat, Phys. Rev. Lett. **81**, 512-515 (1998).
[2] D. E. Kharzeev, L. D. McLerran and H. J. Warringa, Nucl. Phys. A **803**, 227-253 (2008).
[3] H. T. Ding, S. T. Li, J. H. Liu, and X. D. Wang, Phys. Rev. D **105** (3), 034514 (2022).
[4] H. T. Ding, A. Francis, O. Kaczmarek, F. Karsch, E. Laermann and W. Soeldner, Phys. Rev. D **83**, 034504 (2011).
[5] A. Amato, G. Aarts, C. Allton, P. Giudice, S. Hands and J. I. Skullerud, Phys. Rev. Lett. **111** no.17, 172001 (2013).
[6] U. Gursoy, D. Kharzeev and K. Rajagopal, Phys. Rev. C **89** no.5, 054905 (2014).
[7] U. Gursoy, D. Kharzeev, E. Marcus, K. Rajagopal and C. Shen, Phys. Rev. C **98** no.5, 055201 (2018).
[8] S. K. Das, S. Plumari, S. Chatterjee, J. Alam, F. Scardina and V. Greco, Phys. Lett. B **768**, 260-264 (2017).
[9] S. Chatterjee and P. Bozek, Phys. Lett. B **798**, 134955 (2019).
[10] Lucia Oliva, S. Plumari, and V. Greco, JHEP **05**, 034 (2021).
[11] Y. Sun, S. Plumari and V. Greco, Phys. Lett. B **816**, 136271 (2021).
[12] Y. Sun, V. Greco and S. Plumari, Eur. Phys. J. Plus **136** no.7, 726 (2021).
[13] Z.F. Jiang, S.S. Cao, W.J. Xing, X.Y. Wu, C. B. Yang, and B.W. Zhang, Phys. Rev. C **105**, 054907 (2022).
[14] Y. Sun, V. Greco and X.N. Wang, Phys. Lett. B **827**, 136962 (2022).
[15] M. Tanabashi et al. [Particle Data Group], Phys. Rev. D **98** no.3, 030001 (2018).
[16] W. T. Deng and X. G. Huang, Phys. Rev. C **85**, 044907 (2012).

In-medium effects on hidden strangeness production in heavy-ion collisions

Taesoo Song[1,*], *Joerg* Aichelin[2,3], and *Elena* Bratkovskaya[1,4,5]

[1]GSI Helmholtzzentrum für Schwerionenforschung GmbH, Planckstrasse 1, 64291 Darmstadt, Germany
[2]SUBATECH UMR 6457 (IMT Atlantique, Université de Nantes, IN2P3/CNRS), 4 Rue Alfred Kastler, F-44307 Nantes, France
[3]Frankfurt Institute for Advanced Studies, Ruth-Moufang-Strasse 1, 60438 Frankfurt am Main, Germany
[4]Institute for Theoretical Physics, Johann Wolfgang Goethe Universität, Frankfurt am Main, Germany
[5]Helmholtz Research Academy Hesse for FAIR (HFHF), GSI Helmholtz Center for Heavy Ion Physics, Campus Frankfurt, 60438 Frankfurt, Germany

Abstract. We study ϕ meson production in heavy-ion collisions from subthreshold energies of 1.23 A GeV up to RHIC energies within the microscopic Parton-Hadron-String Dynamics (PHSD) transport approach where novel production channels for ϕ mesons based on a coupled channel T-matrix approach are implemented along with the collisional broadening of the ϕ meson spectral width in medium. Since ϕ meson production is closely related to the production of kaons and antikaons, antikaon properties are described via the self-consistent coupled-channel unitarized scheme within a SU(3) chiral Lagrangian (G-matrix) which incorporates explicitly the s− and p− waves of the kaon-nucleon interaction, while the in-medium modifications of kaons are accounted for via a kaon-nuclear potential, which is assumed to be proportional to the local baryon density.

1 Introduction

Recently it has been reported by the HADES collaboration that ϕ mesons are produced in a relatively large quantity in Au+Au collisions at subthreshold energies and the ratio of hidden strangeness to open strangeness reaches values of ≈ 0.5 [1]. The same tendency of an enhanced ϕ production in Ni+Ni and Al+Al collisions at 1.93 A GeV has been reported earlier by the FOPI collaboration [2, 3]. With increasing beam energy this ratio decreases to 0.2 as has been measured recently by the STAR collaboration [4] and at high energies the dependence on the collision energy is mild [5–9].

The goal of this study is to show that the observed 'enhanced' ϕ multiplicity and ϕ/K^- ratio close to threshold can be understood by considering a collisional broadening of the ϕ meson spectral function and accounting for additional multi-step meson-baryon and meson-hyperon reactions for ϕ meson production as predicted by the SU(6) extension of the meson-baryon chiral Lagrangian within a unitary coupled channel T-matrix approach.

*e-mail: T.Song@gsi.de

2 In-medium modification of ϕ meson properties

In order to explore the influence of in-medium effects on the vector-meson spectral function we introduce the collisional broadening by

$$\Gamma^*_\phi(M,|\vec{p}|,\rho) = \Gamma_\phi(M) + \Gamma_{coll}(M,|\vec{p}|,\rho), \tag{1}$$

where M is the mass, $\Gamma_\phi(M)$ the total width of the vacuum spectral function of the ϕ meson, and Γ_{coll} the collisional width approximated as

$$\Gamma_{coll}(M,|\vec{p}|,\rho) = \gamma\,\rho < v\,\sigma^{tot}_{VN} > \approx \alpha_{coll}\,\frac{\rho}{\rho_0}. \tag{2}$$

Here v is the velocity of the ϕ meson in the rest frame of the nucleon current, $\gamma^2 = 1/(1-v^2)$, ρ the nuclear density scaled by $\rho_0 = 0.168$ fm^{-3} (normal nuclear density) and σ^{tot}_{VN} the meson-nucleon total cross section in vacuum. In order to simplify the calculations of $\Gamma_{coll}(\rho)$ we use the linear density approximation [10] with a coefficient α_{coll} which is taken to be 25 MeV [11].

3 ϕ production/absorption within a SU(6) based T-matrix approach

The s-wave scattering amplitude of meson and baryon from the SU(6) chiral effective Lagrangian is written as [12],

$$V^{SIJ}_{ij} = \varepsilon^{SIJ}_{ij}\frac{2\sqrt{s}-M_i-M_j}{4f_if_j}\sqrt{\frac{E_i+M_i}{2M_i}}\sqrt{\frac{E_j+M_j}{2M_j}}, \tag{3}$$

where $i(j)$ indicates the initial (final) meson-baryon scattering states, $M_{i(j)}$ and $E_{i(j)}$ are, respectively, mass and center-of-mass energy of the baryon, $f_{i(j)}$ the decay constant of the meson in the $i(j)$ state, and ε^{SIJ}_{ij} the degeneracy coefficient, corresponding to the scattering channel with $S, I,$ and J being total strangeness, isospin and angular momentum of the collision, respectively [13, 14]. The T-matrix approach can be formulated on the basis of the Born scattering amplitude V^{SIJ}_{ik},

$$T^{SIJ}_{ij} = V^{SIJ}_{ij} + V^{SIJ}_{ik}G^{SIJ}_{kk}T^{SIJ}_{kj}, \tag{4}$$

where k is the intermediate meson-baryon state and the sum is performed over all possible states. G^{SIJ}_{kk} is the product of the meson and baryon propagators of the state k [15], which is renormalized such that $G^{SIJ}_{kk}(s = m_N^2 + m_\pi^2) = 0$ with m_N and m_π being nucleon and pion masses, respectively. The channels considered in this study for ϕ meson production are ηN, $K\Lambda$, $K\Sigma$, ρN, $K\Sigma^*$, $\rho\Delta$, $K^*\Lambda$, $K^*\Sigma$, $K^*\Sigma^* \to \phi N$ for $I = 1/2$ and $K\Sigma$, ρN, $\eta\Delta$, $K\Sigma^*$, $\rho\Delta$, $K^*\Sigma$, $K^*\Sigma^* \to \phi\Delta$ for $I = 3/2$, including their inverse reactions by detailed balance.

4 ϕ meson production in heavy-ion collisions

Fig. 1 shows the PHSD results for the rapidity distribution of reconstructed ϕ mesons from the decay into K^+K^- pairs, compared with the experimental data from the HADES and STAR collaborations. The short dashed orange lines show the PHSD results without including the novel mB channels from the T-matrix approach and without any in-medium modifications of ϕ and $K, \bar{K}$ mesons. The dash-dotted green lines show the results without in-medium modifications of ϕ and $K, \bar{K}$ mesons [17]. The dashed red lines indicate the ϕ rapidity distributions

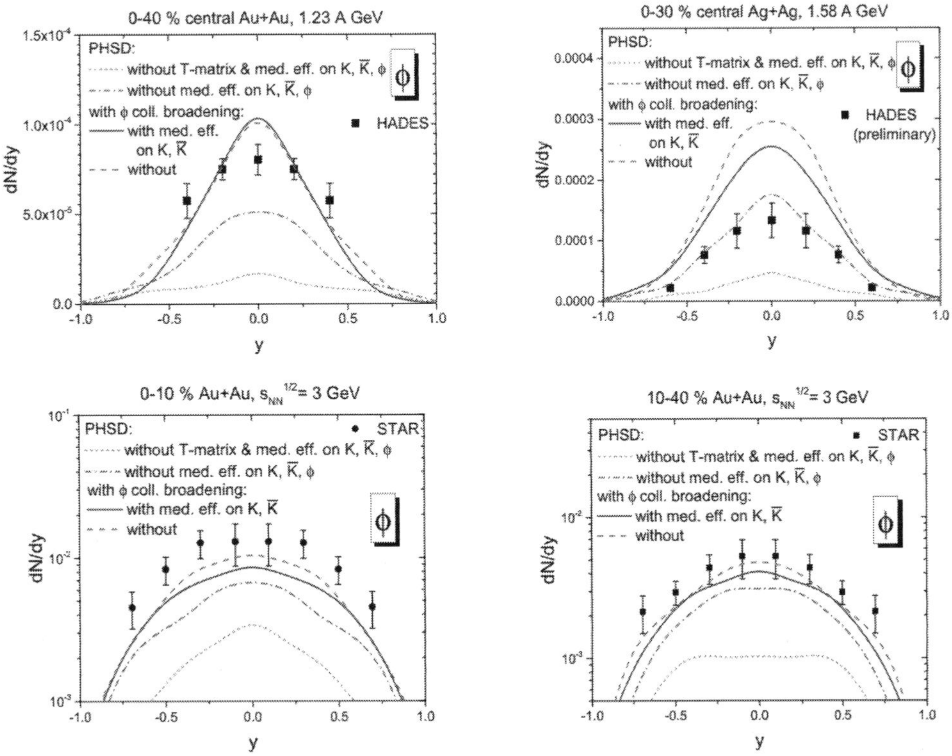

Figure 1. Rapidity distribution of reconstructed ϕ mesons in 0-40 % central Au+Au collisions at E_{kin} = 1.23 A GeV, in 0-30 % central Ag+Ag collisions at E_{kin} = 1.58 A GeV and 0-10 % and 10-40 % central Au+Au collisions at $\sqrt{s_{NN}}$ =3 GeV, compared with experimental data from the HADES and STAR collaborations [1, 4, 16]. Each colored line is explained in the text.

with ϕ collisional broadening, but without in-medium effects for $K, \bar{K}$ mesons. The solid blue lines show the results with collisional broadening for ϕ mesons and with in-medium modifications of $K, \bar{K}$ mesons.The number of ϕ mesons, reconstructed from K^+K^- pairs, is divided by the branching ratio $Br(\phi \rightarrow K^+K^-)$. We note that the rescattering of the K^+ or K^- in the medium reduces the reconstructed ϕ meson to 60-70% in Au+Au reactions at energies between E_{kin} = 1.23 A GeV and $\sqrt{s_{NN}}$ = 3 GeV. In addition, ϕ yield rapidly increases from E_{kin} = 1.23 A GeV to E_{kin} = 1.58 A GeV, because both are sub-threshold energies for ϕ production.

Fig. 2 shows the ϕ/K^- ratio as a function of the collision energy from E_{kin} = 1.23 A GeV to $\sqrt{s_{NN}}$ = 200 GeV. The PHSD results are presented for four different scenarios as in Fig. 1. An inclusion of the in-medium effects for $K, \bar{K}$, which leads to a strong enhancement of the K^- yield [17] and, as a result, to a reduction of the ϕ/K^- ratio.

5 Summary

In this study we have investigated the hidden strangeness (ϕ meson) production in heavy-ion collisions from subthreshold to relativistic energies within the microscopic off-shell PHSD transport approach. We have found that a collisional broadening of the ϕ meson spectral function lead to an enhancement of ϕ meson production, especially at subthreshold energies

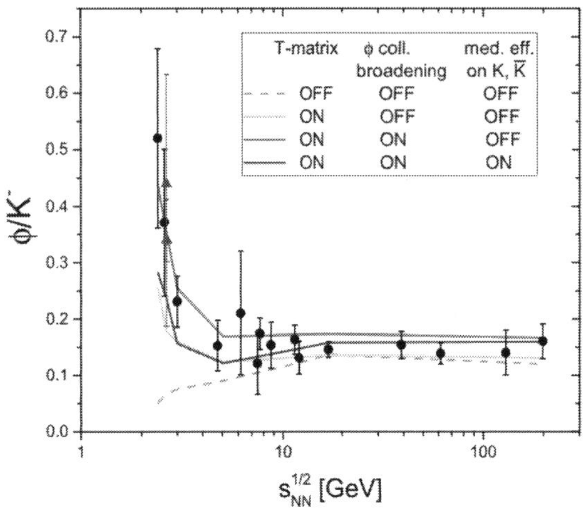

Figure 2. The PHSD results for the ratio ϕ/K^- at midrapidity ($|y| \leq 0.3$) as a function of the collision energy for four different scenarios: with and without novel mB channels for the ϕ meson production from the T-matrix approach and with and without the collisional broadening of the ϕ meson width and in-medium effects on (anti-)kaons (cf. the legend). The solid symbols show the compilation of the experimental data from Refs. [3–9].

and that the novel $mB \to \phi B$ channels from the SU(6) chiral Lagrangian in the T-matrix approach also enhance considerably the ϕ production in heavy-ion collisions.

Acknowledgements

We acknowledge support by the Deutsche Forschungsgemeinschaft (DFG, German Research Foundation): grant BR 4000/7-1 and by the GSI-IN2P3 agreement under contract number 13-70. This project has, furthermore, received funding from the European Union's Horizon 2020 research and innovation program under grant agreement No 824093 (STRONG-2020). Also we thank the COST Action THOR, CA15213. The computational resources have been provided by the LOEWE-Center for Scientific Computing and the "Green Cube" at GSI, Darmstadt.

References

[1] J. Adamczewski-Musch and others, Phys. Lett. B **778**, 403–407 (2018)
[2] N. Herrmann, Nucl. Phys. A **610**, 49C–62C (1996)
[3] K. Piasecki and others, Phys. Rev. C **94**, 014901 (2016)
[4] M. S. Abdallah and others, Phys. Lett. B **831**, 137152 (2022)
[5] S. V. Afanasiev and others, Phys. Rev. C **66**, 054902 (2002)
[6] B. B. Back and others, Phys. Rev. C **69**, 054901 (2004)
[7] J. Adams and others, Phys. Lett. B **612**, 181–189 (2005)
[8] C. Alt and others, Phys. Rev. C **77**, 024903 (2008)
[9] C. Alt and others, Phys. Rev. C **78**, 044907 (2008)
[10] E. L. Bratkovskaya and W. Cassing, Nucl. Phys. A **807**, 214–250 (2008)
[11] T. Song, J. Aichelin and E. Bratkovskaya, arXiv:2205.10251 [nucl-th]
[12] D. Gamermann, C. Garcia-Recio, J. Nieves and L. L. Salcedo, Phys. Rev. D **84**, 056017 (2011)
[13] E. Oset and A. Ramos, Nucl. Phys. A **635**, 99–120 (1998)
[14] D. Cabrera, L. Tolós, J. Aichelin and E. Bratkovskaya, Phys. Rev. C **90**, 055207 (2004)
[15] J. Nieves and E. Ruiz Arriola, Phys. Rev. D **64**, 116008 (2001)
[16] S. Spies, EPJ Web Conf. **259**, 01007 (2022)
[17] T. Song and others, Phys. Rev. C **103**, 044901 (2021)

Baryon Number Transport, Strangeness Conservation and Ω-hadron Correlations

Xiatong Wu[1,*], *Weijie* Dong[2,**], *Xiaozhou* Yu[2,***], *Hui* Li[2], *Gang* Wang[1], *Huan Zhong* Huang[1,2], and *Zi-Wei* Lin[3]

[1]Department of Physics and Astronomy, University of California, Los Angeles, CA 90095, USA
[2]Key Laboratory of Nuclear Physics and Ion-beam Application, Fudan University, Shanghai, China
[3]Department of Physics, East Carolina University, Greenville, NC 27858, USA

Abstract. Although strange quarks are produced in $s\bar{s}$ pairs, the ratio of Ω^- to $\bar{\Omega}^+$ is greater than one in heavy-ion collisions at lower RHIC energies. Thus the produced Ω hyperons must carry net baryon quantum numbers from the colliding nuclei. We present results of K-Ω correlations from AMPT model simulations of Au+Au collisions at $\sqrt{s_{NN}} = 14.6$ GeV, to probe dynamics for baryon number transport to mid-rapidities at this beam energy. We use both the default and string-melting versions to illustrate how hadronization schemes of quark coalescence and string fragmentations could leave imprints on such correlations. Implications on the measurements of these correlations with the STAR experiment at RHIC will also be discussed.

1 Introduction

In heavy-ion collisions at $\sqrt{s_{NN}} < 50$ GeV, the anti-hyperon to hyperon ratios are significantly below one [1], which implies that net hyperons must carry the net baryon number from the incident nuclei. The u or d quarks inside Λ's and Ξ's may originate from the colliding nuclei, and thus carry a portion of the net baryon number. However, there must be a mechanism for net baryon numbers to transport to Ω's without the delivery of u or d quarks, since the three s quarks in Ω are produced in pairs.

Ω production can carry dynamical information from strangeness conservation (SC), baryon number conservation (BNC), and baryon number transport (BNT). A primordial Ω^- is either produced along with three kaons and no anti-baryon (SC+BNT), or with one anti-baryon and some kaons (SC+BNC). In the former (SC+BNT), the s quarks in the kaons may combine with u or d quarks that originate from the colliding nuclei, and thus the Ω^- can effectively obtain baryon numbers from them. This process may also be sensitive to a possible gluonic junction mechanism of baryon number transport [2], where the baryon number is carried by the junction rather than by the valence quarks. The Ω^- arises from the hadronization of the junction by obtaining three s quarks, whereas the kaons can be viewed as the "leading mesons" [3]. In the latter (SC+BNC), all quarks are pair-produced and there is no net baryon number associated with the Ω^-. These two scenarios can be characterized by ΔN_K and $\Delta N_{\bar{B}}$,

*e-mail: maxwoo@physics.ucla.edu
**e-mail: wjdong19@fudan.edu.cn
***e-mail: xzyu18@fudan.edu.cn

the differences in the numbers of K's (K^+ and K^0) and anti-baryons, respectively, between events with an Ω^- and events without any, as shown in Table 1. These values can be counted precisely in models with different hadronization schemes, and provide baselines for more accessible observables in experiments, such as K-Ω and $\bar{B}$-Ω correlations.

Table 1: Expected differences in N_K and $N_{\bar{B}}$ between events with Ω^- and events without any.

	ΔN_K	$\Delta N_{\bar{B}}$
Scenario 1 (SC+BNT)	3	0
Scenario 2 (SC+BNC)	1-3	1

2 Model and Energy Selection

We choose A Multi-Phase Transport (AMPT) model [4] (non-public version v1.25t4cu/v2.25t5cu) to perform simulations for this study, with both the default and string-melting versions. The former includes only the minijet partons in the parton cascade phase and uses the Lund string fragmentation model to convert strings to hadrons. In the latter, both excited strings and minijet partons participate in parton cascade, and a quark coalescence model is used for hadronization. Comparison between these two modes may reveal the difference in the final-state manifestation of the two hadronization schemes. This version of AMPT also satisfies the quantum number (strangeness, electric charge and baryon number) conservation event by event, facilitating the study of the interplay of these quantum numbers in the context of the Ω production.

We have generated about 150 million (50 million default + 100 million string-melting) minimum bias events of Au+Au at $\sqrt{s_{NN}}$ = 14.6 GeV. In the majority of the events, at most one $\Omega^-(\bar{\Omega}^+)$ is produced. Table 2a shows that the difference in the average number of $s\bar{s}$ pairs between events with and without Ω^- is close to 3 in both the default and string-melting versions. This suggests that apart from the Ω^- and the associated hadrons, these two event classes have limited differences in terms of strangeness production. We compare the numbers of strange hadrons and anti-baryons between events with and without Ω^- in Table 2b. With reference to Table 1, both the default and string-melting versions of AMPT show a mixture of the two Ω-production scenarios, while the string-melting mode favors the first scenario.

Table 2: Differences in quantities between events with and without Ω^-.

(a) Average number of $s\bar{s}$ pairs.

	$\Delta N_{pair(s\bar{s})}$
AMPT-SM	3.07 ± 0.03
AMPT-default	3.16 ± 0.03

(b) Numbers of K (including both K^+ and K^0) and $\bar{B}$ (i.e., $\bar{\Lambda}$, $\bar{\Xi}$, $\bar{\Omega}$, $\bar{p}$ and $\bar{n}$).

	ΔN_K	$\Delta N_{\bar{B}}$
AMPT-SM	2.44 ± 0.02	0.119 ± 0.006
AMPT-default	1.76 ± 0.03	0.28 ± 0.01

3 Hadron-Ω Correlation

The results in Table 2b indicate that the string-melting version of AMPT should show a stronger K^+-Ω^- correlation or a weaker $\bar{B}$-Ω^- correlation. Also, since $\bar{\Omega}^+$ cannot carry baryon numbers, a difference in the two "opposite-strangeness-sign" correlations such as K^+-Ω^- and K^--$\bar{\Omega}^+$) may relate the Ω^- production to baryon number transport. The dynamical information about this process could be revealed by the shape of these correlations.

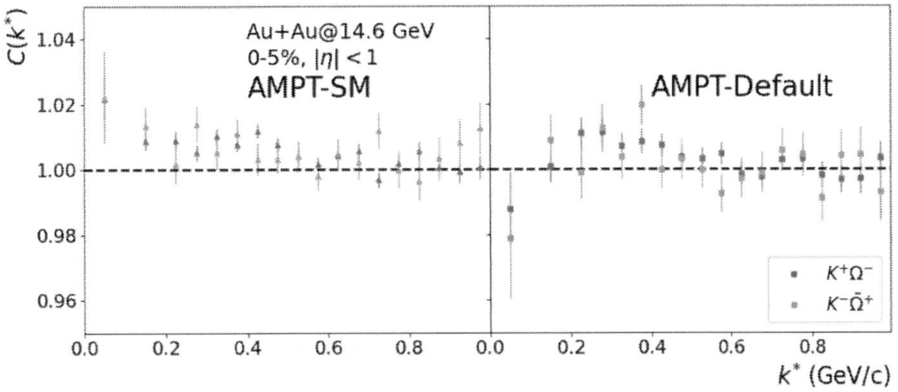

Figure 1: AMPT simulations of "opposite-strangeness-sign" K-Ω correlation functions, $C(k^*) = A(k^*)/B(k^*)$, for Au+Au collisions at 14.6 GeV.

We first use the traditional event-mixing normalization to study the k^* correlations for $K^\pm$-Ω, Λ-Ω and Ξ-Ω, where k^* is half the difference between the pair-rest-frame momenta. The two-particle k^* distribution $A(k^*)$ is divided by the mixed-event distribution $B(k^*)$ after normalizing each distribution at a common k^* interval. For all hadron-Ω correlations under study, there seems to be no significant difference between the two "opposite-strangeness-sign" pair correlations with the current statistics, in either the string-melting or the default version of AMPT. An example of K-Ω correlations is shown in Fig. 1. A potential discrepancy may exist in the first points between the two AMPT modes, but its significance is limited by statistics.

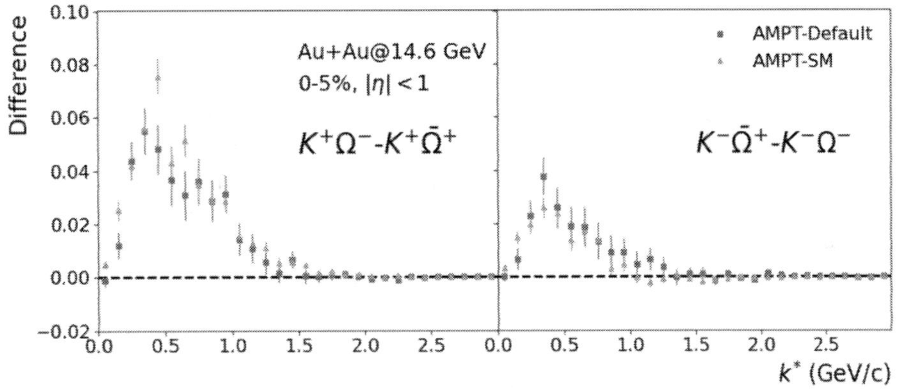

Figure 2: AMPT calculations of two cases of K-Ω correlation differences in the k^* space: (left) K^+-$\Omega^- - K^+$-$\bar{\Omega}^+$ and (right) K^--$\bar{\Omega}^+ - K^-$-Ω^-, for Au+Au collisions at 14.6 GeV.

We also examine the correlation difference between the "opposite-strangeness-sign" and the "same-strangeness-sign" distributions, each normalized by the number of events. In this

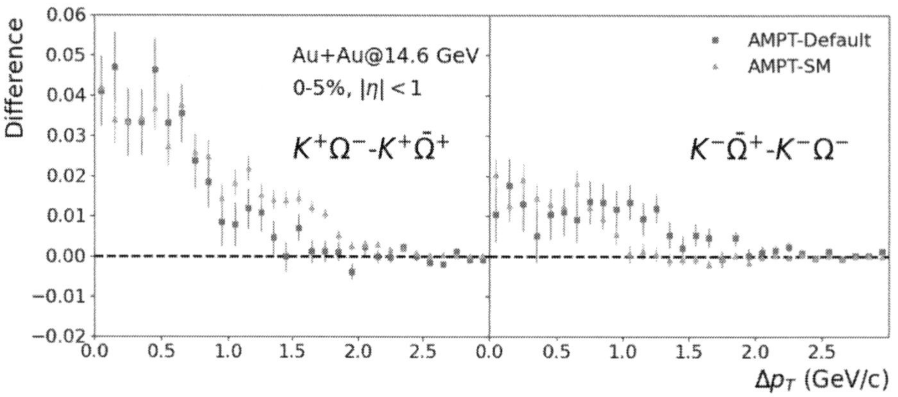

Figure 3: AMPT calculations of two cases of K-Ω correlation differences in the Δp_T space: (left) K^+-Ω^- – K^+-$\bar{\Omega}^+$ and (right) K^--$\bar{\Omega}^+$ – K^--Ω^- , for Au+Au collisions at 14.6 GeV.

definition, the latter is used as a baseline for the former so that they can be compared on the same footing and not affected by the difference between $K^+(\bar{B})$ and $K^-(B)$ multiplicities. These correlation differences reflect the strength and dynamic of the enhancement of $\bar{s}(s)$-carrying hadrons in the presence of an $\Omega^-(\bar{\Omega}^+)$. In the context of the Ω production, these two differences reveal the quantitative effects of SC+BNT and SC only, respectively. As shown in Fig. 2 and Fig. 3, while both the string-melting and default versions of AMPT show transported-quark effects, the spreads of such effects are different in global correlation spaces, such as Δp_T.

4 Conclusion

Ω hyperons can serve as a viable probe to various hadronization and baryon number transport mechanisms, as quark level correlations of $s\bar{s}$ may show different features in $K^+(\bar{B})$-Ω^- and $K^-(B)$-$\bar{\Omega}^+$ correlation due to the presence of baryon number transport dynamics. In these proceedings, we explore two Ω production scenarios through K-Ω correlations using AMPT events of Au+Au collisions at $\sqrt{s_{NN}}$ = 14.6 GeV. With the traditional event-mixing normalization, we find no significant difference between K^+-Ω^- and K^--$\bar{\Omega}^+$ correlation functions $C(k^*)$ with the current statistics, which suggests that $C(k^*)$ may be dominated by effects such as strangeness conservation or strong interaction that are common to both pairs. The correlation difference reveals transported-quark effects and a potential difference in global correlation widths between the two AMPT versions. Further investigation is needed to discern whether such differences arise from different hadronization procedures.

This work is supported by the U.S. DOE under the grant number DE-FG02-88ER40424, the National Science Foundation under Grant No. PHY-2012947, and the Natural Science Foundation of China under the grant number 11835002.

References

[1] J. Adam *et al.* (STAR), Phys. Rev. C **102**, 034909 (2020).
[2] D. Kharzeev, Phys. Lett. B **378**,238 (1996).
[3] H. Z. Huang. Selected Topics on Baryon Transport in Nuclear Collisions. R. Seto (World Scientific, 1999) 3-12.
[4] Z. W. Lin, C. M. Ko, B. A. Li, B. Zhang, and S. Pal, Phys. Rev. C **72(6)**, 064901 (2005).

Results of femtoscopic correlations at CMS

Raghunath Pradhan for the CMS collaboration[1,*]

[1]Indian Institute Of Technology, Madras
Chennai, Tamilnadu, India, 600036

Abstract. The two particle correlations as a function of relative momenta of identified hadrons involving K^0_S and $\Lambda/\overline{\Lambda}$ are measured in PbPb collisions at $\sqrt{s_{NN}} = 5.02$ TeV with the data samples collected by the CMS experiment at the LHC. Such correlations are sensitive to the quantum statistics and possible final state interactions between the particles. The source radii are extracted from $K^0_S K^0_S$ correlations in different centrality regions and found to decrease from central to peripheral collisions. The strong interaction scattering parameters are extracted from $\Lambda K^0_S \oplus \overline{\Lambda} K^0_S$ and $\Lambda\Lambda \oplus \overline{\Lambda\Lambda}$ correlations using the Lednicky-Lyuboshits model, and compared with other experimental and theoretical results. In addition, we present results for the source radii of charged hadrons considering the Lévy type source distributions in PbPb collisions at $\sqrt{s_{NN}} =$ 5.02 TeV.

1 Introduction

Identical and nonidentical particle short-range correlations in relative momentum, known as "femtoscopic" correlations, can be used to study the space-time extent of the particle emitting source created in relativistic heavy ion collisions [1]. The identical particle correlations are sensitive to quantum statistics (QS) and to possible final-state interactions (FSI), while nonidentical particle correlations are only sensitive to final state interactions. The correlations among the neutral K^0_S, Λ, and $\overline{\Lambda}$ particles, collectively referred to as V^0 particles, are of specific interest. They can be used to extract the size of the particle-emitting source, and, in a way complementary to dedicated scattering experiments, the strong-interaction parameters, i.e., the scattering length and the effective range. Because of their heavy mass and absence of Coulomb interactions, femtoscopy based on K^0_S particles complements the more commonly studied charged hadrons femtoscopy. By studying $\Lambda K^0_S(\overline{\Lambda} K^0_S)$ and $\Lambda\Lambda(\overline{\Lambda\Lambda})$ correlations, it is possible to extract the strong interaction scattering parameters for baryon-meson and baryon-baryon systems, which can shed light on the nature of the strong interaction.

Generally, in femtoscopic measurements, Gaussian source distributions are assumed. However, recent high precision correlation measurements at the BNL RHIC facility for gold-gold (AuAu) collisions at $\sqrt{s_{NN}} = 200$ GeV have shown that the Gaussian approximation is not adequately reproduce the measured results [2]. Therefore, Lévy alpha-stable distribution, characterized by the Lévy-exponent α is introduced.

*e-mail: raghunath.pradhan@cern.ch

This note presents the $K^0_S K^0_S$, $\Lambda K^0_S \oplus \overline{\Lambda} K^0_S$, $\Lambda\Lambda \oplus \overline{\Lambda\Lambda}$ and charged hadron femtoscopic correlations in lead-lead (PbPb) collisions at center-of-mass energy per nucleon pair of $\sqrt{s_{NN}}$ = 5.02 TeV using the data recorded by the CMS [3] experiment at the LHC. The $K^0_S K^0_S$ and charged hadron correlations were measured in the extended range of centrality bin (0–60%), where centrality is defined as the fraction of the total nucleus-nucleus cross section, with 0% denoting the maximum overlap of the colliding nuclei. The charged hadron correlations were measured with considering the Lévy type source distributions. The $\Lambda K^0_S \oplus \overline{\Lambda} K^0_S$ and $\Lambda\Lambda \oplus \overline{\Lambda\Lambda}$ correlations were measured in centrality range 0–80%.

2 $K^0_S K^0_S$, $\Lambda K^0_S \oplus \overline{\Lambda} K^0_S$, and $\Lambda\Lambda \oplus \overline{\Lambda\Lambda}$ femtoscopic correlations

The left panel of figure 1 shows the $K^0_S K^0_S$ correlation measurement as a function of relative momenta of the particle pair q_{inv} [1, 4] in 20–30% centrality with $0 < k_T < 2.5$ GeV, where $k_T \equiv |\vec{p}_{T,1} + \vec{p}_{T,2}|/2$ is the average transverse momentum of the pair [4]. The size of the particle-emitting source R_{inv} extracted from $K^0_S K^0_S$ correlation using the Lednicky-Lyuboshits fit [5] together with nonfemtoscopic background [4] for different centrality ranges and plotted in the right panel of figure 1 [4]. It can be seen that R_{inv} decreases from central (0–10%) to peripheral (50–60%) events, as expected from a simple geometric picture of the collisions. The values of R_{inv} as extracted by considering only the QS effect are larger than those found from considering both QS and FSI effects, which suggests that the FSI effect needs to be consider for the accurate measurement of R_{inv}.

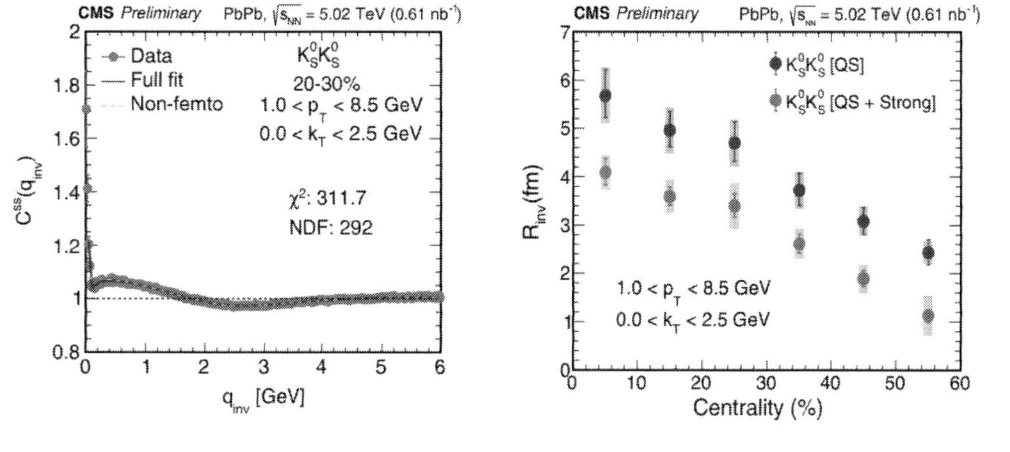

Figure 1. Left: An example of correlation measurement and their fit for $K^0_S K^0_S$ in 20–30% centrality. In this panel, red circles are the experimental results, the blue solid line is the full fit [4], and the green dotted line is the nonfemtoscopic background [4]. Right: R_{inv} as a function of centrality by considering only the QS (blue circles) and both the QS and strong FSI effects (red circles). For each data point, the line and shaded area indicate the statistical and systematic uncertainties, respectively.

Figure 2 shows the $\Lambda K^0_S \oplus \overline{\Lambda} K^0_S$ (left) and $\Lambda\Lambda \oplus \overline{\Lambda\Lambda}$ (right) correlation measurements in $0 - 80\%$ centrality with no restriction on k_T [4]. The strong-interaction scattering parameters: real scattering length ($\Re f_0$), imaginary scattering length ($\Im f_0$), and effective range (d_0), were extracted from the $\Lambda K^0_S \oplus \overline{\Lambda} K^0_S$ and $\Lambda\Lambda \oplus \overline{\Lambda\Lambda}$ correlations using Lednicky-Lyuboshits fit [5]. Figure 3 shows d_0 versus $\Re f_0$ (left) and $\Im f_0$ versus $\Re f_0$ (right). Comparisons to theoretical calculations and results from other experiments are shown [6–10]. The negative value of $\Re f_0$ in $\Lambda K^0_S \oplus \overline{\Lambda} K^0_S$ correlations suggests that the ΛK^0_S ($\overline{\Lambda} K^0_S$) interaction is repulsive while $\Im f_0$

is consistent with zero within uncertainty, preventing us from drawing any conclusion about inelastic processes [4]. A positive $\Re f_0$ value for the $\Lambda\Lambda \oplus \overline{\Lambda\Lambda}$ correlations indicates that the $\Lambda\Lambda(\overline{\Lambda\Lambda})$ interaction is attractive.

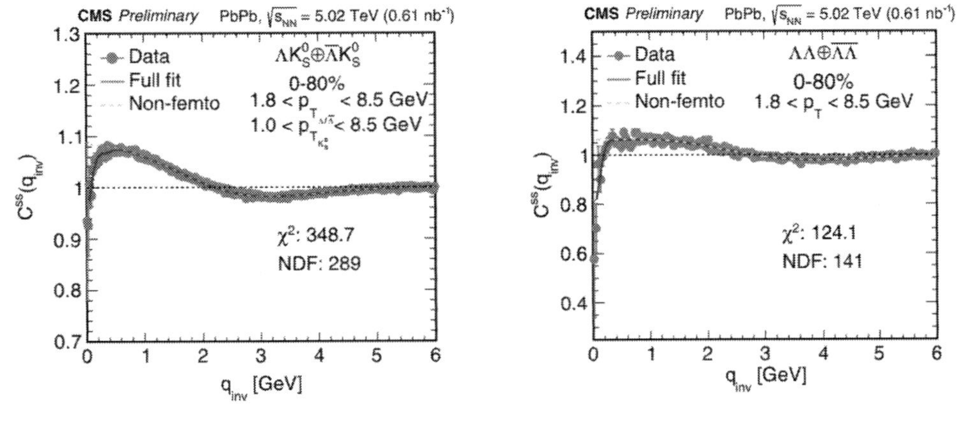

Figure 2. The correlation measurements and their fits for $\Lambda K^0_S \oplus \overline{\Lambda} K^0_S$ (left) and $\Lambda\Lambda \oplus \overline{\Lambda\Lambda}$ (right)in 0–80% centrality. In these plots, red circles are the experimental results, the blue solid line is the full fit, and the green dotted line is the nonfemtoscopic background fit [4]. For each data point, the line indicate the statistical uncertainties.

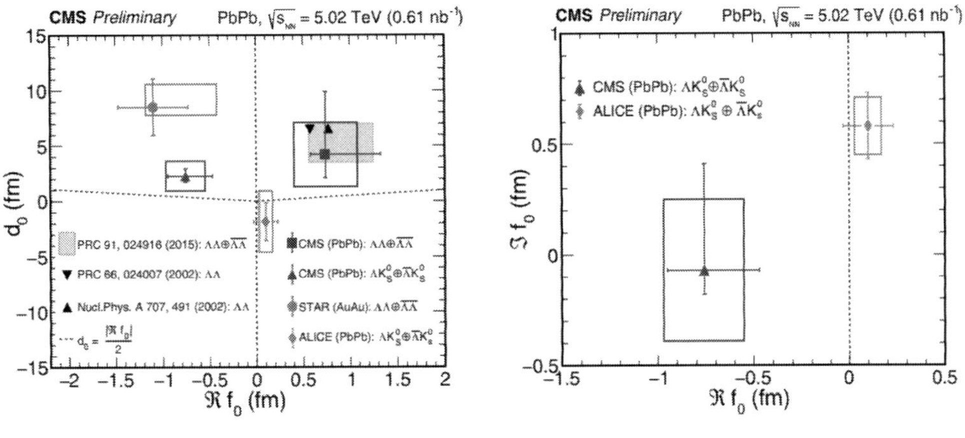

Figure 3. The values of d_0 and $\Re f_0$ (left) and the values of $\Im f_0$ and $\Re f_0$ (right). In the left plot, the blue triangle and square marker are for $\Lambda K^0_S \oplus \overline{\Lambda} K^0_S$ and $\Lambda\Lambda \oplus \overline{\Lambda\Lambda}$ correlations, respectively. The results are compared with the $\Lambda\Lambda \oplus \overline{\Lambda\Lambda}$ results from STAR experiment [7] and $\Lambda K^0_S \oplus \overline{\Lambda} K^0_S$ result from ALICE experiment [6]. A reanalysis of STAR data for $\Lambda\Lambda \oplus \overline{\Lambda\Lambda}$ correlations is shown in the shaded area [8]. Theory calculations of the $\Lambda\Lambda$ interaction parameters are shown as black triangles [9, 10]. In the right plot, the triangle is for $\Lambda K^0_S \oplus \overline{\Lambda} K^0_S$ correlation, and is compared with ALICE $\Lambda K^0_S \oplus \overline{\Lambda} K^0_S$ results [6]. For each data point, the two lines and the box indicate the (one-dimensional) statistical and systematic uncertainties, respectively.

3 Charged hadron femtoscopic correlations

Figure 4 shows the source size $R \equiv R_{\text{inv}}$ versus the transverse mass m_T in different centrality classes for positive (left) and negative (right) hadron pairs considering the Lévy type source distributions. [2]. It can be seen that R decreases with m_T for both positive and negative pairs, which is expected from hydrodynamics [2]. A clear centrality dependence is also visible similar to $K^0_S K^0_S$ pairs in Fig. 1.

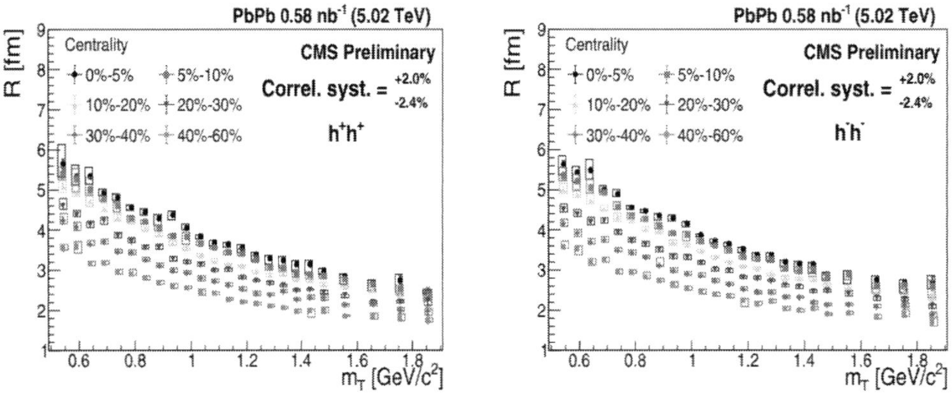

Figure 4. Source size R versus the transverse mass m_T in different centrality classes for positive (left) and negative (right) hadron pairs. The error bars are the statistical uncertainties, while the boxes indicate the systematic uncertainties. The correlated systematic uncertainty is also indicated.

4 Summary

The identified hadrons femtoscopic correlation ($K^0_S K^0_S$, $\Lambda K^0_S \oplus \overline{\Lambda} K^0_S$, and $\Lambda\Lambda \oplus \overline{\Lambda\Lambda}$) and charged hadron femtoscopic correlations are presented in PbPb collisions at a center-of-mass energy per nucleon pair of $\sqrt{s_{\text{NN}}} = 5.02$ TeV, as measured by the CMS experiment at the LHC. This is the first report of $\Lambda\Lambda \oplus \overline{\Lambda\Lambda}$ correlations in PbPb collisions. The source size as a function of centrality as extracted for $K^0_S K^0_S$ and charged hadron correlations show similar behavior. The Lednicky-Lyuboshits model fit to the data suggests that the $\Lambda\Lambda \oplus \overline{\Lambda\Lambda}$ interaction is attractive, whereas the $\Lambda K^0_S \oplus \overline{\Lambda} K^0_S$ interaction is repulsive.

References

[1] M. A. Lisa, S. Pratt, R. Soltz, and U. Wiedeman, Ann. Rev. Nucl. Part. Sci. **55**, 357 (2005)
[2] CMS Collaboration, CMS-PAS-HIN-21-011
[3] CMS Collaboration, JINST **03**, S08004 (2008)
[4] CMS Collaboration, CMS-PAS-HIN-21-006
[5] R. Lednicky and V. Lyuboshits, Sov. J. Nucl. Phys. **35**, 770 (1982)
[6] ALICE Collaboration, Phys. Rev. C **114** 055201 (2021)
[7] STAR Collaboration, Phys. Rev. Lett. **103** 022301 (2015)
[8] K. Morita, T. Furumoto, and A. Ohnishi, Phys. Rev. C **91** 024916 (2015)
[9] E. Hiyama et al., Phys. Rev. C **66** 024007 (2002)
[10] I. N. Filikhin and A. Gal, Nucl. Phys. A **707** 491 (2002)

EPJ Web of Conferences **276**, 03004 (2023) https://doi.org/10.1051/epjconf/202327603004
SQM 2022

A multi-differential investigation of strangeness production in pp collisions with ALICE

Romain Schotter[1,2,3,*], for the ALICE collaboration.

[1]Institut Pluridisciplinaire Hubert Curien, UMR7178, CNRS – Strasbourg, France
[2]Université de Strasbourg – Strasbourg, France
[3]EUR QMat – Strasbourg, France

Abstract. In these proceedings, two multi-differential analyses performed in pp collisions collected by the ALICE collaboration during the LHC Run 2 are presented. One investigates the dependence of strange particle production with multiplicity and *effective* energy, whereas the other clarifies how strangeness enhancement is correlated to the leading jet in the event. The results suggest that strangeness production at the LHC depends strongly on effective energy, and originates dominantly from the transverse region with respect to the leading jet direction.

1 Introduction

By colliding heavy nuclei at extremely high energy, a new state of nuclear matter, in which quarks and gluons are deconfined — the quark–gluon plasma (QGP) —, is formed. One of its historical key signatures is the *strangeness enhancement* [1], which consists in the increase of the relative yields of multistrange particles, such as Ξ(dss) and Ω(sss), to non-strange hadrons. Recently, it has been observed that such yield enhancement also scales smoothly with the charged particle multiplicity in proton–proton (pp) collisions [2]. The presence of this phenomenon in such a small system questions the very foundations of the QGP concept.

Several phenomenological models, based on different approaches and mechanisms, are being developped but none of them has been able to provide an unambiguous explanation so far [3]. Further experimental inputs are required in order to distinguish them, and this is achieved via more multi-differential studies related to the strange hadron production.

One can perform an analysis in order to separate the contribution of initial-state effects on the strangeness enhancement from the final ones. Indeed, the distribution of the charged particle multiplicity is a characteristic of the final state of the collision, but it also depends on the energy effectively available for particle production, which is related to the initial interactions. The idea is to investigate the dependence of Ξ baryon production multi-differentially in multiplicity and effective energy. Similarly, the relative contribution of hard processes — such as jets — and softer processes — occuring in transverse region to the jet axis — to the production of Ξ baryons can be now studied more differentially.

*e-mail: romain.schotter@cern.ch

2 Detector setup and data sample

In these two analyses, the Ξ baryons are studied in their cascade decay channel: $\Xi^{\pm} \rightarrow \pi^{\pm}\Lambda \rightarrow \pi^{\pm}\pi^{\pm}p^{\mp}$ (63.9%). They are reconstructed at midrapidity ($|y| < 0.5$), using the central-barrel detectors of the ALICE experiment [4]: the Inner Tracking System (ITS) and the Time Projection Chamber (TPC) for the track reconstruction and particle identification. At forward pseudorapidity (η), charged particle multiplicity is determined with the V0M estimator, based on the sum of the signal amplitudes measured in the scintillator arrays, V0A ($2.8 < \eta < 5.1$) and V0C ($-3.7 < \eta < -1.7$). The number of clusters found in the two innermost layers of the ITS (SPD) can also provide a multiplicity estimation but at midrapidity ($|\eta| < 0.8$): this is the SPDCluster estimator. Finally, the *ZDC Energy Sum* estimates the energy of particles produced at very forward rapidity ($6.5 < |\eta| < 7.4$ for protons, and $|\eta| > 8.8$ for neutrons) that have been deposited in the two Zero Degree Calorimeters (ZDC).

For the first analysis, the data sample is composed of 129×10^6 minimum bias events at a center-of-mass energy $\sqrt{s} = 13$ TeV, recorded in 2015, 2017 and 2018. Concerning the second study, the dataset contains approximatively 420×10^6 events, coming from pp collisions at $\sqrt{s} = 13$ TeV collected in 2016, 2017, and 2018. Additionally, 920×10^6 pp collisions at $\sqrt{s} = 5$ TeV are used as well for comparison.

3 Dependence of strange particle production with multiplicity and effective energy

3.1 Details on the effective energy and the double differential analysis

In these proceedings, the effective energy is defined as the energy effectively available for particle production in the initial stages of the pp collision, and could be equal in principle to the center-of-mass energy. One has also to consider the baryons emitted in the very forward direction — the so-called forward leading baryon emission [5] —, which carry a large fraction of the incident beam energy and therefore reduce the energy available for particle production at midrapidity. The energy carried by these baryons can be measured by the two ZDCs and thus let us define a proxy for the effective energy as $\sqrt{s} - \langle \text{ZDC energy sum} \rangle$.

As mentionned in Section 1, multiplicity at midrapidity and effective energy are correlated [6]. In order to disentangle contributions from the initial (effective energy) and final (multiplicity) stages of the collision, the analysis is performed differentially in event multiplicity classes based on two estimators: SPDCluster and V0M. It has been observed that:

- Setting the SPDCluster class sharply fixes the multiplicity at midrapidity, and for these classes, further selections on the V0M class allow to vary the effective energy.
- Conversely, fixing the V0M class constrains the effective energy to a narrower range, allowing further SPDCluster selections to classify events with different multiplicity.

3.2 Experimental results

Figure 1 shows the Ξ yield per charged particle when multiplicity at midrapidity is fixed in the double differential analysis. The black diamonds correspond to different V0M multiplicity percentile class. On the left panel, one can observe that the Ξ yield increases with the charged particle multiplicity at midrapidity: this is the strangeness enhancement. On the right panel, the correlation between multiplicity and effective energy is clearly visible from the decrease (or increase) of the Ξ yield with the average ZDC energy sum (or the effective energy).

The double differential analysis indicates that the Ξ yield per charged particle increases at fixed multiplicity in the left panel of Fig. 1. From the right panel, one can see that this

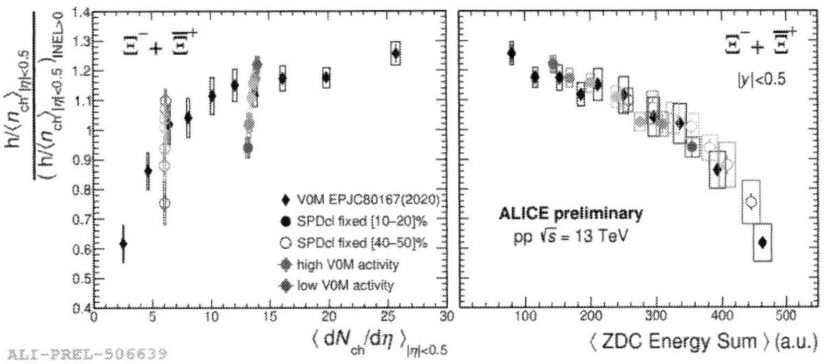

Figure 1. Ξ yield per charged particle self-normalized to INEL>0 as a function of multiplicity at midrapidity on the left, and of the ZDC energy sum on the right. The black diamonds correspond to the analysis performed in V0M classes alone [7]; the colored full (open) squares present the double differential analysis with SPDCluster fixed in [10, 20]% ([40, 50]%).

yield actually increases with effective energy. Moreover, the V0M standalone and the double differential measurement points are compatible within uncertainties, suggesting that the so-known strangeness enhancement with multiplicity is likely a dependence on effective energy.

The situation with constrained effective energy is depicted in Fig. 2. The right panel shows that effective energy has been divided into two ranges: one with a large and another one with a small average effective energy, represented in full and open markers, respectively. As can be seen on the left panel, the Ξ yield shows an almost flat dependence with the charged particle multiplicity at midrapidity for the case of large effective energies. On the contrary, for smaller effective energies the yield increases rapidly at first and then the trend with multiplicity becomes milder. It seems that, with constrained effective energy, the rise of strangeness enhancement with multiplicity is strongly affected.

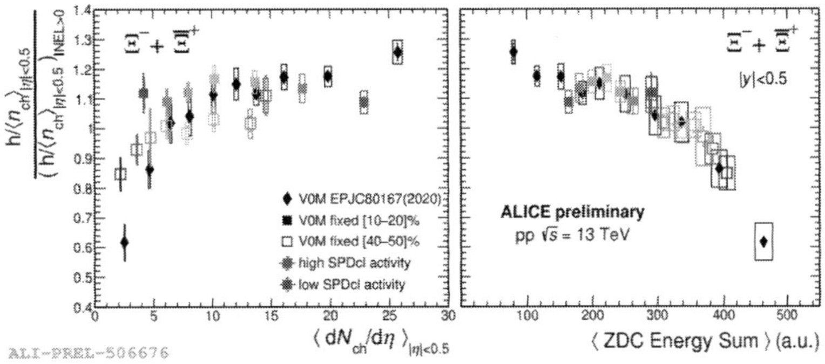

Figure 2. Ξ yield per charged particle self-normalized to INEL>0 as a function of multiplicity at midrapidity on the left, and of the ZDC energy sum on the right. The black diamonds correspond to the analysis performed in V0M classes alone [7]; the colored full (open) circles present the double differential analysis with V0M fixed in [10, 20]% ([40, 50]%).

4 Study of strange hadron production in the regions toward and transverse to the leading jet

4.1 Angular correlation method

To separate strange hadrons produced towards the leading jet from the ones produced transverse to the leading jet, the angular correlation method is applied. It consists in forming pairs between the leading particle in the event, i.e. the primary charged particle with the highest p_T and $p_T > 3$ GeV/c, and the particles of interest, e.g. the Ξ baryons. The angular correlation is provided by the distributions of pairs in relative pseudorapidity ($\Delta\eta$) and azimuthal angle ($\Delta\varphi$) ; strange hadrons produced towards the leading jet originate from small ($\Delta\eta$, $\Delta\varphi$).

4.2 Experimental results

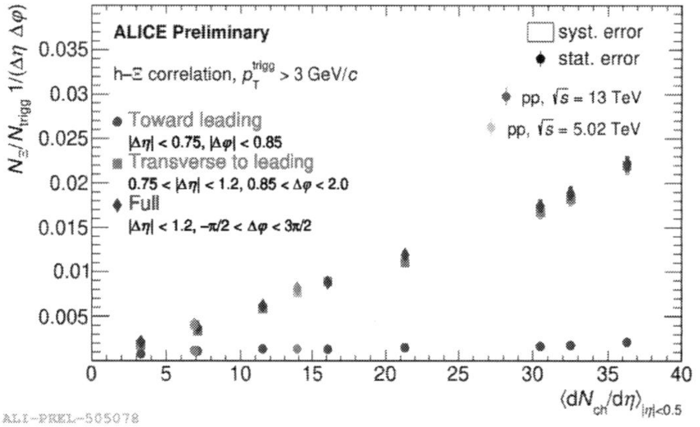

Figure 3. Distribution of yield per trigger particle and per unit of ($\Delta\eta$, $\Delta\varphi$) of Ξ baryons as a function of the charged particle multiplicity at midrapidity: full yield in blue diamonds, toward the leading jet yield in red circles, and transverse to the leading jet yield in green squares. Results for pp at $\sqrt{s}$ = 13 TeV are in dark colors, and pp at $\sqrt{s}$ = 5 TeV in light colors.

Figure 3 shows the Ξ yield density per triggered event — ratios are normalized to the considered ($\Delta\eta$, $\Delta\varphi$) region — extracted in three ($\Delta\eta$, $\Delta\varphi$) regions as a function of the charged particle multiplicity at midrapidity in pp collisions. One can observe that both full and transverse to leading jet yields increase with the charged particle multiplicity at midrapidity, whereas the production in the toward-leading-jet region exhibits a quasi-flat dependence on multiplicity. This means that Ξ are dominantly produced in the transverse region to the leading jet, and therefore the transverse to leading jet processes are the dominant contribution to the strangeness enhancement in pp collisions. It should be also noted that the measured yields in pp at $\sqrt{s}$ = 13 TeV are consistent with the ones at $\sqrt{s}$ = 5 TeV.

5 Conclusion

These two multi-differential analyses have managed to provide new insights on the strangeness production in pp collisions: the initial stage of the collision plays an important role in the strangeness enhancement, and transverse to leading jet processes are the dominant contribution to the strangeness production. These prominent results promise to challenge QCD-inspired models, and both studies are currently being pushed forward in that direction. Moreover, with the large amount of data expected in ALICE with the LHC Run 3 [8], these analyses could be refined, and even extended to triple strange baryons.

References

[1] Rafelski, Springer International Publishing, Cham, 389-400 (2016)
[2] ALICE Collaboration, Nature Physics **13**, 535-539 (2017)
[3] ALICE Collaboration, Eur. Phys. J. C **80**, 693 (2020)
[4] ALICE Collaboration, J. Inst. **3**, S08002 (2008)
[5] Akindinov *et al*, Eur. Phys. J. C **50**, 341-352 (2007)
[6] ALICE Collaboration, J. High Energ. Phys. **2022**, 86 (2022)
[7] ALICE Collaboration, Eur. Phys. J. C **80**, 167 (2020)
[8] ALICE Collaboration, ALICE-PUBLIC-2020-005 (2020)

Midrapidity cluster formation in heavy-ion collisions

Elena Bratkovskaya[1,2,3,*], *Susanne* Glässel[4], *Viktar* Kireyeu[3,5], *Jörg* Aichelin[6,7], *Marcus* Bleicher[2,3,1], *Christoph* Blume[4], *Gabriele* Coci[2,3], *Vadim* Kolesnikov[5], *Jan* Steinheimer[7], and *Vadim* Voronyuk[3,5]

[1]GSI Helmholtzzentrum für Schwerionenforschung GmbH, Planckstr. 1, 64291 Darmstadt, Germany
[2]Institut für Theoretische Physik, Johann Wolfgang Goethe-Universität, Max-von-Laue-Str. 1, 60438 Frankfurt, Germany
[3]Helmholtz Research Academy Hessen for FAIR (HFHF), GSI Helmholtz Center for Heavy Ion Physics. Campus Frankfurt, 60438 Frankfurt, Germany
[4]Institut für Kernphysik, Johann Wolfgang Goethe-Universität, Max-von-Laue-Str. 1, 60438 Frankfurt, Germany
[5]Joint Institute for Nuclear Research, Joliot-Curie 6, 141980 Dubna, Moscow region, Russia
[6]SUBATECH, Université de Nantes, IMT Atlantique, IN2P3/CNRS 4 rue Alfred Kastler, 44307 Nantes cedex 3, France
[7]Frankfurt Institute for Advanced Studies, Ruth Moufang Str. 1, 60438 Frankfurt, Germany

Abstract. We study the production of clusters and hypernuclei at midrapidity employing the Parton-Hadron-Quantum-Molecular-Dynamics (PHQMD) approach, a microscopic n-body transport model based on the QMD propagation of the baryonic degrees of freedom with density dependent 2-body potential interactions. In PHQMD the cluster formation occurs dynamically, caused by the interactions. The clusters are recognized by the Minimum Spanning Tree (MST) algorithm. We present the PHQMD results for cluster and hypernuclei formation in comparison with the available experimental data at relativistic energies. PHQMD allows to study the time evolution of formed clusters and the origin of their production, which helps to understand how such weakly bound objects are formed and survive in the rather dense and hot environment created in heavy-ion collisions. It offers therefore an explanation of the 'ice in the fire' puzzle. To investigate whether this explanation of the 'ice in the fire' puzzle applies only to the MST results we study also the deuterons production by coalescence. We embed MST and coalescence in the PHQMD and UrQMD transport approaches in order to obtain model independent results. We find that both clustering procedures give very similar results for the deuteron observables in the UrQMD as well as in the PHQMD environment. This confirms that our solution for the 'ice in the fire' puzzle is common to MST and coalescence and independent of the transport approach.

1 Introduction

Cluster and hypernuclei production in heavy-ion collisions is presently under active experimental and theoretical investigation. The production mechanisms of hypernuclei may shed light on the theoretical understanding of the dynamical evolution of heavy-ion reactions which cannot be addressed by other probes. In particular, the formation of heavy projectile/target like hypernuclei elucidates the physics at the transition region between spectator and participant matter. Since hyperons are produced in the overlap region, multiplicity as well as rapidity distributions of hypernuclei formed in the target/projectile region depend crucially on the interactions of the hyperons with the hadronic matter, e.g. cross sections and potentials. On the other hand, midrapidity hypernuclei as well as their non-strange counterparts test the phase space distribution of baryons in the expanding participant matter, especially whether the phase space distributions of strange and non-strange baryons are similar and whether they are in thermal equilibrium.

The description of cluster and hypernuclei formation is a challenging theoretical task which requires

I) the microscopic dynamical description of the time evolution of heavy-ion collisions;
II) the modelling of the mechanisms for the cluster formation.

We study this production of hypernuclei and clusters employing the Parton-Hadron-Quantum-Molecular-Dynamics (PHQMD) approach [1, 2], a microscopic n-body transport model based on the QMD propagation of the baryonic degrees of freedom with density dependent 2-body potential interactions. In PHQMD the cluster formation occurs dynamically, caused by the interactions. The clusters are recognized by the Minimum Spanning Tree (MST) algorithm. We present the PHQMD results for cluster and hypernuclei formation in comparison with the available experimental data at AGS, SPS, RHIC-BES and RHIC fixed target energies. We also provide predictions on cluster production for the upcoming FAIR experiment. PHQMD allows to study the time evolution of formed clusters and the origin of their production, which helps to understand how such weakly bound objects are formed and survive in the rather dense and hot environment created in heavy-ion collisions. It offers therefore an explanation of the 'ice in the fire' puzzle.

*invited speaker,e-mail: E.Bratkovskaya@gsi.de

To investigate whether this explanation of the 'ice in the fire' puzzle applies only to the MST results we study also the deuterons production by coalescence [3]. We embed MST and coalescence in the PHQMD and UrQMD transport approaches in order to obtain model independent results [3]. We find that both clustering procedures give very similar results for the deuteron observables in the UrQMD as well as in the PHQMD environment. This confirms that our solution for the 'ice in the fire' puzzle is common to MST and coalescence and independent of the transport approach.

2 Cluster production in the PHQMD and UrQMD approaches

In this section we briefly recall the basic ideas of the PHQMD and UrQMD approaches and cluster identification procedures - coalescence and MST. For further details of the comparison of the PHSD and UrQMD dynamics we refer the readers to the recent review [4] and on comparison of the coalescence and MST - to the Ref. [3].

The PHQMD approach [1, 2] and the Ultra-relativistic-Quantum-Molecular Dynamics (UrQMD) [5, 6] describe the dynamics of heavy-ion collision on a microscopic basis using Quantum Molecular Dynamics (QMD). In QMD approaches [7, 8] the nucleons are presented as Wigner densities of the wave functions where the nucleons interact by mutual density dependent two-body interactions between all nucleons. This allows to account for n-body nucleon correlations (smeared out in mean-field approaches) which are important for the cluster production.

The collision integral and description of QGP and mesonic dynamics in the PHQMD is based on the off-shell microscopic Parton-Hadron-String Dynamics (PHSD) approach [9–12]), based on the Kadanoff-Baym equations [13] in first-order gradient expansion [14]. The description of the QGP phase in PHSD/PHQMD is described in terms of strongly interacting quasiparticles based on the DQPM (Dynamical Quasi-Particle Model) [14]. In the PHQMD the clusters are formed dynamically, by nucleon-nucleon interactions. As introduced in Ref. [1], the cluster identification/recognition in the PHQMD [1] is done either by the Minimum Spanning Tree (MST) procedure [7] (also used in this study) or by the Simulated Annealing Clusterization Algorithm (SACA) (see [15, 16]), as it was previously done in QMD [15, 17, 18] calculations. Clusters are identified with a MST in coordinate space, i.e when the distance between the proton-neutron pair in their rest frame is smaller than r_0 = 4 fm. We note that additional cuts in momentum space do not change much the cluster distributions found by the MST [19].

UrQMD [5, 6]includes hadron-string dynamics. As degrees of freedom it includes the full set of established hadrons and their resonances, summing up to more than 50 baryon species and 40 meson species. The UrQMD is a hybrid approach where the QGP phase can be included (optionally) into the UrQMD dynamics by switching to a hydrodynamical hybrid description [20, 21] for the most dense stages of the reaction. We note that for the results presented here we use UrQMD in hadron-string mode without QGP, i.e. without a transition to the hydro mode. The light cluster formation in the UrQMD is realized by the coalescence procedure. It has been applied to study deuteron production in the UrQMD approach for various energies, see e.g. [22–26].

We note that the coalescence procedure assumes that deuterons are created at freeze-out, i.e. when the last collision of the nucleons of the pair takes place and if at this time the relative distance of the two nucleons in coordinate and momentum space is lower than the coalescence parameters. The MST procedure can be applied any time for finding the nucleon correlations in coordinate space. This allows to follow the history of the cluster production and study their stability. As has been demonstrated in Ref. [3] the MST procedure asymptotically provides results which are similar to a coalescence procedure.

3 Results on cluster production

In this Section we present selected results on the light cluster and hypernuclei production at relativistic energies and discuss the space-time distribution of deuteron production. Since the clusters in the semi-classical QMD approach are not fully stable and nucleons can "spontaneously" evaporate, we show the results at different time in order to demonstrate this uncertainties (cf. Ref. [2] for more details).

3.1 Excitation function

In Fig. 1 we show the midrapidity excitation function of dN/dy of protons (left), antiprotons (middle) and deuterons (right) as a function of $\sqrt{s_{NN}}$ for central Au+Au collisions - 5% most central (protons and antiprotons) and 7% most central (deuterons) - in comparison with the experimental data from the NA49 Collaboration [27]. The PHQMD results are shown as red lines for t = 50 fm/c, blue lines for t = 60 fm/c and green lines for t = 70 fm/c.

One can see that the proton and antiproton as well as the deuteron dN/dy are well reproduced at all energies - from the lowest SPS energies up to the highest RHIC energies. As a consequences, the experimental d/p ratio versus $\sqrt{s_{NN}}$ is also reproduced rather well as follows from Fig. 2, which shows the excitation function of the deuteron to proton d/p (left) and antideuteron to antiproton $\bar{d}/\bar{p}$ (right) ratios from central Au+Au collisions at midrapidity as a function of $\sqrt{s_{NN}}$. The PHQMD results are shown as red lines for t = 50 fm/c, blue lines for t = 60 fm/c and green lines for t = 70 fm/c (similar to Fig. 1). The form of the $\sqrt{s_{NN}}$-dependence of $\bar{d}/\bar{p}$ is well reproduced, however, we overestimate this ratio by about a factor of two, if the yield of antiparticles is taken at the same time as the yield of the particles. As follows from middle part of Fig. 1 the $\bar{p}$ distribution does not depend on time, this enhancement is consequently due to a overpredicted $\bar{d}$ yield.

3.2 Transverse spectra of light clusters

In Figure 3 we show the comparison of the PHQMD results with (preliminary) fixed target STAR data [31] for

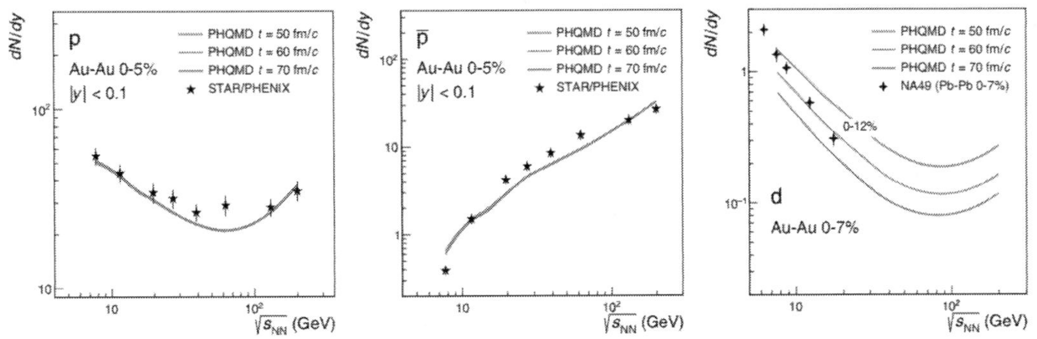

Figure 1. The midrapidity excitation function of dN/dy of protons (left), antiprotons (middle) and deuterons (right) as a function of $\sqrt{s_{NN}}$ for central Au+Au collisions (5% most central (protons and antiprotons) and 7% most central (deuterons) in comparison with the experimental data from the NA49 Collaboration [27] (crosses) where the midrapidity intervals are taken as $-0.4 < y < 0.0$ for $\sqrt{s_{NN}}$ = 6.3 GeV, 7.6 GeV and 8.8 GeV; $-0.6 < y < -0.2$ for 12.3 GeV and $-0.6 < y < -0.4$ for 17.3 GeV, as well as from the STAR and PHENIX Collaborations [28, 29] (stars) for $|y| < 0.1$. The rapidity intervals for the PHQMD results correspond to the experimental ones. The PHQMD results are shown as red lines for t = 50 fm/c, blue lines for t = 60 fm/c and green lines for t = 70 fm/c. The PHQMD results for protons and antiprotons are scaled to account for the protons from weak decay feed-down that are included in the experimental data. The figure is taken from Ref. [2].

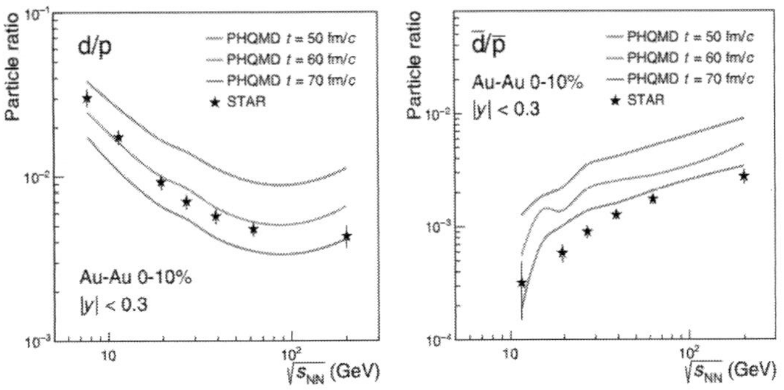

Figure 2. The excitation function of the deuteron to proton (left) and antideuteron to antiproton ratios (right) for central Au+Au collisions as a function of $\sqrt{s_{NN}}$. The experimental data from the STAR Collaboration [30] in the rapidity interval $|y| < 0.3$ are indicated as stars. The PHQMD results are shown as red lines for t = 50 fm/c, blue lines for t = 60 fm/c and green lines for t = 70 fm/c. The figure is taken from Ref. [2].

the p_T distribution of $p, d, t, {}^3He, {}^4He$ from Au+Au central collisions at $\sqrt{s}$ = 3 GeV for different rapidity bins. The PHQMD results are shown for t = 60 fm/c. The right lower plot shows the rapidity distribution (for $y > 0$) of light clusters. One can see that the PHQMD reproduces well the p_T spectra while the corresponding y distribution is underestimated for the light clusters. We note that the STAR data are obtained by integration of the p_T spectra extrapolated to low p_T region. This extrapolation is different from the actual PHQMD calculations at lower p_T.

3.3 Hypernuclei production

The study of hypernuclei production can provide information on the local phase space density of non-strange baryons, as well as the phase space density of the hyperons themselves since the hyperons are produced at the hadronization of the QGP and in elementary hadronic reactions during the heavy-ion collisions and then redistributed over the expanding fireball by collisions or potential interactions with nucleons. Here we show the PHQMD results in comparison to the recent experimental data of STAR collaboration at $\sqrt{s_{NN}}$ = 3 GeV.

Figures 4 and 5 display the comparison of the PHQMD transverse momentum distribution with the preliminary experimental data from the STAR Collaboration. In Fig. 4 we show the results for ${}^3_\Lambda$H, while in Fig. 5 those for ${}^4_\Lambda$H for different rapidity bins are presented. The filled circles indicate the experimental data from the STAR Collaboration [32]. The PHQMD results are taken at the times t = 80 fm/c (red lines), 85 fm/c (blue lines) and 90 fm/c (green lines), but at this low energy the time dependence of the cluster yield is weak.

The PHQMD reproduces well the shape of the experimental p_T spectra. We note that the hypernuclei production is sensitive to the ΛN potential. The PHQMD over-

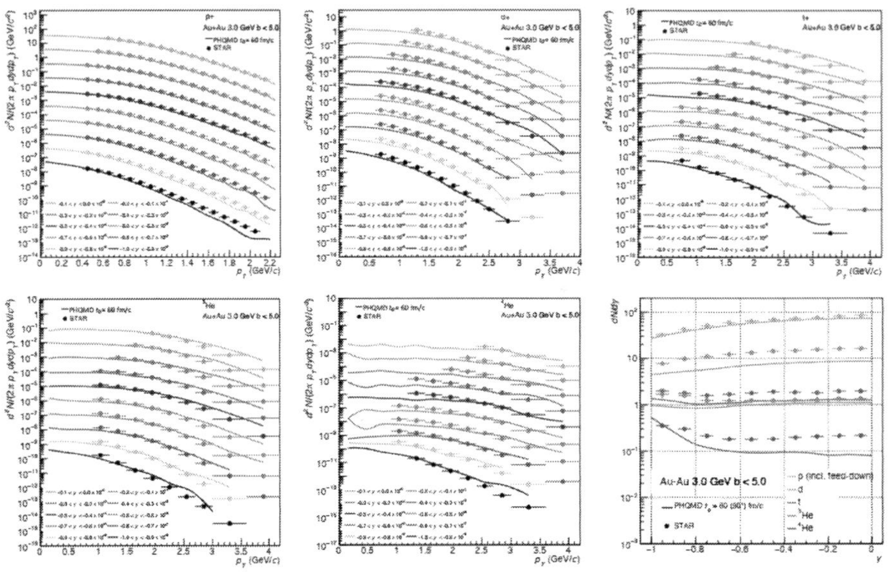

Figure 3. The PHQMD results for the p_T distribution of $p, d, t, {}^3He, {}^4He$ in comparison with (preliminary) STAR fixed target data [31] for Au+Au central collisions at $\sqrt{s} = 3$ GeV for different rapidity bins. The right lower plot shows the rapidity distribution (for $y > 0$) of light clusters.

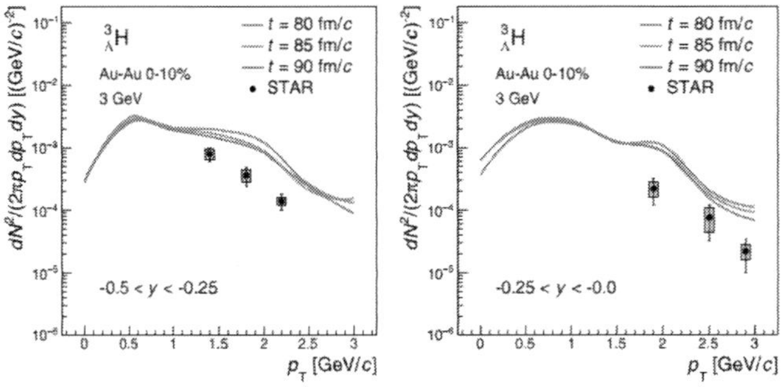

Figure 4. Transverse momentum distribution of $^3_\Lambda$H for different rapidity intervals as indicated in the legends in central Au+Au collisions at $\sqrt{s_{NN}} = 3$ GeV. The filled circles indicate the experimental data from the STAR Collaboration [32]. The PHQMD results are taken at the times $t = 80$ fm/c (red lines), 85 fm/c (blue lines) and 90 fm/c (green lines). The figure is taken from Ref. [2].

estimates the differential yield of $^3_\Lambda H$, but reproduces the yield of $^4_\Lambda H$. This demonstrates that PHQMD is a good starting tool for more extended studies of the hypernucleus production in heavy-ion collisions.

3.4 Space and time distribution of deuteron production

In Ref. [3] we have compared two mechanisms of deuteron identification by the coalescence and MST procedures which have been implemented in two independent microscopic transport approaches - PHQMD and UrQMD. We found that both clustering procedures give very similar results for the deuteron observables in the UrQMD and PHQMD environment. Furthermore, the PHQMD and UrQMD calculations agree well with the NA49 experimental data on deuteron production in Pb+Pb collisions at $\sqrt{s_{NN}} = 8.8$ GeV (selected for the comparison of the MST and coalescence methods and transport models in this study).

Since the microscopic transport approaches allow to study in details the production process, the detailed investigation, performed in Ref. [3], showed that the coordinate space distribution of the produced deuterons differs from that of the free nucleons and other hadrons. Thus, deuterons are not destroyed by additional rescattering.

This finding can be demonstrated by showing the transverse spatial distribution of nucleons and deuterons.

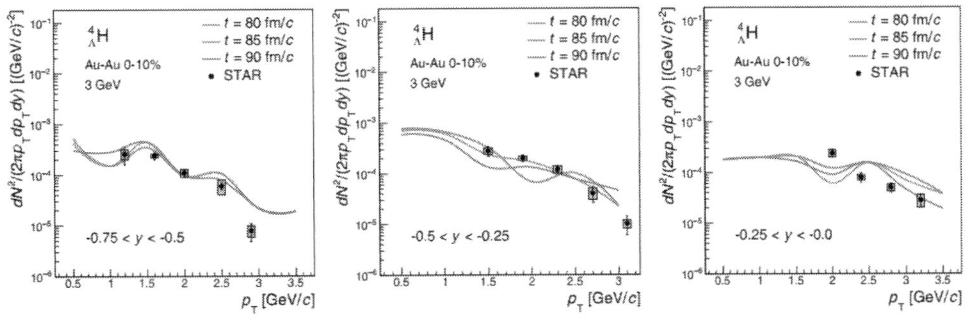

Figure 5. Transverse momentum distribution of $^4_\Lambda$H for different rapidity intervals as indicated in the legends in central Au+Au collisions at $\sqrt{s_{NN}} = 3$ GeV. The filled circles indicate the preliminary experimental data from the STAR Collaboration [32]. The PHQMD results are taken at the times $t = 80$ fm/c (red lines), 85 fm/c (blue lines) and 90 fm/c (green lines). The figure is taken from Ref. [2].

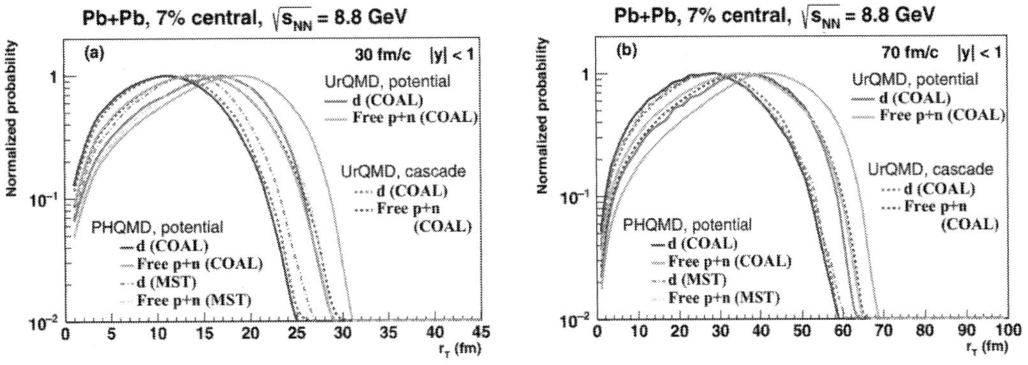

Figure 6. Transverse distance of unbound nucleons ($p + n$) and deuterons at 30 fm/c (a) and at 70 fm/c (b). The solid red line shows the deuterons from UrQMD (potential mode), the solid orange line the free $p + n$ from UrQMD (potential mode), the dashed magenta line represents deuterons from UrQMD (cascade mode), the dashed purple line the free $p + n$ from UrQMD (cascade mode), the blue solid line shows the deuterons from PHQMD (potential mode), the cyan solid line the free $p + n$ from PHQMD (potential mode). Additionally, the dark green dot dashed line shows PHQMD deuterons found by the MST algorithm and the light green dot dot dashed line represents free $p + n$ from PHQMD with MST algorithm. The figure is taken from Ref. [3].

In the left hand side of Fig. 6 we depict the normalized distribution at a fixed center-of-mass time of 30 fm/c and on the right hand side at a fixed center-of-mass time of 70 fm/c. Again, the deuterons are identified by the coalescence procedure. For PHQMD we show additionally the results for the MST procedure, which gives nearly identical results as coalescence.

As follows from Fig. 6 the nucleon and deuteron transverse positions follow the expansion flow with the maxima of the distributions shifting by approximately 20 fm/c in the interval between 30 fm/c to 70 fm/c. Remarkable that the deuterons remain at smaller radial distances than the nucleons; also the average r_T of deuterons is in all transport approaches smaller than that of the free nucleons. Consequently, the deuterons are produced at a smaller distance to the reaction center than the free protons or probably at a later time. This is a hint for a production at the end of the hadronization of the QGP or from nucleons which are closer to the center of the reaction than the average.

4 Summary

In this contribution we recall the selected results on midrapidity light cluster and hypernuclei production with the PHQMD transport approach where the clusters are identified with a minimum spanning tree (MST) in coordinate space. In QMD, being a semi-classical approach, clusters are not fully stable, thus the time at which the clusters are identified, has an influence on the multiplicity, but not on dynamical variables and is almost the same for the energy range investigated here. The excitation function of the light cluster yield as well as the p_T spectra of clusters are well reproduced in a wide energy range of center of mass energies, from a few GeV up to the top RHIC energies.

The recent experimental data by STAR Collaboration on hypernuclei production at an invariant energy of 3 GeV are reasonable well described. Moreover, the cluster identification procedure by the MST has been compared with the coalescence mechanism implemented in two independent transport approaches - the UrQMD and PHQMD - for studying the deuteron production. The coalescence as well as the MST procedure show that the deuterons remain in transverse direction closer to the center of the heavy-ion collision than free nucleons. It offers therefore an explanation of the 'ice in the fire' puzzle: the deuterons follow behind the front of the expanding baryonic fireball, they are spatially separated which might explain why they are not destroyed by collisions with the hadrons of the expanding fireball.

The authors acknowledge inspiring discussions with H. Liu, C.M. Ko, I. Vasiliev. This study is part of a project that has received funding from the European Union's Horizon 2020 research and innovation program under grant agreement STRONG – 2020 - No 824093. Furthermore, we acknowledge support by the Deutsche Forschungsgemeinschaft (DFG, German Research Foundation): grants BR 4000/7-1 and BL 982/3-1 and by the GSI-IN2P3 agreement under contract number 13-70. JS acknowledges support from the BMBF through the ErUM-data-pilot project as well as the Samson AG. Further support was provided by the PPP Program of the DAAD. Computational resources were provided by the Center for Scientific Computing (CSC) of the Goethe University, by the "Green Cube" at GSI, Darmstadt.

References

[1] J. Aichelin, E. Bratkovskaya, A. Le Fèvre, V. Kireyeu, V. Kolesnikov, Y. Leifels, V. Voronyuk, G. Coci, Phys. Rev. C **101**, 044905 (2020), 1907.03860
[2] S. Gläßel, V. Kireyeu, V. Voronyuk, J. Aichelin, C. Blume, E. Bratkovskaya, G. Coci, V. Kolesnikov, M. Winn, Phys. Rev. C **105**, 014908 (2022), 2106.14839
[3] V. Kireyeu, J. Steinheimer, J. Aichelin, M. Bleicher, E. Bratkovskaya, Phys. Rev. C **105**, 044909 (2022), 2201.13374
[4] M. Bleicher, E. Bratkovskaya, Prog. Part. Nucl. Phys. **122**, 103920 (2022)
[5] S.A. Bass et al., Prog. Part. Nucl. Phys. **41**, 255 (1998), nucl-th/9803035
[6] M. Bleicher et al., J. Phys. G **25**, 1859 (1999), hep-ph/9909407
[7] J. Aichelin, Phys. Rept. **202**, 233 (1991)
[8] C. Hartnack, A. Le Fèvre, Y. Leifels, J. Aichelin, 14th International Conference on Nuclear Reaction Mechanisms pp. 115–122 (2015), 1511.02646
[9] W. Cassing, E.L. Bratkovskaya, Phys. Rev. C **78**, 034919 (2008), 0808.0022
[10] W. Cassing, E.L. Bratkovskaya, Nucl. Phys. A **831**, 215 (2009), 0907.5331
[11] E.L. Bratkovskaya, W. Cassing, V.P. Konchakovski, O. Linnyk, Nucl. Phys. A **856**, 162 (2011), 1101.5793
[12] O. Linnyk, E.L. Bratkovskaya, W. Cassing, Prog. Part. Nucl. Phys. **87**, 50 (2016), 1512.08126
[13] L.P. Kadanoff, G. Baym, *Quantum Statistical mechanics* (W. A. Benjamin, Inc., New York, 1962)
[14] W. Cassing, Eur. Phys. J. ST **168**, 3 (2009), 0808.0715
[15] R.K. Puri, C. Hartnack, J. Aichelin, Phys. Rev. C **54**, R28 (1996)
[16] R.K. Puri, J. Aichelin, J. Comput. Phys. **162**, 245 (2000), nucl-th/9811018
[17] P.B. Gossiaux, R. Puri, C. Hartnack, J. Aichelin, Nucl. Phys. A **619**, 379 (1997), nucl-th/9706038
[18] A. Le Fèvre, J. Aichelin, C. Hartnack, Y. Leifels, Phys. Rev. C **100**, 034904 (2019), 1906.06162
[19] V. Kireyeu, Phys. Rev. C **103**, 054905 (2021), 2103.10542
[20] J. Steinheimer, M. Bleicher, H. Petersen, S. Schramm, H. Stocker, D. Zschiesche, Phys. Rev. C **77**, 034901 (2008), 0710.0332
[21] H. Petersen, J. Steinheimer, G. Burau, M. Bleicher, H. Stöcker, Phys. Rev. C **78**, 044901 (2008), 0806.1695
[22] S. Sombun, K. Tomuang, A. Limphirat, P. Hillmann, C. Herold, J. Steinheimer, Y. Yan, M. Bleicher, Phys. Rev. C **99**, 014901 (2019), 1805.11509
[23] P. Hillmann, J. Steinheimer, T. Reichert, V. Gaebel, M. Bleicher, S. Sombun, C. Herold, A. Limphirat, J. Phys. G **47**, 055101 (2020), 1907.04571
[24] V. Gaebel, M. Bonne, T. Reichert, A. Burnic, P. Hillmann, M. Bleicher, Eur. Phys. J. A **57**, 55 (2021), 2006.12951
[25] A. Kittiratpattana, M.F. Wondrak, M. Hamzic, M. Bleicher, C. Herold, A. Limphirat, Eur. Phys. J. A **56**, 274 (2020), 2006.03052
[26] P. Hillmann, K. Käfer, J. Steinheimer, V. Vovchenko, M. Bleicher, Journal of Physics G: Nuclear and Particle Physics (2022), 2109.05972
[27] T. Anticic et al. (NA49), Phys. Rev. C **94**, 044906 (2016), 1606.04234
[28] L. Adamczyk et al. (STAR), Phys. Rev. C **96**, 044904 (2017), 1701.07065
[29] S.S. Adler et al. (PHENIX), Phys. Rev. C **69**, 034909 (2004), nucl-ex/0307022
[30] J. Adam et al. (STAR), Phys. Rev. C **99**, 064905 (2019), 1903.11778
[31] H. Liu et al. (STAR) (2022), 2208.04650
[32] M. Abdallah et al. (STAR), Phys. Rev. Lett. **128**, 202301 (2022), 2110.09513

Strange Hadron Production in Au+Au Collisions at RHIC Beam Energy Scan

Yingjie Zhou[1,*] (for the STAR Collaboration)

[1]Key Laboratory of Quark & Lepton Physics (MOE) and Institute of Particle Physics, Central China Normal University, Wuhan 430079, China

Abstract. Strangeness production has been suggested to be a sensitive probe to the early-time dynamics of the nuclear matter created in heavy-ion collisions. Transverse momentum distributions and yields of strange hadrons provide important information on their production mechanisms and can help us probe the properties of the created medium and its evolution.
Thanks to the high statistics data taken during the STAR BES-II program in 2018-2021, a series of measurements on the properties of strangeness production at low energies are carried out. In these proceedings, the production of K^-, K^0_S, ϕ, Λ, and Ξ^- in Au+Au collisions at $\sqrt{s_{\rm NN}}$ = 3 GeV are presented. The strange hadron transverse momentum spectra, rapidity density distributions, and particle ratios are shown. These results are compared with UrQMD model calculations, and the extracted kinetic freeze-out parameters are discussed and compared with the ones from higher collision energies.

1 Introduction

The main goal of the STAR experiment is to study the properties of the QCD matter under extreme conditions, i.e., high temperature and high density, by colliding heavy ions at ultra-relativistic speed. The yields and particle ratios of strange hadrons provide important information about their production mechanisms in these collisions. The RHIC Beam Energy Scan (BES) program covers a wide range of energies to explore the transition from a hadronic dominated phase to a partonic dominated one. Of particular interest is the high baryon density region which is accessible through the STAR fixed-target (FXT) program covering the energy range from 13.7 GeV down to 3 GeV.

2 Data Analysis

In these proceedings, we focus on results obtained from FXT Au+Au collisions at $\sqrt{s_{\rm NN}}$ = 3 GeV recorded in 2018. In total, approximately 260M minimum bias events are used in this analysis. Particle identification (PID) is performed using the energy loss (dE/dx) information from Time Projection Chamber (TPC) and the particle velocity (β) information from Time of Flight (TOF). Short-lived particles are reconstructed via their hadronic decay channels, using the KF Particle Finder package [1] which is based on the Kalman Filter method. The combinatorial background is estimated with the rotating daughter method, in

*e-mail: yingjiezhou@mails.ccnu.edu.cn

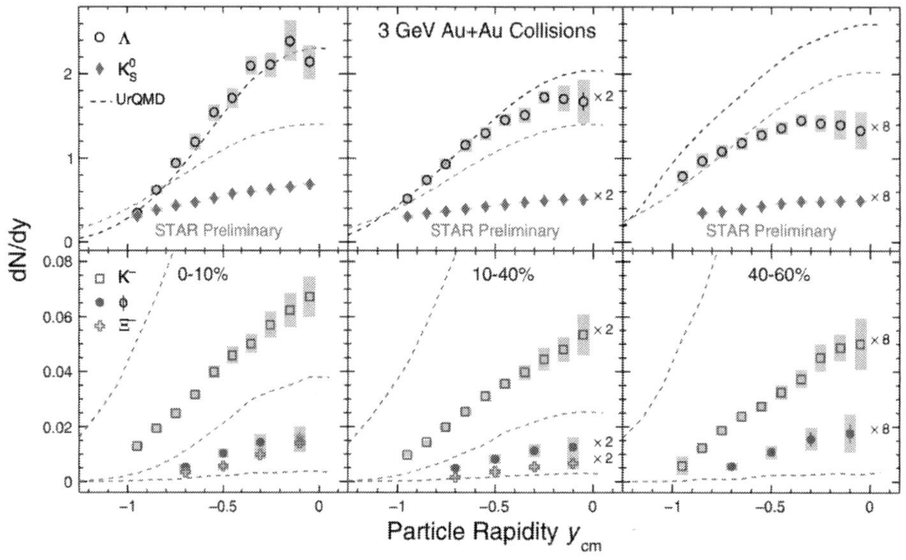

Figure 1. The rapidity dependence of dN/dy of $\Lambda, K^0_S, K^-, \phi$ and Ξ^- in Au+Au collisions at $\sqrt{s_{NN}}$ = 3 GeV in different centralities. The yields at 10-40% and 40-60% are scaled up by a factor of 2 and 8 respectively for better visibility. The vertical lines and bands represent the statistical and systematic uncertainties. The dashed lines are the calculations from UrQMD.

which a daughter track of the K^0_S, Λ and Ξ^- is rotated by a random angle between 150 to 210 degrees in the transverse plane. For ϕ meson, the combinatorial background is estimated by the mixing event technique.

3 Results and Discussion

3.1 Centrality and Rapidity Dependence of Yields of Strange Particles

The transverse momentum (p_T) spectra of Λ and K^0_S are extrapolated to the unmeasured region with several fitting functions (such as blast-wave function, m_T exponential, etc.) to obtain the p_T-integrated yields in different rapidity and centrality regions. The dN/dy of Λ and K^0_S in different centrality intervals are shown in Fig. 1, and are compared with other strange particles, K^-, ϕ, and Ξ^- [2], and UrQMD calculations [3]. For Λ, UrQMD reproduces the measured rapidity distributions in central collisions, but overestimates the data in peripheral collisions. UrQMD significantly overestimates the K^-, K^0_S, and Ξ^- yields, and underestimates the ϕ yields. To gain more insight on the production of strange baryons, the ratio of the Λ yield to the proton yield is presented in Fig. 2, as a function of rapidity and centrality. The Λ/p ratio exhibits clear rapidity and centrality dependence, whose shape can be qualitatively described by UrQMD. An enhancement of Λ/p ratio is observed at mid-rapidity compared to target rapidity, likely due to an increase in hadronic interactions from target to mid-rapidity.

3.2 Dependence of Strangeness Production on Number of Participating Nucleons

The total strange hadron yields divided by the mean number of participants ($\langle N_{part}\rangle$) is shown in Fig. 3 as a function of $\langle N_{part}\rangle$. For comparison, the proton yields are also shown. The

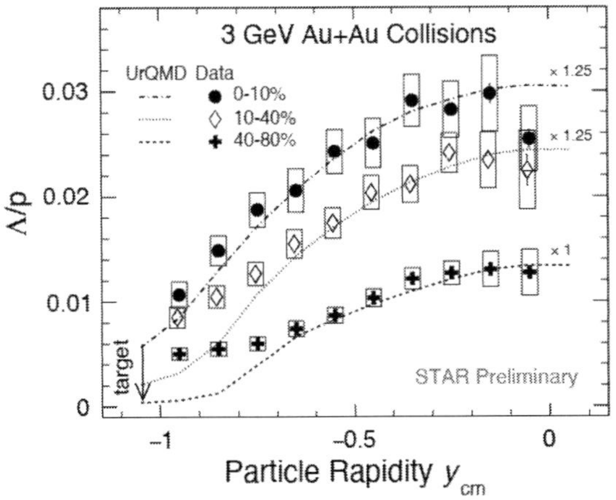

Figure 2. Rapidity dependence of Λ/p for different centrality bins in Au+Au collisions at $\sqrt{s_{NN}}$ = 3 GeV. Vertical lines represent statistical uncertainties, while boxes represent systematic uncertainties. 10-40% centrality is shifted left for better visibility. The curves represent the calculations from UrQMD and are scaled up to compare with data.

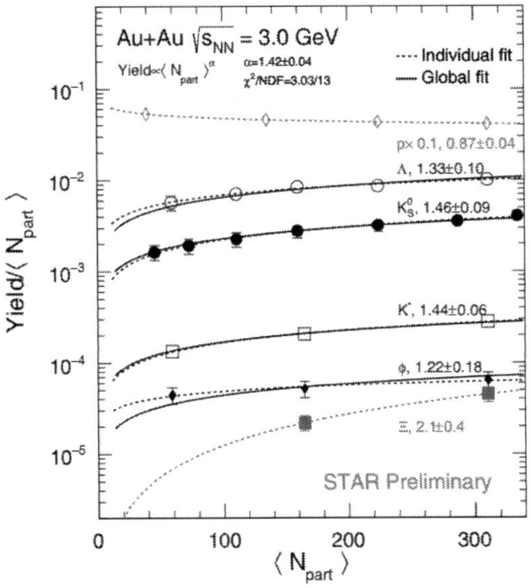

Figure 3. Hadron yields per mean number of participants as a function of $\langle N_{part}\rangle$. All hadron yields are fitted with a function of the form $f \propto C\langle N_{part}\rangle^{\alpha}$ shown as black dash lines, and the fitted α values are shown in the figures. The black solid lines represent a simultaneous fit to all strange hadron yields (K^-, K^0_S, Λ, ϕ).

hadron yields are fitted with the function $dN/dy \propto C\langle N_{part}\rangle^{\alpha}$ and the slope parameter α is extracted for each particle separately. The extracted slope parameters for K^-, K^0_S, Λ, and ϕ are consistent with each other, indicating universal centrality dependence of strangeness production in $\sqrt{s_{NN}} = 3$ GeV collisions. A combined fit to all four strange hadrons gives $\alpha = 1.42 \pm 0.04$. In contrast to strange particles, the yield of protons divided by $\langle N_{part}\rangle$ decreases with centrality. It is also interesting to note here that the slope parameter of Ξ^- is 2.1 ± 0.4, which deviates from the scaling trend. This could be due to the fact that Ξ^- is produced below its *NN* threshold, 3.25 GeV, while the *NN* production thresholds of the other strange particles (K^-, K^0_S, Λ, and ϕ) lie below $\sqrt{s_{NN}} = 3$ GeV. The precise measurement of Ξ^- yield will be carried out with the additional 2 billion Au+Au events on tape.

3.3 Kinetic Freeze-out Properties

The spectra of p, Λ and K^0_S at $\sqrt{s_{NN}} = 3$ GeV are fitted with a boost-invariant blast-wave model [4] which assumes a radially boosted thermal source. The fit parameters, i.e. effective

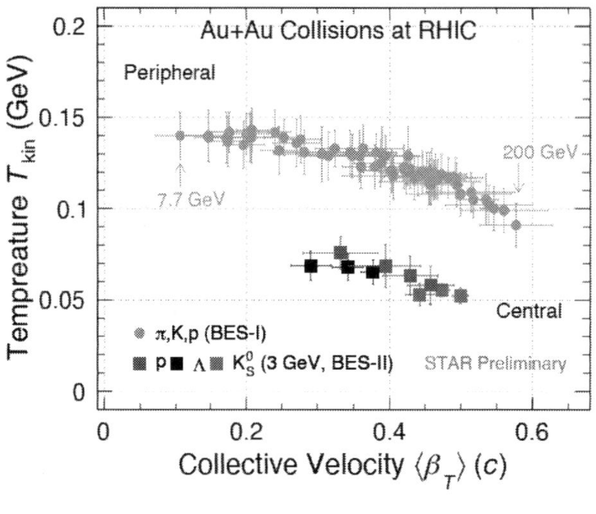

Figure 4. Effective temperature T_{kin} vs collective velocity $\langle\beta_T\rangle$ for p (green), Λ (black) and K_S^0 (blue) extracted from blast-wave fits at $\sqrt{s_{\mathrm{NN}}} = 3\,\mathrm{GeV}$. The fit parameters for (π, K, p) from $\sqrt{s_{\mathrm{NN}}} = 7.7$ to 200 GeV are shown for comparison.

temperature T_{kin} and collective velocity $\langle\beta_T\rangle$, are obtained and shown in Fig. 4. $\langle\beta_T\rangle$ increases from peripheral to central collisions for all three particles. We observe that T_{kin} of Λ is systematically higher than that of K_S^0 at $\sqrt{s_{\mathrm{NN}}} = 3\,\mathrm{GeV}$. The results are compared to those obtained from (π, K, p) spectra in $\sqrt{s_{\mathrm{NN}}} = 7.7$ to 200 GeV Au+Au collisions [5]. The T_{kin} of p, Λ and K_S^0 are significantly lower compared to higher energy collisions.

4 Summary

In summary, measurements on strangeness production in 3 GeV Au+Au collisions have been discussed. Yields of Λ and K_S^0 have been presented as a function of rapidity and centrality. The measured strange particle yields are inconsistent with UrQMD calculations. The yields of all strange particles except for Ξ^- scale with $\langle N_{part}\rangle$ to the power α, where $\alpha = 1.42 \pm 0.04$, indicating universal strangeness production as a function of centrality. The apparent deviation of Ξ^- from this trend could be due to the fact that at $\sqrt{s_{\mathrm{NN}}} = 3\,\mathrm{GeV}$, Ξ^- is produced below its NN production threshold. Finally, blast-wave fits to Λ and K_S^0 spectra result in a lower effective kinetic freeze-out temperature compared to that of π, K, p spectra at $\sqrt{s_{\mathrm{NN}}} = 7.7\,\mathrm{GeV}$ or higher.

Acknowledgement: This work was supported in part by the National Key Research and Development Program of China under Grant No. 2020YFE0202002, and the Fundamental Research Funds for the Central Universities (Nos. CCNU22QN005, CCNU20TS005).

References

[1] K. I., J. Phys. Conf. Ser. 1 p. 602 (2020)
[2] M.S. Abdallah et al. (STAR), Phys. Lett. B **827**, 137003 (2022)
[3] S.A. Bass et al., Prog. Part. Nucl. Phys. **41**, 255 (1998)
[4] E. Schnedermann, J. Sollfrank, U. Heinz, Physical Review C **48**, 2462 (1993)
[5] L. Adamczyk et al. (STAR), Phys. Rev. C **96**, 044904 (2017)

Production of K^{*0} in Au+Au collisions at $\sqrt{s_{NN}}$ = 19.6 GeV from RHIC BES-II

Aswini Kumar Sahoo (for the STAR collaboration)[1,*]

[1]Indian Institute of Science Education and Research, Berhampur, 760010, India

Abstract. We present the study of K^{*0} in Au+Au collisions at $\sqrt{s_{NN}}$ = 19.6 GeV from RHIC BES-II. The ratio of resonanse to non-resonance (K^{*0}/K) is shown as function of centrality and center-of-mass energy, which implies the dominance of hadronic re-scattering over regeneration in central A+A collisions. The lower limit of hadronic phase lifetime ($t_{kin} - t_{chem}$) is also reported using a toy model ansartz. The results are compared with previous RHIC and LHC measurements.

1 Introduction

Relativistic heavy-ion collisions provide a unique opportunity to understand the deconfined state of matter called the Quark-Gluon Plasma. The short-lived hadronic resonances are a very good probe to study the hadron gas phase, which characterises the late-stage evolution of heavy-ion collisions. $K^{*0}(892)$ has a lifetime ~ 4.16 fm/c which is smaller compared to that of fireball (~ 10 fm/c). Due to its short lifetime, the decay daughters may interact with the medium and change the properties of the resonances. In between chemical (CFO) and kinetic (KFO) freeze out, the daughter particles of $K^{*0}(892)$ could undergo in-medium effects like re-scattering and regeneration. Hence the final yield of the $K^{*0}(892)$ depends on the interplay of these effects, and can be used to study the hadronic phase of heavy-ion collisions [1].

2 Data Sets and Analysis Details

In these proceedings, we report the production yield of K^{*0} meson in Au+Au collisions at $\sqrt{s_{NN}}$ = 19.6 GeV, accumulated by the STAR experiment in 2019 as part of the RHIC BES-II program. The $K^{*0}(\overline{K^{*0}})$ is reconstructed via its decay channel K^{*0} $(\overline{K^{*0}}) \rightarrow K^+\pi^-$ $(K^-\pi^+)$. The vertex positions along the beam (V_z) and radial (V_r) directions are required to be within $|V_z| < 145$ cm and $|V_r| < 2$ cm. The daughter particles are identified using both the Time Projection Chamber (TPC) and the Time Of Flight (TOF) detector. In BES-II, the inner part of the TPC has been upgraded for better momentum resolution, wider transverse momentum (p_T) and pseudo-rapidity coverages.

The combinatorial background is estimated using the track rotaion method, where one of the daughter (here π) track momentum is rotated by 180° in transverse plane in order to break the correlation among the pairs originating from same parent resonance (K^{*0}). The K^{*0} signal is obtained after subtracting the combinatorial background from the same event $K\pi$ pairs,

*e-mail: aswinikumar.aks96@gmail.com

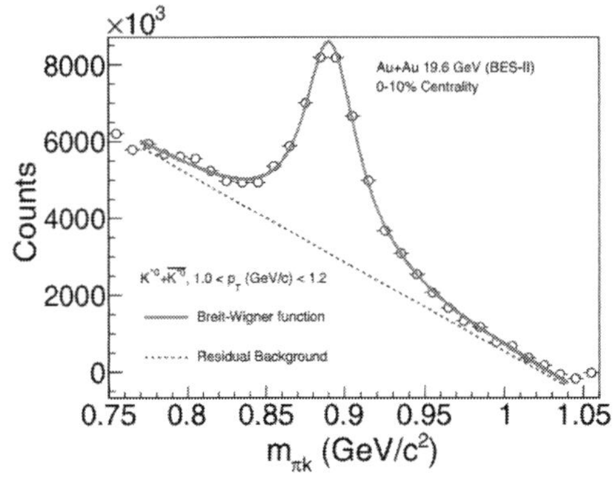

Figure 1. K^{*0} signal for $1.0 < p_T < 1.2$ GeV/c in 0-10% Au+Au collisions at $\sqrt{s_{NN}}$ = 19.6 GeV.

which is then fitted with the Breit-Wigner function and a first-order polynomial representing the residual background. Figure 1 shows an example of K^{*0} signal extraction for $1.0 < p_T < 1.2$ GeV/c in 0-10% Au+Au collisions at $\sqrt{s_{NN}}$ = 19.6 GeV. The yield is estimated by integrating the signal within $0.77 < m_{\pi K} < 1.04$ GeV/c^2 after residual background subtraction.

3 Results

3.1 Particle ratios

If the K^{*0} meson decays before KFO, the decay daughters (π and K) can re-scatter with other hadrons present in the medium, so that their momenta may get changed. Hence, one may not be able to reconstruct the parent resonance. On the other hand pions and kaons present in the medium may regenerate K^{*0} via pseudo-elastic scattering (πK→ K^{*0}). Hence the yield of the resonance is controlled by the relative contribution of these in-medium effects. The resonance to non resonance ratio (K^{*0}/K) can help us probe this phenomenon.

The left panel of Fig. 2 shows the K^{*0}/K and ϕ/K ratio as a function of the avarage number of participating nucleons $\langle N_{part}\rangle$. Here, we observe that the K^{*0}/K ratio decreases from peripheral to central collisions, while ϕ/K remains almost constant throughout all centralities. The thermal model prediction overestimates the K^{*0}/K, whereas the ϕ/K ratio is consistent with the prediction. In right panel of Fig. 2 we present K^{*0}/K as a function of collision energy for both elementary and heavy-ion collisions. Here we observe the ratio in A+A collisions is suppressed as compared to the elementary collisions. All these measurements indicate a dominant hadronic re-scattering effect over regeneration in central A+A collisions. The ϕ meson, which has a long lifetime (~ 46 fm/c), may remain immune to these in-medium effects.

3.2 Hadronic phase lifetime

Here the time difference between CFO and KFO is considered as the hadronic phase lifetime. Since it's span can not be measured directly, we can use the K^{*0}/K ratio to extract the lower limit of hadronic phase lifetime [1] using the following [24] relation:

$$\left(\frac{K^{*0}}{K}\right)_{KFO} = \left(\frac{K^{*0}}{K}\right)_{CFO} \times e^{-\Delta t/\tau_{K^{*0}}}, \quad (1)$$

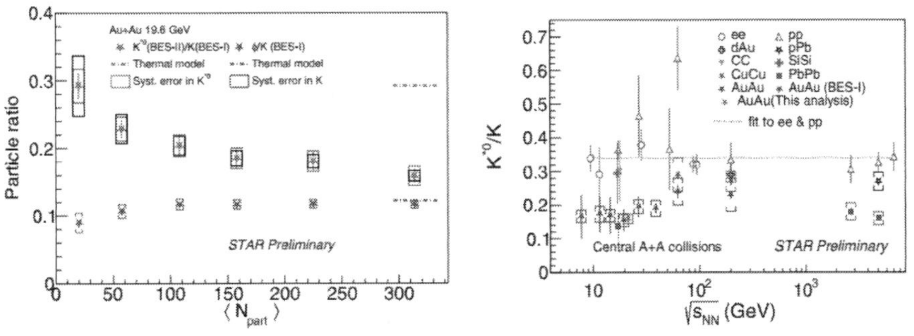

Figure 2. Left panel: Resonance to non-resonance ratio as a function of $\langle N_{part} \rangle$ along with the thermal model predication . Here K^{*0}/K and ϕ/K (BES-I)[2] denotes $(K^{*0} + \overline{K^{*0}})/(K^{+} + K^{-})$ and $2\phi/(K^{+} + K^{-})$ respectively. Here the K^{*0} measurement is based on BES-II data, whereas the measurement of kaons is from BES-I [3]. Hence their systematic uncertainties are denoted by different boxes. Right panel: K^{*0}/K as a function of collision energy [4–23]. The bars and caps (boxes) denote the statistical and systematic uncertainties respectively.

Here the $(K^{*0}/K)_{CFO}$ and $(K^{*0}/K)_{KFO}$ are taken to be the K^{*0}/K ratios measured in elementary and heavy-ion collisions respectively. This method assumes that no K^{*0} regeneration takes place between the chemical and kinetic freeze out, and all K^{*0} that decay before the kinetic freeze out are lost due to the re-scattering effect. The calculated Δt is boosted by the lorentz factor which is estimated as $\sqrt{1 + (\langle p_T \rangle / mc)^2}$.

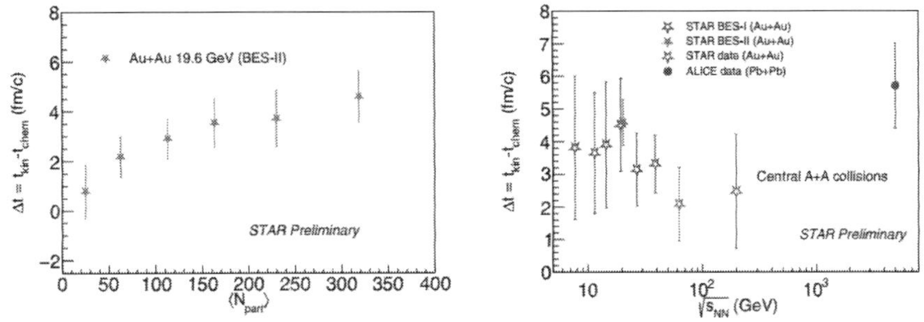

Figure 3. Left panel: Lower limit of hadronic phase lifetime (Δt) as a function of $\langle N_{part} \rangle$. Right panel: Lower limit of hadronic phase lifetime (Δt) as a function of collision energy. The result is compared with previous STAR and ALICE measurements [11, 13, 25]. The error bars are the quadratic sum of statistical and systematic uncertainties.

The left panel of Fig. 3 denotes the lower limit of hadronic phase lifetime as a function of N_{part}. The lifetime seems to increase from central to peripheral collisions. In the right panel of Fig. 3, Δt is plotted as a function of $\sqrt{s_{NN}}$ for central A+A collisions. RHIC measurement seems to be smaller compared to that of LHC, but the measurement uncertainties are large.

4 Summary

Measurement of K^{*0} at mid-rapidity in Au+Au collisions at 19.6 GeV (BES-II) is presented. The K^{*0}/K in central collisions appears to be less than that in peripheral collisions, and also the ratio is suppressed in heavy-ion collisions in comparison with elementary collisions. On the other hand, ϕ/K remains almost independent of centrality. This suggests that the hadronic phase formed in A+A collisions is mostly re-scattering dominated. The lower limit of hadronic phase lifetime is estimated using the K^{*0}/K ratio, which seems to be smaller as compared to LHC measurements in cental heavy-ion collisions, A study with high statistics BES-II data across various energies is needed to draw definitive conclusions.

References

[1] C. Adler et al. (STAR), Phys. Rev. C **66**, 061901 (2002)
[2] J. Adam et al. (STAR), Phys. Rev. C **102**, 034909 (2020)
[3] L. Adamczyk et al. (STAR), Phys. Rev. C **96** (2017)
[4] H. Albrecht et al. (ARGUS), Z. Phys. C **61**, 1 (1994)
[5] Y.J. Pei, Z. Phys. C **72**, 39 (1996)
[6] W. Hofmann, Ann. Rev. Nucl. Part. Sci. **38**, 279 (1988)
[7] K. Abe et al. (SLD), Phys. Rev. D **59**, 052001 (1999)
[8] M. Aguilar-Benitez et al., Z. Phys. C **50**, 405 (1991)
[9] T. Akesson et al. (Axial Field Spectrometer), Nucl. Phys. B **203**, 27 (1982)
[10] A. Adare et al. (PHENIX), Phys. Rev. C **90**, 054905 (2014)
[11] J. Adams et al. (STAR), Phys. Rev. C **71**, 064902 (2005)
[12] B.I. Abelev et al. (STAR), Phys. Rev. C **78**, 044906 (2008)
[13] M.M. Aggarwal et al. (STAR), Phys. Rev. C **84**, 034909 (2011)
[14] T. Anticic et al. (NA49), Phys. Rev. C **84**, 064909 (2011)
[15] A. Aduszkiewicz et al. (NA61/SHINE), Eur. Phys. J. C **80**, 460 (2020)
[16] B.B. Abelev et al. (ALICE), Phys. Rev. C **91**, 024609 (2015)
[17] J. Adam et al. (ALICE), Phys. Rev. C **95**, 064606 (2017)
[18] S. Acharya et al. (ALICE), Phys. Lett. B **802**, 135225 (2020)
[19] B. Abelev et al. (ALICE), Eur. Phys. J. C **72**, 2183 (2012)
[20] S. Acharya et al. (ALICE), Phys. Rev. C **102**, 024912 (2020)
[21] S. Acharya et al. (ALICE), Phys. Lett. B **807**, 135501 (2020)
[22] S. Acharya et al. (ALICE) (2021), `2110.10042`
[23] J. Adam et al. (ALICE), Eur. Phys. J. C **76**, 245 (2016)
[24] S. Singha et al., Int. J. Mod. Phys. E **24**, 1550041 (2015)
[25] S. Acharya et al. (ALICE), Phys. Lett B **802**, 135225 (2020)

Results on system size dependence of strangeness production in the CERN SPS energy range from NA61/SHINE

Piotr Podlaski[1,*] for the NA61/SHINE Collaboration

[1]Faculty of Physics, University of Warsaw, Warsaw, Poland

Abstract. NA61/SHINE is a multipurpose fixed-target facility at the CERN Super Proton Synchrotron. The main goals of the NA61/SHINE strong interactions program are to discover the critical point of strongly interacting matter and study the properties of the onset of deconfinement. To reach these goals, hadron production measurements are performed in the form of a two-dimensional scan by varying collision energy and system size. The Collaboration has recently finished data acquisition for its original program on strong interactions, accumulating broad data samples on hadron production in various systems in the SPS energy range.
In this contribution, the NA61/SHINE results on identified charged kaon and pion production in p+p, Be+Be and Ar+Sc collisions at the SPS energy range ($\sqrt{s_{NN}}$=5.1–17.3 GeV) are presented. The NA61/SHINE measurements of small and intermediate-mass ion collisions establish an interesting system size dependence, showing a rapid change of hadron production properties that starts when moving from Be+Be to Ar+Sc system. In particular, Ar+Sc is the smallest system for which a significant enhancement of K^+/π^+ ratio with respect to $p+p$ collisions is observed. Obtained energy and system size dependence of the measured charged hadron multiplicities are compared with available world data and various theoretical models.

1 Introduction

NA61/SHINE is a large acceptance hadron spectrometer located in the CERN's North Area [1]. Eight large volume Time Projection Chambers (TPC), accompanied by the Time of Flight detectors (TOF) provide tracking and identification of produced particles. The Projectile spectator Detector (PSD), a precise zero-degree hadron calorimeter, measures the energy of projectile spectators, which can be related to the centrality of the collision.
NA61/SHINE performed a unique, two-dimensional scan in system size (p+p, p+Pb, Be+Be, Ar+Sc, Xe+La, Pb+Pb) and momentum of the beam (13A – 150A GeV/c). NA61/SHINE physics program focuses on studies of the phase diagram of strongly interacting matter and searches for the critical point.

2 Studies of the onset of deconfinement

A first-order phase transition between Quark-Gluon Plasma and hadron gas was predicted by the Statistical Model of The Early Stage (SMES) [2, 3] to be located in the CERN SPS energy range. Left panel of Fig. 1 shows the inverse slope parameter of the positive kaon m_T

*e-mail: piotr.podlaski@cern.ch

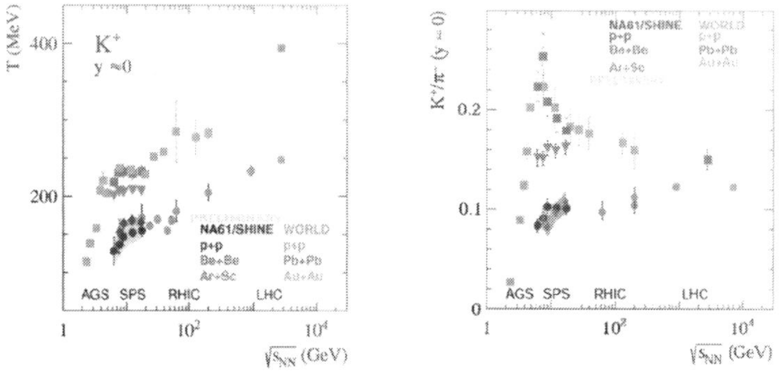

Figure 1. Energy dependence of the inverse slope parameter of K^+ (*left*) and energy dependence of the K^+/π^+ particle yield ratio in mid-rapidity (*right*) for p+p, Be+Be, Ar+Sc and Pb+Pb collisions [4–20].

spectra at midrapidity for different colliding systems as a function of collision energy. Within SMES the characteristic plateau in the energy dependence of the inverse slope parameter for heavy ion collisions (Pb+Pb, Au+Au) can be attributed to the existence of the mixed phase of QGP and hadron gas. The NA61/SHINE results on light and intermediate systems (p+p, Be+Be, Ar+Sc) exhibit a qualitatively similar behavior, with the plateau value growing with the system size. The most prominent signature of the phase transition, predicted within SMES, is a rapid, non-monotonic change of the K^+/π^+ ratio as a function of collision energy, the horn. The right panel of Fig. 1 presents a compilation of results on positive kaon to positive pion multiplicity ratio at midrapidity. The horn structure was not observed in data on central Ar+Sc collisions. Moreover, Ar+Sc results on K^+/π^+ ratio at low SPS energies are located between light (p+p, Be+Be) and heavy (Pb+Pb, Au+Au) systems, while for higher collision energies the K^+/π^+ ratio is following the heavy systems.

Figure 2 presents K^+/π^+ multiplicity ratio and inverse slope parameter of the K^+ transverse mass spectra as a function of the system size for the top SPS energy (150*A* GeV/*c* beam momentum). System size is quantified by the mean number of wounded nucleons in a collision $\langle W \rangle$. Dynamical models (EPOS, UrQMD, SMASH) successfully describe the K^+/π^+ ratio for light systems (p+p and Be+Be) but fail for heavier ones (Ar+Sc, Pb+Pb). On the other hand PHSD, the model with phase transition, reproduces the data for heavy systems but overestimates the K^+/π^+ ratio for lighter ones. Both statistical models taken into consideration tend to overestimate the data, regarding if they include the phase transition (SMES)

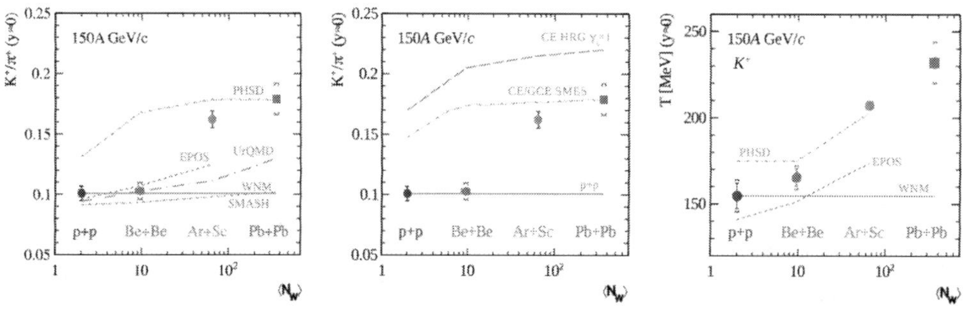

Figure 2. System size dependence of the K^+/π^+ ratio at midrapidity measured at 150*A* GeV/*c* compared with dynamical (*left*) and statistical (*center*) models. *Right*: System size dependence of the inverse slope parameter T of K^+ at the same collision energy [3, 4, 9, 16, 21, 22].

or not (HRG). Both considered quantities (K^+/π^+ ratio and inverse slope parameter T) show qualitatively similar system size dependence, which cannot be fully described by any of the statistical or dynamical models used for comparison.

3 Strangeness production in p+p interactions

The NA61/SHINE has accumulated almost 30 million p+p collisions at 158 GeV/c beam momentum, allowing for precise studies of the strangeness production mechanisms in elementary collisions, as well as obtaining references to A+A systems. The left panel of Fig. 3 shows the measured rapidity spectrum of K^0_S mesons produced in inelastic p+p interactions at 158 GeV/c. The measured mean multiplicity of $\langle K^0_S \rangle$ is $0.162 \pm 0.001 \pm 0.011$. This result, accompanied by the energy scan of K^0_S production in inelastic p+p interactions, will serve as reference measurements for future analyses of collisions of heavier nuclei.

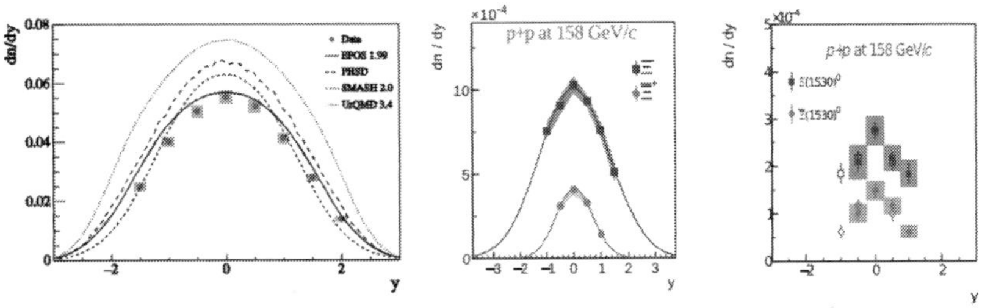

Figure 3. Rapidity spectra of: K^0_S (*left*), Ξ^+ and $\bar{\Xi}^-$ (*center*), $\Xi^0(1530)$ and $\bar{\Xi}^0(1530)$ (*right*) in p+p interactions at 158 GeV/c [23–26].

Unique, high statistics p+p dataset allows for studies of multi-strange baryon production by the NA61/SHINE. Center and right panels of Fig. 3 presents rapidity spectra of Ξ^+, $\bar{\Xi}^-$, $\Xi^0(1530)$ and $\bar{\Xi}^0(1530)$ measured in inelastic p+p collisions at 158 GeV/c. The production of anti-Ξ hyperons is suppressed compared to the production of Ξ hyperons, namely the multiplicity ratios are the following: $\overline{\Xi}^+/\Xi^-$ ratio is $0.24 \pm 0.01 \pm 0.05$. and $\overline{\Xi}^0(1530)/\Xi^0(1530)$ ratio is $0.40 \pm 0.03 \pm 0.05$. Noteworthy, $\Xi^0(1530)$ measurement by the NA61/SHINE is the only one available in the SPS energy range.

The Ξ multiplicities were used to calculate strangeness enhancement factors, E. The E is defined as follows:

$$E = \frac{2}{\langle W \rangle} \frac{\mathrm{d}n/\mathrm{d}y(A+A)}{\mathrm{d}n/\mathrm{d}y(p+p)} \tag{1}$$

where the midrapidity density of Ξ in A+A collision, scaled by the mean number of wounded nucleons, is divided by the corresponding density measured in inelastic p+p collisions (W = 2 for protons). Fig. 4 shows enhancement factors for Ξ^+, $\bar{\Xi}^-$, recalculated using the latest NA61/SHINE results, as a function of the mean number of wounded nucleons. The results shown here provide a new reference for the calculation of strangeness enhancement factors in A+A collisions.

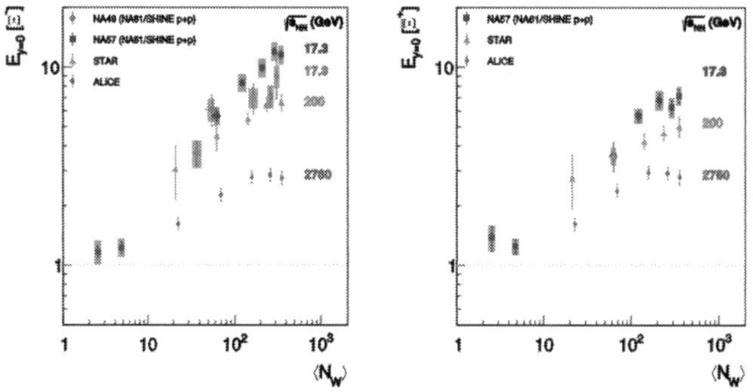

Figure 4. Strangeness enhancement factors calculated using Ξ^{+} and $\bar{\Xi}^{-}$ yields measured by NA61/SHINE [24, 25].

Acknowledgments: This work was supported by the Polish Minister of Education and Science (contract No. 2021/WK/10).

References

[1] N. Abgrall et al. (NA61/SHINE), JINST **9**, P06005 (2014)
[2] M. Gazdzicki, M.I. Gorenstein, Acta Phys. Polon. B **30**, 2705 (1999), hep-ph/9803462
[3] R.V. Poberezhnyuk, M. Gazdzicki, M.I. Gorenstein, Acta Phys. Polon. B **46**, 1991 (2015)
[4] A. Aduszkiewicz et al. (NA61/SHINE), Eur.Phys.J. **C77**, 671 (2017)
[5] B.I. Abelev et al. (STAR), Phys. Rev. C **79**, 034909 (2009)
[6] M. Kliemant, B. Lungwitz, M. Gazdzicki, Phys. Rev. C **69**, 044903 (2004)
[7] K. Aamodt et al. (ALICE), Eur. Phys. J. C **71**, 1655 (2011), 1101.4110
[8] B.B. Abelev et al. (ALICE), Phys. Lett. B **736**, 196 (2014), 1401.1250
[9] A. Acharya et al. (NA61/SHINE), Eur. Phys. J. C **81**, 73 (2021)
[10] L. Adamczyk et al. (STAR), Phys. Rev. C **96**, 044904 (2017)
[11] J. Adams et al. (STAR), Phys. Rev. Lett. **92**, 112301 (2004)
[12] K. Adcox et al. (PHENIX), Phys. Rev. C **69**, 024904 (2004)
[13] S.S. Adler et al. (PHENIX), Phys. Rev. C **69**, 034909 (2004)
[14] I. Arsene et al. (BRAHMS), Phys. Rev. C **72**, 014908 (2005)
[15] L. Ahle et al. (E-802, E-866), Phys. Rev. C **60**, 044904 (1999)
[16] C. Alt et al. (NA49), Phys. Rev. C **77**, 024903 (2008)
[17] T. Anticic et al. (NA49), Phys. Rev. C **86**, 054903 (2012)
[18] S.V. Afanasiev et al. (NA49), Phys. Rev. C **66**, 054902 (2002)
[19] B. Abelev et al. (ALICE), Phys. Rev. Lett. **109**, 252301 (2012), 1208.1974
[20] J. Adamczewski-Musch et al. (HADES), Phys. Lett. B **778**, 403 (2018), 1703.08418
[21] V. Kireyeu, I. Grishmanovskii, V. Kolesnikov, V. Voronyuk, E. Bratkovskaya, Eur. Phys. J. A **56**, 223 (2020), 2006.14739
[22] A. Motornenko, V.V. Begun, V. Vovchenko, M.I. Gorenstein, H. Stoecker, Phys. Rev. C **99**, 034909 (2019), 1811.10645
[23] A. Acharya et al. (NA61/SHINE), Eur. Phys. J. C **82**, 96 (2022), 2106.07535
[24] A. Aduszkiewicz et al. (NA61/SHINE), Eur. Phys. J. C **80**, 833 (2020), 2006.02062
[25] A. Aduszkiewicz et al. (NA61/SHINE), Eur. Phys. J. C **82**, 174 (2022)
[26] A. Acharya et al. (NA61/SHINE), Eur. Phys. J. C **81**, 911 (2021), 2105.09144

Production of pions, kaons, and (anti-)protons in Au+Au collisions at $\sqrt{s_{NN}}$ = 54.4 GeV at RHIC

Krishan Gopal[1,*], for the STAR collaboration

[1]Indian Institute of Science Education and Research (IISER) Tirupati, India

Abstract. We report systematic measurements of bulk properties of the system created in Au+Au collisions at $\sqrt{s_{NN}}$ = 54.4 GeV recorded by the STAR detector at the Relativistic Heavy Ion Collider (RHIC). The transverse momentum spectra of $\pi^{\pm}$, $K^{\pm}$, and $p(\bar{p})$ are studied at mid-rapidity ($|y| < 0.1$) for different centrality classes. We also present mid-rapidity measurement of particle yields (dN/dy) and particle ratios in Au+Au collisions at $\sqrt{s_{NN}}$ = 54.4 GeV. The kinetic freeze-out parameters (T_{kin} and $<\beta>$) are extracted from particle spectra using the blast-wave model, and all the results are compared with previously published experimental results.

1 Introduction

Quantum Chromodynamics (QCD) predicts that at very high energy density and/or very high temperature, a new state of matter is formed where quarks and gluons are deconfined, known as the Quark-Gluon Plasma (QGP) [1]. The study of transverse momentum spectra of produced particles is important for investigating the bulk properties of the QGP, such as integrated yields (dN/dy), average transverse momentum $\langle p_T \rangle$, particle ratios, and freeze-out properties. This information is also useful for understanding particle production mechanisms.

2 Analysis details

We have studied the transverse momentum spectra of $\pi^{\pm}$, $K^{\pm}$, and $p(\bar{p})$ at mid-rapidity ($|y| < 0.1$) for different centrality classes in Au+Au collisions at $\sqrt{s_{NN}}$ = 54.4 GeV, recorded by STAR [2] experiment in the year 2017. In the STAR experiment, charged particle identification is performed utilizing the Time Projection Chamber (TPC) and Time of Flight (TOF) detectors depending on the momentum of the particle. For low momentum particles, the TPC is used, while for particles with intermediate or high momenta ($p_T > 1$ GeV/c), the TOF is used. TPC works on the principle of ionization energy loss of the charged particles passing through TPC's gas volume.

The TPC uses the ionization energy loss (dE/dx) for particle identification, from which the variable z is defined:

$$z_X = ln(\frac{< dE/dx >}{< dE/dx >_X^B})$$

*e-mail: krishangopal@students.iisertirupati.ac.in

where $< dE/dx >^B_X$ is the expected energy loss based on the Bichsel function [3] and X is the particle type ($e^{\pm}$, $\pi^{\pm}$, $K^{\pm}$, p, or $\bar{p}$). A multi-Gaussian fit is carried out on z_X distributions to extract raw yields.

For the TOF detector, we use mass square information, $m^2 = p^2(\frac{c^2T^2}{L^2} - 1)$ where m, p, L, T and c are mass of particle, momentum of the particle, path length, time of travel by the particle and speed of light, respectively. From the above two methods we obtain raw spectra from TPC and TOF. The TPC tracking efficiency and acceptance, energy loss correction, pion and proton background subtraction are applied to calculate corrected spectra. For the TOF analysis, TOF matching efficiency correction is also applied.

3 Results and discussions

3.1 Transverse momentum spectra and particle yields

We calculate the p_T spectra of $\pi^{\pm}$, $K^{\pm}$, and $p(\bar{p})$ in Au+Au collisions at $\sqrt{s_{NN}}$ = 54.4 GeV. Figure 1 shows the p_T spectra of π^+, K^+, and p in different centrality classes from central (0-5%) to peripheral (70-80%) collisions. The p_T spectra of identified hadrons show a clear centrality dependence. We fit the pion ($\pi^{\pm}$), kaon ($K^{\pm}$) spectra with the Levy function and $p(\bar{p})$ spectra with a double-exponential function, to calculate particle yields (dN/dy).

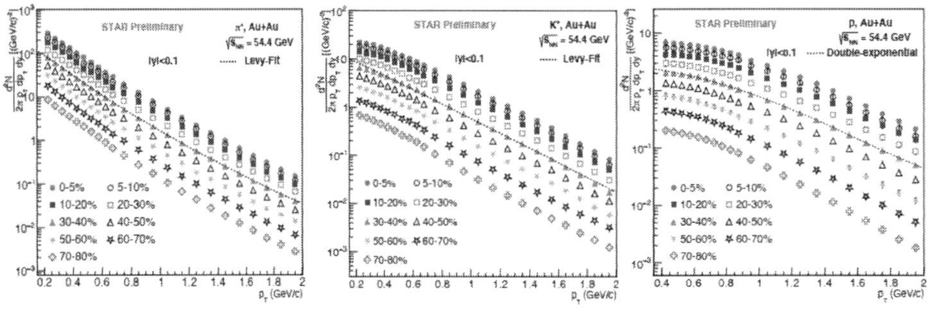

Figure 1. p_T spectra of π^+, K^+, and p measured at mid-rapidity ($|y| < 0.1$) in Au+Au collisions at $\sqrt{s_{NN}}$ = 54.4 GeV in different collision centralities. The statistical and systematic uncertainties are added in quadrature.

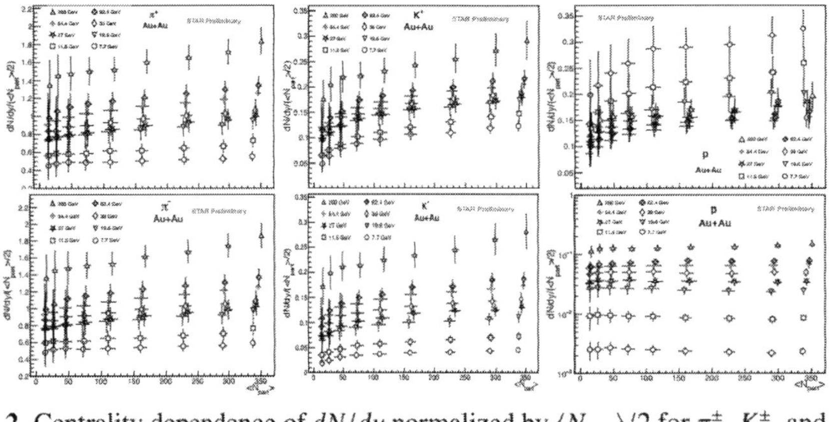

Figure 2. Centrality dependence of dN/dy normalized by $\langle N_{\text{part}}\rangle/2$ for $\pi^{\pm}$, $K^{\pm}$, and $p(\bar{p})$ at mid-rapidity ($|y| < 0.1$) in Au+Au collisions at $\sqrt{s_{NN}}$ = 54.4 GeV. Errors shown are quadrature sums of statistical and systematic uncertainties.

Figure 2 shows the centrality dependence of dN/dy normalized by half of the average number of participating nucleons ($\langle N_{\text{part}}\rangle/2$) for $\pi^{\pm}$, $K^{\pm}$, and $p(\bar{p})$ in Au+Au collisions at $\sqrt{s_{NN}}$ = 54.4 GeV, along with other RHIC's Beam Energy Scan (BES) energies.

The yields of $\pi^{\pm}$, $K^{\pm}$, and $\bar{p}$ decrease with decreasing collision energy, whereas the proton yield is the highest for the lowest energy of 7.7 GeV, which indicates the largest baryon stopping at mid-rapidity at this energy.

3.2 Particle ratios

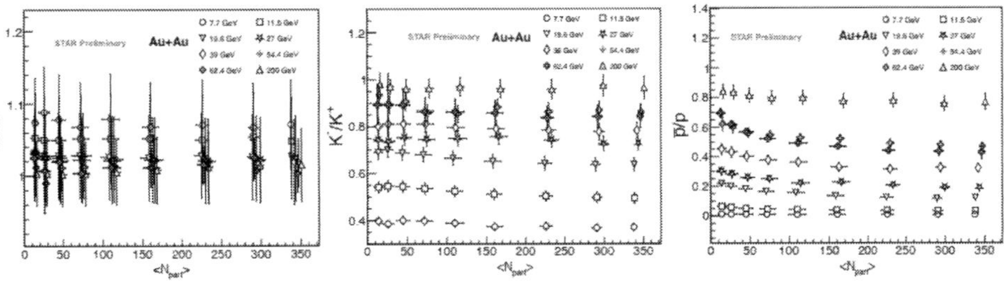

Figure 3. Variation of π^-/π^+, K^-/K^+, and $\bar{p}/p$ ratios as a function of $\langle N_{\text{part}}\rangle$ at mid-rapidity ($|y| < 0.1$) in Au+Au collisions at $\sqrt{s_{NN}}$ = 54.4 GeV.

Particle ratios of $\pi^{\pm}$, $K^{\pm}$, and $p(\bar{p})$ in Au+Au collisions at $\sqrt{s_{NN}}$ = 54.4 GeV are shown in Fig. 3. The π^-/π^+ ratio is close to unity. The K^-/K^+ ratio shows a mild centrality dependence. The $\bar{p}/p$ ratio decreases from peripheral to central collisions. Figure 4 shows the centrality dependence of mixed ratios (K^-/π^-, K^+/π^+, $\bar{p}/\pi^-$, and p/π^+). With increasing energy K^-/π^- shows an increasing behavior. The K^+/π^+ ratio has the maximum value at 7.7 GeV due to the associate production of K^+ at low energy. Also K^+/π^+ increases from peripheral to central collisions. The $\bar{p}/\pi^-$ ratio shows an increasing trend with increasing collision energy and shows a mild centrality dependence. The p/π^+ ratio decreases with increasing energy due to reduced baryon stopping as we go to higher center-of-mass energies.

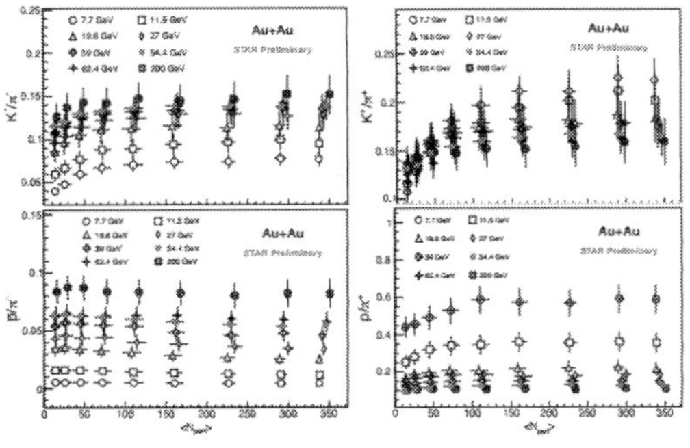

Figure 4. K^-/π^-, K^+/π^+, $\bar{p}/\pi^-$, and p/π^+ ratios as a function of $\langle N_{\text{part}}\rangle$ at mid-rapidity ($|y| < 0.1$) in Au+Au collisions at $\sqrt{s_{NN}}$ = 54.4 GeV.

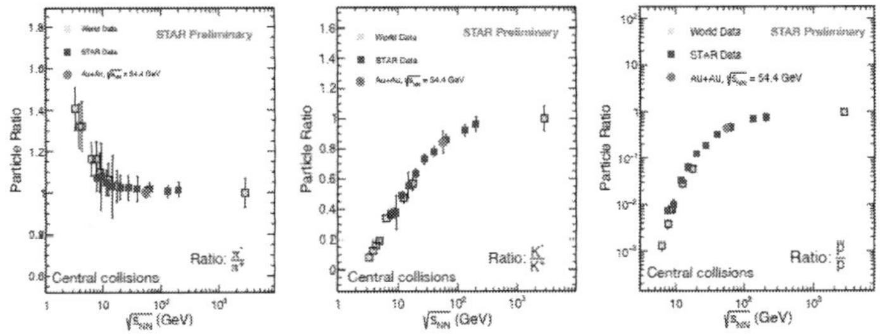

Figure 5. π^-/π^+, K^-/K^+, and $\bar{p}/p$ ratios at mid-rapidity ($|y| < 0.1$) in 0–5% Au+Au collisions at $\sqrt{s_{NN}}$ from 7.7-200 GeV.

Figure 5 shows the collision energy dependence of the particle ratios π^-/π^+, K^-/K^+, and $\bar{p}/p$ in most central collisions at different energies. The latest result from 54.4 GeV follows the trend shown from previous measurements of AGS, SPS, RHIC, and LHC [4] [5]. The displayed uncertainties are statistical and systematic uncertainties added in quadrature for each ratio plots (Fig. 3, Fig. 4, and Fig. 5).

3.3 Kinetic freeze-out parameters

The left panel of Fig. 6 shows the simultaneous fit to the p_T spectra of π, K, p and its anti-particle in 0-5% central collisions in $\sqrt{s_{NN}} = 54.4$ GeV collisions using the blast-wave model [6] [7]. The right panel of Fig. 6 shows the variation of kinetic freeze-out temperature (T_{kin}) with average transverse radial flow velocity ($\langle\beta\rangle$) for the different center-of-mass energies and centralities. $\langle\beta\rangle$ decreases as we go from most central to peripheral collisions, whereas T_{kin} increases from central to peripheral collisions due to the short-lived fireball formed in peripheral collisions.

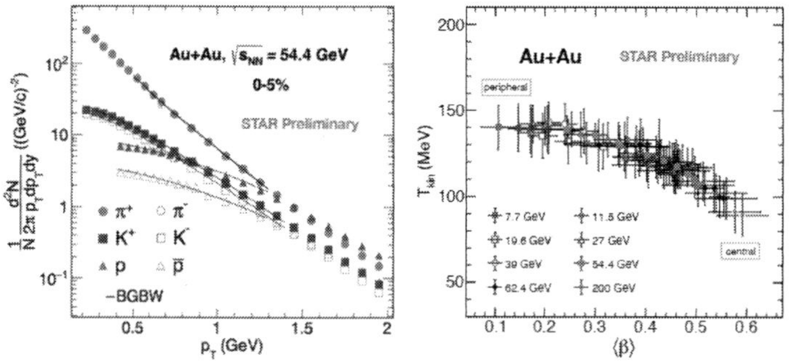

Figure 6. Left panel: Simultaneous Blast-Wave model fits to the p_T-spectra of π, K, $p(\bar{p})$ in 0–5% central Au+Au collisions at $\sqrt{s_{NN}} = 54.4$ GeV. Right panel: Variation of T_{kin} with $\langle\beta\rangle$ for different energies and collision centralities.

4 Conclusions

We presented transverse momentum spectra for $\pi^\pm$, $K^\pm$, and $p(\bar{p})$ in Au+Au collisions at $\sqrt{s_{NN}} = 54.4$ GeV. Particle ratios from Au+Au collisions at $\sqrt{s_{NN}} = 54.4$ GeV follow the trends from measurements at lower and higher energies. The yields of $\pi^\pm$, $K^\pm$ and $\bar{p}$ decrease with decreasing collision energy, whereas the proton yield is the highest at 7.7 GeV. Also the p/π^+ ratio is the largest at $\sqrt{s_{NN}} = 7.7$ GeV and decreases with increasing energy due to the reduced baryon stopping. The kinetic freeze-out temperature and $\langle\beta\rangle$ show an anti-correlation. T_{kin} decreases from peripheral to central collisions indicating a short-lived fireball in peripheral collisions and $\langle\beta\rangle$ increases from peripheral to central collisions indicating a larger radial flow for central collisions.

References

[1] K. Rajagopal and F. Wilczek, arXiv hep-ph/0011333 (2000).
[2] K. Ackermann et al. (STAR Collaboration), Nucl. Instrum. Meth. A 499, 624 (2003).
[3] H. Bichsel, Nucl. Instrum. Meth. A 562, 154 (2006).
[4] B. I. Abelev et al. (STAR Collaboration), Phys. Rev. C 79, 34909 (2009).
[5] B. I. Abelev et al. (STAR Collaboration), Phys. Rev. C 96, 044904 (2017).
[6] E. Schnedermann, J. Sollfrank, and U. Heinz, Phys. Rev. C 48, 2462 (1993).
[7] D. Teaney, J. Lauret, and E.V. Shuryak, arXiv:nucl- th/0110037.

Elliptic flow of strange and multi-strange hadrons in isobar collisions at $\sqrt{s_{\rm NN}}$ = 200 GeV at RHIC

Priyanshi Sinha[1,*] (for the STAR Collaboration)

[1]Indian Institute of Science Education and Research (IISER) Tirupati, India

Abstract. We report the elliptic flow (v_2) of K_s^0, Λ, $\bar{\Lambda}$, ϕ, Ξ^-, and $\overline{\Xi}^+$ at mid-rapidity for Ru+Ru and Zr+Zr collisions at $\sqrt{s_{\rm NN}}$ = 200 GeV. The transverse momentum (p_T) and centrality dependence of v_2 have been studied. The number of constituent quark (NCQ) scaling for all the (multi-)strange hadrons has also been tested. The dependence in the ratio of average v_2 ($\langle v_2 \rangle$) on centrality shows a deviation from unity indicating a difference in nuclear structure and deformation between the isobars. The results show a systematic size dependence when compared to Cu+Cu, Au+Au, and U+U collisions at similar beam energies.

1 Introduction

A dedicated isobar collision run of $^{96}_{44}$Ru+$^{96}_{44}$Ru and $^{96}_{40}$Zr+$^{96}_{40}$Zr at $\sqrt{s_{\rm NN}}$ = 200 GeV was successfully carried out at RHIC in 2018 to measure the charge separation along the magnetic field, called the Chiral Magnetic Effect (CME) [1]. These collisions are considered to be an effective way to minimize the flow-driven background contribution in search for the possibly small CME signal [2, 3]. The recent studies also show probing of nuclear structures via v_2 ratios as well as the v_2-$[p_T]$ correlations in isobar collisions [4, 5]. The deformation parameters are different between the two species and flow measurements are highly sensitive to them. Strange and multi-strange hadrons have a small hadronic cross-section compared to light hadrons, making their elliptic flow an excellent probe for understanding the initial state anisotropies of these isobar collisions.

2 Analysis details

In these proceedings, we report strange and multi-strange hadron v_2 in Ru+Ru and Zr+Zr collisions at $\sqrt{s_{\rm NN}}$ = 200 GeV using the data collected by the STAR experiment. A total of nearly 650M events have been analysed for both the isobar collisions. ϕ-mesons have been reconstructed using the invariant mass technique through its hadronic decay channel: $\phi \rightarrow K^+K^-$ (48.9 ± 0.5%) in midrapidity $|y| < 0.5$ [6]. Event mixing technique is used for combinatorial background estimation. The weakly decaying neutral strange particles K_s^0 and $\Lambda(\bar{\Lambda})$ are reconstructed using invariant mass technique and their weak-decay (V0) topology through the decay channel: $K_s^0 \rightarrow \pi^+ + \pi^-$ (69.20 ± 0.05%) and $\Lambda \rightarrow p + \pi^-$ (63.9 ± 0.5%), respectively [6, 7]. The multi-strange particle $\Xi^-(\overline{\Xi}^+)$ decays into a charged daughter and

*e-mail: priyanshisinha@students.iisertirupati.ac.in

a neutral V0 particle (Λ), which in turn decays into two charged particles. Their reconstruction involves finding two secondary vertices and the various topological selections. The combinatorial background for the weakly decaying particles is constructed using rotational background method [8]. The η-sub event plane method with an η gap of 0.1 has been used to calculate v_2 of these (multi-)strange hadrons [7]. The maximum EP resolution of nearly 48% is achieved for both the collision systems.

3 Results

The left panels of Figs. 1 and 2 show the v_2 of strange and multi-strange hadrons as a function of p_T for minimum bias Ru+Ru and Zr+Zr collisions at $\sqrt{s_{\mathrm{NN}}} = 200$ GeV, respectively. An approximate mass ordering at low p_T and a baryon-meson splitting at intermediate p_T have been observed. All particles and anti-particles tend to follow the number of constituent quark (NCQ) scaling within 10% shown in the right panel of Fig. 1 and 2, indicating the partonic collectivity as well as domination of quark coalescence mechanism during hadronization at intermediate p_T-region. A clear centrality dependence of v_2 has been observed for K_s^0, Λ

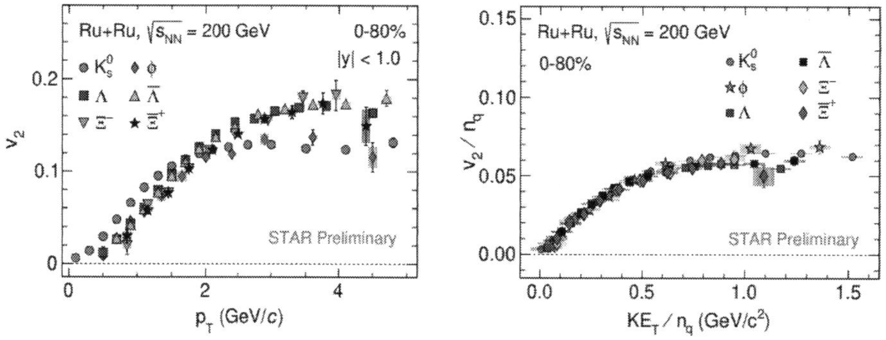

Figure 1. Left panel: v_2 as a function of p_T of (multi-)strange hadrons; Right panel: NCQ-scaled v_2 as a function of transverse kinetic energy for Ru+Ru collisions at $\sqrt{s_{\mathrm{NN}}} = 200$ GeV. The vertical lines and shaded boxes denote statistical and systematic uncertainties, respectively.

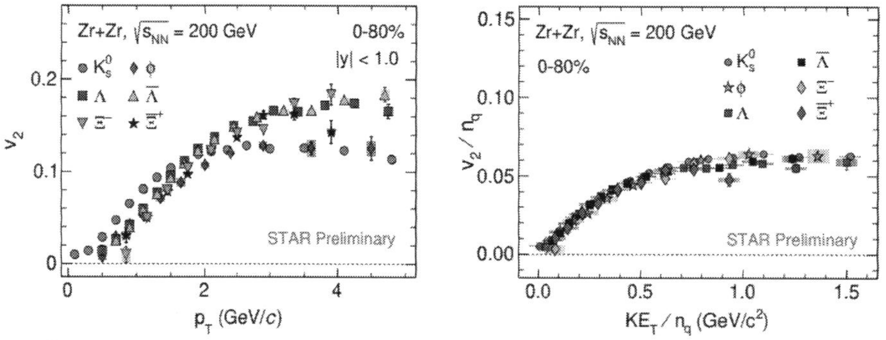

Figure 2. Left panel: v_2 as a function of p_T of strange hadrons; Right panel: NCQ-scaled v_2 as a function of transverse kinetic energy for Zr+Zr collisions at $\sqrt{s_{\mathrm{NN}}} = 200$ GeV. The vertical lines and shaded boxes denote statistical and systematic uncertainties, respectively.

and Ξ^- as shown in Fig. 3 for the isobar collision systems. We also study the p_T-integrated v_2 for strange hadrons as a function of the collision centrality in the isobar collisions as shown in Fig. 4. The ratios of v_2 between the two isobar collisions for K_s^0, Λ, and $\bar{\Lambda}$ are

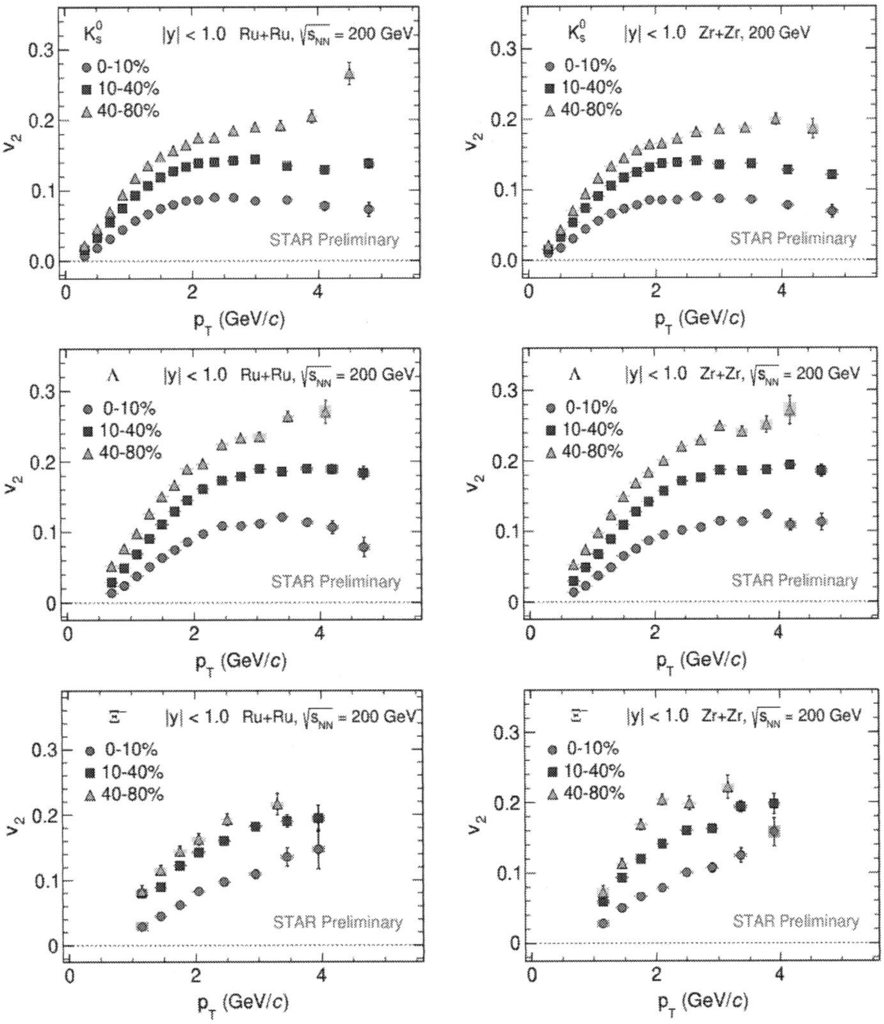

Figure 3. Left panel: Centrality dependence of v_2 of Λ as a function of p_T in Ru+Ru collisions; Right Panel: Same for Zr+Zr collisions at $\sqrt{s_{NN}}$ = 200 GeV. The vertical lines and shaded boxes denote statistical and systematic uncertainties, respectively.

comparable with charged hadron data and show a deviation of nearly 2% from unity in mid-central collisions, indicating a difference in nuclear structure and shape [1]. We studied the system size evolution of v_2 by comparing the minimum bias ${}^{63}_{29}$Cu+${}^{63}_{29}$Cu, ${}^{96}_{44}$Ru+${}^{96}_{44}$Ru, ${}^{96}_{40}$Zr+${}^{96}_{40}$Zr, ${}^{197}_{79}$Au+${}^{197}_{79}$Au collisions at $\sqrt{s_{NN}}$ = 200 GeV, and ${}^{238}_{92}$U+${}^{238}_{92}$U collisions at $\sqrt{s_{NN}}$ = 193 GeV [9–11]. Figure 5 shows an approximate system size dependence of v_2 for $p_T > 1.8$ GeV/c, based on the nuclear size. v_2 in U+U and Au+Au is observed to be higher, whereas that in Cu+Cu is slightly lower than those in isobar collisions.

4 Summary

In summary, we have presented the elliptic flow of K_s^0, Λ, $\bar{\Lambda}$, ϕ, Ξ^-, and $\overline{\Xi}^+$ in Ru+Ru and Zr+Zr collisions at $\sqrt{s_{NN}}$ = 200 GeV. We observed a mass ordering at low p_T and a baryon-meson splitting at intermediate p_T in both isobar species. All the strange particles and anti-particles follow the NCQ scaling indicative of the partonic degrees of freedom and coalescence picture for hadronization. The p_T-integrated v_2 ratio of strange hadrons shows

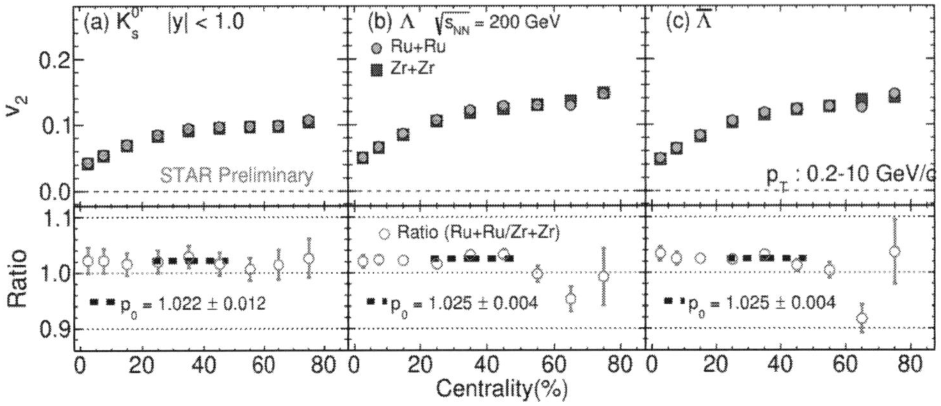

Figure 4. p_T-integrated v_2 as a function of centrality for K^0_s, Λ, and $\bar{\Lambda}$ in Ru+Ru and Zr+Zr collisions at $\sqrt{s_{NN}}$ = 200 GeV. The vertical lines on the ratio includes statistical and systematic uncertainties. The dotted lines denotes the fitting with a constant.

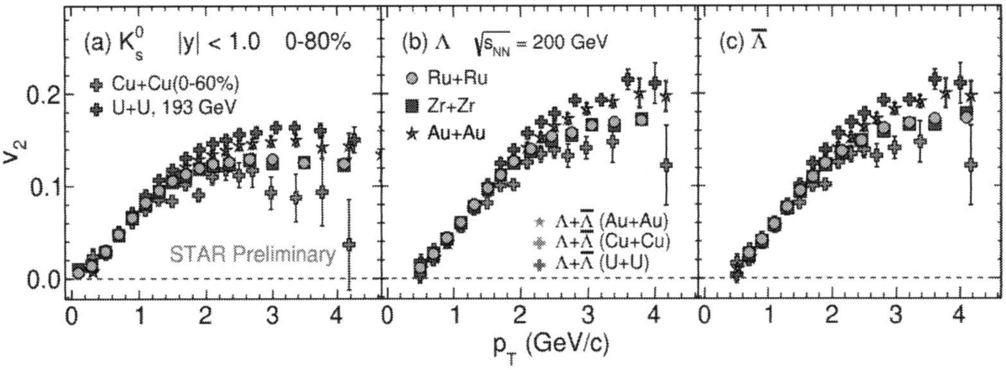

Figure 5. v_2 of strange hadrons in minimum bias Cu+Cu, Ru+Ru, Zr+Zr, Au+Au collisions at $\sqrt{s_{NN}}$ = 200 GeV and U+U collisions at $\sqrt{s_{NN}}$ = 193 GeV [9–11].

a deviation of about 2% from unity. We also observed a system size dependence of v_2 when comparing to different collision systems at similar beam energy. These measurements provide further constraint on the deformation and nuclear density difference in the two isobar nuclei.

References

[1] M. S. Abdallah et. al. (STAR Collaboration), Phys. Rev. C **105**, 14901 (2022).
[2] D. E. Kharzeev, Phys. Lett. B **633**, 260-264 (2006).
[3] D. E. Kharzeev, J. Liao, Nature Rev. Phys. **3**, 55-63 (2021).
[4] C. Zhang and J. Jia, Phys. Rev. Lett. **128**, 022301 (2022).
[5] J. Jia, S. Huang, and C. Zhang, Phys. Rev. C **105**, 014906 (2022).
[6] Particle Data Group, Chinese Phys. C **38**, 090001 (2014).
[7] L. Adamczyk et. al. (STAR Collaboration), Phys. Rev. C **88**, 014902 (2013).
[8] J. Adams et. al. (STAR Collaboration), Phys. Rev. Lett. **95**, 122301 (2005).
[9] B. I. Abelev et al. (STAR Collaboration), Phys. Rev. C **81**, 044902 (2010).
[10] B. I. Abelev et al. (STAR Collaboration), Phys. Rev. C **77**, 054901 (2008).
[11] M. S. Abdallah et al. (STAR Collaboration), Phys. Rev. C **103**, 064907 (2021).

Reconstruction of charged Kaons and $\phi(1020)$ in Ag+Ag collisions at $\sqrt{s_{NN}} = 2.55$ GeV with HADES

Marvin Kohls[1,*] for the HADES Collaboration

[1]Institut für Kernphysik, Goethe University Frankfurt am Main

Abstract. In March 2019, the HADES collaboration recorded $13.7 \cdot 10^9$ Ag(1.58*A* GeV)+Ag events as part of the FAIR Phase-0 program. The analysis steps necessary to extract the phase space information for K^+, K^- and $\phi(1020)$ are presented in this proceeding.

1 Introduction

The **H**igh **A**cceptance **DiE**lectron **S**pectrometer (HADES) is a fixed target experiment connected to the SIS18 synchrotron at GSI in Darmstadt. SIS18 can accelerate heavy ion beams to kinetic energies of up to 2 GeV per nucleon. This realizes center of mass energies of $\sqrt{s_{NN}} \approx 2 - 3$ GeV.

The energetically most favourable process for strange hadron production is the reaction $NN \rightarrow N\Lambda K$ with a threshold energy of 2.55 GeV, leading to the creation of a K^+/K^0 and a Λ^0. Hence, at the investigated collision energy of $\sqrt{s_{NN}} = 2.55$ GeV strangeness is either produced at or below the elementary production threshold. As a consequence, the production of strangeness is driven by collective effects of the created medium rather than isolated NN collisions.

Furthermore, the evaluation of density distributions within neutron star mergers and heavy ion collisions shows comparable values [1]. Consequently, investigating heavy ion collisions at low energies and high baryochemical potentials offers a probe for otherwise purely astronomical events. In particular, it allows the investigation of the microscopic composition of dense nuclear matter, which is hardly possible in astronomical observations. The observed steep rise of their excitation function makes the $K^{\pm}$ and ϕ yield a sensitive probe for the properties of the created medium.

2 The HADES Experiment

The spectrometer setup during the measurement of Ag+Ag collisions at $\sqrt{s_{NN}} = 2.55$ GeV is depicted in Figure 1. Due to the forward boost of secondary particles resulting from the fixed target configuration, the detectors are all situated in the forward direction in the laboratory reference frame.

In the azimuthal direction, the detector setup reaches a nearly total 360° coverage. A range between approximately 18° and 85° is covered in the polar angle direction. This corresponds to a proton rapidity coverage between -0.65 to 0.5 in the center of mass system.

*e-mail: m.kohls@gsi.de

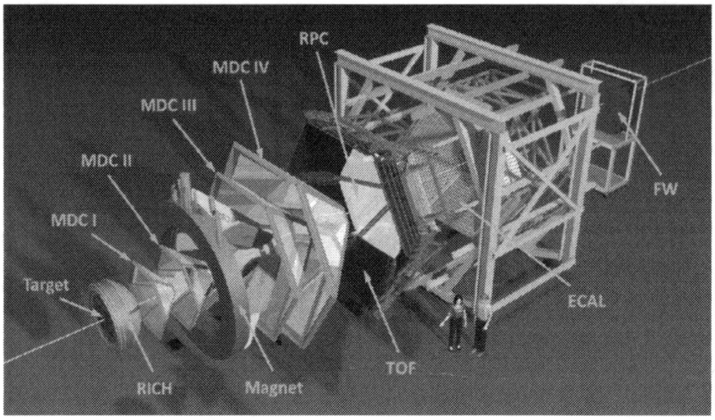

Figure 1. The HADES detector setup during the Ag+Ag beamtime in March 2019 stretched along the beam axis to illustrate the individual subdetectors.

In front of the target, a diamond-based t_0 detector is situated, registering the starting time for each reaction. Afterwards, **M**ultiwire **D**rift **C**hambers (MDCs) form four layers of Ar/CO_2 gas-filled detection volumes with six stereo layers of sensing wires each. With the toroidal magnet between layers II and III, the MDCs perform the charged particle tracking and energy loss measurement before and after the magnetic field. The MDCs are followed by the TOF and **R**esistive **P**late **C**hamber (RPC) time of flight wall, usually referred to as **M**ultiplicity and **E**lectron **T**rigger **A**rray (META) detectors. TOF consists of 384 scintillation rods, so energy loss measurements are possible. META also functions as a trigger for the event multiplicity. More details on the setup can be found in [2].

With the META time of flight information and the bent trajectory reconstructed by the MDCs, the p/q of a particle can be determined. Together with the energy loss information, these are designated PID criteria.

The **R**ing **I**maging **Ch**erenkov Detector (RICH) and the **E**lectromagnetic **Cal**orimeter (ECal) are not used in this analysis. RICH is used for the identification of leptons via the produced Cherenkov light. ECal detects and aids in identifying leptons, $\pi^{\pm}$ and photons. The **F**orward **W**all (FW) in return registers the spectators of the collision in order to determine the event plane.

3 Analysis

As the production of charged Kaons and ϕ-mesons happens at (K^+) or below (K^-,ϕ) the free NN production threshold, the corresponding particle multiplicities are orders of magnitude below those of protons, $\pi^{\pm}$ and light nuclei.

As a consequence, not only track quality criteria are used to increase the significance of the $K^{\pm}$ signals but also the MDC and TOF energy loss information to suppress the p and $\pi^{\pm}$ contributions in the mass spectra. In order to obtain the true signal from the particle spectra, the background is interpolated ($K^{\pm}$) below the peak or estimated by the event mixing technique (ϕ) and then subtracted. This is performed for all mass spectra in the corresponding $(m_t - m_0)$-y intervals, as indicated on the left side of Figure 2, where the mass ($K^{\pm}$) and invariant mass (ϕ) spectra (black) and the corresponding background estimation (red) are displayed. In total about 1.34×10^7 K^+, 1.21×10^5 K^- and 5.67×10^3 ϕ could be reconstructed for differential analyses. For the K^+ mass spectrum the tails of the π^+ and proton mass distributions overlap with the signal and rise steeply towards the edges of the spectra. Similarly the

π^- mass tail overlaps with the K^- mass spectrum. The apparent difference in mass resolution between K^+ and K^- results from the mass spectra being taken from RPC (K^+) and TOF (K^-), which possess a different timing resolution.

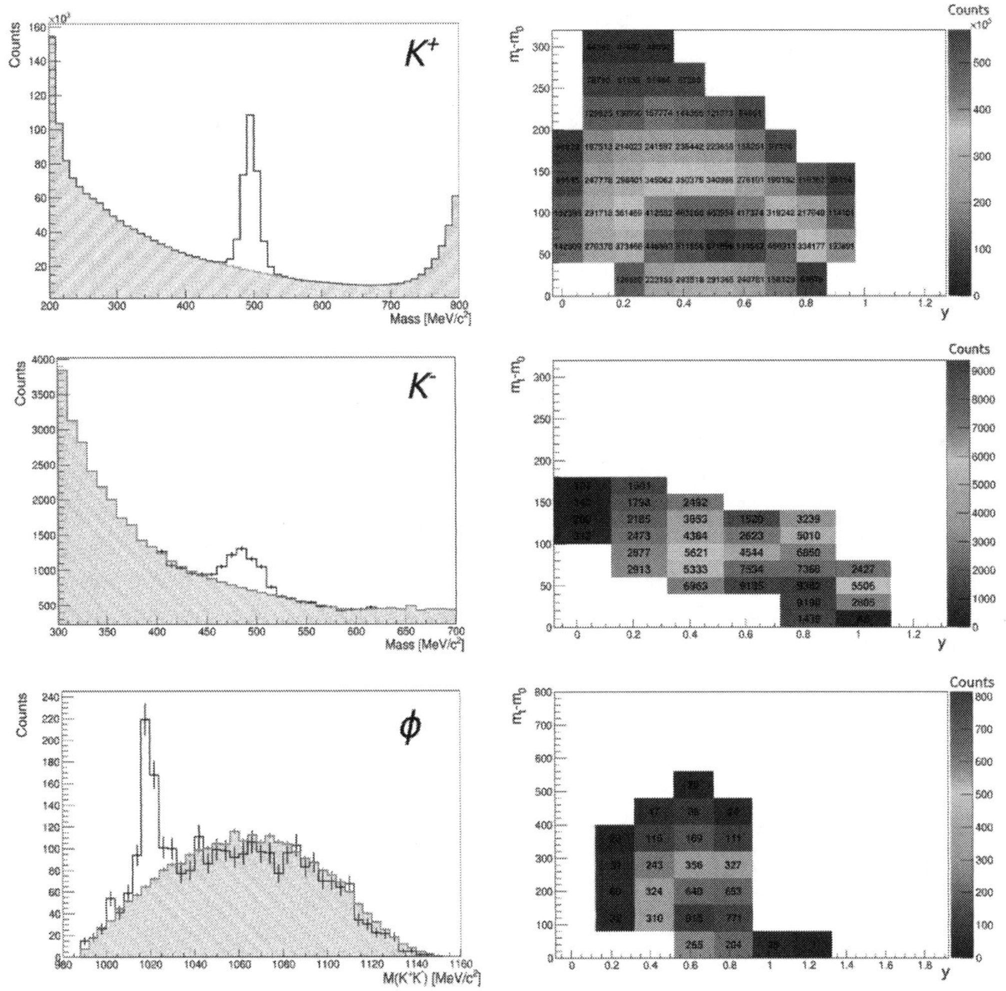

Figure 2. Exemplary differential mass spectra for K^+ , K^- and ϕ (left) and the resulting reconstructed phase space (right) in $0 - 30\%$ centrality. The full collection of spectra can be found in [2].

This results in a measured (raw) yield map of the particles' phase space, as displayed on the right side of Figure 2. The distribution has to be corrected for loss in geometrical acceptance and tracking as well as identification efficiency. UrQMD [3] generated events are transported through a GEANT representation of the HADES detector setup to obtain the relative loss due to those effects. This relative loss is used to correct the reconstructed yield in each phase space bin, resulting in the final $(m_t - m_0)$ spectra for each rapidity interval. More details on the procedure can be found in [2].

The obtained spectra are fitted by a double isotropic statistical distribution [4]

$$\left(\frac{d^2N}{dm_t dy}\right)_{DIS} = \left(\frac{d^2N}{dm_t dy}\right)_{IS} (m_t, y - \eta) + \left(\frac{d^2N}{dm_t dy}\right)_{IS} (m_t, y + \eta) \tag{1}$$

which is the superposition of two radially symmetric thermal sources of the form

$$\frac{d^2N}{dm_t dy} = 2\pi \cdot C \cdot \cosh(y) \cdot e^{-\frac{m_t \cosh(y)}{T_{\mathrm{Eff}}}} \tag{2}$$

each with a distance η to the centre of mass rapidity. The superposition is motivated by the generally non-spherical nature of the collision system. Summing over the available data and integrating the fitted function to extrapolate the yield in the uncovered phase space region, one obtains a final integrated yield for each rapidity interval. Plotting the particle production yield as a function of rapidity, a dN/dy distribution is obtained. Integrating equation 2 over dm_t and fitting the resulting superimposed function onto the dN/dy distribution is used to describe the unmeasured phase space region for the total production yield per event. Besides the yield, T_{Eff} can also be extracted from the fit function for every rapidity bin and will be presented in an upcoming publication.

4 Outlook

This analysis aims to investigate previous results like the observed steep rise of the ϕ/K^- [5], and the ϕ/Ξ^- [6] ratio towards low collision energies. Furthermore, we aim to check the universal scaling of strange hadron production yields according to $\mathrm{Mult}/\left\langle \mathrm{A_{part}} \right\rangle \propto \left\langle \mathrm{A_{part}} \right\rangle^{\alpha}$ as observed in [6] and [5]. In [5] it is found as well, that the lower T_{Eff} for the K^- can be explained by feed-down from the $\phi \to K^+K^-$ channel, which will also be investigated.

Lastly, direct comparisons of HADES results to existing data and hadronic transport models like UrQMD [3] and SMASH [7] as well as thermal models, like presented in [8], will be performed.

References

[1] J. Adamczewski-Musch, J., Arnold, O., Behnke, C. et al., **Nat. Phys. 15**, 1040–1045 (2019)

[2] M. Kohls, *Reconstruction and Analysis of Charged Kaons and φ-Mesons from Ag+Ag-Collisions at 1.58 AGeV*, PhD Thesis, Goethe University Frankfurt, to be published

[3] S. A. Bass, M. Belkacem, M. Bleicher, M. Brandstetter, L. Bravina, C. Ernst, L. Gerland, M. Hofmann, S. Hofmann and J. Konopka, *et al.*, **Prog. Part. Nucl. Phys. 41** (1998), 255-369

[4] S. Spies, *Strange hadron production in Ag+Ag collisions at 1.58A GeV*, PhD Thesis, Goethe University Frankfurt, 2022

[5] H. Schuldes [HADES], **Nucl. Phys. A 967** (2017), 804-807

[6] M. S. Abdallah *et al.* [STAR], **Phys. Lett. B 831** (2022), 137152

[7] H. Petersen, D. Oliinychenko, M. Mayer, J. Staudenmaier and S. Ryu, **Nucl. Phys. A 982** (2019), 399-402

[8] S. Harabasz, W. Florkowski, T. Galatyuk, M. Gumberidze, R. Ryblewski, P. Salabura and J. Stroth, **Phys. Rev. C 102** (2020) no.5, 054903

Interpretation of particle yields in pp interactions at $\sqrt{s}$ = 8.8, 12.3 and 17.3 GeV within statistical hadronization model

Tomasz Matulewicz[1,*] and *Krzysztof* Piasecki[1]

[1]Institute of Experimental Physics, Faculty of Physics, University of Warsaw, Warsaw, Poland

Abstract. The statistical hadronization model ThermalFist was applied to numerous hadron yields measured in p+p collisions at $\sqrt{s}$ = 8.8, 12.3 and 17.3 GeV, including recently published yields of ϕ-mesons, measured by the NA61/SHINE Collaboration. We consistently used the energy-dependent widths of Breit-Wigner mass distributions of hadronic resonances, as this approach was generally found to provide better agreement with experimental data. The well-established experimental ϕ meson yields are consequently accounted for (although neglecting this particle was found to improve the fit quality). The canonical treatment of particles with open strangeness with the grand canonical approach for non-strange particles gave a moderately reasonable agreement with the measured yields, only when the volume of strange particles was allowed to vary freely. In all the studied cases this volume is found to be greater than the canonical one.

1 Introduction

For over three decades the predictions of the statistical hadronization model have been compared to the yields of hadrons emitted from heavy-ion (AA) collisions, leading to the very successful description of experimental results. A spectacular example is the precise description within the Grand Canonical Ensemble (GCE) of Pb+Pb collisions at the LHC [1], covering 9 orders of magnitude in multiplicity of produced particles. To some surprise, the statistical hadronization model also describes quite well elementary interactions like proton-proton collisions [2, 3]. With much lower particle multiplicities in p+p interactions compared to AA collisions, the conservation of the relevant physical quantities is an issue. In particular, strangeness is found to be rarely produced in the beam energy range from SIS18 up to SPS. Special treatment of strangeness is usually applied via: (i) reduction of yields of hadrons containing (anti-)strange quarks, parameterized by factor γ_S, (ii) different volume occupied by strangeness compared to the non-strange bulk, and (iii) strangeness-canonical approach (SC) combining the canonical treatment of strange hadrons and GCE for the rest of matter.

The recent experimental results by the NA61/SHINE collaboration [4] cover also the production of $K(892)^0$ mesons in p+p collisions at $\sqrt{s}$ = 8.8 and 12.3 GeV and encouraged us to expand the analysis performed at $\sqrt{s}$ = 17.3 GeV, where the volume occupied by strange particles was unexpectedly found to be larger than the bulk volume [3, 5].

*e-mail: tomek.matulewicz@cern.ch

2 Analysis

The proton-proton collisions at $\sqrt{s}$ = 17.3 GeV were studied extensively by the NA49 and NA61/SHINE collaborations at the CERN SPS, providing (now) in total 17 yields of hadrons (up to double-strange Ξ). Several hadrons have been measured by two mentioned experiments and the adopted values are tabulated in [3]. Reasonable description of these yields within the statistical hadronization model was possible with $\gamma_s \sim 0.4$ (in agreement with the systematics [6]) and the volume occupied by strange particles was found larger than that of the bulk. This unexpected finding can not be ruled out as femtoscopy of kaon pairs (obtained at higher energies ($\sqrt{s} \in$ [27.4 – 900] GeV)) does not provide sufficient resolution. New yields of strange particles are available now from p+p collisions at $\sqrt{s}$ = 8.8 and 12.3 GeV, allowing to extend our initial observation towards lower energies. The hadron yields available for the statistical hadronization model calculations (see Table 1) show two distinct features of the measurements at $\sqrt{s}$ = 17.3 GeV: (i) the number of measured species is above that at RHIC energy of 200 GeV, and (ii) only here the neutron multiplicity is available, so the initial baryon number B = 2 is reconstructed as B = 1.92±0.11 [3]. One may expect that not yet available yields would be provided by the NA61/SHINE experiment, particularly the Λ baryon at lower beam energies.

Table 1. Summary of experimentally measured total particle yields in pp collisions. The lines separate the following particles: (i) light non-strange mesons, (ii) non-strange baryons (iii) hidden strangeness meson, (iv) open strangeness mesons, (v) $|S|$=1 baryons, (vi) $|S|$=2 baryons, (vii) $|S|$=3 baryons. The results are denoted: □=NA49, ■=NA61/SHINE, □ ⊕ ■=merged NA49 and NA61/SHINE ([3] and references therein), ★=STAR [7] (except π^0, which was measured by PHENIX [8]). Red color denotes recent experimental results

experiment	NA61				NA49&NA61	STAR
$\sqrt{s}$ (GeV)	6.3	7.7	8.8	12.3	17.3	200
π^0						⋈
π^+	■	■	■	■	□⊕■	★
π^-	■	■	■	■	□⊕■	★
p	■	■	■	■	□⊕■	★
$\bar{p}$	■	■	■	■	□⊕■	★
n					□	
ϕ			■	■	□⊕■	★
K_S^0			■	■	■	
K^+	■	■	■	■	□⊕■	★
K^-	■	■	■	■	□⊕■	★
$K(892)^0$			■	■	□⊕■	
$\overline{K(892)}^0$					□	
Λ					■	★
$\bar{\Lambda}$						★
$\Lambda(1520)$					□	
Ξ^-					■	★
$\bar{\Xi}^+$					■	★
$\Xi(1530)^0$					■	
$\overline{\Xi(1530)}^0$					■	
Ω						★
$\overline{\Omega}$						★

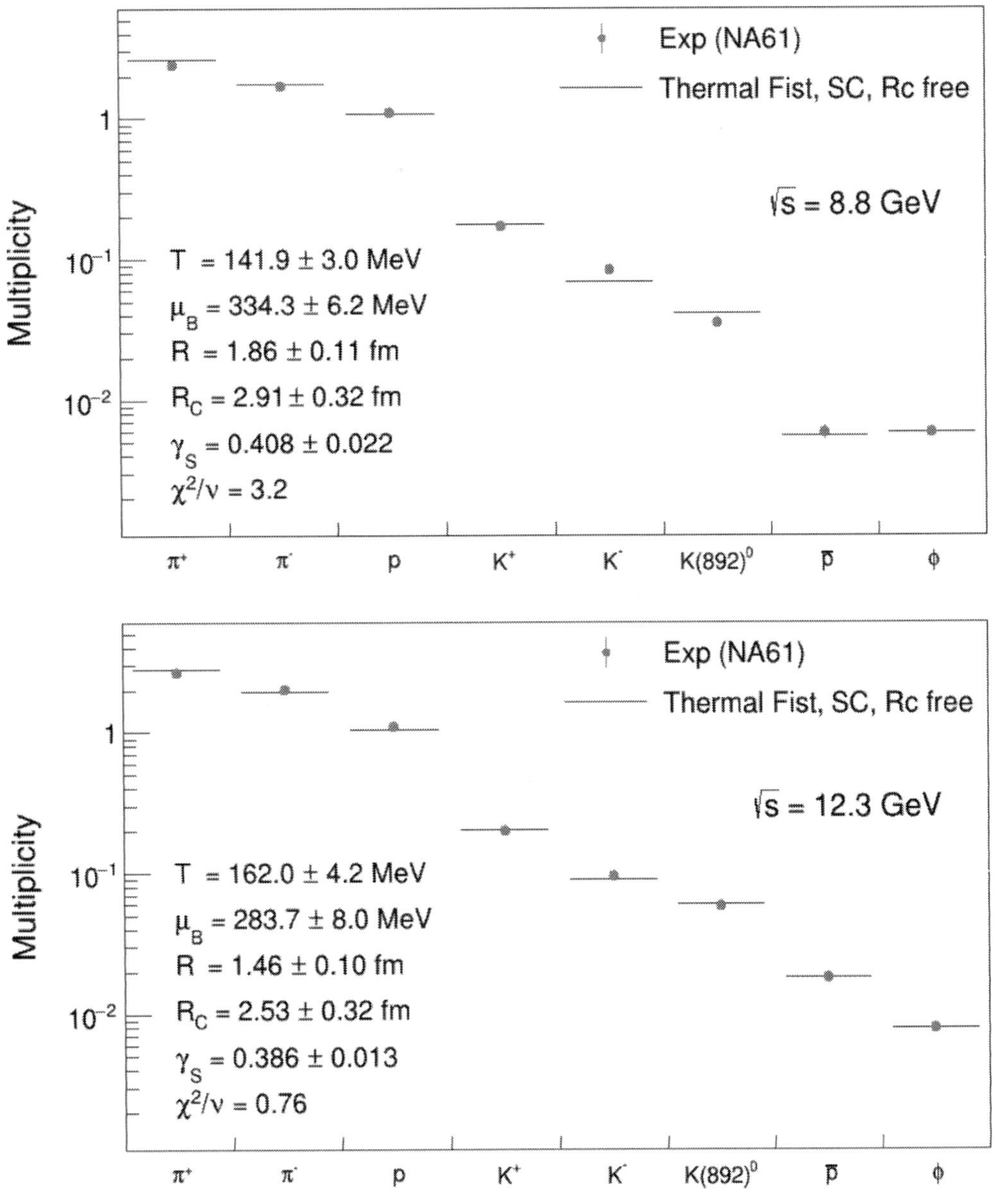

Figure 1. Results of statistical hadronization model fit (lines) to the experimental yields (full dots) of hadrons from p+p at $\sqrt{s} = 8.8$ GeV (upper panel) and 12.3 GeV (lower panel).

For the analysis we used the Thermal-FIST code [9] in the SC version, i.e the canonical treatment of hadrons with open strangeness and grand canonical for the rest of matter. We consequently include the yield of the ϕ-meson to the sample. The list of particles also included the light nuclei up to ^{4}He and their respective hypernuclei, but without charmed hadrons (negligible influence). We use the energy-dependent widths of Breit-Wigner distributions of masses of hadronic resonances which seems to be a more correct description of the mass distributions of hadronic resonances. The volumes are represented in terms of effective radii of spheres (R_C for the strangeness, and R for non-strange matter), which are both free parameters as well as the γ_S undersaturation factor. The resulting particle yields are well described by this version of the model, while the agreement without making R_c free is unsatisfactory. The parameters of the model are listed in Table 2.

Table 2. Values of best fits of SC statistical model to the experimental yields of hadrons from p+p collision at 3 collision energies (see text for details).

$\sqrt{s}$ (GeV)	T (MeV)	μ_B (MeV)	γ_S	R (fm)	R_C (fm)	χ^2/NDF
8.8	141.9±3.0	334.3±6.2	0.408±0.022	1.86±0.11	2.91±0.32	3.2
12.3	162.0±4.2	283.7±8.0	0.386±0.013	1.46±0.10	2.53±0.32	0.76
17.3	173.6±2.7	245.3±3.9	0.418±0.010	1.35±0.06	1.70±0.09	3.4

As in the case of the analysis of 17.3 GeV data [5], the volume of strange particles is found to be greater than of the bulk, and this difference seems to grow (although the accuracy is reduced). The comparison of measured and calculated multiplicities at $\sqrt{s}$ = 8.8 GeV and 12.3 GeV (Figure 1) shows good agreement also for the ϕ meson. Recently, much better agreement of statistical hadronization model with ALICE data was observed thanks to larger value of V_C than the canonical volume V_A [10]. Therefore, it seems that the size of strangeness canonical volume deserves more attention in statistical model calculations.

3 Summary and conclusions

The experimental particle yields in p+p collisions at $\sqrt{s}$ = 8.8 and 12.3 GeV, recently extended by the NA61/SHINE collaboration to include strange $K(892)^0$ mesons, were quite satisfactorily described within the Thermal-Fist hadron resonance gas model treating strange particles according to the canonical ensemble. This description was obtained only when the strangeness canonical volume was allowed to vary freely. As in $\sqrt{s}$ = 17.3 GeV case, this volume is found to be greater than that of the bulk. This effect is not excluded by femtoscopic analysis, because of lacking precision, and is considered also in recent model calculations of nucleus-nucleus collisions at LHC energies.

References

[1] A. Andronic, P. Braun-Munzinger, K. Redlich, J. Stachel, Nature **561**, 321 (2018).
[2] V.V. Begun, V. Vovchenko, M.I. Gorenstein, H. Stoecker, Phys. Rev. C **98**, 054909 (2018).
[3] T. Matulewicz, K. Piasecki, J. Phys. G: Nucl. Part. Phys. 48, 085004 (2021).
[4] NA61/SHINE Collaboration: Eur. Phys. J. **C80**, 199 (2020); **C80**, 460 (2020); **C80**, 833 (2020); **C81**, 911 (2021); **C82**, 96 (2022); **C82**, 322 (2022).
[5] K. Piasecki, T. Matulewicz, POS(PANIC2021)242 (2022).
[6] P. Castorina, S. Plumaria, H. Satz, Int. Jour. Mod. Phys. E **26**, 1750081 (2017).
[7] STAR Collaboration: Phys. Rev. **C75**, 064901 (2007); Phys. Lett. **612B**, 181 (2005).
[8] PHENIX Collaboration: Phys. Rev. Lett. **91**, 241803 (2003).
[9] V. Vovchenko, H. Stoecker, Computer Physics Communications **244**, 295 (2019).
[10] J. Cleymans, P.M. Lo, K. Redlich, N. Sharma, Phys. Rev. **C103**, 014904 (2021).

Event-shape studies of strangeness production in $\sqrt{s}$ = 13 TeV proton–proton collisions with ALICE

Adrian Nassirpour[1]*, on behalf of the ALICE Collaboration

[1]Lund University, Sweden

Abstract.
The ALICE Collaboration has observed that the ratio of strange to non-strange hadron yields increases strongly with multiplicity in small collision systems at LHC energies. The origin of this effect is still not fully understood. Models need to incorporate final-state interactions to accommodate the new ALICE results, and require new measurements to discriminate between the various phenomenological descriptions.
Two different ALICE studies, aiming to further elucidate the underlying mechanisms of strangeness production in high-multiplicity proton–proton (pp) collisions at $\sqrt{s}$ =13 TeV, are presented in this proceeding. First, the production of η and π^0 are studied as a function of multiplicity, combining several analysis techniques, allowing a measurement for the neutral meson production over a vast transverse momentum (p_{T}) range. This is followed by a measurement of π, K, ϕ, and Ξ as a function of the unweighed transverse spherocity $S_{\mathrm{O}}^{p_{\mathrm{T}}=1}$, an event-shape observable which allows us to explore particle production in azimuthal topologies dominated by either hard or soft QCD processes. The results are compared to predictions from PYTHIA and EPOS-LHC.

1 Introduction

The goal of this work is to further study the increased production of strange hadrons relative to π mesons observed in high-multiplicity proton–proton (pp) and proton-lead (p–Pb) collisions at the Large Hadron Collider (LHC). Previously thought to be a unique feature of heavy-ion collisions, it was one of the first proposed signatures of the formation of a quark–gluon plasma (QGP), a strongly interacting medium. Its discovery in smaller collision systems raises questions regarding the physical mechanism of strangeness production. This is compounded by the fact that several theoretical models are able to describe such an effect without modeling a QGP, and can instead reproduce the enhancement by introducing new phenomenological final-state interactions.

The onset of traditional QGP-like features in high-multiplicity small system collisions seems to be incongruent with the idea that proton–proton interactions can be modelled as an almost incoherent superposition of multiple parton–parton interactions (MPIs). PYTHIA 8.3 [1], a QCD-inspired Monte-Carlo (MC) theoretical model where hadronization occurs via "Lund strings", can describe the strangeness enhancement in high-multiplicity collisions by introducing "rope hadronization"; a phenomenological description where several Lund

*e-mail: adrian.nassirpour@hep.lu.se

strings close in phase space overlap and form ropes, effectively increasing string tension and therefore the probability to produce heavier strange quarks.

In contrast, the event generator EPOS-LHC relies on a two-component description to model proton–proton collisions. In this picture, two different systems will form during the initial stages of a proton–proton interaction; a dense core dominated by soft, thermally equilibrated (QGP-like), and a dilute corona, dominated by traditional pp physics. Because the temperature of the core is expected to be close to the strange-quark mass, strangeness is favored to be produced in the core. Events with enhanced strangeness will have a relatively larger contribution from the core versus corona, and vice versa for suppressed strangeness production [2].

In these proceedings, two ALICE measurements performed in high-multiplicity pp collisions at $\sqrt{s}$ = 13 TeV are presented. These results will further help to distinguish which models accurately predict the underlying hadronization processes in high-multiplicity small-systems. The first measurement is of the π^0 and η production as a function of multiplicity, to probe whether the neutral vector meson η, which has a strange–antistrange ($s\bar{s}$) component in the wave function, is enhanced in high-multiplicity events relative to π^0, which has no strangeness. The second study presents measurements on strange particle production as a function of the unweighted transverse spherocity $S_0^{p_T=1}$, aimed at investigating the dependence of the azimuthal event topology for high-multiplicity particle production.

2 π^0 and η production as a function of multiplicity at $\sqrt{s}$ = 13 TeV

η and π^0 mesons are measured through the $X \rightarrow \gamma\gamma$ decay channel. The yield of both mesons are reconstructed by fitting a Gaussian function, convoluted with an exponential low-energy tail, to the invariant mass of photon pairs, $M_{\gamma\gamma}$, after subtracting the combinatorial background. The final yields are measured by performing the signal extraction procedure utilizing different ALICE sub-detector systems, for different p_T intervals. The measurements are then combined into a single multiplicity-dependent particle spectrum for each meson, over a broad p_T range. The primary identification techniques are:

- **Photo-Conversion Method (PCM)**: At low-to-intermediate p_T ranges, it is possible to reconstruct photons that have been converted in the detector material, for $|\eta| < 0.8$ with full azimuthal coverage. This is done by measuring the conversion products via electron identification and a secondary vertex finder.
- **Photon Spectrometer (PHOS)**: At intermediate p_T, photons are reconstructed using the PHOS electromagnetic calorimeter, located at the bottom of ALICE, covering $|\eta| < 0.12$, with 70° of azimuthal coverage.
- **Electromagnetic Calorimeter (EMC)**: The EMC is used to reconstruct photons at very high-p_T, located at $|\eta| < 0.67$, with 106° of azimuthal coverage.

The multiplicity estimation is performed by measuring the deposited charge in the forward V0A and V0C detectors. By taking the mean signal of both detectors (V0M), a multiplicity distribution is created. The multiplicity selection is based on percentiles of the V0M signal.

2.1 Results

The multiplicity-differential η/π^0 ratio as a function of p_T is shown in the left panel of Fig. 1. The ratio suggests that there is little to no multiplicity modification of η production relative to π^0, across the entire p_T range. All presented multiplicity classes are consistent with the minimum bias (0–100%) baseline, even for the top 0–0.01% high-multiplicity events.

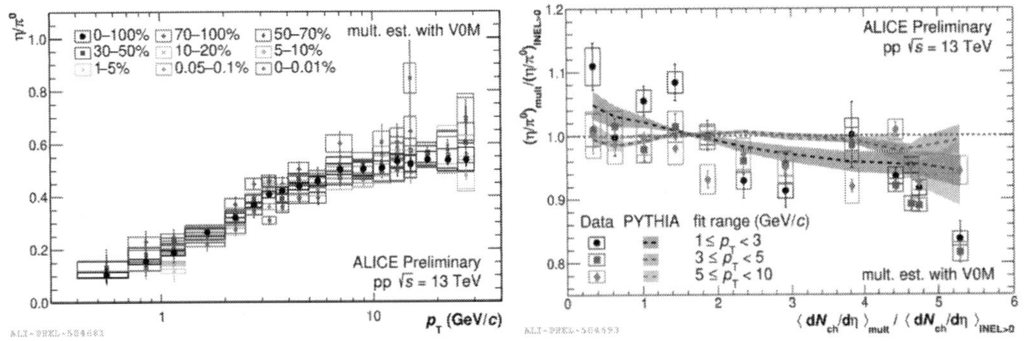

Figure 1: (left): The η/π^0 ratio as a function of p_{T}, in different multiplicity intervals. (right): The double-ratio (DR) between the multiplicity-dependent and minimum-bias η/π^0 ratios, as a function of the self-normalized, average charged-particle density at midrapidity, in three different p_{T} intervals. The results are compared to predictions from PYTHIA8 Monash (dashed lines). Statistical/systematic uncertainties are represented by lines/boxes, respectively.

Further insight can be gained by studying the double ratio (DR), which is a ratio between the multiplicity dependent and minimum-bias η/π^0 ratios. This allows us to study a possible multiplicity modification with high precision. The DR is presented as a function of multiplicity, for three different p_{T} ranges, in the right panel of Fig. 1. The multiplicity selection is the same as presented in the single ratio (left panel), but is expressed in terms of the equivalent self-normalised charged-particle density at midrapidity. These results suggest a small suppression of η production as a function of multiplicity, with a slight enhancement at low multiplicities in the lowest p_{T} range. PYTHIA 8 captures the ordering between three different p_{T} intervals, but is not able to quantitatively describe the trends presented in the data, especially the suppression of η/π^0 at high-multiplicity. Further studies are required to allow for a more quantitative comparison.

3 Particle production as a function of spherocity $S_{\mathrm{O}}^{p_{\mathrm{T}}=1}$

The azimuthal topology of the event can give insight into the very early stages of the collision, and which type of physics drives particle production. This topology can be estimated using the unweighted transverse spherocity, $S_{\mathrm{O}}^{p_{\mathrm{T}}=1}$, defined as

$$S_{\mathrm{O}}^{p_{\mathrm{T}}=1} = \frac{\pi^2}{4} \min_{\hat{n}} \left(\frac{\Sigma_i |\hat{p_{\mathrm{T},i}} \times \hat{n}|}{N_{\mathrm{trks}}} \right), \tag{1}$$

where $\hat{p_{\mathrm{T},i}}$ and $\hat{n}$ are unit vectors. $S_{\mathrm{O}}^{p_{\mathrm{T}}=1}$ is calculated using all charged tracks, N_{trks}, with $p_{\mathrm{T}} > 0.15$ GeV/c. $S_{\mathrm{O}}^{p_{\mathrm{T}}=1}$ can be used as an estimator of the event topology, characterized by two different limits. $S_{\mathrm{O}}^{p_{\mathrm{T}}=1} \rightarrow 1$ represents events with isotropic topologies, where particle production is expected to be driven by several soft processes. In contrast, events where $S_{\mathrm{O}}^{p_{\mathrm{T}}=1} \rightarrow 0$ are likely to be dominated by hard physics processes, with either back-to-back or single-cone jet-like structures.

In these proceedings, high-multiplicity events within the top 10% of the multiplicity distribution are analyzed (requiring also $N_{\mathrm{trks}} > 10$, to ensure that $S_{\mathrm{O}}^{p_{\mathrm{T}}=1}$ is topologically meaningful), in conjunction with the 20% top and bottom percentiles of the $S_{\mathrm{O}}^{p_{\mathrm{T}}=1}$ distribution denoted as Isotropic and Jetty events, respectively. The $S_{\mathrm{O}}^{p_{\mathrm{T}}=1}$ study utilizes a midrapidity ($|\eta| < 0.8$) multiplicity estimation to minimize the implicit multiplicity dependence of $S_{\mathrm{O}}^{p_{\mathrm{T}}=1}$

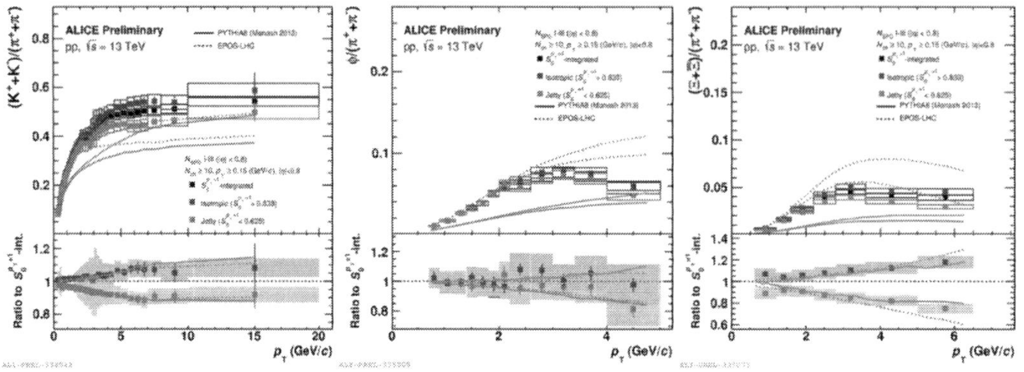

Figure 2: K-to-π (left), ϕ-to-π (middle), and Ξ-to-π (right) ratios as a function of p_{T} in bins of $S_{\mathrm{O}}^{p_{\mathrm{T}}=1}$, for Jetty (red), Isotropic (blue), and $S_{\mathrm{O}}^{p_{\mathrm{T}}=1}$-integrated (black) events. Lower panels show the ratios to $S_{\mathrm{O}}^{p_{\mathrm{T}}=1}$-integrated. The results are compared to predictions from PYTHIA8-Monash (solid lines) and EPOS-LHC (dashed lines).

(events with higher multiplicities will on average be more isotropic), and gain sensitivity to variations in the hardness of the event, see Ref. [3] for more details.

$\pi^{\pm}$ and $K^{\pm}$ are identified via the specific energy loss $\mathrm{d}E/\mathrm{d}x$ measured by the TPC, and the time-of-flight (TOF). ϕ meson Ξ baryon yields are reconstructed by calculating the invariant mass of the identified decay daughters, through the decay channels $\phi \to K^{+}K^{-}$ and $\Xi^{-}[\Xi^{+}] \to \pi^{-}[\pi^{+}] + \Lambda[\bar{\Lambda}]$.

3.1 Results

The p_{T}-differential particle ratios for K-to-π, ϕ-to-π, and Ξ-to-π are presented in Fig. 2, where the DR w.r.t the $S_{\mathrm{O}}^{p_{\mathrm{T}}=1}$-integrated high-multiplicity class are displayed in the lower panels. The K/π and Ξ/π ratios suggest an increased production of K and Ξ in isotropic events, but a suppression in Jetty events. This indicates that strange particle production in high-multiplicity collisions favors isotropic topologies, generally driven by multiple softer collisions, instead of di-jet topologies which are driven by hard processes. The ϕ-to-π highlights that the ϕ meson does not behave like a strange particle w.r.t $S_{\mathrm{O}}^{p_{\mathrm{T}}=1}$, as both Isotropic and Jetty productions of ϕ are consistent with the high-multiplicity average.

The EPOS-LHC and PYTHIA 8 models are able to qualitatively describe the behavior exhibited in the DR, predicting the interplay between $S_{\mathrm{O}}^{p_{\mathrm{T}}=1}$ and high-multiplicity production. However, neither EPOS-LHC nor PYTHIA 8 are able to capture the trends observed in the single ratios. PYTHIA 8 is largely overestimating the yield of pions relative to the production of K, ϕ, and Ξ. While EPOS-LHC overpredicts the high-p_{T} production of ϕ and Ξ, it underestimates the amount of kaons produced relative to what is observed in the data.

4 Summary

These proceedings have highlighted two studies that aim to give further insight into the hadronization processes in high-multiplicity pp collisions. First, the η/π^{0} was presented, whereby combining analyses across several ALICE detector systems, the ratio could be measured over an unprecedented p_{T} range. It was shown that there is little multiplicity-dependence in the single η/π^{0} ratio. However, the double ratio to minimum-bias indicated a small overall suppression of η production at extremely high multiplicities.

Furthermore, utilizing the unweighted transverse spherocity $S_{\mathrm{O}}^{p_{\mathrm{T}}=1}$ to categorize events based on the azimuthal event topology, it was shown that strange particle production is enhanced in isotropic topologies, which are often dominated by multiple soft processes. Remarkably, the ϕ-to-π ratio suggests that ϕ production is insensitive to the azimuthal topology in high-multiplicity events, and in the context of $S_0^{p_T=1}$ behaves like a non-strange particle.

References

[1] T.Sjöstrand et al, *Computer Physics Communications* **191**, 159–177 (2015)
[2] T. Pierog et al, *Phys.Rev.C* **92** (2015) 3, 034906
[3] A. Nassirpour et al, *J.Phys.Conf.Ser.* **1602** (2020) 1, 012007

Constraints on hadron resonance gas interactions via first-principles Lattice QCD susceptibilities

Jamie M. Karthein[1,*], *Volker* Koch[2], *Claudia* Ratti[3], and *Volodymyr* Vovchenko[2,3,4]

[1]Center for Theoretical Physics, Massachusetts Institute of Technology, Cambridge, MA 02139, U.S.A.
[2]Nuclear Science Division, Lawrence Berkeley National Laboratory, Berkeley, CA 94720, U.S.A.
[3]Department of Physics, University of Houston, Houston, TX 77204, U.S.A.
[4]Institute for Nuclear Theory, University of Washington, Seattle, WA 98195, U.S.A.

Abstract. We investigate extensions of the Hadron Resonance Gas (HRG) Model beyond the ideal case by incorporating both attractive and repulsive interactions into the model. When considering additional states exceeding those measured with high confidence by the Particle Data Group, attractive corrections to the overall pressure in the HRG model are imposed. On the other hand, we also apply excluded-volume corrections, which ensure there is no overlap of baryons by turning on repulsive (anti)baryon-(anti)baryon interactions. We emphasize the complementary nature of these two extensions and identify combinations of conserved charge susceptibilities that allow us to constrain them separately. In particular, we find interesting ratios of susceptibilities that are sensitive to one correction and not the other. This allows us to constrain the excluded volume and particle spectrum effects separately. Analysis of the available lattice results suggests the presence of both the extra states in the baryon-strangeness sector and the repulsive baryonic interaction, with indications that hyperons have a smaller repulsive core than non-strange baryons. We note that these results are interesting for heavy-ion-collision systems at both the LHC and RHIC.

1 Introduction

The phase structure of strongly-interacting matter as mapped out in the QCD phase diagram represents highly sought after information on the rich phenomena associated with the underlying theory. Along the temperature axis, i.e. at vanishing baryon chemical potentials, the transition from ordinary nuclear matter to the Quark-Gluon Plasma is known to be a crossover from first-principles Lattice QCD techniques [1]. In addition, the low temperature phase of strongly-interacting matter is well-described by a gas of hadronic resonances [2, 3]. Its agreement with the equation of state from lattice calculations has led to its wide applications, especially in the study of the chemical freeze-out in heavy-ion collisions (HICs) [4–6]. However, more differential observables like susceptibilities of conserved charges have revealed discrepancies between the predictions of the fundamental theory and the hadronic phase treatment [7, 8], specifically in the transition region of the QCD phase diagram. As laid out in Ref. [9], we consider two HRG model extensions: an extended hadronic spectrum and

*e-mail: jmkar@mit.edu

the excluded volume interaction in the baryon sector. The resulting HRG model considerably improves the description of lattice QCD results for fluctuations of conserved charges.

2 Beyond the Ideal HRG Model

In the Boltzmann approximation, the partial pressures in the ideal HRG model for particle species i can be written in the following form:

$$p_i(T, \mu_B, \mu_Q, \mu_S) = d_i \tilde{\phi}(T, m_i) \lambda_i(T, \mu_i), \quad \tilde{\phi}(T, m_i) = \frac{m_i^2 T^2}{2\pi^2} K_2(m_i/T). \tag{1}$$

The full pressure is then given as the sum of the partial pressures of all hadronic sectors i with various BQS quantum numbers:

$$p(T, \mu_B, \mu_Q, \mu_S) = \tilde{\phi}_0(T) + \sum_{i \neq 0} 2\, \tilde{\phi}_i(T) \cosh(\mu_i/T), \tag{2}$$

where $\mu_i = B_i \mu_B + Q_i \mu_Q + S_i \mu_S$ is the chemical potential of the corresponding i^{th} sector and $\tilde{\phi}_0(T)$ is the $i = 0$ term, representing charge neutral species. Each term in Eq. (2) corresponds to the partial pressure associated with the particular set of hadronic quantum numbers. The complementary extensions to the ideal HRG model explored in this work involve supplementing the well-known resonances from the Particle Data Booklet [10] and prohibiting the overlap of baryon states.

Our first modification to the standard HRG model is the incorporation of hadronic states beyond the experimentally well-known ones. The PDG provides an assessment of how well established various measured hadronic states are, with a rating based on a number of stars (*). The **** states are those which are very well known, such as protons or $\Delta(1232)$ resonances. Conversely, the * states are the least established, for example high-mass resonances like $\Delta(1750)$. The three hadronic lists under study here are as follows: PDG2016 – ordinary hadronic list (*** – ****) from the 2016 Particle Data Booklet; PDG2016+ – hadronic list with established and unconfirmed (*–****) states; Quark Model (QM) – the list which incorporates all states predicted by the Quark Model. The latter two lists were introduced and described in detail in [6, 11]. An update to the QM list was provided in Ref. [9] based on the findings in Ref. [12].

The next extension to the HRG model is the excluded volume model. This corresponds to including repulsive interactions between hadrons. Many versions of the EV-HRG model have been considered in the literature. Here we follow the approach introduced in Refs. [3, 13] where EV interactions are included only for baryon-baryon and antibaryon-antibaryon pairs. This corresponds to a minimalistic EV extension that does not affect meson-meson and meson-baryon interactions, which are presumed to be dominated by resonance formation and thus already included in the HRG model. The pressure is partitioned into contributions of non-interacting mesons and interacting baryons and antibaryons: $p = p_M^{\rm id} + p_B^{\rm ev} + p_{\bar{B}}^{\rm ev}$.

$$p_M^{\rm id} = \tilde{\phi}_0(T) + \sum_{i \neq 0,\, i \in M} 2\, \tilde{\phi}_i(T) \cosh(\mu_i/T), \quad p_{B(\bar{B})}^{\rm ev} = \sum_{i \in B} \tilde{\phi}_i(T) \exp(\pm \mu_i/T) \exp\left(\frac{-b\, p_{B(\bar{B})}^{\rm ev}}{T}\right) \tag{3}$$

Here, $i \in M$ corresponds to mesons ($B_i = 0$), $i \in B$ corresponds to baryons ($B_i = 1$), b is the baryon excluded volume parameter, and $\tilde{\phi}(T)$ is given in Eq. (1). Equation (3) can be solved in terms of the Lambert W function:

$$p_{B(\bar{B})}^{\rm ev} = \frac{T}{b} W[\varkappa_{B(\bar{B})}(T, \mu_B, \mu_Q, \mu_S)], \quad \varkappa_{B(\bar{B})}(T, \mu_B, \mu_Q, \mu_S) = b \sum_{i \in B} \tilde{\phi}_i(T) \exp(\pm \mu_i/T). \tag{4}$$

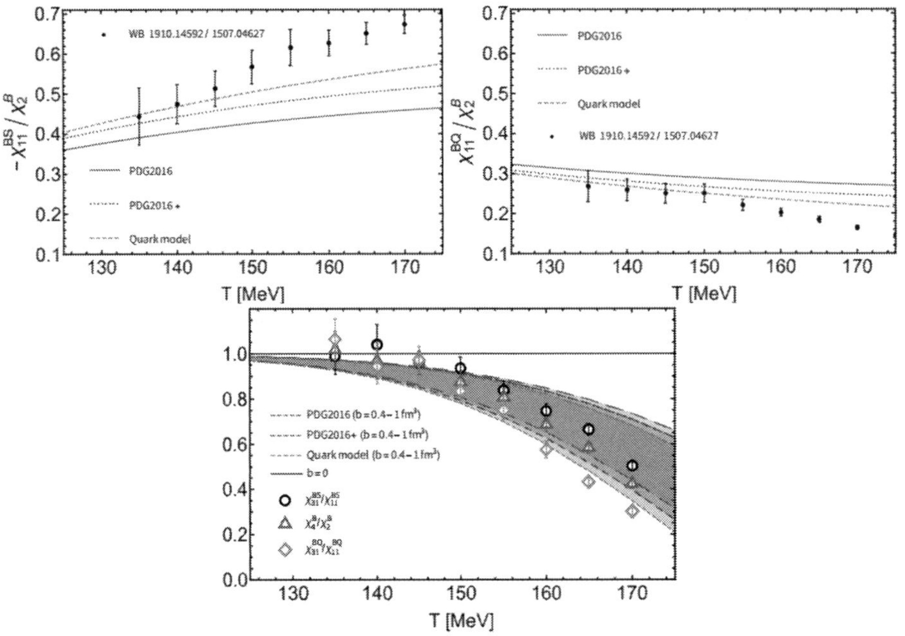

Figure 1. Temperature dependence of susceptibility ratios in the EV-HRG model (colored lines) and from Lattice QCD (grayscale points). Top: second order susceptibility ratios insensitive to the excluded volume parameter b. Bottom: fourth-to-second order susceptibility ratios, insensitive to the hadron spectrum. Here the colored bands correspond to a range of b : $0.4 - 1$ fm.

3 Results

In order to constrain the hadron spectrum, we consider ratios of second order susceptibilities: $\chi_{11}^{BQ}/\chi_2^B, \chi_{11}^{BS}/\chi_2^B$, in which the excluded volume parameter b cancels out. On the other hand, the ratios are sensitive to the particle list via the "partial pressures" $\tilde{\phi}_j$ from Eq. 2. The ratios χ_{11}^{BQ}/χ_2^B and χ_{11}^{BS}/χ_2^B are shown in Fig. 1. The comparison with continuum extrapolated lattice data [14, 15] suggests the need for additional hyperon states from PDG2016+/QM, as previously discussed in Ref. [11].

Next, we can place limits on the excluded volume parameter, b by considering ratios of fourth-to-second order susceptibilities. From Eq. 4, $\varkappa_B = 0$ corresponds to the ideal HRG model, leading to $\frac{\chi_4^B}{\chi_2^B} = \frac{\chi_{31}^{BS}}{\chi_{11}^{BS}} = \frac{\chi_{31}^{BQ}}{\chi_{11}^{BQ}} = 1$ regardless of the inclusion of any additional hadronic states. Thus, these three ratios maintain their equality in the EV-HRG model under consideration and are sensitive to the EV parameter b. Furthermore, given the previous equality within the model, we can estimate the limits of validity of the model as the point where the lattice results of these quantities begin to deviate from one another. Generally speaking, the behavior of the three ratios is very similar over the temperature range, as shown in Fig. 1. Quantitatively, however, we see that when $T \leq 150$ MeV the three ratios sit on top of each other. Where they fall out of agreement, i.e. around $T \sim 160$ MeV, could indicate the breakdown of the model, since these ratios are predicted to be identical in the EV-HRG model.

4 Conclusions

We have implemented additional interactions in the Hadron Resonance Gas model with the motivation to achieve the best agreement with first-principles Lattice QCD data. By utilizing the lattice results on conserved charge susceptibilities, we were able to constrain the free parameters, namely the hadron spectrum and the excluded volume parameter b. We find that the hadron spectrum needs further contribution from strange resonances beyond the well-known PDG states. On the other hand, the excluded volume is constrained to $0.4 \leq b \leq 1\text{fm}^3$ by studying fourth-to-second order ratios. The EV-HRG predicts these ratios to be identical, however, the lattice data reveal statistically significant differences between the three ratios at $T \gtrsim 155$ MeV, which follow a hierarchy $\chi_{31}^{BS}/\chi_{11}^{BS} > \chi_4^B/\chi_2^B > \chi_{31}^{BQ}/\chi_{11}^{BQ}$. We argue that this hierarchy indicates a flavor dependence in the baryon excluded volumes, namely that strange baryons have generally smaller excluded volumes than non-strange baryons.

5 Acknowledgements

This material is based upon work supported by the National Science Foundation under grant no. PHY1654219, PHY2208724 and PHY-2116686, by the U.S. Department of Energy, Office of Science, Office of Nuclear Physics, under contract number DE-AC02-05CH11231231, and within the framework of the Beam Energy Scan Theory (BEST) Collaboration. This work was supported by the National Science Foundation (NSF) within the framework of the MUSES collaboration, under grant number OAC-2103680. J.M.K. is supported by an NSF Ascending Postdoctoral Scholar Fellowship under Award No. 2138063. V.V. acknowledges support via the Feodor Lynen program of the Alexander von Humboldt foundation.

References

[1] Y. Aoki, G. Endrodi, Z. Fodor, S.D. Katz, K.K. Szabo, Nature **443**, 675 (2006), `hep-lat/0611014`

[2] S. Borsanyi, Z. Fodor, C. Hoelbling, S.D. Katz, S. Krieg, C. Ratti, K.K. Szabo (Wuppertal-Budapest), JHEP **09**, 073 (2010), `1005.3508`

[3] V. Vovchenko, M.I. Gorenstein, H. Stoecker, Phys. Rev. Lett. **118**, 182301 (2017), `1609.03975`

[4] V. Vovchenko, V.V. Begun, M.I. Gorenstein, Phys. Rev. C **93**, 064906 (2016), `1512.08025`

[5] R. Bellwied, J. Noronha-Hostler, P. Parotto, I. Portillo Vazquez, C. Ratti, J.M. Stafford, Phys. Rev. **C99**, 034912 (2019), `1805.00088`

[6] P. Alba, V.M. Sarti, J. Noronha-Hostler, P. Parotto, I. Portillo-Vazquez, C. Ratti, J.M. Stafford, Phys. Rev. C **101**, 054905 (2020), `2002.12395`

[7] A. Bazavov et al., Phys. Rev. Lett. **113**, 072001 (2014), `1404.6511`

[8] S. Borsanyi, Z. Fodor, J.N. Guenther, S.K. Katz, K.K. Szabo, A. Pasztor, I. Portillo, C. Ratti, JHEP **10**, 205 (2018), `1805.04445`

[9] J.M. Karthein, V. Koch, C. Ratti, V. Vovchenko, Phys. Rev. D **104**, 094009 (2021)

[10] P. Zyla et al. (Particle Data Group), PTEP **2020**, 083C01 (2020)

[11] P. Alba et al., Phys. Rev. **D96**, 034517 (2017), `1702.01113`

[12] D. Bollweg, J. Goswami, O. Kaczmarek, F. Karsch, S. Mukherjee, P. Petreczky, C. Schmidt, P. Scior (HotQCD), Phys. Rev. D **104** (2021), `2107.10011`

[13] V. Vovchenko, A. Pasztor, Z. Fodor, S.D. Katz, H. Stoecker, Phys. Lett. B **775**, 71 (2017), `1708.02852`

[14] R. Bellwied, S. Borsanyi, Z. Fodor, J.N. Guenther, J. Noronha-Hostler, P. Parotto, A. Pasztor, C. Ratti, J.M. Stafford (2019), `1910.14592`

[15] R. Bellwied, S. Borsanyi, Z. Fodor, S. Katz, A. Pasztor, C. Ratti, K. Szabo, Phys. Rev. D **92**, 114505 (2015), `1507.04627`

https://doi.org/10.1051/epjconf/202327604001

Measurement of the production of (anti)(hyper)nuclei

Chiara Pinto[1,*] for the ALICE Collaboration

[1]Technische Universität München, Physik Department, James-Franck-Str. 1, 85748 Garching bei München

Abstract. In recent years, ALICE has extensively studied the production of light (anti)(hyper)nuclei in different collision systems and center-of-mass energies. Nevertheless, the production mechanisms of light (hyper)nuclei is still unclear and under intense debate in the scientific community. Two classes of models are typically used to describe nuclear production: the statistical hadronisation model and the coalescence ones. In heavy-ion collisions, both models describe well the production yields of light nuclei and their ratios to the yields of hadrons, making it difficult to distinguish between the two. On the contrary, small collision systems, such as pp and p–Pb collisions, are ideal to study the (anti)(hyper)nuclei production mechanisms, thanks to the large separation between the different predictions. In this paper, recent results on light nuclei production measured with ALICE at the LHC in small collision systems are discussed in the context of the nucleosynthesis models, allowing us to exclude some configurations of the SHM and coalescence models in order to set tighter constraints to the available theoretical predictions.

1 Introduction

The production mechanism of light (anti)nuclei in high-energy hadronic collisions is under intense debate in the heavy-ion physics community, despite the large amount of measurements published in recent years. Currently, the experimental data are described by two classes of phenomenological approaches: the thermal-statistical and the coalescence models. In the Statistical Hadronization Model (SHM) [1], hadrons are produced by a thermally and chemically equilibrated source and their abundances are fixed at the chemical freeze-out. This model provides an excellent description of the measured hadron yields in central nucleus–nucleus collisions, corresponding to charged-particle multiplicities of the order of 10^3 [2]. For small systems, such as pp and p–Pb collisions, the production of light nuclei can be described using a different implementation of this model based on the canonical ensemble, where exact conservation of quantum numbers is required [3]. Significant deviations in this case are observed between the experimental data and predictions from the canonical SHM [4]. The mechanisms of hadron production and the propagation of loosely-bound states through the hadron gas phase are not addressed by this model.

On the other hand, the production of light (anti)nuclei can be modelled via the coalescence of protons and neutrons that are close by in phase space at the kinetic freeze-out and match the spin, thus forming a nucleus [5]. The key observable of the coalescence models is the

*e-mail: chiara.pinto@cern.ch

coalescence parameter B_A, which is related to the production probability of a nucleus via this process and can be calculated from the overlap of the nucleus wave function and the phase space distribution of the constituents via the Wigner formalism [6]. Experimentally, B_A is obtained as the ratio between the invariant yield of a nucleus with mass number A and the square of the yield of protons:

$$B_A = \left(\frac{1}{2\pi p_{\mathrm{T}}^{\mathrm{A}}} \frac{\mathrm{d}^2 N_{\mathrm{A}}}{\mathrm{d}y \mathrm{d}p_{\mathrm{T}}^{\mathrm{A}}}\right) \Bigg/ \left(\frac{1}{2\pi p_{\mathrm{T}}^{\mathrm{p}}} \frac{\mathrm{d}^2 N_{\mathrm{p}}}{\mathrm{d}y \mathrm{d}p_{\mathrm{T}}^{\mathrm{p}}}\right)^A, \tag{1}$$

where the labels A and p indicate the nucleus and the proton, respectively, and $p_{\mathrm{T}}^{\mathrm{p}} = p_{\mathrm{T}}^{\mathrm{A}}/A$.

In this paper, recent results on light nuclei production are discussed in the context of the aforementioned nucleosynthesis models. In high multiplicity (HM) pp collisions, by combining the measurements of the production of (anti)nuclei and femtoscopic measurements, the coalescence parameters are compared with parameter-free coalescence predictions. In addition, by comparing the production of in-jet and out-of-jet (anti)deuterons, it is possible to observe an increase of the (anti)deuteron coalescence probability in jets as compared to that out of jets. Additional information can be extracted from the study of very large and extremely loosely bound objects such as the hypertriton ($^3_\Lambda$H). This nucleus has a large wave function, hence its production yield in small collisions is extremely sensitive to the nucleosynthesis models.

2 Testing coalescence models

In high multiplicity pp collisions at a center-of-mass energy $\sqrt{s}$ = 13 TeV, the coalescence models can be tested by combining the measurement of the production of light (anti)nuclei [7] with the precise measurement of the source radius with femtoscopic techniques [8]. Specifically, it is possible to compare the experimentally measured coalescence parameter B_2 with parameter-free coalescence predictions [9], where the only ingredients needed are the emission source size, which is measured for this specific dataset, and the deuteron wave function. As shown in the left panel of Fig. 1, several wave functions for deuterons have been tested. The Gaussian wave function provides the best description of the currently available ALICE data, despite the Hulthen one would be favoured by low-energy scattering experiments. These results show that additional studies in this direction are needed to further investigate the production mechanisms of nuclei.

Further insights into the (anti)nuclei production mechanisms are obtained by measuring their production inside and outside of jets in pp collisions. Hadrons inside the jet cone are closer in phase space compared to hadrons outside of jets. In the coalescence picture, this condition should result in a larger coalescence probability in jets compared to that out of jets. The ALICE Collaboration has recently measured the deuteron production in jets and outside jets in pp collisions at $\sqrt{s}$ = 13 TeV. For this study, the leading particle with the highest transverse momentum ($p_{\mathrm{T}}^{\mathrm{lead}} > 5$ GeV/c) at midrapidity is used as a proxy for the presence of a jet. The (anti)deuteron production is measured in three different azimuthal regions, identified by the relative angular distance with respect to the leading particle. Three equal-size regions, $\pi/3$ wide, are consequently defined: the one around the leading particle (Toward), the one back-to-back to it (Away), and the one transverse to both of them (Transverse). The Toward and Away regions contain contributions from the leading and recoil jets in addition to the underlying event, while the Transverse region is dominated by the underlying event. Consequently, the transverse momentum distribution of deuterons in jets is obtained by subtracting the underlying event (Transverse contribution) from the Towards region. The results are found to be in agreement with those obtained with the two-particle correlation method

[10]. The coalescence parameters B_2 in jets and in the underlying event are obtained as a function of the transverse momentum per nucleon (p_T/A), using Eq. 1. As shown in the right panel of Fig. 1, the coalescence parameter in jets is a factor of about 15 larger than that in the underlying event. This observed enhancement is in agreement with the coalescence picture, and it can be interpreted as due to the reduced distance in phase space between nucleons in jets.

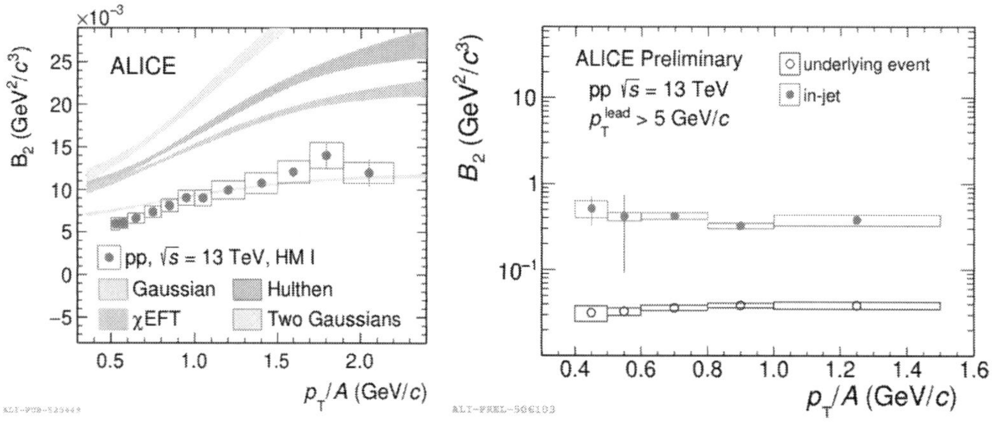

Figure 1. *Left.* Comparison between measurements and theoretical predictions for the coalescence parameter B_2 for (anti)deuterons as a function of p_T/A. Theoretical predictions are obtained using different wave functions to describe nuclei: Gaussian (yellow), Hulthen (blue), χEFT (gray) and two Gaussians (green). *Right.* Coalescence parameter B_2 in jets (full markers) and in the underlying event (empty markers) as a function of p_T/A.

3 Hypertriton measurement in small systems

A key measurement to understand the nuclear production mechanisms in high-energy collisions is the measurement of (anti)hypertritons in small systems [4]. Hypertriton is a bound state of a proton, a neutron and a Λ baryon. This state is characterised by a quite small Λ separation energy, of the order of few hundreds of keV [11], and consequently by a wide wave function that can extend up to a radius of about 10 fm [12]. Therefore, the size of the ${}^3_\Lambda$H wave function is much larger than the hadron emission radius estimated with femtoscopic techniques in small systems (1–2 fm, [13, 14]). For this reason, the ${}^3_\Lambda$H yield in pp and p–Pb collisions predicted by the coalescence models, where the ratio between the nucleus size and the source size directly influences its yield, is suppressed with respect to the statistical hadronisation model expectations, where the nuclear size does not enter explicitly. Hence, the measurement of the ${}^3_\Lambda$H-to-Λ ratio in small systems is crucial to distinguish between the available nucleosynthesis models. In Fig. 2, the ${}^3_\Lambda$H-to-Λ ratio measured in pp, p–Pb [4] and Pb–Pb collisions [15] are compared with the expectations of the canonical statistical hadronization model [16] and with the predicitions of the 2- and 3-body coalescence one [17]. The data show a good agreement with the 2-body coalescence predictions, while some tension with SHM at low charged-particle multiplicity density ($dN_{ch}/d\eta|_{|\eta|<0.5}$) is observed. In particular, the SHM configuration with a correlation volume equal to three units of rapidity $V_C = 3dV/dy$ is excluded by more than 6σ in the low $dN_{ch}/d\eta|_{|\eta|<0.5}$ region.

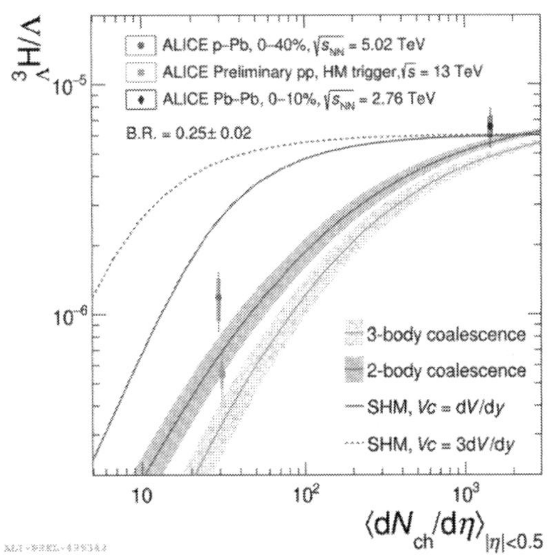

Figure 2. Measurement of the $^3_\Lambda$H-to-Λ ratio in pp, p–Pb [4] and Pb–Pb collisions [15] as a function of the average charged-particle multiplicity density, in comparison with the expectations from the canonical SHM [16] and the 2- and 3-body coalescence model [17], shown as black lines, blue and light blue bands, respectively.

References

[1] A.Andronic, P.Braun-Munzinger, J.Stachel, H.Stöcker, Phys. Lett. B 697 203-207(2011).
[2] A.Andronic, P.Braun-Munzinger, K.Redlich, J.Stachel, Nature 561 321–330 (2018).
[3] V. Vovchenko, B. Doenigus, H. Stoecker, Phys. Lett. B 785 324 (2018).
[4] ALICE Collaboration, Phys. Rev. Lett. 128 25, 252003 (2022).
[5] Butler, S. T. and Pearson, C. A., Phys. Rev. 129 836 (1963).
[6] R. Scheibl and U. W. Heinz, Phys. Rev. C59 1585-1602 (1999).
[7] ALICE Collaboration, JHEP 01 106 (2022).
[8] ALICE Collaboration, Phys. Lett. B 811 135849 (2020).
[9] K. Blum, M. Takimoto, Phys. Rev. C 99 4, 044913 (2019).
[10] ALICE Collaboration, Phys. Lett. B 819 136440 (2021).
[11] D. H. Davis, Nucl. Phys. A 754 (2005).
[12] H. Nemura, Y. Suzuki, Y. Fujiwara, C. Nakamoto, Prog. Theor. Phys. 103 (2000).
[13] ALICE Collaboration, Phys. Rev. C 99 024001 (2019).
[14] ALICE Collaboration, Phys. Rev. Lett. 123 112002 (2019).
[15] ALICE Collaboration, Phys. Lett. B 754 (2016).
[16] V. Vovchenko, B. Doenigus, H. Stoecker, Phys. Lett. B785 (2018).
[17] K.J. Sun, C. M. Ko, B. Doenigus, Phys. Lett. B 792 (2019).

EPJ Web of Conferences **276**, 04002 (2023) https://doi.org/10.1051/epjconf/202327604002
SQM 2022

Hypernuclei at relativistic energies

Benjamin Dönigus[1,*]

[1]Institut für Kernphysik, Goethe Universität Frankfurt, Frankfurt, Germany

Abstract. This article summarizes some of the current theoretical developments and the experimental status of hypernuclei in relativistic heavy-ion collisions and elementary collisions. In particular, the most recent results of hyperhydrogen of mass $A = 3$ and 4 are discussed. The highlight at SQM2022 in this perspective was the discovery of the anti-hyperhydrogen-4 by the STAR Collaboration, in a large data set consisting of different collision systems. Furthermore, the production yields of hyperhydrogen-4 and hyperhelium-4 from the STAR Collaboration can be described nicely by the thermal model when the excited states of these hypernuclei are taken into account. In contrast, the production measurements in small systems (pp and p–Pb) from the ALICE Collaboration tends to favour the coalescence model over the thermal description. New measurements from STAR, ALICE and HADES Collaborations of the properties, e.g. lifetime, of $A = 3$ and 4 hypernuclei give similar results of these properties. Also the anti-hyperhydrogen-4 lifetime is in rather good agreement with previous measurements. Interestingly, the new STAR measurement on the R_3 value, that is connected to the branching ratio, points to a Λ separation energy that is below 100 keV but definitely consistent with the value of 130 keV assumed since the 70s.

1 Introduction

Hypernuclei are composite objects, consisting of nucleons (protons and neutrons) and at least one hyperon. So the theoretical description of these systems needs to take into account the hyperon-nucleon interaction and in cases where there are more than one hyperon also the hyperon-hyperon interaction. The fact that scattering data on their interactions was scarce yet, led basically to the tenor that the main source of information are hypernuclei. Only recently new experiments started (see for instance [1, 2]) that will help to constrain the existing models better, e.g. [3–6]. In addition, the hyperon-nucleon and hyperon-hyperon interaction can also be studied through intensity correlations of the Hanburry Brown and Twiss type [7–9] (nowadays often called femtoscopy, since the experimental technique was used to measure sizes of the emitting source in high-energy collisions leading to values of few femtometer). Examples and more details of how the technique is applied, in particular how that can be used to extract interaction parameters, can be found in [10].

The hyperon-hyperon and hyperon-nucleon interactions might also become important in case higher (energy) densities are reached in neutron stars. The study of the appearance of hyperons in neutron stars led to the so called "hyperon puzzle" [11]. This is connected to

*e-mail: benjamin.doenigus@cern.ch

a weakening of the equation-of-state and thus the often discussed impossibilty to describe the observed neutron star masses of above two solar masses when hyperons are involved. Recently, the solution was seen in the necessity of three-body forces that have been neglected before [12–14]. This could be avoided if the used interactions are modified according to the results from femtoscopy [10].
Of particular interest is the lightest hypernucleus, the hypertriton $^3_\Lambda$H. It consists of a proton,

Symbol	Name	Decay Modes	Mass (GeV/c^2)	B_Λ (MeV)	$\sqrt{\langle r^2 \rangle}$ (fm)
$^3_\Lambda$H	hypertriton	^{3}He + π^- d + p + π^-	2.9911	0.162	4.9
$^4_\Lambda$H	hyperhydrogen-4	^{4}He + π^- ^{3}H + p + π^-	3.9224	2.169	2.0
$^4_\Lambda$He	hyperhelium-4	^{3}He + p + π^-	3.9217	2.347	2.0

Table 1. Compilation of properties of A = 2 and 3 hypernuclei discussed in this article. Decay modes listed are the mesonic decays not containing neutral particles. The mass and binding energies are from [15]. The size in the last column is the rms radius ($\sqrt{\langle r^2 \rangle}$) calculated in Ref. [16].

a neutron and a Λ hyperon. Since the Λ separation energy (B_Λ = 0.162±0.044 MeV [15] is so small it is often imagined as a deuteron core with a Λ with a distance of about $\sqrt{\langle r^2_{\rm d\Lambda} \rangle} \approx 10.6$ fm [17, 18]. If one does a more sophisticated calculation [16, 18, 19] one can also determine the rms radius ($\sqrt{\langle r^2 \rangle}$) of the whole object. Table 1 gives some infromation on the hypernuclei that are discussed here. One clearly sees that the A = 4 hypernuclei are more bound and therefore also slightly more compact than the hypertriton.

1.1 Thermal model

The thermal model description of production yields dN/dy of hadrons works well also for light nuclei (as can be seen from Fig, 1, where the yield dN/dy is shown as function of the collision energy $\sqrt{s_{\rm NN}}$ from thermal model prediction [25], together with (anti-)deuteron measurements from the STAR BES [23] and ALICE results for different nuclei [20–22]). The measured yields in heavy-ion collisions are rather well described by using a grand-canonical ensemble, leading to three free parameters, i.e. chemical freeze-out temperature $T_{\rm ch}$, volume V and baryo-chemical potential $\mu_{\rm B}$. In case of small multiplicities per event one needs to

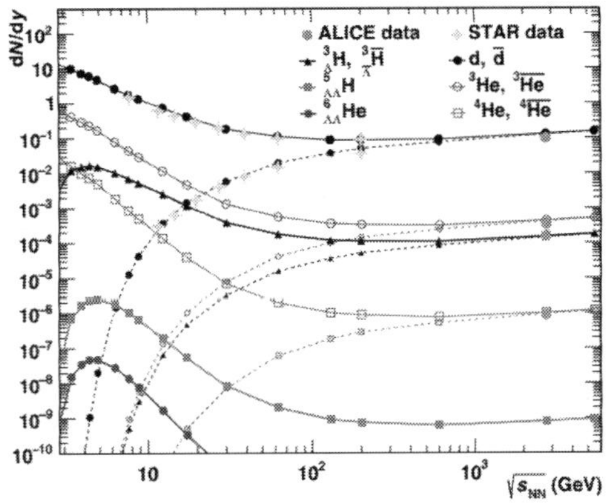

Figure 1. Themal model predictions for the production of several nuclei and hypernuclei, together with experimental results of the ALICE Collaboration [20–22] and the STAR Collaboration [23]. Figure from [24].

use instead a canonical ensemble. There a correlation volume V_c (or correlation radius R_c) is needed in addition, in which the corresponding quantum numbers (typically charge Q, baryon number B and/or strangeness S) are conserved expicitly [26, 27].

At the LHC $\mu_B \approx 0$, therefore often a value of zero is assumed in fits and that works well. Recent studies of yield ratios, also involving the hypertriton measurements, lead to a very precise determination of μ_B, for more details see [28]. The volume V is mainly determined by the pion yields so T_{ch} is the main parameter, which allows to describe the yields of many (anti-)hadrons over ten orders of magnitude [29].

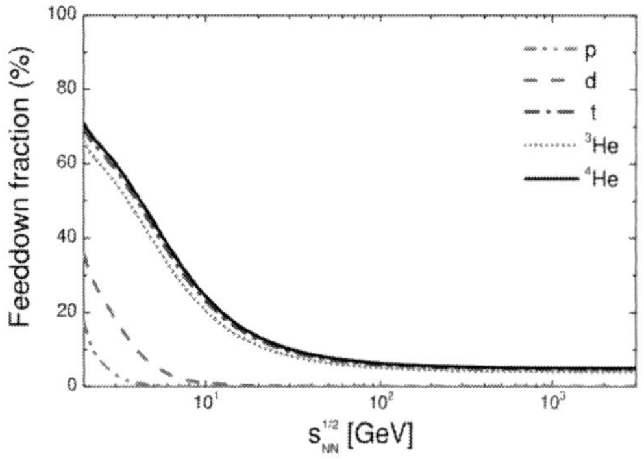

Figure 2. Fractions of final yields of p, d, t, ^{3}He, and ^{4}He coming from decays of the excited nuclei estimated in the thermal model, from [30]

The experimental (final) yields of hadrons suffer typically from large feeddown from higher mass states (resonances, that decay strongly or electromagnetically, since week decays are experimentally separable), that increase the yield from the (primordial) expectation based only on the quantum numbers and mass of the particular state. For the thermal model description of the yields of light nuclei it seems from a first side that there is no feeddown to be taken into account [31, 32]. That is at least for collision energies above $\sqrt{s_{NN}} > 20$ GeV the case, since feed down fractions from excited nuclei into stable ones are below 10% [30]. This can be seen from Fig. 2. At the LHC the feeddown is negligible for deuterons and only about 5% for ^{3}H, ^{3}He and ^{4}He, which is currently covered by the experimental uncertainties.

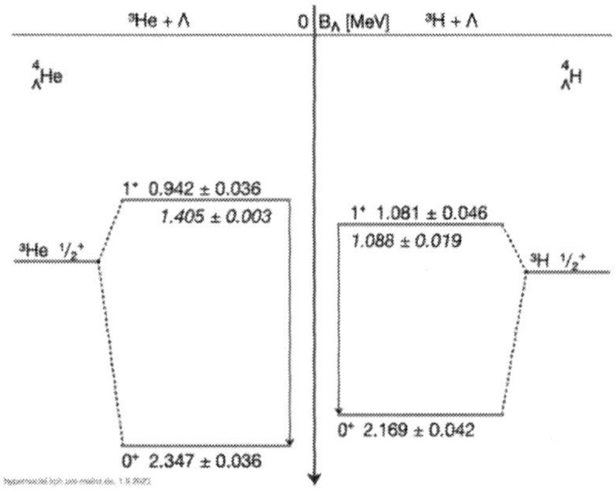

Figure 3. Binding energy level scheme of $A = 4$ hypernuclei. Directly taken from [15]

For the hypertriton there are no excited states known. This is different for $A = 4$ hypernuclei, where both states ($^4_\Lambda$H and $^4_\Lambda$He) have an excited state that is decaying into the ground state by emission of a photon of about one MeV as visible from Fig. 3. That means that the yield of the ground state is significantly higher than expected, without taking the excited state into account. A rough estimate can be given already from the degeneracy (as visible from Fig. 3): with $g = 2J + 1$ one gets $g(^4_\Lambda\mathrm{H}) = 1$ and $g(^4_\Lambda\mathrm{H}^*) = 3$, so the expected yield is four times as high, whereas a factor one comes from the ground state and a factor three from the excited state (the same holds for the $^{4}_{\Lambda}$He).

2 Hypernuclei

2.1 Production

New and important results have been shown and discussed in the context of the production of (anti-)hypernuclei. The highlight at SQM2022 was definitely the observation of the anti-hyperhydrogen-4 $^{4}_{\bar{\Lambda}}\bar{\mathrm{H}}$ by the STAR Collaboration [33] in a huge data set of different collision systems (Au–Au, U–U, Zr–Zr and Ru–Ru) and data taking campaigns. This also showed a nice agreement with the above mentioned factor four enhancement, compared for the case when only the ground is taken into account. That is also in agreement with the results for the production of $^{4}_{\Lambda}$H and $^{4}_{\Lambda}$He in the high statistics data at $\sqrt{s_{\mathrm{NN}}} = 3$ GeV [34].

The biggest tension for the thermal model at the moment comes from the measurements in small systems (pp and p–Pb) by the ALICE Collaboration [35, 36]. These results give a significant tension for the canonical thermal model expectation [27] and seem to favour the coalescence model [37, 38].

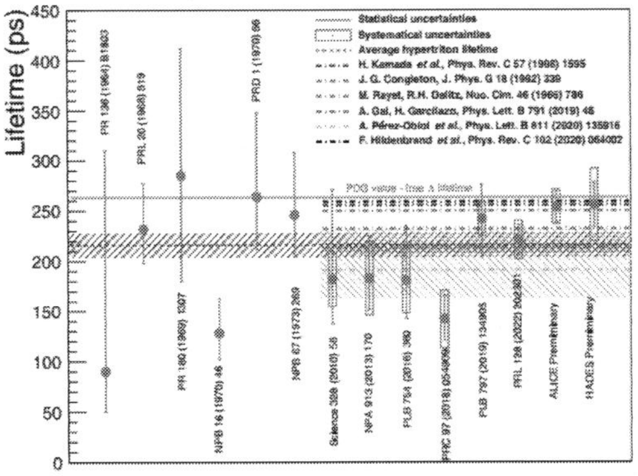

Figure 4. Hypertriton lifetime measured by the ALICE, HADES, HypHI and the STAR Collaborations in heavy-ion collisions, compared with previously published results. The blue dashed line with the shaded band represents the world average of the shown hypertriton lifetime measurements. The magenta full horizontal line indicates the lifetime of the free Λ hyperon as reported by the Particle Data Group [39]. The dashed-dotted lines of different colours are theoretical model expectations of the hypertriton lifetime. Figure updated from [24]

2.2 Lifetime, Branching Ratio and Binding Energy

A very active field from our community is also connected to measure properties of hypernuclei in high precision (in particular since the large data sets contain large numbers of hypernuclei). The first results about ten years ago were on the lifetime of the hypertriton and then several measurements followed becoming more and more precise (as shown in Fig. 4). The first four measurements from heavy-ion collisions (STAR, HypHI, ALICE) aligned well at a central value significantly below the free Λ lifetime. From the aforementioned low Λ separation energy and the corresponding halo-like structure ($\sqrt{\langle r^2_{\mathrm{d}\Lambda}\rangle} \approx 10.6$ fm) the lifetime of the hypertriton is expected to be very close to the free Λ lifetime. One might say that from the current most situation is cleared up, but taking all measurements (also the new preliminary result from HADES [40]) into account and building an average (visible from the blue dashed line in Fig. 4 and the resulting uncertainty in the blue shaded band) one gets a 4.2σ distance even taking the uncertainty of the PDG value of the Λ [39] into account. So from that perspective the "hypertriton puzzle" [17] is not solved yet.

HADES and STAR Collaborations [33, 34, 40] also have shown results on the lifetime of the hyperhydrogen-4 and STAR also on the hyperhelium-4. There is no puzzle for these hypernuclei since the Λ is much tighter bound there, so a deviation from the free Λ lifetime is expected.

There is a strong correlation between the R_3 ratio (defined as the ratio between the charged two-body decay mode to all charged decay modes) and the binding energy. If one uses new model constraints [18] and compares to the world average including recent STAR measurements [34] one gets a value very consitent with the Λ separation energy discussed before. Interestingly, when only the new preliminary value of the STAR Collaboration is used one rather gets a value for B_Λ below 100 keV, which is consistent with the preliminary result for B_Λ from the ALICE Collaboration.

3 Summary and Conclusion

Many new interesting results on (anti-)(hyper)nuclei have been presented recently. In particular nice is that a larger energy range is now available since the HADES Collaboration [40] joined the stage and STAR has large data sets at low and intermediate energy. A definite highlight is the first observation of anti-hyperhydrogen-4 by the STAR Collaboration [33].

The typically discussed production models, i.e. thermal and coalescence, are giving differently good description at different energies. Typically, both models give a rather good description. Using the current descriptions from theory the experimental results at small systems seem to slightly prefer the coalescence model [37, 38], whereas the canonical statistical model is significantly off [27].

The (average) lifetimes of all hypernuclei are significantly below the free Λ lifetime, which is expected for the mass $A = 4$ hypernuclei but not for the hypertriton. In that view, the "hypertriton puzzle" is not yet solved.

It is clear from all new results that more data and even more precise data can be expected in the next years by all involved experiments.

Acknowledgements

The author thanks the ALICE, HADES, HypHI and STAR Collaborations and the theory colleagues involved in this area of research for useful discussions. B.D. acknowledges the support from Bundesministerium für Bildung und Forschung through ErUM-FSP T01 (Förderkennzeichen 05P21RFCA1).

References

[1] K. Miwa et al. (J-PARC E40), Phys. Rev. Lett. **128**, 072501 (2022), `2111.14277`
[2] K. Miwa (J-PARC E40), Few Body Syst. **63**, 31 (2022)
[3] J. Haidenbauer, U.G. Meißner (2022), `2208.13542`
[4] M.M. Nagels, T.A. Rijken, Y. Yamamoto, Phys. Rev. C **102**, 054003 (2020)
[5] V.G.J. Stoks, T.A. Rijken, Phys. Rev. C **59**, 3009 (1999), `nucl-th/9901028`

[6] T.A. Rijken, V.G.J. Stoks, Y. Yamamoto, Phys. Rev. C **59**, 21 (1999), nucl-th/9807082
[7] R. Hanbury Brown, R.Q. Twiss, Nature **178**, 1046 (1956)
[8] R. Hanbury Brown, R.Q. Twiss, Nature **177**, 27 (1956)
[9] G. Goldhaber, S. Goldhaber, W.Y. Lee, A. Pais, Phys. Rev. **120**, 300 (1960)
[10] L. Fabbietti, V. Mantovani Sarti, O. Vazquez Doce, Ann. Rev. Nucl. Part. Sci. **71**, 377 (2021), 2012.09806
[11] L. Tolos, L. Fabbietti, Prog. Part. Nucl. Phys. **112**, 103770 (2020), 2002.09223
[12] D. Lonardoni, A. Lovato, S. Gandolfi, F. Pederiva, Phys. Rev. Lett. **114**, 092301 (2015), 1407.4448
[13] D. Logoteta, Universe **7**, 408 (2021)
[14] E. Friedman, A. Gal (2022), 2204.02264
[15] P. Eckert, P. Achenbach et al., *Chart of hypernuclides — Hypernuclear structure and decay data* (2021), https://hypernuclei.kph.uni-mainz.de
[16] H. Nemura, Y. Suzuki, Y. Fujiwara, C. Nakamoto, Prog. Theor. Phys. **103**, 929 (2000), nucl-th/9912065
[17] P. Braun-Munzinger, B. Dönigus, Nucl. Phys. A **987**, 144 (2019), 1809.04681
[18] F. Hildenbrand, H.W. Hammer, Phys. Rev. C **100**, 034002 (2019), [Erratum: Phys.Rev.C 102, 039901 (2020)], 1904.05818
[19] H. Nemura, Y. Akaishi, Y. Suzuki, Phys. Rev. Lett. **89**, 142504 (2002), nucl-th/0203013
[20] J. Adam et al. (ALICE), Phys. Rev. C **93**, 024917 (2016), 1506.08951
[21] J. Adam et al. (ALICE), Phys. Lett. B **754**, 360 (2016), 1506.08453
[22] S. Acharya et al. (ALICE), Nucl. Phys. A **971**, 1 (2018), 1710.07531
[23] J. Adam et al. (STAR), Phys. Rev. C **99**, 064905 (2019), 1903.11778
[24] B. Dönigus, Eur. Phys. J. A **56**, 280 (2020)
[25] A. Andronic, P. Braun-Munzinger, J. Stachel, H. Stöcker, Phys. Lett. B **697**, 203 (2011), 1010.2995
[26] V. Vovchenko, B. Dönigus, H. Stoecker, Phys. Rev. C **100**, 054906 (2019), 1906.03145
[27] V. Vovchenko, B. Dönigus, H. Stoecker, Phys. Lett. B **785**, 171 (2018), 1808.05245
[28] M. Ciacco (ALICE Collaboration) (2022), these proceedings
[29] A. Andronic, P. Braun-Munzinger, K. Redlich, J. Stachel, Nature **561**, 321 (2018), 1710.09425
[30] V. Vovchenko, B. Dönigus, B. Kardan, M. Lorenz, H. Stoecker, Phys. Lett. **B**, 135746 (2020), 2004.04411
[31] B. Dönigus, Int. J. Mod. Phys. E **29**, 2040001 (2020), 2004.10544
[32] B. Dönigus, G. Röpke, D. Blaschke (2022), 2206.10376
[33] J. Wu (STAR Collaboration) (2022), these proceedings
[34] Y. Ji (STAR Collaboration) (2022), these proceedings
[35] C. Pinto (STAR Collaboration) (2022), these proceedings
[36] S. Acharya et al. (A Large Ion Collider Experiment, ALICE), Phys. Rev. Lett. **128**, 252003 (2022), 2107.10627
[37] K.J. Sun, C.M. Ko, B. Dönigus, Phys. Lett. B **792**, 132 (2019), 1812.05175
[38] T. Reichert et al. (2022), these proceedings
[39] P. Zyla et al. (Particle Data Group), PTEP **2020**, 083C01 (2020), and 2021 update
[40] L. Chlad (HADES Collaboration) (2022), these proceedings

Measurements on the production and properties of light hypernuclei at STAR

Yuanjing Ji[1]*, for the STAR Collaboration

[1]Lawrence Berkeley National Laboratory

Abstract. The hyperon-nucleon (Y-N) interaction, an important ingredient for the nuclear equation-of-state (EoS), remains poorly constrained. Precise measurements of hypernucei intrinsic properties and production yields in heavy-ion collisions are crucial for the understanding of their production mechanisms and the strength of the Y-N interaction. Thanks to the high statistics data taken from the STAR BES II program, a series of hypernuclei measurements are carried out at low energies. In these proceedings, we present the kinematic and centrality dependence of light hypernuclei production yields and strangeness population factor (S_3, S_4) in Au+Au collisions at $\sqrt{s_{\rm NN}}$ = 3 GeV. We also report the energy dependence of $^3_\Lambda$H yields and S_3 at mid-rapidity from 3 to 27 GeV Au+Au collisions. Precise measurements of $^4_\Lambda$He lifetime and $^3_\Lambda$H branching ratio are also reported. These results are compared with model calculations and physics implications are discussed.

1 Introduction

Hypernuclei are bound systems of nucleons and hyperons. They introduce additional degree of freedom in baryon interactions from hyperons. Thus, hypernuclei are regarded as natural laboratory to investigate hyperon-nucleon (Y-N) interactions. Y-N interaction is the important ingredient for understanding the Equation of State (EoS) of neutron stars and the hadronic phase of heavy-ion collisions. Thermal models [1] predict that hypernuclei are abundantly produced in heavy-ion collisions at high baryon density regions. The second phase of the Beam Energy Scan at RHIC, BES-II program, collects high statistical data in Au+Au collisions with center of mass energies ranging from 3-27 GeV. It maps the QCD phase diagram from 200 MeV baryon chemical potential (μ_B) to around 750 MeV. Thus, BES II program provides us an excellent opportunity to investigate hypernuclei physics in heavy-ion collisions.

2 Analysis details

For low energies collisions (3-17 GeV), the collider can be run under the Fixed-target (FXT) mode to significantly enhance the luminosity. Around 2.6×10^8 good events are collected for 3 GeV Au+Au collisions in 2018. The gold target is located at the west side of the STAR detector, while the gold beam comes from the west to east direction. Tracks are reconstructed by Time Projection Chamber (TPC), which covers the pseudo-rapidity range of $-1.5 < \eta <$

*e-mail: yuanjingji@lbl.gov

0 with respect to the target position. Charged particles are identified utilizing the particle energy loss information provided by TPC. In these proceedings, ${}^{3}_{\Lambda}$H is reconstructed via both ${}^{3}_{\Lambda}\mathrm{H} \rightarrow dp\pi^{-}$ and ${}^{3}_{\Lambda}\mathrm{H} \rightarrow {}^{3}\mathrm{He}\pi^{-}$ channels, and ${}^{4}_{\Lambda}$H and ${}^{4}_{\Lambda}$He are reconstructed via ${}^{4}_{\Lambda}\mathrm{H} \rightarrow {}^{4}\mathrm{He}\pi^{-}$ and ${}^{4}_{\Lambda}\mathrm{He} \rightarrow {}^{3}\mathrm{He}p\pi^{-}$. Hypernuclei candidates are reconstructed utilizing KF Particle package [2] to enhance significance.

3 Results and discussions

3.1 Production of hypernuclei in heavy-ion collisions

Figure 1 (left) shows the rapidity dependence of ${}^{3}_{\Lambda}$H and ${}^{4}_{\Lambda}$H production yields in Au+Au collisions at 3 GeV [3]. Both ${}^{3}_{\Lambda}$H and ${}^{4}_{\Lambda}$H show different trends in 0-10% and 10-50% centralities. Dashed lines are calculations from the transport model (JAM) with instant coalescence of all hadrons as an afterburner [4]. This simple coalescence model can qualitatively describe the data with tuned coalescence parameters. Figure 1 (right) shows the light nuclei and hypernuclei average transverse momentum $\langle p_T \rangle$ as a function of particle mass. A linear trend is observed, suggesting the dominance of collective radial motion. Similar phenomena are also observed in light nuclei and hypernuclei directed flow measurements [5]. Those results are qualitatively consistent with that hypernuclei are produced from the coalescence of hyperons and nucleons. To investigate the role of *Y*-*N* interaction in heavy-ion collisions, we calculate

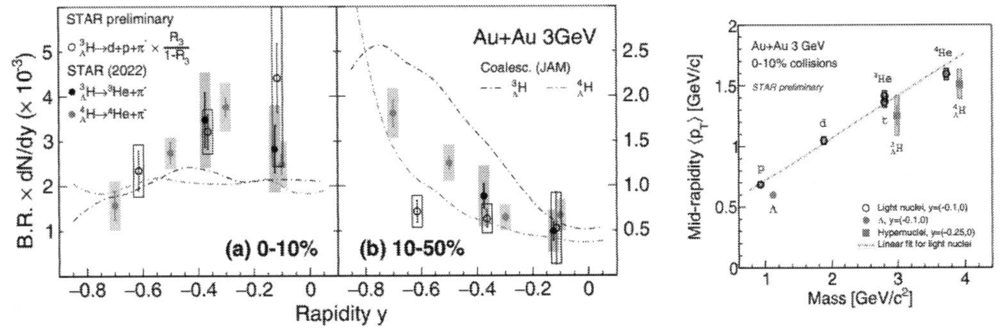

Figure 1. The left plot shows the rapidity dependence of ${}^{3}_{\Lambda}$H (black solid and open circles), ${}^{4}_{\Lambda}$H (red solid circles) $B.R. \times dN/dy$ as a function of rapidity in 0-10% and 10-50% centralities. ${}^{3}_{\Lambda}$H yields via ${}^{3}_{\Lambda}\mathrm{H} \rightarrow dp\pi^{-}$ channel are scaled by a factor of $R_3/(1-R_3)$. Dashed lines are JAM calculations coupled with instant coalescence [4]. The right plot shows the average transverse momentum $\langle p_T \rangle$ of different particle species as a function of particle mass at mid-rapidity in 3 GeV Au+Au collisions.

the strangeness population factor [6] that takes Λ baryon over proton yield ratio as a reference to hypernuclei to light nuclei yield ratio:

$$S_A = \frac{{}^{A}_{\Lambda}\mathrm{H}(A \times p_T)}{{}^{A}\mathrm{He}(p_T) \times \frac{\Lambda}{p}(p_T)} = \frac{B_A({}^{A}_{\Lambda}\mathrm{H})(p_T)}{B_A({}^{A}\mathrm{He})(p_T)} \quad (1)$$

As one can see in Eq. 1, S_A has a direct connection with the ratio of coalescence parameters B_A. Figure 2 shows $S_A(A = 3, 4)$ as a function p_T/A in 0-10% and 10-40% centralities, where A is the mass number. Within the current uncertainties, no obvious p_T, rapidity and centrality dependence of S_A are observed at 3 GeV. The results imply that B_A of light nuclei and hypernuclei might follow similar tendency versus p_T, rapidity and centralities. Suppressed ${}^{3}_{\Lambda}\mathrm{H}/{}^{3}\mathrm{He}$ yield ratios are observed with respect to Λ/p yield ratio in Au+Au collisions at 3 GeV with

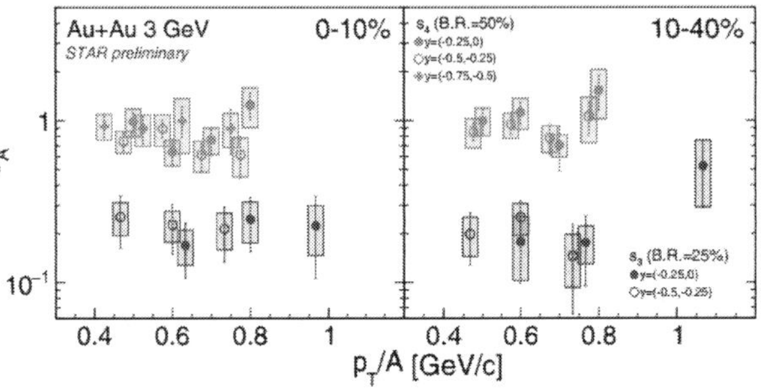

Figure 2. S_3 (blue markers) and S_4 (magenta markers) are shown as a function of p_T/A, where A is the mass number. S_3 and S_4 are presented in 3 rapidity bins at 0-10% and 10-40% centralities.

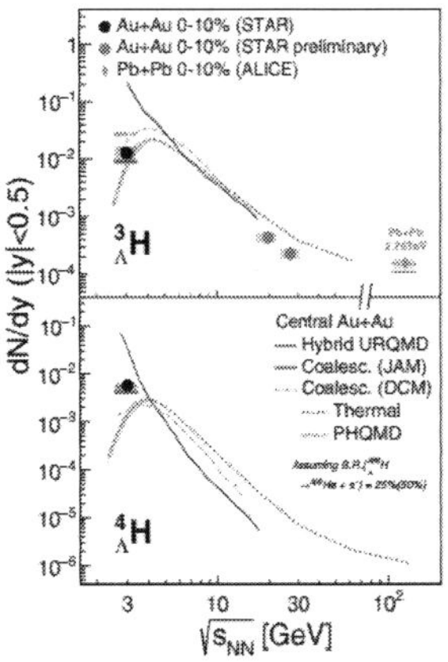

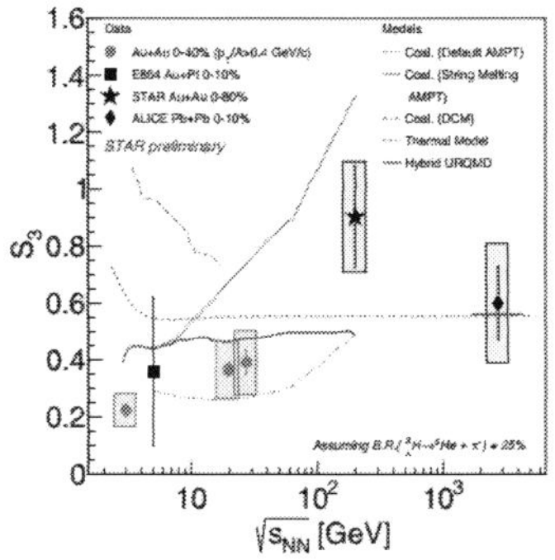

Figure 3. dN/dy of ${}^3_\Lambda$H (top panel) and ${}^4_\Lambda$H (bottom panel) at mid-rapidity as a function of the center-of-mass energy. Our results are compared with the following model calculations: Hybrid UrQMD [7], JAM [4], DCM [7], Thermal [1], PHQMD[8].

Figure 4. The energy dependence of S_3 is shown. Green circles are the STAR new measurements. Model calculations shown in the plot are from: AMPT [6], DCM [7], Thermal [1], Hybrid UrQMD [7].

$S_3 < 1$. ${}^4_\Lambda\text{H}/{}^4\text{He}$ yield ratio is comparable to Λ/p yield ratio, which can be explained by the feed-down contributions from the exited state ${}^4_\Lambda\text{H}^*(J^+ = 1)$.

Figure 3 shows the first energy dependence of ${}^3_\Lambda$H production yields in high μ_B region. An enhanced production of hypernuclei is observed at RHIC BES II energies with respect to LHC energies, which results from the increased baryon density at low energies. The thermal model [1] predicts the trend while can not quantitatively describe the yields. Again, we investigate the strangeness population factor versus collision energy, which removes the absolute difference between Λ and p yields versus beam energy. Figure 4 (right) shows the energy dependence of S_3 from RHIC to LHC energies. A hint of increasing S_3 from $\sqrt{s_{\rm NN}}$= 3 GeV to 2.76 TeV is observed. None of the shown models [1, 6, 7] in Fig. 4 (right) can describe the S_3 data quantitatively. We are looking forward to further developments from theory communities.

3.2 Measurements of hypernuclei lifetimes and branching ratio

The ${}^3_\Lambda$H branching ratio R_3 is defined as $R_3 = \frac{B.R.({}^3_\Lambda\text{H}\rightarrow{}^3\text{He}\pi^-)}{B.R.({}^3_\Lambda\text{H}\rightarrow dp\pi)+B.R.({}^3_\Lambda\text{H}\rightarrow{}^3\text{He}\pi)}$. A recent model calculation [9] predicts that ${}^3_\Lambda$H R_3 is sensitive to its binding energy, B_Λ, which is directly connected to *Y*-*N* interaction strength. Hypernuclei lifetime can be extracted by measuring hypernuclei yields as a function of $L/\beta\gamma$ where L is its decay length [3]. The new measurement on ${}^{3}_{\Lambda}$H R_3 from STAR is highlighted as the red solid star in Fig. 5 (left) with the updated world average value $R_3 = 0.32 \pm 0.03$ shown as the blue band. Figure 5 (right) is the lifetime of ${}^{4}_{\Lambda}$He, where the STAR new result is indicated by the red circle. The updated world average value is consistent with the calculation based on isospin rule [10].

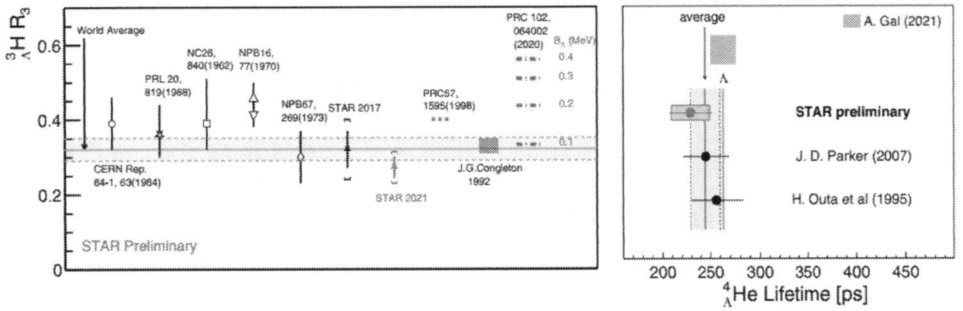

Figure 5. The left plot shows the R_3 of ${}^{3}_{\Lambda}$H, and the right plot shows the lifetime of ${}^{4}_{\Lambda}$He. The red solid star (left) and circle (right) markers are the STAR new measurements. The updated world average values are indicated as blue bands.

4 Summary

In summary, we present a series of measurements on hypernuclei production and intrinsic properties utilizing high statistics data collected during the BES II program at STAR. We report new measurements on ${}^{4}_{\Lambda}$He lifetime and ${}^{3}_{\Lambda}$H R_3. The kinematic and centrality dependence of ${}^{3}_{\Lambda}$H production yields and S_A in 3 GeV Au+Au collisions are presented. Energy dependence of ${}^{3}_{\Lambda}$H yields and S_A in the mid-rapdity from 3-27 GeV are also reported. Our measurements support the coalescence mechanism of hypernuclei production in heavy-ion collisions. Within the uncertainties, no obvious kinematic or centrality dependence of S_3 is observed in 3 GeV Au+Au collisions. Hopefully, our measurements will set strong constraints on hypernuclei internal structures and inspire new insights into the strength of *Y*-*N* interaction.

References

[1] A. Andronic et al., Phys. Lett. B **697**, 203 (2011), 1010.2995
[2] M. Zyzak, Ph.D. thesis, Frankfurt U. (2016)
[3] M. Abdallah et al. (STAR), Phys. Rev. Lett. **128**, 202301 (2022), 2110.09513
[4] H. Liu et al., Phys. Lett. B **805**, 135452 (2020), 1909.09304
[5] C. Hu (STAR), EPJ Web Conf. **259**, 11014 (2022)
[6] S. Zhang et al., Phys. Lett. B **684**, 224 (2010), 0908.3357
[7] J. Steinheimer et al., Phys. Lett. B **714**, 85 (2012), 1203.2547
[8] S. Gläßel et al., Phys. Rev. C **105**, 014908 (2022), 2106.14839
[9] F. Hildenbrand, H.W. Hammer, Phys. Rev. C **102**, 064002 (2020), 2007.10122
[10] A. Gal, EPJ Web Conf. **259**, 08002 (2022), 2108.10179

Rescattering effects on resonances production in small systems with ALICE at the LHC

Antonina Rosano[1,2,*] on behalf of the ALICE Collaboration

[1]Dipartimento MIFT, Università di Messina - Messina, Italy
[2]INFN, Sezione di Catania - Catania, Italy

Abstract. Recent multiplicity-dependent analyses of pp and p–Pb collision data have revealed that particle production shares similar features with that in heavy-ion collisions. Studies using resonances could help to understand the possible onset of collective-like phenomena and the presence of a hadronic phase in small collision systems. Measurements of the differential yields of resonances with different lifetime, mass, quark content, and quantum numbers could enable understanding the mechanisms that influence the shape of particle momentum spectra, lifetime of the hadronic phase, strangeness production, parton energy loss, and collective effects. New ALICE results on various hadronic resonances in small collision systems at Large Hadron Collider (LHC) energies, including the multiplicity dependence measurements of $\Lambda(1520)$ and $K^{*}(892)^{\pm}$ and the production of ϕ-meson pairs are presented here. Results will be also compared with model calculations.

1 Introduction

Hadronic resonances are useful tools to characterize the late stage evolution (hadronic phase) of the system created by heavy-ion collisions at ultra-relativistic energies. In such collisions the critical values of temperature ($T_c \simeq 160$ MeV) and energy density ($\epsilon_c \simeq 1$ Gev/fm^3) can be reached with the consequent formation of a deconfined state of free quarks and gluons, known as quark–gluon plasma (QGP). Resonances with a lifetime comparable to that of the fireball ($\sim$ 1-10 fm/c at LHC energies) may be sensitive to the competitive rescattering and regeneration mechanisms that could lead to a suppression or an enhancement of the resonance measured yield, respectively. The suitable conditions for QGP formation are expected to be reached only in heavy-ion collisions, with data from small systems (pp and p–Pb) collected only as a reference baseline. However recent studies of pp and p–Pb collisions at the LHC for events with high charged-particle multiplicities have shown patterns that are reminiscent of phenomena observed in A–A collisions. For example the strangeness enhancement [1], the hardening of hadron p_{T} spectra [2–5], and the hint of suppression with increasing multiplicity of the ratios of short-lived resonances to long-lived hadrons [5] have been observed even in small collision systems. The presence of typical features of heavy-ion collisions also in pp and p–Pb collisions is an intriguing observation whose interpretation could lead to new perspectives in the theoretical foundations behind QGP formation and to a better understanding of the mechanisms involved.

*e-mail: arosano@unime.it

2 Methodology

ALICE (A Large Ion Collider Experiment [6]) is one of the large experiments installed at the LHC at CERN. A complete description of the ALICE apparatus and its performance can be found in [6, 7]. The main sub-detectors involved in resonance analyses are the Inner Tracking System (ITS), the Time Projection Chamber (TPC), the Time-Of-Flight detector (TOF), and the V0A and V0C scintillators. The ITS and the TPC are used for primary vertex determination, tracking and particle identification (PID) via the measurement of the particle specific energy loss. The TOF detector is designed for PID measuring particle time of flight, and the V0A ($2.8 < \eta < 5.1$) and V0C ($-3.7 < \eta < -1.7$) hodoscopes are used for triggering and selecting events based on the charged particle multiplicity at forward rapidities. Resonances are reconstructed via invariant mass distribution of the hadronic decay products. The shape of the uncorrelated background is estimated by the event-mixing or like-sign technique. After combinatorial background subtraction, the invariant mass distribution is then fitted with a polynomial function to describe the residual background and with a Breit-Wigner, a Voigtian or a Gaussian function to describe the signal. Corrective factors as geometrical acceptance and detector efficiency, branching ratio, trigger selection efficiency, and signal-loss factor are applied to the raw yields in order to estimate the final yields.

3 Results and discussion

ALICE has collected data from several collision systems at different LHC energies, measuring the production of a large set of hadronic resonances [5, 8–15]. The latest results from the multiplicity ($\langle dN_{ch}/d\eta\rangle_{|\eta|<0.5}$) dependent analyses performed at $|y| < 0.5$ of $K^*(892)^{\pm}$ in pp collisions at $\sqrt{s} = 13$ TeV and $\Lambda(1520)$ in pp collisions at $\sqrt{s} = 5.02$ and 13 TeV, and of ϕ-meson pair in minimum bias pp collisions at $\sqrt{s} = 7$ TeV are discussed here.
Both $K^{*\pm}$ and $\Lambda(1520)$ p_T-differential distributions get harder for $p_T \lesssim 5$ GeV/c going from low to high multiplicity classes (Fig. 1). The trend is qualitatively similar to the behaviour observed in Pb–Pb collisions that it is usually attributed to a collective radial expansion of the system (radial flow), although the colour reconnection mechanism seems to mimic collective-like effects in small systems [16].

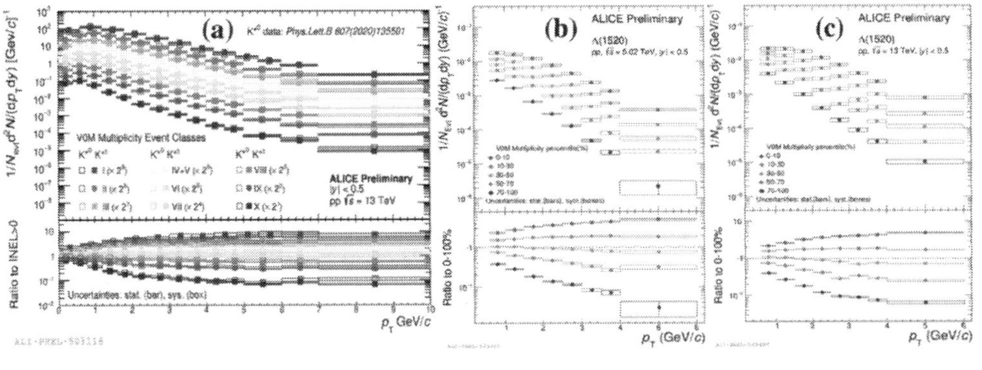

Figure 1. (Colour online) The p_T-differential distributions of $K^{*\pm}$ in pp collisions at $\sqrt{s} = 13$ TeV **(a)**, $\Lambda(1520)$ at $\sqrt{s} = 5.02$ **(b)** and 13 TeV **(c)**. Lower panels show the ratios of the p_T distributions to the inclusive (INEL>0) spectra. The $K^{*\pm}$ data are also compared to K^{*0} measurements [5].

In addition, these results have many qualitative similarities to those reported for longer lived hadrons, where a hardening at low p_T of the p_T-differential distributions have been observed too, and are consistent with previous measurements of K^{*0} and $\Lambda(1520)$ both in pp and p–Pb collisions [2–5, 10, 14].

Figure 2a shows the $\Lambda(1520)$ p_T-integrated yields as a function of the charged particle multiplicity density in pp collisions. The yields exhibit a linear increase with increasing multiplicity and results at $\sqrt{s}$ = 5.02 and 13 TeV are consistent. Therefore, as observed also for other hadron species [2, 5], the particle production rate is independent of collision energy and scales with the event multiplicity. The $K^{*\pm}/K^0_S$ ratios are shown in Fig. 2b. The ratio of resonance p_T-integrated yields to the production of their non-resonant hadronic states, is an important tool to verify the presence of a suppression in resonances production and its dependence on the system size. The $K^{*\pm}/K^0_S$ ratio is suppressed in central Pb–Pb collisions with respect to peripheral Pb–Pb and pp collisions. This is an expected behaviour in Pb–Pb collisions since the measured yield of short-lived resonances as $K^{*\pm}$ ($\tau \simeq 4$ fm/c) can be affected by rescattering effects during the hadronic phase. However an intriguing suppression of the $K^{*\pm}/K^0_S$ ratio is observed from low to high multiplicity pp collision too, which is significant at the 7σ level taking into account the fraction of systematic uncertainties uncorrelated with multiplicity. The latest measurement confirms and improves the precision of previous K^{*0} results [5] due to lower systematic uncertainties. The measured values are compared with several model calculations. Among them, EPOS-LHC [17] without the UrQMD hadronic afterburner and the hadron resonance gas model in partial chemical equilibrium (HRG-PCE) [18] provide the best description for pp and Pb–Pb data respectively, reproducing the decreasing trend.

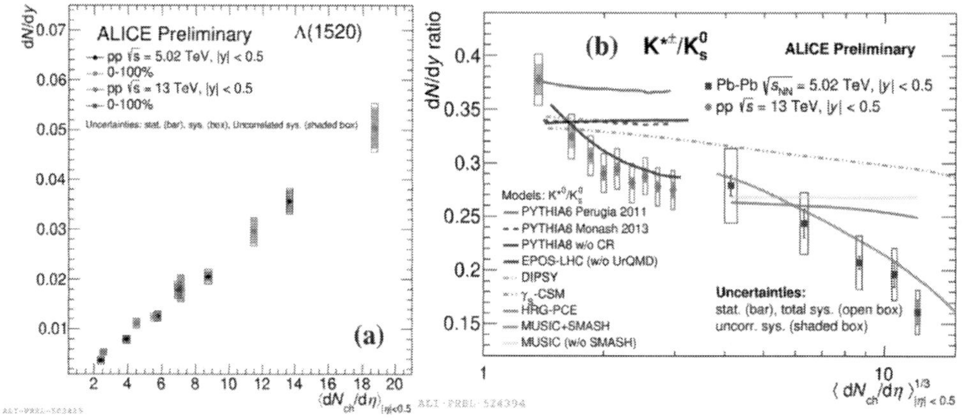

Figure 2. (Colour online) p_T-integrated yields of $\Lambda(1520)$ as a function of $\langle dN_{ch}/d\eta\rangle_{|\eta|<0.5}$ in pp collisions at $\sqrt{s}$ = 5.02 and 13 TeV **(a)**. Ratios of p_T-integrated yields of $K^{*\pm}/K^0_S$ in pp at $\sqrt{s}$ = 13 TeV and in Pb–Pb collisions at $\sqrt{s_{NN}}$ = 5.02 TeV as a function of multiplicity. Measurements are also compared with predictions from common event generators [17–24] **(b)**.

Figure 3 shows the results of the study on the double ϕ production in pp collisions at $\sqrt{s}$ = 7 TeV, performed in order to investigate strangeness production in small systems. In terms of statistical properties of the ϕ-meson production, the average yield ($\mu = \langle Y_\phi\rangle$) and the variance ($\sigma^2 = \langle Y^2_\phi\rangle - \langle Y_\phi\rangle^2$) have been considered. Looking at the variance definition, the second term can be directly measured while the first one can be obtained via the ϕ meson pair production ($\langle Y_{\phi\phi}\rangle$):

$$\langle Y^2_\phi\rangle = 2\langle Y_{\phi\phi}\rangle + \langle Y_\phi\rangle \Rightarrow \sigma^2 = 2\langle Y_{\phi\phi}\rangle + \langle Y_\phi\rangle - \langle Y_\phi\rangle^2 \tag{1}$$

This information can be used to define γ_ϕ, a new parameter that describes the accordance of the production statistic with a Poissonian beahaviour:

$$\gamma_{\phi} = \frac{\sigma^2}{\mu} - 1 = \frac{2\langle Y_{\phi\phi} \rangle}{\langle Y_{\phi} \rangle} - \langle Y_{\phi} \rangle \tag{2}$$

In particular, $\gamma_{\phi} = 0$ is the expected value if the production follows a Poissonian distribution. The obtained results show $\gamma_{\phi} > 0$ indicating that the production distribution is enhanced and not purely Poissonian.

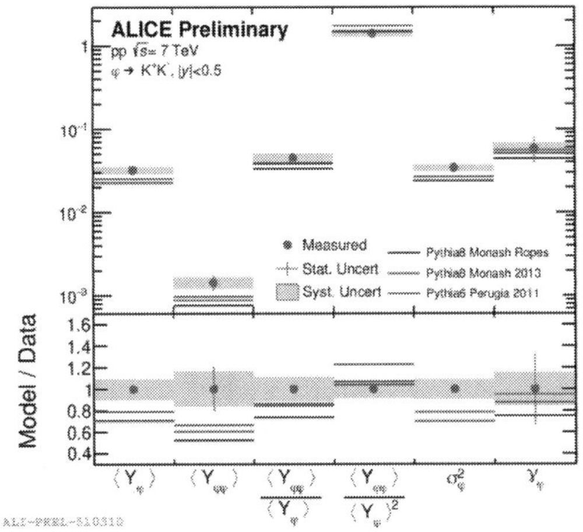

Figure 3. (Colour online). Inclusive ϕ ($\langle Y_{\phi} \rangle$) and double ϕ-meson yields ($\langle Y_{\phi\phi} \rangle$) with their combinations. The key observable γ_{ϕ} is also shown. Results are compared with the most commonly used PYTHIA tuned models (solid colored lines) [19–21]. Lower panel shows the ratios of the model predictions to the measured data.

References

[1] ALICE Collaboration: J. Adam et al., Nat. Phys. **13**, 535 (2017)
[2] ALICE Collaboration: S. Acharya et al., Eur. Phys. J. C **80**, 167 (2020)
[3] ALICE Collaboration: S. Acharya et al., Eur. Phys. J. C **80**, 693 (2020)
[4] ALICE Collaboration: B. Abelev et al., Phys. Lett. B **728**, 25-38 (2014)
[5] ALICE Collaboration: S. Acharya et al., Phys. Lett. B **807**, 135501 (2020)
[6] ALICE Collaboration: K. Aamodt et al., JINST **3**, 8002 (2008)
[7] ALICE Collaboration: K. Aamodt et al., Int. J. Mod. Phys. A **29**, 1430044 (2014)
[8] ALICE Collaboration: S. Acharya et al., Phys. Rev. C, **99**, 064901 (2019)
[9] ALICE Collaboration: S. Acharya et al., Phys. Lett. B, **802**, 135225 (2020)
[10] ALICE Collaboration: J. Adam et al., Eur. Phys. J. C **76**, 245 (2016)
[11] ALICE Collaboration: B. Abelev et al., Eur. Phys. J. C **75**, 1 (2015)
[12] ALICE Collaboration: D. Adamova et al., Eur. Phys. J. C **77**, 389 (2017)
[13] ALICE Collaboration: K. Aamodt et al., arXiv:2205.13998 [nucl-ex] (2022)
[14] ALICE Collaboration: S. Acharya et al., Eur. Phys. J. C **80**, 160 (2020)
[15] ALICE Collaboration: S. Acharya et al., Phys. Rev. C **99**, 024905 (2019)
[16] A. Ortiz Velasquez et al., Phys. Rev. Lett. **111**, 042001 (2013)
[17] T. Pierog et al., Phys. Rev. C **92**, 034906 (2015)
[18] A. Motornenko et al., Phys. Rev. C **102**, 024909 (2020)
[19] P. Skands, Phys. Rev. D **82**, 074018 (2010)
[20] P. Skands et al., Eur. Phys. J. C **74**, 3024 (2014)
[21] R. Acconcia et al., Phys. Rev. D **97**, 036010 (2018)
[22] C. Flensburg et al., J. High Energy Phys. **8**, 103 (2011)
[23] V. Vovchenko et al., Phys. Rev. C **100**, 054906 (2019)
[24] B. Schenke et al., Phys. Rev. C **82**, 014903 (2010)

Exploring the hadronic phase of relativistic heavy-ion collisions with resonances in ALICE

Dukhishyam Mallick[1,*]

[1]for the behalf of ALICE Collaboration,
National Institute of Science Education and Research, HBNI, Jatni - 752050, India and Homi Bhabha National Institute, Training School Complex, Anushaktinagar, Mumbai 400094, India

Abstract.
Hadronic resonances, having short lifetimes, are useful for studying the hadron-gas phase that characterizes the late-stage evolution of high-energy nuclear collisions. Indeed, regeneration and rescattering processes occurring in the hadron gas modify the measured yields of hadronic resonances and can be studied by measuring resonance yields as a function of system size and comparing them to model predictions with and without hadronic interactions. Measurements of the differential yields of resonances with different lifetime, mass, quark content, and quantum numbers help in understanding particle production mechanisms, the lifetime of the hadronic phase, strangeness production, parton energy loss, rapidity yield asymmetry, and collective effects. With its excellent tracking and particle identification capabilities, the ALICE experiment measured a comprehensive set of mesonic and baryonic resonances. We present recent results on resonance production in pp, p–Pb, Xe–Xe and Pb–Pb collisions at various center-of-mass energies, highlighting new results of $K^{*\pm}$, $\Lambda(1520)$, and $\Sigma^{*\pm}(1385)$. The results are also compared to lower energy measurements and model calculations.

1 Introduction

High-energy heavy-ion (AA) collisions allow studying the deconfined quark-gluon plasma (QGP) and its properties created in such collisions. The hot and dense medium created in heavy-ion collisions evolves with time and cools down, it reaches a temperature or time such that the deconfined quarks and gluons combine to form hadrons, after that hadron resonance gas is formed. The hadronic resonance production serves as a unique tool to understand the properties of the hadronic phase, i.e, the phase between chemical freeze-out (when inelastic collisions among the constituents cease) and kinetic freeze-out (when elastic collision ceases) created in heavy-ion collisions. Due to their short lifetimes ($\sim$ fm/c), resonance yields and particle ratios are expected to be modified due to the interaction of their decay products within the hadronic medium via regeneration and rescattering processes. Recent measurements of the particle ratios of resonance to stable hadron, such as K^{*0}/K show a decreasing trend with charged particle multiplicity in heavy-ion collisions. The K^{*0}/K yield ratio at most central collisions is suppressed in comparison to peripheral heavy-ion collisions. It is understood

*e-mail: dukhishyam.mallick@cern.ch

that K^{*0} has a short lifetime ($\sim$ 4 fm/c), so rescattering processes mainly modify the yield of K^{*0} [1]. Recently, a theoretical study suggests the lifetime of resonance is not the only key indicator to understanding the suppression of hadronic resonances; the final yield of resonances depends on other factors such as mean free path, cross section of decay products, and chemical freeze-out temprature [2]. The systematic comparison of measurements related to various resonances with different lifetimes, quark content, and varying mass may enable us to investigate the dynamics of the hadronic phase and study the lifetime of the hadronic phase.

2 Analysis details

Due to their short lifetime, resonances cannot be detected directly. They are reconstructed from their hadronic decay products using the invariant mass method. A detailed description of the ALICE detector setup and its performance is available at [3]. The sub-detectors relevant for these studies are the Time Projection Chamber (TPC), the Time-of-Flight detector (TOF), the Inner Tracking System (ITS), covering pseudorapidity window of $|\eta| < 0.9$, and the V0A ($2.8 < \eta < 5.1$) and V0C ($3.7 < \eta < 1.7$) detectors [4]. The ITS and TPC are used for charged-particle tracking and primary vertex reconstruction. Particle identification is done by measuring specific energy loss in the TPC gas and the time-of-flight information using TOF. The V0 detectors are used for triggering and centrality measurement. The measurements of hadronic resonance production are carried out at midrapidity ($|y| < 0.5$) for heavy-ion collisions. Depending on the analysis, the combinatorial backgrounds have been estimated using event mixing or like-sign method. For strange and multi-strange hadrons, a set of topological selections are applied. The raw yields are extracted from the signal distribution after subtracting the combinatorial background. After the combinatorial background subtraction, the distribution is fitted with a fit function (Breit-Wigner or Voitgian) for the signal and polynomial function of different order for the residual background distribution. Further, the raw p_T spectrum is corrected for the detector acceptance $\times$ efficiency and branching ratio.

3 Results and discussion

Figure 1 (left) shows the transverse momentum (p_T) spectra of K^{*0} and ϕ compared with blast wave model predictions (upper) and the ratio of data to model predictions (below) for most central (0-10%) in Pb–Pb collisions at $\sqrt{s_{NN}} = 5.02$ TeV. The Blast Wave model predictions are obtained using parameters from fits to pion, kaon, and (anti)proton p_T distributions [5]. It is observed that at low p_T (< 3 GeV/c), the data to model predictions ratio for K^{*0} is suppressed, whereas ϕ, the blast wave prediction describes the p_T spectrum well in the measured p_T range. It is expected that the ϕ yield is not significantly affected modified as its larger lifetime ($\sim$ 42.6 fm/c) and is consistent with the Blast Wave model because it does not have a rescattering effect. Figure 1 (right) shows the ratio of resonance yields to those of long-lived particles as a function of the cubic root of the average charged-particle multiplicity ($\langle dN_{ch}/d\eta \rangle^{1/3}$ is proportional to the radius of the particle-emitting source) for various resonances in different collision systems and energies measured by the ALICE [1] and STAR [6] Collaborations. A decreasing trend with multiplicity in the ratios ρ/π, $K^{0,*\pm}/K$, and $\Lambda(1520)/\Lambda$ is observed when going from peripheral to central collisions, whereas the $\Sigma^{*\pm}/\Lambda$, Ξ^{*0}/Ξ, and ϕ/K ratios are nearly constant across all systems and centrality classes. The new measurement of $K^{*\pm}$ is consistent with previous results of K^{*0} in Pb–Pb collisions at $\sqrt{s_{NN}} = 5.02$ TeV. These results suggest the dominance of rescattering effects over regeneration in the hadronic phase. However, the new measurements of $\Lambda(1520)$ is carried out

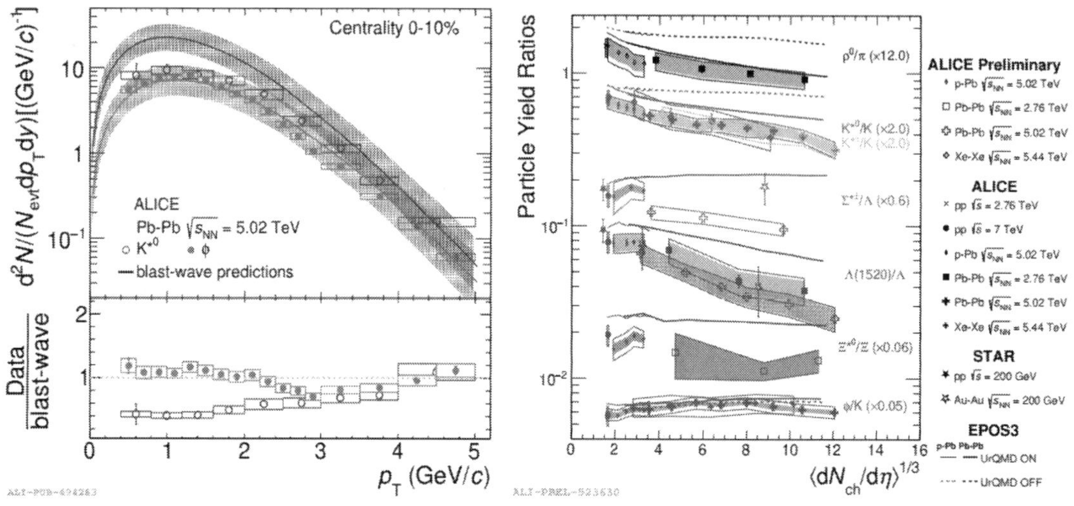

Figure 1. Left: transverse momentum distributions of K^{*0} (blue) and ϕ (red) resonances in Pb–Pb collisions at $\sqrt{s_{NN}}$ = 5.02 TeV along with blast wave predictions. The shaded bands indicate the uncertainties in the fit parameters of the model distributions. The lower panels show the ratios of the measured distribution to the values from the model. Right: summary of p_T-integrated yield ratios of different resonances to their ground state particles (ρ/π, $K^{0,*\pm}/K$, $\Sigma^{*\pm}/\Lambda$, $\Lambda(1520)/\Lambda$, Ξ^{*0}/Ξ and ϕ/K) as a function of $\langle dN_{ch}/d\eta\rangle^{1/3}$ in pp, p–Pb, Xe–Xe and Pb–Pb collisions along with model predictions from EPOS3 (with and without URQMD) and STAR data.

in Pb–Pb collisions at $\sqrt{s_{NN}}$ = 5.02 TeV with high statistics, it is found that $\Lambda(1520)/\Lambda$ is more suppressed than $K^{0,*\pm}/K$ as a function of $\langle dN_{ch}/d\eta\rangle^{1/3}$ even though its lifetime is about 3 times longer than $K^{0,*\pm}$. The ratios are compared to the EPOS3 with UrQMD [7], which qualitatively describes the measurements.

The measured suppression in particle ratios (resonance to stable hadron yield), is most probably considered one of the causes due to the rescattering of decay products in the hadronic phase. The timespan between chemical and kinetic freeze-out ($\tau_{kin} - \tau_{chem}$) is estimated using the exponential decay law as $r_{kin} = r_{chem} \times e^{(\tau_{kin}-\tau_{chem})/\tau_{res}}$ under assumption of i) negligible regeneration effects and ii) simultaneous freeze-out occur for all particle species from particle ratios [1]. In this calculation, the measured yield ratio in pp collisions is used as a proxy for r_{chem} at chemical freeze-out, r_{kin} is the measured yield ratio in Pb-Pb collisions, and τ_{res} is the proper lifetime of resonance [1].

Results for the estimated duration of hadronic phase as a function of $\langle dN_{ch}/d\eta\rangle^{1/3}$ is shown in Fig. 2 for various resonances particles in Pb–Pb collisions at $\sqrt{s_{NN}}$ = 2.76 and 5.02 TeV. It increases with increasing system size for all particles, which is the mirror image of decreasing trend of particle ratios as a function of system size. The time duration obtained from ρ/π, $K^{0,*\pm}/K$ ratios are lower than the time calculation obtained from Λ^*/Λ. It may be due to the simplest assumption of the exponential decay model, neglecting regeneration effects and sudden freeze-out for all particles.

4 Conclusions

ALICE has measured several resonances with varying lifetimes (from 1–50 fm/c), quark content, and mass in different colliding systems and energies. A suppression in p_T-differentail

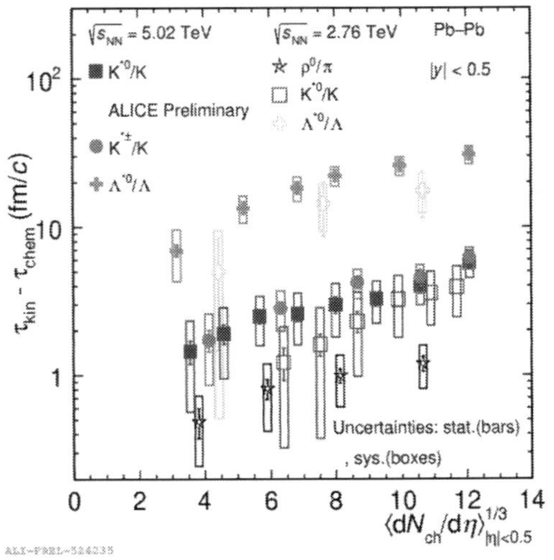

Figure 2. Lower limit on the hadronic phase lifetime between chemical and kinetic freeze-out as a function of $\langle dN_{ch}/d\eta\rangle^{1/3}$ for different resonances in Pb-Pb collisions at LHC energies. The statistical and systematic errors are represented by vertical bars and boxes around the data points.

yield of K^{*0} in compared to blast wave predictions at low p_T in most central heavy-ion collisions suggests that rescattering effects play a dominant role at low p_T. The suppression in p_T-integrated yield ratios of ρ/π, $K^{0,*\pm}$ /K is observed in central heavy-ion collisions, suggesting that rescattering is the dominant process over regeneration in the hadronic phase. However, yield ratios of long-lived resonances (ϕ/K) remain flat with multiplicity, not significantly affected by rescattering and regeneration processes. Recently, the new measurement of $\Lambda(1520)/\Lambda$ ratio shows stronger suppression than $K^{0,*\pm}$/K. The simplest exponential decay model without considering the regeneration effects is used to estimate the lower limit of the hadronic phase lifetime for ρ, $K^{0,*\pm}$, Λ^{*0} in Pb–Pb collisions $\sqrt{s_{NN}} = 5.02$ TeV. The estimated lifetime of the hadronic phase increases with system size, and values of Λ^* is higher than ρ, and $K^{0,*\pm}$. It suggests that the final yield of resonances depends not only on the resonance lifetime but also on factors such as the mean free path of resonances, cross section of decay products, and chemical freeze-out temperature.

In the future, precise and new measurements in Run 3, along with more realistic parameterization for the modification of resonance yield, will help us further understand the hadronic phase and its lifetime in heavy-ion collisions at LHC energies.

References

[1] ALICE Collaboration, S. Acharya et al., Phys. Lett. B 802, 135225 (2020).
[2] Dmytro Oliinychenko, Chun Shen, arXiv:2105.07539 (2021).
[3] ALICE Collaboration, B.B. Abelev et al., Int. J. Mod. Phys. A 29, 1430044 (2014).
[4] ALICE Collaboration, E. Abbas et al., JINST 8, P10016 (2013).
[5] ALICE Collaboration, S. Acharya et al., Phys. Rev. C 101, 044907 (2020).
[6] STAR Collaboration, M.M. Aggarwal et al., Phys. Rev. C 84 034909, (2011).
[7] T. Pierog, K. Werner, Nucl. Phys. B Proc. Suppl. 196, (2009).

Collective flow of light nuclei and hyper-nuclei in Au+Au collisions at $\sqrt{s_{NN}}$ = 3, 14.6, 19.6, 27, and 54.4 GeV using the STAR detector

Rishabh Sharma[1,*], for the STAR Collaboration

[1]Indian Institute of Science Education and Research (IISER) Tirupati

Abstract. Light nuclei and hyper-nuclei are produced in abundance in heavy-ion collisions. The production mechanism of these species in heavy-ion collisions still remains to be understood. In these proceedings, we report the transverse momentum and centrality dependence of elliptic flow (v_2) of d, t, and ^{3}He in Au+Au collisions at $\sqrt{s_{NN}}$ = 14.6, 19.6, 27, and 54.4 GeV. Mass number scaling of $v_2(p_T)$ of light nuclei is shown. We also report the first observation of hyper-nuclei $^3_\Lambda$H and $^4_\Lambda$H directed flow (v_1) in $\sqrt{s_{NN}}$ = 3 GeV mid-central (5-40%) Au+Au collisions in the fixed target mode.

1 Introduction

The production and interaction of light nuclei and hyper-nuclei in high-energy heavy-ion collisions have been a focus of theoretical and experimental interests for a long time. The production of light nuclei in heavy-ion collisions can be explained by the coalescence of produced or transported nucleons [1–5]. Due to their low binding energies, it is more likely that they are formed at later stages of the evolution of the fireball. Therefore, studying the collective flow of light nuclei and hyper-nuclei can provide insights into their production mechanism in heavy-ion collisions. Further, the study of the collective flow of hyper-nuclei will shed light on the hyperon-nucleon (YN) interaction in dense nuclear medium [6, 7].
In the following sections, elliptic flow (v_2) of d, t, and ^{3}He in Au+Au collisions at $\sqrt{s_{NN}}$ = 14.6, 19.6, 27, and 54.4 GeV and the directed flow (v_1) of Λ, $^3_\Lambda$H, and $^4_\Lambda$H in mid-central Au+Au collisions at $\sqrt{s_{NN}}$ = 3 GeV are discussed.

2 Analysis details

The data presented here was collected in the phase 2 of the Beam Energy Scan (BES-II) program by the STAR experiment at RHIC. Particle identification is performed using the Time Projection Chamber (TPC) and Time of Flight (TOF) detectors. The TPC is the primary tracking detector in the STAR experiment which uses specific ionisation energy loss (dE/dx) in a large gas volume to detect trajectories of charged particles. TOF detector allows the identification of particles of interest using a restriction on the mass-square (m^2) of the particle. It helps improve particle identification in the intermediate transverse momentum region. The hyper-nuclei $^3_\Lambda$H and $^4_\Lambda$H are reconstructed from the following decay channels: $^3_\Lambda\text{H} \rightarrow {}^3\text{He}$ +

*e-mail: rishabhsharma@students.iisertirupati.ac.in

π^-, ${}^3_\Lambda\text{H} \rightarrow d + p + \pi^-$, and ${}^4_\Lambda\text{H} \rightarrow {}^4\text{He} + \pi^-$. In this study, v_2 measurements of light nuclei are performed in Au+Au collisions in the collider mode whereas v_1 measurements of hyper-nuclei are performed in the fixed-target (FXT) mode. In the fixed target mode, the gold beam of 3.85 GeV/u is bombarded on a thin gold target with 1% interaction probability. The target is located at 200 cm along the beam direction from the center of the TPC. In Au+Au collisions at $\sqrt{s_{NN}}$ = 14.6, 19.6, 27, and 54.4 GeV, the second order event plane angle is constructed using the TPC to measure the v_2 of d, t, and ^{3}He. The v_1 of hyper-nuclei in Au+Au collisions at $\sqrt{s_{NN}}$ = 3 GeV is measured using the first order event plane determined by the Event Plane Detector (EPD) [8]. EPD is designed to measure the azimuthal patterns of charged particles emitted in backward and forward direction relative to the beam direction.

3 Results

3.1 Elliptic flow of light nuclei

Figure 1 shows v_2 as a function of p_T in minimum bias Au+Au collisions at $\sqrt{s_{NN}}$ = 14.6, 19.6, 27, and 54.4 GeV. A monotonous increase in v_2 of light nuclei with p_T across all four center-of-mass energies is observed.

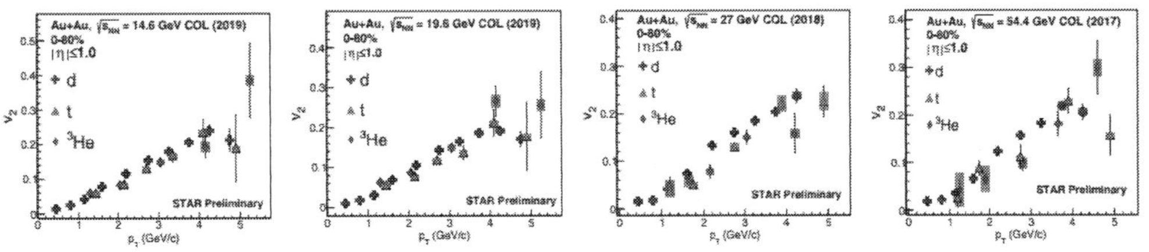

Figure 1. $v_2(p_T)$ of light nuclei (d, t, and ^{3}He) in minimum bias Au+Au collisions at $\sqrt{s_{NN}}$ = 14.6, 19.6, 27, and 54.4 GeV. Vertical lines and shaded area at each marker represent statistical and systematic uncertainties, respectively.

Centrality dependence of v_2 of d is shown in Fig. 2. The nuclei v_2 is measured in two centrality ranges 0-30% and 30-80% for Au+Au collisions at $\sqrt{s_{NN}}$ = 19.6, 27, and 54.4 GeV. $v_2(p_T)$ of d shows a strong centrality dependence for all the three energies. It is observed that peripheral collisions have a higher v_2 compared to more central collisions. This can be explained by the fact that more peripheral collisions have a greater spatial anisotropy compared to the central collisions.

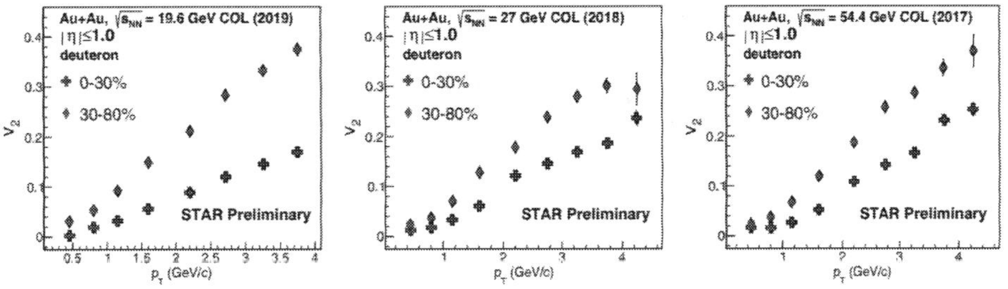

Figure 2. Centrality dependence of v_2 of d as a function of p_T in Au+Au collisions at $\sqrt{s_{NN}}$ = 19.6, 27, and 54.4 GeV. Vertical lines and shaded area at each marker represent statistical and systematic uncertainties, respectively.

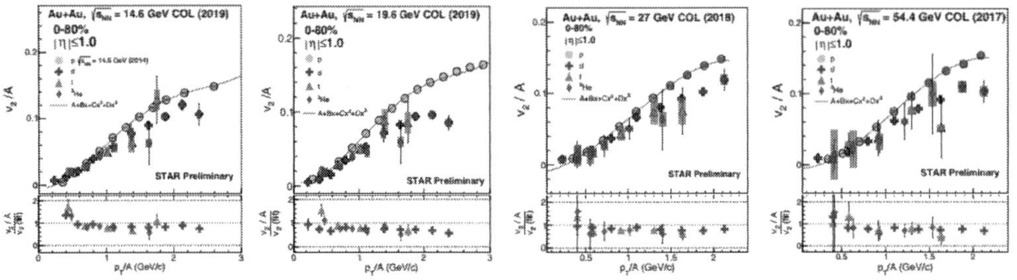

Figure 3. Atomic mass number scaling v_2/A of light nuclei as a function of p_T/A in minimum bias Au+Au collisions at $\sqrt{s_{NN}}$ = 14.6, 19.6, 27, and 54.4 GeV. Proton v_2 has been fitted with a third-order polynomial. The bottom panel in each plot shows the ratio between the v_2/A of light nuclei and the fit to proton v_2. Vertical lines and shaded area at each marker represent statistical and systematic uncertainties, respectively.

Coalescence model predicts that if a composite nuclei is produced by coalescence of n number of nucleons, very close to each other in phase-space, then the v_2 of the composite nuclei will be n-times the v_2 of the individual nucleons [9, 10]. This is called as mass number scaling. Figure 3 shows the comparison of light nuclei v_2/A as a function of p_T/A (where A is the mass number of the nuclei) with proton v_2/A (where A = 1). Light nuclei v_2 has been observed to follow the mass number scaling within 20-30%.

3.2 Directed flow of hyper-nuclei

Figure 4 shows v_1 as a function of rapidity (y) for Λ, ${}^{3}_{\Lambda}$H, and ${}^{4}_{\Lambda}$H in Au+Au collisions at $\sqrt{s_{NN}}$ = 3 GeV in 5-40% centrality. Results of hyper-nuclei v_1 are compared to those of light nuclei p, d, t, ^{3}He, and ^{4}He. Linear fits (yellow-red lines) to Λ, ${}^{3}_{\Lambda}$H, and ${}^{4}_{\Lambda}$H v_1 are compared with the linear fits (gray lines) to light nuclei v_1. These fit results have been extrapolated to the positive y region. It is observed that the linear fit to the Λ hyperon v_1 is consistent with that of proton, and the fit results of hyper-nuclei v_1 are also similar to those of the corresponding light nuclei with the similar mass number within statistical uncertainties.

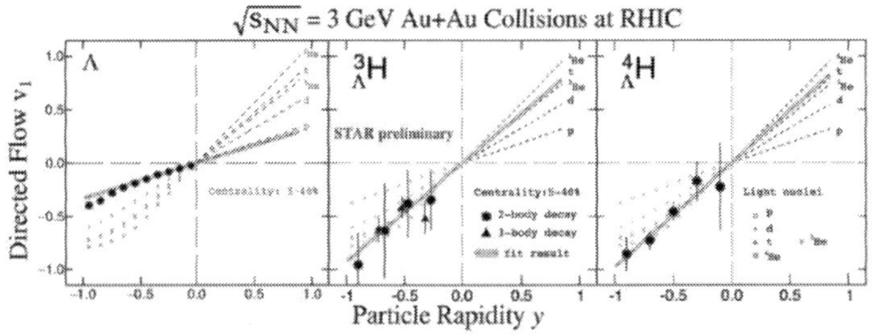

Figure 4. v_1 of Λ, ${}^{3}_{\Lambda}$H, and ${}^{4}_{\Lambda}$H as a function of rapidity from mid-central (5-40%) Au+Au collisions at $\sqrt{s_{NN}}$ = 3 GeV. Vertical lines at each marker represent statistical uncertainty.

Mid-rapidity hyper-nuclei slopes of v_1 were extracted from the linear fits as shown in Fig. 4. Mass dependence of these v_1 slopes (dv_1/dy) is shown in Fig. 5. The dv_1/dy of light nuclei p, d, t, ^{3}He, and ^{4}He from the same dataset and centrality is also shown as open circles. Linear fit to the light nuclei dv_1/dy is shown as yellow-red line. It is observed that

https://doi.org/10.1051/epjconf/202327604006

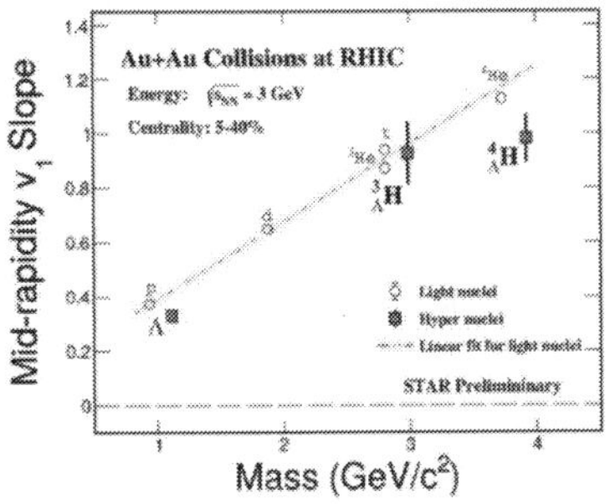

Figure 5. Mass dependence of the mid-rapidity v_1 slope $dv_1/dy|_{y=0}$ for Λ, ${}^3_\Lambda$H, and ${}^4_\Lambda$H, from mid-central (5-40%) Au+Au collisions at $\sqrt{s_{NN}}$ = 3 GeV. Vertical lines at each marker represent statistical uncertainty.

the v_1 slopes of hyper-nuclei are consistent with those of light nuclei with similar mass within statistical uncertainties. This mass dependence of the v_1 slope might suggest that coalescence could be the dominant mechanism of hyper-nuclei production in heavy-ion collisions.

4 Summary

In summary, we have studied the v_2 of d, t, and ^{3}He in Au+Au collisions at $\sqrt{s_{NN}}$ = 14.6, 19.6, 27, and 54.4 GeV. A monotonic rise of light nuclei v_2 with p_T is observed. The v_2 of d is observed to show a strong centrality dependence being higher for more peripheral collisions compared to more central collisions. Light nuclei v_2 is also found to follow the mass number scaling within 20-30%. In addition, we have also reported the v_1 of ${}^3_\Lambda$H and ${}^4_\Lambda$H from Au+Au collisions at $\sqrt{s_{NN}}$ = 3 GeV. The rapidity dependence of hyper-nuclei v_1 and dv_1/dy is compared to those of the light nuclei. It is observed that within statistical uncertainties, the results of hyper-nuclei v_1 and dv_1/dy are quite close to that of corresponding light nuclei with similar mass. These observations may suggest that coalescence of nucleons and hyperons is the dominant mechanism of hyper-nuclei production in such collisions.

References

[1] S. T. Butler and C. A. Pearson, Phys. Rev. Lett., **129**, 836 (1963).
[2] A. Schwarzschild and A. Zupancic, Phys. Rev. Lett., **129**, 854–862 (1963).
[3] H. H. Gutbrod et al., Phys. Rev. Lett., **37**, 667 (1976).
[4] H. Sato and K. Yazaki, Phys. Lett. B, **98**, 153–157 (1981).
[5] E. A. Remler, Annals of Physics, **136**, 293–316 (1981).
[6] D. Gerstung, N. Kaiser, and W. Weise, Eur. Phys. J., **A56**, 175 (2020).
[7] D. Lonardoni et al., Phys. Rev. Lett., **114**, 092301 (2015).
[8] J. Adams et al. (STAR Collaboration), Nucl. Instrum. Meth. A **968**, 163970 (2020).
[9] T. Z. Yan, Y. G. Ma, X. Z. Cai et al., Phys. Lett. B, **638**, 50–54 (2006).
[10] Y. Oh and C. M. Ko, Phys. Rev. C, **76**, 054910 (2007).

Measurements of global and local polarization of hyperons in isobar collisions at 200 GeV from STAR

Xingrui Gou[1,*], for the STAR Collaboration

[1]Institute of Frontier and Interdisciplinary Science & Key Laboratory of Particle Physics and Particle Irradiation (Ministry of Education), Shandong University, Qingdao, Shandong, 266237, China

Abstract. In heavy-ion collisions, the observation of the global and local polarization of hyperons has revealed the existence of large vorticities perpendicular to reaction plane due to system's orbital angular momentum and along beam direction due to collective velocity field, respectively. With the high-statistics data from isobar collisions of Ru+Ru and Zr+Zr at $\sqrt{s_{NN}}$ = 200 GeV collected by the STAR experiment, we present differential measurements of global polarization for $\Lambda/\bar{\Lambda}$ as a function of centrality. These measurements allow us to study the possible magnetic field driven effects through the polarization difference between Ru+Ru and Zr+Zr, owing to a larger magnetic field in the former. Furthermore, the first measurements of Λ hyperon local polarization along beam direction relative to the third-order event plane as well as the second-order event plane are presented. A comparison of results from isobar and Au+Au collisions provides important new insights into the collision system size dependence of the vorticities in heavy-ion collisions.

1 Introduction

In non-central heavy-ion collisions, the produced system has large orbital angular momentum and may have a strong vortical structure, which leads to the global spin polarization of hyperons through the spin-orbital interaction [1]. Due to the nature of the weak decay, Λ hyperon's polarization can be determined through the angular distribution of decay daughter proton in parent's rest frame [1].

Global polarization has been observed for Λ and $\bar{\Lambda}$ hyperons in Au+Au collisions from $\sqrt{s_{\rm NN}}$ = 7.7 to 200 GeV by the STAR experiment [2, 4]. Significant global polarization of $\Lambda(\bar{\Lambda})$ observed from low to high energy is consistent with the expectation from the spin-orbit coupling picture, which can be described by hydrodynamic models. Some models predict a system size dependence of global polarization [3, 5]. Experimentally, the system size dependence can be studied by comparing results in isobar collisions with those in Au+Au collisions. Futhermore, the magnetic field effects may cause a splitting between Λ and $\bar{\Lambda}$ global polarization, and initial magnetic field difference between Ru+Ru and Zr+Zr collision may lead to different Λ global polarization in the two systems.

On the other hand, STAR has measured the local polarization with respect to the second-order event plane in Au+Au collisions at $\sqrt{s_{\rm NN}}$ = 200 GeV [7]. The local polarization as a function of azimuthal angle relative to the second-order event plane shows a sine modulation,

*e-mail: xgou@rcf.rhic.bnl.gov;xingruigou@mail.sdu.edu.cn

as expected from quadrupole structure of vorticity along the beam direction [7]. With high statistics isobar data, measurements of local polarization in smaller systems and relative to higher harmonic event planes can provide new insights into polarization phenomena.

In these proceedings, we report $\Lambda(\bar{\Lambda})$ global and local polarization as a function of centrality in Ru+Ru and Zr+Zr collisions at $\sqrt{s_{\rm NN}} = 200$ GeV using the data collected by STAR experiment.

2 Global polarization results

In the STAR experiment, the first-order event plane can be determined by Zero Degree Calorimeters with Shower Maximum Detectors (ZDC SMD), the second-order and third-order event planes are determined by the Time Projection Chamber detector (TPC). $\Lambda(\bar{\Lambda})$ hyperons have been reconstructed through its decay channel: $\Lambda \rightarrow \pi^- + p$ ($\bar{\Lambda} \rightarrow \pi^+ + \bar{p}$). The residual background under the mass peak is smaller than 3%. The global polarization is determined by

$$P_\Lambda = \frac{8}{\alpha\pi}\frac{1}{A^0}\frac{< \sin(\Psi_1 - \phi_{\rm p}^*) >}{{\rm Res}(\Psi_1)}, \quad (1)$$

where α is the decay parameter, A^0 is the acceptance correction factor, ϕ_p^* is the azimuthal angle of decay proton in Λs rest fame and ${\rm Res}(\Psi_1)$ is the first-order event plane resolution [2].

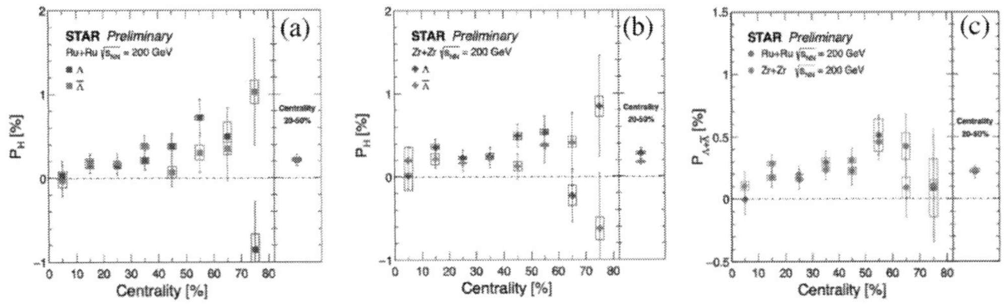

Figure 1: Global polarization of Λ and $\bar{\Lambda}$ as a function of centrality in Ru+Ru(a), Zr+Zr(b) collisions at $\sqrt{s_{\rm NN}} = 200$ GeV. Panel (c) shows Λ+$\bar{\Lambda}$ global polarization results in isobar collisions. Open boxes and vertical lines represent systematic and statistical uncertainties.

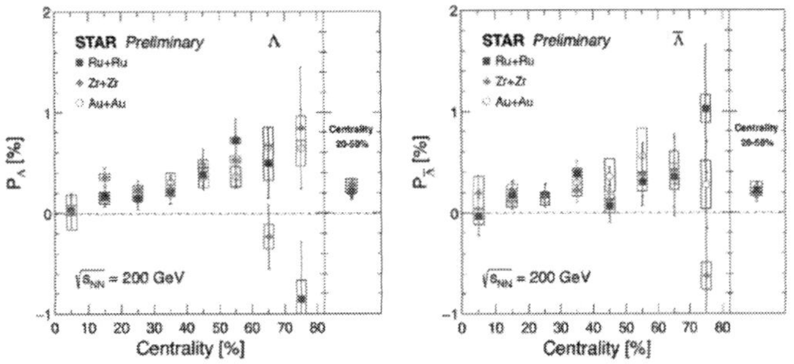

Figure 2: Λ(left) and $\bar{\Lambda}$(right) global polarization as a function of centrality in Ru+Ru, Zr+Zr, and Au+Au collisions at $\sqrt{s_{\rm NN}} = 200$ GeV.

Figure 1 (a) and (b) show Λ and $\bar{\Lambda}$ global polarization $P_{\Lambda/\bar{\Lambda}}$ as a function of centrality in Ru+Ru and Zr+Zr collisions. The trend of the results increase from central to peripheral centrality. For better precision, we also combine 20-50% centrality results. No significant difference between Λ and $\bar{\Lambda}$ global polarization in Ru+Ru and Zr+Zr collisions has been observed which indicates that no magnetic field effects on the hyperon polarization is observed in isobar collisions within current statistical limitation.

Figure 1 (right) shows Λ+$\bar{\Lambda}$ global polarization $P_{\Lambda+\bar{\Lambda}}$ as a function of centrality in Ru+Ru and Zr+Zr collisions. The results are consistent in each centrality between Ru+Ru and Zr+Zr collisions.

Figure 2 shows Λ and $\bar{\Lambda}$ global polarization comparison between isobar and Au+Au collisions. The results are consistent between isobar and Au+Au collisions for the whole centrality range, indicating there might be little collision system size dependence.

3 Local polarization results

The component of the polarization along the beam direction can be measured by

$$\langle \cos\theta^*_{\rm p} \rangle = \int \frac{dN}{d\Omega^*} \cos\theta^*_{\rm p} d\Omega^*, \tag{2}$$

where θ^*_p is the polar angle of the daughter proton in the Λ rest frame [7].

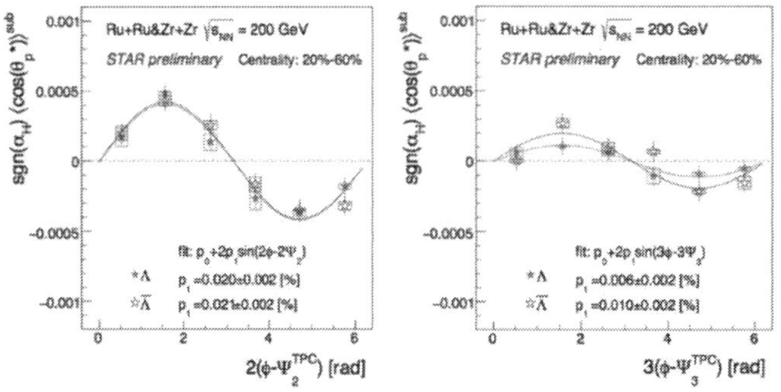

Figure 3: $\langle \cos\theta^*_{\rm p} \rangle$ of Λ and $\bar{\Lambda}$ hyperons as a function of azimuthal angle ϕ relative to the second-order event plane Ψ_2(left) and third-order event plane Ψ_3(right) for 20% − 60% centrality in isobar collisions at $\sqrt{s_{\rm NN}}$ = 200 GeV. $\langle\rangle^{sub}$ denotes the subtraction of the acceptance effect. Solid lines show the fit with the sine function.

Figure 3 shows $\langle \cos\theta^*_{\rm p} \rangle$ of Λ and $\bar{\Lambda}$ hyperons as a function of azimuthal angle ϕ relative to the second-order event plane Ψ_2(left) and third-order event plane Ψ_3(right) for 20% − 60% centrality respectively. The solid lines are the fits to the results with $\rm p_0 + 2p_1 \sin(n\phi - n\Psi_n)$. The signal on Figure 3 (left) shows a clear sine modulation, as expected from quadrupole structure of vorticity along the beam direction. The trend is similar to that in Au+Au collisions. Figure 3 (right) shows the first measurements of $\langle \cos\theta^*_{\rm p} \rangle$ with respect to the third-order event plane Ψ_3. The results also show a sine modulation for both Λ and $\bar{\Lambda}$, indicating a $\rm v_3$ driven polarization.

Figure 4 (left) presents the centrality dependence of the second and third Fourier sine coefficients of the local polarizaiton $\langle \rm P_z \sin[n(\phi - \Psi_n)] \rangle$. The increase of the results with centrality is in line with the increasing of elliptic flow magnitude towards peripheral collisions. A significant local polarization with respect to the third-order event plane has been observed which increases with centrality. The results show no significant difference between the second-order and third-order local polarization within uncertrainties. The hydrodynamic model with a shear term [6] reasonably describes the data for central collisions, but not for the peripheral ones.

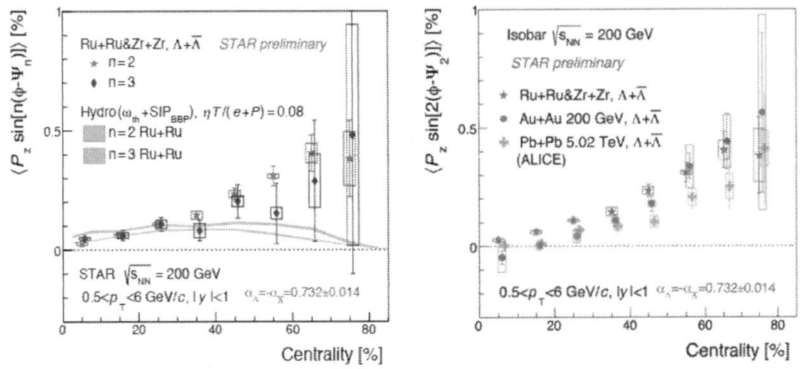

Figure 4: Left: the second and third Fourier sine coefficients of the local polarization of $\Lambda+\bar{\Lambda}$ as a function of the collision centrality in isobar collisions at $\sqrt{s_{\rm NN}}$ = 200 GeV. Right: the comparison of the second Fourier sine coeffcient of $\Lambda+\bar{\Lambda}$ local polarization among isobar, Au+Au collisions at $\sqrt{s_{\rm NN}}$ = 200 GeV and Pb+Pb collisions at $\sqrt{s_{\rm NN}}$ = 5.02 TeV.

Figure 4 (right) shows the $\langle P_z \sin[2(\phi - \Psi_2)]\rangle$ of $\Lambda+\bar{\Lambda}$ local polarization with respect to the second-order event plane as a function of the collision centrality in isobar, Au+Au, and Pb+Pb collisions [8]. A hint of system size dependence has been observed comparing isobar and Au+Au collisions at $\sqrt{s_{\rm NN}}$ = 200 GeV, while the energy dependence is not obvious between $\sqrt{s_{\rm NN}}$ = 200 GeV Au+Au collisions and $\sqrt{s_{\rm NN}}$ = 5.02 TeV Pb+Pb collisions.

4 Summary

The global and local polarizations of Λ and $\bar{\Lambda}$ have been measured in Ru+Ru and Zr+Zr collisions at $\sqrt{s_{\rm NN}}$ = 200 GeV. For global polarization, Λ and $\bar{\Lambda}$ results are consistent, showing that the magnetic field effects on global polarization are not observed in isobar collisions within current statistical limitation. Global polarization results are consistent between Ru+Ru, Zr+Zr, and Au+Au collisions. No obvious collision system size dependence is observed. Significant local polarization signals with respect to the second-order and third-order event plane are observed in isobar collisions at $\sqrt{s_{\rm NN}}$ = 200 GeV. A slight hint of collision system size dependence has been observed, while energy dependence is not obvious.

Acknowledgements

The author is supported partially by the National Natural Science Foundation of China under No.11890710, No.11890713.

References

[1] Z. T. Liang and X. N. Wang, Phys. Rev. Lett. 94,102301 (2005); 96, 039901(E) (2006).
[2] L. Adamczyk et al., (STAR Collaboration), Nature 548, 62 (2017).
[3] I. Karpenko and F. Becattini, Eur. Phys.J. C 77, 213 (2017).
[4] J. Adam et al., (STAR Collaboration), Phys. Rev. C 98, 014910 (2018).
[5] S. Alzhrani, S. Ryu, and C. Shen, Phys. Rev. C 106, 014905 (2022).
[6] B. C. Fu, et al., Phys. Rev. Lett. 127, 142301 (2021).
[7] J. Adam et al., (STAR Collaboration), Phys. Rev. Lett. 123, 132301(2019)
[8] S. Acharya et al., (ALICE Collaboration), arXiv:2107.11183.

Light hypernuclei in heavy-ion collisions

Tom Reichert[1,2,*], *Jan* Steinheimer[3], *Volodymyr* Vovchenko[4,5,3], *Benjamin* Dönigus[6], and *Marcus* Bleicher[1,2,7]

[1]Institut für Theoretische Physik, Goethe Universität Frankfurt, Max-von-Laue-Str. 1, D-60438 Frankfurt am Main, Germany
[2]Helmholtz Research Academy Hesse for FAIR (HFHF), GSI Helmholtzzentrum für Schwerionenforschung GmbH, Campus Frankfurt, Max-von-Laue-Str. 12, 60438 Frankfurt am Main, Germany
[3]Frankfurt Institute for Advanced Studies, Ruth-Moufang-Str. 1, D-60438 Frankfurt am Main, Germany
[4]Institute for Nuclear Theory, University of Washington, Seattle, WA 98195-1550, USA
[5]Nuclear Science Division, Lawrence Berkeley National Laboratory, 1 Cyclotron Road, Berkeley, CA 94720, USA
[6]Institut für Kernphysik, Goethe Universität Frankfurt, Max-von-Laue-Str. 1, D-60438 Frankfurt am Main, Germany
[7]GSI Helmholtzzentrum für Schwerionenforschung GmbH, Planckstr. 1, D-64291 Darmstadt, Germany

Abstract. Prediction for hyper nuclei multiplicities from GSI to LHC energies from the Ultra-relativistic Quantum Molecular Dynamics (UrQMD) model combined with a final state coalescence approach is presented and compared to the thermal model. The influence of the coalescence radius on the collision energy and centrality dependence of the ${}^{3}_{\Lambda}H/\Lambda$ ratio is discussed.

1 Introduction

Light (hyper)nuclei production in nuclear collisions has become an active field of research in recent years. The precise understanding of the mechanisms creating those nuclei is crucial to understand the strong interaction in more detail. Furthermore, especially hypernuclei are a very important tool to investigate the nuclear Equation-of-State (EoS) at high densities. In the literature, there are basically two models frequently used to describe (hyper)nuclei production: (i) A (grand) canonical thermal model [1, 2] or (ii) a final-state coalescence approach [3]. Here, we aim to compare both methods and explore light-nuclei as well as hypernuclei production from SIS up to LHC energies.

2 Coalescence and the thermal model

Coalescence. The Ultra-relativistic Quantum Molecular Dynamics (UrQMD) [5–7] is a relativistic transport model providing an effective solution of the Boltzmann equation with hadrons as degrees of freedom. The scattering term is implemented via a geometric interpretation of the cross section, and further via string fragmentation and resonance excitation. An additional Equation-of-State can be used as well as a switch to a hybrid mode with intermediate hydrodynamic evolution suited for higher energies.

*e-mail: treichert@itp.uni-frankfurt.de

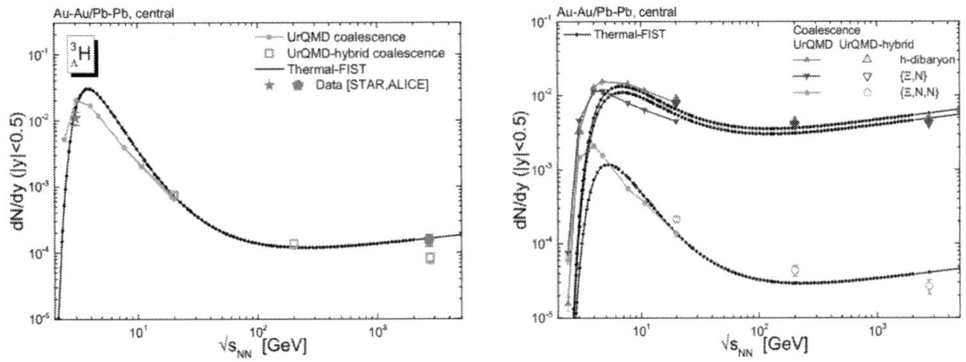

Figure 1. Mid-rapidity yields of the hyper triton (left panel) and hitherto undiscovered states (right panel) in central collisions as a function of the collision energy. The thermal model (Thermal-FIST) prediction is compared with the coalescence results of the UrQMD model and available data from ALICE [14] and STAR [15].

The coalescence approach relates the invariant momentum distribution of a cluster to the invariant spectra of its components at kinetic freeze-out (defined as last scattering point in the UrQMD simulations) via a phase-space volume $\rho_{AB}(\Delta x, \Delta p)$.

$$\frac{\mathrm{d}N}{\mathrm{d}^3k} = g \int \mathrm{d}p_1^3 \mathrm{d}p_2^3 \mathrm{d}x_1^3 \mathrm{d}x_2^3 f_A(p_1, x_1) f_B(p_2, x_2) \rho_{AB}(\Delta x, \Delta p) \delta(k - (p_1 + p_2)) \tag{1}$$

In the following, we practically implement the phase-space coalescence volume as a box in phase-space which is defined by two parameters $\Delta x_{\max}$ and $\Delta p_{\max}$, which have to be determined for each (hyper)nucleus. For a more detailed description of the procedure, see [8–10].

Thermal model. In the thermal model approach the light nuclei are incorporated into the partition function of hadronic matter at chemical freeze-out as separate degrees of freedom and their abundances are given by corresponding thermal distribution functions, incorporating the feed down contributions from excited nuclei. Here we employ the open source thermal model Thermal-FIST [11] to estimate the (hypernuclei) multiplicities. The thermal parameters are taken from fits to stable hadron multiplicities, the so-called freeze out curve. For small systems the Thermal-FIST model can also incorporate the canonical treatment of the baryon number, charge and strangeness within a given correlation volume V_c [12, 13].

3 Results

The midrapidity multiplicity of the hypertriton ${}^3_\Lambda$H, in central heavy-ion collisions, as a function of the collision energy is shown in the left panel of Fig. 1. The UrQMD model calculations are shown by orange lines and symbols, the Thermal-FIST results as a black line and the experimental data points are shown as red symbols. The parameters were fixed to describe the data points and yield $\Delta x_{\max}^{pn\Lambda} = 9.5$ fm (the estimated size of the wave function) and $\Delta p_{\max}^{pn\Lambda} = 0.135$ GeV. The same parameter set is further used to predict the multiplicity of hitherto undiscovered particles. In the right panel of Fig. 1 we show the energy dependence of the midrapidity yields of a possible H-dibaryon state (red), a bound ΞN (blue) and a bound ΞNN (orange) configuration.

Recently, it has been argued that the hypertriton multiplicity is highly sensitive to the system size due to the small Λ separation energy and therefore large spatial separation [16]. In such a scenario, the parameters Δx and Δp can be used to study the source volume at

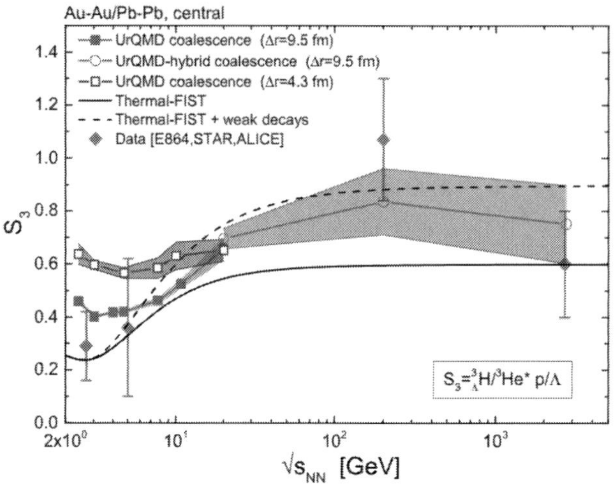

Figure 2. Energy dependence of the double ratio $S_3 = {}^3_\Lambda\mathrm{H}/{}^3\mathrm{He} \cdot p/(\Lambda + \Sigma^0)$. Here the UrQMD results are shown as green (set I) or blue (set II) lines with error bands, the thermal model (Thermal-FIST) results are shown by black lines and the available data points are shown by red symbols.

kinetic freeze out [10]. Thus, a second parameter set for the hypertriton, $\Delta x^{pn\Lambda}_{\max} = 4.3$ fm (same parameter as for the triton [9]), can be used to analyze the system size dependence. The differences can be clearly seen in the ratio $S_3 = {}^3_\Lambda\mathrm{H}/{}^3\mathrm{He} \cdot p/(\Lambda + \Sigma^0)$ shown as a function of energy in Fig. 2. Here, the UrQMD results are shown as green lines (parameter set I) and blue lines (parameter set II) with symbols and error bands, the Thermal-FIST results are shown by black lines and the available data points are shown by red symbols. It should be pointed out that it is not clear from the experimental results to what degree the feed down from the Λ weak decay may be taken into account. Therefore, the thermal model results are shown for two extreme scenarios, i.e. the solid line corresponds to results where feed down from the weak decay is not included in the proton number while in the dashed line it is fully included. The thermal model results with decay of excited nuclei describe the experimental data well. The UrQMD model results with parameter set I also describe the experimental data remarkably well.

In the following the centrality dependence of (hyper)nuclei at LHC energies is further discussed. Varying the centrality at fixed energy allows to analyze the volume dependence without changing the chemical composition of the system. We start with a discussion of the deuteron to proton ratio in the left panel of Fig. 3. The UrQMD model calculations are shown as green lines and the thermal model results are shown as a black line. The ALICE data points are shown as red symbols. The data clearly shows an increase of the ratio d/p from ultra-peripheral to semi-peripheral collisions and a further decrease to very central collisions (also observed in [8]). This phenomenon has been understood as an annihilation effect which affects the deuteron more pronounced, while the rest of the structure arises from volume scaling and canonical effects. One might argue that this effect should be visible for the ratio ${}^3\mathrm{He}/p$ as well. In the right panel of Fig. 3 we therefore show the ratio ${}^3\mathrm{He}/p$ from UrQMD as a green line, from Thermal-FIST as a black line and the same ratio measured by ALICE as magenta symbols. The data and both models qualitatively agree on the canonical effects visible at large centralities. For central collisions the data points do not suggest a decrease, but the available data may not be sufficient to draw final conclusions. However, the UrQMD model simulation clearly predicts a maximum at semi-peripheral collisions. Can this effect be

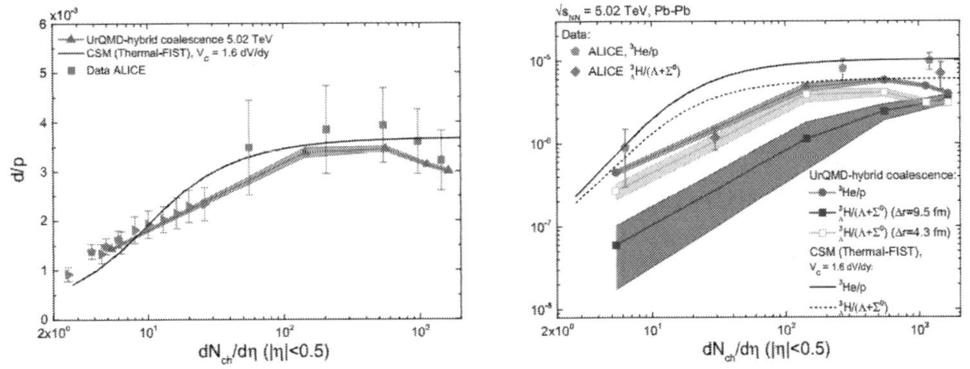

Figure 3. Left panel: Centrality dependence of the d/p ratio from models at $\sqrt{s_{\rm NN}}$ = 5.02 compared to experimental data from the ALICE experiment. Right panel: The centrality dependence of the ^{3}He to proton and $^3_\Lambda$H to Λ ratio for Pb-Pb collisions at $\sqrt{s_{\rm NN}}$ = 5.02 TeV. In the case of the $^3_\Lambda$H to Λ ratio we show two scenarios corresponding to two different choices of the Δr parameter (Δr = 9.5 fm or Δr = 4.3 fm). The predictions from the canonical thermal model fits are shown as black lines. ALICE data are shown as symbols [14, 17].

seen in the ratio $^3_\Lambda$H/(Λ + Σ^0) as well? Fig. 3 also shows this ratio as a function of centrality as calculated in the UrQMD model with parameter set I (blue line) and parameter set II (light blue line), in the thermal model (dashed line) and measured by ALICE (red symbols) [14, 17]. The coalescence model using a larger coalescence parameter, motivated from the wave function, shows a much faster drop towards peripheral collisions and no peak at more central collisions. In contrast, the coalescence model using the small Δr reveals a similar drop off as the 3-Helium to proton ratio. However, more precise measurements are needed to draw a final conclusion.

4 Conclusion

We have employed the coalescence (UrQMD) and thermal (Thermal-FIST) models to predict light cluster and hypernuclei yields from SIS to LHC energies. In UrQMD, cluster production is employed via a coalescence mechanism while in the thermal model the clusters are explicitly included in the partition sum as degrees of freedom. We found a good agreement of measured multiplicities over a broad range of energies both with the transport model and the thermal model. Differences arise in the centrality dependence of the ratio $^3_\Lambda$He/(Λ + Σ^0). Here, the UrQMD results are sensitive to the values of coalescence parameters corresponding to the coalescence source size. A larger size leads to a much faster drop towards peripheral collisions. More data is required to draw further conclusions.

M.B. acknowledges support by the EU–STRONG 2020 network. The computational resources for this project were provided by the Center for Scientific Computing of the GU Frankfurt and the Goethe–HLR.

References

[1] A. Andronic, P. Braun-Munzinger, K. Redlich, and J. Stachel, Nature **561** 321-330 (2018)

[2] P. Braun-Munzinger, K. Redlich, and J. Stachel, Quark-Gluon Plasma 3, World Scientific Publishing (2003)

[3] S.T. Butler, and C.A. Pearson, C. A., Phys. Rev. **129** 836-842 (1963)

[4] J.I. Kapusta, Phys. Rev. C **21** 1301-1310 (1980)

[5] S. A. Bass, M. Belkacem, M. Bleicher, M. Brandstetter, L. Bravina, C. Ernst, L. Gerland, M. Hofmann, S. Hofmann and J. Konopka, *et al.* Prog. Part. Nucl. Phys. **41**, 255-369 (1998)

[6] M. Bleicher, E. Zabrodin, C. Spieles, S. A. Bass, C. Ernst, S. Soff, L. Bravina, M. Belkacem, H. Weber and H. Stoecker, *et al.* J. Phys. G **25**, 1859-1896 (1999)

[7] M. Bleicher and E. Bratkovskaya, Prog. Part. Nucl. Phys. **122**, 103920 (2022)

[8] S. Sombun, K. Tomuang, A. Limphirat, P. Hillmann, C. Herold, J. Steinheimer, Y. Yan and M. Bleicher, Phys. Rev. C **99**, no.1, 014901 (2019)

[9] P. Hillmann, K. Käfer, J. Steinheimer, V. Vovchenko and M. Bleicher, J. Phys. G **49**, no.5, 055107 (2022)

[10] T. Reichert, J. Steinheimer, V. Vovchenko, B. Dönigus and M. Bleicher, to be published.

[11] V. Vovchenko and H. Stoecker, Comput. Phys. Commun. **244**, 295-310 (2019)

[12] V. Vovchenko, B. Dönigus, and H. Stoecker, Phys. Lett. B **785**, 171-174 (2018)

[13] V. Vovchenko, B. Dönigus, and H. Stoecker, Phys. Rev. C **100** 054906 (2019)

[14] J. Adam *et al.* [ALICE], Phys. Lett. B **754**, 360-372 (2016)

[15] M. Abdallah *et al.* [STAR], Phys. Rev. Lett. **128**, 202301 (2022)

[16] K. J. Sun, C. M. Ko and B. Dönigus, Phys. Lett. B **792**, 132-137 (2019)

[17] S. Acharya *et al.* [ALICE], Phys. Rev. Lett. **128**, 252003 (2022)

Development of Future Electromagnetic Calorimeter Technologies and Applications for the Electron-Ion Collider with GEANT4 Simulations

Zhaozhong Shi[1,*], *Craig* Woody[2], *Ian* Delk[3], and *John* Lajoie[3]

[1]Los Alamos National Laboratory, P.O. Box 1663, Bikini Atoll Road, SM-30, Los Alamos, New Mexico, USA 87545
[2]Brookhaven National Laboratory, PO Box 5000, Upton, NY USA 11973
[3]Iowa State University, 701 Morrill Road, Ames, IA USA 50011

Abstract. The Electron-Ion Collider (EIC) is a future collider planned to be built at BNL in about a decade. It will provide physicists with high luminosity and highly polarized beams with a wide range of nuclei species at different energies, covering an extensive kinematic range. The EIC physics goals include measuring the Generalized Parton Distribution (GPD) from Deeply Virtual Compton Scattering (DVCS) and Deeply Virtual Meson Production (DVMP) experiments, performing precision 3D imaging of the nuclei structure, studying color confinement and hadronization mechanisms, and understanding the spin structure of the proton. In order to meet the physics goals of EIC, a high-resolution electromagnetic calorimeter (EMCAL) is required to measure electrons and photons and to achieve good particle identification. We propose to develop a tungsten/shashlik (W/shashlik) EMCAL with better readout configuration to achieve better energy and position resolution. In this work, we will present the GEANT4 detector simulation results of W and Pb shashlik EMCAL to study π^0 merging probability as a function of π^0 energy and the performance of position and energy resolutions of the EMCAL for ECCE design.

1 Introduction

It is arguably believed that Quantum Chromodynamics (QCD) is the last frontier of the Standard Model. So far, there is still no first principle calculations yet available to precisely describe the basic properties of proton such as mass, spin, and charge radius. Therefore, the EIC, providing high luminosity and highly polarized beams with a variety of nuclei species, will be built to perform 3D tomography of the nucleons and nuclei by probing the transverse momentum distribution and generalized parton distribution function (GPD). Experimentally, GPDs could be accessed in DVCS and DVMP processes [?]. According to the EIC Yellow Report [?], in order to probe the kinematic regime of $10^{-5} < x < 0.7$, 1 GeV$^2 < Q^2 < 1000$ GeV^2, and $0 < |t| < 1.6$ GeV^2, the π^0 produced from DVMP should be reconstructed up to at least 50 GeV. Hence, due to the limited installation space, EIC experiments are required to have compact EMCALs with excellent energy resolution, excellent electron-photon separation performance, sufficient radiation hardness, and capabilities of detecting and identifying DVCS single photon and photon pairs from π^0 decays [?].

*e-mail: zhaozhongshi@lanl.gov

2 Shashlik EMCAL Design

We proposed to apply novel technologies in the development of the next generation shashlik sampling EMCAL for the EIC experiments. In the shashlik design, EMCAL plates with absorber material and scintillators are connected by wavelength shifting readout fibers. There are two design options: W and Pb shashlik EMCALs. The Pb shashlik design has the advantages of lower cost and easier to machine. One existing example is the PHENIX EMCAL [**?**]. We can refurbish and reuse it by adding more silicon photomultiplier (SiPM) readout to increase its segmentation. Moreover, the ECCE Collaboration proposed a Pb shashlik EMCAL design with very fine granularity to the call for the EIC Detector 1 [**?**].

On the other hand, W absorber material offers shorter radiation length and Moliére radius, allowing for compact EMCAL design, which is important for EIC experiments. However, W is generally more expensive and hard to machine. Thus, a softer alloy made of 80% W and 20% Cu as the absorber material of the EMCAL prototype was employed by the UTFSM group in Chile [**?**] and will be studied in this paper. The WCu EMCAL design with high granularity and individual fiber readout is a novel calorimeter technology. The technical details of the shashlik EMCAL design can be found in reference [**?**].

3 Detector Simulation Setup

In our design, the EMCAL towers are optically isolated and equipped with wavelength shifting fibers that read out the visible energy. We utilize *TracePro* [**?**], a commercial software that models the optics of light rays, to simulate the light collection efficiency of individual fibers within a tower through position scans with fine steps.

Using the GEANT4 simulation framework, we setup a testbed for the shashlik EMCAL in the EIC hadron going direction to model test team conditions. The EMCAL is a solid circular wall placed about three meters from the beam origin. We use a particle gun to generate the γ and π^0 and simulate photon EM showers in the PHENIX Pb, WCu, and ECCE Pb shashlik EMCALs with different readout granularities. In particular, since the ECCE EMCAL has a 2 $\times$ 2 fiber readout for each tower, we define several readout options: individual fiber readout (green), individual tower readout (blue), 2 $\times$ 2 towers (red), 3 $\times$ 3 towers (orange) and 4 $\times$ 4 towers (magenta), and 6 $\times$ 6 towers (cyan). The simulation details are documented in reference [**?**].

4 Analysis Results

We performed single photon simulations to study the position and energy resolutions of the WCu and ECCE EMCAL shown below in Figure **??**.

The ECCE EMCAL readout with individual fiber, individual tower, and 2 $\times$ 2 towers all have the same position resolution performance. As the granularity decreases to 3 $\times$ 3 towers readout, the position resolution is significantly worse due to position bias correction. For 4 $\times$ 4 towers readout, the position resolution is only slightly worse than 3 $\times$ 3 towers readout. The position resolution continues to drop as the readout size increases to 6 $\times$ 6 towers, which is larger than the Moliére radius of the ECCE EMCAL. The position resolution of the WCu EMCAL readout by tower (black) has similar performance to the ECCE tower.

The energy resolution does not depend on granularity. The ECCE EMCAL has a smaller constant term in energy resolution than the WCu EMCAL because its light collection efficiency is more uniform than the WCu one.

In addition, we studied the π^0 reconstruction and merging probability of the PHENIX and WCu EMCALs shown in Figure **??**.

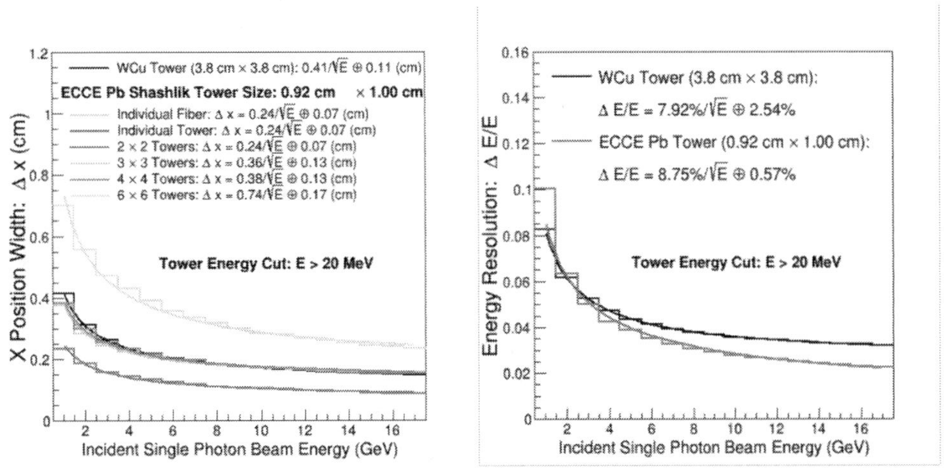

Figure 1. The position resolution of the WCu and ECCE EMCALs with different readout schemes (left) and the energy resolution of the WCu and ECCE EMCALs as well as their fits to the function $y = a/\sqrt{E} \oplus b$ are shown above. Individual fiber (green), individual tower (blue), and 2×2 towers (red) are overlapping with each other and only the 2×2 towers data points are seen.

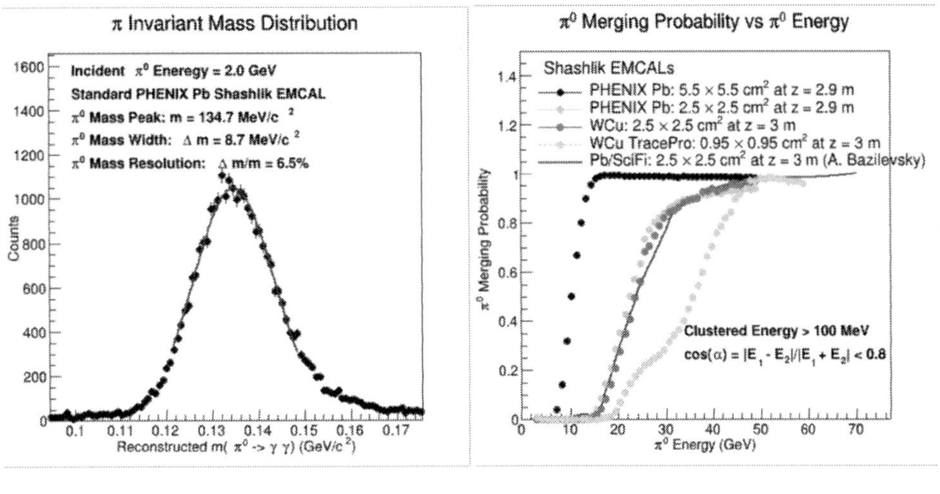

Figure 2. The invariant mass of π^0 reconstructed from the two photon (left) with the EMCAL and π^0 merging probability as a function of π^0 energy with different EMCAL designs (right) are shown above. Only cluster counting information is used to determine the π^0 merging probability. We did not look into the shower profiles of the clusters to distinguish a single photon or two photons.

After the single photon calibration, the shashlik EMCAL is able to reconstruct π^0 with the correct invariant mass (about 135 MeV/c^2) and good resolution. However, the PHENIX EMCAL (black) can only reconstruct π^0 up to about 10 GeV. When we increase the PHENIX EMCAL segmentation by a factor of 2 (orange), the EMCAL pushes the π^0 energy up to about 40 GeV. However, if we change the absorber material from Pb (orange) to WCu (red), which has a smaller Moliére radius, but keep the same granularity, the WCu EMCAL is only slightly better than the PHENIX Pb EMCAL. For reference, we compare our shashlik EMCALs to

Pb/SciFi EMCAL (blue) [**?**] with similar granularity and find that they have similar π^0 decay photon separation performance.

5 Summary

In summary, we have explored different Pb and WCu shashlik sampling EMCAL designs and studied their performance including the case that tower size is smaller than the Moliére radius. We also utilized the *TracePro* software to model the optics inside the tower and obtain the light collection efficiency maps.

A strong dependence of the readout granularity on the position resolution and π^0 merging probability is observed and found the significant improvement of the position resolution and the π^0 merging probability as the granularity increases. However, the dependence of Moliére radius of the EMCAL is weak because the position resolution of WCu EMCAL with individual towers readout (3.8×3.8 cm^2) and ECCE Pb shashlik EMCAL with 4×4 towers readout (3.68×4.0 cm^2) and the π^0 merging probability of WCu and PHENIX EMCALs at 2.5×2.5 cm^2 tower granularity are both similar. We are able to reconstruct π^0 up to 50 GeV with the WCu EMCAL if we read out individual fibers using only cluster counting without utilizing the shower shape information.

We also find that due to the lack of sensitivity to the Moliére radius in the ECCE EMCAL design, the tower readout size could be increased by a factor of 2, which would reduce the readout channels by a factor of 4, thus saving on cost. In addition, the energy resolution of ECCE EMCAL is better than the WCu EMCAL thanks to the better uniformity of its light efficiency collection map.

6 Acknowledgement

This work is supported by the United States Department of Energy Office of Science Graduate Student Research (SCGSR) Award and Los Alamos National Laboratory Laboratory Directed Research & Development (LDRD). We would also like to thank SQM 2022 conference organizers for providing us with opportunities to present this work.

References

[1] A. Ferrero (for the Compass Collaboration), "Study of DVCS and DVMP processes at COMPASS", J. Phys. Conf. Ser. 295, 012039 (2011)
[2] R. A. Khalek et. al., "Science Requirements and Detector Concepts for the Electron-Ion Collider: EIC Yellow Report", BNL-220990-2021-FORE, JLAB-PHY-21-3198, LA-UR-21-20953, March 8, 2021
[3] L. Aphecetche et. al. (PHENIX Collaboration), "PHENIX calorimeter", Nucl. Instrum. Meth. A 499, 521-536 (2003)
[4] ECCE Proto-Collaboration, "ECCE Calorimetry Note", ecce-note-det-2021-02, December 1, 2021
[5] C. Woody, "Report on W/SciFi and W/Shashlik Calorimeter R&D", EIC R&D Committee Meeting, July 11, 2019
[6] *TracePro* User Manual, https://www.lambdares.com/wp-content/uploads/TraceProDownload/TracePro_User_Manual.pdf
[7] Z. Shi, C. Woody, I. Delk, and J. Lajoie, "Development of Future Electromagnetic Calorimeter Technologies and Applications for the Electron-Ion Collider with GEANT4 Simulations", Deep Inelastic Scattering 2022 Conference Proceedings, https://doi.org/10.5281/zenodo.6856618, July 18 2022

[8] A. Brazilevsky, “EIC Calorimetry: YR Summary”, 1st EECE Workshop, February 11, 2021

Physics programme of the ALICE 3 experiment for the LHC Runs 5 and 6

Raphaelle Bailhache[1,*]

[1]Goethe-Universität, Max-von-Laue-Straße 1, 60438 Frankfurt am Main, Germany

Abstract. Ultrarelativistic heavy-ion collisions are used to study the physics of strongly interacting matter under extreme conditions, similar to those of the early universe. In such collisions a deconfined state of quarks and gluons, the quark-gluon plasma (QGP), is formed. Nuclear collisions at the LHC provide access to the highest-temperature, longest-lived experimentally accessible QGP. After three years of Long Shutdown and intensive installation of detector and accelerator upgrades, ALICE is about to take data at a peak Pb–Pb collision rate of 50 kHz to further characterize the properties of the QGP. Even after the ambitious scientific programme for the upcoming Runs 3 and 4, open questions will remain. Therefore, a next-generation LHC heavy-ion experiment ALICE 3 is proposed for the 2030s. It should give access to next-level measurements of electromagnetic probes and heavy-flavour hadrons, including multi-charm states and exotic hadrons, inaccessible in the LHC Runs 3 and 4.

1 Introduction

ALICE (A Large Ion Collider Experiment) is a general-purpose heavy-ion experiment at the CERN LHC. Its main goal is to study the properties of the quark-gluon plasma (QGP) created in such collisions. After a major upgrade of the detector to ALICE 2, the LHC Run 3, starting in 2022, will give further insight into the properties of the QGP [1]. Despite the rich planned scientific programme, fundamental questions will still remain open beyond Run 4, like the QGP properties driving its constituents to equilibration, the hadronisation mechanisms of the medium, the partonic equation of state and its temperature dependence, and the underlying dynamics of chiral symmetry restoration. To answer such questions, precision differential measurements of dileptons, systematic measurements of (multi-)heavy-flavoured hadrons down to low transverse momentum (p_T), and hadron interaction and fluctuation measurements are crucial. They require qualitative improvements in detector performance and statistics, calling for a next-generation heavy-ion experiment ALICE 3 [2].

2 ALICE 3 detector concept

A sketch of ALICE 3 is shown in Fig. 1. It consists of a compact all-silicon tracker with high-resolution vertex detector placed in a superconducting magnet system. Additional detectors are included to identify γ, $e^\pm$, $\mu^\pm$, $K^\pm$ and $\pi^\pm$ over a large pseudo-rapidity acceptance ($-4 < \eta < 4$). The full set-up could be installed inside of the current ALICE L3 magnet and

*e-mail: rbailhache@ikf.uni-frankfurt.de

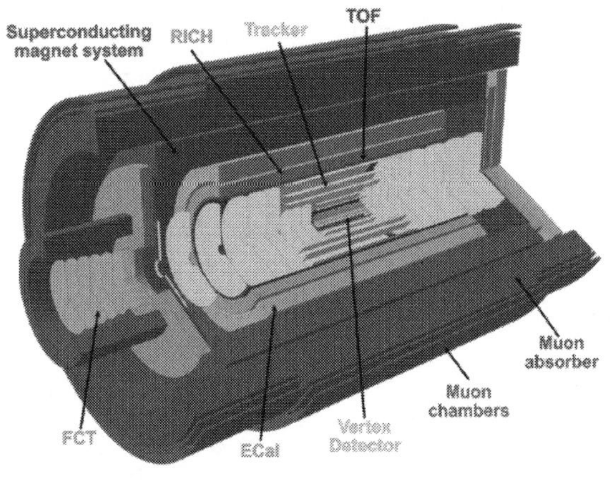

Figure 1. Sketch of the ALICE 3 set-up [2].

collect data during 6 years with one month per year of heavy-ion data taking. The inner tracker is made of three layers of curved wafer-scaled ultra-thin silicon sensors positioned inside the beam pipe [2], featuring an unprecedented low material budget. To provide the required pointing resolution, the first layer is placed at a radius of 5 mm from the beam axis during data taking, thanks to a retractable vertex detector. R&D challenges concern mechanical supports, cooling and radiation tolerance. To illustrate the overall improvement in detector capabilities, the left panel of Fig. 2 shows the expected detector pointing resolution and effective expected statistics of heavy-ion collisions for ALICE 1 (Runs 1 and 2), ALICE 2 (Run 3), ALICE 2.1 (Run 4) and ALICE 3 (Runs 5 and 6). Time-of-flight (TOF) detectors, ring-imaging Cherenkov (RICH) detectors, calorimeters (ECal), muon chambers and a forward conversion tracker (FCT) are foreseen to identify particles. Up to a few GeV/c, $e^{\pm}$/pi$^{\pm}$ separation will be provided with the TOF detectors, made of silicon sensors with $\approx$ 20 ps timing resolution, whereas for higher momentum the RICH detectors, using aerogel as radiator material, will complete the particle identification capabilities (see right panel of Fig. 2). The overall large total silicon surface requires cost-effective sensor development and automatisation of the detector production. More details on the foreseen technologies, the detector system and the R&D plans can be found in [2].

3 Physics goals and expected performance

3.1 Electromagnetic radiation

The hot medium created in heavy-ion collisions emits black-body radiation in form of real and virtual photons. The latter produce dilepton pairs. In Fig. 3, the expected invariant mass (m_{ee}) spectrum of thermal e^+e^- pairs produced in central Pb–Pb collisions at a center-of-mass energy per nucleon pair, $\sqrt{s_{NN}}$, of 5.02 TeV and measured with ALICE 3 after one month of data taking is shown [2]. To achieve such statistical and systematic precision, very good electron identification is needed down to low p_T ($\geq$ 80 MeV) with a small detector material budget in order to reduce the combinatorial background from γ conversion and π^0 Dalitz decays. Moreover, the excellent pointing resolution of ALICE 3 is decisive to suppress efficiently the very large background from heavy-flavour hadron decays at LHC energies, in particular for $m_{ee} \geq 1$ GeV/c^2 (see right panel of Fig. 3). In this m_{ee} region, the slope of the m_{ee} spectrum provides direct information about the temperature (T) of the QGP. Whereas

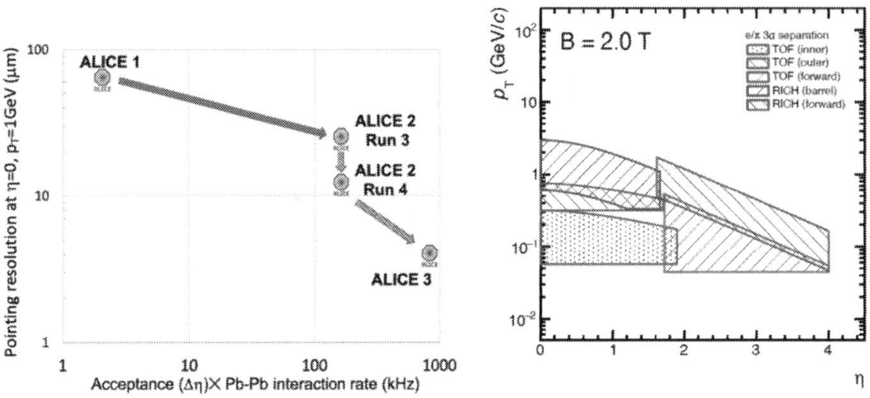

Figure 2. Left: Pointing resolution and effective expected statistics in heavy-ion collisions for ALICE 1, 2, 2.1 and 3 [2]. Right: 3σ separation between $e^{\pm}$ and $\pi^{\pm}$ with the sub-detectors in p_{T} and η [2].

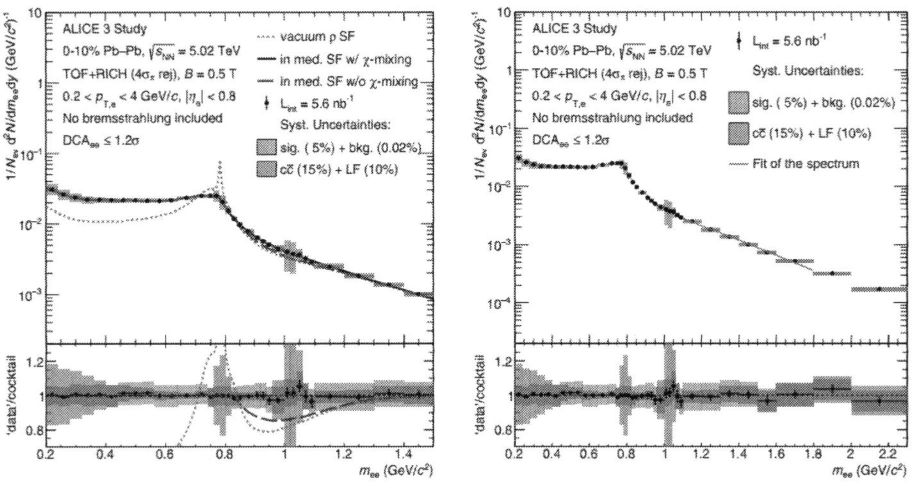

Figure 3. Simulated raw invariant mass spectrum of thermal dielectron from the QGP and hadron gas measured with ALICE 3 in central Pb–Pb collisions at $\sqrt{s_{\mathrm{NN}}} = 5.02\,\mathrm{TeV}$ after one month of data taking according to theoretical predictions from [3, 4] at low (left) and higher (right) masses [2].

first T measurements are expected with ALICE 2 (Runs 3 and 4), ALICE 3 will open up the possibility to probe the time dependence of T by extracting T as a function of mass or pair transverse momentum $p_{\mathrm{T,ee}}$, as shown on the left panel of Fig. 4. Similarly, the measurement of the thermal dielectron elliptic flow (labelled excess dielectron v_2 in the right panel of Fig. 4) would be for the first time statistically achievable with ALICE 3 as a function of m_{ee} and $p_{\mathrm{T,ee}}$. Such measurements probe the time evolution of the flow field and provide decisive inputs to the so-called photon puzzle, i.e. simultaneous understanding of the yield and v_2 of direct real photons also part of the ALICE 3 programme. Finally, the m_{ee} spectrum of thermal dielectrons at smaller m_{ee} ($m_{\mathrm{ee}} \leq 1.2\,\mathrm{GeV}/c^2$) is dominated by thermal production of ρ mesons. Their spectral function (SF) is modified in the hot medium where chiral symmetry restoration (CSR) is predicted. The expected high precision measurement of the m_{ee} spectrum

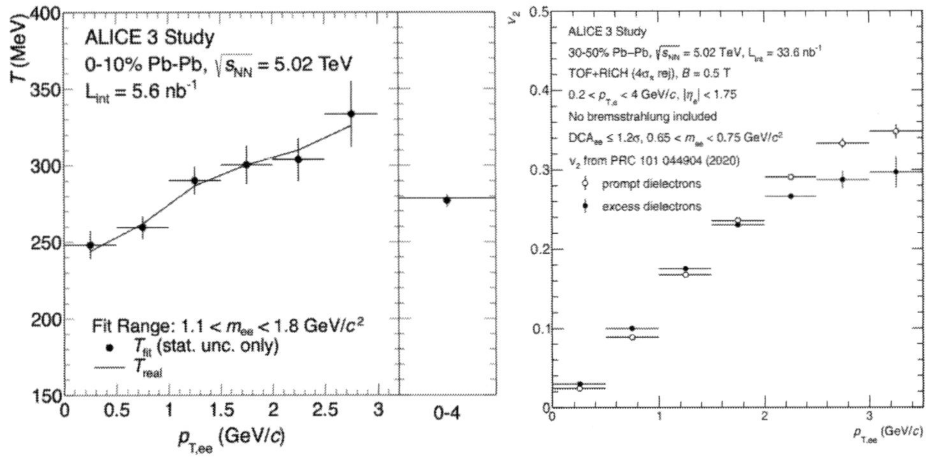

Figure 4. Left: Extracted temperature parameter T_{fit} from the simulated m_{ee} dielectron spectrum for different $p_{\text{T,ee}}$ intervals in central Pb–Pb collisions after one month of data taking with ALICE 3 [2]. Right: Projections for prompt and thermal dielectron v_2 measurements in $0.65 \le m_{\text{ee}} \le 0.75\,\text{GeV}/c^2$ with ALICE 3 in semi-central Pb–Pb collisions after six months of data taking [2].

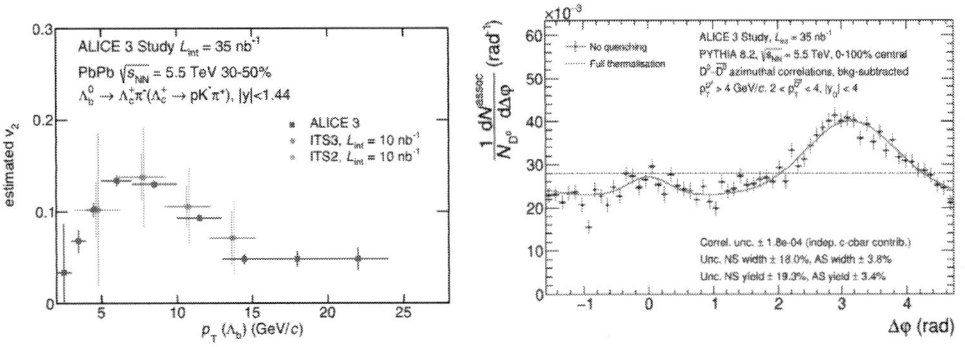

Figure 5. Left: Expected v_2 measurements of the Λ_b baryon with ALICE for Run 3 (ITS2), Run 4 (ITS3) and Runs 5 and 6 (ALICE 3) in semi-central Pb–Pb collisions [2]. Right: Simulated $D^0\bar{D}^0$ azimuthal correlation measurements with ALICE 3 in minimum-bias Pb–Pb collisions [2].

in central Pb–Pb collision with ALICE 3 (see left panel of Fig. 3) would allow the study of CSR mechanisms like $\rho - a_1$ mixing (denoted χ-mixing).

3.2 Heavy-quark transport and correlations

The (heavy-)quark transport properties in the QGP, i.e. the diffusion coefficients, can be extracted from precise nuclear modification factor and v_2 measurements of different charm and beauty hadrons down to low p_T to disentangle effects at the partonic level from those due to the hadronisation process. The left panel of Fig. 5 shows the expected performance for Run 3 (ITS2), Run 4 (ITS3) and Runs 5 and 6 (ALICE 3) for the Λ_b v_2 measurement. The particle identification capabilities, pointing resolution and acceptance of ALICE 3 give an unique access to the low p_T range.

The angular correlation of heavy-flavour hadron pairs is sensitive to the energy loss mechanisms and degree of heavy-quark thermalisation in the QGP, particularly at low p_T. In the

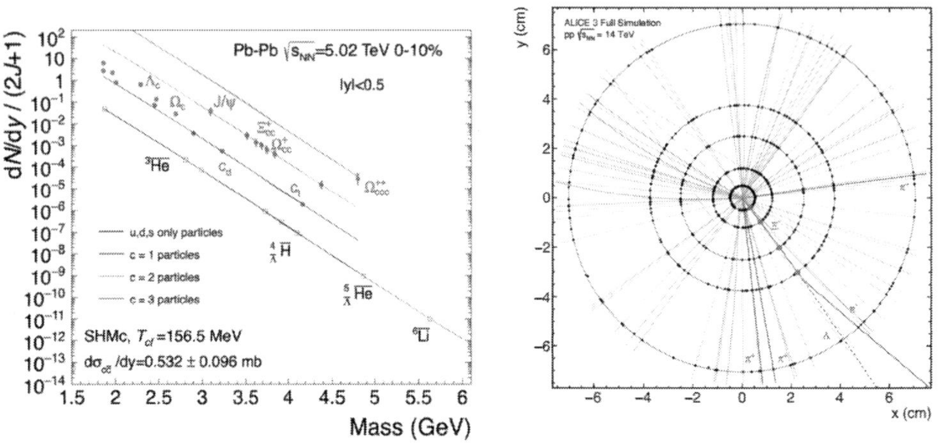

Figure 6. Left: Statistical-thermal model predictions for (anti-)(hyper-)nuclei (black) and (multi-)charm states (red) in central Pb–Pb collisions at $\sqrt{s_{\mathrm{NN}}}$ = 5.02 TeV [5]. Right: ALICE 3 strangeness tracking with the Ξ_{cc}^{++} decay into $\Xi_c^+ + \pi^+$ and the successive decay $\Xi_c^+ \rightarrow \Xi^- + 2\pi^+$ [2].

right panel of Fig. 5 the projection for $D^0\bar{D}^0$ azimuthal correlation measurement in minimum-bias Pb–Pb with ALICE 3 is presented for six years of data taking. Such measurements will not be in reach in heavy-ion collisions after Runs 3 and 4, since they require a high D^0 purity and reconstruction efficiency, as well as a large rapidity coverage.

3.3 Hadron formation

Multi-charm baryons are unique probes of the hadron formation mechanisms since they are expected to be produced essentially by combination of uncorrelated charm quarks in heavy-ion collisions. The yields of multi-charm baryons relative to the number of produced charm quarks are predicted to be significantly enhanced in heavy-ion relative to proton-proton collisions. In the statistical hadronisation model [5], the computed yield of hadron at a given mass scales with g_c^n, where g_c ($\approx$ 30) is the charm fugacity factor and n the number of constituent charm quarks in the state, as shown for central Pb–Pb collisions at the LHC on the left panel of Fig. 6. With ALICE 2, the measurement of (anti)-hypernuclei up to atomic number $A = 4$ and single-charm states should be in reach. A test of the full physical picture will be only possible with ALICE 3 by measuring additional states, i.e. hypernuclei with $A > 4$ and multi-charm states. With a first tracking layer at only 5 mm of the beam axis at mid-rapidity, strange baryons like Ξ^- can be directly tracked before they decay. This increases significantly the pointing resolution of the reconstructed strange baryon trajectory to the primary vertex, improving the ability to distinguish weak decays from primary or secondary sources. The latters are characteristic of multi-charm baryons, as depicted in the right panel of Fig. 6 with the tracking of the non-prompt Ξ^- strange baryon in the Ξ_{cc}^{++} decay chain. In the left panel of Fig. 7, the expected significances for Ξ_{cc}^{++} and Ω_{cc}^{+} in central Pb–Pb collisions are shown.

The nature of recently discovered exotic bound states, like T_{cc}^{+} [6], is still unclear. Molecular states or more compact objects like tetraquarks are both not excluded. Two particle momentum correlation measurement, e.g. between D^0 and D^{*+} for T_{cc}^{+}, can be used to search for possible DD bound states. The projections for such measurements with ALICE 3 in pp and Pb–Pb collisions for six years of data taking are shown in the right panel of Fig. 7. The very good pointing resolution and large acceptance of the detector would give the unique op-

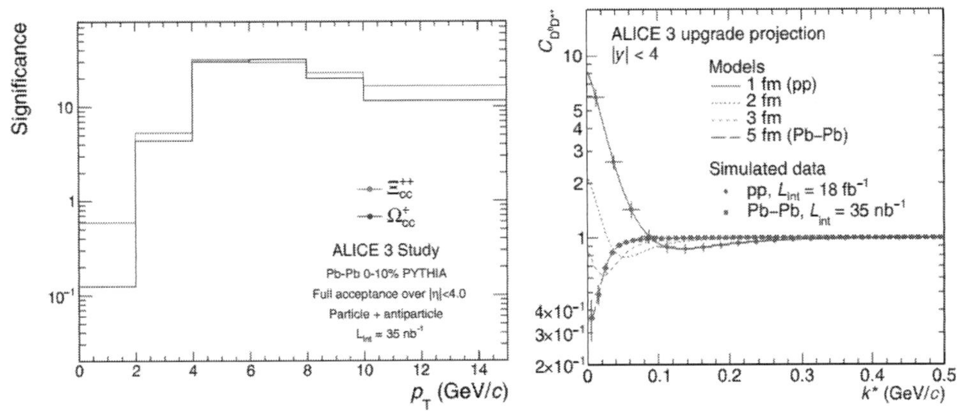

Figure 7. Left: Expected significance of Ξ_{cc}^{++} and Ω_{cc}^{+} measurements in central Pb–Pb collisions for six years of data taking with ALICE 3 [2] as a function of p_T; Right: Expected D^0D^{*+} momentun correlation measurements in Pb–Pb and pp collisions with ALICE 3 after six years of data taking in the presence of a bound state [2].

portunity to perform such studies and determine the properties of the DD interaction potential by looking at the behaviour of the correlation function in the different colliding systems.

3.4 Beyond QGP physics

More physics topics, not related to the QGP, are part of the ALICE 3 programme [2]. The detector and accelerator set-up would open up the possibility to search for axion-like particles (ALP) in ultra-peripheral Pb–Pb collisions with a mass between 0.05 and 5 GeV/c^2 [2]. In hadronic collisions, ultra-soft real and virtual photon productions remain not understood. Although the cross sections are computable using the Low-theorem, large excesses are observed by several experiments at lower $\sqrt{s}$, see [2] for an overview. A systematic study of very-soft γ production is proposed with ALICE 3 using a forward conversion tracker.

4 Conclusions

ALICE 3 is a next-generation multipurpose detector at the LHC, proposed as a follow-up to the present ALICE experiment. Its main goal is to unravel the remaining unknown microscopic dynamics of the strongly-interacting matter produced in heavy-ion collisions. An innovative detector concept is required to allow high precision measurements of electromagnetic probes and heavy-flavour hadrons including multi-charm states and exotic hadrons inaccessible in Runs 3 and 4.

References

[1] Z. Citron *et al*, CERN-LPCC-2018-07 (2018)
[2] ALICE Collaboration, CERN-LHCC-2022-009 (2022)
[3] R. Rapp, Adv. High Energy Phys. **2013**, 148253 (2013)
[4] P.M. Hohler and R. Rapp, Phys. Lett. B**731**, 103 (2014)
[5] A. Andronic *et al*, JHEP **07**, 035 (2021)
[6] LHCb Collaboration, Nature Physics **18**, 751 (2022)

Joint ATLAS/CMS ZDC upgrade project for the High Luminosity LHC

Riccardo Longo[1,*], on behalf of the ATLAS and CMS collaborations

[1]Department of Physics, University of Illinois, 1110 W. Green St., Urbana IL 61801-3080, USA

Abstract. The High Luminosity LHC (HL-LHC) provides the opportunity to study heavy ion, proton-nucleus, photon-nucleus and photon-photon collisions with unprecedented luminosities at the TeV scale. The LHC heavy ion community has mapped out an extensive range of physics measurements at the HL-LHC that will push forward our understanding of both QCD, QED and even electroweak physics. The measurement of forward neutrons and photons in Zero Degree Calorimeters (ZDCs) is essential for event classification and triggering. In order to reach the required luminosities, the LHC interaction regions will be redesigned, necessitating the need to build new ZDCs that will be both narrower and much more radiation tolerant. This challenge motivated the formation of a joint project between ATLAS and CMS to build new ZDCs for Run 4, JZCaP. The ZDCs are based on radiation-hard fused silica rods that produce Cherenkov light. These rods have been developed by Heraeus Quartzglas in collaboration with JZCaP and the LHC BRAN and FLUKA groups. The Run 4 ZDCs (HL-ZDCs) are the first joint detector project between CMS and ATLAS. This talk will present the capabilities of the new ZDCs and recent R&D highlights.

1 Introduction

Zero Degree Calorimeters (ZDCs) played a central role in the success of the Heavy Ion (HI) physics programs of the ATLAS [1] and CMS [2] experiments during LHC Run 1 and 2. The ZDCs detect far-forward neutral particles produced in A+A and p+A collisions, which are mostly spectators from the interactions, enabling the possibility to distinguish between different classes of physics processes by looking at forward neutron topologies. Central and mid-central A+A collisions are characterized by the detection of several spectator neutrons in both arms of the ZDC (XnXn topology). In peripheral and ultra-peripheral (UPC) A+A collisions and p+A collisions, the absence of neutrons in one (Xn0n or 0nXn) or both of the ZDCs (0n0n) typically indicates that the corresponding nucleus likely did not interact hadronically and, instead, emitted a coherent photon or coherent pomeron. The measurement of the single-neutron peak is the primary tool used to calibrate the energy scale of the calorimeter (with two- and three-neutron peaks complementing the calibration). Therefore, the ability to resolve the 1n and subsequent peaks is essential for achieving a well-controlled energy scale.

In both the ATLAS (IP1) and CMS (IP5) interaction regions, the ZDC is installed in a dedicated slot in the Target Absorber for Neutrals (TAN). The TANs are radiation absorbers

*e-mail: riccardo.longo@cern.ch

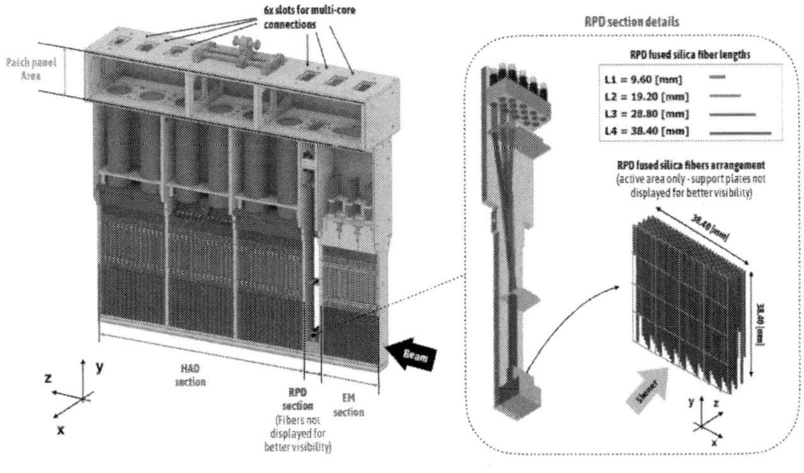

Figure 1. Left: HL-ZDC detector design. Right: detailed view of the RPD section.

aimed at shielding the downstream superconducting magnets from forward neutral particles emerging from the IP. The high luminosity upgrade of the LHC, HL-LHC, will necessitate substantial modifications to the LHC optics. The recombination chamber will be moved 13 m closer to IP1 and IP5 and, as a consequence, the existing TAN will be replaced by a new absorber, the TAXN [3], that will require a substantially-reduced separation between the beam pipes, resulting in only 4.6 cm available in the transverse direction for the implementation of the upgraded ZDC (HL-ZDC). The higher collision rate and the closer position to the IP will also increase the radiation levels in the TAXN. Thanks to detailed simulations provided by the CERN FLUKA group, it was possible to estimate the peak dose that will be accumulated by the HL-ZDC. Taking into account all the HI-related data taking foreseen in Run 4, a peak dose of 4.5 MGy is expected to be registered in the detector, in correspondence with the active area. Therefore, the identification of a radiation-hard Cherenkov radiator is a crucial step in the finalization of the HL-ZDC design.

The harsh radiation environment of the HL-LHC also introduces radio-protection (RP) implications to be taken into account while designing the HL-ZDC and planning its installation. After running dedicated simulations, the CERN RP team has estimated an ambient dose equivalent rate of $\sim$ 800 μSv/h near the TAXN after a week of technical stop [4], the typical cool-down time preceding the ZDC installation. Such a high dose equivalent rate imposes a detector design that is easy to handle and connect, in order to minimize the exposure for personnel involved in the operations.

The HL-ZDC design will also include a common Reaction Plane Detector (RPD) section. By mapping the transverse profile of the neutron showers, the RPD will be capable of measuring the direction of the reaction plane on an event-by-event basis, enabling a set of collective flow measurements carried out using the same detector in both ATLAS and CMS.

2 HL-ZDC design

A three-dimensional view of the current design of the HL-ZDC is shown in Figure 1. Contrary to the existing ZDCs, the HL-ZDC is characterized by a single-unit structure to facilitate efficient detector handling during the installation procedure. Starting from the interaction point (IP) side and proceeding along z, one can identify three different sections:

- Electromagnetic (EM) section.

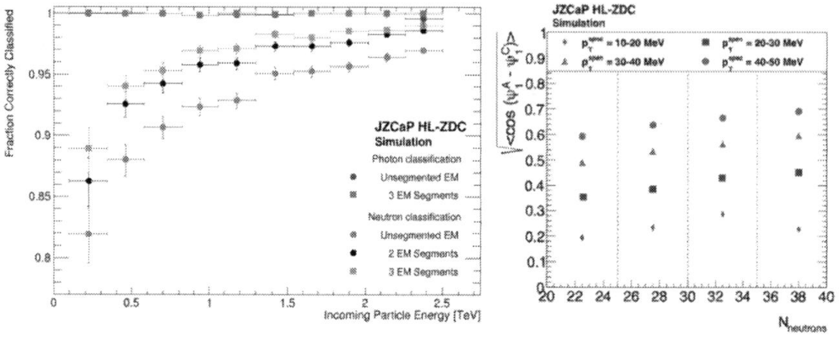

Figure 2. Left: HL-ZDC detector design. Right: detailed view of the RPD section.

- Reaction Plane Detector (RPD) section.
- Hadronic (HAD) section.

In all three functional components, detection is based on Cherenkov light emission in high-purity, ultra-radiation-hard fused silica material. The EM and HAD sections are calorimeters with different sampling ratios, while the RPD consists of an array of fused-silica fibers of different lengths that map the transverse profile of the shower produced within the EM module.

The EM section is the first part of the HL-ZDC encountered by particles originating in the IP. The active region consists of layers of fused silica rods (25 each of 1.5 mm diameter and 275 mm height), alternating with 4 mm thick W plates. The EM is equipped with 25 W plates, corresponding to $\sim$ 30 radiation lengths, and 1 nuclear interaction length. Above the tungsten plates, the fused silica rods are guided by stainless steel plates located atop the tungsten. The HL-ZDC EM module light signal is read out in 4 horizontal and 3 longitudinal segments. The segmentation is implemented via Winston-Cone-shaped [5] air light guides that collect and re-direct the light generated in each element toward the corresponding Photomultiplier tube (PMT). The segmentation in the longitudinal direction enhances photon-neutron discrimination, as shown in Figure 2 left panel. By contrast, the horizontal sections enable locating the beam position in the case of horizontal crossing angles.

The RPD is sandwiched between the EM and HAD section, and is characterized by a "Pan Flute" geometry (see Figure 1, right panel), proposed to optimize the light-collection mechanism and to have a detector core that is fully radiation-hard. This implementation is analogous to the ATLAS Run 3 RPD, recently constructed and tested with the beam at the CERN SPS. The RPD active region is comprised of 256 radiation-hard fused silica fibers arranged as a virtual 4×4 array of square tiles, each with an area of 9.60 mm^2. Since the structure of the RPD signals can be interpreted as pixels of an image, machine learning (ML) based pattern recognition techniques are planned to reconstruct the RP. The RPD will provide large data samples characterized by high dimensionality and complex dependencies between inputs and outputs. These conditions are suitable for applying deep learning techniques, such as convolutional neural network (CNN). The right panel in Figure 2 shows the RP resolution achievable by applying CNN reconstruction algorithms to the RPD signals, estimated by means of MC simulations. The RP reconstruction performance is analyzed as a function of the neutron multiplicity detected in the ZDC and of the transverse momentum kick ($p_T^{\rm spec}$) imparted to the spectators during the collision. In all the cases considered, it will be possible to achieve a meaningful measurement of the RP.

The majority of the HL-ZDC length ($\sim$ 530 mm) is dedicated to the HAD section, comprised of a total of 4.5 nuclear interaction lengths divided into three sub-modules. Each sub-module consists of 15 layers of 1.5 mm diameter fused silica rods sandwiched between

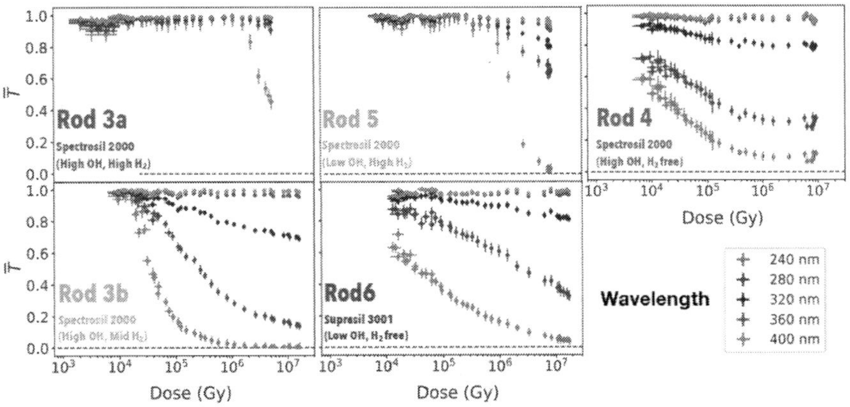

Figure 3. Optical transmission of different fused silica materials as a function of dose received in the TAN. Each panel corresponds to a fused silica material characterized by different levels of OH and H_2 dopants. Results are displayed from 240 nm (lower limit of the measurement) to 400 nm. As one can note, negligible losses in transmittance are observed for all the materials when approaching this wavelength. For this reason, results from 400 nm to 1500 nm (upper limit of the measurement) are not included at this point. Rod 3a, characterized by high concentrations of OH and H_2 dopants, exhibits a mostly perfect optical transmission up to the MGy scale.

10 mm tungsten plates. The fused silica rods terminate at trapezoidal air light guides that direct the Cherenkov photons to the closest PMT. Each HAD sub-module is equipped with two light guides and two PMTs, resulting in a total of six PMT channels to be read out. Thanks to the greater depth compared to the existing detectors (5.5 vs 4.5 λ_{int}), the energy resolution of the HL-ZDC is expected to be better than those characterizing the Run 3 detectors [3], despite the narrower transverse dimensions.

3 R&D highlights

The unprecedented collision rate that will be reached by the HL-LHC poses radiation hardness challenges that are common for all forward detectors installed in the TAXN, such as the ZDC and the BRAN luminosity monitor [6]. For this reason, JZCaP has started an R&D effort with the CERN BRAN and FLUKA teams to carry out radiation hardness studies on fused silica materials irradiated in the LHC. The first irradiation campaign took place in Run 2, when seven 40 cm long fused-silica rods, characterized by different material compositions, were inserted in a Run 4 BRAN prototype installed in the TAN. Thanks to detailed FLUKA simulations transporting collision products from the center of the ATLAS detector hosted in the experimental cavern up to the TAN [7], it was possible to estimate the dose accumulated in the BRAN rods, which spans over four orders of magnitude. Therefore, by cutting the rods into 1 cm segments, it was possible to form sub-sets of 40 samples, each characterized by the same material composition and different irradiation levels. Figure 3 shows preliminary results of this analysis which correlate transmittance over a wide wavelength range with the dose accumulated in each sample. The results, currently being drafted for publication, already informed the choice of material for the ATLAS Run 3 ZDC refurbishing. For the sake of the same refurbishing campaign, a dedicated polishing procedure was developed to improve the transmission of light from the rods to the light guide, see [8] for more details. The refurbished ATLAS ZDC was operated successfully for the first time during the 2021 pilot run [9].

Acknowledgements

This work is supported by the National Science Foundation, Grant no. PHY-2111046.

References

[1] ATLAS Collaboration, JINST **3** (2008), S08003
[2] CMS Collaboration, JINST **3** (2008), S08004
[3] P. Santos Díaz, F. Sanchez Galan, R. De Maria, E. Kepes, P. Steinberg, R. Longo, S. Mazzoni, F. Cerutti, M. Sabate Gilarte and A. Infantino, *et al.* Phys. Rev. Accel. Beams **25** (2022) no.5, 053001
[4] A. Infantino, EDMS 2331769 (public), 2020
[5] Winston, R. and Welford, W. T. New York: Academic Press, 193-196, 1978
[6] H. S. Matis, M. Placidi, A. Ratti, W. C. Turner, E. Bravin and R. Miyamoto, Nucl. Instrum. Meth. A **848** (2017), 114-126 [arXiv:1612.01238 [physics.acc-ph]].
[7] S. Yang, M. S. Gilarte, A. Tate, N. Santiago, R. Longo, S. Mazzoni, F. Cerutti, E. Bravin, M. G. Perdekamp and G. Lerner, *et al.* [arXiv:2204.01937 [physics.acc-ph]].
[8] ATLAS Collaboration,ATL-PHYS-SLIDE-2022-071, https://cds.cern.ch/record/2806543
[9] ATLAS Collaboration, ATL-COM-FWD-2022-006, https://cds.cern.ch/record/2805122?

Status and Performance of sPHENIX Experiment

Hideki Okawa[1]* for the sPHENIX Collaboration

[1]Institute of Modern Physics and Key Laboratory of Nuclear Physics and Ion-beam Application (MOE), Fudan University, Yangpu, Shanghai, China

Abstract. sPHENIX is a state-of-the-art experiment at the Relativistic Heavy Ion Collider. Hard probes are to be measured in $p+p$, $p+$Au and Au+Au collisions to understand the properties of the quark-gluon plasma. sPHENIX covers a wide range of physics programs: jet correlations and substructure, upsilon spectroscopy, open heavy flavor and cold quantum chromodynamics. It provides unique opportunities in the low transverse momentum region and also offer kinematic overlap with the Large Hadron Collider experiments. In these proceedings, the scientific mission of sPHENIX, detector design and key performance parameters are described, and expected projections of some measurements under consideration are presented.

1 Introduction

The ultra-relativistic nuclear collisions at the Relativistic Heavy Ion Collider (RHIC) and the Large Hadron Collider (LHC) have revealed the existence of a new form of matter called the quark-gluon plasma (QGP) [1]. While the macroscopic properties of the QGP have been measured in detail, such properties are not apparent from the fundamental interactions governed by the quantum chromodynamics (QCD). The sPHENIX experiment will provide a crucial insight to one of the two highest priorities in Hot QCD: to understand the inner workings of the QGP via measurements of hard probes in heavy-ion collisions at RHIC [2].

sPHENIX was proposed as the upgrade and replacement of the PHENIX experiment in 2010. Detailed physics cases and detector design proposals have been developed in the following years [3, 4]. The sPHENIX Collaboration was formed in early 2016, and has more than 360 members from 82 institutions in 14 countries as of 2022. The detector construction is nearly complete, and the first data taking will take place in early 2023.

2 sPHENIX Physics

The scientific mission of sPHENIX is to probe the QGP at multiple length scales and can be achieved with three years of operation. Table 1 presents an overview of data that are expected to be obtained in the first three years [5]. The run plan is consistent with the currently envisioned schedule of the Electron Ion Collider (EIC).

More than 50% of the Year-1 (2023) data are planned to serve for the commissioning of the detector and its operation as well as to validate the calibration and reconstruction. Year-2

*e-mail: hideki.okawa@cern.ch

Table 1. Summary of the sPHENIX Beam Use Proposal for years 2023–2025. The values in parentheses correspond to the 28 cryo-week scenario.

Year	Species	$\sqrt{s_{NN}}$ [GeV]	Cryo Weeks	Physics Weeks	Rec. Lum. $\|z\| < 10$ cm	Samp. Lum. $\|z\| < 10$ cm
2023	Au+Au	200	24 (28)	9 (13)	3.7 (5.7) nb^{-1}	4.5 (6.9) nb^{-1}
2024	$p^{\uparrow}p^{\uparrow}$	200	24 (28)	12 (16)	0.3 (0.4) pb^{-1} [5 kHz] 4.5 (6.2) pb^{-1} [10%-*str*]	45 (62) pb^{-1}
2024	$p^{\uparrow}$+Au	200	–	5	0.003 pb^{-1} [5 kHz] 0.01 pb^{-1} [10%-*str*]	0.11 pb^{-1}
2025	Au+Au	200	24 (28)	20.5 (24.5)	13 (15) nb^{-1}	21 (25) nb^{-1}

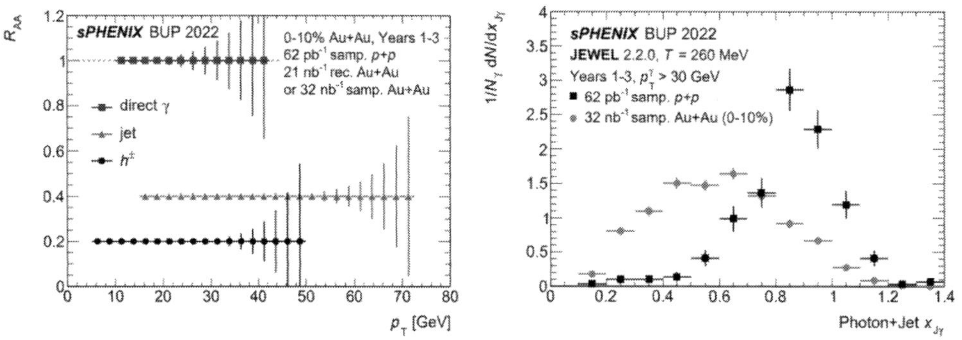

Figure 1. Statistical projections of the nuclear modification factor R_{AA} for photons, jets and charged hadrons (left) and the jet-photon momentum balance $x_{J\gamma}$ (right).

(2024) will collect large datasets of p+p and p+Au collisions, which are the critical references for the heavy ion physics and will also offer new opportunities for the cold QCD. Year-3 (2025) will collect a significantly large dataset of Au+Au collisions, which will provide unprecedented statistical precision in jet and heavy flavor measurements. With the three years of operation, a total of 141 billion Au+Au events is expected to be recorded. If any opportunity arises, additional runs would continue to recoup benefits from the sPHENIX investment [5].

sPHENIX has four main physics programs: (1) jet structure to probe various momentum and angular scales, (2) upsilon spectroscopy to vary the size of the probe, (3) open heavy flavor to probe with various momentum scales and parton masses, and (4) cold QCD to study proton spin, transverse momentum, and cold nuclear effects. The high data rates, precision trackers and hermetic calorimeters offer access to a wide p_T range of up to around 50 GeV for hadrons, 40 GeV for photons and 70 GeV for jets (Figure 1 (left)). sPHENIX will offer kinematic overlap with the LHC, which is possible for the first time at RHIC.

2.1 Jet structure

The jet-photon momentum balance $x_{J\gamma}$ is a “flag-ship” observable at sPHENIX (Figure 1 (right)). It is expected to have a dramatic difference between the LHC and RHIC energies [6], and such a comparison would only be possible with sPHENIX.

The jet substructure will provide another probe to measure the QGP properties and an insight to the connection to the fundamental QCD. For example, the groomed momentum sharing will allow for exploring the parton shower development in the QGP.

sPHENIX has unique opportunities for the jet measurements, being able to precisely measure down to the low-p_T region, which is challenging at the LHC. The dependence of the

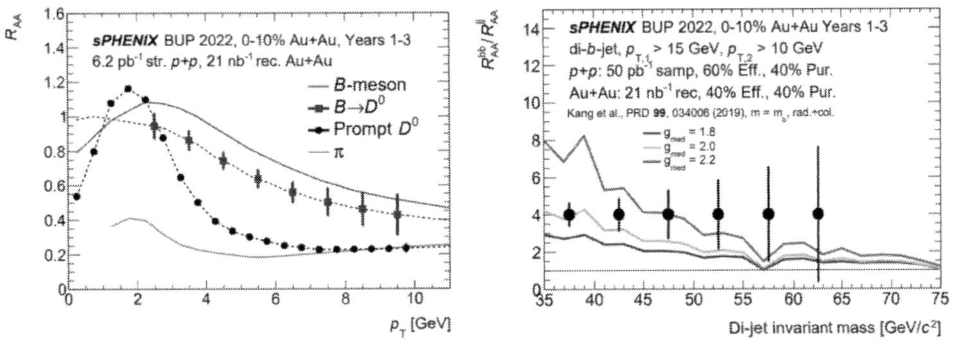

Figure 2. Statistical projections for the R_{AA} measurements of prompt/non-prompt D^0 mesons as a function of p_T (left) and of the b-jet-light-jet-super-ratio as a function of the dijet invariant mass (right).

nuclear modification factor R_{AA} on the jet distance parameter R^{jet} will probe the interplay of the out-of-cone energy loss and the medium response. Currently, there is a tension between the ALICE and ATLAS measurements. The elliptic flow v_2 is a crucial observable to understand the properties of the QGP, but modeling it simultaneously with the suppression has not been successful so far in most cases. Its precision measurement at sPHENIX may provide an insight in further understanding the QGP.

2.2 Open heavy flavor & parton energy loss

The heavy-flavor quarks are unique probes as their initial productions can be predicted from the perturbative QCD, and they are conserved from the initial hard scattering, experiencing the full evolution of the QGP.

A precise measurement of non-prompt D^0 suppression will be possible, thanks to the performance of the Monolithic Active Pixel Sensor Vertex Detector (MVTX). Furthermore, the streaming readout allows for collecting an enormous amount of minimum bias data for unbiased heavy-flavor mesons down to almost p_T=0 as shown in Figure 2 (left). The determination of b-quark R_{AA} and v_2 will provide clean access to the diffusion at RHIC.

The MVTX and full calorimeter implementation will offer the first b-jet tagging at RHIC. The current b-jet tagging algorithms exploiting the track distance of closest approach (DCA) and the secondary vertices respectively, show compatible performance to CMS [7, 8]. sPHENIX provides outstanding precision in the low-jet-p_T region, which is challenging at the LHC. In addition, making use of the di-b-jet mass for the R_{AA} measurement enhances the sensitivity to the transport property, and taking the ratio to the light-jet-pair mass will particularly offer strong sensitivity to the parton mass effect (Figure 2 (right)). Sufficient statistics at sPHENIX will also allow for measuring the b-jet substructure [5, 9], which is another approach to quantify the role of the parton mass.

2.3 Upsilon spectroscopy

Measuring the centrality and p_T dependence of the nuclear modification factor of the upsilons is crucial to compare with the LHC. sPHENIX can reconstruct the upsilon states with excellent mass resolution $\delta M < 125$ MeV using the electrons. Recently, the Υ(3S) state has been observed for the first time in the Pb-Pb collisions by CMS [10]. sPHENIX has the unique opportunity to precisely measure the three separated states at RHIC (Figure 3 (left)).

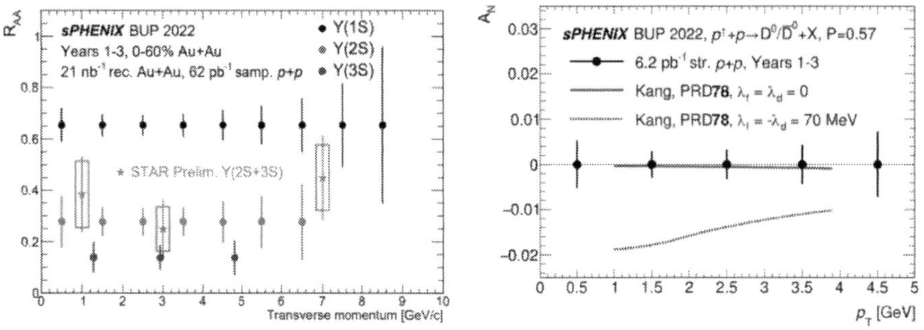

Figure 3. Statistical projections for the upsilon R_{AA} as a function of $p_{\rm T}$ (left) and the transverse spin asymmetry (A_N) for the D^0 mesons (right).

2.4 Cold QCD

The capability of sPHENIX to pursue measurements with jets, photons and heavy-flavor particles provides exciting programs for the cold QCD as well. In p+p collisions, the transverse single spin asymmetry (TSSA) can be accessed via prompt photons and D^0 mesons (Figure 3 (right)), which would deepen understanding about the gluon dynamics in transversely polarized nucleons with the tri-gluon correlations. In p+Au collisions, measuring the nuclear dependence of TSSA will offer insight to its origin with much improved precision from the PHENIX experiment.

3 sPHENIX detector

sPHENIX is the first new detector at RHIC at Brookhaven National Laboratory in over 20 years of operation (Figure 4). It has the capability to utilize the full delivered luminosity of RHIC with the high data rates of 15 kHz for all the subdetectors and large hermetic coverage of 2π in azimuth and pseudo-rapidity of $|\eta|<1.1$. The trigger capability includes the streaming readout, which will be described in Section 3.4. The sPHENIX detector consists of a precision tracking system, electromagnetic and hadronic calorimeters, and minimum bias and event plane detectors. The 1.4 T superconducting solenoid from the BaBar experiments surrounds the tracking system, electromagnetic and inner hadronic calorimeters. The sPHENIX detector will allow for full, unbiased jet reconstruction, b-jet tagging, and measurements of the three upsilon states separating in both Au+Au and p+p collisions.

3.1 Tracking Detectors

The tracking system consists of four components: MVTX, Intermediate Silicon Tracker (INTT), Time Projection Chamber (TPC) and TPC Outer Tracker (TPOT). The MVTX, the inner-most tracking detector (2.3 < radius < 3.9 cm), provides the precision vertexing. It has three layers of MAPS using the ALICE Pixel Detector (ALPIDE) [11]. The DCA resolution in xy and z reaches below 40 μm at $p_{\rm T}$>0.5 GeV (Figure 5 (left)), which is crucial for the open heavy flavor program. The INTT consists of two layers of silicon strips with 86 μm in pitch. It has a fast integration time of O(100ns), which can resolve one beam crossing and provide the pileup suppression. The TPC is based on Gas Electron Multiplier and has 48 layers with gateless and continuous readout. It has an outer radius of 78 cm, leading to only 1/30 of the ALICE TPC volume. The tracking efficiency is around 90% for the p+p collisions (Figure 5

Figure 4. The engineering drawing of the sPHENIX detector with its support structure.

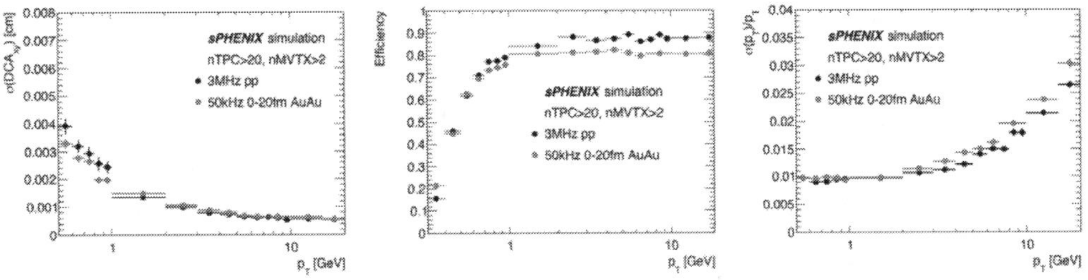

Figure 5. Tracking performance for the DCA resolution in the transverse direction (xy) (left), reconstruction efficiency (middle), and the p_{T} resolution (right) as a function of track p_{T}.

(middle)) and the p_{T} resolution is below 2% for p_{T}<10 GeV (Figure 5 (right)), which allows for measuring and discriminating the three upsilon states. The TPOT has eight modules of Micromegas inserted between the TPC and the electromagnetic calorimeter (EMCal). It offers calibration of beam-induced space charge distortions.

3.2 Calorimeters

The calorimeters consist of the EMCal, inner hadronic calorimeter (IHCal), and outer hadronic calorimeter (OHCal) with a common light collection and readout using silicon photomultipliers. The EMCal and IHCal are housed inside the solenoid.

The EMCal provides the electron identification for the upsilons and the photon measurements. It adopts a Tungsten-scintillating fiber sampling technique with a sampling fraction of 2.3%. It has a highly granular segmentation of $\Delta\eta \times \Delta\phi = 0.025 \times 0.025$. The energy resolution has been measured in testbeams to be less than $16\%/\sqrt{E} \oplus 5\%$ [12, 13].

The Inner (Outer) HCal exploits aluminum (steel) absorber plates and scintillating tiles with embedded wavelength shifting fibers. The OHCal is also used as a return yoke for the solenoidal magnetic field. Both the IHCal and OHCal have a granular segmentation of $\Delta\eta \times \Delta\phi = 0.1 \times 0.1$. The energy resolution of the overall HCal has been measured in a testbeam [12] and meets the requirement for the jet energy resolution to be around $20\%/\sqrt{E}$.

3.3 Minimum Bias and Event Plane Detectors

The Minimum Bias Detector (MBD) is a reuse of the PHENIX Beam-Beam Counter, consisting of 3 cm-thick quartz radiator on the mesh dynode photomultipliers. It has the timing resolution of 120 ps.

The sPHENIX Event Plane Detector (sEPD) exploits 1.2-cm-thick scintillators with wavelength shifting fibers. It consists of two wheels of 12 sectors with 31 optically-isolated tiles. The sEPD provides significant improvement in the event plane resolution: $\mathrm{Res}(\Psi_2) = 0.14$ (MBD only) and 0.55 (with sEPD).

3.4 Hybrid Data Acquisition Structure

sPHENIX adopts a hybrid system of a streaming readout from the inner trackers (MVTX, INTT, and TPC) and the calorimeter triggers for the high-p_T objects. The streaming readout is a triggerless configuration recording 10% of all collisions. It will increase the amount of minimum bias data in the $p+p$ and $p+$Au collisions by orders of magnitude. It is crucial for track-only measurements in open heavy-flavor physics as well as cold QCD.

4 Conclusions

sPHENIX is a state-of-the-art experiment at RHIC. Hard probes are to be measured in $p+p$, $p+$Au and Au+Au collisions to understand the properties of the QGP and cover a wide range of physics: jet correlations and substructure, upsilon spectroscopy, open heavy flavor and cold QCD. sPHENIX provides unique opportunities in low-p_T measurements and also offer kinematic overlap with the LHC experiments.

sPHENIX has entered the detector construction phase since September 2019. The detector construction and installation as well as the data taking preparation are on schedule. The detector will be ready for commissioning with beam in early 2023, and sPHENIX will complete the scientific mission of RHIC.

References

[1] W. Busza, K. Rajagopal, W. van der Schee, Ann. Rev. Nucl. Part. Sci. **68**, 339 (2018)
[2] A. Aprahamian *et al.*, Reaching the horizon: The 2015 long range plan for nuclear science (2015)
[3] A. Adare *et al.*, sPHENIX Science Proposal (2014), `arXiv:1501.06197`
[4] sPHENIX Collaboration, sPHENIX Technical Design Report (2019), `https://indico.bnl.gov/event/7081`
[5] sPHENIX Collaboration, sPHENIX Beam Use Proposal, sPH-TRG-2022-001 (2022), `https://indico.bnl.gov/event/15845`
[6] W. Dai, I. Vitev, B.W. Zhang, Phys. Rev. Lett. **110**, 142001 (2013)
[7] sPHENIX Collaboration, sPHENIX MVTX proposal, sPH-HF-2018-001 (2018), `https://indico.bnl.gov/event/4072`
[8] CMS Collaboration, Phys. Rev. Lett. **113**, 132301 (2014)
[9] H.T. Li, I. Vitev, Phys. Lett. B **793**, 259 (2019)
[10] CMS Collaboration, CMS Note, CMS-PAS-HIN-21-007 (2022)
[11] ALICE Collaboration, J. Phys. G: Nucl. Part. Phys. **41**, 087002 (2014)
[12] C.A. Aidala *et al.*, IEEE Trans. Nucl. Sci. **65**, 2901 (2018)
[13] C.A. Aidala *et al.*, IEEE Trans. Nucl. Sci. **68**, 173 (2021)

The physics program of the NA60+ experiment

Alessandro De Falco[1,*] for the NA60+ Collaboration.

[1]Università/INFN Cagliari, Italy

Abstract.
The NA60+ experiment, which has been proposed as a fixed target experiment at the CERN SPS, is designed to study the phase diagram of the strongly interacting matter at high baryochemical potential by performing precision studies of thermal dimuons, heavy quark and strangeness production in Pb-Pb collisions at center-of-mass energies ranging from 5 to 17 GeV. The progress of the R&D and the key points of the NA60+ physics program will be described.

1 Introduction

One of the key points in the understanding of the QCD under extreme conditions is the exploration of its phase diagram. Lattice QCD calculations predict a cross-over transition from the hadronic matter to the Quark-Gluon Plasma (QGP) at vanishing baryochemical potential μ_B around a critical temperature $T_c \approx 155$ MeV, while at large μ_B values a first order phase transition is expected. A strong experimental program is being carried out since two decades at the CERN-SPS, BNL-RHIC and CERN-LHC at top energies to characterize the region around $\mu_B \sim 0$, showing that a deconfined state of matter is formed, with properties consistent to the predictions from lattice QCD. More recently, interest in the experimental study of the large μ_B region increased. The search of a critical point in the phase diagram, the order of the phase transition and of the properties of the medium at large μ_B, the chiral symmetry restoration effects and the temperature at which the onset of the deconfinement takes place constitute fundamental issues in the understanding of the phase diagram properties.

The target of the NA60+ experiment is to study electromagnetic and hard probes in the range $200 < \mu_B < 400$ MeV via an energy scan at the SPS in the range $6 < \sqrt{s} < 17$ GeV. Electromagnetic probes, and in particular dileptons, will allow to determine the temperature of the system via a measurement of the thermal dimuon mass spectrum. Also chiral symmetry restoration effects related to the modification spectral function of the ρ and its chiral partner a_1 can be investigated. Hard probes bring information on the onset of deconfinement via the measurement of J/ψ suppression versus CM energy, while the production of open charm hadrons can provide information on the transport properties of the QGP.

2 The experimental apparatus

The detector concept of NA60+ is based on the design of NA60. The layout of the experiment is shown in figure 1. The two main components of the apparatus are a muon spectrometer

*e-mail: alessandro.de.falco@ca.infn.it

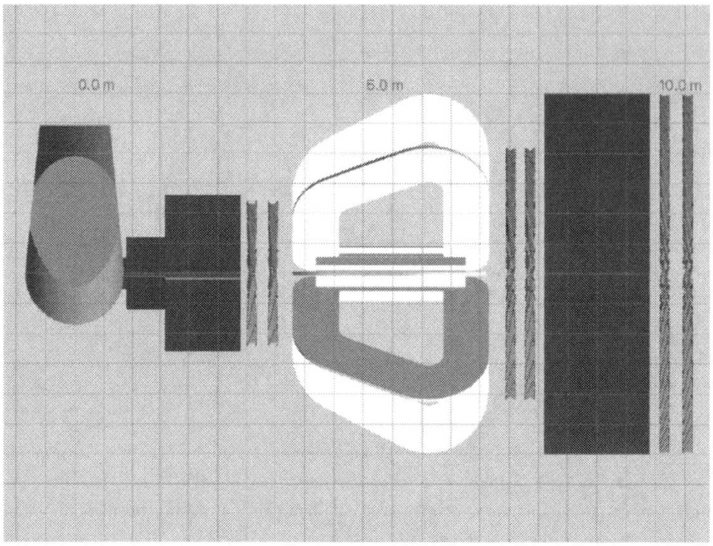

Figure 1. Conceptual design of the NA60+ experimental apparatus. The figure represents the set-up adapted to low-energy collisions, with a thinner hadron absorber and the muon spectrometer relatively closer to the target.

(MS) and a vertex spectrometer (VS), separated by a thick hadron absorber made of BeO and graphite, which acts as a muon filter.

The MS is composed of a set of six tracking stations (plotted as green boxes in the figure) and a large aperture toroidal magnet. Two tracking stations are located upstream the magnet, which is followed by two other tracking stations. A thick graphite wall (brown rectangle in the figure) filters out possible residual hadrons. Finally, a set of two more tracking stations is placed downstream of the graphite wall. The relatively low rates after the hadron absorber, $\sim$ 2 kHz for a Pb beam rate of 10^{-6} s^{-1}, allows to adopt the GEM or MWPC technologies for the tracking stations. A prototype of trapezoidal MWPC is currently being constructed and will be tested at the CERN SPS in autumn 2022. The prototype detector constitutes a module that would be replicated several times and arranged in such a way to cover the designed geometry in each tracking station. As an alternative option, triple GEM modules will also be studied. The toroidal magnet, which is being designed for NA60+, is foreseen to provide a magnetic field of 0.5 T over a volume of 120 m^3. A small-scale prototype (1:5) was constructed and tested. The simulations of the magnetic field were found to be in agreement with the results of the tests within 3%.

The MS can be moved on rails, to cover the midrapidity region at different collision energies. At top SPS energies, the absorber thickness will be increased.

The hadron absorber, while providing muon identification, deteriorates the muon momentum resolution, due to multiple scattering and fluctuations in the energy loss. This loss in resolution is recovered by matching the tracks measured in the MS with those reconstructed in the VS, both in coordinates and momentum space.

The VS (figure 2) consists of a set of five to ten planes of ultra-thin, large area Monolithic Active Pixel Sensors (MAPS) for tracking, embedded in the gap of a dipole magnet. MAPS sensors are composed of 25 mm long units, which are replicated through a stitching technique to cover an area of 15$\times$15 cm^2. Each plane contains four MAPS sensors. The sensor thickness is of about 0.1% X_0, reducing the effect of multiple scattering in the VS. The spatial resolution is expected to be 5 μm or better, with an improvement of a factor of two with respect to the hybrid technology. The magnet considered for the VS is the CERN MEP48 dipole, that provides a field up to $B = 1.47$ T.

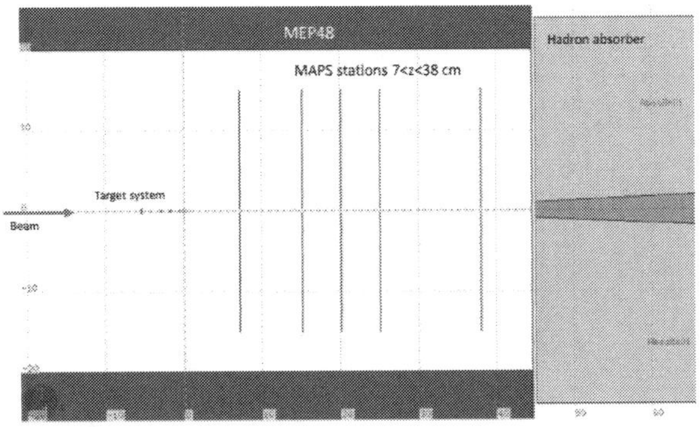

Figure 2. Schematic layout of the silicon planes inside MEP48.

3 Physics performances

The physics performances of NA60+ were studied executing fast simulations of the signals with semi-analytical tracking based on the Kalman filter, while the simulation of the background was performed with FLUKA. The obtained opposite sign dimuon mass spectrum is shown in figure 3 (left). The combinatorial background and the fake matches contribution can be evaluated using event mixing techniques and subtracted to the opposite sign mass spectrum. The signal mass spectrum resulting after background subtraction is dominated by the hadronic cocktail for $M < 1.5$ GeV/c^2. In this region, a precision measurement of the ρ spectral function can be performed, complementing the NA60 measurement in In-In at top SPS energy with results at lower energy.

In the region $1 < M < 1.5$ GeV/c^2, a dimuon enhancement due to the chiral mixing between the ρ and its chiral partner a_1 via 4π states can be observed [1]. With the foreseen accuracy of the measurement, a 20-30% enhancement, expected in case of full mixing, can be detected by NA60+.

A study of the thermal dimuon mass spectrum can be performed for $M > 2$ GeV/c^2. A fit with the form $dN/dM \propto M^{-3/2}\exp(-M/T_S)$ allows to extract the parameter T_S, which represents a space-time average of the thermal temperature over the fireball evolution and can be determined with a precision of the order of 10 MeV. The determination of the evolution of T_S vs CM energy for $\sqrt{s_{NN}} < 10$ GeV may allow to discover a plateau in the caloric curve (fig. 3, right) that would be present in case of a phase transition of the first order [2].

Charmonium suppression in Pb-Pb collisions was extensively studied at the top SPS energy by the NA50 collaboration [3]. NA60+ aims to extend the measurements at lower energies, down to $E_{\mathrm{lab}}/A = 40$ GeV, in order to search for the onset of the deconfinement. A statistics ranging from 10^4 to 10^5 reconstructed J/ψ, depending on the beam energy, can be collected at the foreseen Pb beam intensity, allowing for a precise measurement of the suppression. Also data taking with p-A collisions is foreseen, in order to determine precisely cold nuclear matter effects.

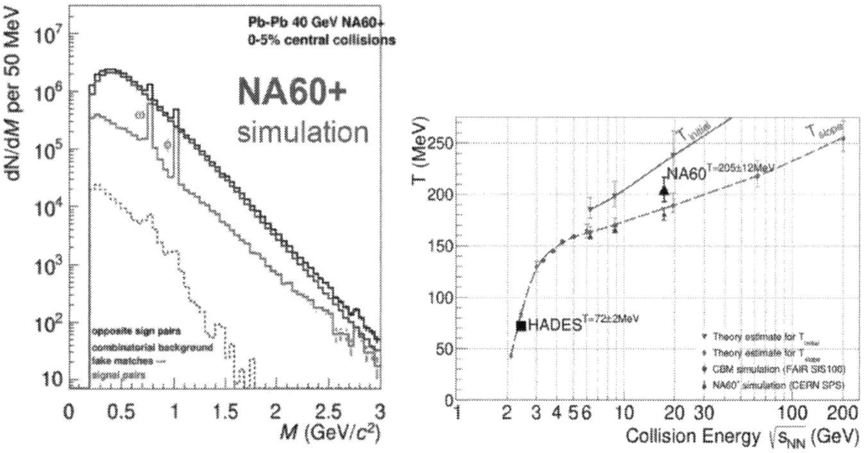

Figure 3. Left: simulated dimuon mass spectrum in central Pb-Pb collisions at $E = 40$ GeV per nucleon. Right: medium temperature evolution vs $\sqrt{s_{NN}}$ in central Pb-Pb collisions.

Open charm measurements will be performed using the vertex telescope as a stand-alone detector, reconstructing the decays of charmed hadrons into two or three charged hadrons. The huge combinatorial background can be reduced by applying geometrical selections on the displaced decay vertex topology. The MAPS technology can provide a signal to background ratio $\sim$ 10 times higher than the corresponding value obtainable with hybrid pixel sensors. No open charm measurement are currently available below the top SPS energy. In one month data taking, a measurement of the D^0 yield in central Pb-Pb collisions with a statistical precision much better than 1% can be obtained, allowing for a precise determination of the yield and v_2 as a function of p_T, rapidity and centrality. At $\sqrt{s_{NN}} = 10.6$ GeV due to the lower production cross section, a reduction of the statistics by an order of magnitude is expected. Still, the measurement will be feasible with a statistical precision at the level of the percent.

Finally, strangeness measurements in the hadronic decay channels will allow to explore the low multiplicity region to have a complete view of strangeness enhancement with multiplicity. A large statistics is expected, allowing for a high p_T-reach, studies in narrow centrality bins and an extension to multistrange hadrons of the elliptic flow measurements.

4 Timeline

The EoI [4] was submitted in 2019 at the CERN SPSC. A Letter of Intent is being prepared and will be submitted at the end of 2022. If the experiment will be approved, the goal is to have the detector ready for data taking after the LHC Long Shutdown 3 and run for at least 5-6 years, with $\sim$ one month data taking for each Pb beam incident energy, complemented by a few weeks of data taking with proton beams for cold nuclear matter studies.

References

[1] M. Dey, V.L. Eletsky, B.L. Ioffe, Phys. Lett. B **252**, 620 (1990)
[2] R. Rapp, H. van Hees, Phys. Lett. B **753**, 586 (2016), `1411.4612`
[3] B. Alessandro et al. (NA50), Eur. Phys. J. C **39**, 335 (2005), `hep-ex/0412036`
[4] T. Dahms, E. Scomparin, G. Usai (NA60+ Collaboration), Tech. rep., CERN, Geneva (2019), `https://cds.cern.ch/record/2673280`

Simulation study of Dual-Readout Calorimeter for a forward calorimeter at the Electron-Ion Collider

Yongjun Kim[1,*] On behalf of the Korea DRC team for the EIC

[1]Pusan National University, Busan, South Korea

Abstract. The Electron-Ion Collider (EIC) is a future particle accelerator to be built at the Brookhaven National Laboratory, and the primary purpose of experiments at the EIC is to resolve the question of partonic structure of nucleons and nuclei. To achieve the physics goals of the EIC, a hadron calorimeter of high energy resolution is required at forward rapidity. A Dual-readout Calorimeter (DRC) which has been developed for future collider experiments is considered as an upgrade option of the forward hadron calorimeter for the ECCE experiment at the EIC. The DRC consisting of two types of optical fiber, Cherenkov and Scintillation fibers, can achieve high energy resolution by measuring a fraction of electromagnetic shower in a hadronic shower. A performance study of DRC for the EIC such as geometry, material, and energy resolution is ongoing based on the existing simulation framework for high energy experiments, and the DRC simulation details will be transported to the EIC simulation framework. In this presentation, we will introduce the simulation study of the DRC for the EIC.

1 Introduction

The primary purpose of the Electron-Ion Collider (EIC) is to study the partonic structure of protons and nuclei more accurately [1]. It is required to be equipped high-quality detectors to achieve precise measurements. At forward rapidity where hadrons are going, electromagnetic (EM) and hadronic calorimeters will be located to measure hadrons and jets, and the required hadronic energy resolution is $50\%/\sqrt{E} \pm 10\%$.

Dual-Readout Calorimeter (DRC), proposed as the main calorimeter in the IDEA detector at the Future Circular Collider [2], can also be utilized for experiments at the EIC. The DRC is a sampling calorimeter consisting of absorber material and two types of optical fiber; Cherenkov and Scintillation fibers. The left panel of Fig. 1 shows a cross-sectional view of the DRC tower, and fibers of yellow and blue color are two different types of fibers. The different responses of EM particles and hadrons in two fiber types can provide information to determine the EM fraction of each hadron shower. The right panel of Fig. 1 shows a correlation of calibrated energy in Cherenkov and Scintillation fibers, data points from hadrons are located around the straight line. The corrected energy is calculated by applying a rotation matrix with the angle θ, and a better energy resolution can be achieved. See the recent review for more details on the DRC [3].

*e-mail: yongjun.kim@cern.ch

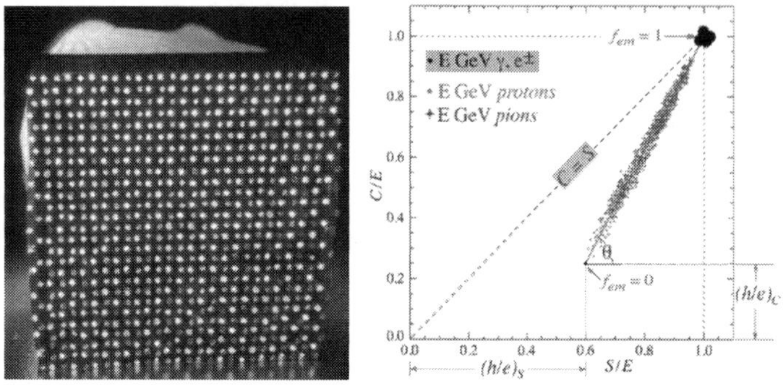

Figure 1. Left: Cross-sectional view of the DRC tower, Right: Correlation of energy in two channels for photons, electrons, and hadrons.

2 Performance study with GEANT4 simulation

A simulation framework for the DRC with GEANT4 has been developed for experiments with future e^+e^- colliders. We utilize the framework and use towers located at the endcap region for the simulation study. We change the tower length from 2.5 m to 1.25 m for the EIC design. Copper is used as an absorber material for the reference design, and performance with other materials like iron and tungsten is also studied. Electrons, charged pions, and jets with various energies are used to evaluate the detector performance.

2.1 Single EM particle

Energy calibration is performed separately for Cherenkov and Scintillation fibers using simulations with 20 GeV electrons. The results of the calibration are shown on the left panel of Fig. 2. The middle panel of Fig. 2 shows the energy resolution for electrons as a function of energy, and the resolution of the two channels is similar. We use a linear combination of two channels for the final energy measurement, and the energy resolution is $11\%/\sqrt{E} + 0.6\%$. In terms of energy linearity shown on the right panel of Fig. 2, the ratio between measured energy and beam energy is within 1% in a wide range of beam energy.

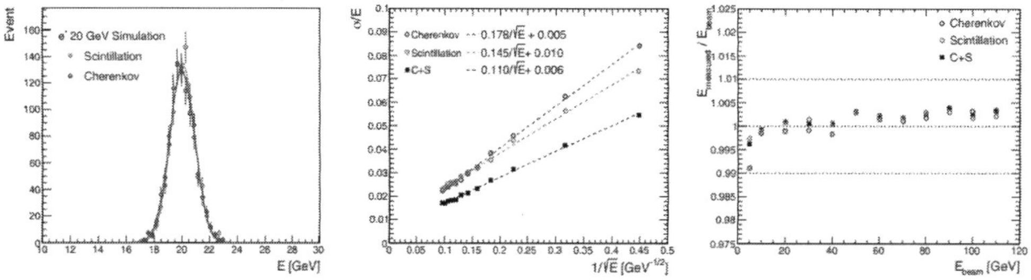

Figure 2. Left: Energy distribution of two channels for 20 GeV electrons. Center: Energy resolution for electrons with each channel and combined channels. Right: Energy linearity (ratio of measured energy to beam energy) for electrons.

2.2 Single hadron

For the performance of hadrons, we simulate positively charged pion (π^+) with various energies. Figure 3 shows the distribution of measured energy in Cherenkov and Scintillation channel and dual-readout corrected energy, and the left (right) panel shows the simulation results using 1.25 m (2.5 m) DRC towers of copper absorber. There are tails on the left-hand side in distributions with shorter towers, which indicates a longitudinal shower leakage when using 1.25 m towers for the EIC design. Note that a small fraction of events passing through the entire detector without initiating a hadron shower is removed. To account for the non-Gaussian shape due to shower leakage, we use a standard deviation of the histogram for the energy resolution instead of the Gaussian width obtained from fit to the distribution.

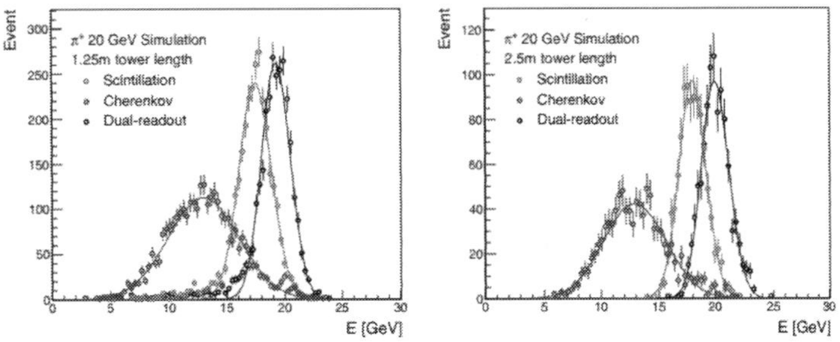

Figure 3. Left: Energy distribution of 20 GeV π^+ in 1.25 m DRC towers. Right: Energy distribution of 20 GeV π^+ in 2.5 m DRC towers.

Figure 4 shows the energy resolution and linearity obtained from π^+ simulations with the EIC design. In the left plot, the energy resolution is compared with the default length results. The energy resolution with 1.25 m towers worsens as the beam energy increases because of shower leakage. The magenta line represents the requirement of the EIC detector, and the energy resolution with 1.25 m towers still satisfies the requirement below 70 GeV. Note that a single hadron of 70 GeV is expected to be rarely produced at the EIC. Regarding the energy linearity shown on the right side, we can obtain more accurate hadron energy with the dual-readout correction method than with two single channel.

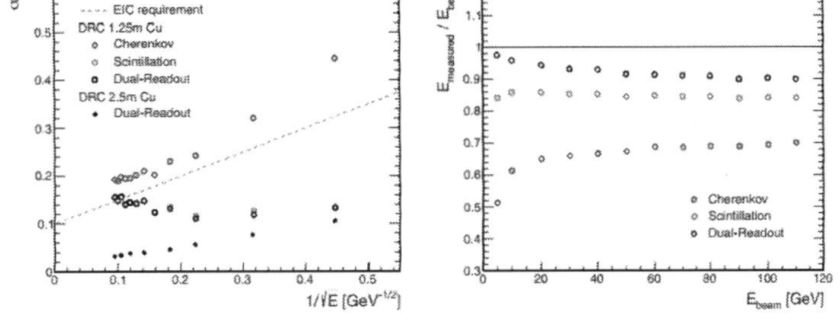

Figure 4. Left: Energy resolution for π^+ with different lengths of towers compared with the EIC requirement. Right: Energy linearity for π^+ with 1.25 m DRC towers.

2.3 Jet

We also study the performance of jet energy measurement with the EIC design of the DRC. The PYTHIA8 is utilized to generate hadrons from a single quark fragmentation, and the anti-k_T algorithm with a resolution parameter $R = 0.8$ is used to reconstruct jets. Not to use jets fragmented out of the endcap region, we select jets at least 80% of the initial quark energy is within the $R = 0.8$ range. Jets with Cherenkov and Scintillation channels are reconstructed separately, and a dual-readout correction is applied later. Figure 5 shows simulation results of jets with the DRC. The energy resolution with the EIC design is worse than the default design due to hadron shower leakage. The resolution becomes smaller as the jet energy increases, unlike in the single hadron case. This is because the energy of hadrons from jet fragmentation is not large enough to suffer from the shower leakage significantly. The magenta line represents the requirement of the EIC detector, and the energy resolution with 1.25 m towers satisfies the requirement. Regarding the energy linearity shown in the right plot, more accurate jet energy can be obtained with the dual-readout correction method.

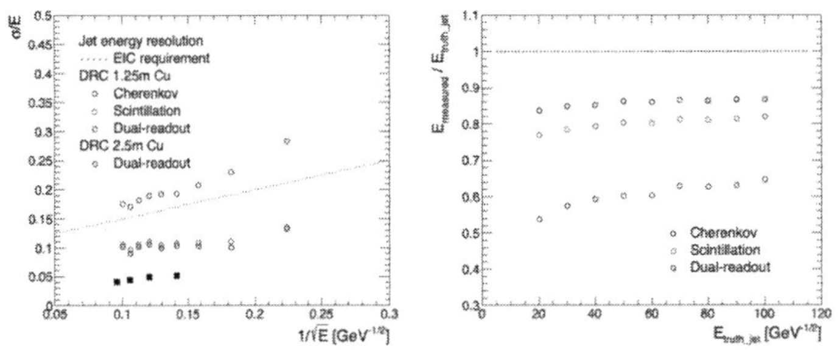

Figure 5. Left: Energy resolution for jets with different lengths of towers compared with the EIC requirement. Right: Energy linearity for jets with 1.25 m DRC towers.

3 Summary and Outlook

We have performed a simulation study of the Dual-Readout Calorimeter for experiments at the EIC. Based on the simulation framework developed for future collider experiments, a modified design with a shorter length (1.25 m) is used for the simulation of single electrons, hadrons, and jets. Although there is a longitudinal leakage of hadron showers with the EIC design, the obtained energy resolution of hadrons and jets satisfies the EIC requirement. For the plan, we will implement detailed features of the DRC simulation to the EIC simulation framework, such as optical photon propagation, light attenuation, and readout electronics.

References

[1] R. Abdul Khalek and others, *Science Requirements and Detector Concepts for the Electron-Ion Collider: EIC Yellow Report*, arXiv:2103.05419

[2] A. Abada and others, *FCC-ee: The Lepton Collider: Future Circular Collider Conceptual Design Report Volume 2*, Eur. Phys. J. ST 228 (2019) 2, 261

[3] S. Lee and others, *Dual-Readout Calorimetry*, Rev. Mod. Phys. 90, 025002 (2018)

The LHCspin project

A polarised gas target at the Large Hadron Collider

Marco Santimaria[1,*], *Vittore* Carassiti[2], *Giuseppe* Ciullo[2,3], *Pasquale* Di Nezza[1], *Paolo* Lenisa[2,3], *Saverio* Mariani[4,5], *Luciano Libero* Pappalardo[2,3], and *Erhard* Steffens[6]

[1]INFN - Laboratori Nazionali di Frascati, via Enrico Fermi 54, 00044 Frascati (RM), Italy
[2]INFN - Ferrara, via Saragat 1, 44122 Ferrara (FE), Italy
[3]Department of Physics, University of Ferrara, via Saragat 1, 44122 Ferrara (FE), Italy
[4]INFN - Firenze, via Sansone 1, 50019 Sesto Fiorentino (FI), Italy
[5]Department of physics, University of Firenze, via Sansone 1, 50019 Sesto Fiorentino (FI), Italy
[6]Physics department, FAU Erlangen-Nürnberg, Staudtstr. 7 / B2, 91058 Erlangen, Germany

Abstract. The goal of LHCspin is to develop, in the next few years, innovative solutions and cutting-edge technologies to access spin physics in polarised fixed-target collisions at high energy, exploring the unique kinematic regime offered by LHC and exploiting new final states by means of the LHCb detector.

1 Introduction

The LHC delivers proton and lead beams with an energy of 7 TeV and 2.76 TeV per nucleon, respectively, with world's highest intensity. Fixed-target proton-gas collisions occur at a centre-of-mass energy per nucleon of up to 115 GeV in the case of a hydrogen target. The large centre-of-mass boost offers an unprecedented opportunity to investigate partons carrying a large fraction of the target nucleon momentum, i.e. large Bjorken−x values.

The LHCb detector [1] is a general-purpose forward spectrometer specialised in detecting hadrons containing c and b quarks, and the only LHC detector able to collect data in both collider and fixed-target mode. It is fully instrumented in the $2 < \eta < 5$ region with a vertex locator (VELO), a tracking system, two Cherenkov detectors, electromagnetic and hadronic calorimeters and a muon detector.

The fixed-target physics program of LHCb is active since the installation of the SMOG device [2], enabling the injection of noble gases in the beam pipe section crossing the VELO detector at a pressure of $O(10^{-7})$ mbar.

With the SMOG2 upgrade [3], an openable gas storage cell, shown in the left panel of Fig. 1, has been installed in 2020 in front of the VELO. The cell boosts the target areal density by a factor of 8 to 35 depending on the injected gas species, and creates a localised beam-gas collision region which is well detached from the proton-proton vertices. As shown in the right panel of Fig. 1, full tracking efficiency is expected in the beam-gas region, despite its upstream position with respect to the VELO. Due to these developments, fixed-target data will be collected in the upcoming Run 3 with a novel reconstruction software allowing for the simultaneous data-taking of beam-gas and beam-beam collisions. SMOG2 will offer a

*e-mail: marco.santimaria@cern.ch

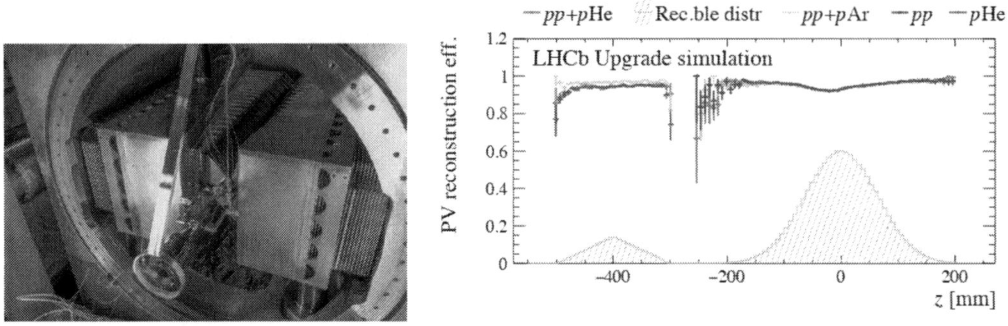

Figure 1. Left: picture of the SMOG2 storage cell after installation. Right: simulated vertex reconstruction efficiency for simultaneous beam-gas and beam-beam collisions [4].

rich physics program for the Run 3 and, at the same time, will also allow to investigate the dynamics of the novel beam-target system, setting the basis for future developments.

The LHCspin project [5, 6] aims at extending the fixed-target physics program in Run 4 and Run 5 with the installation of a polarised gas target, bringing spin physics at LHC for the first time. The experimental setup of LHCspin is discussed in Sec. 2, while a selection of physics opportunities is presented in Sec. 3.

2 Experimental setup

The LHCspin experimental setup is in R&D phase and calls for the development of a new generation polarised target. The starting point is the setup of the HERMES experiment at DESY [7] and comprises three main components: an Atomic Beam Source (ABS), a Target Chamber (TC) and a diagnostic system. The ABS consists of a dissociator with a cooled nozzle, a Stern-Gerlach apparatus to focus the wanted hyperfine states, and adiabatic RF-transitions for setting and switching the target polarisation between states of opposite sign. The ABS injects a beam of polarised hydrogen or deuterium into the TC, which is located in the LHC primary vacuum. The TC hosts a T-shaped openable storage cell, sharing the SMOG2 design, and a dipole holding magnet ($B = 300$ mT), as shown in Fig. 2.

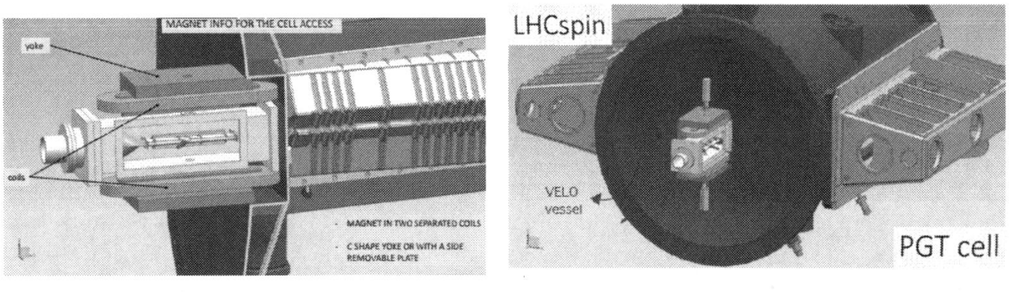

Figure 2. Two views of the TC with the magnet coils (orange) and the iron return yoke (blue) enclosing the storage cell. The VELO vessel and detector box are shown in green and grey, respectively.

The diagnostic system continuously analyses gas samples drawn from the TC and comprises a target gas analyser to detect the molecular fraction, and thus the degree of dissociation, and a Breit-Rabi polarimeter to measure the relative population of the injected hyperfine states. An instantaneous luminosity of $O(10^{32})$ $cm^{-2}s^{-1}$ is foreseen for p-H collisions during Run 4.

3 Physics case

The physics case of LHCspin covers three main areas: exploration of the wide physics potential offered by unpolarised gas targets, investigation of the nucleon spin and heavy-ion collisions.

Unpolarised gas targets

By means of the SMOG2 gas feed system, LHCspin will allow the injection of several species of unpolarised gases: H_2, D_2, He, N_2, O_2, Ne, Ar, Kr and Xe with negligible impact on the LHC beam lifetime. This gives an excellent opportunity to investigate parton distribution functions (PDFs) in both nucleons and nuclei in the large-x and intermediate Q^2 regime, which is especially affected by lack of experimental data and impact several fields from basic QCD tests to astrophysics. For example, the large acceptance and high reconstruction efficiency of LHCb on heavy flavour states enables the study of gluon PDFs, which are a fundamental input for theoretical predictions [8]. Searches for an intrinsic charm component in the proton [9] and antiproton production in p-He collisions [10] are two other high-profile example measurements that have already been pioneered at LHCb. With the large amount of data to be collected, nuclear PDFs can also be investigated in greater detail, helping to shed light on the intriguing anti-shadowing effect [11], which is expected to be dominant in the x range covered with LHCspin.

Spin physics

Beside standard collinear PDFs, LHCspin will offer the opportunity to probe polarised quark and gluon distributions by means of proton collisions on polarised hydrogen and deuterium. For example, measurements of transverse-momentum dependent PDFs (TMDs) provide a map of parton densities in 3-dimensional momentum space. Light quark TMDs, especially in the high-x regime, can be accessed by measuring transverse single spin asymmetries (TSSAs) in Drell-Yan processes, while gluon densities, such as the gluon Sivers function, can be probed via heavy-flavour production. Such asymmetry can be large (left panel of Fig. 3) and well in reach with just 1 month of LHCspin data, as shown in simulated $J/\psi \rightarrow \mu^+\mu^-$ events (right panel of Fig. 3).

It is also attractive to go beyond a 3-dimensional description by building observables which are sensitive to Wigner distributions [13] and to measure the elusive transversity PDF, whose knowledge is currently limited to valence quarks at the leading order [14], as well as its integral, the tensor charge, which is of direct interest in constraining physics beyond the Standard Model [15].

Heavy-ion collisions

Thermal heavy-flavour production is negligible at the typical temperature of few hundreds MeV of the system created in ultra-relativistic heavy-ion collisions. Quarkonia states ($c\bar{c}$, $b\bar{b}$) are instead produced on shorter timescales, and their energy change while traversing the medium represents a powerful way to investigate Quark-Gluon Plasma (QGP) properties.

LHCb capabilities allow to both cover the aforementioned charmonia and bottomonia studies and to extend them to bottom baryons as well as exotic probes. QGP phase diagram exploration at LHCspin can be performed with a rapidity scan, complementing RHIC's

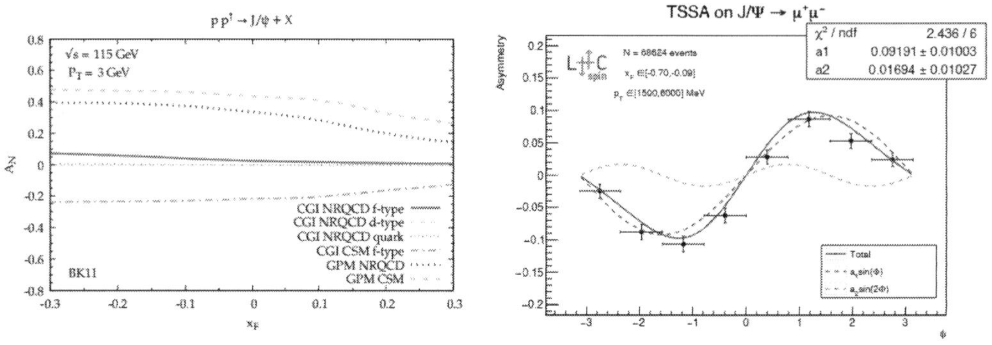

Figure 3. Left: theoretical predictions for A_N in inclusive J/ψ production [12]. Right: simulated $J/\psi \to \mu^+\mu^-$ azimuthal asymmetries with a fit curve superimposed made of two Fourier terms.

beam-energy scan, while flow measurements will greatly benefit from the excellent particle identification performance of LHCb on charged and neutral light hadrons. An interesting topic joining heavy-ion collisions and spin physics is the dynamics of small systems which can be probed via ellipticity measurements in lead-ion collisions on polarised deuterons [16].

References

[1] A.A. Alves, Jr. et al. (LHCb), JINST **3**, S08005 (2008)
[2] R. Aaij et al. (LHCb), JINST **9**, P12005 (2014), 1410.0149
[3] Tech. rep., CERN, Geneva (2019), https://cds.cern.ch/record/2673690
[4] *SMOG2 Velo tracking efficiency* (2019), https://cds.cern.ch/record/2688875
[5] C.A. Aidala et al. (2019), 1901.08002
[6] M. Santimaria, V. Carassiti, G. Ciullo, P.D. Nezza, P. Lenisa, S. Mariani, L.L. Pappalardo, E. Steffens, SciPost Phys. Proc. p. 50 (2022), 2105.10012
[7] A. Airapetian et al. (HERMES), Nucl. Instrum. Meth. A **540**, 68 (2005), physics/0408137
[8] C. Hadjidakis et al. (2018), 1807.00603
[9] R. Aaij et al. (LHCb), Phys. Rev. Lett. **122**, 132002 (2019), 1810.07907
[10] R. Aaij et al. (LHCb), Phys. Rev. Lett. **121**, 222001 (2018), 1808.06127
[11] K.J. Eskola, P. Paakkinen, H. Paukkunen, C.A. Salgado, Eur. Phys. J. C **77**, 163 (2017), 1612.05741
[12] U. D'Alesio, L. Maxia, F. Murgia, C. Pisano, S. Rajesh, Phys. Rev. D **102**, 094011 (2020)
[13] S. Bhattacharya, A. Metz, J. Zhou, Phys. Lett. B **771**, 396 (2017), [Erratum: Phys.Lett.B 810, 135866 (2020)], 1702.04387
[14] M. Radici, A. Bacchetta, Phys. Rev. Lett. **120**, 192001 (2018), 1802.05212
[15] A. Courtoy, S. Baeßler, M. González-Alonso, S. Liuti, Phys. Rev. Lett. **115**, 162001 (2015), 1503.06814
[16] W. Broniowski, P. Bożek, Phys. Rev. C **101**, 024901 (2020), 1906.09045

Hadron production from heavy ion collisions

Che Ming Ko[1,*]

[1]Cyclotron Institute and Department of Physic and Astronomy, Texas A&M University, College Station, Texas 77843-3366, USA

Abstract. A brief review of some topics in hadron production from heavy ion collisions is given. They include charged pion ratio as a probe of nuclear symmetry energy, in-medium effects on pion production, enhanced Λ_c/D^0 ratio, Λ local polarization, and $X(3872)$ production.

1 Introduction

Hadron production from heavy ion collisions has been a topic of continuous interest since earlier experiments at Bevalac on pion [1, 2] and kaon [3] production. Following the demonstration in Ref. [4] that kaon production in heavy ion collisions at energy below its production threshold of 1.56 GeV in a nucleon-nucleon collision is sensitive to the stiffness of nuclear equation of state at high density, the comparison of transport model results with the experimental data from Au + Au collisions at SIS has led to the conclusion that the nuclear equation of state is soft with a compressibility of $K \sim 200$ MeV at the nuclear matter saturation density [5], similar to that extracted from the flow observables in heavy ion collisions [6]. Hadron production from heavy ion collisions has since been extensively studied at SIS, AGS, SPS, RHIC and the LHC. The present talk gives a brief review on some topics from these studies, which include charged pion ratio as a probe of nuclear symmetry energy, in-medium effects on pion production, enhanced Λ_c/D^0 ratio, Λ hyperon local polarization, and X(3872) production.

2 Charged pion ratio and the nuclear symmetry energy

As for subthreshold kaon production, studies based on isospin-dependent transport models have shown that the π^-/π^- ratio in heavy ion collisions with neutron-rich nuclei at near-threshold energies is sensitive to the density dependence of the nuclear symmetry energy [7], which is the energy needed per nucleon to convert a symmetric nuclear matter to a pure neutron matter, with the soft one giving a larger value than the stiff one. Dedicated experiments have recently been carried out by the SπRIT Collaboration to study pion production from ^{108}Sn+^{112}Sn, ^{112}Sn+^{124}Sn, and ^{132}Sn+^{124}Sn collisions at $E/A = 270$ MeV [8], and the measured π^-/π ratios have been compared with the predictions from seven transport codes using both a stiff and a soft symmetry energy of $\sim 46 - 56$ MeV and $\sim 120 - 150$ MeV, respectively, for the

*e-mail: ko@comp.tamu.edu

density slope parameter L at the nuclear matter saturation density. Although the difference between the predicted single π^-/π^- ratio from ^{132}Sn+^{124}Sn and ^{108}Sn+^{112}Sn collisions for soft and stiff symmetry energies in some codes is smaller than the errors in experimental measurements, it is larger in other codes. The differences among predictions from the different transport codes are thus too large to allow for the extraction of reliable constraints on the nuclear symmetry energy from the data. This finding may explain previous contradictory conclusions on symmetry energy constraints obtained from the pion data in Au+Au system by the FOPI Collaboration [9]. These new results call for a better understanding of the differences among transport codes, which has been pursued for many years by the transport model evaluation project (TMEP) Collaboration [10], and new observables that may be more sensitive to the density dependence of the symmetry energy. However, a subsequent study of the SπRIT π^-/π^- ratio data [11] based on the dcQMD model [12], which includes the implementation of total energy conservation for the colliding system besides at the two-body level due to the momentum and isospin asymmetry dependence of nuclear interactions, which is neglected in most transport codes, has deduced the value $42 < L < 117$ MeV from ratios of measured high transverse momentum charged pions. This value is slightly lower but consistent with that deduced from the measured neutron skin thickness $R_n - R_p = 0.283 \pm 0.071$ fm of ^{208}Pb by the PREX-2 Collaboration [13] but contradicts the expected smaller L from the smaller $R_n - R_P = 0.121 \pm 0.026(\mathrm{exp}) \pm 0.024(\mathrm{model})$ fm of ^{48}Ca [14] measured by the CEREX Collaboration. It is also inconsistent with the value $L = 58 \pm 19$ MeV based on constraints from neutron star masses, radii, and tidal deformability as well as the predictions from chiral effective theory [15, 16].

3 In-medium effects on pion production

In high energy heavy ion collisions without the formation of a quark-gluon plasma, pions are mainly produced from the reactions $NN \leftrightarrow N\Delta$ and $\Delta \leftrightarrow N\pi$ involving the Δ resonance. This is the case in the SπRIT experiments on Sn+Sn collisions at $E/A = 270$ MeV and also in the HADES experiments at SIS on Au+Au collisions at $E/A = 1.23$ GeV or the center-of-mass energy of $\sqrt{S_{NN}} = 2.4$ GeV [17]. Using information for these reactions extracted in free space, results from five popular transport models not only fail to describe the data by giving too large pion yields but also disagree among themselves. Since the cross section for the reaction $NN \leftrightarrow N\Delta$ is known to become smaller in nuclear medium [18–20], a recent study has included this effect in the RVUU model by introducing the density-dependent cross section $\sigma^{\mathrm{med}}_{NN\to N\Delta} = \sigma^{\mathrm{free}}_{NN\to N\Delta} e^{-\alpha(\rho/\rho_0)^{1.5}}$ [21]. With $\alpha = 0.39$ for Δ^{++} and Δ^+ production and $\alpha = 0.7$ for Δ^0 and Δ^-, a good description of the measured collision centrality dependence of π^+ and π^- multiplicities, rapidity distributions, and transverse momentum spectra is obtained. Remaining small differences between the theoretical results and the experimental data, particularly for the pion transverse momentum spectra, may indicate the need of the pion mean-field potentials, which are at present absent in RVUU.

4 Enhanced Λ_c/D^0 ratio at RHIC and the LHC

Because of the attractive color-spin interaction between quarks, color nonsinglet diquark $[qq]$ states may be present in QGP [22, 23]. In vacuum, the color-spin interaction between two quarks gives the $[ud]$ diquark mass as $m_{[\mathrm{ud}]} = m_{\mathrm{u}} + m_{\mathrm{d}} - C\vec{s}_{\mathrm{u}} \cdot \vec{s}_{\mathrm{d}}/m_{\mathrm{u}}m_{\mathrm{d}}$.

With the coonstituent quark mass $m_{\mathrm{u}} = m_{\mathrm{d}} = 0.3$ GeV, the spin operator $\vec{s}_i$ ($i = \mathrm{u,d}$), and a constant $C/m_{\mathrm{u}}^2 = 0.193$ GeV fitted to the $N - \Delta$ mass splitting [24], a maximum binding energy of 0.145 GeV or mass $m_{[\mathrm{ud}]} = 0.455$ GeV is obtained for the diquark. Expecting decreasing coupling in QGP, the diquark mass could be increased to $m_{[\mathrm{ud}]} = 0.6$ GeV but still be present, which would then lead to enhanced production of Λ_c and Λ_b in relativistic heavy ion collisions at RHIC and the LHC through the coalescence of c and b quarks with bound diquarks near the phase transition [25, 26]. Experiments have recently been carried out by the STAR Collaboration for Au+Au collisions at 200 GeV [27] and the ALICE Collaboration for Pb+Pb collisions at 5.02 TeV [28] to measure the transverse momentum dependence of the Λ_c/D^0 yield ratio. These results can be successfully described by various theoretical studies based on the quark coalescence model without diquarks present in the produced QGP but with additional assumptions, such as a unit coalescence probability for charm quarks of zero momentum [29], the collective flow effect in the coalescence process [30, 31], and the contribution of missing resonances to Λ_c production [32] as in the study of deeply subthreshold Ξ^- and ϕ meson [33] production in heavy ion collisions at SIS. Because of these very different assumptions, to decisively exclude the presence of diquarks in QGP require further studies.

5 Λ local polarization

The Λ hyperon global polarization in the direction perpendicular to the reaction plane of non-central Au+Au collisions measured by the STAR Collaboration [34] can be successfully described by both hydrodynamic [35] and coarse-grained transport [36] models with the assumption that Λ hyperons are in thermal equilibrium with the vorticity field generated during collisions. These models [37, 38] failed, however, to describe the measured Λ local polarization along the beam direction [39] by giving a wrong sign for its azimuthal angle dependence in the transverse plane of the collisions. On the other hand, the nonequilibrium chiral kinetic approach [40], which takes into account the effect of the vorticity field on the equations of motion and scatterings of quarks and antiquarks, can describe not only the Λ global polarization [41], when Λ hyperons are formed from the coalescence of polarized strange quarks with u and d quarks, but can also give the correct sign in the azimuthal dependence of the Λ local polarization [42] due to the redistribution of axial charges in the transverse plane. The latter result was confirmed by a covariant angular-momentum-conserved chiral transport model study [43]. This dynamic effect in the chiral kinetic approach can be included in the hydrodynamic approach by adding the effect from the thermal shear field [44–46]. The study based on the 3+1 MUSIC hydrodynamic model then indicates that strange quarks show a similar local spin polarization as that of measured Λ [47]. In another study based on the 3+1 viscous eHLLe and EVHO-QGP hydrodynamic models, the experimental data is found to be consistent with the isothermal local equilibrium of Λ spin polarization in thermal vorticity and shear fields at temperature of 150 MeV [48]. Both Λ global and local spin polarizations decrease, however, with temperature as the hadronic matter cools [49]. The fact that the experimental data is consistent with Λ spin polarizations at the hadronization temperature indicates that they freeze out or decouple early from the hadronic matter. This would be the case if the $\Lambda - \pi$ scattering is through the spin 3/2, parity positive $\Sigma^*(1358)$ resonance since the ratio of Λ spin non-flip to flip probabilities in this scattering is 3.5.

6 Exotic X(2872) production

Besides allowing the study of the properties of QGP, heavy ion collisions at relativistic energies also provides a new method for studying the structure of exotic hadrons [50–52], which is a topic of the most active areas of research in hadron physics. Based on the coalescence model for hadron production, it was shown in Refs. [50–52] that compared to the case of a nonexotic hadron with normal quark numbers, the yield of an exotic hadron is typically an order of magnitude smaller when it is a compact multiquark state and a factor of 2 or more larger when it is a loosely bound hadronic molecule. More detailed studies based on the kinetic approach have been carried out to study the hadronic effects on the production of the exotic hadron $X(3872)$, which was first observed by the Belle Collaboration [53] at KEK from the exclusive decay process $B^\pm \to K^\pm\pi^+\pi^- J/\Psi$ as a narrow charmonium-like state that has a mass of 3872 MeV near the $M_{D^0} + M_{D^{*0}}$ mass threshold and decays into $\pi^+\pi^- J/\Psi$. With its quantum numbers $J^{PC} = 1^{++}$ determined recently by the LHCb Collaboration at the LHC [54], $X(3872)$ can thus be either a compact $c\bar{c}d\bar{d}$ tetraquark state or a $D^0\bar{D}^{*0}$ ($\bar{D}^0D^{*0}$) molecule. In Ref. [55], $X(2872)$ production from Au+Au collisions at 200 GeV was studied in the kinetic approach with the initial $X(3872)$ abundance given by the coalescence model for the tetraquark or the hadronic molecule scenario, including the reactions $X\rho \leftrightarrow D^0\bar{D}^{*0}$ and $X\rho \leftrightarrow \bar{D}^0D^{*0}$ is found not to affect much its initial abundance in both scenarios, with the conclusion of an order-of-magnitude larger yield for the tetraquark than the hadronic molecule scenario as in Refs. [50–52]. A factor of four reduction of $X(3872)$ yield in the tetraquark scenario is, however, reported in Ref. [56] after including an anomalous vertex in the reactions $X\rho \leftrightarrow D^0\bar{D}^{*0}$ and $X\rho \leftrightarrow \bar{D}^0D^{*0}$ as it enhances their cross sections by more than two order-of-magnitude to about 1 mb. In a more recent study [57], the initial $X(3872)$ abundance in Pb+Pb collisions at 5 TeV for the tetraquark scenario is taken from the statistical hadronization model, and it remains essentially unchanged during the hadronic evolution by using cross sections of about 50 mb for the reactions $X\rho \leftrightarrow D^0\bar{D}^{*0}$ and $X\rho \leftrightarrow \bar{D}^0D^{*0}$ estimated from the $X(3872)$ radius of 1.3 fm. For the hadronic molecule scenario in this study, $X(3872)$ is assumed to be produced from the above reactions with its final abundance depending on the temperature below which $X(3872)$ is bounded, with its final yield always less than that in the tetraquark scenario, resulting in a conclusion that is opposite to that of Ref. [55]. In another study [58] based on a schematic coalescence of the quarks or charmed mesons from the AMPT model [59], it is again found that the yield of $X(3872)$ is larger for the hadronic molecule than for the tetraquark scenario. However, instead of a factor of ten in the yield ratio for the two scenarios, a factor of 200 is reported due to the unrealistic application of the coalescence model only in the coordinate space.

Experimentally, $X(3872)$ has recently been unambiguously identified from the $J/\Psi\pi^+\pi^-$ invariant mass spectrum in Pb+Pb collisions at 5.02 TeV by the CMS Collaboration [60]. The yield ratio $X(3872)/\Psi(2S)$ measured at $15 < p_T < 50$ GeV/c is about one. Because of the large momentum, the observed $X(3872)$ is likely formed from hard quarks or charmed mesons from initial hard scatterings in heavy ion collisions instead of the thermal ones in the produced QGP considered in above theoretical studies. According to Ref. [61], the ratio $X(3872)/\Psi(2S)$ at $p_T = 10$ GeV from initial hard scatterings is about 0.1 as in pp collisions, which is about an order-of-magnitude smaller than that measured in CMS experiments. Understanding such a large enhancement of high momentum $X(3872)$ production in relativistic heavy ion collisions is an interesting question and may also provide a means to understand the structure of $X(3872)$. Of course, it is important to also measure low momentum $X(3872)$ in these collisions and compare the experimental data with the theoretical predictions.

7 Summary

Since the first experiments at Bevalac on pion and kaon production from heavy ion collisions, a lot of progress has been made in the study of hadron production in heavy ion collisions from energies near the pion production threshold in a nucleon-nucleon collision to ultrarelativistic energies at the LHC. These studies have led to a better understanding of the equation of state of both symmetric and asymmetric nuclear matter at high density as well as the in-medium effects on hadron production. Although the enhanced Λ_c/D^0 at intermediate momentum measured at RHIC and the LHC can be described without diquarks in the QGP, different assumptions have been introduced in the coalescence description of Λ_c production. Besides the vorticity field generated in non-central collisions, which is responsible for the Λ global polarization, the shear field is needed to better understand the measured Λ local polarization. The study of the production of exotic hadrons, whether they are multiquark or hadronic molecules, in heavy ion collisions provides a new means to understand their structures. However, there remain many puzzles such as the role of diquarks in heavy baryon production and the large $X(3872)/\Psi(2S)$ ratio at high momentum. Many more progresses and puzzles, which are not discussed here, have made the study of hadron production from heavy ion collisions an exciting topics in high energy nuclear physics, which is expected to continue in the future.

ACKNOWLEDGEMENTS

This work was supported in part by the US Department of Energy under Award No. DE-SC0015266.

References

[1] I. Tanihata et al., Phys. Lett. B **87**, 349 (1979)
[2] K.L. Wolf et al., Phys. Rev. Lett. **42**, 1448 (1979)
[3] S. Schnetzer et al., Phys. Rev. Lett. **49**, 989 (1982)
[4] J. Aichelin, C.M. Ko, Phys. Rev. Lett. **55**, 2661 (1985)
[5] C. Fuchs, A. Faessler, E. Zabrodin, Y.M. Zheng, Phys. Rev. Lett. **86**, 1974 (2001)
[6] P. Danielewicz, R. Lacey, W.G. Lynch, Science **298**, 1592 (2002)
[7] B.A. Li, L.W. Chen, C.M. Ko, Phys. Rept. **464**, 113 (2008)
[8] G. Jhang et al. (SπRIT, TMEP), Phys. Lett. B **813**, 136016 (2021)
[9] W. Reisdorf et al. (FOPI), Nucl. Phys. A **781**, 459 (2007)
[10] H. Wolter et al. (TMEP), Prog. Part. Nucl. Phys. **125**, 103962 (2022)
[11] J. Estee et al. (SπRIT), Phys. Rev. Lett. **126**, 162701 (2021)
[12] M.D. Cozma, M.B. Tsang, Eur. Phys. J. A **57**, 309 (2021)
[13] B.T. Reed et al., Phys. Rev. Lett. **126**, 172503 (2021)
[14] D. Adhikari et al. (CREX), Phys. Rev. Lett. **129**, 042501 (2022)
[15] R. Essick, I. Tews, P. Landry, A. Schwenk, Phys. Rev. Lett. **127**, 192701 (2021)
[16] Y. Lim, J.W. Holt, Phys. Rev. Lett. **121**, 062701 (2018)
[17] J. Adamczewski-Musch et al. (HADES), Eur. Phys. J. A **56**, 259 (2020)

[18] A.B. Larionov, U. Mosel, Nucl. Phys. A **728**, 135 (2003)
[19] Q. Li, Z. Li, Physics Letters B **773**, 557 (2017)
[20] Y. Cui, Y. Zhang, Y. Tian, Z. Li, Int. J. of Mod. Phy. E **30**, 2150069 (2021)
[21] K. Godbey, Z. Zhang, J.W. Holt, C.M. Ko, Phys. Lett. B **829**, 137134 (2022)
[22] J.F. Donoghue, K.S. Sateesh, Phys. Rev. D **38**, 360 (1988)
[23] E.V. Shuryak, I. Zahed, Phys. Rev. C **70**, 021901 (2004)
[24] S.H. Lee, S. Yasui, W. Liu, C.M. Ko, Eur. Phys. J. C **54**, 259 (2008), 0707.1747
[25] S.H. Lee et al., Phys. Rev. Lett. **100**, 222301 (2008)
[26] Y. Oh, C.M. Ko, S.H. Lee, S. Yasui, Phys. Rev. C **79**, 044905 (2009)
[27] J. Adam et al. (STAR), Phys. Rev. Lett. **124**, 172301 (2020)
[28] S. Acharya et al. (ALICE), Phys. Lett. B **793**, 212 (2019)
[29] S. Plumari et al., Eur. Phys. J. C **78**, 348 (2018)
[30] S. Cho, K.J. Sun, C.M. Ko, S.H. Lee, Y. Oh, Phys. Rev. C **101**, 024909 (2020)
[31] S. Cao et al., Phys. Lett. B **807**, 135561 (2020)
[32] M. He, R. Rapp, Phys. Rev. Lett. **124**, 042301 (2020)
[33] J. Steinheimer, M. Bleicher, J. Phys. **G43**, 015104 (2016)
[34] L. Adamczyk et al. (STAR), Nature **548**, 62 (2017)
[35] I. Karpenko, F. Becattini, Eur. Phys. J. C **77**, 213 (2017)
[36] H. Li, L.G. Pang, Q. Wang, X.L. Xia, Phys. Rev. **C96**, 054908 (2017)
[37] F. Becattini, I. Karpenko, Phys. Rev. Lett. **120**, 012302 (2018)
[38] X.L. Xia et al., Phys. Rev. C **98**, 024905 (2018), 1803.00867
[39] T. Niida (STAR), Nucl. Phys. A **982**, 511 (2019)
[40] Y. Sun, C.M. Ko, Phys. Rev. C **95**, 034909 (2017)
[41] Y. Sun, C.M. Ko, Phys. Rev. **C96**, 024906 (2017)
[42] Y. Sun, C.M. Ko, Phys. Rev. C **99**, 011903 (2019)
[43] S.Y.F. Liu, Y. Sun, C.M. Ko, Phys. Rev. Lett. **125**, 062301 (2020)
[44] S.Y.F. Liu, Y. Yin, JHEP **07**, 188 (2021)
[45] F. Becattini, M. Buzzegoli, A. Palermo, Phys. Lett. B **820**, 136519 (2021)
[46] C. Yi, S. Pu, D.L. Yang, Phys. Rev. C **104**, 064901 (2021)
[47] B. Fu, S.Y.F. Liu, L. Pang, H. Song, Y. Yin, Phys. Rev. Lett. **127**, 142301 (2021)
[48] F. Becattini et al., Phys. Rev. Lett. **127**, 272302 (2021)
[49] Y. Sun, Z. Zhang, C.M. Ko, W. Zhao, Phys. Rev. C **105**, 034911 (2022)
[50] S. Cho et al. (ExHIC), Phys. Rev. Lett. **106**, 212001 (2011)
[51] S. Cho et al. (ExHIC), Phys. Rev. C **84**, 064910 (2011)
[52] S. Cho et al. (ExHIC), Prog. Part. Nucl. Phys. **95**, 279 (2017)
[53] S.K. Choi et al. (Belle), Phys. Rev. Lett. **91**, 262001 (2003)
[54] R. Aaij et al. (LHCb), Phys. Rev. D **92**, 011102 (2015)
[55] S. Cho, S.H. Lee, Phys. Rev. C **88**, 054901 (2013)
[56] L.M. Abreu et al., Phys. Lett. B **761**, 303 (2016)
[57] B. Wu et al., Eur. Phys. J. A **57**, 122 (2021)
[58] H. Zhang et al., Phys. Rev. Lett. **126**, 012301 (2021)
[59] Z.W. Lin, C.M. Ko, B.A. Li, B. Zhang, S. Pal, Phys. Rev. C **72**, 064901 (2005)
[60] A.M. Sirunyan et al. (CMS), Phys. Rev. Lett. **128**, 032001 (2022)
[61] A. Andronic, P. Braun-Munzinger, M.K. Köhler, K. Redlich, J. Stachel, Phys. Lett. B **797**, 134836 (2019)

Causality violations in realistic simulations of relativistic nuclear collisions

Christopher Plumberg[1,2,*], Dekrayat Almaalol[1,**], Travis Dore[1,***], Jorge Noronha[1,****], and Jacquelyn Noronha-Hostler[1,†]

[1]Illinois Center for Advanced Studies of the Universe, Department of Physics, University of Illinois at Urbana-Champaign, Urbana, IL 61801, USA
[2]Department of Physics, Pepperdine University, Malibu, CA 90263, USA

Abstract. We show that relativistic causality is violated in the early stages of state-of-the-art heavy-ion hydrodynamic simulations of nuclear collisions. Up to 75% of the initial fluid cells violate nonlinear causality constraints, while superluminal propagation is observed by up to 15% the speed of light. Only after 2-3 fm/c of evolution, do 50% of the fluid cells become definitely causal. Inclusion of pre-equilibrium evolution significantly reduces the number of acausal cells, but it does not eliminate them. Relativistic causality thus imposes constraints on the model parameter space of heavy-ion collision simulations.

1 Introduction

Relativistic viscous hydrodynamics is vital for the phenomenological modeling of ultrarelativistic heavy ion-collisions and provides strong evidence for the formation of a fluid-like state of matter known as the quark-gluon plasma (QGP) in high-energy nuclear collisions. Comparisons to experimental data require modeling all the stages of a heavy-ion collision: the initial stages, the pre-equilibrium stages, relativistic hydrodynamics, and hadronic interactions. Relativistic viscous fluid-dynamics is currently determined by equations of motion for an extended set of dynamical variables which include both equilibrium and non-equilibrium quantities.

Traditionally, hydrodynamics is only expected to accurately describe the behavior of systems close to equilibrium. In the far-from-equilibrium domain, dissipative contributions to the energy-momentum tensor of the system can become comparable to the equilibrium pressure P. Recently, new constraints have been derived [1] involving the magnitude of the viscous currents and the QGP's transport coefficients, which ensure that causality holds in the nonlinear regime of the class of second order hydrodynamic equations of motion [2–4] used in heavy-ion simulations. These constraints define the physically allowable space of out-of-equilibrium corrections to the initial state, providing new theoretical guidance for relativistic viscous hydrodynamics.

*e-mail: christopher.plumberg@pepperdine.edu
**e-mail: almaalol@illinois.edu
***e-mail: tdore2@illinois.edu
****e-mail: jn0508@illinois.edu
†e-mail: jnorhos@illinois.edu

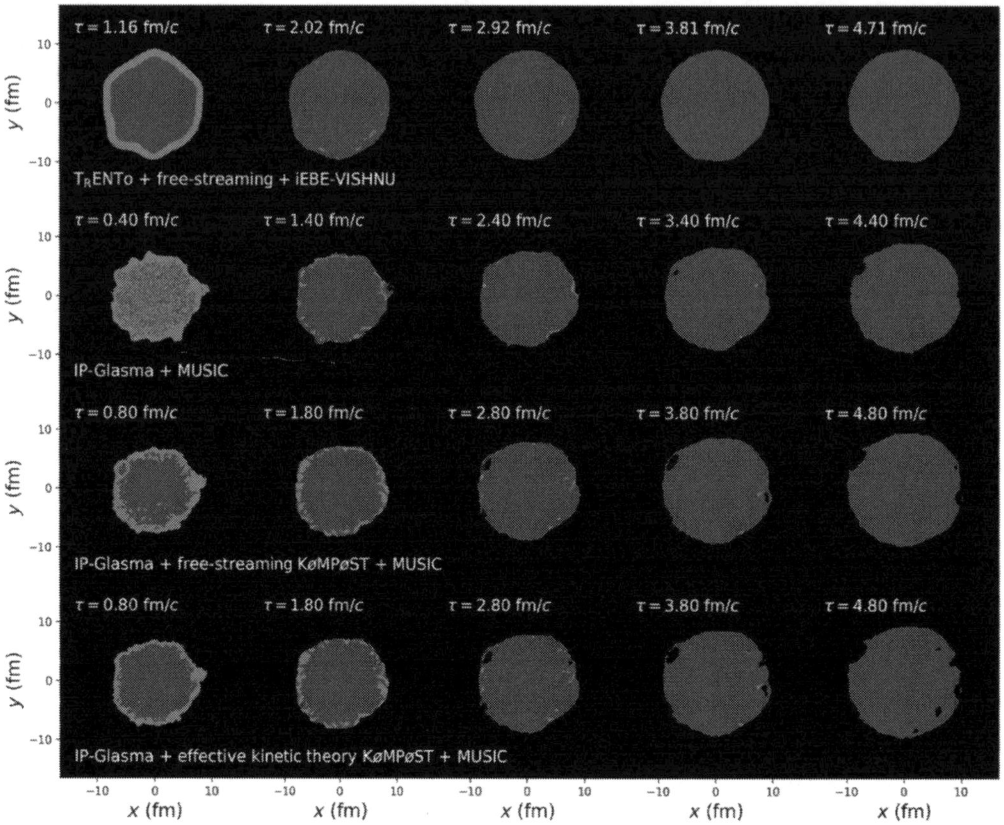

Figure 1. From top to bottom: the TFV scenario and the three IKM scenarios (no KøMPøST, free-streaming KøMPøST, EKT KøMPøST). Colors correspond to the following cell classifications: certainly causal (blue), certainly acausal (red), and purple (indeterminate). Taken from Ref. [5].

In this work we investigate these causality constraints for the most well-behaved scenario in heavy ions collisions simulations: central LHC Pb+Pb collisions. Two state-of-the-art open-source frameworks are used in our study: The first couples T_RENTo+free-streaming+VISHNU ("TFV") and the second framework couples IP-Glasma+(KøMPøST)+MUSIC ("IKM"). Further details about the individual components of each framework are provided in Ref. [5]. Both frameworks have constrained parameters through a Bayesian analysis and have been extensively compared to experimental data.

2 Modeling

In the TFV framework, we adopt the Bayesian tune to LHC p+Pb and Pb+Pb data which combines T_RENTo initial conditions with a conformal, pre-hydrodynamic free-streaming phase, a boost-invariant hydrodynamic phase, and a hadronic afterburner UrQMD. We use the maximum-likelihood parameters for the transport coefficients. In the IKM framework, the initial conditions are from IP-Glasma, coupled to classical Yang-Mills evolution, followed by a boost-invariant hydrodynamics (MUSIC), with a possible intervening phase of pre-hydrodynamic dynamics using KøMPøST: "FS" free-streaming or "EKT" effective kinetic theory. Three different scenarios are considered: (i) IP-Glasma + MUSIC; (ii) IP-Glasma + KøMPøST (FS) + MUSIC; (iii) IP-Glasma + KøMPøST (EKT) + MUSIC. Further

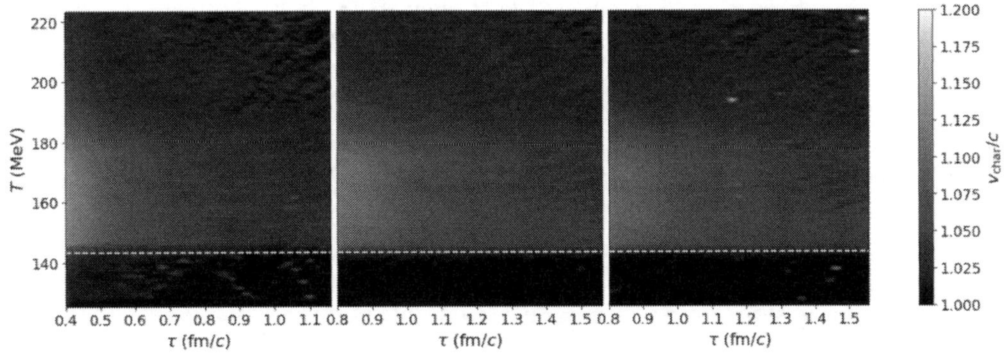

Figure 2. Characteristic velocities for PbPb collision in the IKM framework: Left: No KøMPøST, Middle: Free Streaming, Right: KøMPøST EKT. Taken from Ref. [5].

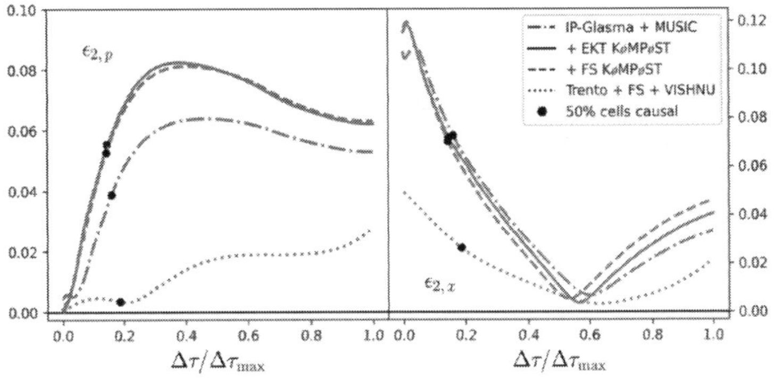

Figure 3. The momentum anisotropy $\epsilon_{2,p}$ (left) and spatial eccentricity $\epsilon_{2,x}$ (right) versus time. Black dots represent the points at which exactly half of the hydrodynamic cells (with $\varepsilon \geq \varepsilon_{FO}$) become explicitly causal. Taken from Ref. [5].

details of our simulations are provided in [5]. Both frameworks are used to generate a single, $\sqrt{s_{NN}} = 2.76$ TeV central Pb+Pb event to which the nonlinear causality constraints are then applied.

The constraints from [1], which apply to the Israel-Stewart-like [2–4] equations of motion used in both frameworks, provide separate sets of *necessary* and *sufficient* conditions for causality. Both sets of conditions correspond to simple inequalities involving transport coefficients and viscous currents. For the explicit expressions of the constraints, see [1, 5]. On the basis of these constraints, we sort grid points in the simulations into three different categories: (i), points at which the sufficient conditions (and consequently the necessary conditions) hold, hence causality is certainly respected; (ii), points at which one or more necessary conditions (and consequently sufficient conditions) are violated, hence causality is unquestionably violated; and (iii), points at which all necessary conditions are satisfied but one or more sufficient conditions fail, hence the analysis cannot determine if causality is violated.

3 Results

The time evolution of the causality analysis for a typical Pb+Pb event is shown for the TFV framework and all scenarios of the IKM framework in Fig. 1. Note that only the fluid cells that have not yet frozen out are plotted. The hydrodynamic simulations are all characterized by pervasive violations of causality, particularly in the first 1-2 fm/c of the collision. For the TFV framework, most of the severe causality violation occurs near the edge of the system where Knudsen and inverse Reynolds numbers become large, though still above freeze out. For the IKM framework, without KøMPøST, approximately 75% of cells in the initial state violate causality. However, the inclusion of KøMPøST pre-equilibrium evolution significantly reduces the causality violation present in the IP-Glasma initial state, bringing it down to approximately 1/3 of fluid cells. EKT offers only a modest improvement over FS.

In Fig. 2, we show the times and temperatures over which characteristic velocities were found to propagate faster than the speed of light in the three cases of the IKM framework. The details concerning the calculation of these velocities are given in [1, 5]. These show that the characteristic velocities were calculated to be around 15% greater than the speed of light, most pervasive both at early times and near the transition temperature. One can note that while the calculated super-luminal speeds seem to decrease as pre-equilibrium is turned on, its duration in time is increased. Also, we note that the scenario with KøMPøST EKT has small regions of very high superluminal characteristic velocities.

We show in Fig. 3 how the momentum anisotropy $\epsilon_{2,p}$ and the spatial eccentricity $\epsilon_{2,x}$[1] evolve with time in different scenarios. The black dots indicate the point in time for each scenario when half of the fluid cells are certainly causal. Whereas most of the final $\epsilon_{2,p}$ in the TFV framework is built up after the majority of the system has become causal, in the IKM scenarios the majority of the $\epsilon_{2,p}$ anisotropy is built up at early times (up to 20 – 30% of evolution time) and nearly half of the final anisotropy is built up when most of the system either explicitly violates causality or the sufficient conditions are not met. Although we cannot at this stage remove the effects of causality violations entirely from our simulations, these results suggest that both collective dynamics and spatial geometry will be affected once causality constraints are taken into account.

4 Conclusions

In this study we showed that there are sizable causality violations in state-of-the-art simulations of heavy-ion collisions. The TFV and IKM frameworks, with parameters constrained by experimental data, yield up to 75% of fluid cells explicitly violating causality in the earliest stages of central Pb+Pb collisions at the LHC. A pre-equilibrium phase prior to the hydrodynamic evolution significantly reduces the amount of causality violation, but it does not fully eliminate it.

Acknowledgements

J.N.H, T.D., D.A., and C.P. are supported by the US-DOE Nuclear Science Grant No. DE-SC0020633. J.N. is partially supported by the U.S. Department of Energy, Office of Science, Office for Nuclear Physics under Award No. DE-SC0021301.

References

[1] F.S. Bemfica, M.M. Disconzi, V. Hoang, J. Noronha, M. Radosz, Phys. Rev. Lett. **126**, 222301 (2021)
[2] W. Israel, J.M. Stewart, Annals Phys. **118**, 341 (1979)
[3] R. Baier, P. Romatschke, D.T. Son, A.O. Starinets, M.A. Stephanov, JHEP **04**, 100 (2008), 0712.2451
[4] G.S. Denicol, H. Niemi, E. Molnar, D.H. Rischke, Phys. Rev. D **85**, 114047 (2012), [Erratum: Phys.Rev.D 91, 039902 (2015)], 1202.4551
[5] C. Plumberg, D. Almaalol, T. Dore, J. Noronha, J. Noronha-Hostler, Phys. Rev. C **105**, L061901 (2022)

[1]The definitions for these quantities are provided in Ref. [5].

Observation of the $\gamma\gamma \rightarrow \tau^+\tau^-$ production in ultraperipheral PbPb collisions with the CMS experiment

Georgios K. Krintiras[1,*] (on behalf of the CMS Collaboration)

[1]Department of Physics and Astronomy, The University of Kansas, Malott Hall, 1251 Wescoe Hall Dr., Lawrence, KS 66045, US

Abstract. We report the observation of the photon-induced $\gamma\gamma \rightarrow \tau^+\tau^-$ production based on a data sample of 404 μb^{-1} collected by the CMS experiment at a per nucleon center-of-mass energy of 5.02 TeV. The cross section is measured in a fiducial phase space region via the decay process involving one muon and three charged hadrons, and is found to be $\sigma(\gamma\gamma \rightarrow \tau^+\tau^-) = \mathbf{4.8 \pm 0.6}$ (stat) $\pm$ **0.5** (syst) μb, in agreement with leading-order quantum electrodynamics predictions. The $\sigma(\gamma\gamma \rightarrow \tau^+\tau^-)$ measurement is used to determine the anomalous magnetic moment of the τ lepton a_τ, which is currently poorly constrained.

1 Introduction

Ultraperipheral collisions (UPC) of heavy ions provide an extremely clean environment to probe the τ lepton anomalous magnetic moment a_τ at the LHC, hence offering ample room for searching for physics beyond the standard model. Here, we present [1] the observation of the photoproduction of pairs of τ leptons in lead-lead (PbPb) collisions PbPb $(\gamma\gamma) \rightarrow$ $Pb^{(*)}$+$Pb^{(*)}$ $\tau^+\tau^-$ (hereafter referred to as $\gamma\gamma \rightarrow \tau^+\tau^-$) by analyzing a data sample collected by the CMS experiment [2] at a per nucleon center-of-mass energy of $\sqrt{s_{NN}} = 5.02$ TeV in 2015, and corresponding to an integrated luminosity of 404 μb^{-1}. The τ leptons are reconstructed in a decay process involving in the final state one muon and three particles ("prongs") assume to be charged π mesons, and the cross section $\sigma(\gamma\gamma \rightarrow \tau^+\tau^-)$ is measured in a fiducial phase space region defined by the muon and pion transverse momenta (p_T) and their pseudorapidities (η). The ATLAS Collaboration has also reported a measurement of $\gamma\gamma \rightarrow \tau^+\tau^-$ using a larger PbPb data sample with an integrated luminosity of 1.44 nb^{-1} [3].

2 Event selection, signal and background estimation

The exchange of photons in UPC does not primarily disassociate the lead ions but scatters them at small angles close to the beam direction, such that the final state is uniquely characterized by low track multiplicity. We thus filter events from UPC by requiring in real time ("online") a single muon with no p_T threshold requirement and minimum event activity above the noise threshold in the forward hadron (HF) calorimeter ($3.0 < |\eta| < 5.2$). A further event selection is applied offline for the "τ_μ" and "τ_{3prong}" candidates, which at the same time defines the $\sigma(\gamma\gamma \rightarrow \tau^+\tau^-)$ fiducial phase space region as summarized in Table 1.

*e-mail: gkrintir@cern.ch

A dedicated $\gamma\gamma \rightarrow \tau^+\tau^-$ [4] Monte Carlo (MC) sample is generated with MadGraph5_amc@nlo (v2.6.5) [5], decayed and hadronized with pythia8 (v2.1.2) [6], and simulated by Geant4 [7] to study the detector effects. The background contamination is estimated in a data-driven way using uncorrelated phase space regions defined based on the track multiplicity and the HF activity. In all cases, good agreement is observed between the measured distributions and the sum of the signal simulation and background estimation, e.g., as shown in Fig. 1 (left) for the invariant mass of the $\tau_{3\mathrm{prong}}$ candidate.

Table 1. Summary of the kinematic event selection, which also defines the fiducial phase space region for the $\sigma(\gamma\gamma \rightarrow \tau^+\tau^-)$ measurement [1]. Additional quality criteria are applied for the τ_μ candidate and all selected tracks.

τ_μ selection	$p_T > 3.5$ GeV for $\lvert\eta\rvert < 1.2$
	$p_T > 2.5$ GeV for $1.2 < \lvert\eta\rvert < 2.4$
Track selection	$p_T > 0.5$ GeV for the leading-p_T
	$p_T > 0.3$ GeV for the (sub-)subleading-p_T
	$\lvert\eta\rvert < 2.5$
$\tau_{3\mathrm{prong}}$ selection	$p_T^{\mathrm{vis}} > 2$ GeV and $m_\tau^{\mathrm{vis}} < 1.5$ GeV

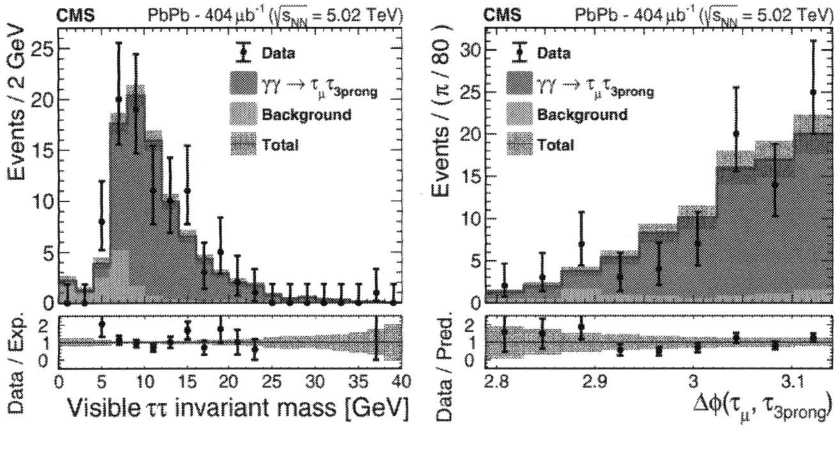

Figure 1. Left: Invariant mass of the three pions forming the $\tau_{3\mathrm{prong}}$ candidate [1]. Right: Difference in azimuthal opening angle between the τ_μ and $\tau_{3\mathrm{prong}}$ candidates [1]. The lower panels show the ratios of data to the signal-plus-background prefit expectation (left) or postfit prediction (right).

3 Results

A binned maximum likelihood method is used for the signal extraction. The difference in opening azimuthal angle between the τ_μ and $\tau_{3\mathrm{prong}}$ leptons, $\Delta\phi(\tau_\mu, \tau_{3\mathrm{prong}})$, is used as the final discriminant. The sum in quadrature of systematic uncertainties, which may affect the normalization and the shape of the $\Delta\phi(\tau_\mu, \tau_{3\mathrm{prong}})$ distribution, is found to be 9.7%, with the statistical component contributing with an almost equal amount. To compare the compatibility of the data with the background-only and signal plus background hypotheses, where the signal is allowed to be scaled by some factor r, we construct a test statistic based on the profile likelihood ratio. The best fit value of the signal strength is $r = 0.99^{+0.16}_{-0.14}$, with the number of signal events of $N_{\text{sig}} = 77 \pm 12$. The post-fit $\Delta\phi(\tau_\mu, \tau_{3\text{prong}})$ distribution is shown in Fig. 1 (right).

The cross section is found to be $\sigma(\gamma\gamma \to \tau^+\tau^-) = 4.8 \pm 0.6\,(\text{stat}) \pm 0.5\,(\text{syst})\,\mu\text{b}$, and it is measured in the fiducial phase space region, following the kinematic requirements listed in Table 1. The result, summarized in Fig. 2 (left), is compared to leading-order quantum electrodynamics (QED) predictions [4, 8]. Further assuming the correction factor of Ref. [4] to extrapolate the fiducial cross section measurement to the full phase space region, we then use the dependence of the total $\sigma(\gamma\gamma \to \tau^+\tau^-)$ as a function of a_τ to extract the model-dependent value of $a_\tau = 0.001^{+0.055}_{-0.089}$. The comparison of the world best a_τ constraints [9] to the ones extracted from this analysis and its expected performance [10] in the High-Luminosity LHC program is given in Fig. 2 (right).

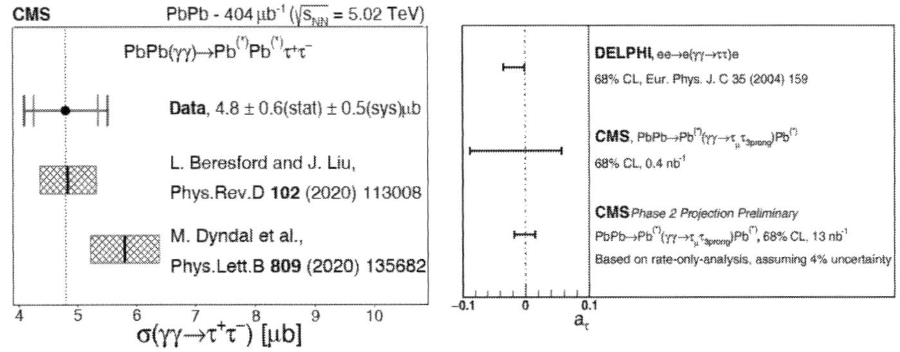

Figure 2. Left: The cross section, $\sigma(\gamma\gamma \to \tau^+\tau^-)$, measured in a fiducial phase space region at $\sqrt{s_{_{\text{NN}}}} = 5.02$ TeV [1], and compared to theoretical predictions at leading-order accuracy in QED [4, 8]. Right: Comparison between the world best a_τ constraints [9] and the ones extracted from this analysis [1] and its performance [10] using the integrated PbPb luminosity expected from the High-Luminosity LHC program.

4 Summary

In summary, an observation of the τ lepton pair production in ultraperipheral nucleus-nucleus collisions is reported. Events with one muon and three particles identified as charged pions in the final state are reconstructed from a lead-lead data sample collected by the CMS experiment at $\sqrt{s_{_{\text{NN}}}} = 5.02$ TeV in 2015, and corresponding to an integrated luminosity of $404\,\mu\text{b}^{-1}$. This measurement introduces a novel experimental strategy using heavy ion data recorded by the LHC in order to measure the τ lepton magnetic moment a_τ. This approach can be used to extract constraints on a_τ that surpass the current best from previous lepton-lepton colliders.

Acknowledgements

GKK is supported by the Office of Nuclear Physics in the Department of Energy (DOE NP) under grant number DE-FG02-96ER40981.

References

[1] CMS Collaboration, "Observation of τ lepton pair production in ultraperipheral lead-lead collisions at $\sqrt{s_{\text{NN}}} = 5.02$ TeV", (2022). arXiv:2206.05192. Accepted by *Phys. Rev. Lett.*

[2] CMS Collaboration, "The CMS Experiment at the CERN LHC", *JINST* **3** (2008) S08004, .

[3] ATLAS Collaboration, "Observation of the $\gamma\gamma \to \tau\tau$ process in PbPb collisions and constraints on the τ lepton anomalous magnetic moment with the ATLAS detector", (2022). arXiv:2204.13478. Accepted by *Phys. Rev. Lett.*

[4] L. Beresford and J. Liu, "New physics and τ $g-2$ using LHC heavy ion collisions", *Phys. Rev. D* **102** (2020) 113008, , arXiv:1908.05180. [Erratum: 10.1103/PhysRevD.106.039902].

[5] J. Alwall et al., "The automated computation of tree-level and next-to-leading order differential cross sections, and their matching to parton shower simulations", *JHEP* **07** (2014) 079, , arXiv:1405.0301.

[6] T. Sjöstrand et al., "An introduction to PYTHIA 8.2", *Comput. Phys. Commun.* **191** (2015) 159, , arXiv:1410.3012.

[7] GEANT4 Collaboration, "GEANT4—a simulation toolkit", *Nucl. Instrum. Meth. A* **506** (2003) 250, .

[8] M. Dyndal, M. Klusek-Gawenda, M. Schott, and A. Szczurek, "Anomalous electromagnetic moments of τ lepton in $\gamma\gamma \to \tau^+\tau^-$ reaction in PbPb collisions at the LHC", *Phys. Lett. B* **809** (2020) 135682, , arXiv:2002.05503.

[9] DELPHI Collaboration, "Study of τ pair production in photon-photon collisions at LEP and limits on the anomalous electromagnetic moments of the τ lepton", *Eur. Phys. J. C* **35** (2004) 159, , arXiv:hep-ex/0406010.

[10] CMS Collaboration, "Snowmass White Paper Contribution: Physics with the Phase-2 ATLAS and CMS Detectors", CMS Physics Analysis Summary CMS-PAS-FTR-22-001, 2022.

Understanding the initial state effects by the measurement of the Drell-Yan process in pPb collisions with CMS

Hyunchul Kim[1,*] on behalf of the CMS Collaboration

[1]Chonnam National University

Abstract. The Drell-Yan process is considered as an essential probe to understand initial-state protons, usually described by the parton distribution functions (PDFs) for stand-alone protons and by the nuclear PDFs (nPDFs) for protons confined in the nucleus. In the LHC era, Z and W boson production in pPb and PbPb collisions have been used to investigate the initial-state effects. In this presentation, we report results for the Drell-Yan process in pPb collisions at a center-of-mass energy of 8.16 TeV with the CMS detector. Differential cross sections are presented as functions of dimuon p_T, rapidity and ϕ^* in a wider dimuon mass region which includes not only the Z boson mass range but also the lower mass region down to 15 GeV. In addition, the forward-backward asymmetries are shown in both mass regions, and the uncertainties are found to be smaller than in the model calculations. The results in the Z mass region are the most precise to date, while the measurements in the lower mass region allow access to a new phase space of nPDF studies at a lower longitudinal momentum fraction x and lower energy scale Q^2. The results are compared to the EPPS16 and nCTEQ15WZ nPDFs, and the free-proton PDF CT14. An improved understanding of the nuclear PDF and the modeling of pPb collisions is achieved.

1 Introduction

The parton distribution function (PDF) is defined as the probability density for finding partons i.e. quarks and gluons inside the proton as a function of the momentum fraction x of the parton at energy scale Q^2 [1]. The nuclear PDF (nPDF) can be modified because the nucleons exist in the nucleus, compared to that in a free proton [2]. These (n)PDFs are extracted from global fits to experimental data [3, 4].

Drell-Yan (DY) process, proposed by Sidney Drell and Tung-Mow Yan in 1970 to explain the production of lepton-antilepton pairs in high energy collisions [5], is an important probe for nPDF studies as they are created in the initial stages of the collisions and are free from final state interactions.

In the DY process, quark and antiquark pairs from colliding nucleons are annihilated and decayed into lepton pairs with opposite charges by exchanging a virtual γ or a Z boson. A Z boson has been used as the clean signal to test nPDFs because of the higher mass and small backgrounds. In addition, the measurement through a virtual γ allows access to a wider x range.

*e-mail: hyunchul.kim@cern.ch

In this contribution, the study of DY dimuon production from proton-lead (pPb) collisions at $\sqrt{s_{NN}} = 8.16$ TeV with the CMS detector is reported. The detailed information about the analysis can be found in [6].

2 Event Selection

The differential cross section of the DY process via dimuon decay channel is measured within the dimuon invariant mass range $15 < m_{\mu\mu} < 600$ GeV. The transverse momentum (p_T), the absolute value of the rapidity in the nucleon-nucleon center-of-mass frame (y_{CM}), and the ϕ^* dependence are measured in two dimuon mass ranges for $15 < m_{\mu\mu} < 60$ GeV and $60 < m_{\mu\mu} < 120$ GeV, targeting to lower mass range and Z boson mass, respectively. The variable ϕ^* is defined as $\tan\left(\frac{\pi-\Delta\phi}{2}\right)\sin(\theta^*_\eta)$, where $\Delta\phi$ is the opening angle between decaying muons [7–9]. The θ^*_η is related to the pseudorapidities of the muons by the relation $\cos(\theta^*_\eta) = \tanh(\Delta\eta/2)$, where $\Delta\eta$ is the difference in pseudorapidity between the two muons. The ϕ^* value are measured by the angular variables with better precision than p_T, especially at lower p_T values. The forward-backward ratios (R_{FB}) are defined as the ratio of the cross sections in the positive rapidity range to that in the negative rapidity range and measured in the same dimuon mass ranges.

For the analysis, the proton-lead ion collision data at a nucleon-nucleon center-of-mass energy of $\sqrt{s_{NN}} = 8.16$ TeV, with a total integrated luminosity of 173.4 ± 6.1 nb^{-1} are used. At least two muons in the pseudorapidity in a laboratory frame with opposite charges are required. The main event and dimuon candidate selection conditions are that the requiring p_T of the leading muon and another muon is larger than 15 and 10 GeV, respectively. The other detailed selection conditions can be found in [6].

3 Results

At the beginning of the analysis, various backgrounds are estimated by their source. The backgrounds involving two isolated muons, such as $Z/\gamma^* \rightarrow \tau^+\tau^-$, $t\bar{t}$, tW, or dibosons are estimated from the "eμ method" with simulated samples. The contribution from heavy-flavor meson decays is only a small fraction and corrected from same-sign eμ events. The contributions with one or more muons in jets are subtracted by the "misidentification rate method" [10, 11] After subtracting the sum of background contributions from data, the domination of the DY signal is measured, and the agreement between data and the expectation is shown. Following the background estimation, detector resolution and final state radiation are corrected for. The acceptance and efficiency correction is considered to the corrected differential cross section. The differential fiducial cross sections without acceptance correction are also measured.

Figure 1 shows the fiducial cross sections as the function of dimuon rapidity in the $15 < m_{\mu\mu} < 60$ GeV and $60 < m_{\mu\mu} < 120$ GeV ranges. The experimental data are compared with the expectations from POWHEG, using the CT14 only proton PDF [12] and CT14+EPPS16 nPDF sets [3]. The measurement in the lower mass range $15 < m_{\mu\mu} < 60$ GeV is allowed to access the lower x region. In the Z boson mass range $60 < m_{\mu\mu} < 120$ GeV, EPPS16 nPDF shows better agreement with data than CT14 PDF alone. But the concrete conclusions in favor of nPDF expectation are prevented by imperfect modeling in POWHEG, from p_T or ϕ^* differential results in the same mass region.

The R_{FB} results in the $15 < m_{\mu\mu} < 60$ GeV and $60 < m_{\mu\mu} < 120$ GeV are shown in Fig. 2. The results from the data are compared with CT14 proton PDF and EPPS16 and nCTEQ15WZ nPDF predictions [4] based on CT14. In the Z boson mass region, R_{FB} is

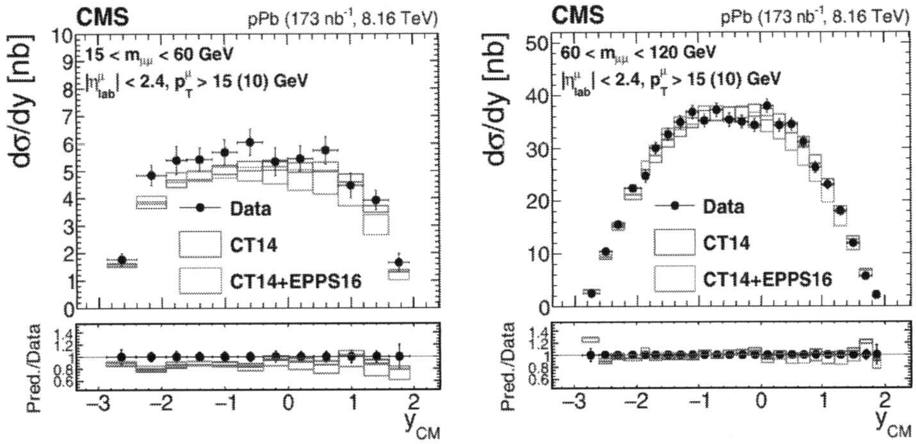

Figure 1. Differential fiducial cross section for the DY process measured in the dimuon channel, as a function of the rapidity in the center-of-mass frame for $15 < m_{\mu\mu} < 60$ GeV (left) and $60 < m_{\mu\mu} < 120$ GeV (right). The vertical error bars on the data represent the quadratic sum of the statistical and systematic uncertainties. Model predictions from CT14 [12] protn PDF (blue) or CT14+EPPS16 [3] (red) nPDF sets are shown. The boxes show the 68% confidence level (n)PDF uncertainty on these predictions. The ratios of predictions over data are shown in the lower panels, where the data and (n)PDF uncertainties are shown separately, as error bars around one and as colored boxes, respectively.

smaller than unity as shown in the experimental data and the expectations from nPDFs in contrast to proton PDF only. The data points that have smaller uncertainties compared to expectation can also contribute to further improving the models.

Results are consistent with previous W boson in pPb measurements at the same collision energy [13] and Z boson measurements in pPb measurements at $\sqrt{s_{\rm NN}} = 5.02$ TeV [14]. Recently the DY measurement contributed to the update of nPDF models such as nNNPDF3.0 [15]. In addition, more contributions to further nPDF developments by the constraints are expected.

4 Summary

Measurements of cross sections and forward-backward ratios for the Drell-Yan process are reported for the first time in proton-ion collisions with the CMS detector. The results are presented in two mass regions, in the range below the Z boson mass, and around the Z boson mass. The measurements in the lower mass range allow access to a new phase space for (n)PDF studies. These novel results provide strong constraints to theoretical calculations and are expected to become more essential for the predictions and analysis of future LHC Run3 data.

References

[1] A.D. Martin, W.J. Stirling and R.G. Roberts, Phys. Rev. D **50**, 6734 (1994)
[2] J. J. Ethier and E. R. Nocera, Annu. Rev. Nucl. Part. Sci. **70**, 43 (2020)
[3] K.J. Eskola, P. Paakkinen, H. Paukkunen and C.A. Salgado, Eur. Phys. J. C **77**, 163 (2017)

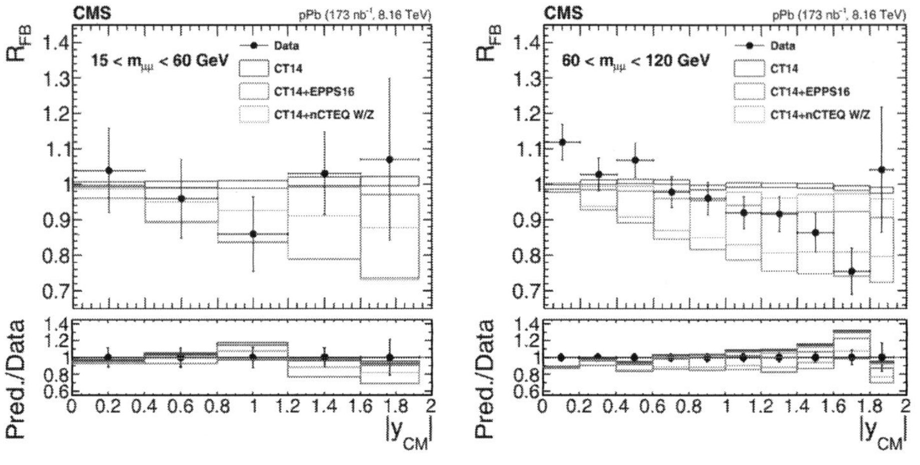

Figure 2. Forward-backward ratios for 15 – 60 GeV (left) and 60 – 120 GeV (right). The vertical error bars on the data points represent the quadratic sum of the statistical and systematic uncertainties. The theory predictions from the CT14 [12] proton PDF (blue), CT14+EPPS16 [3] (red), or CT14+nCTEQ15WZ [4] (green) nPDF sets. The boxes show the 68% confidence level (n)PDF uncertainty in these predictions. The ratios of predictions over data are shown in the lower panels, where the data and the (n)PDF uncertainties are shown separately, as error bars around one and as coloured boxes, respectively.

[4] A. Kusina et al., Eur. Phys. J. C **80**, 968 (2020)
[5] S.D. Drell and T.-M. Yan, Phys. Rev. Lett. **25**, 316 (1970)
[6] CMS Collaboration, JHEP **05**, 182 (2021)
[7] A. Banfi, S. Redford, M. Vesterinen, P. Waller and T.R. Wyatt, Eur. Phys. J. C **71**, 1600 (2011)
[8] A. Banfi, M. Dasgupta, S. Marzani and L. Tomlinson, Phys. Lett. B **715**, 152 (2012)
[9] S. Marzani, EPJ Web Conf. **49**, 14007 (2013)
[10] CMS Collaboration, JHEP **08**, 063 (2022)
[11] CMS Collaboration, JHEP **04**, 025 (2015)
[12] S. Dulat et al., Phys. Rev. D **93**, 033006 (2016)
[13] CMS Collaboration, Phys. Lett. B **800**, 135048 (2020)
[14] CMS Collaboration, Phys. Lett. B **759**, 36 (2016)
[15] R. A. Khalek, R. Gauld, T. Giani, E. R. Nocera, T. R. Rabemananjara, J. Rojo, arXiv:2201.12363 (2022)

Scaling properties of background- and chiral-magnetically-driven charge separation: implications for the chiral magnetic effect in heavy ion collisions

Roy Lacey[1,*]

[1]Depts. of Chemistry & Physics, Stony Brook University, Stony Brook, New York 11794, USA

Abstract. The scaling properties of the $R_{\Psi_2}(\Delta S)$ correlator and the $\Delta\gamma$ correlator are used to investigate a possible chiral-magnetically-driven (CME) charge separation in p+Au, d+Au, Ru+Ru, Zr+Zr, and Au+Au collisions at $\sqrt{s_{\rm NN}}$ = 200 GeV, and in p+Pb ($\sqrt{s_{\rm NN}}$ = 5.02 TeV) and Pb+Pb collisions at $\sqrt{s_{\rm NN}}$ = 5.02 and 2.76 TeV. The results for p+Au, d+Au, p+Pb, and Pb+Pb collisions, show the $1/\mathrm{N_{ch}}$ scaling for background-driven charge separation. However, the results for Au+Au, Ru+Ru, and Zr+Zr collisions show scaling violations which indicate a CME contribution in the presence of a large background. In mid-central collisions, the CME accounts for approximately 27% of the signal + background in Au+Au and roughly a factor of two smaller for Ru+Ru and Zr+Zr, which show similar magnitudes.

Metastable domains of gluon fields with non-trivial topological configurations can form in the magnetized chiral relativistic quark-gluon plasma (QGP) [1] produced in collisions at RHIC and the LHC. The colliding ions generate the magnetic field ($\vec{B}$) at early times [2]. The interaction of chiral quarks with the gluon fields can drive a chiral imbalance resulting in an electric current $\vec{J}_V = \frac{N_c e\vec{B}}{2\pi^2}\mu_A$, along the $\vec{B}$-field, i.e., perpendicular to the reaction plane; N_c is the color factor, and μ_A is the axial chemical potential that quantifies the imbalance between right- and left-handed quarks. The resulting final-state charge separation, termed the chiral magnetic effect (CME) [1], is of great experimental and theoretical interest. However, its experimental characterization has been hampered by significant background.

The charge separation can be quantified via the P-odd sine term a_1, in the Fourier decomposition of the charged-particle azimuthal distribution [3]:

$$\frac{dN_{\rm ch}}{d\phi} \propto 1 + 2\sum_n (v_n \cos(n\Delta\phi) + a_n \sin(n\Delta\phi) + ...) \tag{1}$$

where $\Delta\phi = \phi - \Psi_{\rm RP}$ gives the particle azimuthal angle with respect to the reaction plane (RP) angle, and v_n and a_n denote the coefficients of the P-even and P-odd Fourier terms, respectively. A direct measurement of a_1, is not possible due to the strict global $\mathcal{P}$ and $\mathcal{CP}$ symmetry of QCD. However, their fluctuation and/or variance $\tilde{a}_1 = \left\langle a_1^2 \right\rangle^{1/2}$ can be measured with charge-sensitive correlators such as the γ-correlator [3] and the $R_{\Psi_2}(\Delta S)$ correlator [4].

The γ-correlator measures charge separation as: $\gamma_{\alpha\beta} = \left\langle \cos\left(\phi_\alpha + \phi_\beta - 2\Psi_2\right)\right\rangle$, $\Delta\gamma = \gamma_{\rm OS} - \gamma_{\rm SS}$, where Ψ_2 is the azimuthal angle of the 2nd-order event plane which fluctuates

*e-mail: Roy.Lacey@Stonybrook.edu

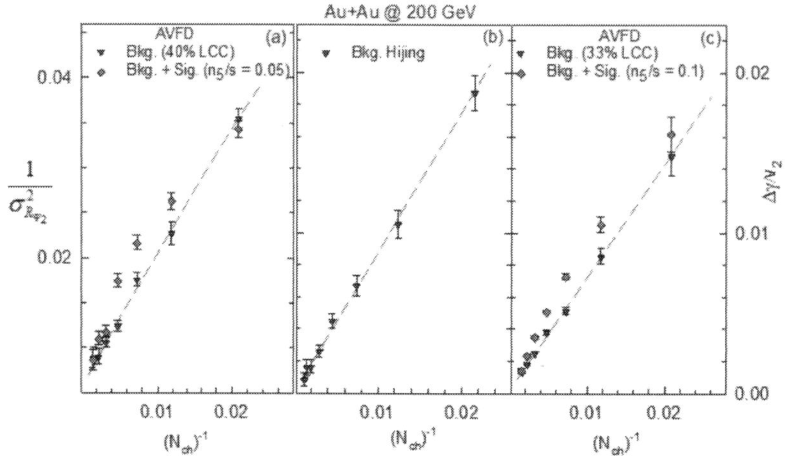

Figure 1. $\sigma^{-2}_{R_{\Psi_2}}$ vs. $1/N_{ch}$ (a) and $\Delta\gamma/v_2$ vs. $1/N_{ch}$ (b) and (c) for simulated Au+Au collisions at $\sqrt{s_{NN}} = 200$ GeV. The results for $\sigma^{-2}_{R_{\Psi_2}}$ and $\Delta\gamma/v_2$ from the AVFD model are shown for background and for signal + background as indicated. The Hijing model results are only shown for the background. The dashed lines represent linear fits to the background values.

about the RP, ϕ denote the particle azimuthal emission angles, α, β denote the electric charge (+) or (−) and SS and OS represent same-sign (++, −−) and opposite-sign (+ −) charges. Measurements of the quotient $\Delta\gamma/v_2$ with the 2nd-order anisotropy coefficient v_2, are usually employed to aid quantification of the background-driven charge separation.

The $R_{\Psi_2}(\Delta S)$ correlator measures charge separation relative to Ψ_2 via the ratio: $R_{\Psi_2}(\Delta S) = C_{\Psi_2}(\Delta S)/C^{\perp}_{\Psi_2}(\Delta S)$, where $C_{\Psi_2}(\Delta S)$ and $C^{\perp}_{\Psi_2}(\Delta S)$ are correlation functions that quantify charge separation ΔS, approximately parallel and perpendicular (respectively) to the $\vec{B}$-field. The charge-shuffling procedure used to construct the correlation functions ensures identical properties for their numerator and denominator, except for the charge-dependent correlations, which are of interest [4]. $C_{\Psi_2}(\Delta S)$ measures both CME- and background-driven charge separation while $C^{\perp}_{\Psi_2}(\Delta S)$ measures only the background. After correcting the $R_{\Psi_2}(\Delta S)$ distributions for the effects of particle-number fluctuations and the event-plane resolution, their inverse variance $\sigma^{-2}_{R_{\Psi_2}}$ are used to quantify the charge separation [4].

In this work, we use model simulations to chart the scaling properties of $\sigma^{-2}_{R_{\Psi_2}}$ and $\Delta\gamma/v_2$ for the background and signal + background, respectively, in A+A collisions. We then leverage these scaling properties to identify and characterize a possible CME-driven charge separation using previously published data for p+Au, d+Au, Ru+Ru, Zr+Zr and Au+Au collisions at RHIC [5–10], and p+Pb and Pb+Pb collisions at the LHC [11–14].

Figure 1 shows the results for $\sigma^{-2}_{R_{\Psi_2}}$ and $\Delta\gamma/v_2$ obtained with the AVFD and Hijing models for Au+Au collisions. Note that these models emphasize different sources for the charge-dependent non-flow background; the initial axial charge density n_5/s and the degree of local charge conservation (LCC) regulate the magnitude of the CME- and background-driven charge separation in the AVFD model. The solid triangles in Fig. 1 show that the background scales as $1/N_{ch}$ – the expected trend for the charge-dependent non-flow correlations. By contrast, the signal (Sig.) + background values (solid diamonds) indicate positive deviations from the background scaling [16, 17]. This dependence can be represented as; $\Delta\gamma/v_2 = a + b/(N_{ch})^{1-c}$, and $\sigma^{-2}_{R_{\Psi_2}} = a' + b'/(N_{ch})^{1-c'}$, for the small values of n_5/s indicated in Fig. 1. Here, a, b and c are parameters; c characterizes the degree of the scaling violation.

Note that for $c \sim 0$ the $1/\mathrm{N_{ch}}$ scaling for the background is retrieved, as demonstrated with the AVFD model in Fig. 1.

The scaling violation gives a direct signature of the CME-driven contributions to the charge separation (Figs. 1 (a) and (c)). It can be quantified via the fraction: $f^{\Delta\gamma}_{\mathrm{CME}} = [\Delta\gamma/v_2(Sig. + Bkg.) - \Delta\gamma/v_2(Bkg.)]/[\Delta\gamma/v_2(Sig. + Bkg.)]$, $f^{R}_{\mathrm{CME}} = [\sigma^{-2}_{R_{\Psi_2}}(Sig. + Bkg.) - \sigma^{-2}_{R_{\Psi_2}}(Bkg.)]/[\sigma^{-2}_{R_{\Psi_2}}(Sig. + Bkg.)]$. The scaling patterns in Fig. 1 suggest that the observation of $1/\mathrm{N_{ch}}$ scaling for the experimental $\sigma^{-2}_{R_{\Psi_2}}$ and $\Delta\gamma/v_2$ measurements would strongly indicate background-driven charge separation with little room for a CME contribution. However, observing a violation of this $1/\mathrm{N_{ch}}$ scaling would indicate the CME-driven contribution. Figs. 1 (a) and (c) also indicate comparable background and signal + background $\sigma^{-2}_{R_{\Psi_2}}$ and $\Delta\gamma/v_2$ values in central and peripheral collisions, suggesting that the background dominates over that of the CME-driven contributions in these collisions. Note the reduction of $\vec{B}$ in central collisions and the enhanced de-correlation between the event plane and the $\vec{B}$-field in peripheral collisions. Since the background dominates in central and peripheral collisions, the $\sigma^{-2}_{R_{\Psi_2}}$ and $\Delta\gamma/v_2$ measurements for these collisions can be leveraged with $1/\mathrm{N_{ch}}$ scaling to obtain a quantitative estimate of the background over the entire centrality span (cf. Fig. 1). Here, an important proviso is to experimentally establish that the background in $p(d)$+A and A+A collisions scale over the full centrality span.

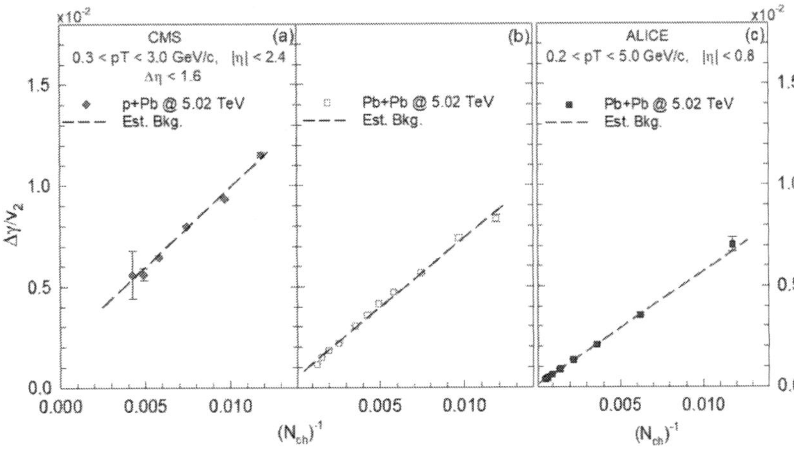

Figure 2. $\Delta\gamma/v_2$ vs. $1/\mathrm{N_{ch}}$ for p+Pb (a) and Pb+Pb [(c) and (d)] collisions at $\sqrt{s_{\mathrm{NN}}} = 5.02$ TeV. The dashed lines indicate an estimate of the background. The data are taken from Refs. [12, 13, 15].

The v_2 and $\Delta\gamma$ values reported for p+Au, d+Au, Ru+Ru, Zr+Zr and Au+Au collisions at RHIC [5–10], and p+Pb and Pb+Pb collisions at the LHC [12–15] were used to investigate the scaling properties of $\Delta\gamma/v_2$. Fig. 2 shows the results for p+Pb and Pb+Pb collisions at $\sqrt{s_{\mathrm{NN}}} = 5.02$ TeV. They indicate that $\Delta\gamma/v_2$ essentially scales as $1/\mathrm{N_{ch}}$ ($c \approx 0$), suggesting negligible CME contributions in these collisions. They also confirm that the combined sources of background (LCC, resonances, back-to-back jets, ...), which should be substantial, especially for p+Pb, scale as $1/\mathrm{N_{ch}}$. Note as well that the CME contribution is negligible in $p(d)$+A collisions because of significant reductions in $\vec{B}$, and the sizable de-correlation between the event plane and the $\vec{B}$-field [12]. Thus, the scaling patterns of $\Delta\gamma/v_2$ for these systems' sizable backgrounds give a direct experimental constraint on the validity of $1/\mathrm{N_{ch}}$ scaling of the background.

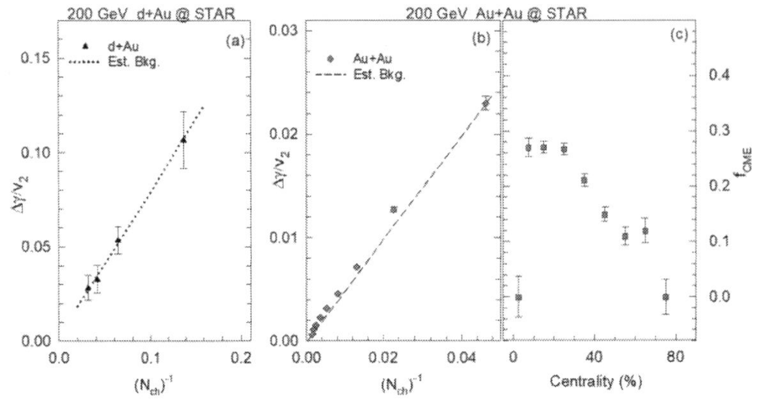

Figure 3. $\Delta\gamma/v_2$ vs. $1/N_{ch}$ [(a) and (b)] and f_{CME} vs. centrality (c) for *d*+Au and Au+Au collisions at $\sqrt{s_{NN}}$ = 200 GeV. The dotted and dashed lines indicate an estimate of the background contributions. The data are taken from Refs. [7, 9**?**].

The scaling results for Au+Au collisions at $\sqrt{s_{NN}}$ = 200 GeV are shown in Fig. 3. The $1/N_{ch}$ scaling apparent for *d*+Au collisions (Fig. 3 (a)) confirms the expectation that the CME is negligible in these collisions. It also confirms that the combined sources of background (LCC, resonances, back-to-back jets, ...), which could be substantial in *d*+Au collisions, show $1/N_{ch}$ scaling. In contrast to *d*+Au, the results for Au+Au (Fig. 3(b)) show visible indications of a violation ($c > 0$) to the $1/N_{ch}$ scaling observed for background-driven charge separation in *p*(*d*)+A collisions. Similar violations were observed for Ru+Ru and Zr+Zr [17]. The scaling violation is similar to that observed for signal + background in Figs. 1 (a) and (c), suggesting an unambiguous non-negligible CME contribution to the measured $\Delta\gamma/v_2$ in Au+Au, Ru+Ru, and Zr+Zr collisions. The estimates of the background for all three systems are obtained by leveraging the $\Delta\gamma/v_2$ measurements for peripheral and central collisions with $1/N_{ch}$ scaling [17]. Here, it is noteworthy that the simulated results from the AVFD and HIJING models, as well as the measurements presented in Figs. 2 and 3(a), provide strong constraints that the combined sources of background, scale as $1/N_{ch}$ over the full centrality span. The background estimates were used to extract f_{CME} values for Au+Au, (Fig. 3 (c)) Ru+Ru and Zr+Zr collisions respectively. They indicate non-negligible f_{CME} values that vary with centrality. In mid-central collisions, f_{CME} ~ 27% for Au+Au collisions, which is roughly a factor of two larger than the values for Ru+Ru and Zr+Zr. Within the uncertainties, no significant difference between the values for Ru+Ru and Zr+Zr was observed, suggesting that $\Delta\gamma/v_2$ is sensitive to CME-driven charge separation in Ru+Ru and Zr+Zr collisions but may be insensitive to the signal difference between them [17].

In summary, the scaling properties of the $R_{\Psi_2}(\Delta S)$ and the $\Delta\gamma$ correlators have been used to characterize the CME in several colliding systems at RHIC and the LHC. The results for *p*+Au and *d*+Au collisions at $\sqrt{s_{NN}}$ = 200 GeV and *p*+Pb ($\sqrt{s_{NN}}$ = 5.02 TeV) and Pb+Pb collisions at $\sqrt{s_{NN}}$ = 5.02 and 2.76 TeV, scales as $1/N_{ch}$ consistent with background-driven charge separation. However, the results for Au+Au, Ru+Ru and Zr+Zr collisions ($\sqrt{s_{NN}}$ = 200 GeV) show scaling violations which indicate a CME-driven contribution in the presence of significant background. In mid-central collisions, f_{CME} ~ 27% for Au+Au collisions and approximately a factor of two smaller in Ru+Ru and Zr+Zr collisions but with similar magnitudes for the two isobars.

References

[1] D. Kharzeev, Phys. Lett. B **633**, 260-264 (2006)
[2] M. Asakawa, A. Majumder and B. Muller, Phys. Rev. C **81** (2010), 064912
[3] S. A. Voloshin, Phys. Rev. C **70** (2004), 057901
[4] N. Magdy *et al.*, Phys. Rev. C **97** (2018) no.6, 061901
[5] B. I. Abelev *et al.* [STAR], Phys. Rev. C **81** (2010), 054908
[6] B. I. Abelev *et al.* [STAR], Phys. Rev. Lett. **103** (2009), 251601
[7] L. Adamczyk *et al.* [STAR], Phys. Rev. C **88** (2013) no.6, 064911
[8] L. Adamczyk *et al.* [STAR], Phys. Rev. Lett. **113** (2014), 052302
[9] J. Adam *et al.* [STAR], Phys. Lett. B **798** (2019), 134975
[10] M. Abdallah *et al.* [STAR], Phys. Rev. C **105** (2022) no.1, 014901
[11] B. Abelev *et al.* [ALICE], Phys. Rev. Lett. **110** (2013) no.1, 012301
[12] V. Khachatryan *et al.* [CMS], Phys. Rev. Lett. **118** (2017) no.12, 122301
[13] A. M. Sirunyan *et al.* [CMS], Phys. Rev. C **97** (2018) no.4, 044912
[14] S. Acharya *et al.* [ALICE], Phys. Lett. B **777** (2018), 151-162
[15] S. Acharya *et al.* [ALICE], JHEP **09** (2020), 160
[16] R. A. Lacey, N. Magdy, P. Parfenov and A. Taranenko, [arXiv:2203.10029 [nucl-ex]].
[17] R. A. Lacey and N. Magdy, [arXiv:2206.05773 [nucl-ex]].

Electroweak-boson production from small to large systems with ALICE at the LHC

Guillaume Taillepied[1,*] for the ALICE Collaboration

[1]Research Division and ExtreMe Matter Institute EMMI,
GSI Helmholtzzentrum für Schwerionenforschung GmbH, Darmstadt, Germany

Abstract. Results on the recent measurements of the electroweak-boson production performed by the ALICE Collaboration are reported. The $W^{\pm}$-boson production was measured at midrapidity in pp collisions at $\sqrt{s} = 13$ TeV, via the electronic decay of the boson. It was also measured in association with a hadron emitted back-to-back with respect to the electron. In heavy-ion collisions, the $W^{\pm}$- and Z^{0}-boson production was measured at forward rapidity in p–Pb collisions at $\sqrt{s_{\rm NN}} = 8.16$ TeV and Pb–Pb collisions at $\sqrt{s_{\rm NN}} = 5.02$ TeV, in their muonic decay channel. Measurements in pp collisions shed light on multiple parton interactions and colour reconnection mechanisms. In heavy-ion collisions the measurements of electroweak bosons allow to probe the initial state of the collision.

1 Introduction

One of the important observations in the LHC era is that the self-normalised production of hadrons increases faster than linearly as a function of the self-normalised charged-particle multiplicity [1]. This enhancement in high-multiplicity events is not fully understood, but calculations including multiple parton interactions (MPI) and colour reconnection (CR) are able to reproduce the observed trend. The measurement of the multiplicity-dependent production of the $W^{\pm}$ boson, which is colourless, with associated hadron in proton–proton (pp) collisions can then provide further understanding of the MPI and CR mechanisms.

In heavy-ion collisions (HIC), electroweak bosons are valuable probes of the initial phase of the collision, on which a precise knowledge is required to disentangle QGP-induced phenomena from other nuclear effects. Being based on hard processes, the production of the $W^{\pm}$ and Z^{0} bosons is highly sensitive to the initial state, especially the modifications of the Parton Distribution Functions (PDF) in the nucleus with respect to that in the proton. The bosons decay before the typical time of creation of the QGP and can be measured via their leptonic decays, providing a final state insensitive to the strong force. The whole process is then medium-blind, carrying the information from the initial state to the detector where it can be recorded.

The measurements presented here were performed using data collected with the ALICE detector [2]. In pp collisions at $\sqrt{s} = 13$ TeV, electrons from $W^{\pm}$-boson decays were measured at midrapidity, in the rapidity interval $|y| < 0.6$. Electrons coming from $W^{\pm}$-boson

*e-mail: g.taillepied@cern.ch

decays typically have a large transverse momentum (p_T) and are well isolated from other particles. They were thus identified by considering electrons in the range $30 < p_T^e < 60$ GeV/c, then defining an isolation cone surrounding the electron in which the total energy is requested to be below a certain threshold. The associated hadrons, produced together with $W^\pm$ bosons, were detected away-side in azimuth with respect to the electron coming from the decay of the boson, applying a lower minimum p_T selection of 5 GeV/c.

Measurements in HIC were performed in proton–lead (p–Pb) collisions at $\sqrt{s_{NN}}$ = 8.16 TeV and lead–lead (Pb–Pb) collisions at $\sqrt{s_{NN}}$ = 5.02 TeV. The data were collected from muon-triggered events in the muon forward spectrometer, covering the rapidity interval $2.5 < y < 4$. The Z^0-boson candidates were reconstructed by combining high-p_T muons ($p_T^\mu > 20$ GeV/c) in pairs of opposite charge, considering the pairs with an invariant mass in the range $60 < m_{\mu^+\mu^-} < 120$ GeV/c^2. The number of W^- and W^+ bosons was extracted via a template fit to the single-muon p_T distribution, accounting for the various contributions to the inclusive spectrum. As the low-p_T region features a very low signal-to-background ratio, the measurements were performed on muons with $p_T > 10$ GeV/c.

The available measurements of electroweak bosons performed by the ALICE Collaboration at forward rapidity can be found in Refs. [3–5].

2 Results

2.1 Measurements in pp collisions

Figure 1 shows the self-normalised multiplicity-dependent yield of electrons from $W^\pm$ boson decays (combining W^- and W^+ for increased precision), and that of hadrons associated with a $W^\pm$ boson, measured in pp collisions at $\sqrt{s}$ = 13 TeV. The yield of electrons from $W^\pm$-boson decays is consistent with a linear increase as a function of the charged-particle multiplicity, while the yield of associated hadron shows a faster-than-linear increase. This measurement thus suggests a significant correlation of the associated hadron production with the event multiplicity, and the trend is well reproduced by PYTHIA 8 [6] calculations including MPI and CR mechanisms. This analysis also constitutes the first measurement of electroweak bosons at midrapidity with ALICE, and will serve as reference for similar future measurements in HIC.

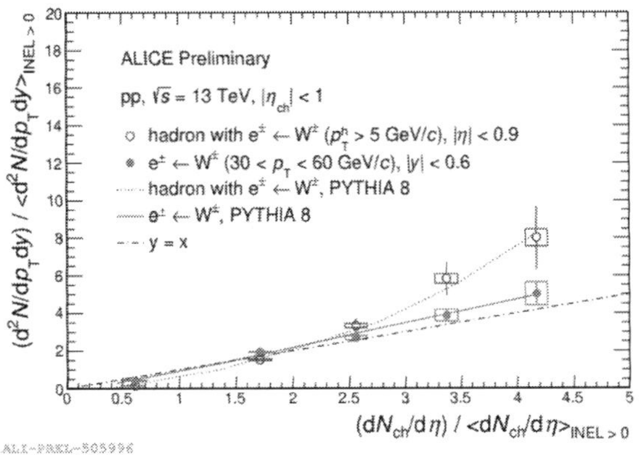

Figure 1. Self-normalised yield of electron from $W^\pm$ decays (red), and hadron associated with $W^\pm$ bosons (blue), as a function of the charged-particle multiplicity. The coloured lines correspond to calculations with PYTHIA 8 [6] including MPI and CR.

2.2 Measurements in p–Pb collisions

The measurements of the Z^0 and W^+ bosons as a function of rapidity in p–Pb collisions at $\sqrt{s_{NN}}$ = 8.16 TeV are shown in the left and right panels of Fig. 2, respectively. The larger number of events available for the W^+ boson allowed splitting the acceptance into several rapidity intervals. The Z^0-boson measurement is compared with predictions from the EPPS16 [7] and nCTEQ15 [8] nuclear PDFs (nPDFs), as well as predictions from the CT14 [9] free PDF accounting for the isospin effect but without nuclear modifications. The models are in good agreement with the measurement, although no strong conclusion can be derived on the nuclear modifications due to statistical limitations.

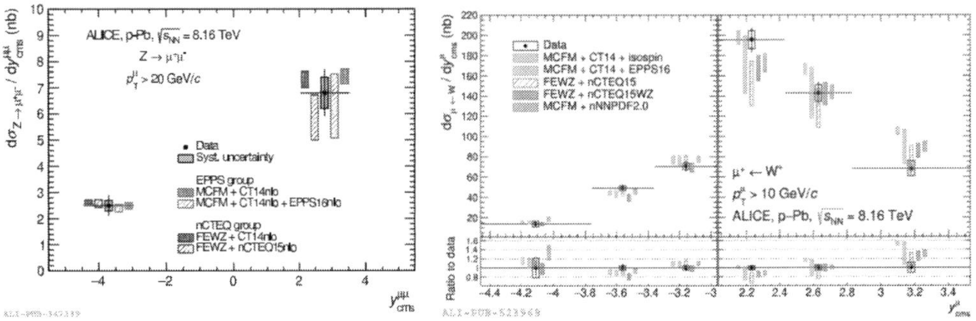

Figure 2. Left: cross section of dimuons from Z^0 decays measured in p–Pb collisions at $\sqrt{s_{NN}}$ = 8.16 TeV, compared with theoretical calculations with and without nuclear modifications of the PDF. **Right**: cross section of muons from W^+ decays in the same collisions systems, compared with model predictions from various PDF and nPDF sets.

The W^+ cross section shown in the right panel of Fig. 2 is compared with the same models, as well as two recent nPDF sets: nCTEQ15WZ [10], an improvement of nCTEQ15 with the addition of electroweak-boson measurements from the LHC for the nPDF determination; and nNNPDF2.0 [11], a new model obtained following a methodology based on machine learning. The models including nuclear modifications are in agreement with one another, and provide a good description of the data. Calculations without nuclear modifications, on the contrary, overestimate the production at large positive rapidity. The resulting 3.5σ deviation is the strongest observation of nuclear modification of the PDF measured by the ALICE Collaboration from electroweak-boson measurements. This measurement helps constraining the nPDF models in the very low Bjorken-x region (down to about 10^{-4}), where the available constraints are scarce.

2.3 Measurements in Pb–Pb collisions

The results obtained in Pb–Pb collisions at $\sqrt{s_{NN}}$ = 5.02 TeV are illustrated in Fig. 3. In the left panel, the yield of dimuons from Z^0 decays, scaled with the average nuclear overlap function $\langle T_{AA} \rangle$, is compared with a previously published result obtained from the 2015 data only, as well as predictions with and without nuclear modifications. The combination of the 2015 and 2018 data periods for this new measurement improved the available integrated luminosity by a factor three, allowing for a significant reduction of the uncertainty. The various nPDFs provide a good description of the measured yields, while a deviation by 3.4σ from the CT14-only calculation is observed, constituting another strong sign of nuclear modifications.

In the right panel of Fig. 3, the $\langle T_{AA} \rangle$-scaled yield of electrons and positrons from $W^\pm$ decays are shown as a function of the collision centrality. They are compared with HG-PYTHIA [12] calculations of the nuclear modification factor of hard scatterings, scaled with the value measured in 0–90% centrality. The scaled calculations are in good agreement with

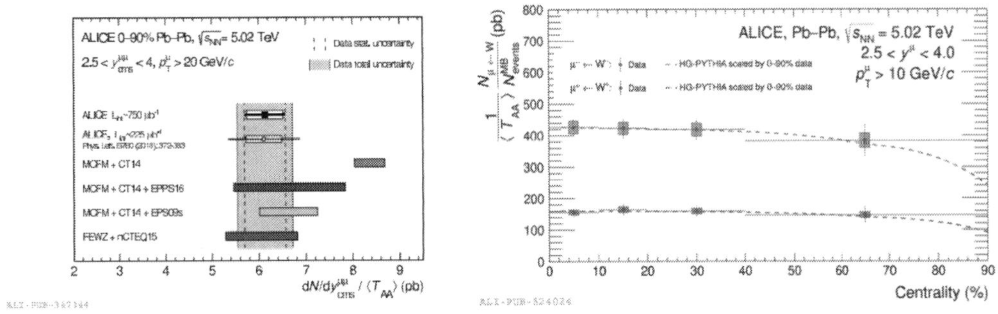

Figure 3. Left: $\langle T_{AA}\rangle$-scaled yield of dimuons from Z^0-boson decays in Pb–Pb collisions at $\sqrt{s_{NN}}$ = 5.02 TeV, compared with theoretical calculations with and without nuclear modifications of the PDF. **Right**: $\langle T_{AA}\rangle$-scaled yield of muons from $W^{\pm}$-boson decays as a function of centrality, in Pb–Pb at $\sqrt{s_{NN}}$ = 5.02 TeV. The measured distribution is compared with HG-PYTHIA [12] predictions of the nuclear modification factor of hard scatterings, scaled to the measured value in 0–90% centrality.

the data. They predict a sizeable drop for the most peripheral collisions, but statistical limitations prevent to have enough granularity in this region in the measurement, such that no experimental conclusion can be drawn. Nonetheless, this measurement is the first measurement of the $W^{\pm}$-boson production in Pb–Pb collisions at forward rapidity, where low values of Bjorken-x are attained.

3 Conclusion

The recent measurements of electroweak-boson production show their usefulness in the characterization of key features of QCD. The measurement at midrapidity in pp collisions is well described by calculations including MPI and CR. In heavy-ion collisions, the measurements are generally well described by nuclear models, while significant deviations from free-PDF calculations can be observed. These measurements can thus be used to help constraining the nPDF models.

References

[1] **ALICE** Collaboration, JHEP **09**(2015)148, arXiv:1505.00664.
[2] **ALICE** Collaboration, JINST **3**(2008)S08002.
[3] **ALICE** Collaboration, JHEP **02**(2017)077, arXiv:1611.03002.
[4] **ALICE** Collaboration, JHEP **09**(2020)076, arXiv:2005.11126.
[5] **ALICE** Collaboration, arXiv:2204.10640.
[6] C. Bierlich et al., arXiv:2203.11601.
[7] K. J. Eskola et al., EPJC **77**(2017)163, arXiv:1612.05741.
[8] K. Kovarik et al., PRD **93**(2016)085037, arXiv:1509.00792.
[9] S. Dulat et al., PRD **93**(2016)033006, arXiv:1506.07443.
[10] A. Kusina et al., EPJC **80**(2020)968, arXiv:2007.09100.
[11] R. A. Khalek et al., JHEP **09**(2020)183, arXiv:2006.14629.
[12] C. Loizides and A. Morsch, PLB **773**(2017)408–411, arXiv:1705.08856.

A physics-based neural network reconstruction of the dense matter equation of state from neutron star observables

Shriya Soma[1,2,3,*], *Lingxiao* Wang[1,3], *Shuzhe* Shi[4], *Horst* Stöcker[1,2,5], and *Kai* Zhou[1,**]

[1]Frankfurt Institute for Advanced Studies (FIAS), D-60438 Frankfurt am Main, Germany

[2]Institut für Theoretische Physik, Goethe Universität, D-60438 Frankfurt am Main, Germany

[3]Xidian-FIAS International Joint Research Center, D-60438 Frankfurt am Main, Germany

[4]Center for Nuclear Theory, Department of Physics and Astronomy, Stony Brook University, Stony Brook, New York, 11794, USA.

[5]GSI Helmholtzzentrum für Schwerionenforschung GmbH, D-64291 Darmstadt, Germany

Abstract. We introduce a novel technique that utilizes a physics-driven deep learning method to reconstruct the dense matter equation of state from neutron star observables, particularly the masses and radii. The proposed framework involves two neural networks: one to optimize the EoS using Automatic Differentiation in the unsupervised learning scheme; and a pre-trained network to solve the Tolman–Oppenheimer–Volkoff (TOV) equations. The gradient-based optimization process incorporates a Bayesian picture into the proposed framework. The reconstructed EoS is proven to be consistent with the results from conventional methods. Furthermore, the resulting tidal deformation is in agreement with the limits obtained from the gravitational wave event, GW170817.

1 Introduction

Neutron stars (NSs) are unique laboratories for studying the properties of strongly interacting dense matter. NS observables like masses and radii are direct probes to the underlying equation of state (EoS). The recent increase in data from electromagnetic and gravitational wave observations of neutron stars and their mergers has resulted in an immense wave of studies aimed at constraining the dense matter EoS [1–10]. Past attempts to reconstruct the neutron star EoS include conventional methods like the Bayesian inference [11, 13, 20], or the more recent supervised machine learning inference [14–17]. In this work, we introduce a novel physics-based deep learning approach to constrain the neutron star EoS. The proposed method involves an unsupervised learning algorithm in the Automatic Differentiation (AD) framework. We adopt a deep neural network representation for the EoS. This allows for a versatile and unbiased characterization of the EoS, e.g., the occurrence of a strong first order phase transition.

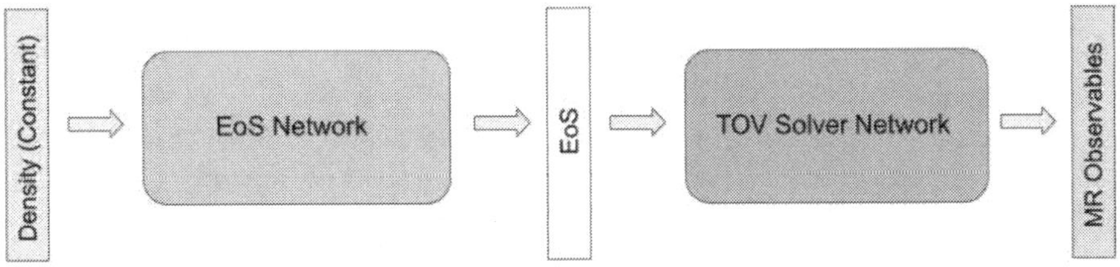

Figure 1. A schematic representation of the proposed algorithm for reconstructing the neutron star EoS via automatic differentiation.

2 Automatic Differentiation

We present a schematic illustration of the proposed mechanism in Fig. 1. The `EoS Network`, as the name suggests, is used to flexibly represent an EoS. It uses density (ρ_i) as the input, and outputs the corresponding pressure (P_i). The `TOV Solver Network`, on the other hand operates as an efficient emulator for the TOV equations. It is a pre-trained network that outputs the mass-radius (M-R) curve, given any input EoS (further details on the network structure, its training procedure, etc, can be found in [18]). The `EoS Network` is combined with the well-trained `TOV-Solver Network` network and optimized in an unsupervised learning scheme. As a result, the trainable parameters in the optimization procedure are the weights of the `EoS Network` alone. Furthermore, we constrain the weights of the `EoS Network` to be non-negative. This ensures a monotonically non-decreasing output for the EoS, so as to fulfill thermodynamic stability. We therefore optimize the `EoS Network` to output an EoS, which in turn reproduces an M-R curve that best fits the NS observations. In order to account for the observational uncertainty, we sample several sets of M-R curves from a Gaussian distribution of the given observations and train the network correspondingly. The optimization process uses a gradient-based algorithm within the AD framework to minimize the loss function, χ^2,

$$\chi^2 = \sum_{i=1}^{N_{\rm obs}} \frac{(M_i - M_{{\rm obs},i})^2}{\Delta M^2_{{\rm obs},i}} + \frac{(R_i - R_{{\rm obs},i})^2}{\Delta R^2_{{\rm obs},i}}. \tag{1}$$

Here, $(M_{{\rm obs},i}, R_{{\rm obs},i})$ are the M-R observations, $(\Delta M_{{\rm obs},i}, \Delta R_{{\rm obs},i})$ are their respective uncertainties, and (M_i, R_i) represent the output of the `TOV-Solver Network`. However, the observational data is limited and not uniformly distributed across the M-R curve. Moreover, the uncertainties of the observations used in this study vary by large factors due to different measurement techniques. As a consequence, there is a possibility of a disordering in the M-R pairs sampled from the normal distribution of the respective observations. To tackle the induced complication, we implement the 'closest approach' optimization [20]. Therefore, the loss in each iteration during the training is,

$$\chi^2 = \sum_{i=1}^{N_{\rm obs}} \frac{(M(\rho_{ci}) - M_{{\rm obs},i})^2}{\Delta M^2_{{\rm obs},i}} + \frac{(R(\rho_{ci}) - R_{{\rm obs},i})^2}{\Delta R^2_{{\rm obs},i}}, \tag{2}$$

where the central density, ρ_{ci}, for every $i^{\rm th}$ observation is updated as,

$$\rho_{ci} = \arg\min_{\rho_c} \frac{(M(\rho_c) - M_{{\rm obs},i})^2}{\Delta M^2_{{\rm obs},i}} + \frac{(R(\rho_c) - R_{{\rm obs},i})^2}{\Delta R^2_{{\rm obs},i}}. \tag{3}$$

*Presenter, e-mail: soma@fias.uni-frankfurt.de
**e-mail: zhou@fias.uni-frankfurt.de

We implement Eq. (3) to determine the central densities of $(M_{\text{obs}}, R_{\text{obs}})$, such that they result in the least distance between the M-R observations and the M-R curve obtained from the `TOV-Solver Network`. We further discard any reconstructed EoS that does not comply with the causal condition or fails to support a $1.9 M_{\odot}$ star. The procedure described above was first tested on several mock M-R data and has been proven to work efficiently [18]. In the next section, we present the results of the reconstructed EoS using the current NS observational data.

3 Results

In order to reconstruct the NS EoS, we use the existent M-R observational data [8, 10, 21–23], which also include the recent NICER measurements. We approximate the uncertainty of each M-R observational point to fit a 1D normal distribution for both mass and radius, individually [15]. We sample 1000 M-R curves from the fitted distributions and reconstruct the EoS from these 18 M-R observations. The resulting EoS is plotted in the left panel of Fig. 2. The pink shaded curve depicts the 95% confidence interval (CI) of the reconstructed EoS from the method discussed above. The mean of the reconstructed EoS curve is marked by the dashed red line. Furthermore, the results are compatible with EoS reconstructions from previous works that include alternate methods like Bayesian inference [24, 25] and machine learning inference [15]. The M-R curve corresponding to the EoS in Fig. 2 is plotted in the right panel of the same figure. Furthermore, the deduced tidal deformability of a $1.4 M_{\odot}$ neutron star, $\Lambda_{1.4 M_{\odot}}$, from the reconstructed EoS, using the present method is evaluated at $\Lambda_{1.4} = 224^{+107.3}_{-107.3}$ (95% CI). This value falls within the estimated range of $\Lambda_{1.4 M_{\odot}} = 190^{+390}_{-120}$, obtained from the gravitational wave event, GW170817 [26].

The next generation telescopes and gravitational-wave detectors provide scope for higher precision on NS observables. With the findings of this work, we therefore conclude that in future, there is hope for a finer reconstruction of the dense matter EoS.

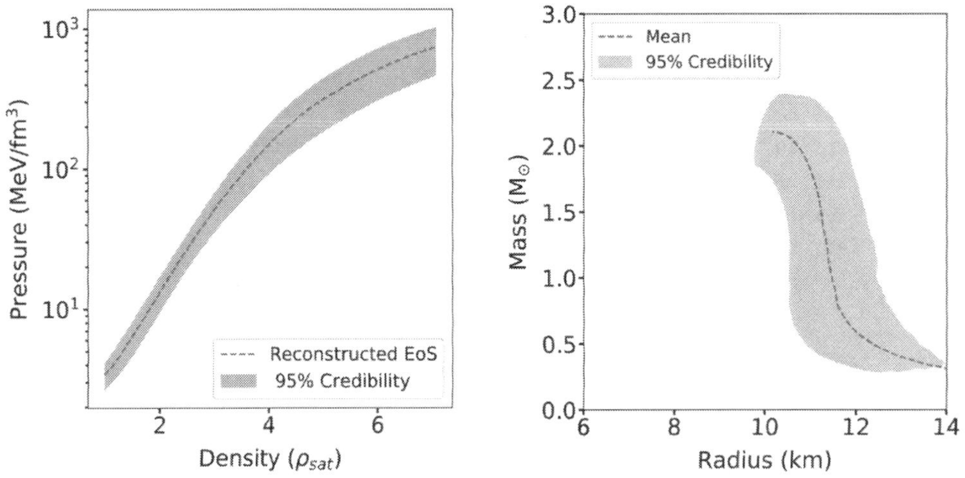

Figure 2. The left panel depicts the 95% CI of the reconstructed EoS from the proposed algorithm. The corresponding M-R curve is plotted in the right panel.

The authors acknowledge support from HGS-HIRe for FAIR, Deutscher Akademischer Austausch-dienst (DAAD), GSI-F&E, BMBF, SAMSON AG, Xidian-FIAS International Joint Research Center (XFIJR), INFN, U.S. Department of Energy, Office of Science, Office of Nuclear Physics, grant No. DE-FG88ER40388 and the Walter Greiner Gesellschaft.

References

[1] J. Antoniadis, P. C. C. Freire, N. Wex, *et al.* Science **340** (2013), 6131
[2] E. Fonseca, H. T. Cromartie, T. T. Pennucci, *et al.* Astrophys. J. Lett. **915** (2021) 1, L12
[3] R. W. Romani, D. Kandel, A. V. Filippenko, *et al.* Astrophys. J. Lett. **908** (2021) 2, L46
[4] B. P. Abbott *et al.* [LIGO Scientific and Virgo], Phys. Rev. X **9** (2019) 1, 011001
[5] L. Rezzolla, E. R. Most and L. R. Weih, Astrophys. J. Lett. **852** (2018) 2, L25
[6] E. R. Most, L. R. Weih, L. Rezzolla and J. Schaffner-Bielich, Phys. Rev. Lett. **120** (2018) 26, 261103
[7] M. Shibata, E. Zhou, K. Kiuchi and S. Fujibayashi, Phys. Rev. D **100** (2019) 2, 023015
[8] M. C. Miller, F. K. Lamb, A. J. Dittmann, *et al.* Astrophys. J. Lett. **887** (2019) 1, L24
[9] S. Soma and D. Bandyopadhyay, Astrophys. J. **890** (2020), 139
[10] T. E. Riley, A. L. Watts, P. S. Ray, *et al.* Astrophys. J. Lett. **918** (2021) 2, L27
[11] A. W. Steiner, J. M. Lattimer and E. F. Brown, Astrophys. J. **722** (2010), 33-54
[12] C. A. Raithel, F. Özel and D. Psaltis, Astrophys. J. **844** (2017) 2, 156
[13] S. Traversi, P. Char and G. Pagliara, Astrophys. J. **897** (2020), 165
[14] F. Morawski and M. Bejger, Astron. Astrophys. **642** (2020), A78
[15] Y. Fujimoto, K. Fukushima and K. Murase, JHEP **03** (2021), 273
[16] M. Ferreira and C. Providência, JCAP **07** (2021), 011
[17] P. G. Krastev, Galaxies **10** (2022) no.1, 16
[18] S. Soma, L. Wang, S. Shi, H. Stöcker and K. Zhou, JCAP **08** (2022), 071
[19] F. Özel and D. Psaltis, Phys. Rev. D **80** (2009), 103003
[20] C. A. Raithel, F. Özel and D. Psaltis, Astrophys. J. **844** (2017) 2, 156
[21] F. Özel, D. Psaltis, T. Güver, *et al.* Astrophys. J. **820** (2016) no.1, 28
[22] J. Nättilä, M. C. Miller, A. W. Steiner, *et al.* Astron. Astrophys. **608** (2017), A31
[23] D. Gonzalez-Caniulef, S. Guillot and A. Reisenegger, Mon. Not. Roy. Astron. Soc. **490** (2019) no.4, 58
[24] A. W. Steiner, J. M. Lattimer and E. F. Brown, Astrophys. J. Lett. **765** (2013), L5
[25] F. Özel and P. Freire, Ann. Rev. Astron. Astrophys. **54** (2016), 401-440
[26] B. P. Abbott *et al.* [LIGO Scientific and Virgo], Phys. Rev. Lett. **121** (2018) 16, 161101

Measurement of exclusive vector meson photoproduction in pPb collisions with the CMS experiment

Subash Chandra Behera for the CMS Collaboration[1,*]

[1]Department of Physics, Indian Institute of Technology Madras, Chennai, 600036

Abstract. The exclusive photoproduction of vector mesons provides a unique opportunity to constrain the gluon distribution function within protons and nuclei. Measuring vector mesons of various masses over a wide range of rapidity and as a function of transverse momentum provides important information on the evolution of the gluon distribution within nuclei. A variety of measurements, including the exclusive J/ψ, ρ, and Υ meson production in pPb (at nucleon-nucleon center of mass energies of 5.02 and 8.16 TeV) and PbPb (5.02 TeV) collisions, are presented as a function of squared transverse momentum and the photon-proton center of mass energy. Finally, compilations of these data and previous measurements are compared to various theoretical predictions.

1 Introduction

This article presents the measurement of the exclusive photoproduction of Υ and ρ^0 mesons in pPb collisions at nucleon-nucleon centre-of-mass energy of $\sqrt{s_{\mathrm{NN}}}$ = 5.02 TeV with the CMS detector [1]. Ultraperipheral collisions (UPCs) of proton-lead ions or lead-lead ions happen when the collision impact parameter is larger than the sum of their radii. Therefore the hadronic interaction is largely suppressed [2]. In UPCs, a boosted ion is the source of a quasi-real photon, which converts into a vector meson state through the interaction with two gluons of no net color emitted from the other ion. In this process, the incoming hadrons remain intact. Such a process is called "exclusive". The study of exclusive quarkonia photoproduction can provide a unique probe that targets the hadron structure, with the large mass of the J/ψ and Υ mesons providing a hard scale for calculations based on perturbative quantum chromodynamics (pQCD) [3]. The study of the photoproduction of Υ and ρ^0 mesons from a proton is sensitive to generalized parton distributions (GPDs), which can be approximated by the square of the gluon density in the proton. Exclusive vector meson photoproduction is interesting because the Fourier transform of the t distribution is related to the two-dimensional distribution of the partons in the transverse plane. Here, t represents the squared four-momentum transfer at the proton vertex. This note presents the $|t| \approx p_{\mathrm{T}}^2$ dependence for the Υ and ρ^0.

In this article we discuss the exclusive photoproduction of ρ^0 meson [4] that decay to $\pi^+\pi^-$ and Υ [5] decay to $\mu^+\mu^-$ process in ultra-peripheral pPb collisions at $\sqrt{s_{\mathrm{NN}}}$ = 5.02 TeV. The cross section is measured as a function of $W_{\gamma p}$ and t. This discussion is organized as

*e-mail: subash.chandra.behera@cern.ch

follows. In section 2 we discuss the invariant mass of the vector meson. Section 3 presents the differential cross section as a function of rapidity, $|t|$ and the photon–proton centre-of-mass energy $W_{\gamma p}$. Finally, in section 4 we discuss the summary of this measurement.

2 Invariant mass

Figure 1 shows the invariant mass distribution for dimuons in the range between 8 and 12 GeV that satisfies the selection criteria described in Ref. [5]. To fit the spectrum, an unbinned likelihood fit is performed using RooFIT [52] with a linear function to describe the QED $\gamma + \gamma \rightarrow \mu{+}\mu^-$ continuum background, where the background exponential slope parameter is fixed to the value obtained from a STARLIGHT [6] $\gamma + \gamma \rightarrow \mu{+}\mu^-$ simulation, with three Gaussian functions for the three Υ signal slope parameter of peaks.

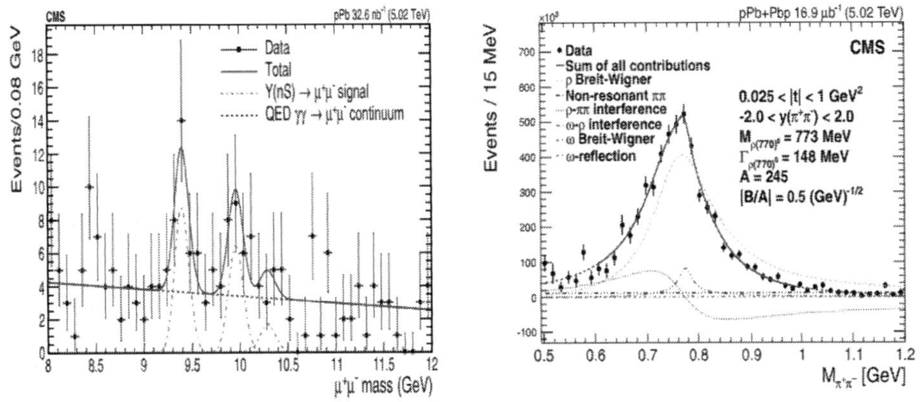

Figure 1. The left-side figure shows the invariant mass distribution of the exclusive dimuon candidates in the range $8 < m_{\mu^+\mu^-} < 12$ GeV that satisfy all the selection criteria, fitted to a linear function for the two-photon QED continuum (blue dashed line) plus three Gaussian distributions corresponding to the $\Upsilon(1S)$, $\Upsilon(2S)$, $\Upsilon(3S)$ mesons (dashed-dotted-red curves). The right-side figure shows the unfolded $\pi^+\pi^-$ invariant mass distribution [4] in the pion pair rapidity interval $|y_{\pi^+\pi^-}| < 2$ fitted with the Söding model [7]. The green dashed line indicates resonant $\rho(770)^0$ production, the red dotted line represents the interference term, the magenta dash-dotted line corresponds to the background from $\omega \rightarrow \pi^+\pi^-\pi^0$, the black dash-dotted line represents the non-resonant contribution, and the blue solid line represents the sum of all these contributions.

The right plot of Figure 1 presents the fit of the unfolded distribution with the modified Söding modell [7]. A least squares fit is performed for the interval $0.6 < M_{\pi^+\pi^-} < 1.1$ GeV, with the quantities, $M_{\rho(770)^0}$, $M_{\omega(783)}$, $\Gamma\rho(770)^0$, $\Gamma_{\omega(783)}$, A, B, C, $\phi_{\omega(783)}$ treated as free parameters.

3 Differential cross section measurement

The differential cross-section of the exclusive $\Upsilon(1S)$ is measured as a function of $W_{\gamma p}$ and y over $|y| < 2.2$, are shown in Figure 2. The p_T^2-differential cross section is fitted with an exponential function in the region $0.01 < p_T^2 < 1.0$ GeV^2, using a χ^2 goodness-of-fit minimization technique. The slope parameter $b = 6.0 \pm 2.1$ (stat) ± 0.3 (syst) GeV^{-2} is extracted, and in agreement with $b = 4.3^{+2.0}_{-1.3}$ (stat) $^{+0.5}_{-0.6}$ (syst) GeV^{-2}, which is measured by the ZEUS Collaboration in the photon–proton centre-of-mass energy in the range $60 < W_{\gamma p} < 220$ GeV. The measured results are consistent with the predictions of pQCD-based models.

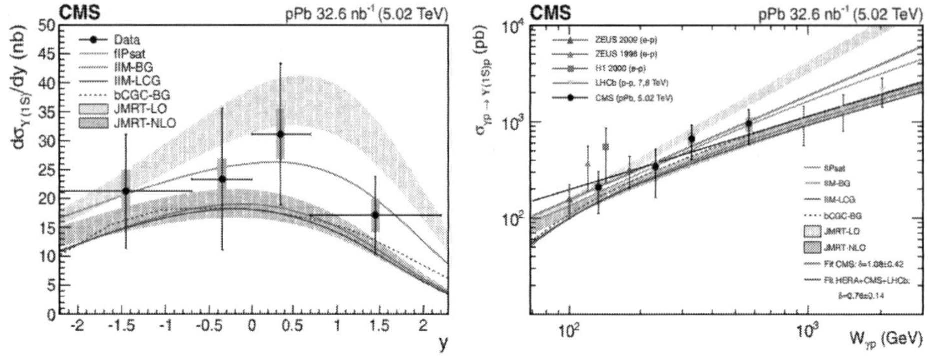

Figure 2. Left plot is the background substracted differential cross section as a function of rapidity measured in pPb collisions. Right plot is the $W_{\gamma p}$ dependence of cross section for the exclusive photo production [5].

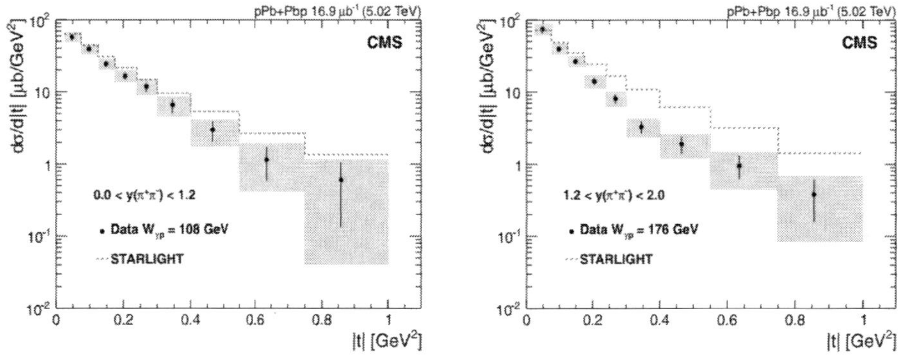

Figure 3. Differential cross section ($d\sigma/dt$) is measured for ρ^0 as a function of rapidity bin [4]. The dashed dotted line indicates the predictions from STARLIGHT. The error bars show the statistical uncertainty, and the shaded green areas represent the systematic uncertainties of that bin.

Figure 2 (left plot) shows the rapidity distribution of the $\Upsilon(1S)$. Our results are compared with the various theoretical predictions, consistent with the data within the relatively large experimental uncertainties. The JMRT-LO [8] results are systematically above the data points and all the other calculations. The data are compared to the various theoretical predictions. A fit with the power-law function in the entire $W_{\gamma p}$ range of data yields $\delta = 1.30$ and $\delta = 0.84$ for the LO and NLO calculations, respectively. We significantly reduced the uncertainty compared to ZEUS [9] and cover a wide range of $W_{\gamma p}$.

The differential cross section of ρ^0 is measured as a function of $|t|$ for different rapidity range as shown in figure 3. The $|t|$ distribution is fitted in the region $0.025 < |t| < 0.5$ GeV^2. The data results are compared with the STARLIGHT prediction and it is observed that the STARLIGHT prediction is comparatively higher than the data in the high t region, an effect that becomes more significant as $W_{\gamma p}$ increases [4].

The cross-section measured as a function of $W_{\gamma p}$ is shown in figure 4. The data results are fitted with $\sigma = \alpha_1 W_{\gamma p}^{\delta_1} + \alpha_2 W_{\gamma p}^{\delta_2}$. δ_1 and δ_2 parameters are extracted from the fit and yields the values $\delta_1 = -0.81 \pm 0.04$ (stat) ± 0.09 (syst) and $\delta_2 = 0.36 \pm 0.07$ (stat) ± 0.05 (syst). Both

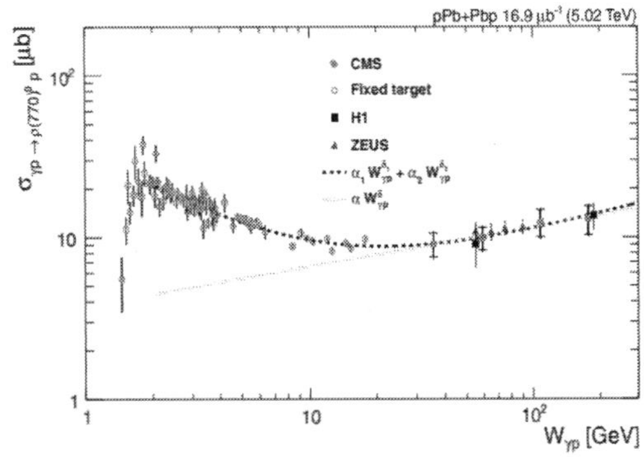

Figure 4. Exclusive ρ meson cross sections measured as a function of $W_{\gamma p}$ [4]. The inner and outer error bars show the statistical uncertainty and systematic uncertainty. The dashed black line shows the result of the Regge fit [10].

the CMS and HERA data together fitted with $\alpha W^{\delta}_{\gamma p}$ and yields $\delta = 0.24 \pm 0.13$ (stat) ± 0.04 (syst).

4 Summary

The cross section of exclusive photoproduction of the Υ and $\rho(770)^0$ is measured in UPC pPb collisions at $\sqrt{s_{\mathrm{NN}}} = 5.02$ TeV with the CMS experiment. The differential cross section as a function of rapidity, p_T^2, and $W_{\gamma p}$ are calculated and compared with previous experimental results from H1, LHCb, and ZEUS. Our results are compared with the previous measurements and are qualitatively consistent within uncertainties. The results discussed in this article provide a wider range of $W_{\gamma p}$ than previous experiments and comparable uncertainties to that of ZEUS. The data is consistent with various pQCD approaches that model the behaviour of the low-x gluon density.

References

[1] CMS Collaboration, *JINST* **9** (2014) P10009, doi = 10.1088/1748-0221/9/10/P10009.
[2] A.J. Baltz et al.,, *Physics Reports* **1** (2008) 458, doi=10.1016/j.physrep.2007.12.001.
[3] Stanley J. Brodsky and L. Frankfurt and J. F. Gunion and A. H. Mueller and M. Strikman, *Physical Review D* **50** (1994) 3134, doi=10.1103/PhysRevD.50.3134.
[4] CMS Collaboration, *Eur. Phys. J. C* **79** (2019) 702, doi = 10.1140/epjc/s10052-019-7202-9.
[5] CMS Collaboration, *The European Physical Journal C* **277** (2019) 79, doi = 10.1007/JHEP11(2013)085.
[6] Klein, Spencer R. and Nystrand, *Computer Physics Communications* **212** (2017) 258, doi = 10.1016/j.cpc.2016.10.016.
[7] P.Söding, *Physics Letters* **19** (1966) 702, doi = 10.1016/0031-9163(66)90451-3.
[8] Jones, S. P. and Martin, A. D. and Ryskin, M. G. and Teubner, T., *JHEP* **11** (2013) 085. doi = 10.1007/JHEP11(2013)085.
[9] A. M. Sirunyan and et al, *Physics Letters B* **708** (2012) 14, doi = 10.1016/j.physletb.2012.01.009.
[10] K. Goulianos, *Physics Reports* **101** (1983) 169, doi = 10.1016/0370-1573(83)90010-8.

(Anti)nucleosynthesis in heavy-ion collisions and (anti)nuclei as "baryonmeter" of the collision

Mario Ciacco[1,*] for the ALICE Collaboration

[1]Dipartimento DISAT del Politecnico and Sezione INFN, Turin, Italy

Abstract. The production mechanism of light (anti)nuclei in heavy-ion collisions has been extensively studied experimentally and theoretically. Two competing (anti)nucleosynthesis models are typically used to describe light (anti)nuclei yields and their ratios to other hadrons in heavy-ion collisions: the statistical hadronization model (SHM) and the nucleon coalescence model. The possibility to distinguish these phenomenological models calls for new experimental observables.
Given their large baryon number, light (anti)nuclei have a high sensitivity to the baryon chemical potential (μ_{B}) of the system created in the collision.
In this talk, the first measurement of event-by-event antideuteron number fluctuations in heavy-ion collisions is presented and compared with expectations of the SHM and coalescence model. In addition, the antinuclei-to-nuclei ratios are used to obtain a measurement of μ_{B} in heavy-ion collisions with unprecedented precision.

1 Introduction

(Anti)nucleosynthesis is one of the open questions of heavy-ion physics. While it is yet to be fully understood how loosely bound objects such as light (anti)nuclei can survive the extreme conditions reached in high-energy nuclear collisions, the general properties of their production can be described by two different phenomenological approaches, namely the Statistical Hadronisation Model (SHM) [1] and the coalescence model [2]. New observables such as (anti)nuclei event-by-event fluctuations, which are characterised by a high discriminating power between these two models, can help to shed light on the nuclear formation mechanisms [3]. Moreover, among all species produced in collisions, light nuclei are the most sensitive to antiparticle-particle balance, which is encoded by the baryon chemical potential μ_B. As a consequence, they provide a precise constraint on the value of μ_B.

2 Antideuteron event-by-event fluctuations

The properties of the (anti)nuclei multiplicity distributions are directly connected to the assumed (anti)nucleosynthesis mechanism [3]: in particular, a Poisson distribution is predicted by the Grand Canonical (GC) SHM [4]. Conservation laws and correlations among the particles produced in the collision can cause deviations from the Poisson baseline. This is the case for the canonical SHM and the coalescence model. The event-by-event fluctuations

*e-mail: mario.ciacco@cern.ch

can be experimentally characterised by the m-th order cumulant κ_m: in this study, the ratio between the cumulants κ_2 and κ_1, which represent the variance and the average of the distribution respectively, is considered. Two versions of the coalescence model are considered: in model A, the multiplicities of the coalescing (anti)protons and (anti)neutrons are assumed to be correlated, while in model B, (anti)protons and (anti)neutrons are emitted by independent sources.

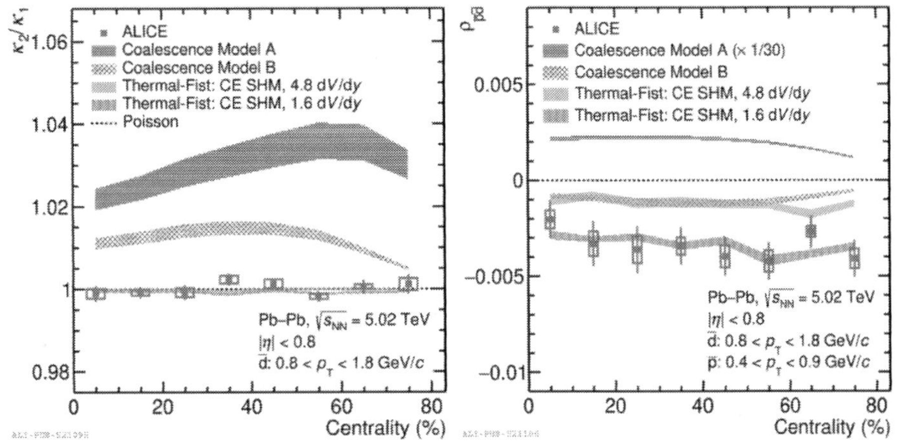

Figure 1: (Left) Ratio of cumulants κ_2 and κ_1 of the antideuteron multiplicity distribution as a function of the event centrality. (Right) Antideuteron-antiproton correlation $\rho_{\bar{p}\bar{d}}$ as a function of the event centrality. In both plots, the ALICE measurements are shown as red points, with statistical and systematic uncertainties represented as vertical lines and boxes, respectively. The predictions obtained from both the coalescence and the statistical model are shown as coloured bands [5].

The κ_2/κ_1 ratio of antideuterons measured by the ALICE Collaboration in Pb–Pb collisions at $\sqrt{s_{NN}} = 5.02$ TeV is shown in the left panel of Fig. 1 as a function of the event centrality [5]. Antideuterons are measured in a narrow transverse momentum (p_T) range, $0.8 < p_T < 1.8$ GeV/c, where the outstanding particle identification (PID) capabilities of ALICE can be fully exploited to select a high-purity antideuteron sample [6]. Moreover, only the antimatter component is considered for this measurement as its observed yield is not contaminated by the products of spallation interactions of primary particles in detector materials. Predictions of coalescence models A (violet) and B (blue), and those of the Canonical Ensemble (CE) SHM with baryon-number correlation volume $V_C = 4.8$ dV/dy (yellow) and $V_C = 1.6$ dV/dy (green) are each compared with the measured points in the left panel of Fig. 1. It can be observed that the measured κ_2/κ_1 ratio is consistent with unity, i.e. with the Poissonian limit, across centrality: the CE SHM is able to reproduce this result, as antideuterons constitute only a small fraction of the antibaryons produced in the collision, while both version A and B of the coalescence model overpredict the observed ratio.

Another variable considered in this study is the antiproton-antideuteron correlation, $\rho_{\bar{p}\bar{d}}$: the results obtained by the ALICE Collaboration as a function of the event centrality are shown in the right panel of Fig. 1. An anticorrelation of about −0.5% is observed across centrality: this result can be succesfully described by the CE SHM with $V_C = 1.6$ dV/dy, i.e. with a baryon-number correlation volume smaller than the one obtained in studies of net-proton fluctuations [7]. The observed anticorrelation can be also qualitatively described by the coalescence model B, while model A is ruled out by the comparison with data, as it predicts a large positive correlation of about 5% between antiprotons and antideuterons. The CE SHM predictions are also in good agreement with $\rho_{\bar{p}\bar{d}}$ as a function of the acceptance window used for the measurement [5].

3 The baryon chemical potential

As shown in section 2, (anti)nucleus production can be successfully described by statistical models: consequently, light (anti)nuclei can be used in turn to determine the model parameters, specifically the baryon chemical potential μ_B. The μ_B measurement described in the following is based on the precise evaluation of antiparticle-to-particle ratios of different species ($\bar{h}/h$), which are connected to the value of μ_B via a relation obtained from the GC SHM under the assumption of strangeness neutrality [8, 9]:

$$\bar{h}/h \propto \exp\left[-2\left(B + \frac{S}{3}\right)\frac{\mu_B}{T} - 2I_3\frac{\mu_{I_3}}{T}\right] \tag{1}$$

where B, S and I_3 are the baryon number, strangeness and isospin third component of the species h, respectively, while μ_{I_3} is the isospin chemical potential. From Eq. 1 it can be seen that the species characterised by a large baryon content are more sensitive to μ_B; moreover, there is also a dependence on strangeness. The μ_B measurement presented here is based on the antiparticle-to-particle ratios of species reconstructed by ALICE that carry the highest values of B, i.e. protons, ^{3}He and hypertriton ($^3_\Lambda$H). The ratio of charged pion yields is also taken into account to put a tight constraint on μ_{I_3}. In order to extract μ_B, the chemical freeze-out temperature T can be fixed to the value $T = 156.2 \pm 1.5$ MeV obtained from previous studies [10], since the dependence of ratios on T in Eq. 1 is negligible for $\mu_B \sim O(1\ \text{MeV})$.

The data sample analysed consists of about 3×10^8 Pb–Pb collisions at $\sqrt{s_{\text{NN}}} = 5.02$ TeV recorded during the 2018 LHC run: the large sample size allows us to perform a centrality-differential study in three centrality classes, namely 0-5%, 5-10% and 30-50%. The analysis of pions, protons and helium relies on standard rectangular selections, while the hypertriton measurement is based on machine learning selections [11]. For all the analysed particles, the contributions due to spallation and weak decays are subtracted to account only for the primary particle yields in the final results.

The ratios obtained for the three analysed centrality classes are shown in Fig. 2. A fit based on Eq. 1 with μ_B and μ_{I_3} as free parameters is performed separately in each centrality interval: it can be observed that the baryon-number hierarchy predicted by the GC SHM is also present in the measured points. The μ_B values obtained are then shown as a function of the average number of participating nucleons $\langle N_{\text{part}} \rangle$ in Fig. 3: the μ_B results obtained for the most central collisions are compatible with previous studies based on SHM fits of antiparticle and particle yields of different species [10], while showing an improvement in precision by about one order of magnitude; in addition, a possible centrality-dependence due to different baryon stopping at different centralities is excluded with the current precision.

4 Conclusions

The study of antideuteron event-by-event fluctuations offers a powerful probe of (anti)nucleosynthesis: the results shown in this contribution favour the CE SHM predictions over the coalescence ones. The large data samples that will be collected during LHC Runs 3 and 4 will allow us to perform more precise measurements of this observable. In addition, the most precise μ_B measurement ever performed is reported here: this study will be further extended by testing the strangeness and isospin dependence of ratios via the strange baryons Λ and Ω, and triton.

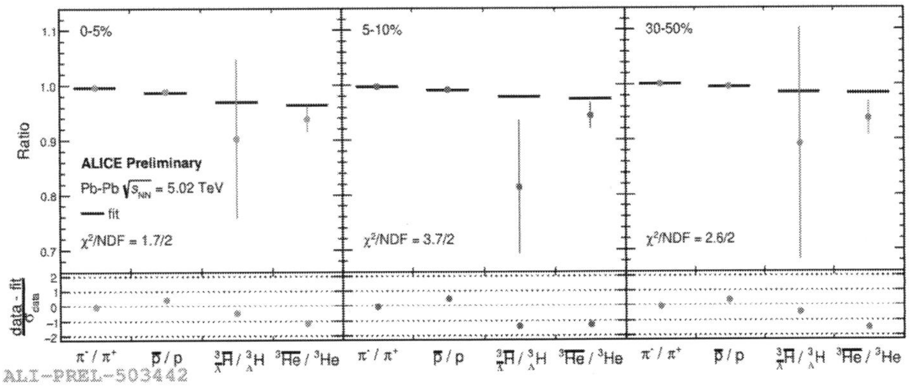

Figure 2: Results of the SHM fit (black lines) to the measured antiparticle-to-particle ratios in the 0-5% (orange), 5-10% (blue) and 30-50% (green) centrality classes. The vertical lines show the sum in quadrature of statistical and p_T- and centrality-uncorrelated systematic uncertainties. The difference between the fit results and the measured ratios normalised to the total uncertainty is shown in the lower panel.

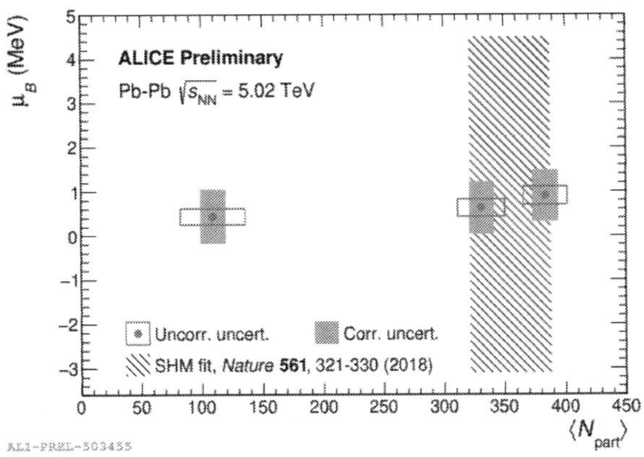

Figure 3. Baryon chemical potential μ_B as a function of the average number of participating nucleons $\langle N_{\text{part}} \rangle$. The uncorrelated uncertainties are shown as open boxes, while the correlated ones as coloured rectangles. The latest published measurement of the baryon chemical potential at the LHC, $\mu_B = 0.7 \pm 3.8$ MeV [10], is represented as a black shaded area.

References

[1] A. Andronic, P. Braun-Munzinger, J. Stachel, H. Stöcker, Phys. Lett. B **697**, 203–207 (2011)

[2] S.T. Butler, C.A. Pearson, Phys. Rev. **129**, 836–842 (1963)

[3] Z. Fecková, J. Steinheimer, B. Tomášik, M. Bleicher, Phys. Rev. C **93**, 054906 (2016)

[4] P. Braun-Munzinger, B. Friman, K. Redlich, A. Rustamov, J. Stachel, arXiv:2007.02463 [nucl-th] (2020)

[5] ALICE Collaboration, arXiv:2204.10166 [nucl-ex] (2022)

[6] J. Adam, et. al (ALICE Collaboration), Phys. Rev. C **93**, 024917 (2015)

[7] S. Acharya, et al. (ALICE Collaboration), Phys. Lett. B **807**, 135564 (2020)

[8] J. Cleymans, I. Kraus, H. Oeschler, K. Redlich, S. Wheaton, Phys. Rev. C **74**, 034903 (2006)

[9] J. Cleymans, H. Satz, Z. Phys. C **57**, 135–147 (1993)

[10] A. Andronic, P. Braun-Munzinger, K. Redlich, J. Stachel, Nature **561**, 321–330 (2018)

[11] S. Acharya, et. al (ALICE Collaboration), arXiv:2107.10627v1 [nucl-ex] (2021)

Initial electromagnetic field dependence of photon-induced production in isobaric collisions at STAR

Kaifeng Shen[1,2,*] (for the STAR Collaboration)

[1]Department of Modern Physics, University of Science and Technology of China, Hefei 230026, China
[2]State Key Laboratory of Particle Detection and Electronics, University of Science and Technology of China, Hefei 230026, China

Abstract. In these proceedings, we present the measurements of e^+e^- pair production at very low transverse momentum in $^{96}_{44}$Ru+$^{96}_{44}$Ru and $^{96}_{40}$Zr+$^{96}_{40}$Zr collisions at $\sqrt{s_{\rm NN}}$ = 200 GeV by the STAR experiment. The electromagnetic field dependence of photon-induced production is studied for the first time, via comparisons between the measurements in isobaric collisions to the published results in Au+Au and U+U collisions. Taken the initial electromagnetic field dependence into account, the excess yield of e^+e^- pair production shows an impact parameter dependence. The results are also compared to model calculations based on EPA-QED, which describe the observed excess yields for different collision species.

1 Introduction

Quark gluon plasma (QGP), a deconfined state of partonic matter, can be produced in ultra-relativistic heavy-ion collisions. Traditionally, di-leptons, one of probes used to study the properties of the QGP, refer to those produced via the violent hadronic interactions in the nuclear overlap region. However, di-letpons can also be generated by interactions of strong electromagnetic fields induced by the colliding nuclei with velocities close to the speed of light. Due to the strong Lorentz contraction and the large number of charge (Z) carried by heavy nuclei, the induced strong electromagnetic fields can be viewed as large flux of quasi-real photons, which leads to production of di-leptons from photon-photon process ($\gamma\gamma \to l^+l^-$). The cross section of such photon-photon process is proportional to Z^4. Recently, significant enhancements of e^+e^- pair at very low transverse momentum (p_T) have been observed in peripheral hadronic heavy-ion collisions [1], which can be reasonably explained by this photon-induced production [2–4]. In order to further prove that photon-induced interactions are the cause of the very low-p_T e^+e^- excess yield observed in heavy-ion collisions, it is essential to measure its production yield in various collision systems with varying Z.

In 2018, the STAR experiment has collected a large sample of $^{96}_{44}$Ru+$^{96}_{44}$Ru and $^{96}_{40}$Zr+$^{96}_{40}$Zr collisions at $\sqrt{s_{\rm NN}}$ = 200 GeV to search for the Chiral Magnetic Effect signals [5] since the electromagnetic fields in Ru+Ru and Zr+Zr are expected to be different while the hadronic backgrounds should be similar. Since photon-induced processes are sensitive to the electromagnetic field, they can be used to experimentally study the difference of the initial electromagnetic fields between Ru+Ru and Zr+Zr collisions. The isobaric collisions also provide a

*e-mail: skfwyl@mail.ustc.edu.cn

unique opportunity to test the electromagnetic field dependence of photon-induced production.

2 Experiment and Analysis

The STAR experiment has recorded around 4 billion minimum-bias Ru+Ru and Zr+Zr collision events in 2018. Electrons with $p_T^e > 0.2$ GeV/c and $|\eta^e| < 1$ are selected by combining the ionization energy loss measured by the Time Projection Chamber (TPC) [6] and the flight time measured by the Time-of-Flight (TOF) detector [7]. The raw signal of e^+e^- pairs at very low p_T is obtained by subtracting background from the unlike-sign pair distribution. A Monte Carlo (MC) simulation is used to estimate the contributions from known hadronic sources, usually referred to the hadronic cocktail, which needs to be subtracted in order to obtain the excess yields.

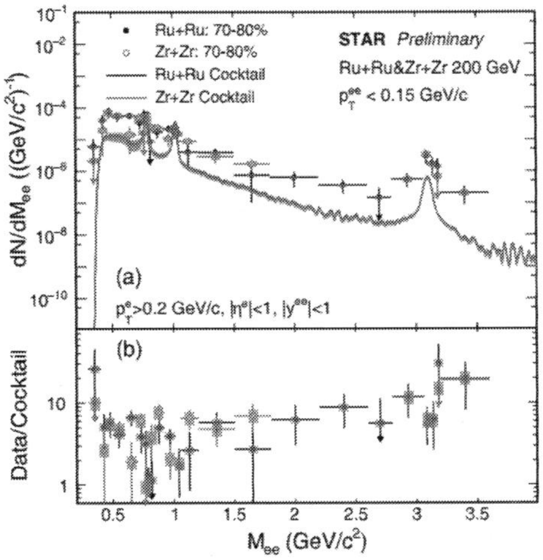

Figure 1: (a) The e^+e^- invariant mass spectra within the STAR acceptance from Ru+Ru and Zr+Zr collisions for pair $p_T^{ee} < 0.15$ GeV/c in 70-80% centrality. The hadronic cocktails are shown as solid lines of different colors. (b) The corresponding ratios of data over cocktail.

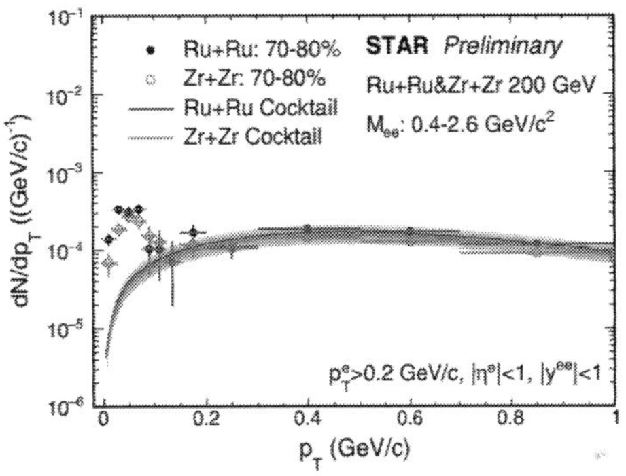

Figure 2: The e^+e^- pair p_T distribution within the STAR acceptance in 0.4-2.6 GeV/c^2 mass region in 70-80% centrality, compared to cocktails.

3 Results

In Fig. 1(a), the e^+e^- invariant mass distributions corrected for detector inefficiency in Ru+Ru and Zr+Zr collisions are shown as solid and open circles respectively in 70-80% centrality for pair $p_T < 0.15$ GeV/c within STAR acceptance ($p_T^e > 0.2$ GeV/c, $|\eta^e| < 1$, and $|y^{ee}| < 1$). The corresponding enhancement factors are defined as ratios of data over hadronic cocktail in the same mass regions, which are displayed in Fig. 1(b). The enhancement factors reach local minima around M_ϕ before rising towards larger mass, similar to the trends seen in Au+Au and U+U collisions [1].

The p_T distributions of e^+e^- pairs in the mass region from 0.4-2.6 GeV/c^2 are shown in Fig. 2 for 70-80% isobaric collisions. While for $p_T^{ee} > 0.15$ GeV/c, data are consistent with hadronic cocktails, significant excesses above hadronic cocktails are seen at $p_T^{ee} < 0.15$ GeV/c.

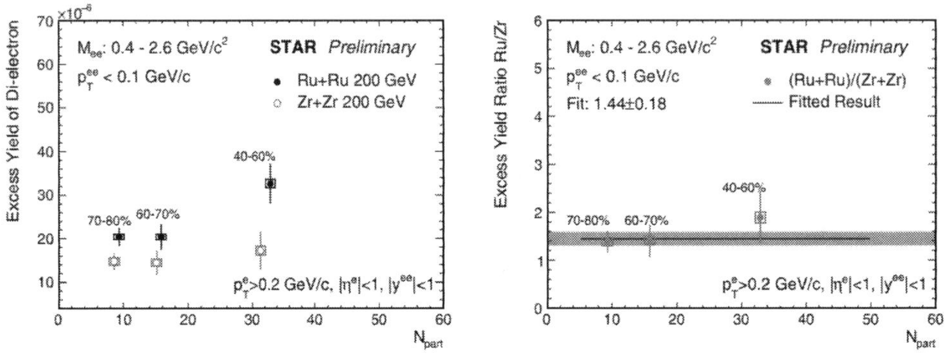

Figure 3: *Left panel*: The centrality dependence of integrated excess yields at $p_T^{ee} < 0.1$ GeV/c in the mass region of 0.4-2.6 GeV/c^2 in Ru+Ru and Zr+Zr collisions within the STAR acceptance. *Right panel*: The centrality dependence of the ratios of integrated low-p_T excesses between Ru+Ru and Zr+Zr collisions. The solid line is the fitted result to data points by a constant function.

After subtracting the hadronic cocktail, the integrated low-p_T excess yields of e^+e^- pairs as a function of average number of participating nucleons $\langle N_{\text{part}} \rangle$ are shown in the left panel of Fig. 3. The integrated excesses yields in Ru+Ru collisions are systematically higher than those in Zr+Zr collisions. The right panel of Fig. 3 is the centrality dependence of excess yield ratios between Ru+Ru and Zr+Zr collisions, and a constant function is utilized to fit the ratios. The fitted result is about 2.4σ higher than unity, which hints at the initial electromagnetic field dependence between the two collision systems.

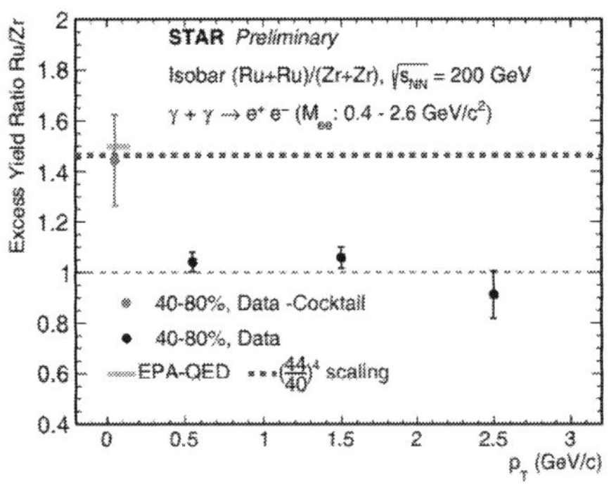

Figure 4: The p_T dependence of e^+e^- yield ratios between Ru+Ru and Zr+Zr collisions in 40-80% centrality. The statistical and systematical uncertainties are combined and shown as the vertical bars. The green solid line is the prediction at low-p_T based on EPA-QED [8] and the blue dash line is the $(\frac{44}{40})^4$ scaling.

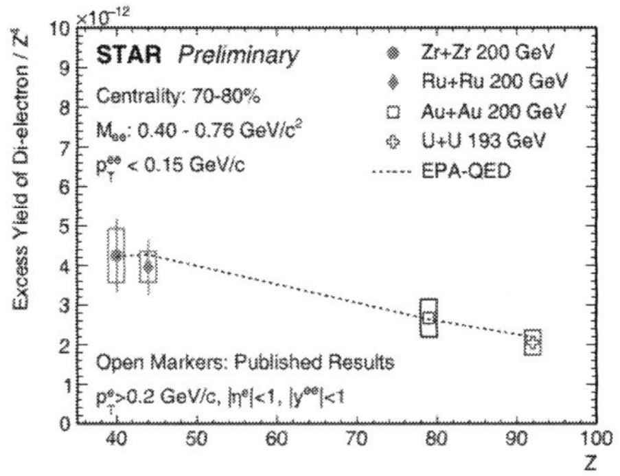

Figure 5: The collision system dependence of the integrated e^+e^- excess yield within the STAR acceptance, scaled with Z^4, in 70-80% centrality. The dash line is the EPA-QED prediction [9].

Figure 4 shows the p_T dependence of the e^+e^- yield ratios. For $p_T > 0.1$ GeV/c, the hadronic cocktail is not subtracted, and the yield ratio is consistent with unity, indicating the hadronic contribution dominates in this p_T range. The two collision systems' distinct nuclear structures could be the cause of the small deviation from unity. At $p_T < 0.1$ GeV/c, the hadronic cocktail is subtracted in order to study the collision system dependence of the photon-induced production. The excess yield ratio at low-p_T can be described by the EPA-QED calculation [8] shown as the green line and also follows the $(\frac{44}{40})^4$ scaling.

Figure 5 shows the integrated excess yields in the mass region of 0.4-0.76 GeV/c^2 at p_T < 0.15 GeV/c in 70-80% centrality scaled by Z^4 as function of Z for different collisions. The Z^4 scaled yield shows a clear collision system dependence, likely originating from the impact parameter dependence of photon-induced interactions. The decreasing trend can be described by the EPA-QED calculations [9] taking such impact parameter dependence into account.

4 Summary

In this contribution, the electromagnetic field and collision species dependence of e^+e^- pair production by photon-photon process from STAR are presented. We observe a hint of initial electromagnetic field dependence in peripheral Ru+Ru and Zr+Zr collisions. Taken the initial electromagnetic field dependence into account, the excess yield of e^+e^- shows a clear impact parameter dependence. The results can be well described by calculations of photon-photon interactions based on EPA-QED, pointing to the existence of photon-photon interactions in hadronic heavy-ion collisions.

Acknowledgements

This work was supported in part by the National Natural Science Foundation of China under Grant Nos. 11720101001, 12175223 and 12005220, Anhui Provincial Natural Science Foundation under Grant Nos. 1908085J02.

References

[1] J. Adam et al. (STAR Collaboration), Phys. Rev. Lett. **121**, 132301 (2018)
[2] W. Zha, L. Ruan, Z. Tang et al., Physics Letters B **781**, 182 (2018)
[3] S.R. Klein, Phys. Rev. C **97**, 054903 (2018)
[4] W. Zha, S.R. Klein, R. Ma et al., Phys. Rev. C **97**, 044910 (2018)
[5] M.S. Abdallah et al. (STAR Collaboration), Phys. Rev. C **105**, 014901 (2022)
[6] M. Anderson, J. Berkovitz, W. Betts et al., Nucl. Instrum. Methods Phys. Res., Sect. A **499**, 659 (2003)
[7] W. Llope, Nucl. Instrum. Methods Phys. Res., Sect. A **661**, S110 (2012)
[8] W. Zha, L. Ruan, Z. Tang et al., Phys. Lett. B **789**, 238 (2019), 1810.02064
[9] W. Zha, J.D. Brandenburg, Z. Tang et al., Physics Letters B **800**, 135089 (2020)

Thermal radiation and direct photon production measurements with dielectrons in Pb–Pb and pp collisions

Hikari Murakami[1,*] for the ALICE Collaboration

[1]Center for Nuclear Study, the University of Tokyo, Japan

Abstract. The latest ALICE results on dielectron measurements in Pb–Pb collisions at $\sqrt{s_{\rm NN}}$ = 5.02 TeV, and in minimum-bias (MB) and high-multiplicity (HM) pp collisions at $\sqrt{s}$ = 13 TeV with LHC Run 2 data are reported. The results are compared to the predictions of dielectron yields from known hadronic sources and thermal radiation from the medium. The contribution of virtual direct photons is extracted from the data by fitting the measured invariant mass distributions. We present the direct to inclusive photon ratio, as well as the direct photon yield in both pp and central Pb–Pb collisions.

1 Introduction

Electromagnetic probes such as photons and dileptons are well suited to study the whole space–time evolution of the hot and dense matter called quark–gluon plasma (QGP), created in ultra-relativistic heavy-ion collisions. They are produced during all stages of the collision with negligible final-state interactions. In the intermediate mass region (IMR : 1.1 GeV/c^2 < $m_{\rm ee}$ < 2.7 GeV/c^2), thermal radiation from the QGP carries information from the partonic phase of the collision. In the low mass region (LMR : $m_{\rm ee}$ < 1.1 GeV/c^2) , thermal radiation from the hadron gas contributes to the dielectron spectrum mainly through the decay of rho mesons whose spectral function is sensitive to chiral-symmetry restoration [1]. At $m_{\rm ee} \ll p_{\rm T,ee}$, the ratio of direct to inclusive photons can be extracted from the dielectron spectrum. Dielectron production in minimum-bias (MB) pp collisions serves as a vacuum baseline for the studies in Pb–Pb collisions. Recently, pp collisions with high charged-particle multiplicities (HM) were found to exhibit surprisingly similar phenomena, i.e. anisotropic flow [2–4] as observed in heavy-ion collisions. The search for thermal photons in small systems may help to better understand the underlying dynamics in such collisions. The analysis in pp collisions uses the full Run 2 data, with and integrated luminosity of $L_{\rm int,MB} \simeq 30.3$ nb^{-1} for MB and $L_{\rm int,HM} \simeq 6.08$ pb^{-1}. This is increased by a factor of 3.8 and 4.4, for the MB and HM datasets with respect to the previous publication [5], respectively. The Pb–Pb data was collected in 2018, and the analysis uses both the MB and centrality triggers with an integrated luminosity of 83 μb^{-1}. The tracking and identification of electrons are done using the Inner Tracking System (ITS), the Time Projection Chamber (TPC), and the Time-Of-Flight detector at midrapidity ($|\eta_e| < 0.8$) [6].

2 Dielectrons in pp collisions at $\sqrt{s}$ = 13 TeV

Figure 1 shows the invariant mass spectra measured in MB and HM pp collisions at $\sqrt{s}$ = 13 TeV compared with the expectation from known hadronic decays, which we also refer

*e-mail: hikari.murakmi@cern.ch

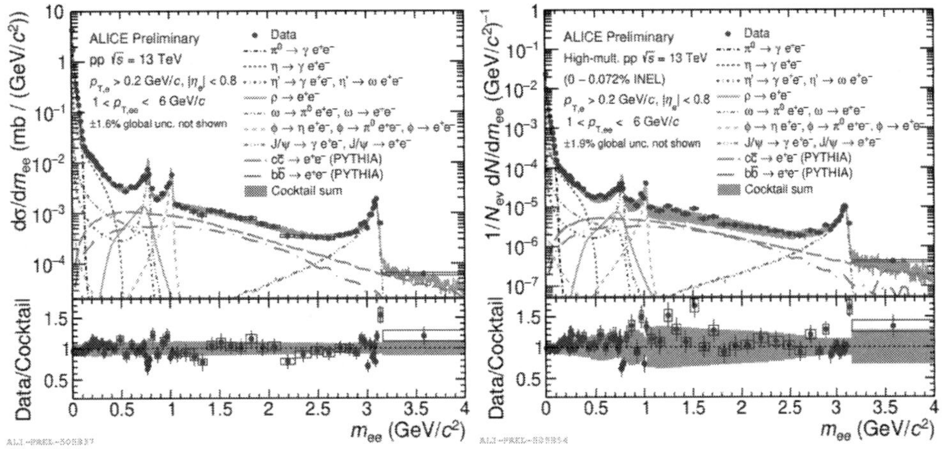

Figure 1. Dielectron production cross section (left) and yields (right) as a function of invariant mass in MB and HM pp collisions at $\sqrt{s}$ = 13 TeV, respectively.

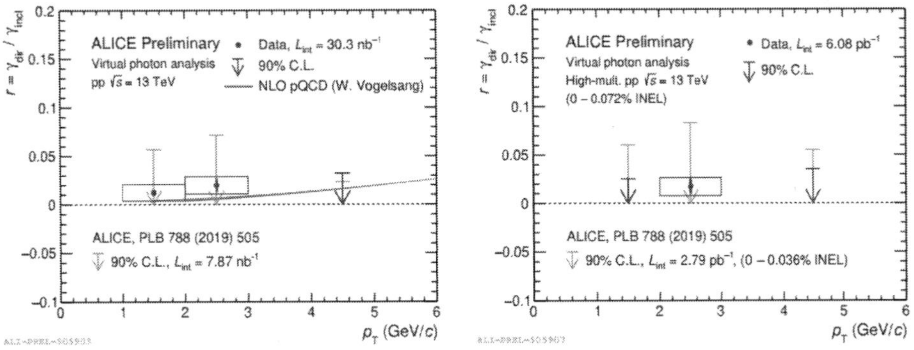

Figure 2. Direct photon fraction r as a function of p_T in MB (left) and HM (right) pp collisions at $\sqrt{s}$ = 13 TeV.

to as the hadronic cocktail. The π^0 and η spectra were measured for MB and HM pp collisions*, and used as input to the cocktail calculations. Dielectrons from semileptonic decays of heavy-flavor (HF) hadrons in MB pp collisions are estimated using the PYTHIA6 event generator. For the HM pp cocktail, an additional p_T-dependent scaling factor is applied for the HF and J/ψ contributions to take into account the multiplicity dependence of heavy-flavor and J/ψ production [7, 8]. The MB data is well described by the cocktail, which establishes the vacuum baseline. In HM pp collisions, the data and the cocktail are consistent and no sign of thermal radiation is observed in the IMR.

The direct virtual photon is measured via the internal conversion technique. The relation between virtual photons and dielectrons is given by the Kroll–Wada formula [9]. The different expected mass shapes of the virtual direct photon contribution and of e^+e^- pairs from light-meson Dalitz decays makes it possible to separate them. The direct photon fraction r defined as the ratio of direct photon yield to the inclusive photon yield $r = \gamma_{\mathrm{dir}}/\gamma_{\mathrm{incl}} = \gamma^*_{\mathrm{dir}}/\gamma^*_{\mathrm{incl}}\big|_{m_{ee}=0}$, is extracted from the data by fitting the dielectron invariant mass spectra with a three component function including the light-flavor decay (f_{LF}), direct virtual photon (f_{dir}), and heavy-flavor decay (f_{HF}) contributions: $f_{\mathrm{fit}} = r f_{\mathrm{dir}} + (1-r) f_{\mathrm{LF}} + f_{\mathrm{HF}}$. The fraction r as a function of p_T is shown in Fig. 2 for MB pp and HM pp collisions. It can be seen that statistical and system-

*Presented at SQM2022 - The 20th International Conference on Strangeness in Quark Matter

atic uncertainties are significantly reduced compared to the previous publication [5]. In MB pp collisions, r is in good agreement with the pQCD calculations and in HM pp collisions, no sign of an increase of r with respect to MB pp collisions is observed.

3 Dielectrons in Pb–Pb collisions at $\sqrt{s_{\rm NN}}$ = 5.02 TeV

The left panel of Fig. 3 shows the mass spectrum of dilepton pairs in the 0–10% central Pb–Pb collisions at $\sqrt{s_{\rm NN}}$ = 5.02 TeV. In the LMR, the results show a hint of an excess above the hadronic cocktail compatible with the prediction of thermal radiation from the hadron gas [10, 11]. The data are compared with different hadronic cocktails, based on two different assumptions for the HF contribution. In the first approach, the HF contribution is estimated using the measured charm and beauty cross sections at midrapidity in pp collisions at $\sqrt{s}$ = 5 TeV [12] scaled with the number of binary collisions. In the second approach, on top of the binary scaling, the HF contribution is weighted by the measured nuclear modification factor $R_{\rm AA}$ of single HF electrons [13]. Due to the large HF background in the IMR, the extraction

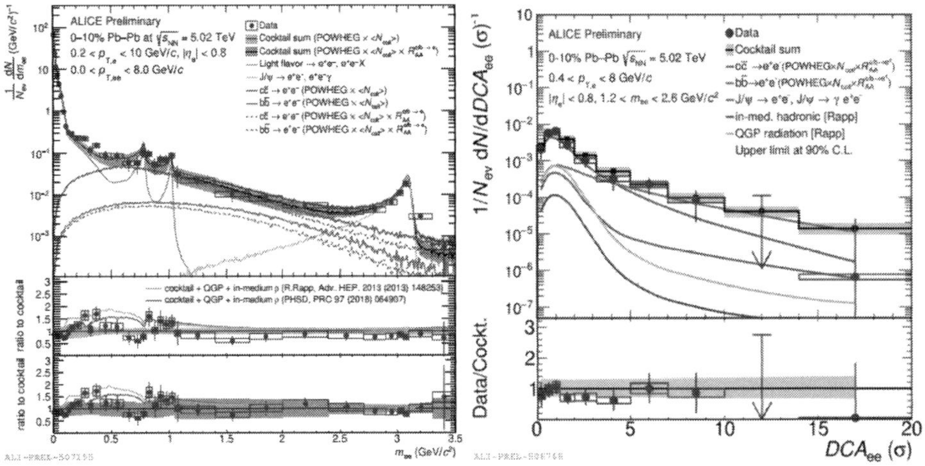

Figure 3. Left: Dielectron $m_{\rm ee}$ spectrum compared with different hadronic cocktails. Right: Dielectron yields as a function of $\rm DCA_{ee}$

of a thermal contribution from the QGP is challenging. In order to separate prompt (thermal) and non-prompt (HF) sources, the distance-of-closest approach to the primary vertex of dielectrons ($\rm DCA_{ee}$) is used. It is expected that prompt thermal dielectrons have smaller $\rm DCA_{ee}$ compared to the one from HF decays. Figure 3 (right) shows the $\rm DCA_{ee}$ distribution of dielectrons at IMR and the comparison with cocktail calculations. The data is well described by the cocktail which takes into account the e^+e^- HF suppression and thermal radiation from the QGP. At high $\rm DCA_{ee}$, the e^+e^- dominant contribution comes from beauty hadron decays. By fixing the beauty contribution to reproduce the measured $\rm DCA_{ee}$ spectra at high $\rm DCA_{ee}$, the charm and prompt (thermal) components can be extracted from the data. The results are compatible with HF suppression, more pronounced for charm than beauty, and a contribution of thermal radiation from the QGP similar to that predicted by models [10].
Finally, Figure. 4 (right) shows the measured ratio $R_\gamma = \gamma_{\rm incl}/\gamma_{\rm decay}$ = 1 / (1 - r) extracted in central Pb–Pb collisions. The result is in agreement with that of the real photon analysis via the gamma conversion method (PCM)*.The virtual photon results show smaller (larger) systematic (statistical) uncertainties compared to the ones from PCM. The direct photon yield can be constructed using the inclusive photon spectrum from PCM via $\gamma^{\rm dir} = \gamma^{\rm incl}_{\rm PCM} \times r$, as

*Presented at QM22 - XXIIXth International Conference Quark Matter 2022

shown in Fig. 4 (right). The result is compared with theory calculations including thermal, pQCD and pre-equilibrium photons,[14–17]. It can be seen that all models agree with the data but some tend to overestimate them at low p_{T}.

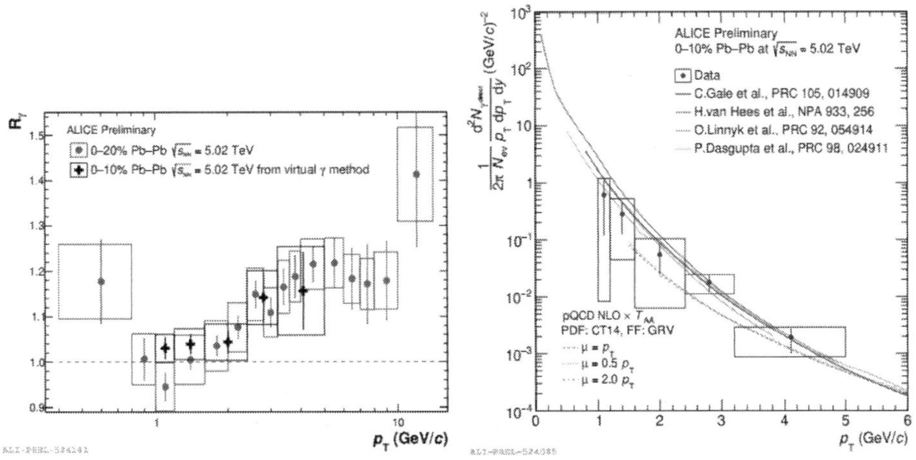

Figure 4. Left: Direct photon excess ratio R_{γ} as a function of p_{T} measured with virtual (black) and real (red) photons. Right: Direct photon spectrum in most central (0–10%) Pb–Pb collisions compared with theory and pQCD calculation.

4 Summary and outlook

The direct photon fraction measured in MB pp collisions at $\sqrt{s} = 13$ TeV is consistent with pQCD calculations and no significant increase of it is seen in HM pp collisions. The direct photon yield measured in most central Pb–Pb collisions at $\sqrt{s_{\mathrm{NN}}} = 5.02$ TeV can be reproduced by the available theory calculations, although the models tend to overestimate the data at low p_{T}. Furthermore, we presented the first results using a $\mathrm{DCA}_{\mathrm{ee}}$ analysis that separates the thermal signal from the HF background.

References

[1] R. Rapp *et al*, Landolt-Bornstein **23**, 134 (2010)
[2] ALICE Collaboration, Phys. Lett. B **719**, 29 - 41 (2013)
[3] ATLAS Collaboration, Phys. Rev. Lett. **110**, 182302 (2013)
[4] CMS Collaboration, JHEP **09**, 091 (2010)
[5] ALICE Collaboration, Phys. Lett. B **788**, 505 (2019)
[6] ALICE Collaboration, Int. J. Mod. Phys. A **29**, 1430044 (2014)
[7] ALICE Collaboration, JHEP **09**, 148 (2015)
[8] ALICE Collaboration, Phys. Lett. B **810**, 135758 (2020)
[9] N.M. Kroll and W. Wada, Phys. Rev. **98**, 1355 - 1359 (1955)
[10] R. Rapp, Adv. High Energy Phys. **2013**, 148253 (2013)
[11] T. Song *et al*, Phys. Rev. C **97**, 064907 (2018)
[12] ALICE Collaboration, Phys. Rev. C **102**, 055204 (2020)
[13] ALICE Collaboration, Phys. Lett. B **804**, 135377 (2020)
[14] C.Gale *et al*, Phys. Rev. C **105**, 014909 (2022)
[15] H.van Hees *et al*, Nucl. Phys. A **933**, 256 (2015)
[16] O. Linnyk *et al*, Phys. Rev. C **92**, 054914 (2015)
[17] P. Dasgupta *et al*, Phys. Rev. C **98**, 024911 (2018)

Spin polarization and alignment in heavy-ion collisions

Subhash Singha[1]*

[1]Institute of Modern Physics Chinese Academy of Sciences, Lanzhou

Abstract. We report the measurements of global spin polarization of hyperons (P_Λ) from FAIR, RHIC and LHC energies. The P_Λ continues to follow an increasing trend at $\sqrt{s_{NN}} < 7.7$ GeV indicating that enormous vorticity prevails even in hadronic dominant region. New measurements of local spin polarization ($P_{z,2}$) from isobar collisions follow the same pattern observed in previous RHIC and LHC energies. An observable motivated by predicted baryonic Spin Hall Effect, the net local polarization ($P^{\rm net}_{z,2}$), show a negative value in Au+Au collisions at $\sqrt{s_{NN}} = 19.6$ and 27 GeV. The first P_z measurement with respect to third order harmonic in 200 GeV RHIC isobar collisions shows a non-zero sine modulation, indicating the presence of complex vortical structure in the medium. Surprising new patterns of spin alignment (ρ_{00}) for various vector mesons (K^{*0}, $K^{*\pm}$, ϕ and J/ψ) emerged at both RHIC and LHC energies. The large deviation in ρ_{00} of several species poses challenge to the theory.

1 Introduction

In non-zero impact parameter heavy-ion collisions, a very large orbital angular momentum (OAM $\sim 10^5 - 10^7\hbar$) is expected in the centroid of the participant matter [1]. The part of OAM transferred to the Quark Gluon Plasma (QGP) can polarize quarks (anti-quarks) due to "spin-orbit" coupling [2]. Further, the polarization of quarks are translated into the polarization of produced hadrons with non-zero spin along the OAM via hadronization processes. In the initial stage, we also expect a large magnetic field ($B \sim 10^{18}$ Gauss) [3], which can induce different spin polarization for quarks (anti-quarks) with different magnetic moments. Thus, through the polarization measurements we can search for the signatures of initial OAM interactions and B-field. These measurements help understand the response of QGP medium under these extreme initial conditions and also provide a unique opportunity to probe the spin dynamics of QGP.

2 Spin polarization of hyperons

Using the parity violating weak decay of hyperons, the global polarization ($P_{\rm H}$) can be measured from the angular distribution of the daughter baryon in parent hyperon's rest frame [4]. In 2007, the STAR experiment set an upper limit on P_Λ (< 2%) in 200 GeV Au+Au collisions [5]. A breakthrough happened when STAR reported the first evidence of positive Λ and $\bar{\Lambda}$ spin polarization in heavy ion collisions from the RHIC Beam Energy Scan (BES) program. Under the assumption of local thermal equilibrium and based on hydrodynamic model,

*e-mail: subhash@impcas.ac.cn

the polarization data averaged over all the BES energies resulted in a vorticity of about $(9 \pm 1) \times 10^{21}\ /s$. This suggests the most vortical fluid ever created in a laboratory experiment [6]. In the BES measurements, $P_{\bar{\Lambda}}$ is consistently larger than P_{Λ} across the beam energies. Due to opposite sign of their magnetic moment, such a splitting in polarization is expected from the B-field. Nevertheless, the precision is not sufficient for a clear conclusion. Subsequently, a precise non-zero P_{Λ} along with differential measurements of P_{Λ} as a function of collision centrality, transverse momentum and rapidity are reported at the top-RHIC energy [7]. The increase of P_{Λ} from central to peripheral collisions is consistent with expectation from the vorticity of the medium. Later, the ALICE experiment at LHC reported P_{Λ} consistent with zero within uncertainties [8]. Recently, the P_{Λ} measurements are extended in the lower beam energy by the fixed target set up at STAR [9] and HADES [10]. A significant P_{Λ} of about 5% is observed at these energies. The P_{Λ} follows a monotonically increasing trend from 5.02 TeV down to 2.4 GeV as shown in Fig. 1. Various transport and hydrodynamic models can successfully capture this energy dependence trend. The monotonic increase of P_{Λ} is understood as an interplay of shear flow and baryon stopping. There are scenarios, e.g, migration of polarization to forward rapidity and lifetime of the QGP etc can also contribute to P_{Λ}. Recent model calculations suggested that $P_{\Lambda} \sim 0$ at $\sqrt{s_{NN}} = 2m_N$. The monotonic rise of P_{Λ} at $\sqrt{s_{NN}} < 7.7$ GeV and a clear increase of P_{Λ} from central to peripheral collisions, in a hadronic dominant matter, is indeed surprising. It is an open question to the community, how the hadron gas can support such enormous vorticity? More experimental measurements and inputs from theory are needed for a better understanding.

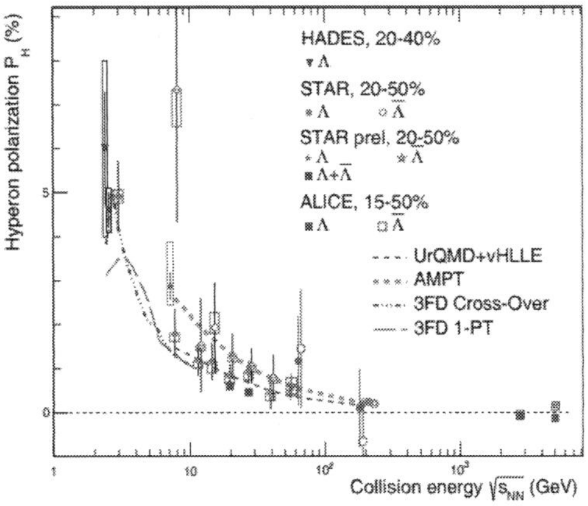

Figure 1. Beam energy dependence of global spin polarization of Λ and $\bar{\Lambda}$ hyperons [5–10]. Experimental measurements are compared to several models.

2.1 Global spin polarization in isobar collisions

The recent high statistics 200 GeV isobar collisions ($^{96}_{44}$Ru+$^{96}_{44}$Ru and $^{96}_{40}$Zr+$^{96}_{40}$Zr) [11] at RHIC offer new opportunities for polarization measurements. Due to larger proton numbers, the Ru+Ru is expected to have larger initial B-field than the Zr collisions. Thus the measurement P_{Λ} and $P_{\bar{\Lambda}}$ polarization in isobar collisions can probe the initial B-field. Furthermore, several model calculations predicted a collision system size dependence in P_{Λ} ($P_{\Lambda}^{\mathrm{O+O}} > P_{\Lambda}^{\mathrm{Ru+Ru}} > P_{\Lambda}^{\mathrm{Au+Au}}$) [12]. The isobar data can be used to test such dependencies.

The STAR collaboration reported P_Λ and $P_{\bar{\Lambda}}$ from isobar collisions [17]. As shown in Fig. 2, the P_Λ ($P_{\bar{\Lambda}}$) in isobar collisions followed same centrality dependence observed in previous Au+Au collisions. No polarization difference is observed between Λ and $\bar{\Lambda}$ or between the isobar collisions species within the current precision. However, the results from RHIC BES-II are expected to provide a precision to probe the B-field. No obvious system size dependence in P_Λ ($P_{\bar{\Lambda}}$) is observed when isobar results are compared to large system Au+Au collisions at a same collision centrality. Results from a smaller system collision (O+O) data can provide more insights into the system size dependence of P_Λ. The STAR reported a first time non-zero polarization for Ξ and Ω hyperons in 200 GeV Au+Au collisions, which reinforces the global nature of spin polarization in heavy-ion collisions [13]. More precise measurements of Ξ and Ω polarization in future RHIC runs can help testing the species dependent splitting of polarization due to different spin and magnetic moments.

2.2 Local spin polarization of hyperons in isobar collisions

Recently it is understood that the collective flow in conjunction with the vorticity of the medium can induce a novel longitudinal polarization (P_z) along the beam direction [14]. The first non-zero sine modulation of P_z for Λ ($\bar{\Lambda}$) hyperons with respect to second order flow harmonics (Ψ_2) is observed in 200 GeV Au+Au collisions by STAR [15]. Although many hydrodynamic and transport models do excellent job in capturing the beam energy dependence of the global hyperon polarization, they failed to explain the pattern in P_z with proper sign, which is termed as a "sign-puzzle". Later the same pattern in P_z is also seen in 5.02 TeV Pb+Pb collisions at LHC by ALICE [16]. Recently there are many theoretical developments in addressing this puzzle. It is understood that the addition of a "shear term" can explain the correct sign of P_z [18].

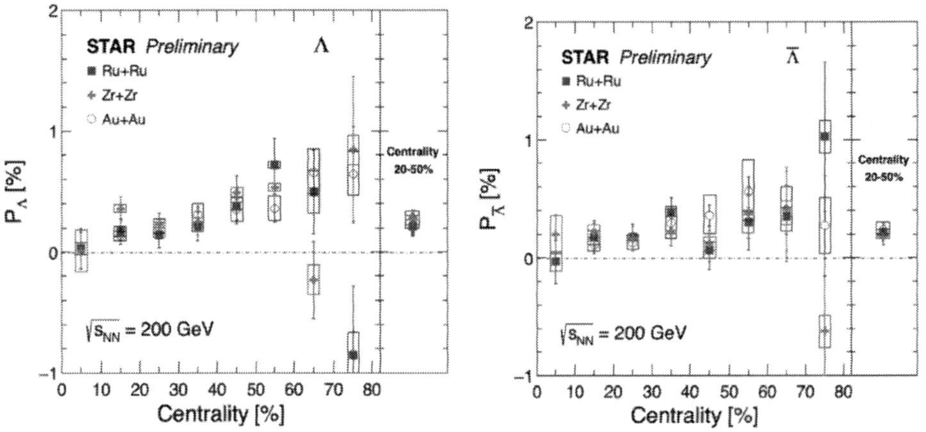

Figure 2. Centrality dependence of global spin polarization of Λ (left) and $\bar{\Lambda}$ (right) in isobar (Ru+Ru and Zr+Zr) collisions. Results are compared with the same from Au+Au collisions [17].

STAR reported P_z with respect to second order harmonics ($P_{z,2}$) from the isobar collision [17]. The left panel in Fig. 3 presents the comparison of isobar results with the same from Au+Au collisions at 200 GeV [15] and Pb+Pb collisions at 5.02 TeV [16]. There are some hints of system size dependence but no obvious energy dependence in $P_{z,2}$ is observed. These measurements in isobar collisions are also extended with respect to third order harmonics ($P_{z,3}$) for the first time and a significant non-zero sine modulation in polarization is observed [17]. Both $P_{z,2}$ and $P_{z,3}$ show a similar centrality dependence trend, however $P_{z,3}$ is systematically smaller than $P_{z,2}$ in peripheral collisions. These local polarization measurements can provide informations on the complex vortical structures, constrain on initial conditions and transport parameters in many models.

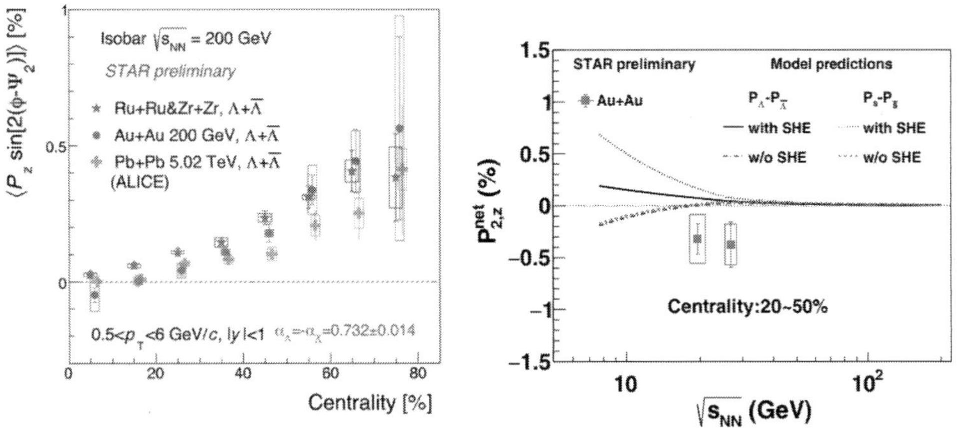

Figure 3. Left: Centrality dependence of local spin polarization ($P_{z,2}$) of Λ hyperons in Ru+Ru, Zr+Zr, Au+Au and Pb+Pb collisions. Right: Beam energy dependence of net local polarization ($P^{\text{net}}_{z,2}$) of Λ and $\bar{\Lambda}$. Results are compared with model predictions.

Recently, it is predicted that the gradient of baryonic chemical potential ($\nabla\mu_B$) can cause a difference in P_z for Λ and $\bar{\Lambda}$ ($P^{\text{net}}_{z,2}$). Having similarity with a spin phenomena in condensed matter physics, it is termed as baryonic Spin Hall Effect (SHE) [19]. It is predicted that the effect of SHE could be more prominent at lower beam energy (larger baryonic chemical potential) and it could produce a positive monotonically increasing $P^{\text{net}}_{z,2}$ with decrease in beam energy. On the contrary, absence of SHE is predicted to generate a similar monotonic pattern but with a negative sign in $P^{\text{net}}_{z,2}$. Abundant production of Λ ($\bar{\Lambda}$) and a large baryonic chemical potential provide an ideal environment to look for such novel effects at the RHIC BES range. The STAR experiment reported the net local polarization $P^{\text{net}}_{z,2}$ in mid-central Au+Au collisions at 19.6 and 27 GeV as shown in right panel of Fig. 3 [20]. The sign of $P^{\text{net}}_{z,2}$ is negative and consistent with absence of SHE. The P_z measurements from lower beam energies ($\sqrt{s_{NN}}$ < 19.6 GeV) will be useful in testifying the presence of SHE.

3 Spin alignment of vector mesons

The spin alignment of vector mesons can provide informations complementary to hyperon spin polarization. The global spin alignment is quantified by the diagonal element of the unity trace Hermitian spin density matrix. Among diagonal elements, only the 00th component (called ρ_{00}) can be measured from the angular distribution (θ^*) of the decay daughter of the vector meson with respect to the polarization axis. The ρ_{00} is expected to be $\frac{1}{3}$ in absence of spin alignment. Any deviation of ρ_{00} from $\frac{1}{3}$ indicates that there is a net spin alignment. In 2008, the measurements from 200 GeV Au+Au collisions by STAR resulted a null ρ_{00} for K^{*0} and ϕ mesons with a large uncertainty [21]. Recent measurements by ALICE in mid-central Pb+Pb collisions at LHC energies indicate that the K^{*0} ρ_{00} significantly smaller than $\frac{1}{3}$ in the low p_T region, while the same for ϕ mesons is consistent with $\frac{1}{3}$ within uncertainty [22]. At low p_T, there is a clear centrality dependence observed for both K^{*0} and ϕ. While the same measurements in p+p collisions at LHC yielded null deviations in ρ_{00}. The ρ_{00} in Pb+Pb collisions is observed to be an order of magnitude larger than the naive expectation from Λ polarization. The STAR collaboration reported of ϕ and K^{*0} ρ_{00} from the Beam Energy Scan [23]. The BES data reveals a surprising pattern. At mid-central collisions, while the ϕ ρ_{00} is significantly larger than $\frac{1}{3}$, the K^{*0} ρ_{00} is consistent with $\frac{1}{3}$ within uncertainties. The large signal for ϕ mesons can not be accommodated by conventional sources of polarization, e.g. vorticity of the medium, electromagnetic field etc. Polarization by a fluctuating vector meson strong force field can explain the large deviation for ϕ mesons in mid-central collisions [24]. The high statistics isobar data from STAR allows a precision measurement of ρ_{00} for K^{*0} and $K^{*\pm}$ and presented in left panel in Fig. 4. Due to a factor of five difference in magnetic moments of K^{*0} and $K^{*\pm}$ one naively expects K^{*0} ρ_{00} being larger than $K^{*\pm}$ due to its coupling with B-field. But surprisingly $K^{*\pm}$ ρ_{00} is observed to be larger than K^{*0} in isobar collisions. The behind such an ordering is not understood yet and require more inputs from theory.

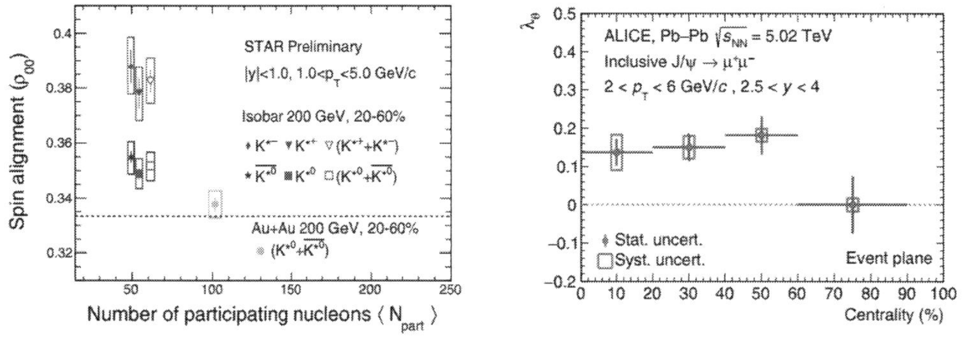

Figure 4. Left: Spin alignment (ρ_{00}) of K^{*0} ($\overline{K}^{*0}$) and $K^{*\pm}$ vector mesons in isobar (Ru+Ru and Zr+Zr) collisions. Results are compared with the same from Au+Au collisions. Right: Centrality dependence of spin polarization ($\lambda_\theta \propto (3\rho_{00} - 1)/(1 - \rho_{00})$) of J/$\psi$ in Pb+Pb collisions at 5.02 TeV [25].

Recently, the ALICE reported the first observation of J/Ψ spin polarization in di-muon channel at forward rapidity. The right panel in Fig. 4 presents the spin polarization parameter λ_θ ($\propto (3\rho_{00} - 1)/(1 - \rho_{00})$) as a function of centrality in Pb+Pb collisions at 5.02 TeV [25]. On the contrary to ρ_{00} of ϕ and K^{*0} near mid-rapidity, the J/ψ ρ_{00} (translated from λ_θ) at forward rapidity is larger than $\frac{1}{3}$. This is surprising and inputs from theory are needed for an explanation.

4 Summary

In summary, the spin polarization of hyperons across FAIR, RHIC and LHC energies established the global nature of hyperon spin polarization in heavy-ion collisions. More precise and differential measurements will provide more insights and constrain various models. Future P_Λ measurements at the low energy will reveal whether or not there is any vanishing polarization. A precise measurements of local spin polarization (P_z) of hyperons are carried out from RHIC to LHC. These results pose challenge to many theories and provide informations on the complex structure of vortices in heavy ion collisions. These measurements are generating discussions of emerging spin phenomena (such as Shear Induced Polarization, baryonic Spin Hall Effect) and help develop spin hydrodynamics, spin kinetic theories etc.

Measurements of spin alignments of vector mesons across RHIC and LHC is very surprising and puzzling. The observation of opposite sign of ρ_{00} deviations among different species requires more inputs from theory for better understanding of the data.

5 Acknowledgement

SS acknowledges support from the Strategic Priority Research Program of Chinese Academy of Sciences (Grant XDB34000000).

References

[1] F. Becattini, *et al.*, Phys. Rev. **C 77** 024906 (2008)
[2] Z. Liang and X. Wang, Phys. Rev. Lett. **94** 102301 (2005), [Erratum: Phys. Rev. Lett. **96** 039901 (2006)]; Phys. Lett. **B 629** 20–26 (2005)
[3] D. Kharzeev, *et al.*, Nucl. Phys. **A 803** 227-253 (2008)
[4] K. Schilling, *et al.*, Nucl. Phys. **B 15** 397–412 (1970)
[5] B. Abelev, *et al.*, [STAR Collaboration], Phys. Rev. **C 76** 024915 (2007); [Erratum: Phys. Rev. **C 95** 039906 (2017)]
[6] L. Adamczyk, *et al.*, [STAR Collaboration], Nature **548** 62–65 (2017)
[7] J. Adam, *et al.*, [STAR Collaboration], Phys. Rev. **C 98** 014910 (2018)
[8] S. Acharya, *et al.*, [ALICE Collaboration], Phys. Rev. **C 101** 044611 (2021); [Erratum: Phys. Rev. **C 105** 029902 (2022)]
[9] M. Abdallah, *et al.*, [STAR Collaboration], Phys. Rev. **C 104** L061901 (2021)
[10] R. Yassine, *et al.*, [HADES Collaboration], arXiv: 2207.05160
[11] M. Abdallah, *et al.*, [STAR Collaboration], Phys. Rev. **C 105** 014901 (2022)
[12] S. Shi, *et al.*, Phys Lett. **B 788** 409413 (2019); S. Alzhrani, *et al.*, arXiv: 2203.15718
[13] J. Adam, *et al.*, [STAR Collaboration], Phys. Rev. Lett. **126** 162301 (2021)
[14] F. Becattini and I. Karpenko, Phys. Rev. Lett. **120** 012302 (2018);
[15] J. Adam, *et al.*, [STAR Collaboration], Phys. Rev. Lett. **123** 132301 (2019)
[16] S. Acharya, *et al.*, [ALICE Collaboration], Phys. Rev. Lett. **128** 172005 (2022)
[17] X. Gou, [for STAR Collaboration], *these proceedings.*
[18] B. Fu, *et al.*, Phys. Rev. Lett. **127** 142301 (2021); F. Becattini, *et al.*, Phys. Rev. Lett. **127** 272302 (2021)
[19] B. Fu, *et al.* arXiv:2208.00430
[20] Q. Hu, [for STAR Collaboration], *SQM 2022 poster.*
[21] B. Abelev, *et al.*, [STAR Collaboration], Phys. Rev. **C 77** 061902 (2008)
[22] S. Acharya, *et al.*, [ALICE Collaboration], Phys. Rev. Lett. **125** 012301 (2020)
[23] M. Abdallah, *et al.*, [STAR Collaboration], arXiv:2204.02302
[24] X. Sheng, *et al.*, Phys. Rev. **D 101** (2020) 096005; Phys. Rev. **D 102** (2020) 056013; arXiv 2206.05868;
[25] [ALICE Collaboration], arXiv:2204.10171

Estimate of a nonflow baseline for the chiral magnetic effect in isobar collisions at RHIC

Yicheng Feng[1,*] for the STAR Collaboration

[1]Department of Physics and Astronomy, Purdue University, West Lafayette, IN 47907, USA

Abstract. Recently, STAR reported the isobar ($^{96}_{44}$Ru + $^{96}_{44}$Ru, $^{96}_{40}$Zr + $^{96}_{40}$Zr) results for the chiral magnetic effect (CME) search. The Ru+Ru to Zr+Zr ratio of the CME-sensitive observable $\Delta\gamma$, normalized by elliptic anisotropy (v_2), is observed to be close to the inverse multiplicity (N) ratio. In other words, the ratio of the $N\Delta\gamma/v_2$ observable is close to the naive background baseline of unity. However, nonflow correlations are expected to cause the baseline to deviate from unity. To further understand the isobar results, we decompose the nonflow contributions to $N\Delta\gamma/v_2$ (isobar ratio) into three terms and quantify each term by using the nonflow in v_2 measurement, published STAR data, and HIJING simulation. From these studies, we estimate a nonflow baseline of the isobar ratio of $N\Delta\gamma/v_2$ for the CME and discuss its implications.

1 Introduction

Quantum chromodynamics (QCD) predicts vacuum fluctuations, rendering nonzero topological charge in a local domain. As a result, there would be more particles with one certain chirality than the other, which is called the chirality anomaly. This effect may happen in heavy-ion collisions. If there is a strong magnetic field created by the spectator protons in heavy-ion collisions, the particles would have their spins aligned either parallel or anti-parallel to the magnetic field direction depending on their charges. Then, with the same chirality preference but opposite spins, the positive and negative charged particles would have opposite momenta. This charge separation phenomenon is called the chiral magnetic effect (CME).

To search for the CME, STAR conducted the isobar collisions in 2018, and recently published their isobar results [1]. The two isobar species, $^{96}_{44}$Ru and $^{96}_{40}$Zr, were expected to have similar background in CME sensitive observables due to the same nucleon number, while Ru was expected to have larger CME signal due to more protons and therefore stronger magnetic field. If so, this CME observable $\Delta\gamma$ over elliptic flow v_2 would be measured to be larger in Ru+Ru than Zr+Zr. However, from the recent STAR isobar paper, the Ru+Ru/Zr+Zr ratio of this observable is below unity, which is on the contrary to the initial expectation. The main reason is the multiplicity difference between the two isobars. If we include the multiplicity scaling, namely $N\Delta\gamma/v_2$, then the background baseline would be naively unity. The isobar results with the multiplicity scaling are above unity. In order to conclude whether this is an evidence for CME, one needs to consider other background contributions to $N\Delta\gamma/v_2$ from nonflow effects.

*e-mail: feng216@purdue.edu

2 $\Delta\gamma$ observable and its nonflow backgrounds

The CME-sensitive observable $\Delta\gamma \equiv C_3/v_2^*$ is defined by

$$C_{3,\text{os}} = \langle\cos(\phi_\alpha^\pm + \phi_\beta^\mp - 2\phi_c)\rangle, \quad C_{3,\text{ss}} = \langle\cos(\phi_\alpha^\pm + \phi_\beta^\pm - 2\phi_c)\rangle, \quad C_3 = C_{3,\text{os}} - C_{3,\text{ss}}, \tag{1}$$

where α, β indicate particles of interest (POI), and c is a reference particle as an estimate of the event plane whose resolution is equal to the elliptic flow v_2^*. We use an asterisk (*) on v_2 to indicate that it is the measured v_2 containing nonflow. The superscripts $+, -$ indicate the sign of particle charge. The subscripts OS, SS stand for opposite-sign and same-sign pairs, respectively. Their difference is taken to cancel charge-independent backgrounds.

The v_2^* measurement contains flow and nonflow, respectively:

$$v_2^{*2} = v_2^2 + v_{2,\text{nf}}^2, \quad \epsilon_\text{nf} \equiv v_{2,\text{nf}}^2/v_2^2, \tag{2}$$

where ϵ_nf denotes their ratio. C_3 is composed of a flow-induced background (major), a three-particle nonflow correlation (minor), and the possible CME signal [2]. Since we mainly focus on the backgrounds, the CME signal term is not written out:

$$C_3 = \frac{N_\text{2p}}{N^2} C_\text{2p} v_{2,\text{2p}} v_2 + \frac{N_\text{3p}}{2N^3} C_\text{3p} = \frac{v_2^2 \epsilon_2}{N} + \frac{\epsilon_3}{N^2}. \tag{3}$$

$C_\text{2p} \equiv \langle\cos(\phi_\alpha + \phi_\beta - 2\phi_\text{2p})\rangle$ is the two-particle (2p) nonflow correlations, such as resonance decay daughters with respect to their parent azimuthal angle (ϕ_2p). $C_\text{3p} \equiv \langle\cos(\phi_\alpha + \phi_\beta - 2\phi_c)\rangle_\text{3p}$ is the three-particle (3p) nonflow correlations, such as jets where all the three particles are correlated. Thus,

$$\frac{N\Delta\gamma}{v_2^*} = \frac{NC_3}{v_2^{*2}} = \frac{\epsilon_2}{1+\epsilon_\text{nf}} + \frac{\epsilon_3}{Nv_2^2(1+\epsilon_\text{nf})} = \frac{\epsilon_2}{1+\epsilon_\text{nf}}\left(1 + \frac{\epsilon_3/\epsilon_2}{Nv_2^2}\right), \tag{4}$$

where $\epsilon_2 \equiv \frac{N_\text{2p} v_{2,\text{2p}}}{N v_2} C_\text{2p}$ and $\epsilon_3 \equiv \frac{N_\text{3p}}{2N} C_\text{3p}$ are short-hand notations. With approximations to leading order, we get the nonflow contributions to the isobar ratio:

$$\frac{(N\Delta\gamma/v_2^*)^\text{Ru}}{(N\Delta\gamma/v_2^*)^\text{Zr}} \equiv \frac{(NC_3/v_2^{*2})^\text{Ru}}{(NC_3/v_2^{*2})^\text{Zr}} = \frac{\epsilon_2^\text{Ru}}{\epsilon_2^\text{Zr}} \cdot \frac{(1+\epsilon_\text{nf})^\text{Zr}}{(1+\epsilon_\text{nf})^\text{Ru}} \cdot \frac{\left[1 + \epsilon_3/\epsilon_2/(Nv_2^2)\right]^\text{Ru}}{\left[1 + \epsilon_3/\epsilon_2/(Nv_2^2)\right]^\text{Zr}}$$
$$\approx 1 + \frac{\Delta\epsilon_2}{\epsilon_2} - \frac{\Delta\epsilon_\text{nf}}{1+\epsilon_\text{nf}} + \frac{\epsilon_3/\epsilon_2/(Nv_2^2)}{1+\epsilon_3/\epsilon_2/(Nv_2^2)}\left(\frac{\Delta\epsilon_3}{\epsilon_3} - \frac{\Delta\epsilon_2}{\epsilon_2} - \frac{\Delta N}{N} - \frac{\Delta v_2^2}{v_2^2}\right). \tag{5}$$

In the above expression the quantities with Δ are the differences between Ru+Ru and Zr+Zr, while the others refer to those from Zr+Zr (or similarly Ru+Ru). We need those ϵ-terms for an improved background estimate.

3 Nonflow estimates

To get the nonflow in v_2^* measurement, ϵ_nf, we perform fit on the two-particle ($\Delta\eta$, $\Delta\phi$) 2D distribution for mid-central collisions (Fig. 1). The fit function is given by:

$$\begin{aligned} f(\Delta\eta, \Delta\phi) = & A_1 G_{\text{NS},W}(\Delta\eta) G_{\text{NS},W}(\Delta\phi) + A_2 G_{\text{NS},N}(\Delta\eta) G_{\text{NS},N}(\Delta\phi) + A_3 G_{\text{NS},D}(\Delta\eta) G_{\text{NS},D}(\Delta\phi) \\ & + \frac{B}{2-|\Delta\eta|}\text{erf}\left(\frac{2-|\Delta\eta|}{\sqrt{2}\sigma_{\Delta\eta,\text{AS}}}\right) G_\text{AS}(\Delta\phi \pm \pi) + D G_\text{RG}(\Delta\eta) \\ & + C[1 + 2V_1\cos(\Delta\phi) + 2V_2\cos(2\Delta\phi) + 2V_3\cos(3\Delta\phi)], \end{aligned} \tag{6}$$

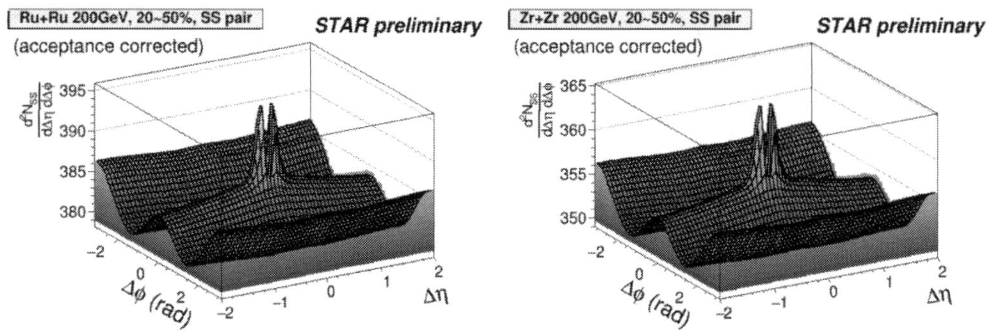

Figure 1. The two-particle $(\Delta\eta, \Delta\phi)$ distributions of SS pairs (left: Ru+Ru; right: Zr+Zr). The POI are from $0.2 < p_T < 2.0$ GeV/c and $|\eta| < 1$. The centrality range is 20–50%, which is defined by cutting on the POI multiplicity. The acceptance effect is corrected by mixed-event technique.

Table 1. Fit parameters and nonflow calculations.

	STAR preliminary	Ru+Ru	Zr+Zr
SS	fit parameter C	381.651 ± 0.011	351.988 ± 0.009
	fit parameter $V_2 = v_2^2$	0.0029716 ± 0.0000029	0.0028668 ± 0.0000025
	$\langle\cos(2\Delta\phi)\rangle_{\text{ss}}$ $(|\Delta\eta| > 0.05)$	0.0035968 ± 0.0000010	0.0034930 ± 0.0000010
inclusive	$\langle\cos(2\Delta\phi)\rangle = v_2^{*2}$ $(|\Delta\eta| > 0.05)$	0.0037161 ± 0.0000007	0.0036088 ± 0.0000007
	nonflow $U = \langle\cos(2\Delta\phi)\rangle - V_2$	0.0007446 ± 0.0000030	0.0007420 ± 0.0000026
	$\epsilon_{\text{nf}} = U/V_2$	(25.06 ± 0.10)%	(25.88 ± 0.09)%

where $G(x)$ is Gaussian functions. The first three terms in Eq. (6) are 2D Gaussian functions (A-terms) for the nearside (NS) correlations. The fourth term is for the awayside (AS) correlations (B-term), whose $\Delta\eta$ projection is an error function divided by $(2 - |\Delta\eta|)$, and whose $\Delta\phi$ projection is a Gaussian with mean value fixed to $\pm\pi$. The fifth term (D-term) is a Gaussian centering at $\Delta\eta = 0$ and independent of $\Delta\phi$ (referred to as the ridge, RG). The sixth term is the flow pedestal (C-term), where $V_n = v_n^2$ is the squared "true flows" assumed to be η-independent. Some of the fit results are listed in Table 1. By comparing the "true flow" from fit (V_2) with the inclusive measurement ($\langle\cos(2\Delta\phi)\rangle$ with a η-gap), we estimate ϵ_{nf} to be approximately 25%. The dominant contribution is from the A_1-term; we take half of this term as the systematic uncertainty for ϵ_{nf}, and obtain $\Delta\epsilon_{\text{nf}} = (-0.82 \pm 0.13 \mp 0.30)\%$, $-\Delta\epsilon_{\text{nf}}/(1+\epsilon_{\text{nf}}) = (0.65 \pm 0.11 \pm 0.22)\%$, $\Delta v_2^2/v_2^2 = \Delta V_2/V_2 = (3.7 \pm 0.1 \mp 0.3)\%$.

Due to the large η gap between TPC and ZDC, there is no 3p nonflow correlation between POI and ZDC event plane, so the ϵ_2 can be obtained from ZDC measurements [1]: $\epsilon_2 = \frac{N\Delta\gamma\{\text{ZDC}\}}{v_2\{\text{ZDC}\}} \approx 0.57 \pm 0.04 \pm 0.02$ (where the tracking efficiency is assumed to be ~80%). For the

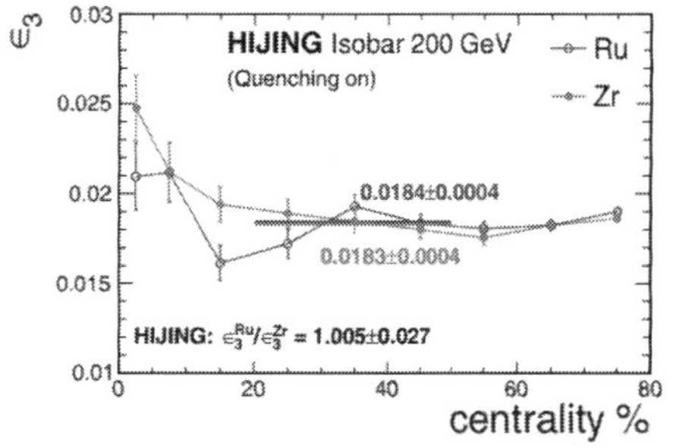

Figure 2. HIJING simulation estimates of ϵ_3.

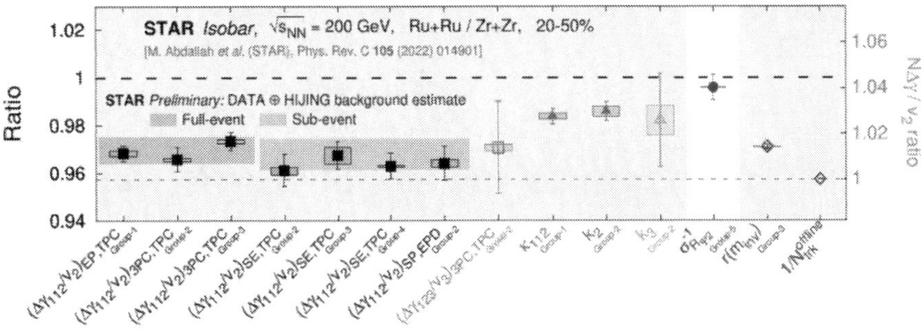

Figure 3. Background estimates with total uncertainties (bands) on the isobar ratio of $\Delta\gamma/v_2$ [1] .

isobar difference, however, the precision of measurements with the ZDC is too poor: $\Delta\epsilon_2/\epsilon_2 \approx (2.3 \pm 9.2)\%$. If we assume two isobars have similar C_{2p}, then $\epsilon_2 \propto Nr$, where the pair multiplicity difference $r \equiv \frac{N_{os}-N_{ss}}{N_{os}}$ is precisely measured [1]. Thus, the isobar difference can be estimated as $\Delta\epsilon_2/\epsilon_2 = \Delta r/r + \Delta N/N = (-2.95 \pm 0.08)\% + 4.4\% = (1.45 \pm 0.08)\%$, which has better precision.

The 3p nonflow ϵ_3 is hard to estimate from data, so we use HIJING model. HIJING does not have flow, so the only term left should be just the 3p nonflow. From this simulation, we get $\epsilon_3 \approx (1.84 \pm 0.04 \pm 0.92)\%$, while the isobar difference is small (consistent with zero) $\Delta\epsilon_3/\epsilon_3 = (0.5 \pm 2.7)\%$ (Fig. 2). We assumed 50% systematic uncertainty for ϵ_3.

With all the estimates above, we finally get by using Eq. (5):

$$\begin{aligned}
\frac{(N\Delta\gamma/v_2^*)^{\mathrm{Ru}}}{(N\Delta\gamma/v_2^*)^{\mathrm{Zr}}} &\approx 1+\frac{\Delta\epsilon_2}{\epsilon_2}-\frac{\Delta\epsilon_{\mathrm{nf}}}{1+\epsilon_{\mathrm{nf}}}+\frac{\epsilon_3/\epsilon_2/(Nv_2^2)}{1+\epsilon_3/\epsilon_2/(Nv_2^2)}\left(\frac{\Delta\epsilon_3}{\epsilon_3}-\frac{\Delta\epsilon_2}{\epsilon_2}-\frac{\Delta N}{N}-\frac{\Delta v_2^2}{v_2^2}\right) \\
&= 1 + (1.45 \pm 0.08)\% + (0.65 \pm 0.11 \pm 0.22)\% \\
&\quad + (0.094 \pm 0.007 \pm 0.048)[(0.5 \pm 2.7) - (1.45 \pm 0.08) - 4.4 - (3.7 \pm 0.1 \pm 0.3)]\% \\
&= 1 + (1.45 \pm 0.08)\% + (0.65 \pm 0.11 \pm 0.22)\% - (0.85 \pm 0.26 \pm 0.44)\% \\
&= 1.013 \pm 0.003 \pm 0.005. \qquad (7)
\end{aligned}$$

All the above estimates are for the full-event method [1]. Following the above procedure for sub-events, we estimate the equivalent of Eq. (7) as $(1.011 \pm 0.005 \pm 0.005)$. We plot those nonflow background estimates together with the STAR isobar data in Fig. 3.

4 Summary

This study estimates nonflow contribution in v_2 by fitting two-particle $(\Delta\eta, \Delta\phi)$ distribution, and 2p nonflow in C_3 using STAR isobar data [1]. The 3p nonflow in C_3 is evaluated by HIJING simulation. Using the above information, we obtain an improved background estimate of isobar ratio $\frac{(N\Delta\gamma/v_2^*)^{\mathrm{Ru}}}{(N\Delta\gamma/v_2^*)^{\mathrm{Zr}}} \approx (1.013 \pm 0.003 \pm 0.005)$ for full-event, and $(1.011 \pm 0.005 \pm 0.005)$ for sub-event.

References

[1] M. Abdallah *et al.* (STAR Collaboration), Phys. Rev. C **105**, 014901 (2022)
[2] Y. Feng, J. Zhao, H. Li, H. j. Xu and F. Wang, Phys. Rev. C **105**, 024913 (2022)
[3] D. E. Kharzeev, L. D. McLerran and H. J. Warringa, Nucl. Phys. A **803**, 227-253 (2008)
[4] S. Choudhury *et al.* Chin. Phys. C **46**, 014101 (2022)

Transverse single-spin asymmetry of very forward neutral pion

Hee-Jin Kim[1,*], *Samson* Clymton[1,**], and *Hyun-Chul* Kim[1,2,***]

[1]Department of Physics, Inha Univesity, Incheon 22212, Republic of Korea

[2]School of Physics, Korea Institute for Advenced Study (KIAS), Seoul 02455, Republic of Korea

Abstract. We present in this talk a recent work on the transverse single-spin asymmetry of the very forward neutral pion in polarized $p+p$ collisions at $\sqrt{s}$ = 510 GeV. The triple-Regge formalism remarkably well desribes the RHICf data at $p_T < 1$ GeV. We found that the neutral pion production at low p_T is interpreted as a diffractive one.

1 Introduction

Recently, the transverse single-spin asymmetry(TSSA) of the neutral pion in the inclusive polarized proton - proton collision was measured by the RHICf Collaboration [1]. The TSSAs in the high $p_T(\gtrsim 2$ GeV) have been understood via pQCD-based models such as the transverse momentum-dependent functions (TMDs) [2–4] or twist-3 factorization [5–12]. However, the produced pions in this measurement have large pseudorapidities($\eta > 6$) and small transverse momenta, so that those approaches may not be vaild in this kinematic region.

The Regge approach has been successfully applied to soft processes, where the center of mass energy s is very large and the transverse momentum is small. In 1970, Mueller found the generalized optical theorem [14], which allows one to approximate the two-body scattering in the high energies to a relevant Reggeon exchange processes of the three-body reaction. The triple-Regge process is one of the kinematic boundaries of the three-body Mueller amplitude which can be applied to the kinematic range of the RHICf experiment.

In the present talk, we show how to compute the TSSA of the neutral pion in the very forward direction by using the triple-Regge exchange diagram. The N, $N^*(1520)$, $\Delta(1232)$ and $\Delta(1600)$ baryon trajectories are introduced. The interefence between the trajectories yield the very foward TSSA. It might indicate that the TSSA of the neutral pion in the very forward direction can be interpreted as a diffractive one.

[*]e-mail: heejin.kim@inha.edu
[**]e-mail: sclymton@inha.edu
[***]e-mail: hchkim@inha.ac.kr

2 The differential cross section in the $p^{\uparrow} + p \to \pi^0 + X$ reaction in the triple-Regge limit

The differential cross section in the inclusive $pp^{\uparrow} \to \pi^0 X$ collision at high-energies is given as

$$d\sigma^{\uparrow} = \frac{1}{s} \sum_X \left| A^{\text{tot}}_{pp^{\uparrow}\to\pi^0 X}(M_X^2) \right|^2, \tag{1}$$

where M_X is the missing mass which is defined as $M_X^2 = (p_p + p_{p^{\uparrow}} - p_{\pi^0})^2$ and $\sum_X$ denotes the summation over the phase space of the X. Since the RHICf energy is sufficiently large, $\sqrt{s} = 510$ GeV, the scattering amplitude can be described in terms of Reggon exchanges. The square of the two-body Reggeon exchange process is then analytically continued to the $3 \to 3$ reaction in the forward direction by Mueller's generalized optical theorem [14]:

$$d\sigma^{\uparrow} = \frac{1}{s} \sum_{i,j} \sum_{\lambda} \beta^i_{\uparrow\lambda}(t) \beta^j_{\uparrow\lambda}(t) \mathcal{P}_i(t) \mathcal{P}^*_j(t) \text{Disc}\, A_{ip\to jp}(M_X^2), \tag{2}$$

where $\beta^i_{\uparrow\lambda}$ is a residue function in the Reggeized amplitude and λ is the helicity of the baryon. Since the Regge approach does not provide any information on the vertex structure, we adopt the effective Lagrangians to compute the residue function. The propagator of the Regge tajectory is defined as [15, 16]

$$\mathcal{P}_i(t) \equiv \alpha'_i \xi_i(t)^{\pm} \Gamma(J_i - \alpha_i(t))(1 - x_F)^{-\alpha_i(t)}, \tag{3}$$

where $\alpha_i(t)$ stands for the Regge trajectory of the i particle and x_F is the Feynman variable which is defined as a longitudinal momentum fraction between the polarized proton and its fragment π^0.

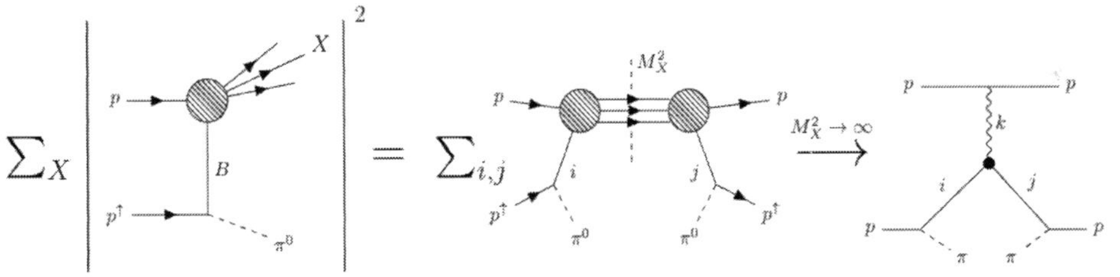

Figure 1. The triple-Regge approximation of the differential cross section. i, j and k indicate the Regge trajectories.

In the $M_X^2 \to \infty$ limit, the discontinuity on the M_X^2 plane also can be replaced by an appropriate Regge trajectory as shown in Fig. 1. Then the $d\sigma^{\uparrow}$ is written as the sum of the triple-Regge diagrams,

$$\text{Disc}\, A_{ip\to jp}(M_X^2) = \sum_k G^k_{ij}(t) \left(\sum_{\lambda'} \beta^k_{\lambda'\lambda'}(t_k = 0) \right) \left(\frac{M_X^2}{s_0} \right)^{\alpha_k(0)}, \tag{4}$$

where s_0 is an energy scaling factor and fixed as 1 GeV^2. One could calculate the ppk vertex on the top side of the triple-Regge diagram by using the effective Lagrangian method. However, since we utilize the generalized optical theorem to derive the triple-Regge exchange process, k trajectories do not carry the momentum ($t_k = 0$). Thus the residue function of this vertex can be written as $\gamma^k_{pp} = \sum_{\lambda'} \beta^k_{\lambda'\lambda'}$. The coupling of three tracjectories G^k_{ij} are parametrized as a function of t.

Table 1. Parameter values in A_N.

	NN^*	N^*N^*	$\Delta\Delta$	$\Delta\Delta^*$	$\Delta^*\Delta^*$
g^k_{ij}	0.028	0.10	0.022	−0.018	0.079
b^k_{ij}	0.2	-	-	-	-

3 Transverse sinlge-spin asymmetry of the very forward π^0

The transverse single-spin asymmetry is given as a ratio of the spin-dependent($d\Delta\sigma$) and spin-averaged($d\sigma$) differential cross sections. In the course of the calculation, one can see that the ij diagonal terms do not contribute to the spin-dependent cross section. It implies that the interference between i and j exchange (more specifically between the signature factors of them) yield the TSSA in the very forward direction.

$d\sigma$ can be simplified by the parity invariance of the residue function β [17]. Firstly, $d\sigma^\uparrow$ vanishes when the trajectory k is unnatural parity state. So we introduce the Pomeron for k exchange as a leading trajectory, because it has the largest Regge intercept $\alpha(0)$ among the natural parity states. Secondly, $d\Delta\sigma$ vanishes if i and j trajectories have opposite naturalities. The leading contributions are N-$N^*(1520)$ and Δ-$\Delta(1600)$ interference for natural and unnatural part, respectively. One define the triple-Regge coupling as

$$G^k_{ij}(t) = \begin{cases} G^k_{ii}(0)e^{-B^k_{ii}|t|} & \text{for } i = j \\ G^k_{ij}(0)\frac{\sqrt{|t|}}{m_\pi}e^{-B^k_{ij}|t|} & \text{for } i \neq j. \end{cases} \tag{5}$$

In order to fit the RHICf data, we define the following new parameters as

$$g^k_{ij} = G^k_{ij}(0)/G^k_{NN}(0), \quad b^k_{ij} = B^k_{ij}(0) - B^k_{NN}(0). \tag{6}$$

Finally, we obtain the TSSA in terms of the triple-Regge amplitude as

$$A_N = \frac{\sum_\lambda \left[\beta^N_{+\lambda}\beta^{N^*}_{-\lambda}\mathrm{Im}\,\mathcal{P}_N\mathcal{P}^*_{N^*}(\sqrt{t}/m_\pi)g^P_{NN^*} + \beta^\Delta_{+\lambda}\beta^{\Delta^*}_{-\lambda}\mathrm{Im}\,\mathcal{P}_\Delta\mathcal{P}^*_{\Delta^*}(\sqrt{t}/m_\pi)g^P_{\Delta\Delta^*}e^{-b^P_{\Delta\Delta^*}|t|}\right]}{\sum_\lambda \left[\sum_i(\beta_{+\lambda})^2|\mathcal{P}_i|^2 g^P_{ii}e^{-b^P_{ii}|t|} + \sum_{i\neq j}\beta^i_{+\lambda}\beta^j_{-\lambda}\mathrm{Re}\,\mathcal{P}_i\mathcal{P}^*_j(\sqrt{t}/m_\pi)g^P_{ij}e^{-b^P_{ij}|t|}\right]}, \tag{7}$$

where P indicates the Pomeron. We list the values of the parameters in Table. 1.

4 Results and Discussion

Our main interest lies in describing the A_N for the soft diffractive π^0 production in the the large x_F region, employing the tripe-Regge formalism. The numerical results show a remarkable agreement with the RHICf experimental data up to $p_T \simeq 0.8$ GeV/c, as shown in Fig. 2.

The very forward π^0 A_N exhibits a rising trend as p_T increase. In Fig. 2, A_N is almost saturated in the region $0.25 \lesssim p_T \lesssim 0.45$ GeV/c. The $\sqrt{|t|}$ factor in the triple-Regge coupling can explain this data point, since t becomes zero at the point between third and fourth points. For the smaller p_T, A_N comes from NN^* interference term. As p_T increases, the contribution of NN^* becomes negative and large $\Delta\Delta^*$ term compensates it, so that the total contributions yield positive values at $p_T > 0.5$ GeV/c. In addition, A_N starts to increase rapidly and reachs over 20 % at about $p_T = 0.8$ GeV/c. The next leading pole contributions moderate A_N at higher p_T. N^*N^* contribution is not sufficient to suppress the large value of A_N, so $\Delta^*\Delta^*$ is required to describe the RHICf data.

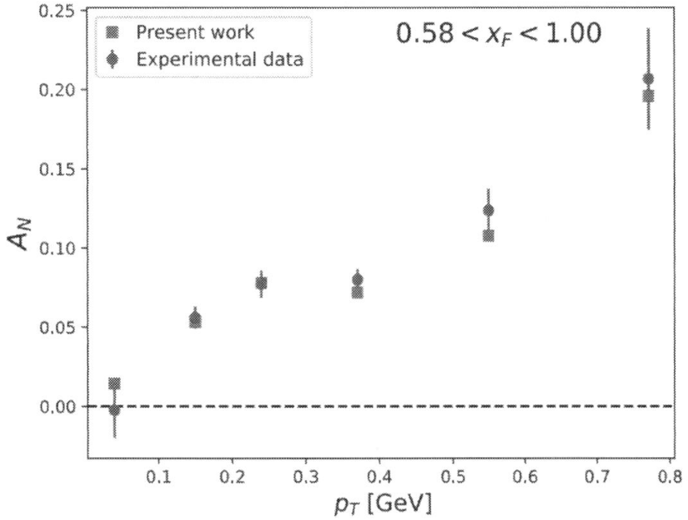

Figure 2. p_T distribution of A_N with $0.58 < x_F < 1$. The square symbol denotes the numerical result for the present work. The circles and solid lines are the RHICf experiment results and errors.

5 Summary and Conclusions

We investigate the transverse single-spin asymmetry for the neutral pion in the very forward direction via triple-Regge exchange processes. The numerical result with the various Regge trajectories show an excellent agreement with the data from the RHICf experiment at large x_F. It indicates that the most of the A_N in the $p_T < 1$ Gev$/c$ and $x_F > 0.58$ is produced by interferences between diffractive scattering processes.

References

[1] M. H. Kim *et al.* [RHIC-f], Phys. Rev. Lett. **124**, no.25, 252501 (2020).
[2] D. W. Sivers, Phys. Rev. D **41**, 83 (1990).
[3] J. C. Collins, Nucl. Phys. B **396**, 161-182 (1993).
[4] J. C. Collins, S. F. Heppelmann and G. A. Ladinsky, Nucl. Phys. B **420**, 565-582 (1994) doi:10.1016/0550-3213(94)90078-7 [arXiv:hep-ph/9305309 [hep-ph]].
[5] A. V. Efremov and O. V. Teryaev, Sov. J. Nucl. Phys. **36**, 140 (1982).
[6] H. Eguchi, Y. Koike and K. Tanaka, Nucl. Phys. B **763**, 198-227 (2007).
[7] J. w. Qiu and G. F. Sterman, Nucl. Phys. B **378**, 52-78 (1992).
[8] J. w. Qiu and G. F. Sterman, Phys. Rev. Lett. **67**, 2264-2267 (1991) doi:10.1103/PhysRevLett.67.2264
[9] L. Gamberg, Z. B. Kang and A. Prokudin, Phys. Rev. Lett. **110**, no.23, 232301 (2013).
[10] K. Kanazawa, Y. Koike, A. Metz and D. Pitonyak, Phys. Rev. D **89**, no.11, 111501 (2014).
[11] L. Gamberg, Z. B. Kang, D. Pitonyak and A. Prokudin, Phys. Lett. B **770**, 242-251 (2017).
[12] L. Gamberg, M. Malda, J. A. Miller, D. Pitonyak, A. Prokudin and N. Sato, [arXiv:2205.00999 [hep-ph]].

[13] J. Cammarota *et al.* [Jefferson Lab Angular Momentum], Phys. Rev. D **102**, no.5, 054002 (2020).

[14] A. H. Mueller, Phys. Rev. D **2**, 2963-2968 (1970).

[15] P. D. B. Collins *An Introduction to Regge Theory and High Energy Physics*, Cambridge University Press, (1977).

[16] A. Donnachie, H. G. Dosch, P. V. Landshoff and O. Nachtmann *Pomeron Physics and QCD*, Cambridge University Press, (2002)

[17] F. E. Paige and D. P. Sidhu, Phys. Rev. D **14**, 2307 (1976).

Exotic Particles and heavy ion collisions

Su Houng Lee[1,*], *Sungsik* Noh[1,**], *Daeho* Park[1,***], and *Hyung-Ok* Yoon[1,****]

[1]Department of Physics and Institute of Physics and Applied Physics, Yonsei University, Seoul 03722, Korea

Abstract. We discuss the structures of exotic candidates and why it is interesting to measure them in heavy ion collisions. We take the $X(3872)$ and T_{cc} to illustrate our point.

1 Introduction

Observations of exotic hadrons opened a new ear in hadronic physics. This started with the observation of the $X(3872)$[1] and then followed by a series of exciting discoveries such as the pentaquark states[2, 3], the tetraquark states X(5568)[4], the T_{cc}[5] and the recent strange pentaquark states. Of particular interest is the flavor exotic T_{cc} state with quark content $(cc\bar{u}\bar{d})$, which suggests that a meson can be composed of two quarks and two anti-quarks. Despite the excitement, we are still left with the pressing question of whether their structures are molecular types composed of loosely bound hadrons or compact multiquark states where the quarks and anti-quarks are confined inside a compact configurations with the typical hadron size. Discriminating their structures will give us important clues to understanding confinement phenomena.

Previously, we have pointed out that for exotic hadrons to be strongly bound and compact, they have to include multiple heavy quarks and therefore could be easily produced in relativistic heavy ion collisions[6, 7]. We have then formed the ExHIC theory collaboration and through the collaborative work found that one could discriminate whether an exotic is a molecular configuration or a compact multiquark state, by studying their production characteristics in a heavy ion collision[8–10]. The CMS collaboration at CERN has recently measured the $X(3872)$ for the first time in a heavy ion collision, and found that its production ratio to $\psi(2S)$ is anomalously larger than that measured in pp collisions[11]. We have shown in a recent work[12], that the ratio of the total yield of molecular configuration of $X(3872)$ to the total yield of $\psi(2S)$ in the statistical hadronization model with charm (SHMc) is 0.806 ± 0.234. The experimental measurement is at high transverse momentum, nevertheless, considering suppression of $\psi(2S)$ compared to the SHMc, the observed ratio seems consistent with a molecular picture for $X(3872)$.

In this talk, we first argue why the $X(3872)$ can not be a compact configuration and thus should be a molecular configuration while T_{cc} could either be a compact multiquark configuration or a molecular configuration. The argument is based on short distance properties using

*e-mail: suhoung@yonsei.ac.kr
**e-mail: sungsiknoh@yonsei.ac.kr
***e-mail: olmu100@yonsei.ac.kr
****e-mail: mero0819@yonsei.ac.kr

a quark model[13] combined with expectations from the pion exchanged model[14, 15] for long range interactions. We then predict the transverse momentum distribution of the molecular configuration of the $X(3872)$ and T_{cc} in Pb-Pb collisions at $\sqrt{s_{NN}}$ =5.02 TeV using the coalescence formula that well reproduces the deuteron data at similar collision energies.

2 Short distance interaction from a quark model

A compact multiquark state that is composed of two hadrons can form a stable multiquark configuration if the short distance interaction between the two composing hadrons is attractive. As is well known, the short distance nucleon-nucleon interaction is repulsive and thus we do not expect to have a dibaryon state with the deuteron quantum number. On the other hand, it is well known that when extended to SU(3), baryon-baryon interaction is attractive in the flavor singlet channel leading to the possibility of a stable H-dibaryon.

To understand the short distance interaction and what happens when two hadrons approach each other within a quark model perspective, let us consider the following typical nonrelativistic Hamiltonian for n quarks that contains the confinement and hyperfine potential for the color and spin interaction.

$$H = \sum_{i=1}^{n}(m_i + \frac{\mathbf{p}_i^2}{2m_i}) - \sum_{i<j}^{n} \lambda_i^c \lambda_j^c V_{ij}^C(r_{ij}) - \sum_{i<j}^{n} \frac{(\lambda_i^c \lambda_j^c)(\sigma_i \sigma_j)}{m_i m_j} V_{ij}^{SS}(r_{ij}). \tag{1}$$

Here, m_i, $\lambda_i^c/2$ and σ_i are the mass, color and spin operators of the i'th quark, respectively. V_{ij}^C and V_{ij}^{SS} are the confinement and hyperfine potential, respectively with r_{ij} being the distance between quarks. The above model is known to well reproduce the properties of the ground state hadrons[16]. Let us now consider what happens to each contribution in Eq. (1) as we bring two hadrons respectively with n_1 and n_2 quarks into a compact configuration of n quarks.

The first terms are the kinetic terms and show that the compact configuration has an additional term corresponding to the extra kinetic energy that one has to overcome to bring the two hadrons into a compact configuration. Assuming the compact configuration to be the size of a typical hadron (~ 0.64fm) and the typical reduced mass μ ~ 0.5 GeV, the extra term roughly amounts to the following.

$$\frac{p_{n_1,n_2}}{2\mu} \sim \frac{1}{1\mathrm{GeV}(0.64\mathrm{fm})^2} \sim 100\ \mathrm{MeV}. \tag{2}$$

This suggests that the attraction has to be sufficiently larger than this value for the configuration to be compact.

The color-color attraction appearing in the second set of terms in Eq. (1) can not give such a large attraction. This can be seen by considering the compact configuration where all the quarks occupy a similar size. Then, the effects of V_{ij}^C are the same for all quarks pairs so that the overall strength is given by the color-color factor. However, assuming that both the compact configuration and individual hadrons are color singlets, we find that the overall factor should be the same as the sum of color spin factor of the two hadrons.

$$\begin{aligned} \sum_{i<j}^{n}(\lambda_i^c \lambda_j^c) &= \frac{1}{2}\Big((\lambda_1^c + ... + \lambda_n^c)^2 - \lambda_1^2 - .. - \lambda_n^2\Big) \\ &= 0 - \frac{8}{3}(n_1 + n_2) = \sum_{i<j}^{n_1}(\lambda_i^c \lambda_j^c) + \sum_{i<j}^{n_2}(\lambda_i^c \lambda_j^c). \end{aligned} \tag{3}$$

Hence, the additional interaction between quarks coming from different hadrons should be small.

Now, let us consider the color spin interaction. To identify the important attractive channel, we again assume that all quarks occupy the same configuration and consider the prefactor, which we define as

$$K = -\sum_{i<j}^{n} (\lambda_i^c \lambda_j^c)(\sigma_i \sigma_j). \tag{4}$$

This factor is responsible for many of the characteristics of hadron interaction at short distance. First of all, the nucleon delta mass difference comes from the difference in the spin orientations, which lead to different K factors. For the nucleon and delta the K factors are, respectively, −8 and +8 leading to the $\Delta K = 16$ which corresponds to the mass difference of 290 MeV. Assuming that the quarks occupy similar size as those inside a nucleon, one can conclude that a unit factor of K corresponds to about 18 MeV. The ratio of the positive K factors between a dibaryon states in the (spin,isospin) $(0, 1)$ and $(1, 0)$ channels is 1.29, which is similar to the ratio between the corresponding nucleon-nucleon repulsions at short distance <0.3 fm observed in a lattice calculation. In SU(3) the ratio between flavor 27 and flavor 1 is −3, which is also consistent with the lattice result at short distance. In a previous publication[13], we have shown that the short distance part of the baryon-baryon interactions for various quantum numbers from the recent lattice calculation can be well reproduced using a constituent quark model with color-spin interaction. In fact, the full constituent quark model calculations are found to be comparable to the simple estimates based on the K factors among the multiquark configurations[13].

Therefore, we expect that a compact multiquark configuration will only exist if there is a strong negative short range interaction between composing hadrons which means a large negative K factor of the compact configuration. The negative K factor was the original motivation for a possible H-dibaryon state[17]. One could also use this K factor to search for possible pentaquark configurations[18].

2.1 The $X(3872)$ and T_{cc}

Let us consider the K factor to the first and latest observed tetraquark states. The quantum number of $X(3872)$ is $I^G(J^{PC}) = 0^+(1^{++})$. This leads to two allowed color-spin wave functions which are most conveniently written in the quark-antiquark basis $(c\bar{c}) \otimes (q\bar{q})$ with c, q being the charm and light quarks. In such basis, the two states are $(1 \otimes 1)_{C=1}(V \otimes V)_{J=1}$ and $(8 \otimes 8)_{C=1}(V \otimes V)_{J=1}$, where the first brackets denote two color singlets (1) or octets (8) combined into color singlet and the second two spin 1 (V) into spin 1. The K factor, after subtracting the lowest $D\bar{D}^*$ threshold, in the above basis becomes

$$K_{X(3872)} - K_D - K_{\bar{D}^*} = \begin{pmatrix} \frac{16}{3m_c^2} + \frac{16}{3m_q^2} + \frac{32}{3m_q m_c} & 0 \\ 0 & -\frac{2}{3m_c^2} - \frac{2}{3m_q^2} - \frac{4}{3m_q m_c} \end{pmatrix}. \tag{5}$$

The factor $\frac{32}{3m_q m_c}$ is due to the subtracted threshold. One notes that the second diagonal component coming from the color octet element gives an attractive contribution. However, noting that the K factor coming from the delta nucleon mass difference is $\frac{16}{m_q^2}$, which gives a mass difference of around 290 MeV, we find that using a contitutent quark mass of $m_c = 1500$ MeV and $m_q = 300$ MeV, the attraction only amounts to about 17.4 MeV assuming that all quarks occupy the size of a nucleon. Considering the additional kinetic energy needed to bring the two mesons together, one can conclude that the attraction is not strong enough to confine the $X(3872)$ into a compact multiquark configuration. A detailed quark model calculation confirms this expectation. This result furthermore implies that the attraction between $D\bar{D}^*$ in this channel is small at short distance.

Assuming that $J^P = 1^+$ for the T_{cc}, a better way to describe the color-spin wave function is using the diquark-antidiquark $(ud)\otimes(\bar{c}\bar{c})$ basis, for which the two states are $(\bar{3}\otimes 3)_{C=1}(S\otimes V)_{J=1}$ and $(6\otimes\bar{6})_{C=1}(V\otimes S)_{J=1}$, where the first and second brackets show the color combination and spin with S denoting the spin zero scalar state, respectively. Using these bases, the K factor for the T_{cc} is given as follows.

$$K_{T_{cc}} - K_D - K_{D^*} = \begin{pmatrix} -\frac{8}{m_q^2} + \frac{8}{3m_c^2} + \frac{32}{3m_q m_c} & \frac{8\sqrt{2}}{m_c m_q} \\ \frac{8\sqrt{2}}{m_c m_q} & -\frac{4}{3m_q^2} + \frac{4}{m_c^2} + \frac{32}{3m_q m_c} \end{pmatrix}. \tag{6}$$

One now notes that with the same parameters as above, the attraction from the first diagonal component is found to be around 104.4 MeV, similar to what was first noted in Ref [19]. Diagonalizing the matrix, the attraction is expected to be larger. Detailed constituent quark model calculations agree on the deeply bound compact configurations for the T_{bb}, but are still controversial for T_{cc} [16, 20, 21].

Using the same analysis, one can estimate the attraction between $D - D^*$ at relative distance estimated within that configuration. Using a simple Gaussian wave function, one finds that the attraction is around 200 MeV at $r_{D-D^*} = 0.3$ f m, which is similar to what has been estimated in the lattice gauge calculation[22]. Hence, we will add the short distance attraction to the long range pion exchange potential and probe the possibility of a weakly bound molecular configuration of the T_{cc}.

3 Perspectives from the pion exchange

For molecular configurations, the most important physics is the pion exchange potential. We will use the pion exchange potentials between $D - \bar{D}^*$ and $D - D^*$ for the $X(3872)$ and T_{cc} quantum numbers, respectively, that were used in Ref.[14, 15].

$$V(r) = V_{short}(r) - \gamma V_0 \left[\begin{pmatrix} 1 & 0 \\ 0 & 1 \end{pmatrix} C_\pi(r) + \begin{pmatrix} 0 & -\sqrt{2} \\ -\sqrt{2} & 1 \end{pmatrix} T_\pi(r) \right], \tag{7}$$

where $V_0 = 1.3$ and $\gamma = 3(-3)$ for $X(3872)$ (T_{cc}) channel[14]. $C_\pi(r)$ is small, so that the effect of $T_\pi(r)$ is important to the possible binding [14]. We have also added the V_{short}, which we take to be zero for $X(3872)$ but attractive for the T_{cc} as discussed before. For the $X(3872)$, the attractive diagonal component of the tensor potential and the D-wave mixing coming from the off diagonal component in the tensor potential lead to a loosely bound molecular configuration [14]. On the other hand, if we do not include V_{short}, because of the repulsion in the tensor part, T_{cc} is not bound despite of the common D-wave mixing Eq. (7) which lowers the ground state energy[14, 15]. Now with the additional attraction, which we parametrize as $V_{\rm short}(r) = V_\omega e^{-\mu_\omega^2 r^2}$ with $\mu_\omega^2 = m_\omega^2 - (m_V - m_P)^2$ to reproduce the r dependence in the lattice result [22] and normalized with an overall constant $V_\omega = -448$ MeV to reproduce the attraction extracted from the constituent quark model discussed previously, we obtain an weakly bound molecular configuration. The results, together with the result for the deuteron within the same formalism[14], are summarized in Table 1

Our analysis shows that taking into account the short range attraction, the molecular structures of the $X(3872)$ and T_{cc} composed of heavy mesons are similar to that of the deuteron composed of the nucleons.

Table 1. Molecular structures obtained using Eq. (1)

Molecule	Content	V_{short}	V_π (S wave)	V_π (mixing)	$\langle r \rangle$ (fm)	E_b (MeV)
d	pn	repulsive	attractive	attractive	1.9	2.2
$X(3872)$	$D\bar{D}^*$	negligible	negligible	attractive	3.0	0.33
T_{cc}	DD^*	attractive	negligible	attractive	2.2	0.65

4 Production of molecular configurations in heavy ion collisions

Since both the molecular tetraquarks and the deuteron are weakly bound broad molecular states, their productions in heavy ion collisions should be formed and determined by the distributions of their respective constituents at the kinetic freeze-out point. Such descriptions can be well encoded in the coalescence model.

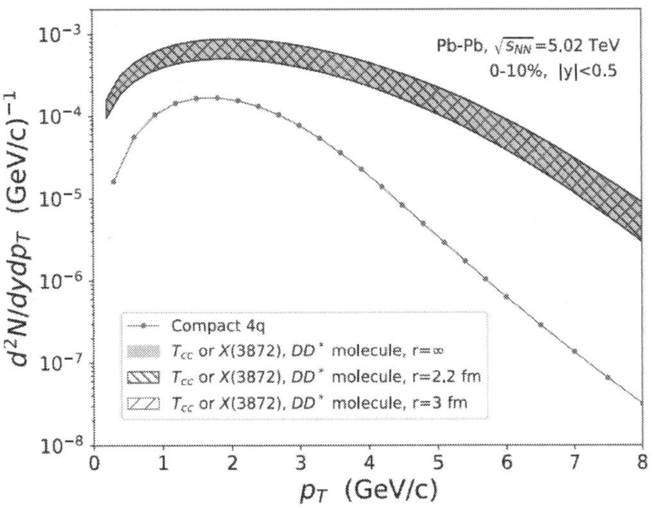

Figure 1. The transverse momentum distribution of $X(3872)$ and T_{cc} (0–10% event). The bands for molecular configurations are due to the uncertainties in the input data.

Using the same formula, we have parametrized the observed transverse momentum distribution of the D^0 and $\bar{D}^{*0}$ at centrality 0–10% for Pb–Pb collisions at $\sqrt{s_{NN}} = 5.02$ TeV[23]. Fig. 1 shows coalescence model result for the transverse momentum distribution of T_{cc} and $X(3872)$ assuming a molecular configuration. Since the chemical potential can be almost neglected, a molecular configuration of T_{cc} will have almost identical transverse momentum distribution of $X(3872)$ as given in Fig. 1. The band in Fig. 1 shows the statistical uncertainty of the input D, D^* data transferred into the coalescence result. One also notes that for the radii given in Table 1, the results are almost identical to the result given in the $\sigma \to \infty$ limit.

We also plot, the expected transverse momentum distribution as of T_{cc} given in Ref. [24], obtained assuming that it has a compact multiquark configuration. The result in Ref. [24] given for collision energy at 2.76 TeV, has been multiplied by 1.63 to take into account the difference in the charm quark number squared at 5.02 TeV as given in [10] to compare with the result for the molecular configuration calculated here. As can be seen in the figure, the expected transverse momentum distributions for two possible configurations of the T_{cc}

are markedly different in addition to the difference in the total yields. Integrating over the transverse momentum, we find that the total yields of molecular (compact) configuration is about a factor of 2.47 (0.625) times[12] that of the statistical model prediction[25]. Therefore, once the transverse momentum distributions of these tetraquarks are measured, we will be able to confirm/discriminate their structures.

Acknowledgements

The work was supported by Samsung Science and Technology Foundation under Project Number SSTF-BA1901-04.

References

[1] S. K. Choi *et al.* [Belle], Phys. Rev. Lett. **91**, 262001 (2003).
[2] R. Aaij *et al.* [LHCb], Phys. Rev. Lett. **115**, 072001 (2015).
[3] R. Aaij *et al.* [LHCb], Phys. Rev. Lett. **122**, no.22, 222001 (2019).
[4] V. M. Abazov *et al.* [D0], Phys. Rev. Lett. **117**, no.2, 022003 (2016).
[5] R. Aaij *et al.* [LHCb], Nature Phys. **18**, no.7, 751-754 (2022).
[6] S. H. Lee, S. Yasui, W. Liu and C. M. Ko, Eur. Phys. J. C **54**, 259-265 (2008).
[7] S. H. Lee in "Proceedings, Workshop on Heavy Ion Collisions at the LHC: Last Call for Predictions: Geneva, Switzerland, May 14 - June 8, 2007," J. Phys. G **35**, 054001 (2008) [arXiv:0711.0974 [hep-ph]].
[8] S. Cho *et al.* [ExHIC], Phys. Rev. Lett. **106**, 212001 (2011).
[9] S. Cho *et al.* [ExHIC], Phys. Rev. C **84**, 064910 (2011).
[10] S. Cho *et al.* [ExHIC], Prog. Part. Nucl. Phys. **95**, 279-322 (2017).
[11] A. M. Sirunyan *et al.* [CMS], Phys. Rev. Lett. **128**, no.3, 032001 (2022).
[12] H. O. Yoon, D. Park, S. Noh, A. Park, W. Park, S. Cho, J. Hong, Y. Kim, S. Lim and S. H. Lee, [arXiv:2208.06960 [hep-ph]].
[13] A. Park, S. H. Lee, T. Inoue and T. Hatsuda, Eur. Phys. J. A **56**, no.3, 93 (2020).
[14] N. A. Tornqvist, Z. Phys. C **61**, 525-537 (1994).
[15] N. A. Tornqvist, Phys. Lett. B **590**, 209-215 (2004).
[16] S. Noh, W. Park and S. H. Lee, Phys. Rev. D **103**, 114009 (2021).
[17] R. L. Jaffe, Phys. Rev. Lett. **38**, 195-198 (1977) [erratum: Phys. Rev. Lett. **38**, 617 (1977)].
[18] W. Park, S. Cho and S. H. Lee, Phys. Rev. D **99**, no.9, 094023 (2019).
[19] S. Zouzou, B. Silvestre-Brac, C. Gignoux and J. M. Richard, Z. Phys. C **30**, 457 (1986).
[20] W. Park and S. H. Lee, Nucl. Phys. A **925**, 161-184 (2014).
[21] W. Park, S. Noh and S. H. Lee, Nucl. Phys. A **983**, 1-19 (2019).
[22] Y. Ikeda, B. Charron, S. Aoki, T. Doi, T. Hatsuda, T. Inoue, N. Ishii, K. Murano, H. Nemura and K. Sasaki, Phys. Lett. B **729**, 85-90 (2014).
[23] S. Acharya *et al.* [ALICE], JHEP **01**, 174 (2022).
[24] S. Cho and S. H. Lee, Phys. Rev. C **101**, no.2, 024902 (2020).
[25] A. Andronic, P. Braun-Munzinger, M. K. Köhler, A. Mazeliauskas, K. Redlich, J. Stachel and V. Vislavicius, JHEP **07**, 035 (2021).

Studies of low-x phenomena with the LHCb detector

Thomas Boettcher[1,*] on behalf of the LHCb collaboration

[1]University of Cincinnati

Abstract. With a unique geometry covering the forward rapidity region, the LHCb detector provides unprecedented kinematic coverage at low Bjorken-x down to $x \sim 10^{-5}$ or lower. The excellent momentum resolution, vertex reconstruction and particle identification allow precision measurements down to very low hadron transverse momentum. In this contribution we present the latest studies of the relatively unknown low-x region using the LHCb detector, including recent measurements of charged and neutral hadron production, as well as direct photon and hadron correlations in proton-proton and proton-lead collisions. Comparisons to various theoretical model calculations are also discussed.

The LHCb detector is a general purpose detector fully instrumented at forward rapidity at the LHC [1]. The LHCb detector's forward acceptance allows it to study kinematic regions inaccessible to other LHC experiments. LHCb is sensitive to iteractions between low- and high-Bjorken-x partons. In proton-lead collisions, LHCb is sensitive to nuclear partons with x as small as 10^{-6}, and LHCb data provides some of the strongest available constraints on nuclear parton distribution functions (nPDFs) for $x < 10^{-3}$ [2, 3]. As a result, LHCb data could be sensitive to novel phenomena such as parton saturation, which is expected to be significant at low x and momentum transfer squared Q^2.

LHCb's sub-detectors include a high-precision Vertex Locator (VELO), which provides a vertex position resolution of $10 - 50\ \mu$m transverse to the beamline [4]. In addition, ring imaging cherenkov detectors allow for charged hadron identification, and the forward spectrometer configuration provides excellent charged-particle momentum resolution. These subdetectors were originally designed to study heavy flavor hadron decays in pp collisions, but they have also allowed LHCb to produce precise measurements of heavy flavor hadron production in pPb collisions. The LHCb experiment recently measured prompt D^0 production at $\sqrt{s_{\mathrm{NN}}} = 8.16$ TeV [5]. The prompt D^0 nuclear modification factor $R_{p\mathrm{Pb}}$ was measured over a wide rapidity range for $0 < p_{\mathrm{T}} < 10$ GeV. The measured $R_{p\mathrm{Pb}}$ is shown in Fig. 1. At forward rapidities, the measured $R_{p\mathrm{Pb}}$ agrees well with nPDF predictions [6–8], as well as with color glass condensate (CGC) predictions, which account for parton saturation effects [9–11]. At backward rapidities, the LHCb measurement shows some tension with nPDF predictions. This tension is particularly clear in the forward-backward ratio (R_{FB}), which is shown in Fig. 2. In addition, modifications due to fully coherent eneryg loss (FCEL) in the nucleus alone are unable to describe the data.

The LHCb experiment's measurements of heavy flavor production are currently being used in state-of-the-art nPDF fits and provide strong constraints in the low-x region [2, 3]. However, the high mass of heavy-flavor hadrons limits the sensitivity of these measurements

*e-mail: boettcts@ucmail.uc.edu

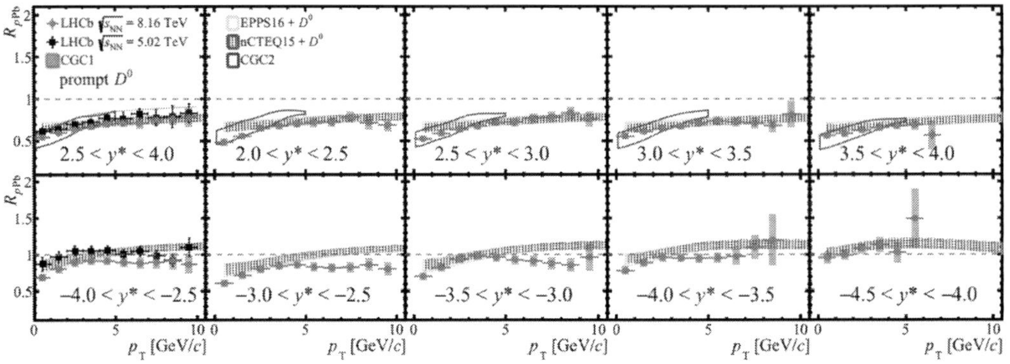

Figure 1. LHCb measurement of the D^0 nuclear modification factor in pPb collisions at $\sqrt{s_{\mathrm{NN}}}$ = 8.16 TeV [5].

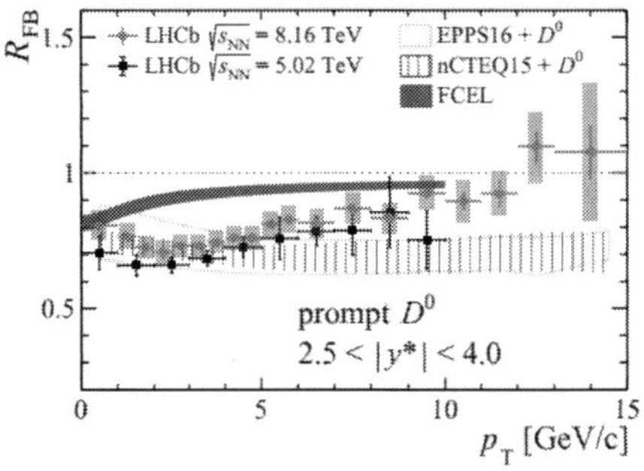

Figure 2. LHCb measurement of the D^0 forward-backward ratio in pPb collisions at $\sqrt{s_{\mathrm{NN}}}$ = 8.16 TeV [5].

to $Q^2 \gtrsim 4\mathrm{GeV}^2$. Measurements of light hadron production can probe lower Q^2, potentially providing greater sensitivity to parton saturation effects. The LHCb experiment has measured inclusive charged-particle production in pPb collisions at $\sqrt{s_{\mathrm{NN}}}$ = 5.02 TeV [12]. The charged-particle $R_{p\mathrm{Pb}}$ was measured as a function of both p_{T} and rapidity. Results are shown in Fig. 3. The forward rapidity data are compared to nPDF and CGC predictions [7, 13]. The measurement agrees well with the nPDF predictions, although the nPDF uncertainties are large. The forward data disagree with the CGC prediction, although subsequent next-to-leading order CGC calculations better describe the data [14]. The backward $R_{p\mathrm{Pb}}$ data show a much larger excess than the nPDF predictions. The backward data also disagrees with a perturbative quantum chromodynamics (pQCD) calculation that include FCEL [15], which successfully describes a similar enhancement observed by the PHENIX collaboration [16]. The failure of available models to describe the LHCb data suggests that other nuclear effects contribute to the enhancement, such as radial flow or final-state recombination [17, 18].

Measurements of $R_{p\mathrm{Pb}}$ of identified particles are necessary to determine the origin of the LHCb charged-particle enhancement. Because charged particles produced in LHC collisions are mostly pions, measurements of π^0 production probe similar kinematics and production mechanisms as inclusive charged-particle measurements. Additionally, π^0s are reconstructed from photon pairs, so the systematic uncertainties in charged-particle and π^0 production measurements are mostly independent. LHCb has measured the π^0 $R_{p\mathrm{Pb}}$ at forward and backward

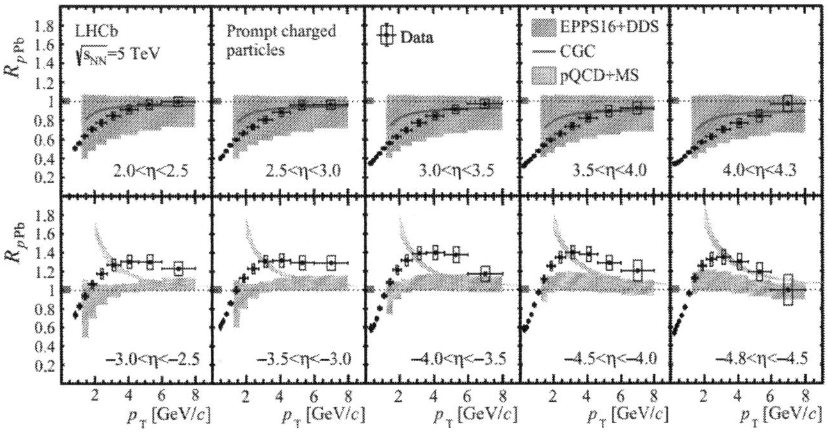

Figure 3. LHCb measurement of the charged-particle nuclear modification factor in *p*Pb collisions at $\sqrt{s_{\mathrm{NN}}} = 5.02$ TeV [12].

rapidity at $\sqrt{s_{\mathrm{NN}}} = 8.16$ TeV [19]. The π^0 candidates were reconstructed from photon pairs. To avoid the effects of overlapping clusters in the electromagnetic calorimeter, one of the photons is required to convert in the detector material. A proton-proton reference sample is constructed by interpolating between measurements at $\sqrt{s} = 5.02$ and 13 TeV. The resulting $R_{p\mathrm{Pb}}$ measurements are shown in Fig. 4. The measurements are compared to pQCD predictions computed using nPDFs that include LHCb measurements of D^0 production at $\sqrt{s_{\mathrm{NN}}} = 5.02$ TeV [20–22]. The forward measurement agrees well with the nPDF prediction, as well as with the LHCb charged-particle $R_{p\mathrm{Pb}}$ measurement. The backward measurement shows a slightly larger enhancement than the nPDF prediction, but smaller than that observed in the LHCb charged-particle data. This suggests that the charged-particle excess could be due to a mass-dependent effect such as radial flow, or due to a baryon-enhancing effect like final state recombination. Additional studies of identified particles such as p, η, and η' could help untangle the contributions from these effects.

Since the LHCb experiment began recording *p*Pb collisions in 2013, it has provided the strongest available constraints on nPDFs at low x. As a result, the low-x gluon nPDF in particular has quickly gone from almost entirely unconstrained to well known in just the last few years [2, 3]. The study of low-x nPDFs has suddenly become a high-precision field. With recent LHCb measurements, low-x nPDFs will soon be overconstrained, and emerging discrepancies between measurements and predictions could point to novel QCD phenomena.

Acknowledgements

The author is supported by the U.S. National Science Foundation.

References

[1] R. Aaij et al. (LHCb), Int. J. Mod. Phys. A **30**, 1530022 (2015), `1412.6352`

[2] R. Abdul Khalek, R. Gauld, T. Giani, E.R. Nocera, T.R. Rabemananjara, J. Rojo, Eur. Phys. J. C **82**, 507 (2022), `2201.12363`

[3] K.J. Eskola, P. Paakkinen, H. Paukkunen, C.A. Salgado, Eur. Phys. J. C **82**, 413 (2022), `2112.12462`

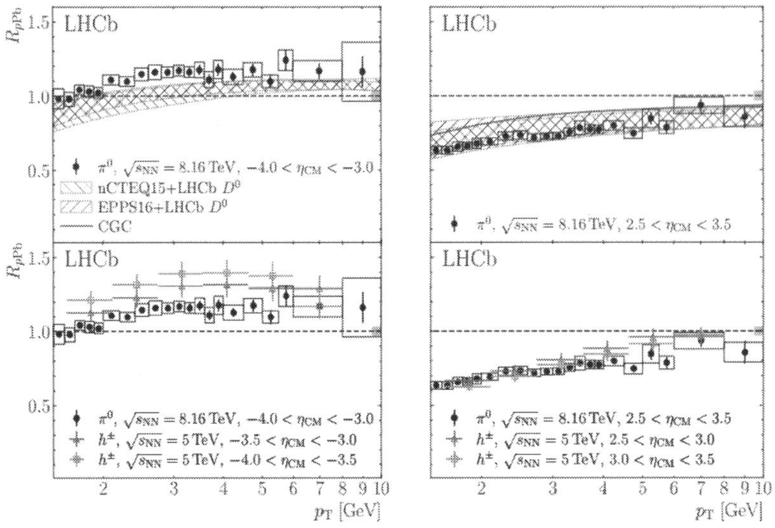

Figure 4. LHCb measurement of the π^0 nuclear modification factor in pPb collisions at $\sqrt{s_{\mathrm{NN}}} =$ 8.16 TeV [19].

[4] R. Aaij et al., JINST **9**, P09007 (2014), 1405.7808
[5] LHCb (2022), 2205.03936
[6] J.P. Lansberg, H.S. Shao, Eur. Phys. J. C **77**, 1 (2017), 1610.05382
[7] K.J. Eskola, P. Paakkinen, H. Paukkunen, C.A. Salgado, Eur. Phys. J. C **77**, 163 (2017), 1612.05741
[8] K. Kovarik et al., Phys. Rev. D **93**, 085037 (2016), 1509.00792
[9] B. Ducloué, T. Lappi, H. Mäntysaari, Phys. Rev. D **91**, 114005 (2015), 1503.02789
[10] B. Ducloué, T. Lappi, H. Mäntysaari, Nucl. Part. Phys. Proc. **289-290**, 309 (2017), 1612.04585
[11] Y.Q. Ma, P. Tribedy, R. Venugopalan, K. Watanabe, Phys. Rev. D **98**, 074025 (2018), 1803.11093
[12] R. Aaij et al. (LHCb), Phys. Rev. Lett. **128**, 142004 (2022), 2108.13115
[13] T. Lappi, H. Mäntysaari, Phys. Rev. D **88**, 114020 (2013), 1309.6963
[14] Y. Shi, L. Wang, S.Y. Wei, B.W. Xiao, Phys. Rev. Lett. **128**, 202302 (2022), 2112.06975
[15] F. Arleo, F. Cougoulic, S. Peigné, JHEP **09**, 190 (2020), 2003.06337
[16] C. Aidala et al. (PHENIX), Phys. Rev. C **101**, 034910 (2020), 1906.09928
[17] K.S. Lee, U.W. Heinz, E. Schnedermann, Z. Phys. C **48**, 525 (1990)
[18] R.C. Hwa, C.B. Yang, Phys. Rev. Lett. **93**, 082302 (2004), nucl-th/0403001
[19] LHCb (2022), 2204.10608
[20] I. Helenius, K.J. Eskola, H. Paukkunen, JHEP **09**, 138 (2014), 1406.1689
[21] D. de Florian, R. Sassot, M. Epele, R.J. Hernández-Pinto, M. Stratmann, Phys. Rev. D **91**, 014035 (2015), 1410.6027
[22] K.J. Eskola, I. Helenius, P. Paakkinen, H. Paukkunen, JHEP **05**, 037 (2020), 1906.02512

EPJ Web of Conferences **276**, 06017 (2023) https://doi.org/10.1051/epjconf/202327606017
SQM 2022

SQM2022: Theoretical Summary

*Berndt*Müller[1,2,*]

[1]Department of Physics, Duke University, Durham, NC 27705, USA
[2]Wright Laboratory, Yale University, New Haven, CT 06520, USA

Abstract. Written version of the theoretical summary lecture presented at the Strangeness in Quark Matter 2022 conference.

1 Introduction

In his Theoretical Overview, J. Kapusta [1] described a large array of topics and open questions that fall within the purview of this conference series. In my summary lecture I will focus on a reduced subset of topics where I believe significant progress was presented at SQM2022. I apologize in advance to those whose beautiful work I have no time and space to cover. The absence of a mention does not reflect negatively on their contributions. Above all, it is worth emphasizing the following points:

- Remarkable progress continues to be made.
- Unexpected new insights continue to be generated.
- The topics of "SQM" have expanded far beyond what was imaginable 40 years ago.
- The huge amount of data of a wide variety being generated by the experiments is both, a benefit and a curse.
- It is vital that the community stays focused on its core goals.

My top-level assessment of the status of the field is:

- We have an accepted "standard" reaction model for the central rapidity region in collisions at $\sqrt{s_{NN}} > 100$ GeV.
- We do not have a widely accepted and data tested model for collisions at $\sqrt{s_{NN}} < 20$ GeV.
- Theoretical tools for the baryon-rich region remain inadequate.
- The mechanisms for hadron emission remain insufficienty understood.
- Probes of chiral symmetry restoration remain at best tenuous.

In the following, I will discuss newly discovered possible flaws in the standard reaction model, core-corona models, the physics of collisions at the energies of the RHIC beam energy scan, critical fluctuations, anomalous chiral phenomena, hadron emission, heavy quarkonia, strangeness, and Bayesian model-to-data comparison.

*e-mail: mueller@phy.duke.edu

2 Reaction Dynamics

The standard reaction model for high-energy collisions relies on a multi-stage approach beginning with some model of the initial energy deposition and followed by, in sequence, hydrodynamic expansion of the quark-gluon plasma (QGP) fluid, Cooper-Frye hadronization, and hadronic Boltzmann transport. All these stages need to be carefully checked for intrinsic consistency and for consistency of the transition between them. Many careful investigations have addressed this issue in the past and validated important aspects of the framework.

At this conference, C. Plumberg [2] reported recent work that takes a careful look at the validity of using second-order viscous relativistic hydrodynamics during the early phase of the hydrodynamic stage. He applied a set of criteria to ascertain whether the evolution of all fluid cells is causal or not. Causality requires that the propagation velocity v satisfies the condition $0 \leq v^2 \leq c^2$, which means that the system of differential equations is hyperbolic and propagation is not superluminal. These requirements result in 6 necessary and 8 sufficient conditions that can be checked for each fluid cell.

The implementation he studied involved energy deposition in the IP-glasma model, possibly followed by a period of free streaming or kinetic theory transport. What he found for central Pb+Pb collisions at $\sqrt{\mathrm{NN}} = 2.76$ TeV is remarkable: (a) When fluid dynamics is initiated at $\tau = 0.4$ fm/c directly from the IP-glasma initial conditions, almost all fluid cells strongly violate causality. (b) When fluid dynamics is preceded either by free-streaming or kinetic transport and initialized at $\tau = 0.8$ fm/c, the necessary, but not the sufficient, conditions for causality are met over a large part of the volume, but causality remains strongly violated near the surface. The violation of causality persists over a period of approximately 20% of the duration of the hydrodynamic stage, during which more than half of the final elliptic flow is generated.

It is currently unclear what a suitable remedy might be that retains the indispensable features of the fluctuating nature of the initial state. The causality violations are associated with the large initial density fluctuations that drive higher Fourier components of the collective flow. Is it perhaps possible to work directly with moments of the initial density fluctuations instead of the full fluctuating density distribution by performing a Fourier decomposition of the fluid in azimuthal angle?

The strong causality violations near the outer edge of the fireball indicate that it is preferable to treat the core and the corona of the fireball differently, applying fluid dynamics to the core and some form of kinetic transport to the corona. In her talk Y. Kanakubo [3] presented a dynamical core-corona model that allows quasiparticles (partons) in the corona to thermalize and become part of the core. The results appear encouraging: While the overwhelming part of the fireball joins the fluid core in Pb+Pb collisions, the corona constitutes most of the fireball in p+p collisions, and even in the most central p+p collisions approximately half of the fireball remains in the corona phase. The transition to core dominance occurs at $dN_{\mathrm{ch}}/d\eta \approx 20$ for both systems. Some interesting questions arise: Can core-corona models of this type for midrapidity be formulated as a rigorous effective kinetic theory of QCD similar to the kinetic equilibration scenarios? What would be the limits for such a formulation? Would a kinetic theory avoid causality violations and provide a more realistic description of the entire mid-rapidity fireball?

While energy deposition at high energies can be described as an instantaneous quench owing to the high degree of Lorentz contraction of the two colliding nuclei, auch an approach fails at the lower energies of the RHIC beam energy scan. In the target rest frame a modified form of this picture may survive because the projectile nucleus is still highly Lorentz contracted to $2R/\gamma_{\mathrm{Lab}}$. The collision then ptakes the form of a time-staggered local quench as the sheet-like projectile sweeps over the target nucleus, and a space-like Cauchy hypersurface for hydrodynamics initial conditions $t_i(z)$ can still be defined? Does this allow to develop a modified "standard model" for the reaction?

In the center-of-mass frame on the other hand, both nuclei are only mildly Lorentz contracted to thickness $2R/\gamma_{\rm cm}$, and the collision cannot be treated as a quench because energy and baryon number deposition occur over extended period of time. As T. Mendenhall [4] showed in his talk, this limits the attainable maximum energy and baryon number density. For example for a central Au+Au collision at $\sqrt{s_{\rm NN}} = 5$ GeV the maximum energy density of slightly more than 1 GeV/fm^3 and baryon number density $n_B \approx 0.7$ fm^{-3} is reached only at $\tau \approx 4.5$ fm/c. This implies that the fireball conditions (T, μ_B) at the time of highest density are rather close to the QCD phase boundary when also a formation time $\tau_{\rm F} \approx 1$ fm/c is taken into account.

3 Critical point and anomalous chiral phenomena

An extension of the new parametrization of the lattice equation of state at non-zero baryon chemical potential to nonvanishing strangeness chemical potential was presented by P. Parotto [5]. The new parametrization produces an *ab initio* equation of state for conditions prevailing in heavy ion collisions (strangeness neutrality) for $\mu_B/T \leq 3.5$ with reasonable uncertainties and covers a large part of the QCD phase diagram explored in the RHIC beam energy scan (down to $\sqrt{s_{\rm NN}} \approx 6$ GeV).

Much theoretical progress has been made on modeling the dynamics near the conjectured critical point in the QCD phase diagram. In her talk at SQM2022 M. Pradeep [6] explored the off-equilibrium critical dynamics by treating the critical mode as an independent degree of freedom in the framework of fluid dynamics (the "Hydro+" approach [7]). The delayed response of the critical mode avoids the singularity that occurs at the critical point but leads to enhanced remnant baryon number fluctuations at freeze-out. For reasonable parameter values the remnant fluctuations are found to depend only weakly on the freeze-out temperature.

In his talk on anomalous chiral phenomena E. Finch [8] discussed the results from the RHIC isobar (^{96}Zr – ^{96}Ru) comparison run [9]. The results, which were obtained by means of a blind analysis, have been re-assessed for various isobar specific corrections that were not understood at the time of the analysis. The STAR collaboration has concluded that these reduce the expected ratio (Ru/Zr) for various putative signatures of the chiral magnetic effect (CME) by approximately 3% from unity to 0.97. The measured values for all signature CME observables agree with this revised baseline within 0.5%.

Efforts to search for other manifestations of anomalous chiral phenomena will surely continue. However, I believe that it is now time for theorists to embark on a rigorous analysis what these impressive results mean. What are the implications of the absence of CME signatures at detectable levels? There are three main drivers of the magnitude of the CME observables: (1) the degree of chiral symmetry restoration in the QGP, (b) the size of QCD winding number fluctuations in the QGP, and (c) the magnitude of the remnant magnetic field in the QGP at late times. Theorists will need to ask which of these quantities are constrained by other observables.

It seems to me that the lattice provides compelling evidence of chiral symmetry restoration even in the absence of direct experimental confirmation. We have experimental bounds on the remnant magnetic field at hadronization from hyperon polarization data [10] which are not far from state-of-the-art theoretical predictions [11]. If this assessment is correct, the experimental data can set an upper bound on winding number fluctuations over the course of the heavy ion collision. How valuable would this information be? Given the tremendous effort invested in the isobar experiment a commensurate theoretical effort is called for.

4 Hadronization

Because the dynamics of QGP hadronization is neither described perturbatively nor by lattice simulations, it is among the largest unknowns in models of relativistic heavy ion collisions. W. Zhao [12] discussed a phenomenological framework that treats bulk fluid hadronization, quark coalescence, and fragmentation consistently within an end-to-end dynamical model of the collision. He concluded that data from Pb+Pb collisions at LHC and Au+Au collisions at RHIC both require the presence of all three hadronization mechanisms in order to reproduce the measured valence quark number scaling. A similar conclusion was reached by V. Minissale [13] who presented work by the Catania group on D^0 and Λ_c production as measured by STAR and ALICE. He suggested that recombination may even be required to describe the Λ_c/D^0 ratio in p+p collisions at LHC for $p_T < 5$ GeV/c.

Quite a few talks discussed new data that probe aspects of the statistical hadronization model (SHM). C. Pinto [14] showed that the SHM describes particle yields in Pb+Pb collisions at $\sqrt{s_{\mathrm{NN}}} = 2.76$ GeV over 9 orders of magnitude from light hadrons to alpha particles with a common formation temperature $T_{\mathrm{chem}} \approx 156$ MeV. This temperature coincides remarkably well with the pseudocritical temperature T_c "measured" on the lattice. He pointed out that the hypertriton ${}^3_\Lambda$H with its intrinsic radius $a > 5$ fm provides for an extreme test case.

The common expectation is that ${}^3_\Lambda\mathrm{H}/\Lambda$ yield ratios for SHM only vary with system size because of exact strangeness neutrality of the fireball. In a coalescence scenario the ratio depends on the size of the ${}^3_\Lambda$H bound state and falls off precipitously with shrinking fireball size. Recent data for this ratio from p+Pb and p+p collisions clearly favor two-body coalescence – a reasonable assumption since the ${}^3_\Lambda$H can be thought of as a loose d+Λ bound state – over the SMH prediction with strangeness neutrality.

I believe that this conclusion is erroneous, because the relative sizes of the emitting source and the emitted particle also comes into play when one performs a careful evaluation in the SHM. This is illustrated in Fig. 1. The left panel depicts the case when the emitted particle is much smaller than the emission volume ($a \ll R$), which is typically the case for standard hadrons such as the Λ. The central panels depicts the opposite case when the emitted particle is much larger than the emission volume ($a \gg R$), which is realized for small fireballs such as those formed in p+Pb and p+p and the ${}^3_\Lambda$H.

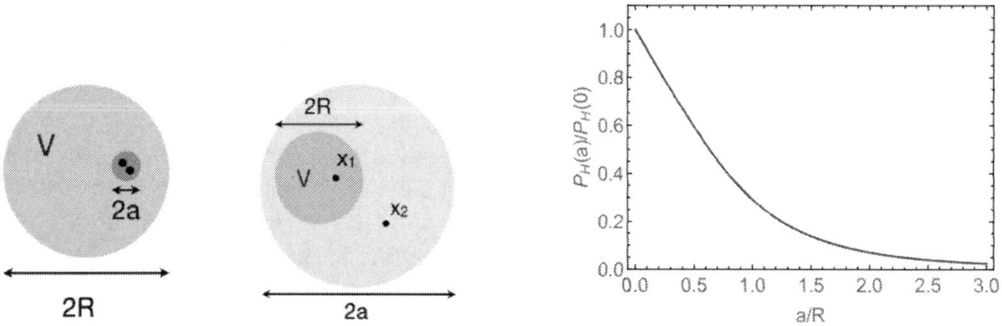

Figure 1. Left panel: The emitted particle (radius a) is much smaller than the fireball (radius R). Middle panel: The emitted particle is much larger than the fireball. Right panel: Ratio of the particle yield calculated from Eq. (2) for a Gaussian wave function as a function of the ratio a/R.

When one calculates the yield of a hadron in the thermal model, one needs to evaluate the expression $P = \mathrm{Tr}[e^{-H/T}\,\theta_V]$, where $\theta_V = 1$ inside the thermal volume and $\theta_V = 0$ outside, assuming the fireball is surrounded by vacuum. One can then integrate out the internal wave function $\psi(x)$ of the emitted particle and obtains the usual SHM result:

$$P_0 \sim \int d^3p\, e^{-E_p/T} \int d^3x_1\, d^3x_2\, |\psi(x_1 - x_2)|^2 \theta_V(x_1)\theta_V(x_2) \approx V \int d^3p\, e^{-E_p/T}. \quad (1)$$

On the other hand, when $a \gg R$ emission of the hadron requires that both (or generally all) constituents of the emitted particle to emerge from the hot fireball, and one finds:

$$P(a) \sim \int d^3p\, e^{-E_p/T} \int d^3x_1\, d^3x_2\, |\psi(0)|^2 \theta_V(x_1)\theta_V(x_2) \approx V^2|\psi(0)|^2 \int d^3p\, e^{-E_p/T}. \quad (2)$$

In general, $P(a)/P_0$ is a function of the ratio a/R, which is one at $a/R = 0$ and tends to zero for $a/R \to \infty$. Using a Gaussian internal wave function one obtains the plot shown in the right-most panel of Fig. 1. Inserting a reasonable estimate of $a = 5$ fm for the radius of the hypertriton and scaling the radius R of the fireball with the measured charged particle multiplicity $dN_{ch}/d\eta$ one finds that the size-corrected prediction of the SHM closely tracks the prediction for two-body coalescence and agrees within uncertainties with the measurements for p+Pb and p+p, as shown in Fig. 2.

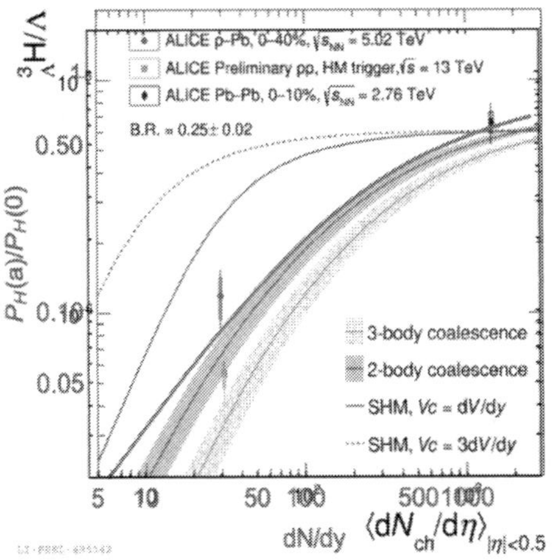

Figure 2. Yield ratio ${}^3_\Lambda$H/Λ as a function of $dN_{ch}/d\eta$ measured in Pb+Pb, p+Pb, and p+p collisions at LHC. The red curve shows the prediction for the SHM with size correction, Eq. (2). The curves with uncertainty bands show the predictions of coalescence models. The solid and dotted curves show the predictions of the SHM with strangeness neutrality for two different values of the strangeness correlation volume V_C.

5 Other topics

Other topics that were widely discussed in the conference were heavy quarkonia and strangeness. B. Gossiaux [15] gave a thorough overview of the theoretical developments relating to quarkonium suppression. The main take-aways were:

- The theoretical description of Υ suppression has reached a mature state with general agreement about the formal framework based on open quantum systems and NRQCD [16]. What is now needed are high statistics data. These will be forthcoming in the next few years from RHIC and LHC. It remains to be seen whether the effect of color screening (Debye mass) on the data be isolated. Υ suppression at $p_T >$ few GeV requires more theoretical work (perhaps by adding SCET to the theoretical framework?).
- Charmonium suppression is an extraordinarily complex process. Many mechanisms contribute: nuclear shadowing (nPDFs) and other cold nuclear matter effects, color screening and thermal ionization; regeneration, especially at LHC, influenced by charm thermalization; and gluon fragmentation, which dominates at high p_T and is affected by parton energy

loss. A similarly comprehensive and rigorous theoretical framework like the one for the Υ states does not yet exist for charmonium. The problem is that the charm quark mass is not large enough for a clean separation of scales.

There exists a veritable wealth of data on the production of strange hadrons; this area is not data limited as L. Bianchi's [17] survey talk made abundantly clear. It seems, however, that there has been only limited theoretical progress in recent years, possibly due to the success of the thermal resonance gas model. The effect of strangeness conservation in the fireball, often called "canonical suppression", is clearly observed STAR data at low collision energies ($\sqrt{s_{\rm NN}} = 2 - 3$ GeV) by looking e. g., at the ϕ/K^- ratio.

Finally, there is continued progress in Bayesian model-to-data comparisons, which are becoming ever more ambitious. A new analysis by the Jyväskylä group, presented by D. J. Kim [18], which used a much expanded set of observables, showed evidence of increasing discriminating power. These analyses require very substantial computing resources which will strain the resources available to the theoretical community. Clever approaches that reduce these requirements without sacrificing rigor are being explored.

6 Final words

There are many other exciting developments which I have not covered. One example is the growing confluence between heavy ion collisions at lower energies, where the fireball contains a large surplus of baryons, and novel astrophysical observations for neutron stars (using γ-rays) and neutron star mergers (using gravitational waves). Both types of data have the ability to constrain the equation of state of cold, dense and hot, dense nuclear matter and connect these with analyses from heavy ion collisions. Clearly, the field remains vibrant and presents many opportunities for young scientists to make discoveries or to find answers to open questions.

Last, but not least, I congratulate the SQM2022 organizers for conducting an inspiring conference that brought the interested scientific community together, either in person or remotely via video link. It was not at all easy to organize the conference under the conditions imposed by the pandemic, and we all owe them much gratitude for their efforts.

References

[1] J. I. Kapusta, talk at this conference.
[2] C. Plumberg, talk at this conference.
[3] Y. Kanakubo, talk at this conference.
[4] T. Mendenhall, talk at this conference.
[5] P. Parotto, talk at this conference.
[6] M. Pradeep, talk at this conference.
[7] M. Stephanov and Y. Yin, Phys. Rev. D **98**, 036006 (2018).
[8] E. Finch, talk at this conference.
[9] M. Abdallah *et al.* [STAR], Phys. Rev. C **105**, 014901 (2022).
[10] J. Adam *et al.* [STAR], Phys. Rev. Lett. **126**, 162301 (2021).
[11] C. Grayson, M. Formanek, J. Rafelski and B. Müller, Phys. Rev. D **106**, 014011 (2022).
[12] W. Zhao, talk at this conference.
[13] V. Minissale, talk at this conference.
[14] C. Pinto, talk at this conference.
[15] B. Gossiaux, talk at this conference.

[16] A. Rothkopf, Phys. Rept. **858**, 1 (2020).
[17] L. Bianchi, talk at this conference.
[18] D. J. Kim, talk at this conference.

Experimental Status of the Chiral Magnetic Effect

Evan Finch[1,*]

[1]Southern Connecticut State University

Abstract. Experimental searches for definitive evidence of the Chiral Magnetic Effect in heavy-ion collisions have become increasingly ingenious over the past decade, but a clear separation of any CME signal from flow-related backgrounds remains elusive. Although the isobar analysis by STAR does not appear to show a signal given the current preliminary background estimate, there is a tantalizing statistically significant signal in Au-Au at top RHIC energy using a spectator plane versus participant plane analysis. I discuss the status of these works in some detail, and discuss the possible advantage of an analysis that takes advantage of the expected correlation between the parity-odd event-by-event CME signal and other parity-odd measures such as the net hyperon helicity.

1 CME and Local Parity Violation

Within the rich field of heavy-ion physics, one of the exciting possibilities is that we may be able to observe the consequences of the creation of metastable regions in which parity and charge-parity symmetries are violated by the strong interaction [1]. This "Local Parity Violation" may manifest itself in various ways in heavy-ion collisions, but one of the most interesting is that this effect in the presence of the enormous magnetic fields generated by the colliding nuclei would cause a separation of electric charges along the direction of the magnetic field [2]-this is the "Chiral Magnetic Effect". If an observation of the CME could be clearly established in heavy-ion collisions, it would imply the existence of these $\mathcal{CP}$-violating regions, the restoration of approximate chiral symmetry in the Quark Gluon Plasma medium, and the action of an ultra-strong magnetic field on the collision region (see [3] for a review).

As illustrated in Figure 1 (taken from [4]), each parity-violating region can be characterized by an integer winding number Q_w which gives equal chirality imbalance (difference in left- and right-handed quarks) among every light quark flavor $N_L^f - N_L^f = 2Q_w$. The sign of Q_w is random region-to-region, with an equal chance of $Q_w > 0$ (for which negative electric charge flows in the direction of magnetic field) and $Q_w < 0$ (positive charge flows in the magnetic field direction). So the CME can be directly searched for only through correlation or fluctuation analyses, essentially comparing the behavior of charge flow along the B-field direction to charge flow perpendicular to that, as is done explicitly with the observable $\gamma \equiv \langle \cos[\varphi_\alpha + \varphi_\beta - 2\psi_2] \rangle$ [5]. Here, ϕ_α and ϕ_β are the azimuthal angles of charged particles in the event, ψ_2 is the second-order flow plane (which is roughly perpendicular to the magnetic field), and the average is taken over all particles in an event. γ quantifies the tendency of same- (or opposite-) charged particles to go in the same direction along the magnetic field as compared to perpendicular to the field. $\Delta\gamma \equiv \gamma^{OS} - \gamma^{SS}$ then makes a further comparison

*e-mail: FinchL3@southernCT.edu

of opposite-signed correlations minus same-signed correlations. For a CME-like effect, we expect a positive value for $\Delta\gamma$, particularly in mid-central collisions where the magnetic field generated by spectator protons is large.

Another possible avenue for a search is to look for other parity odd effects. For example: *if* the strange quark can be treated as light in this context, it has the same chirality imbalance as both the up and down quarks. This means that regions with $Q_w > 0$, for example, give an excess of left-handed strange quarks, which would then be expected to cause a helicity imbalance among Λ hyperons. It has been suggested to look directly at fluctuations in Λ net helicity event-to-event [6, 7]; this may be quite challenging in practice because there are other potential sources of fluctuations, from physics and from detector effects, that are difficult to establish precisely. I suspect that a more promising avenue is to look for a correlation between this effect and the CME [8]-I'll come back to this!

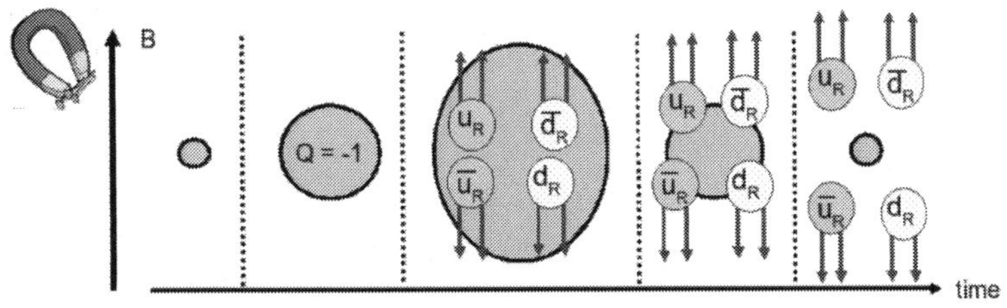

Figure 1: Cartoon representation (from [4]) of the CME effect with charge separation along the direction of the event magnetic field. A given winding number ($Q_w = -1$ in this example) has the same effect on all light quark flavors; up and down as pictured here, but potentially strange quarks as well! This would lead to charge separation along the magnetic field in addition to other parity-odd effects.

2 Isobars: blind analysis and subsequent work

Initial measurements of $\Delta\gamma$ at RHIC [9] and LHC [10] showed a signal qualitatively consistent with expectations from CME, but production of 2-particle opposite-signed clusters in which the clusters have a positive elliptic flow gives a signal in $\Delta\gamma$ (and every other CME-sensitive observable) that is very difficult to distinguish from a CME signal. Previous efforts to separate this flowing-cluster background, perhaps most notably using small-system and event-shape-engineering studies ([11] and [12] give nice overviews) indicate that the CME signal is at most a small fraction of the measured $\Delta\gamma$ signal. To be useful, then, efforts to identify CME need to be able to sort out a CME fraction (f_{CME}) of something like 10% of the measured $\Delta\gamma$.

A significant effort to increase the sensitivity for separating signal from background was undertaken with the RHIC isobar run, for which STAR analyzed approximately 2 billion events each of Zr^{96}_{40}+Zr^{96}_{40} and Ru^{96}_{44}+Ru^{96}_{44} collisions, the central idea being that with more protons, *Ru* collisions produce a higher magnetic field than *Zr* collisions (calculations [13] give a 15%-20% higher value for B^2, which is the quantity that the CME signal should scale with). To the extent that the nuclei are the same shape, the backgrounds should be the same (and to the extent they they have slight differences, those can be largely removed by scaling $\Delta\gamma$ by v_2). So the idea is straightforward: look for any excess in $\Delta\gamma/v_2$ in *Ru* compared to *Zr* and attribute that to CME. The results of the blind analysis[14] for the ratio $(\Delta\gamma/v_2)_{Ru}/(\Delta\gamma/v_2)_{Zr}$ are reproduced in Figure 2 along with some additions that are explained below. The blind analysis results clearly show no excess signal in *Ru*, and in fact even show a small excess in *Zr*, for all the varieties of the $\Delta\gamma/v_2$ measurements (i.e., the leftmost seven points of Figure 2).

An important caveat is that there is a multiplicity (N) difference between the two species at matching centrality, and the $\Delta\gamma$ observable has a trivial $1/N$ statistical dependence for the case where the signal comes dominantly from flowing 2-particle clusters (which is true here). A more informative first-order comparison, then, would be to compare these leftmost seven points not to unity, but instead to the ratio of N^{-1} for the two species (that is the point at the far right of Figure 2); using this baseline would indicate a small CME signal above background. However, there are important further considerations to that comparison; the first is a small non-flow correction to the 1/N scaling based on the numbers of observed same-sign and opposite-signed clusters in the two species, the second is the expected difference between the isobars in contributions to $\Delta\gamma$ from direct 3-particle correlations (e.g. jets). These relatively small corrections are now crucial for understanding whether there is any evidence for CME signal in isobar collisions. Preliminary work by STAR, reported at [15] and discussed at more length in [16], quantifies these two contributions through studies with data and the HIJING model (used to study the 3-particle contributions). This work gives the conclusion that the sum of these non-flow differences in the isobars would have the net effect of moving the comparison baseline for the $\Delta\gamma/v_2$ up from the 1/N line by a factor of 1.0125. Given the current uncertainty in these non-flow contributions, that baseline becomes the tan- and green-colored bands shown around the leftmost seven points in Figure 2. That these data points are consistent with these bands indicates that in the context of these *preliminary* calculations, there is no significant CME signal observed above background in these isobar collisions.

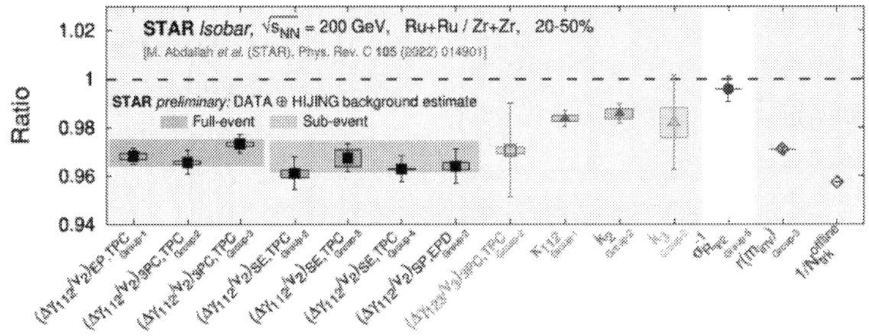

Figure 2: Results of the blind isobar analysis from STAR with some post-blinding additions. The left-most seven points are measurements of the *Ru/Zr* ratio of $\Delta\gamma/v_2$, four of which use sub-event analysis to suppress non-flow contributions. As discussed in the text, a good 1st-order baseline for those seven points is the ratio of inverse multiplicities at the far right, but a more complete *preliminary* baseline estimate is given by the tan and aqua bands that overlay those seven points. Other points in the plot are also discussed briefly in the text.

A couple more words about this figure: The green points represent a method of removing some of the v_2-related contributions to $\Delta\gamma$ by dividing it by the average reaction-plane independent two-particle correlations times the overall v_2. Although this first-order correction will not remove all backgrounds, it is interesting to note that these points do not lie above unity (which is their proper first-order baseline for comparison), so no CME signal is indicated by this method either.

An important study[17] was carried out to complement the isobar analysis, with the purpose of benchmarking the sensitivities of different observables using the analysis code from the isobar analysis to analyze simulated data. Of particular note, this study compared $\Delta\gamma$ to a measurement involving the width of the R correlator [18] and found these methods to have very similar sensitivities to a CME signal; this conclusion was bolstered by analytical work in [17] comparing these observables. I conclude that such additional observables are good tools to double-check the results from $\Delta\gamma$, but we should not expect any qualitatively different conclusions or better sensitivity to come from them. Results of the Ru/Zr ratio for the R correlator inverse width are shown as the blue point in Figure 2 - that this measured ratio is below 1 is consistent with the finding of no measurable CME signal; another useful result derived from [17] is that the first-order baseline for comparison of the R-correlator ratio when constructed independently from two subevents is not unity but rather the ratio of v_2 of the isobars (roughly 1.02 for this case, though non-flow will also contribute).

3 Recent Au+Au full energy results

In my view, the other big step forward for CME measurements in the last year is a recently published paper [19] by STAR using 2.4B events of 200 GeV $Au + Au$ data. The event plane can be determined using the azimuthal angles of spectator nucleons (the "spectator" plane) - that plane should be quite closely correlated with the direction of the event B-field which is mostly generated by the spectator protons, so measuring $\Delta\gamma$ relative to this plane should maximize the contribution from the CME. Alternatively, the event plane can be determined using produced particles near mid-rapidity (the "participant" plane). $\Delta\gamma$ measured with respect to this plane should maximize the contribution of flow-related background effects. Using $\Delta\gamma$ measurements from both of these planes, one can extract the fraction of the $\Delta\gamma$ that comes from CME under the assumption that the sole contributions are CME and v_2-related background. Results from this STAR measurement[19] are shown in Figure 3 - each column of the figure represents the CME fraction (f_{CME}) extracted from a slightly different version of the measurement, with each red point representing a result in mid-central collisions where we may expect the CME fraction to be maximized, and the blue points representing results in peripheral collisions for reference. A statistically significant f_{CME} of around 10% is extracted in mid-central collisions, but there are concerns that non-flow may still contribute. To reduce that effect, the two right-most columns use a sub-event method in which the participant plane is found using one-half of the TPC and the particles of interest for $\Delta\gamma$ from the other half. It is important to note that some non-flow effects can cause f_{CME} to appear *smaller* than its "true" value, so it is not obvious which direction non-flow will drive the f_{CME}. Still, it is crucial to continue to study these results to understand if this is truly a signal!

We should then ask: can we reconcile a possible 10% f_{CME} signal in $Au + Au$ with the apparent lack of signal from the isobar run? STAR met its sensitivity goal in the isobar run to achieve a sensitivity of 0.5% for the Ru/Zr ratio, which implies that a 10% f_{CME} signal would yield a 4σ signal in the isobar analysis-that signal is clearly not observed. So the question becomes: is it reasonable to expect a significantly smaller f_{CME} in the isobar systems than in $Au + Au$? There are simple arguments that make a reasonable case for "yes": the B-field should scale roughly as $A^{1/3}$, and the dominant "flowing clusters" background should roughly scale as multiplicity, or $1/A$, which together argue for a factor of ≈ 3 smaller f_{CME} in isobar than $Au + Au$. A calculation [20] using the AVFD model, reaches a similar conclusion, and while there are many assumptions involved, it does appear reasonable that these results can be consistent with one another.

4 Correlations of charge separation with other parity-odd effects

Looking to the future, one idea that I find very exciting is the possibility of correlating a *parity-odd* CME measurement, event by event, with other signals of Local Parity Violation (LPV). If the strange quark can be considered a light quark, then it will have the same net chirality in each event as the up and down quarks. Some of these strange quarks will go

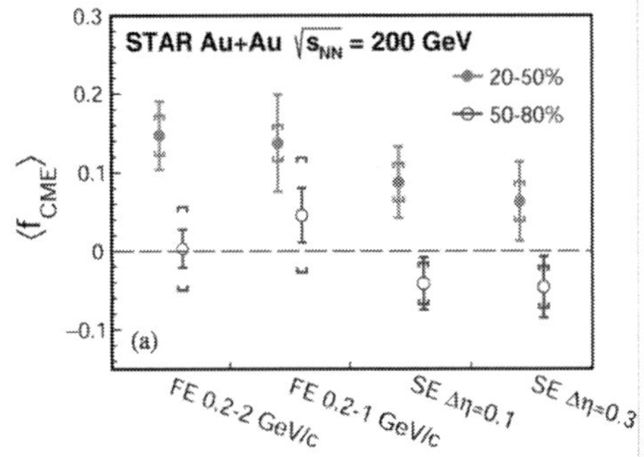

Figure 3: f_{CME}, the fraction of the $\Delta\gamma$ that is due to CME, as determined by an analysis that compares correlations with the participant plane to those with the spectator plane[19]. Sub-event analyses are performed to reduce the contribution of non-flow, as discussed in the text

into the formation of Λ hyperons, which will then acquire a net helicity in each event caused by LPV. Importantly, the handedness of CME and the net lambda (and anti-lambda) helicity should be the same in every event, so that events in which positive charges flow in the direction of the magnetic field vector(i.e., Q_w is negative, as depicted in the Figure 1 cartoon) will have an excess of right-handed (helicity) Λs, and likewise an excess of *left-handed* Λs should be expected in events with positive charges flowing *opposite* to the B-field. To distinguish charge flow "along" the B-field from charge flow "opposite to" the B-field, the first-order event plane is needed for each event. Using ψ_1, the signed charge flow can be characterized by $\Delta a_1 =< \sin(\phi_+ - \psi_1) > - < \sin(\phi_- - \psi_1) >$. In the presence of LPV/CME, we expect a negative correlation between Δa_1 and $\Delta N_\Lambda = N^L_\Lambda - N^R_\Lambda$.

A *preliminary* measurement of the covariance between these two quantities has been made by STAR using about 400M events of 27 GeV $Au + Au$ data. This is shown in Figure 3 as a function of centrality. The signature for LPV would be a negative covariance in this plot for the "on-peak" signal. Nothing statistically significant is observed in this exploratory measurement, but this method will be repeated in larger full-energy data sets and has the advantage that the backgrounds to these two observables are largely (completely?) uncorrelated by background sources so that the expected negative covariance signal should be very resistant to false LPV/CME signals.

5 Summary and Acknowledgements

With the current preliminary background estimates by STAR, there is not a significant CME signal in the isobar data. There **is** a statistically significant signal that implies a CME fraction of around 10% from (in my opinion) the best current analysis of top energy $Au + Au$ data, using both spectator plane and participant plane information to disentangle the background. That signal could be consistent with a non-observation in the isobar data with justifiable assumptions about how the CME signal may scale with system size. The $Au + Au$ analysis is, however, still not free of non-flow backgrounds; more work to understand these backgrounds (both "2-particle + flow" and "direct 3-particle") in detail is still crucial to understanding whether a signal is present.

For the future, it is expected that full energy $Au + Au$ runs of RHIC in 2023 and 2025 will increase the available statistics by roughly an order of magnitude, which will allow a much more thorough study of non-flow backgrounds to pin down whether the spectator/partcipiant plane analysis is seeing a true signal. Other CME-only observables may provide some additional, but not radically different, information, but taking advantage of the parity-odd behavior

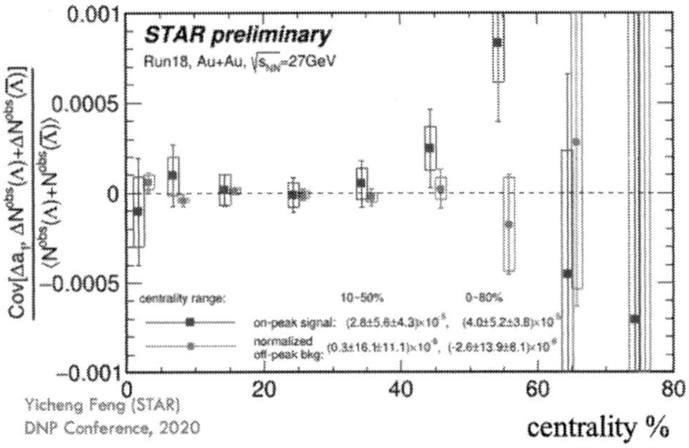

Figure 4: Event-by-event covariance between charge separation along the magnetic field (using first-order event plane information from the STAR EPD detector) and the net handedness of Λs + anti-Λs. A signal for LPV/CME would appear as a negative covariance for the blue points in mid-central collisions. No signal is seen in this 27 GeV data but the method can be used on forthcoming large top energy Au+Au datasets.

of the CME by correlating its signal event-by-event with other parity odd observables, such as net lambda helicity, opens up a qualitatively different way of searching that I consider quite exciting!

This work was supported by the U.S. Department of Energy, Office of Science, under Award Number DE-SC0022076.

References

[1] D E Kharzeev, R D Pisarski, M H G Tytgat, Phys Rev Lett **81**, 512 (1998)
[2] K Fukushima, D E Kharzeev, H J Warringa, Phys Rev D **78**, 074033 (2008)
[3] D E Kharzeev, J Liao, S A Voloshin, and G Wang, Prog Part Nucl Phys **88**, 1 (2016)
[4] D E Kharzeev, L D McLerran, H J Warringa, Nucl Phys A **803**, 227 (2008)
[5] S A Voloshin, Phys Rev C **70**, 057901 (2004)
[6] F Du, L E Finch and J Sandweiss, Phys Rev C **78**, 044908 (2008)
[7] F Becattini, M Buzzegoli, A Palermo, and G Prokhorov, Phys Lett B **822**, 012301 (2021)
[8] L E Finch and S J Murray, Phys Rev C **96**, 044911 (2017)
[9] B I Abelev et al (STAR Collaboration), Phys Rev Lett **103**, 251601 (2009)
[10] B I Abelev et al (ALICE Collaboration), Phys Rev Lett **110**, 251601 (2013)
[11] W Li and G Wang, Ann Rev Nucl Part Sci, **70**, 293 (2020)
[12] F Wang, 2022 Quark Matter Proceedings, arXiv:2207.10044-nucl-ex
[13] W T Deng, X G Huang, G L Ma, and G Wang, Phys Rev C**94** 041901 (2016)
[14] M S Abdallah et al (STAR Collaboration), Phys Rev C **105**, 014901 (2022)
[15] Y Hu, 2022 Quark Matter Proceedings, arXiv:2208.09069-nucl-ex
[16] Y Feng, 2022 Strange Quark Matter Proceedings
[17] S Choudhury et al, Chin Phys C **46** 014101 (2022)
[18] N Magdy, S Shi, J Liao, N Ajitanand, and R Lacey, Phys Rev C **97**, 061901(R) (2018)
[19] M S Abdallah et al (STAR Collaboration), Phys Rev Lett **128**, 092301 (2022)
[20] Y Feng, Y Lin, J Zhao, and F Wang, Phys Lett B **820**, 136549 (2021)

Recent Results from PHENIX

Ron Belmont[1,*] for the PHENIX Collaboration

[1]University of North Carolina at Greensboro

Abstract. We present results from the heavy-ion physics program of the PHENIX experiment at RHIC. These results include jet substructure, open heavy flavor, charmonium, strangeness, electromagnetic probes, femtoscopy, and flow.

1 Introduction

Since ending operations in the summer of 2016, the PHENIX Collaboration has continued to data analysis of the wide variety of high-quality data collected in the last few years of operations. These proceedings represent a subset of recent results, as presented at the Strangeness in Quark Matter conference held in Busan, Korea in June of 2022. These results include measurements of jet substructure, open heavy flavor, charmonium, strangeness, electromagnetic probes, femtoscopy, and flow.

2 Jet Substructure

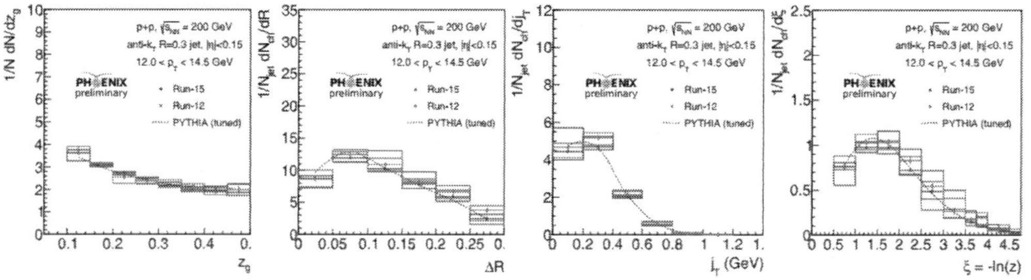

Figure 1. New preliminary results on jet substructure measurements in p+p collisions at $\sqrt{s} = 200$ GeV, from left to right: jet splitting function, radial profile, transverse fragmentation, longitudinal fragmentation.

Shown in Figure 1 is a variety of new preliminary results on jet substructure measurements in p+p collisions at $\sqrt{s} = 200$ GeV. A single jet p_T bin, 12.0 GeV/c < p_T < 14.5 GeV/c, is shown; additional p_T bins have been measured and will be presented in a forthcoming publication. The jets are reconstructed using the anti-k_T algorithm with a jet radius of $R = 0.3$. From left to right, shown are: the jet splitting function dN/dz_g as a function of z_g; the radial profile $dN_{\rm ch}/dR$ as a function of ΔR; the transverse fragmentation $dN_{\rm ch}/dj_T$ as

*e-mail: rjbelmon@uncg.edu

a function of j_T; the longitudinal fragmentation $dN_{\rm ch}/d\xi$ as a function of ξ. These measurements provide important detailed information about the substructure of jets in p+p collisions and also provide the baseline for future measurements in p+Au.

3 Open Heavy Flavor Results

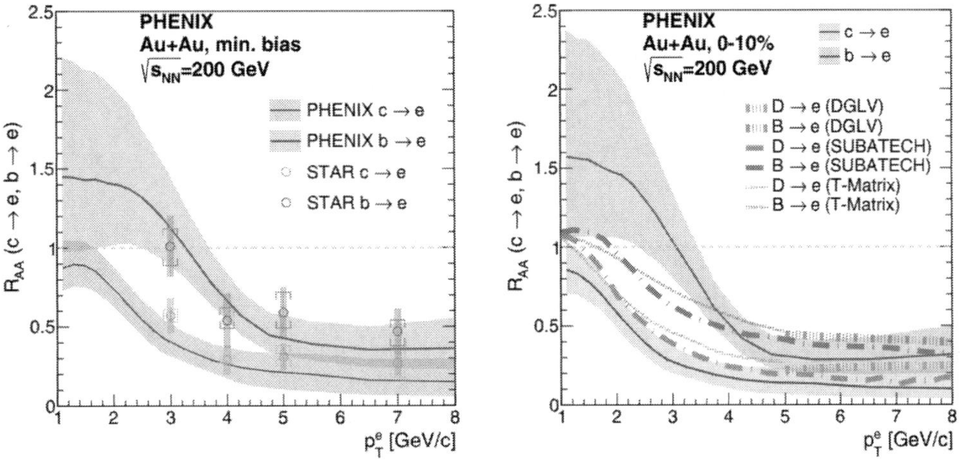

Figure 2. R_{AA} of $c \to e$ and $b \to e$ for minimum bias Au+Au collisions at $\sqrt{s_{NN}}$ = 200 GeV [1] with comparisons to STAR measurements (left panel) and model calculations (right panel).

Some open heavy flavor results are shown in Figure 2, R_{AA} of $c \to e$ and $b \to e$ for minimum bias Au+Au collisions at $\sqrt{s_{NN}}$ = 200 GeV [1]. The left panel shows a comparison to STAR measurements and demonstrates good agreement between the experiments. The right panel shows a comparison to various theoretical models, demonstrating good agreement for charm at all p_T as well as good agreement for bottom at higher p_T.

4 Charmonium

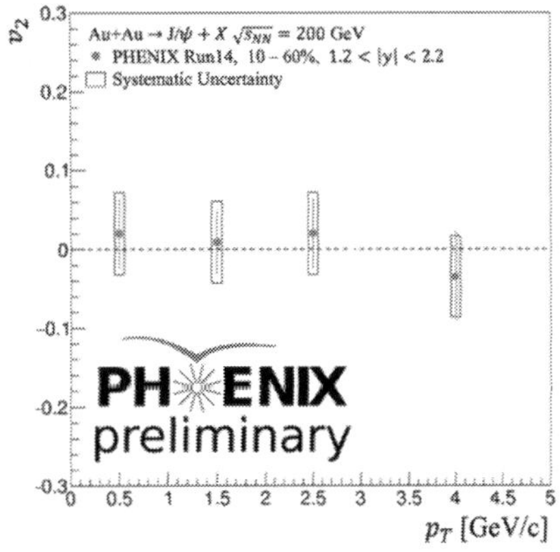

Figure 3. Anisotropy coefficient v_2 of J/ψ for $1.2 < |\eta| < 2.2$ in Au+Au collisions at $\sqrt{s_{NN}}$ = 200 GeV.

 https://doi.org/10.1051/epjconf/202327606019

Shown in Figure 3 is the v_2 of J/ψ for $1.2 < |\eta| < 2.2$ in Au+Au collisions at $\sqrt{s_{NN}} = 200$ GeV. The v_2 of J/ψ at forward/backward rapidity at RHIC energies is consistent with zero. This is contrast with the non-zero v_2 found at mid-rapidity at RHIC energies, as well as the non-zero v_2 found at both mid- and forward/backward rapidity at LHC energies. This may indicate the absence of regeneration at forward/backward rapidity at RHIC energies.

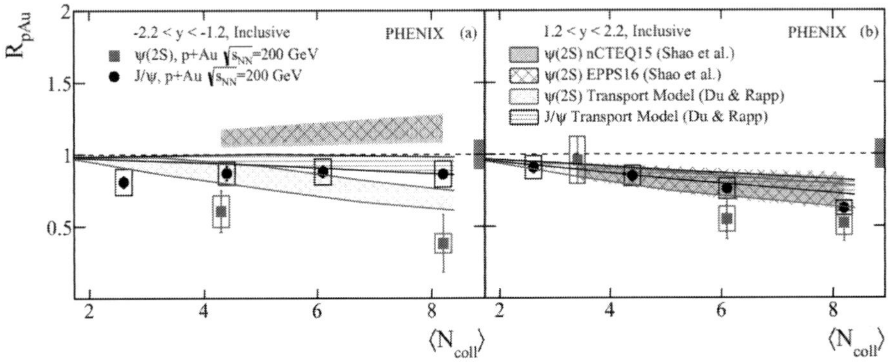

Figure 4. nuclear modification factor R_{AB} of J/ψ and ψ(2S) for $1.2 < |\eta| < 2.2$ in p+Au collisions at $\sqrt{s_{NN}} = 200$ GeV [2].

Figure 4 shows the nuclear modification factor R_{AB} of J/ψ and ψ(2S) at forward and backward rapidity ($1.2 < |\eta| < 2.2$) in p+Au collisions at $\sqrt{s_{NN}} = 200$ GeV [2].

The J/ψ modification is consistent with initial state effects alone at forward and backward rapidity. Likewise is true for the ψ(2S) at forward rapidity, but contrariwise, the ψ(2S) modification indicates presence of final state effects at backward rapidity. The final state effects could be the presence of co-movers, or perhaps an indication of the presence of the quark-gluon plasma (QGP).

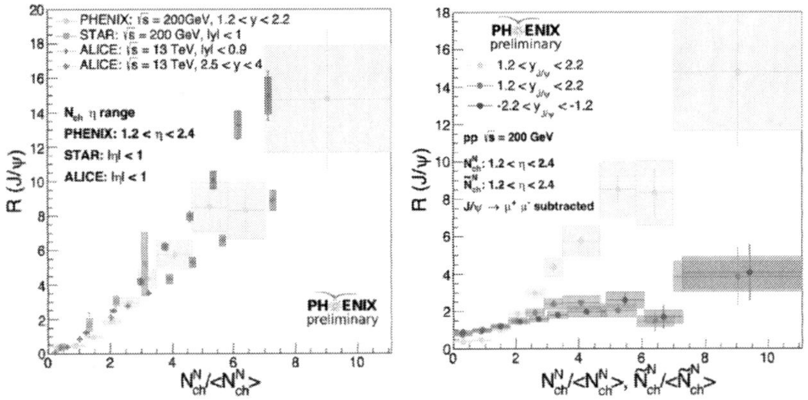

Figure 5. Relative yield of J/ψ in p+p collisions at $\sqrt{s_{NN}} = 200$ GeV. In the left panel, the multiplicity is determined in the same kinematic region as the J/ψ is reconstructed. In the right panel, the multiplicity is determined in a different kinematic region, and compared to the result when the multiplicity is reconstructed in the same region but when the J/ψ contribution to the multiplicity is subtracted off.

Figure 5 shows the relative yield of J/ψ in p+p collisions at $\sqrt{s_{NN}} = 200$ GeV. Relative yield of J/ψ in p+p collisions at $\sqrt{s_{NN}} = 200$ GeV. In the left panel, the multiplicity is determined in the same kinematic region as the J/ψ is reconstructed. Both forward and backward

 https://doi.org/10.1051/epjconf/202327606019

rapidity selections are compared. In the right panel, the multiplicity is determined in a different kinematic region, and compared to the result when the multiplicity is reconstructed in the same region but when the J/ψ contribution to the multiplicity is subtracted off.

The dramatic reduction in J/ψ enhancement when either selecting multiplicity in a different kinematic region, or when adjusting the multiplicity to account for the J/ψ itself, may have considerable implications for the rule of multi-parton interactions in J/ψ production.

5 Strangeness

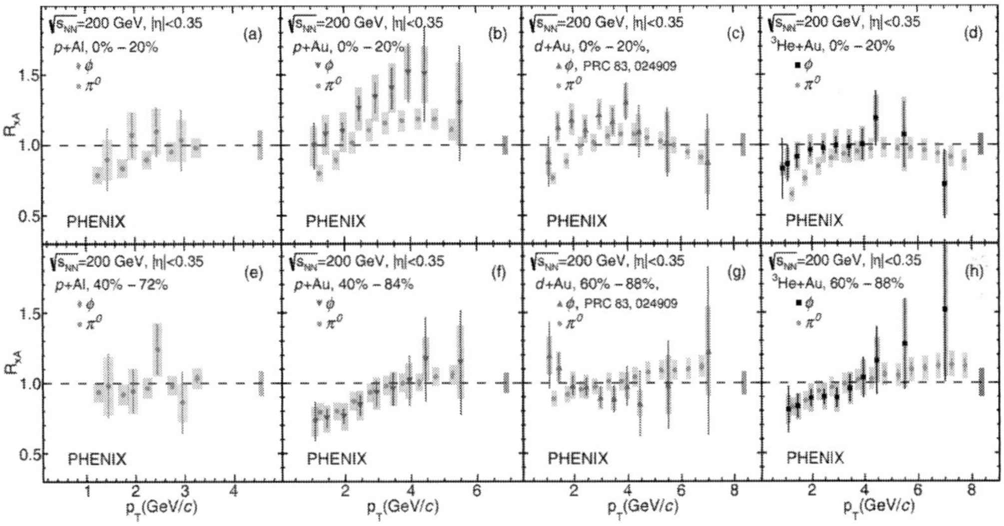

Figure 6. Nuclear modification factor R_{AB} of ϕ mesons in p+Al, p+Au, d+Au, and ^{3}He+Au collisions at $\sqrt{s_{NN}}$ = 200 GeV [3]. The upper row shows the most central collisions and the lower row shows the most peripheral collisions.

Figure 6 shows nuclear modification factor R_{AB} of ϕ mesons in p+Al, p+Au, d+Au, and ^{3}He+Au collisions at $\sqrt{s_{NN}}$ = 200 GeV [3]. The ϕ is mostly consistent with the π^0 in nearly all cases, perhaps suggesting a lack of apparent strangeness enhancement in small systems. However, allowing for the relatively large experimental uncertainties, there is some slight indication of enhancement in central collisions for p+Au, d+Au, and ^{3}He+Au collisions.

6 Electromagnetic Probes

Figure 7 shows nuclear modification factor R_{AB} of π^0 mesons in p+Al, p+Au, d+Au, and ^{3}He+Au collisions at $\sqrt{s_{NN}}$ = 200 GeV [4]. The centrality selection is 0–10% for all four. Seen at intermediate p_T is the usual Cronin enhancement. This enhancement is smaller when the target nucleus is lighter, viz. the R_{AB} of π^0 is smaller in p+Al than it is in p+Au. Additionally, the enhancement is also smaller when the projectile is heavier, as seen in the decreasing trend of the enhancement from p+Au to d+Au and then to ^{3}He+Au.

Shown in Figure 8 is the R_{AB} of π^0 determined using the standard MC Glauber N_{coll} (left panel), the R_{AB} of direct photons determined using the standard MC Glauber N_{coll} (center panel), and the R_{AB} of π^0 determined using experimentally determined N_{coll} (right panel).

It is known direct photons are not modified by nuclear matter, with a mean free path approximately 50 times larger than the nuclear size. Because of this, the R_{AB} of direct photons

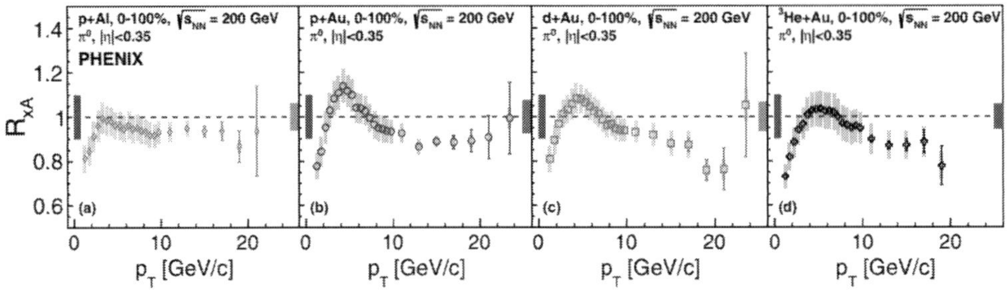

Figure 7. Nuclear modification factor R_{AB} of π^0 in p+Al, p+Au, d+Au, and ^{3}He+Au collisions at $\sqrt{s_{NN}}$ = 200 GeV [4]. Minimum-bias 0–100% is shown for all four collision species.

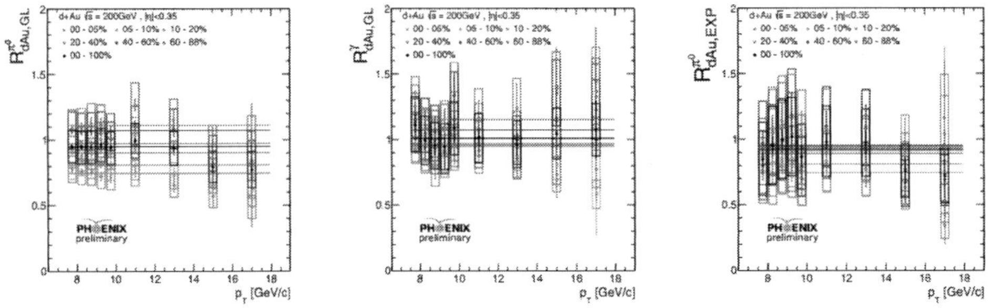

Figure 8. Nuclear modification factor R_{AB} of π^0 (left, right) and direct photon (center). in d+Au collisions at $\sqrt{s_{NN}}$ = 200 GeV. Various centralities are shown. For the left and center panels, the standard MC Glauber determination of $N_{\rm coll}$ is used. On the right, experimentally determined $N_{\rm coll}$ is used instead.

should be exactly unity. The apparent deviation from unity is an indication of event selection bias. The experimentally determined $N_{\rm coll}$ adjusts for this bias by enforcing the direct photon R_{AB} to be exactly unity. When using these new values of $N_{\rm coll}$ to determine the R_{AB} of π^0, the apparent enhancement seen in the left panel is removed in the right panel. This addresses the long-standing mystery of the peripheral enhancement as not being due to some novel physics effect but rather an event selection bias. Contrariwise, the slight suppression of the R_{AB} of π^0 persists when correcting for this bias. The origin of this effect is not yet understood, but could be due to the EMC effect in the nuclear PDFs and/or an indication of suppression caused by the QGP.

7 Flow

The azimuthal triangular anisotropy v_3 of charged hadrons in p+Au, d+Au, and ^{3}He+Au collisions at $\sqrt{s_{NN}}$ = 200 GeV [5] is shown in Figure 9.

The agreement between the 3×2PC method and the event plane method indicates the robustness of the measurement against sensitive experimental effects like the detector alignment and beam optics.

Although results are not shown here due to length restrictions, we want to call your attention to another recent PHENIX publication on a systematic study of the centrality and system size dependence of v_2 in p+p, p+Au, d+Au, and ^{3}He+Au collisions at $\sqrt{s_{NN}}$ = 200 GeV [6]

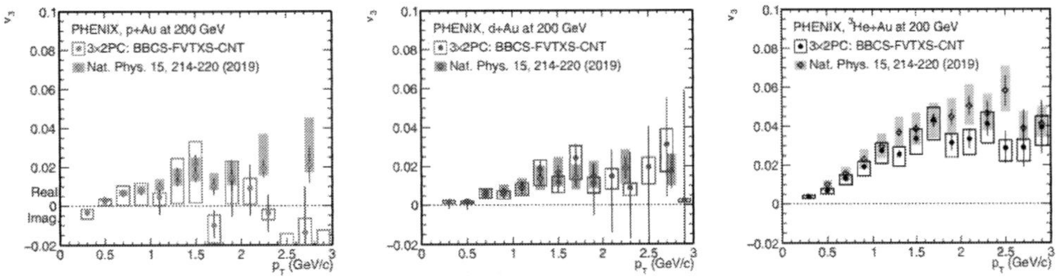

Figure 9. Azimuthal triangular anisotropy v_3 of charged hadrons in p+Au, d+Au, and ^{3}He+Au collisions at $\sqrt{s_{NN}}$ = 200 GeV. The most central 0–5% is shown for all three collision species. Data from Ref. [5]

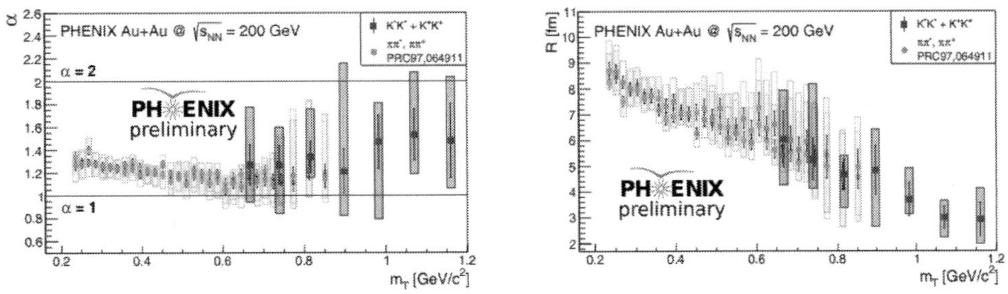

Figure 10. HBT with kaons. Source shape parameter α vs pair m_T is shown in the left panel. Radius parameter R vs pair m_T is shown in the right panel.

8 Femtoscopy

Figure 10 shows the source shape parameter α (left panel) and the radius parameter R (right panel) as a function of the pair m_T. By using a Lévy function, the source shape can be better understood in comparison to a typical Gaussian function. When $\alpha = 1$ the source is equivalent to a Cauchy distribution, and to a Gaussian when $\alpha = 2$. The measurement indicates a source shape somewhere in between but closer to a Cauchy distribution. The results using $\pi^{\pm}$ and $K^{\pm}$ show similar trends, with the charged kaon results continuing to higher pair m_T due to the larger mass of the individual particles.

9 Summary

PHENIX continues to produce high-quality physics results on a variety of topics, including but not limited to jet substructure, open heavy flavor, charmonium, strangeness, electromagnetic probes, femtoscopy, and flow as shown in these proceedings. Further results and publications are forthcoming, and we are looking forward to sharing them at the next SQM.

References

[1] PHENIX Collaboration, arXiv:2203.17058, submitted to Phys. Rev. C (2022)
[2] PHENIX Collaboration, Phys. Rev. C **105, 064912 (2022)**
[3] PHENIX Collaboration, Phys. Rev. C 106, 014908 (2022)
[4] PHENIX Collaboration, Phys. Rev. C 105, 064902 (2022)
[5] PHENIX Collaboration, Phys. Rev. C 105, 024901 (2022)
[6] PHENIX Collaboration, arXiv:2203.09894, submitted to Phys. Rev. C (2022)

EPJ Web of Conferences **276**, 06020 (2023) https://doi.org/10.1051/epjconf/202327606020
SQM 2022

Probing neutron skin and symmetry energy with relativistic isobar collisions

Hao-jie Xu[1,*]

[1]School of Science, Huzhou University, Huzhou, Zhejiang 313000, China

Abstract. In these proceedings, we present the three proposed observables to probe the neutron skin and symmetry energy with relativistic isobar collisions, namely, the isobar ratios of the produced hadron multiplicities (N_{ch}), the mean transverse momenta ($\langle p_\perp \rangle$), and the net charge multiplicities (ΔQ). Our findings suggest potentially significant improvement to neutron skin and symmetry energy determination over traditional low energy methods.

1 Introduction

Recently, the STAR collaboration has published the first result of the chiral magnetic effect (CME) search in isobar $^{96}_{44}$Ru+$^{96}_{44}$Ru and $^{96}_{40}$Zr+$^{96}_{40}$Zr collisions at nucleon-nucleon center-of-mass energy $\sqrt{s_{NN}} = 200$ GeV [1]. While no evidence is found for the CME, the experimental data show sizeable differences in multiplicity and elliptic flow between isobar collisions, indicating significant differences in their nuclear structures, such as the neutron skin thickness and nuclear deformation.

The neutron skin thickness $\Delta r_{np} \equiv \sqrt{\langle r_n^2 \rangle} - \sqrt{\langle r_p^2 \rangle}$, the root mean square difference between neutron and proton distributions, is strongly correlated to the density slope parameter of the symmetry energy. The symmetry energy and its density dependence are crucial to our understanding of the masses and drip lines of neutron-rich nuclei and the equation of state (EOS) of nuclear and neutron star matter. The Δr_{np} has traditionally been measured by low-energy electron and hadron scatterings off nuclei [2–4], and the symmetry energy slope parameters L and L_c are extracted at the nuclear saturation density ρ_0 and critical density ρ_c, respectively [3, 5–8]. Because of the inevitable uncertainties in modeling the strong interaction of the scattering processes in quantum chromodynamics (QCD) [9], large uncertainties on the L and L_c persist. The parity-violating electroweak scattering measurement on the ^{208}Pb by the Lead Radius Experiment (PREX-II), although void of QCD uncertainties, has relatively large statistical uncertainty, $\Delta r_{np} = 0.283 \pm 0.071$ fm [10, 11]. It leads to $L = 105 \pm 37$ MeV [12], at a slight tension with $L = 75 \pm 25$ MeV [13] from traditional scattering experiments.

The sensitivity of relativistic isobar collisions to neutron skin, predicted [14, 15] and confirmed by the STAR data [1], offers a new opportunity to complement low-energy measurements with completely different systematics. Specifically, the ratios between isobar collisions of the produced hadron multiplicities (N_{ch}) [16], the mean transverse momenta ($\langle p_\perp \rangle$) [17], and the net charge multiplicities (ΔQ) [18] are found to be sensitive to the neutron skin

*e-mail: haojiexu@zjhu.edu.cn

difference between the isobar nuclei. Measurements of those ratios can, in turn, offer an unconventional and perhaps more precise means to probe the neutron skin.

2 Symmetry energy and density functional theory

The symmetry energy encodes the energy related to neutron-proton asymmetry in the nuclear matter EOS. It is conventionally defined in the binding energy per nucleon, approximately expressed as [19] $E(\rho, \delta) = E_0(\rho) + E_{\rm sym}(\rho)\delta^2 + O(\delta^4)$, where $\rho = \rho_n + \rho_p$ is the nucleon number density and $\delta = (\rho_n - \rho_p)/\rho$ is the isospin asymmetry with ρ_p (ρ_n) denoting the proton (neutron) density. The symmetry energy can be obtained as $E_{\rm sym}(\rho) = \frac{1}{2}\frac{\partial^2 E(\rho,\delta)}{\partial\delta^2}\Big|_{\delta=0}$. At a given reference density ρ_r, the $E_{\rm sym}(\rho)$ can be expanded in $\chi_r = (\rho-\rho_r)/3\rho_r$ as $E_{\rm sym}(\rho) = E_{\rm sym}(\rho_r) + L(\rho_r)\chi_r + O(\chi_r^2)$, where $L(\rho_r) = 3\rho_r \frac{dE_{\rm sym}(\rho)}{d\rho}\Big|_{\rho=\rho_r}$ is the density slope parameter [19]. Especially, for $\rho_r = \rho_0 \approx 0.16$ fm^{-3} and $\rho_c = 0.11\rho_0/0.16 \approx 0.11$ fm^{-3}, one has $L \equiv L(\rho_0)$ and $L_c \equiv L(\rho_c)$ which characterizes the density dependence of the $E_{\rm sym}(\rho)$ around ρ_0 and ρ_c. A strong constraint $L(\rho_c) = 47.3 \pm 7.8$ MeV is obtained from analyzing the data on the electric dipole polarizability in ^{208}Pb [20]. Generally, it is found that the $L(\rho_c)$ displays a particularly strong positive correlation with the $\Delta r_{\rm np}$ of heavy nuclei.

We use two different nuclear energy density functionals to describe nuclear matter EOS and the properties of finite nuclei, namely, the standard Skyrme-Hartree-Fock (SHF) model (see, e.g., Ref. [21]) and the extended SHF (eSHF) model [22]. Compared to SHF, the eSHF model contains additional momentum and density-dependent two-body forces to simulate the momentum dependence of the three-body forces effectively [22]. Fitting to data using the strategy in Ref. [23], we obtain an interaction parameter set (denoted as Lc47) within eSHF by fixing $L(\rho_c) = 47.3$ MeV [20] with $E_{\rm sym}(\rho_c) = 26.65$ MeV [24]. We also construct two more interaction parameter sets (denoted as Lc20 and Lc70) with $L(\rho_c) = 20$ MeV and 70 MeV, respectively, keeping the same $E_{\rm sym}(\rho_c)$ [24], to explore the effects of the symmetry energy (and neutron skin) variations. For the SHF calculations, we use the well-known interaction set SLy4 [25, 26]. The four interaction parameter sets give similar proton rms radius r_p for ^{96}Zr and ^{96}Ru which are experimentally well constrained, but the neutron radius r_n increases with $L(\rho_c)$ and L, leading to a positive correlation between $\Delta r_{\rm np}$ and $L(\rho_c)$ (and L) as expected. Both DFT calculations predict a halo-type neutron skin thickness for ^{96}Zr, which is crucial to our understanding of non-trivial bump structures in the ratios of the multiplicity distributions and the elliptic flows in non-central isobar collisions [1, 14, 15, 27]

3 Probes in isobar collisions

3.1 Total charge multiplicity $N_{\rm ch}$

The event multiplicity produced in heavy-ion collisions is sensitive to the density distributions of the colliding nuclei, and thus the L. The absolute $N_{\rm ch}$ values are subject to significant model dependence because particle production in heavy ion collisions is generally hard to model precisely. The shape of the $N_{\rm ch}$ distribution is, on the other hand, more robust. It is determined by the interaction cross-section as a function of the impact parameter (b). We use four typical, well developed, commonly used models for relativistic heavy ion collisions to simulate the particle production: the Hijing (Heavy ion jet interaction generator, v1.411) model, the AMPT (A Multi-Phase Transport, v1.26, v2.26) model with string fragmentation (v1.26) and string melting (v2.26), and the UrQMD (Ultra relativistic Quantum Molecular Dynamics, v3.4) model. Except for the input nuclear density distributions, all parameters are set to default.

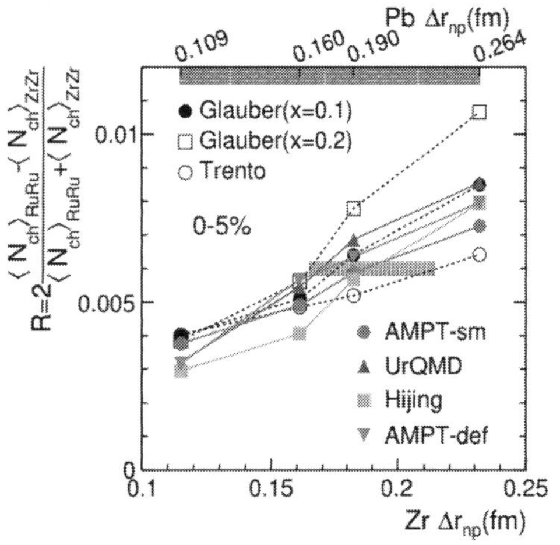

Figure 1. (Color online). The relative $\langle N_{\rm ch}\rangle$ ratio $R_{\langle N_{\rm ch}\rangle}$ as a function of the Zr neutron skin thickness. The four sets of data points in order from left to right are from Lc20, SLy4, Lc47, Lc70 densities. The results from AMPT-SM, UrQMD, Hijing, and AMPT-Def are connected by solid lines. The results from Glauber and Trento models are connected by dashed lines. The figure is taken from Ref. [16].

The $N_{\rm ch}$ distributions show splittings with different L_c. We use the relative $\langle N_{\rm ch}\rangle$ difference between Ru+Ru and Zr+Zr, $R_{\langle N_{\rm ch}\rangle} = 2\frac{\langle N_{\rm ch}\rangle_{\rm RuRu}-\langle N_{\rm ch}\rangle_{\rm ZrZr}}{\langle N_{\rm ch}\rangle_{\rm RuRu}+\langle N_{\rm ch}\rangle_{\rm ZrZr}}$, to quantify the splitting of the $N_{\rm ch}$ tails. Experimental measurements of $N_{\rm ch}$ are affected by tracking inefficiency, usually multiplicity dependent [28]. While this effect mostly cancels out in $R_{\langle N_{\rm ch}\rangle}$, it is better to use only central collisions, e.g. top 5% centrality, where the tracking efficiency is constant to a good degree.

The $R_{\langle N_{\rm ch}\rangle}$ in each model must depend on how much the Ru and Zr nuclear density distributions differ, which can be characterized by $\Delta r_{\rm np}$ of the Zr (or Ru) nucleus. We therefore plot in Fig. 1 the $R_{\langle N_{\rm ch}\rangle}$ in the top 5% centrality against $\Delta r_{\rm np}$ of the Zr nucleus from the eSHF (SHF) calculations with Lc20, Lc47, and Lc70 (SLy4). It is found that $R_{\langle N_{\rm ch}\rangle}$ monotonically increases with $\Delta r_{\rm np}$. This is because, with increasing $\Delta r_{\rm np}$, the difference between Ru and Zr densities increases. This results in an increasing difference in $N_{\rm ch}$ between Ru+Ru and Zr+Zr collisions. The value of $R_{\langle N_{\rm ch}\rangle}$ has a relatively weak model dependence, including Hijing whose tail distribution is significantly narrower than the other models, which can already be discriminated by data measurements. The intuitive geometrical models, namely, the Glauber and Trento models give a similar trend as the above dynamical models. In addition, $R_{\langle N_{\rm ch}\rangle}$ is a relative measure between Ru+Ru and Zr+Zr collisions, and much of the experimental effects are canceled. We thus expect a precise constraint on symmetry energy with our proposed observable $R_{\langle N_{\rm ch}\rangle}$ in relativistic isobar collisions [16].

The $R_{\langle N_{\rm ch}\rangle}$ observable has been measured in top 2% centrality by the STAR collaboration [29]. Based on the Monte Carlo Glauber model simulations with the density distributions obtained from eSHF calculations, the symmetry energy slope parameter is extracted to be $L(\rho_c) = 53.8 \pm 1.7(\text{stat.}) \pm 7.8(\text{syst.})$ MeV, where the systematic uncertainty is dominated by those on nuclear deformations.

3.2 Mean transverse momentum $\langle p_\perp\rangle$

Transverse momentum ($p_\perp$) generation in relativistic heavy ion collisions is sensitive to the initial geometry and the final-state bulk evolution. In hydrodynamics, the $\langle p_\perp\rangle$ values are sensitive to the medium bulk properties. To investigate the effects of bulk properties, we calculate the $\langle p_\perp\rangle$ using the Lc47 densities with three values of shear viscosity [$(\eta/s)_{\rm min}$ = 0.04, 0.08 and 0.16] and with three values of bulk viscosity [$(\zeta/s)_{\rm max}$ = 0.025, 0.081 and 0.1] in the iEBE-VISHNU (an event-by-event (2+1)-dimensional viscous hydrodynamics, together with the hadron cascade model UrQMD) model simulations. The middle values are

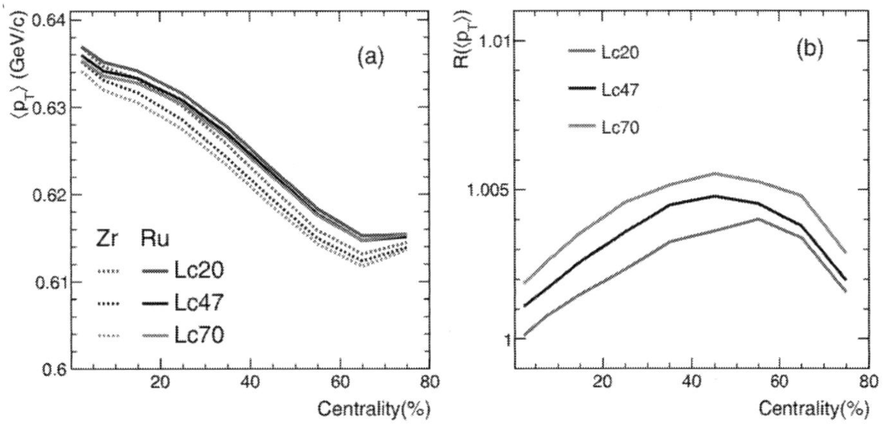

Figure 2. (Color online). (a) The mean transverse momentum $\langle p_\perp \rangle$ in Zr+Zr collisions, and (b) the Ru+Ru/Zr+Zr ratio $R_{\langle p_\perp \rangle}$ as functions of centrality, calculated by the iEBE-VISHNU model with Lc20, Lc47, and Lc70 spherical nuclear densities. Figures are taken from Ref. [17].

typical values used in hydrodynamic simulations [30]. Our findings strongly indicate while the magnitude of $\langle p_\perp \rangle$ depends on the bulk properties, the ratio $R_{\langle p_\perp \rangle} \equiv \frac{\langle p_\perp \rangle^{\text{Ru+Ru}}}{\langle p_\perp \rangle^{\text{Zr+Zr}}}$ is insensitive to them and hence their uncertainties.

Figure 2(a) presents the $\langle p_\perp \rangle$ as functions of centrality in both Ru+Ru and Zr+Zr collisions from the iEBE-VISHNU simulations with various DFT-calculated spherical densities for the isobars. A larger $L(\rho_c)$ gives thicker neutron skin, and results in smaller $\langle p_\perp \rangle$ at each centrality, as expected. On the other hand, the Ru+Ru/Zr+Zr ratio $R_{\langle p_\perp \rangle}$, shown in Fig. 2(b), increases with $L(\rho_c)$. This is because the neutron skin effect is larger in ^{96}Zr than in ^{96}Ru and this effect increases with $L(\rho_c)$. The splittings shown in the figure indicate the sensitivity of L_c on $R_{\langle p_\perp \rangle}$, providing a novel method to constrain symmetry energy in relativistic isobar collisions [17]. The centrality dependence of $R_{\langle p_\perp \rangle}$ is non-trivial and can reach as large as 0.5% above unity.

This method has recently been applied to the isobar data by the STAR collaboration [29]. Based on the $R_{\langle p_\perp \rangle}$ values in the top 5% centrality and state-of-the-art hydrodynamic simulations, the symmetry energy slope parameter has been extracted. The value is $L(\rho_c) = 56.8 \pm 0.4(\text{stat.}) \pm 10.4(\text{syst.})$ MeV, where the systematic uncertainty is dominated by those on nuclear deformations. This result is consistent with that extracted from $R_{\langle N_{\text{ch}} \rangle}$.

3.3 Net charge ΔQ

We note that the final state N_{ch} and $\langle p_\perp \rangle$ do not distinguish between initial neutron or proton participants. Proton-proton (pp), proton-neutron (pn), and neutron-neutron (nn) collisions at relativistic energies produce practically the same $\langle N_{\text{ch}} \rangle$ and $\langle p_\perp \rangle$. It is sensitive only to the overall nucleon density, and therefore indirectly sensitive to the neutron density (and neutron skin) given that the proton density is well determined. But if nuclei had proton skin instead of neutron skin, our study would yield the same result. Ultra-peripheral collisions, where the nuclei are only grazing each other, must have very different mixture of participant protons and neutrons, and therefore likely yield significantly different net-charge numbers (ΔQ).

The ΔQ ratio in Ru+Ru over Zr+Zr collisions, under the superimposition assumption, is $R_{\Delta Q} \equiv \frac{\Delta Q_{\text{RuRu}}}{\Delta Q_{\text{ZrZr}}} = \frac{q_{\text{RuRu}}+\alpha/(1-\alpha)}{q_{\text{ZrZr}}+\alpha/(1-\alpha)}$, where $\alpha \equiv \Delta Q_{nn}/\Delta Q_{pp}$ is the ΔQ ratio in nn to pp interactions and q_{AA} is the fraction of protons among the participant nucleons that can be obtained from

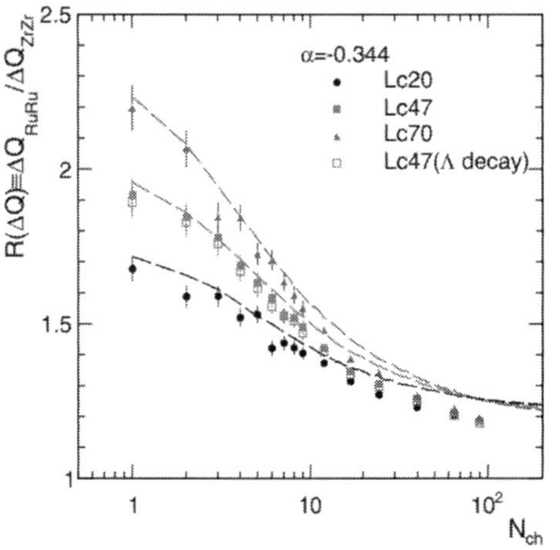

Figure 3. (Color online). $R_{\Delta Q}$ for $b \in [7, 20]$ fm as function of $N_{\rm ch}$ ($|\eta| < 0.5$) simulated by UrQMD with DFT nuclear densities from eSHF (Lc20, Lc47, Lc70). The open red squares show a calculation for the Lc47 case including Λ-hyperon decays. The curves are from superimposition prediction with $\alpha = -0.344$ from UrQMD NN interactions and the q_{AA} from Trento simulation. Figure is taken from Ref. [18].

Trento simulations. Pythia(version 8.240) gives $\alpha \simeq -0.352$ with acceptance cuts $|\eta| < 1$ and $0.2 < p_T < 2$ GeV/c excluding the (anti-)protons with $p_T < 0.4$ GeV/c. The overall $q_{\rm RuRu}$ and $q_{\rm ZrZr}$ values for the whole nuclei are 44/96 and 40/96, respectively; they would give $R_{\Delta Q} \simeq 1.267$. Of course, the simple superimposition assumption breaks down in non-peripheral collisions because of nuclear effects. However, the assumption should be good for grazing AA collisions. The general idea to probe $\Delta r_{\rm np}$ by $R_{\Delta Q}$ is that a sizeable $\Delta r_{\rm np}$ will make the q_{AA} decrease dramatically with increasing impact parameter (b) in those grazing collisions. The $\Delta r_{\rm np}$ of ^{96}Zr is significantly larger than that of ^{96}Ru, so the $R_{\Delta Q}$ ratio amplifies the $\Delta r_{\rm np}$ sensitivity. The $\Delta r_{\rm np}$ of both nuclei are controlled by the L_c parameter, thus a measurement of $R_{\Delta Q}$ can determine its value.

We simulate ~450 million events within $b \in [7, 20]$ fm using UrQMD model. The same acceptance cuts have been applied as done in Pythia simulations. Figure 3 shows $R_{\Delta Q}$ as a function of $N_{\rm ch}$. Using $\alpha = -0.344$, the predicted curves from superimposition assumption are depicted in Fig. 3. The curves can fairly well describe the UrQMD data. This indicates that the grazing collisions in UrQMD, with $N_{\rm ch} \lesssim 10$, are indeed simple superimposition of NN interactions. This is not surprising as only a few nucleons participate in such a grazing AA collision, so any nuclear effect would be negligible. At higher $N_{\rm ch}$ the UrQMD data points deviate from the curves, presumably because those collisions are not simple NN superimpositions anymore. It may also be viewed that the effective α in central AA collisions, because of nuclear effects, is very different from the one calculated using single NN interactions. The splittings shown in the figure indicate the sensitivity of L_c on $R_{\Delta Q}$, providing a novel method to constrain symmetry energy in relativistic isobar collisions [18].

4 Summary

The isobar $^{96}_{44}$Ru+$^{96}_{44}$Ru and $^{96}_{40}$Zr+$^{96}_{40}$Zr collisions at $\sqrt{s_{\rm NN}} = 200$ GeV provide novel means to probe the neutron skins of the isobar nuclei. The neutron skin thickness can be determined, with the help of DFT calculations, from the isobar ratios of the produced hadron multiplicities ($R_{\langle N_{\rm ch}\rangle}$), the mean transverse momenta ($R_{\langle p_\perp\rangle}$), and the net charge multiplicities ($R_{\Delta Q}$). Due to the rather weak dependence of these ratios to the details of QCD, our proposed methods can be used to determine the density slope parameter of symmetry energy with a precision that may be comparable to or even exceed those achieved by traditional low-energy nuclear experiments. Our measurements complement, with different systematics, those low-energy experiments in probing the symmetry energy. The preliminary results on the extracted slope parameter $L(\rho_c)$ from $R_{\langle N_{\rm ch}\rangle}$ and $R_{\langle p_\perp\rangle}$ have been reported by the STAR collaboration [29], with the values of 53.8 ± 1.7 ± 7.8 MeV and 56.8 ± 0.4 ± 10.4 MeV, respectively. These values are consistent with world-wide data from traditional nuclear scattering experiments with comparable precision [29].

This work is supported by National Natural Science Foundation of China (NSFC) under Grants No. 12275082, 12035006, 12075085, 11909059.

References

[1] M. Abdallah *et al.* [STAR], Phys. Rev. C **105**, no.1, 014901 (2022)
[2] B. Frois and C. N. Papanicolas, Ann. Rev. Nucl. Part. Sci. **37**, 133-176 (1987)
[3] M. B. Tsang, J. R. Stone, F. Camera *et al.*, Phys. Rev. C **86**, 015803 (2012)
[4] C. M. Tarbert, D. P. Watts, D. I. Glazier *et al.*, Phys. Rev. Lett. **112**, 242502 (2014)
[5] L. W. Chen, C. M. Ko and B. A. Li, Phys. Rev. C **72**, 064309 (2005)
[6] X. Roca-Maza, M. Centelles, X. Vinas and M. Warda, Phys. Rev. Lett. **106**, 252501 (2011)
[7] C. J. Horowitz, E. F. Brown, Y. Kim *et al.*, J. Phys. G **41**, 093001 (2014)
[8] Y. Wang, Z. Gao, H. Lü and Q. Li, Phys. Lett. B **835**, 137508 (2022)
[9] L. Ray, G. W. Hoffmann and W. R. Coker, Phys. Rept. **212**, 223-328 (1992)
[10] S. Abrahamyan, Z. Ahmed, H. Albataineh *et al.*, Phys. Rev. Lett. **108**, 112502 (2012)
[11] D. Adhikari *et al.* [PREX], Phys. Rev. Lett. **126**, 172502 (2021)
[12] B. T. Reed, F. J. Fattoyev, C. J. Horowitz and J. Piekarewicz, Phys. Rev. Lett. **126**, 172503 (2021)
[13] M. Centelles, X. Roca-Maza, X. Vinas and M. Warda, Phys. Rev. Lett. **102**, 122502 (2009)
[14] H. j. Xu, X. Wang, H. Li *et al.*, Phys. Rev. Lett. **121**, 022301 (2018).
[15] H. Li, H. j. Xu, J. Zhao *et al.*, Phys. Rev. C **98**, 054907 (2018).
[16] H. Li, H. j. Xu, Y. Zhou *et al.*, Phys. Rev. Lett. **125**, 222301 (2020).
[17] H. j. Xu, W. Zhao, H. Li *et al.*, [arXiv:2111.14812 [nucl-th]].
[18] H. j. Xu, H. Li, Y. Zhou *et al.*, Phys. Rev. C **105**, L011901 (2022).
[19] B. A. Li, L. W. Chen and C. M. Ko, Phys. Rept. **464**, 113-281 (2008)
[20] Z. Zhang and L. W. Chen, Phys. Rev. C **90**, 064317 (2014)
[21] E. Chabanat, J. Meyer, P. Bonche, R. Schaeffer and P. Haensel, Nucl. Phys. A **627**, 710-746 (1997)
[22] Z. Zhang and L. W. Chen, Phys. Rev. C **94**, 064326 (2016)
[23] Y. Zhou, L. W. Chen and Z. Zhang, Phys. Rev. D **99**, 121301 (2019)
[24] Z. Zhang and L. W. Chen, Phys. Lett. B **726**, 234-238 (2013)
[25] E. Chabanat, P. Bonche, P. Haensel, J. Meyer and R. Schaeffer, Nucl. Phys. A **635**, 231-256 (1998) [erratum: Nucl. Phys. A **643**, 441-441 (1998)]
[26] X. B. Wang, J. L. Friar and A. C. Hayes, Phys. Rev. C **94**, 034314 (2016)
[27] H. j. Xu, H. Li, X. Wang, C. Shen and F. Wang, Phys. Lett. B **819**, 136453 (2021)
[28] B. I. Abelev *et al.* [STAR], Phys. Rev. C **79**, 034909 (2009)
[29] H. Xu [STAR], Acta Phys. Polon. Supp. **16**, 1-A30 (2023)
[30] J. E. Bernhard, J. S. Moreland and S. A. Bass, Nature Phys. **15**, 1113-1117 (2019)

AUTHOR INDEX

AUTHOR INDEX